AIR POLLUTION IN THE 21ST CENTURY
Priority Issues and Policy

Studies in Environmental Science 72

AIR POLLUTION IN THE 21ST CENTURY
Priority Issues and Policy

Edited by

T. Schneider

1998
ELSEVIER
Amsterdam – Lausanne – New York – Oxford – Shannon – Tokyo

ELSEVIER SCIENCE B.V.
Sara Burgerhartstraat 25
P.O. Box 211, 1000 AE Amsterdam, The Netherlands

ISBN 0-444-82799-4

∞ The paper used in this publication meets the requirements of ANSI/NISO Z39.48-1992 (Permanence of Paper).

This book is printed on acid-free paper.

Printed in The Netherlands

Studies in Environmental Science

Other volumes in this series

1 **Atmospheric Pollution 1978** edited by M.M. Benarie
2 **Air Pollution Reference Measurement Methods and Systems** edited by T. Schneider, H.W. de Koning and L.J. Brasser
3 **Biogeochemical Cycling of Mineral-Forming Elements** edited by P.A. Trudinger and D.J. Swaine
4 **Potential Industrial Carcinogens and Mutagens** by L. Fishbein
5 **Industrial Waste Management** by S.E. Jørgensen
6 **Trade and Environment: A Theoretical Enquiry** by H. Siebert, J. Eichberger, R. Gronych and R. Pethig
7 **Field Worker Exposure during Pesticide Application** edited by W.F. Tordoir and E.A.H. van Heemstra-Lequin
8 **Atmospheric Pollution1980** edited by M.M. Benarie
9 **Energetics and Technology of Biological Elimination of Wastes** edited by G. Milazzo
10 **Bioengineering,Thermal Physiology and Comfort** edited by K. Cena and J.A. Clark
11 **Atmospheric Chemistry. Fundamental Aspects** by E. Mészáros
12 **Water Supply and Health** edited by H. van Lelyveld and B.C.J. Zoeteman
13 **Man under Vibration. Suffering and Protection** edited by G. Bianchi, K.V. Frolov and A. Oledzki
14 **Principles of Environmental Science and Technology** by S.E. Jørgensen and I. Johnsen
15 **Disposal of Radioactive Wastes** by Z. Dlouhý
16 **Mankind and Energy** edited by A. Blanc-Lapierre
17 **Quality of Groundwater** edited by W. van Duijvenbooden, P. Glasbergen and H. van Lelyveld
18 **Education and Safe Handling in Pesticide Application** edited by E.A.H. van Heemstra-Lequin and W.F. Tordoir
19 **Physicochemical Methods for Water and Wastewater Treatment** edited by L. Pawlowski
20 **Atmospheric Pollution 1982** edited by M.M. Benarie
21 **Air Pollution by Nitrogen Oxides** edited by T. Schneider and L. Grant
22 **Environmental Radioanalysis** by H.A. Das, A. Faanhof and H.A. van der Sloot
23 **Chemistry for Protection of the Environment** edited by L. Pawlowski, A.J. Verdier and W.J. Lacy
24 **Determination and Assessment of Pesticide Exposure** edited by M. Siewierski
25 **The Biosphere: Problems and Solutions** edited by T.N. Veziroğlu
26 **Chemical Events in the Atmosphere and their Impact on the Environment** edited by G.B. Marini-Bettòlo
27 **Fluoride Research 1985** edited by H. Tsunoda and Ming-Ho Yu
28 **Algal Biofouling** edited by L.V. Evans and K.D. Hoagland
29 **Chemistry for Protection of the Environment 1985** edited by L. Pawlowski, G. Alaerts and W.J. Lacy
30 **Acidification and its Policy Implications** edited by T. Schneider
31 **Teratogens: Chemicals which Cause Birth Defects** edited by V. Kolb Meyers
32 **Pesticide Chemistry** by G. Matolcsy, M. Nádasy and Y. Andriska
33 **Principles of Environmental Science and Technology (second revised edition)** by S.E. Jørgensen and I. Johnsen
34 **Chemistry for Protection of the Environment 1987** edited by L. Pawlowski, E. Mentasti, W.J. Lacy and C. Sarzanini
35 **Atmospheric Ozone Research and its Policy Implications** edited by T. Schneider, S.D. Lee, G.J.R. Wolters and L.D. Grant

36 **Valuation Methods and Policy Making in Environmental Economics** edited by H. Folmer and E. van Ierland
37 **Asbestos in Natural Environment** by H. Schreier
38 **How to Conquer Air Pollution. A Japanese Experience** edited by H. Nishimura
39 **Aquatic Bioenvironmental Studies: The Hanford Experience, 1944–1984** by C.D. Becker
40 **Radon in the Environment** by M. Wilkening
41 **Evaluation of Environmental Data for Regulatory and Impact Assessment** by S. Ramamoorthy and E. Baddaloo
42 **Environmental Biotechnology** edited by A. Blazej and V. Privarová
43 **Applied Isotope Hydrogeology** by F.J. Pearson Jr., W. Balderer, H.H. Loosli, B.E. Lehmann, A. Matter, Tj. Peters, H. Schmassmann and A. Gautschi
44 **Highway Pollution** edited by R.S. Hamilton and R.M. Harrison
45 **Freight Transport and the Environment** edited by M. Kroon, R. Smit and J.van Ham
46 **Acidification Research in The Netherlands** edited by G.J. Heij and T. Schneider
47 **Handbook of Radioactive Contamination and Decontamination** by J. Severa and J. Bár
48 **Waste Materials in Construction** edited by J.J.J.M. Goumans, H.A. van der Sloot and Th.G. Aalbers
49 **Statistical Methods in Water Resources** by D.R. Helsel and R.M. Hirsch
50 **Acidification Research: Evaluation and Policy Applications** edited by T.Schneider
51 **Biotechniques for Air Pollution Abatement and Odour Control Policies** edited by A.J. Dragt and J. van Ham
52 **Environmental Science Theory. Concepts and Methods in a One-World, Problem-Oriented Paradigm** by W.T. de Groot
53 **Chemistry and Biology of Water, Air and Soil. Environmental Aspects** edited by J. Tölgyessy
54 **The Removal of Nitrogen Compounds from Wastewater** by B. Halling-Sørensen and S.E. Jørgensen
55 **Environmental Contamination** edited by J.-P. Vernet
56 **The Reclamation of Former Coal Mines and Steelworks** by I.G. Richards, J.P. Palmer and P.A. Barratt
57 **Natural Analogue Studies in the Geological Disposal of Radioactive Wastes** by W. Miller, R. Alexander, N. Chapman, I. McKinley and J. Smellie
58 **Water and Peace in the Middle East** edited by J. Isaac and H. Shuval
59 **Environmental Oriented Electrochemistry** edited by C.A.C. Sequeira
60 **Environmental Aspects of Construction with Waste Materials** edited by J.J.J.M. Goumans, H.A. van der Sloot and Th. G. Aalbers.
61 **Caracterization and Control of Odours and VOC in the Process Industries** edited by S. Vigneron, J. Hermia, J. Chaouki
62 **Nordic Radioecology. The Transfer of Radionuclides through Nordic Ecosystems to Man** edited by H. Dahlgaard
63 **Atmospheric Deposition in Relation to Acidification and Eutrophication** by J.W. Erisman and G.P.J. Draaijers
64 **Acid Rain Research: do we have enough answers?** edited by G.J. Heij and J.W. Erisman
65 **Climate Change Research. Evaluation and Policy Implications** edited by S. Zwerver, R.S.A.R. Rompaey, M.T.J. Kok and M.M. Berk
66 **Global Environmental Biotechnology** edited by D.L. Wise
67 **Municipal Solid Waste Incinerator Residues** by A.J. Chandler, T.T. Eighmy, J. Hartlen, O. Hjelmar, D.S. Kosson, S.E. Sawell, H.A. van der Sloot and J. Vehlow
68 **Freshwater and Estaurine Radioecology** edited by G. Desmet, R.J. Blust, R.N.J. Comans, J.A. Fernandez, J.Hilton, and A. de Bettencourt
69 **Acid Atmospheric Deposition and its Effects on Terrestrial Ecosystems in The Netherlands** edited by G.J. Heij and J.W. Erisman
70 **Harmonization of Leaching/Extraction Tests** edited by H.A. van der Sloot, L. Heasman and Ph. Quevauviller
71 **Waste Materials in Construction. Putting Theory into Practice** edited by J.J.J.M. Goumans, G.J. Senden and H.A. van der Sloot

CONTENTS

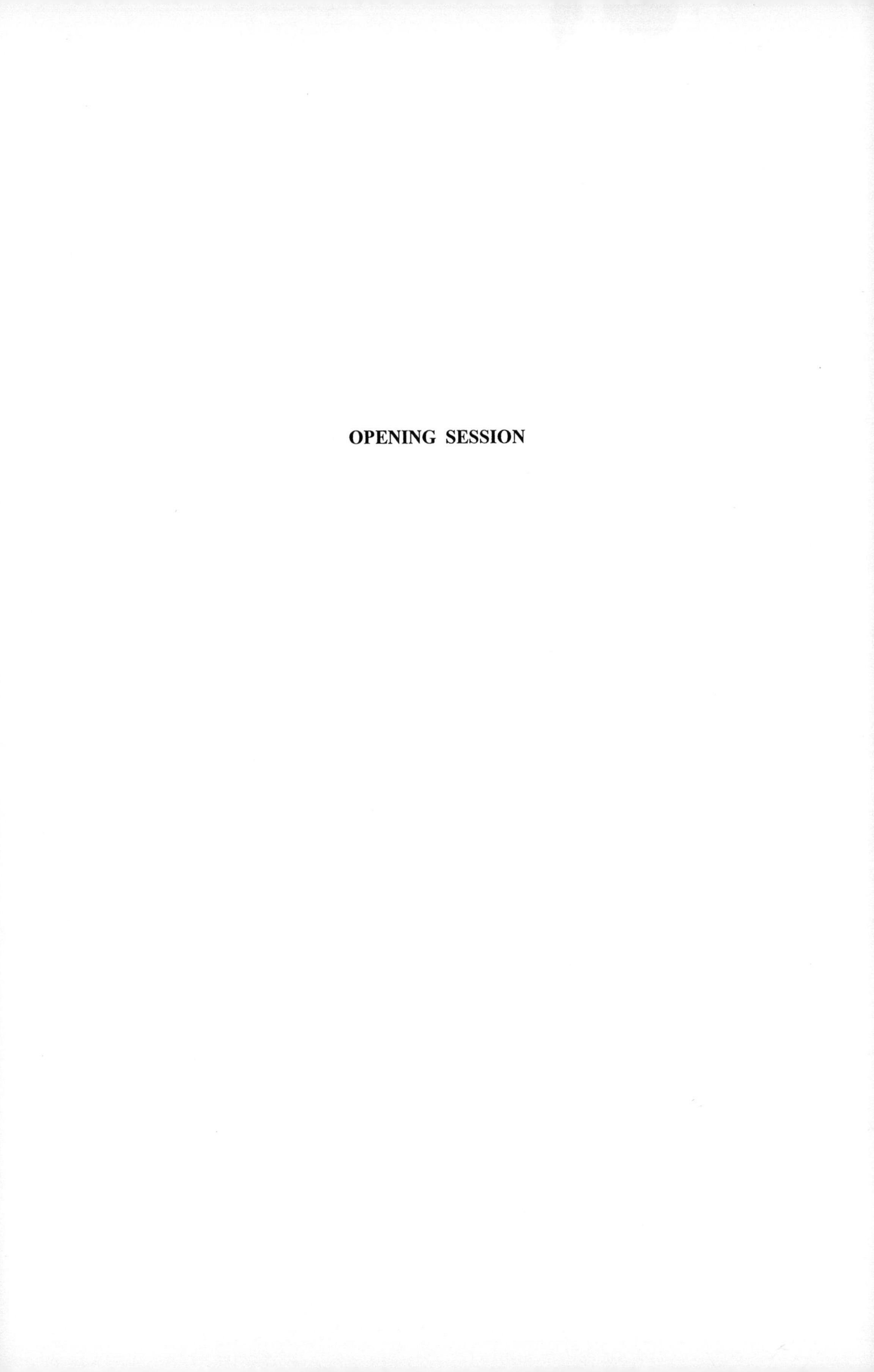

OPENING SESSION

Facing the Air Pollution Agenda for the 21st Century

Ms. M. de Boer
The Netherlands Minister for Housing, Spatial Planning and the Environment
Air Pollution in the 21st Century: priority issues and policy trends.

Your Majesty,
Mr Ambassador,
Ladies and gentlemen,

It is my very great pleasure to welcome you here today to the Fifth US-Dutch Symposium. It is my privilege to welcome Your Majesty. We are very honored that Your Majesty has accepted our invitation to attend the symposium this morning. Your Majesty's interest and presence here today and at the first of these US-Dutch International Symposia in 1982 underlines the importance of developing sustainable solutions to air pollution problems which will challenge us in the next century.
I wish to extend a very cordially welcome to the United States Ambassador to The Netherlands. This symposium is jointly organized by the United States Environmental Protection Agency and my ministry, The Netherlands Ministry of Housing, Spatial Planning and the Environment. The co-operation on environmental policy development between these two organizations is based on a Memorandum of Understanding signed 20 years ago between our two countries. I consider this co-operation to be very valuable. Over the years, contacts have been established at both policy and scientific level. And certainly in the area of air pollution, many fruitful contacts have been made.
The Netherlands wants to play a leading role in environment in the European Union. And so we have involved as many of the other EU Member States as possible in these symposia. I am pleased that the European Commission is represented here today by Mr

Enthoven from the directorate-general which is responsible for environmental matters.

Ladies and gentlemen,

Over the last 30 years, we have made progress in reducing air pollution.

In The Netherlands, a wide range of policy instruments has been formulated which have achieved the desired effect. These are the Air Pollution Act, emission guidelines, covenants with industry and various economic instruments. Since 1975, we have reduced sulphur dioxide emissions by 70% and lead emissions by 90%. Similar successes have been achieved in reducing CFC's and dioxins. However, emission reduction figures for many other substances are more modest. For example, we have only succeeded in reducing hydrocarbons emissions by about 30% compared with 1980. Yet, it should not be forgotten that these reductions have been achieved in spite of continued economic growth. Many air pollution problems persist because much progress in countering these problems is nullified by economic growth and especially growth in traffic. This indicates the nature of the challenge we are facing.

What are we doing this for? This is an easy question to answer. We are doing it for the health of people and of our planet. This is what motivates you as scientists and policy makers, and me as a politician to continue working to improve the quality of the air.

Clean air is a basic condition for health. Health is at issue especially with regard to ozone and particulate matter. Air pollution aggravates respiratory problems, and leads to an increase in sickness absenteeism, an increase in the use of health care services - more medication and hospital admissions - and even to an increase in pre-mature mortality. There are indications that air pollution is one of the contributing factors to the development of chronic obstructive pulmonary disease and an increase in the number of people with these problems, for example, through chronic effects on the respiratory system and through interaction with the body's immune system. In The Netherlands, chronic obstructive pulmonary disease is the third major cause of death after heart disease and cancer, and is in second place in terms of the incidence of disease and disorders. Because of these observations, air pollution is under intensive discussion in the

United States and in Europe.

In addition, clean air is necessary for a healthy environment and for maintaining the biodiversity. Acidification and photochemical air pollution are threatening our forests and vegetation. The presence of various persistent organic pollutants may very well threaten the biodiversity.

Another important target is the prevention of climate change. The international community is more or less in agreement that the increasing concentration of greenhouse gases in the atmosphere since the industrial revolution has led to a gradual increase in the earth's temperature. We do not know yet what the consequences will be. That is the worst of it. But we do know that we cannot allow this to continue. That would be irresponsible, for example, for the countries that are not well protected from the sea, and also for future generations. In terms of the environmental consequences and social implications, the greenhouse problem surpasses all other air quality problems.

But before going into this problem, I want to say something about the immediate problem of air pollution in our cities, and especially the problem of increasing motor traffic.

I hope that during the present Netherlands presidency of the European Union, we can reach agreements on cleaner vehicles and cleaner fuels. This concerns particularly trucks, buses and cars which have an adverse effect on the air quality in our cities and towns.

Technologically, quite a lot can be done to improve the situation. We can use less fuel, we can introduce cleaner fuels, and we can do more about exhaust gases.

In the European Union, we will have to set new targets which industry will have to meet. In the last few months, I have heard from Ministers from countries with large automobile industries that they will support this. I hope that this attitude will gain wide spread acceptance in these countries and that this will lead to effective agreements at the Council of Ministers in June. Industry will then know what requirements they will have to meet in the future, for example in 2005. This will give certainty and also allow enough time to introduce the essential changes and adjustments.

In The Netherlands, we have decided that cars that pollute less will be taxed less.

Taxation will be reduced on cars which run on the most update and cleanest LPG installations. At the same time, we are working on stimulating a change over to LPG for city buses. Because diesel buses have an adverse effect on air quality and cause odor nuisance.

I am pleased that very soon the first electric cars for town use will be introduced to the market. Such technology can also count on favorable financial treatment. These cars are cleaner and therefore it is commendable that the automobile industry is going in this direction, even though slowly. In my opinion, government must be more active in stimulating these developments.

A few words about acidification and smog.
In Europe, we are developing a strategy to reduce acidification and photochemical air pollution. An air emission ceiling for each country in the European Union is being agreed, based on the carrying capacity of the environment and the contribution of each of the Member States to pollution. This is being done so that everywhere in Europe, a high degree of protection is offered against the lowest possible total cost to the European Union as a whole. This strategy is already being applied in the framework of the United Nations Treaty on cross-border pollution. The European Ministers have decided that the summer smog episodes with high ozone peaks must stop in Europe by 2005.
The problem of the greenhouse effect is top of my agenda. I am not dissatisfied with what we have achieved in the last ten years. Thanks to the efforts of scientists, we have a large degree of certainty about the reasons for the increasing greenhouse effect. A number of principle agreements have been reached, and most industrialized countries have started to adapt their energy policy accordingly.
In the area of climate change, there is good co-operation between the United States, The Netherlands and other EU Member States in the ongoing global negotiations. Such co-operation is fruitful as shown by the outcome of the first Conference of the Parties to the Framework Convention on Climate Change in Berlin in 1995. On the other hand: The United States and The Netherlands differ of opinion on the speed and contents of measures to be taken.

The European Union has an important task to take a leading role in global decision making about the greenhouse effect. In December this year, a decision will be made in Kyoto on the target for reducing greenhouse gas emissions after the year 2000.
This is an important step forward because up until now, the target has been only to bring emissions back to 1990 levels.

Views on environment, however, differ greatly between the wealthy industrialized countries and the developing countries. The industrialized countries still consume by far the most but barely muster the political fist and effort to implement essential changes to protect the environment. The developing countries want more prosperity for their people, and rightly so. But in doing so, they must use fossil fuels. I see that the European Union has an important role to play in bringing these two extremes closer together.
At Kyoto, the European Union will pursue a reduction in greenhouse emissions of 15% by 2010 in relation to 1990 for the industrialized countries. A realistic, achievable and necessary goal. The United States also has a requirement that an increasing group of countries pledge themselves to limiting and reducing greenhouse gas emissions. I look forward with interest to the proposal of the United States to commit individually or jointly to a negotiating position similar to that of the European Union.
But we are not pessimistic about this. We all know new fuel systems are very interesting, both technologically and economically. Most oil companies would agree with me on this. The problem is that such investments do not lead to a short-term return. Thus government needs to become involved to enable new technologies to be introduced more quickly. This can be done through development subsidies but also through adjusting fuel prices.
Higher energy prices, however, do not mean that consumers have less money to spend. Such measures need to be introduced in combination with a reduction in income tax. We have already introduced a modest fuel tax in The Netherlands. The European Commission is also proposing such a tax on an European scale. It is a modest but nevertheless an important start.
We are now at the start of a new movement. In the last century economies and societies developed through increasing human productivity. In the next century they must develop

through increasing the productivity of fuel and natural resources.
It is economic growth which is stopping us from getting the problem of air quality under control.
But economic growth is essential. This is certainly so when we see hunger, sickness and poverty in a large part of the world. The wealthy countries are fortunate enough to be able to make choices about the direction of economic growth - whether to consume more and to increase mobility, or whether to act to benefit the quality of the living environment, health and climate!
Economic growth leads to extra income for government, for example, from the sale of natural gas and oil, company taxation, and duties and taxes on fuel. It is logical that we use some of this income on environmental policy. In this way, we can link the cause of the problem directly with its solution. In making this choice we are thinking not only of our own prosperity but also that of the generations to come.

Ladies and gentlemen,
The previous four US-Dutch International Symposia have made a valuable contribution to establishing a sound foundation for environmental policy development.
I trust that this symposium will provide a stimulus for taking up the challenges of the 21st century, and that it will contribute to continuing co-operation between the United States and The Netherlands - on environmental policy development.

In this expectation, I have pleasure in opening this symposium.

Thank you.

Air Quality: The European Perspective

Marius Enthoven, Director-General
DGXI: Environment, Nuclear Safety and Civil Protection European Commission
200 Rue de la Loi, 1049 Brussels, Belgium

1. INTRODUCTION

In the 1970's and 1980's the European Community introduced several pieces of legislation pertaining to air quality objectives and emission reduction (Figure 1).

Figure 1.

Summary of earlier initiatives relating to air quality and emission reduction

- Directive 70/220/EEC relating to measures to be taken against air pollution by emissions from motor vehicles.
- Directive 80/779/EEC on air quality limit values and guide values for sulphur dioxide and suspended particulates.
- Directive 82/884/EEC on a limit value for lead in the air.
- Directive 85/203/EEC on air quality standards for nitrogen dioxide.
- Directive 88/609/EEC on the limitation of emissions of certain pollutants into the air from large combustion plants.

Air quality standards were established for sulphur dioxide and particulate material (black smoke), nitrogen dioxide and lead. Emission controls were introduced for combustion plants: vehicle emission standards were introduced on a voluntary basis as early as 1970. However, it would be fair to say that these initiatives were not particularly structured, systematic or coordinated. The 5th Environmental Action Program which was adopted by the Commission in 1992, set down specific objectives with regard to a range of environmental issues including air quality and the reduction of atmospheric emissions (Figure 2).

Figure 2.

Objectives for acidification/transboundary atmospheric pollution set down in the 5th Environmental Action Program

Acidification/Transboundary Pollution		
	Objective	Targets up to 2000
NOx	No exceeding ever of critical loads and levels	30% reduction
SOx	idem	35% reduction
General VOC's	idem	30% reduction

These objectives have recently been confirmed/reinforced in the Review of the 5th Environmental Action program which was adopted by the Commission in 1996. The 5th Environmental Action Program provides a coherent and consistent framework which serves as the Commission's blueprint for future action/initiatives on the Environment.

With regard to air quality, Community policy operates in the following manner. Air quality objectives are established and agreed at the level of the Community. However, these standards are minimum standards: Member States are fully at liberty to pursue a more aggressive policy and many do so. However, the air quality standards agreed at the level of the Community must be respected everywhere throughout the 15 Member States. These standards are the legal guarantee to the European Citizens that they will have the basic right of clean air to breathe.

Clearly, air quality standards on their own will not automatically produce cleaner air: they must be linked to parallel initiatives to reduce polluting emissions. The Community clearly has a role to play in reducing emissions. However, the approach which we have developed is built on the principle of partnership and burden sharing: the Community's emission reduction policy consists of a number of interlocking and complementary actions at the level of the Community, the Member States, the Regions, the Cities and the individual citizens. Together these interlocking and complementary measures must result as a minimum in the achievement of the Community air quality standards throughout all the Member States (Figure 3).

Figure 3.

How Community, national and local measures
combine to achieve air quality objectives

Community measures + National measures + Local measures

↓

↓

Together must as a minimum achieve

↓

↓

Air quality standards adopted at the level of the Community

It must however, be clear that the existence of a clear set of air quality policy objectives at the level of the Community does not and cannot, imply that the solution to the challenge of air pollution resides exclusively with the Community institutions. The most effective level for action is and will continue to be at the level of the Member States and the cities.

2. NEW AIR QUALITY STANDARDS FOR THE NEW MILLENNIUM

As I mentioned at the start of my speech, air quality standards for a number of key pollutants have existed for many years. However, in January of this year the European Council adopted a new directive on the management and assessment of ambient air quality. This directive will provide the framework for the development of future policy on air quality over the next twenty years (Figure 4).

Figure 4.

Directive on ambient Air Quality

Objectives

- Define and establish objectives for ambient air quality (AAQ).
- Asses AAQ on common basis.
- Obtain information on AAQ and make it available to public.
- Maintain/improve AAQ.

Under the framework directive it is foreseen that new air quality standards will be laid down for a wide range of pollutants and that common procedures will be developed for monitoring and assessing air quality as well as providing comprehensible and timely information to the general public.

The framework directive sets down a very tough time schedule for the Commission to come forward with proposals dealing with individual pollutants (Figure 5).

Figure 5.

Air Quality Pollutants / Calendar					
1)	SO_2	31/12/96	5)	Pb	31/12/96
2)	NO_2	31/12/96	6)	O_3	01/04/98
3)	Fine Part.	31/12/96	7)	Benzene	31/12/97
4)	SPM	31/12/96	8)	CO	31/12/97

By June of this year the Commission will come forward with proposals for SO_2, NO_2, Particulate material and lead. Proposals on CO and Benzene will follow by the end of the year. Ozone will be the subject of a proposal in the Spring of 1998.

In developing these proposals for daughter directives on individual pollutants, the Commission has worked very closely with the experts in the Member States. There has also been very close contact with the EPA. Representatives of the EPA have participated in meetings of the management group responsible for overseeing the development of the daughter directives. The starting point for the development of the air quality standards for the individual pollutants are the recommendations from the WHO Regional Office for Europe.

I understand from reading the international press that our colleagues from the USA are currently involved in a major battle with regard to the revision of the air quality standards for tropospheric ozone and particulate matter. I would like to take this opportunity to wish our colleagues from the EPA the strength and fortitude to see this fight through to a successful conclusion despite the very negative campaign which is being fought against them.

Turning now to recent and forthcoming action on reducing atmospheric emissions.

2. ROAD TRANSPORT EMISSIONS

One of the most important source of emissions is road transport. In June 1996, the Commission agreed a strategy for the control of road transport emissions into the next century. The strategy foresees that by 2010 emissions of the major pollutants from the road transport sector will be reduced by between 65 and 70% as compared to 1990 (Figure 6).

Figure 6.

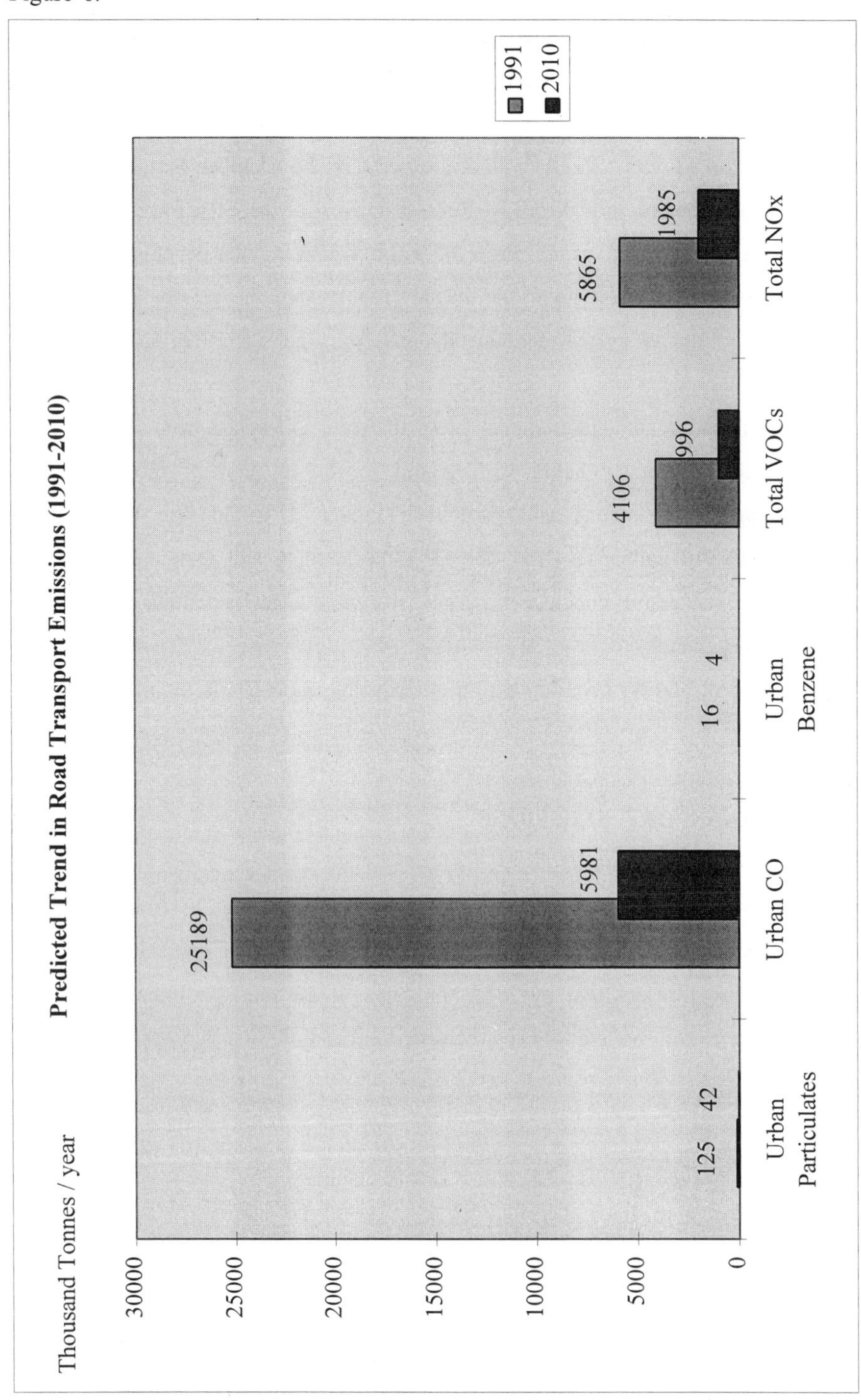
Predicted Trend in Road Transport Emissions (1991-2010)
Thousand Tonnes / year
30000
25000
20000
15000
10000
5000
0
1991
2010
125
42
25189
5981
16
4
4106
996
5865
1985
Urban Particulates
Urban CO
Urban Benzene
Total VOCs
Total NOx

The Commissions policy on road transport emissions is based on a technical assessment carried out together with the European Oil and Automobile manufacturers and known as the Auto Oil program. The Commission has already brought forward three of the legislative proposals necessary to implement its strategy on road transport emissions (Figure 7).

Figure 7.

Drawing on the results of the Auto-Oil Program the Commission has Decided upon a Package of Legislative Proposals:

- a proposal on a moderate reformulation of petrol and diesel fuels;
- proposals to strengthen the existing emission limits for passenger cars, light duty vehicles and heavy duty vehicles;
- a proposal to strengthen legislation on inspection and maintenance.

With regard to passenger car emission standards the Commissions proposals for indicative standards for 2005 are of a comparable severity to the ULEV standards in California (Figure 8).

Figure 8.

Comparison of Proposed EU Emission Standards with USA and Japan (figures are in g/km)

	CO		HCs		NOx		NOx plus HCs		PM
	P	D	P	D	P	D	P	D	D
EU-2000	2.3	0.64	0.2	-	0.15	0.5	-	0.56	0.05
EU 'stage 2' 2005	1.0	0.5	0.1	-	0.08	0.25	-	0.3	0.025
US Tier 2- 2003	1.54	1.54	0.1	0.1	0.15	0.15	-	0.25	-
CARB: ULEV	1.53	1.53	0.03	0.03	0.14	0.14	-	0.17	0.03

Vehicle emissions standards are becoming of increasing importance internationally. It is clearly in everyone's interest that developing economies such as China and India take advantage of the clean technologies as they develop their vehicle fleets. With this objective in mind the Community and the USA in the context of the Transatlantic Business Dialogue have identified the need to move towards commonly agreed test procedures and emission standards at an international level.

4. NON-ROAD MOBILE MACHINERY

On both sides of the Atlantic there has been an increasing recognition that non-road mobile machinery such as bulldozers, tractors, backhoes and graders represent a significant source of atmospheric emissions. The Commission has worked very closely with colleagues from the EPA in Ann Arbor, Michigan to develop compatible approaches to the control of emissions from non-road mobile machinery. In the Community, it is expected that the Council and the Parliament will adopt a directive within the next few months.

5. STATIONARY SOURCES

With regard to emissions from stationary sources, I would highlight a number of recent or ongoing initiatives (Figure 9).

Figure 9.

Recent en planned initiatives on stationary emission sources

- Directive 96/61/EC concerning integrated pollution prevention and control.
- Proposal for a Directive on limitation of emissions of volatile organic compounds due to the use of organic solvents in certain industrial activities.
- Revision of the 1988 large combustion plants Directive.

In 1996, the Community adopted the directive on Integrated Pollution Prevention and Control (IPPC). This directive provides for all major industrial installations to be permitted and to control their emissions according to Best Available Technology.
In 1996, the Commission also put forward a proposal for a directive to control the losses of Volatile Organic Compounds associated with the use of solvents in industrial processes. Finally, during the course of 1997, the Commission will put forward a proposal to revise and strengthen the controls applied to emissions of NOx and SO_2 generated by combustion plants.

6. TRANSBOUNDARY POLLUTION-ACIDIFICATION AND TROPOSPHERIC OZONE

Atmospheric pollution which has a significant transboundary component such as acidification and tropospheric ozone represent a particular challenge requiring a concerted and coherent response. In March 1997, the Commission adopted a strategy for a Community strategy to combat acidification. This strategy includes a number of specific actions in order to reduce emissions of SO_2, NOx and NH_3 (Figure 10).

Figure 10.

The main elements of the proposed Acidification Strategy:

- the establishment of national emission ceilings;
- ratification of the 1994 Protocol to further reduce sulphur emissions;
- a Directive limiting the sulphur content of heavy fuel oils;
- revision of the 1988 large combustion plants Directive;
- the designation of the Baltic Sea and the North Sea as 'SO_2 Emission control Areas'.

One element in the proposed strategy to which I would like to draw your particular attention is the emission of SO_2 and NOx by shipping in the Baltic Sea, the North Sea and the English Channel. In order to reduce the burden of acidifying pollutants in the Community it is important that the Baltic Sea and the North Sea/Channel are recognized as special areas where ships would be required to burn bunker fuel with a sulphur content of 1.5% or less. The Commission will be pushing for these provisions to be introduced into the revision of the IMO Marpol Convention due to be completed in fall of this year. We hope we can count on the support of our US colleagues in this initiative.

With regard to tropospheric ozone, the Commission should before March 1998, come forward with a proposal for a Community strategy to combat this type of pollution. This strategy will clearly be linked to the air quality standards for tropospheric ozone which will be put forward at the same time.

One of the central elements in the Commission's future policy with regard to atmospheric pollution will be an agreement on national emission ceilings for NOx, SO_2 NH_3 and VOC's compatible with the attainment of the environmental targets for acidification and tropospheric ozone. So far we have only completed a provisional analysis of the emission ceilings which will be necessary in order to achieve the interim targets for acidification (Figure 11).

Figure 11.

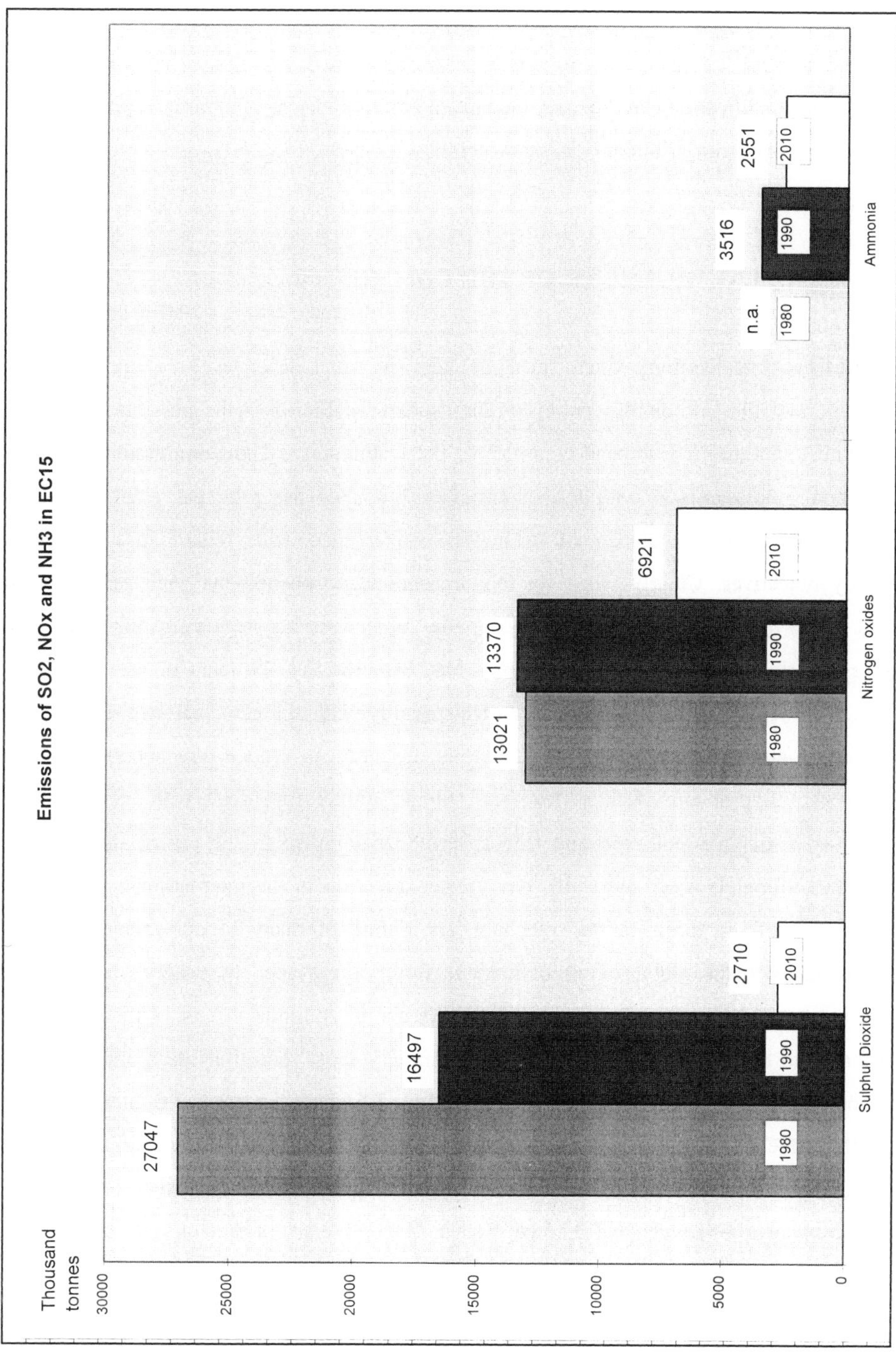

Emissions of SO2, NOx and NH3 in EC15
Thousand tonnes
30000
25000
20000
15000
10000
5000
0
27047
16497
2710
13021
13370
6921
n.a.
3516
2551
1980
1990
2010
Sulphur Dioxide
Nitrogen oxides
Ammonia

You can see on the basis of these provisional figures that the Community's emission reduction policy is extremely ambitious.

When we have completed an integrated assessment of the emission ceilings associated with our acidification and ozone objectives we will bring forward a proposal for a directive establishing national emission ceilings for NOx, SO_2, NH_3 and VOC's.

7. THE ENLARGEMENT OF THE COMMUNITY

A further consideration which must be kept in mind when we are considering the evolution of the Community policy on atmospheric pollution is the possible accession of a number of Eastern European countries to the Community. The Commission is currently engaged in discussions with ten countries and one of the principal elements in these discussions is the challenge of achieving convergence with regard to environmental policy objectives. Clearly when we talk transboundary phenomena such as acidification and ozone the need to develop common policy objectives becomes even more important.

In conclusion, the European Community has, over the last 25 years played a significant role with regard to the improvement of air quality with the establishment of air quality standards and the introduction of numerous measures to reduce atmospheric emissions. However, as problems such as lead and sulphur dioxide are dealt with, new challenges such as particulate pollution and tropospheric ozone move to the top of the agenda. In the next millennium we must also pay greater attention to the international aspects of air pollution both in terms of the transboundary nature of many of our problems but also with regard to the globalization of markets for products such as vehicles and fuels and the need to develop common standards. If we are to continue to ensure that our citizens are able to breathe clean air, we in the Community must share our ideas and experiences in particular with our colleagues from the United States who are confronted with many of the same problems. Meetings such as the present symposium provide the fora within which we can learn from each other and identify opportunities for further collaboration I congratulate the Dutch and United States Governments in having the foresight and commitment to initiate and sustain these meetings over the last twenty years...

Adressing Airborn Environmental Risks: Integration of Science and Policy

W.H. Farland

National Center for Environmental Assessment
Office of Research and Development
U.S. Environmental Protection Agency

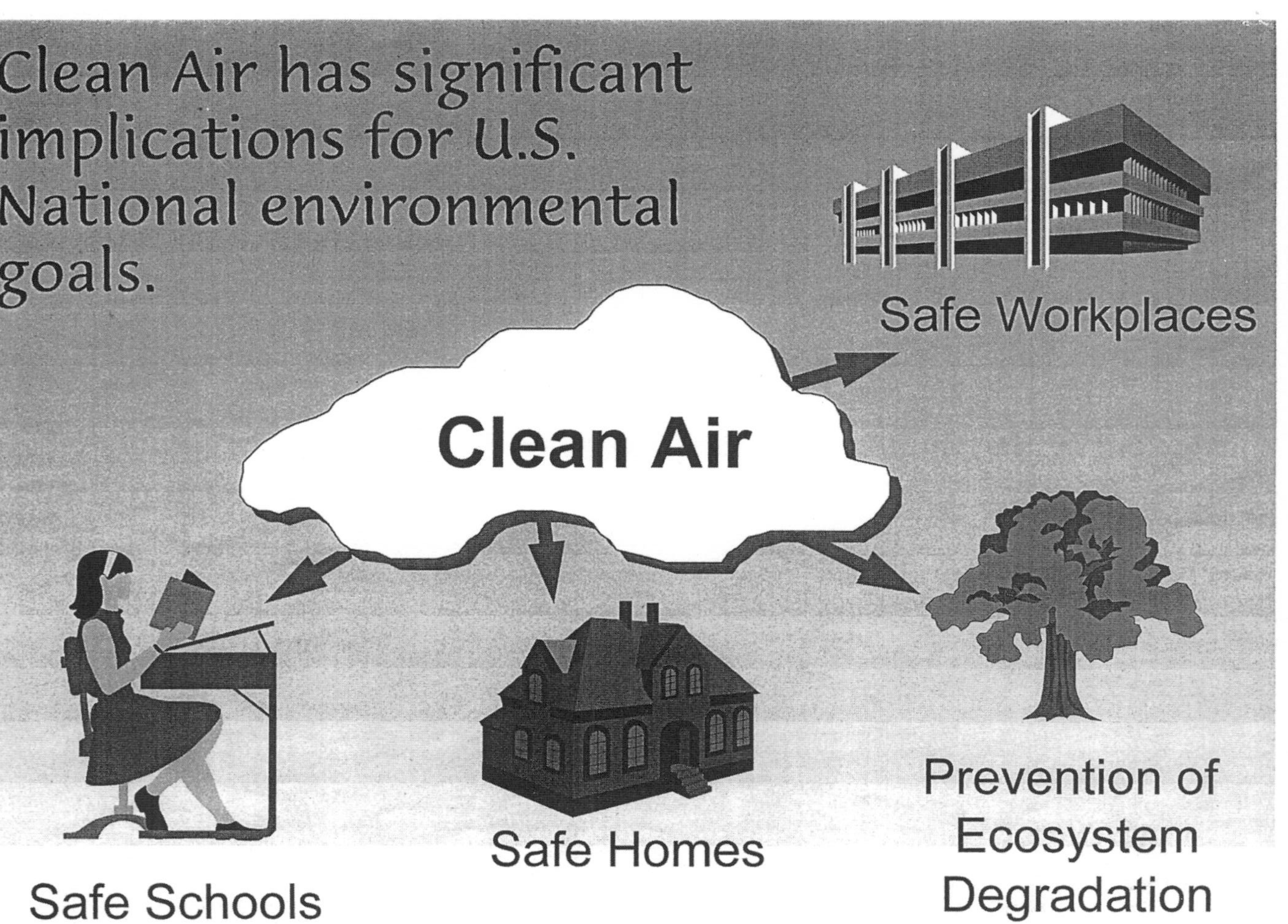
Clean Air has significant implications for U.S. National environmental goals.
Clean Air
Safe Workplaces
Safe Schools
Safe Homes
Prevention of Ecosystem Degradation

What is the nature of the Air Pollution problem?

- U.S. industrial base annually emits 2.7 million tons per year of toxic chemicals into the air.
- Mobile sources emissions are estimated to account for as much as 50% of the national volatile organic compound (VOC) emissions.
- An estimated 50 million people live within 6.5 miles of monitored sites or 1.25 miles of modeled facilities where concentration of one or more chemicals exceeded the health reference level.
- Evidence shows that children or individuals with pre-existing disease may be at particular risk due to increased sensitivity or greater exposure.
- Pollution stresses the health and sustainability c ecosystems.

UNITED STATES · ENVIRONMENTAL PROTECTION AGENCY

Past approaches to Assessment/ Regulation have focused on single pollutant or simple classes such as:

- Air Toxics
  - → Benzene
  - → Lead
  - → PAHs
  - → Mercury
- Particulate Matter
- Tropospheric Ozone
- Indoor Air

Future Direction -- One Atmosphere Approach

Concerns to be addressed:

- Health and Ecological Effects of air pollution mixtures
- Integrated Modeling/Measurement of exposure, air quality characterization, and fate → transport → transformation
- Optimization of multi-pollutant risk management approaches

Example: Impacts of National Ambient Air Quality Standards (NAAQS) for O_3 and $PM_{2.5/10}$ and national reduction programs for acid rain and air toxics

National Research Council (NRC) Perspective

"The quality of risk analysis will improve as the quality of input improves. As we learn more about biology, chemistry, physics, and demography, we can make progressively better assessments of the risks involved. Risk assessment evolves continually, with re-evaluation as new models and data become available."

Science and Judgment in Risk Assessment (NRC, 1994)

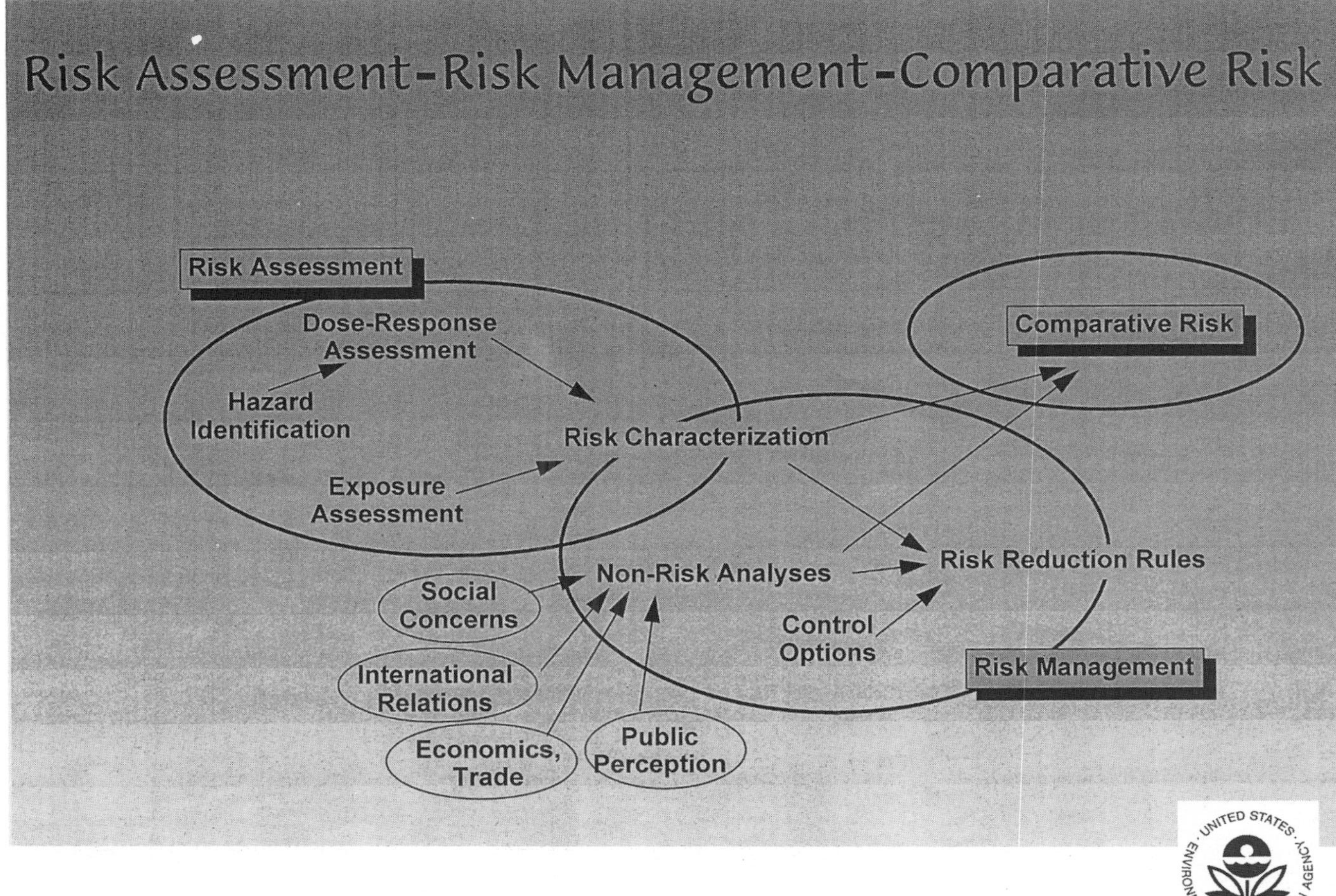

Risk Assessment-Risk Management-Comparative Risk
Risk Assessment
Dose-Response Assessment
Hazard Identification
Exposure Assessment
Risk Characterization
Comparative Risk
Non-Risk Analyses
Risk Reduction Rules
Control Options
Risk Management
Social Concerns
International Relations
Economics, Trade
Public Perception
UNITED STATES · ENVIRONMENTAL PROTECTION AGENCY

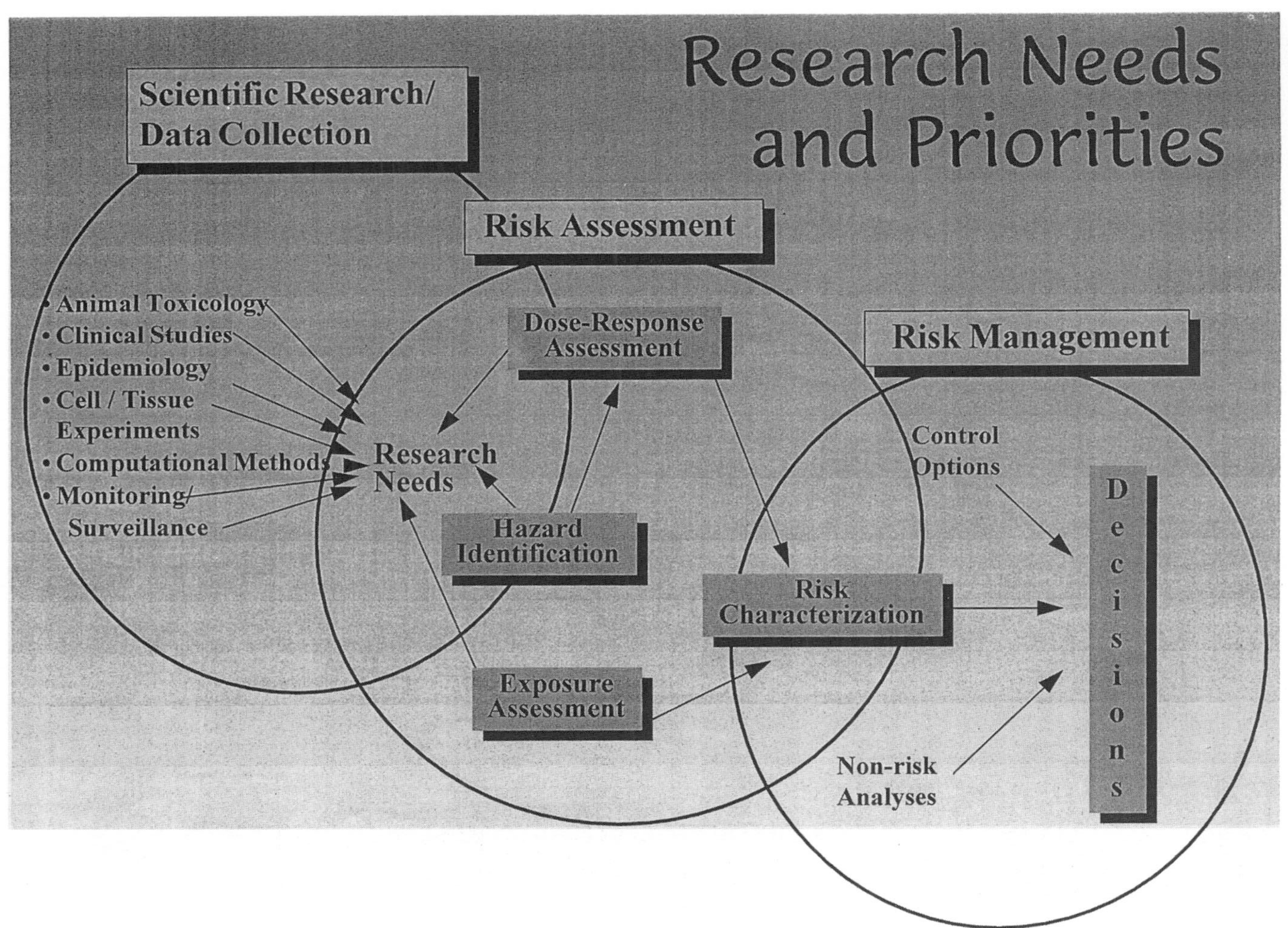

Research Needs and Priorities
Scientific Research/ Data Collection
Risk Assessment
Risk Management
• Animal Toxicology
• Clinical Studies
• Epidemiology
• Cell / Tissue Experiments
• Computational Methods
• Monitoring/ Surveillance
Research Needs
Dose-Response Assessment
Hazard Identification
Exposure Assessment
Risk Characterization
Control Options
Non-risk Analyses
Decisions

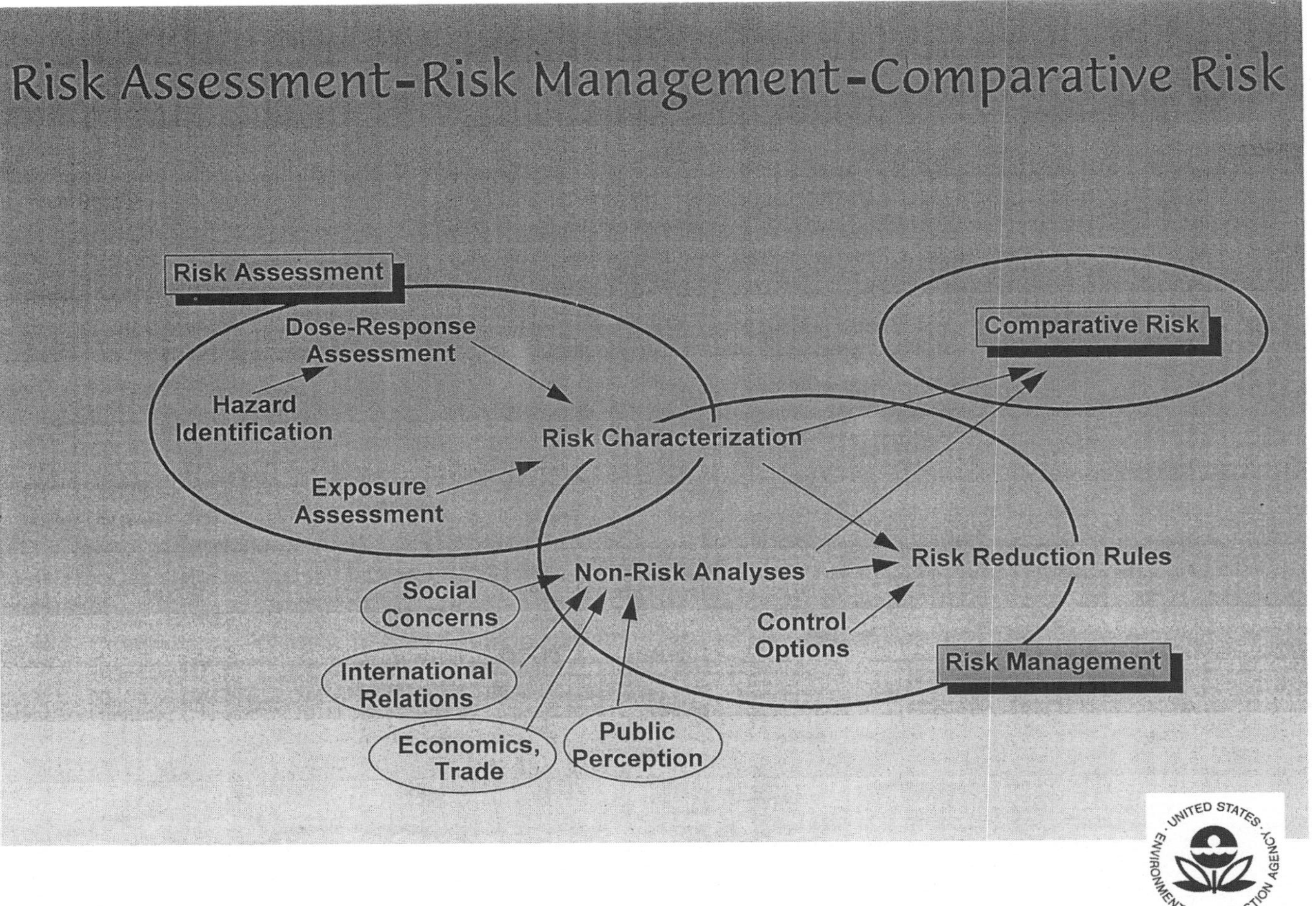
Risk Assessment-Risk Management-Comparative Risk
Risk Assessment
Dose-Response Assessment
Hazard Identification
Risk Characterization
Exposure Assessment
Comparative Risk
Risk Reduction Rules
Non-Risk Analyses
Social Concerns
International Relations
Economics, Trade
Public Perception
Control Options
Risk Management
UNITED STATES · ENVIRONMENTAL PROTECTION AGENCY

Environmental Risk

Fields of Analysis

Risk Assessment

Nature of effects

Potency of agent

Exposure

Population at risk

- Average risk
- High-end risk
- Sensitive groups

Uncertainties of science

Uncertainties of analysis

Identify

Describe

Measure

Risk Management

Social importance of risk

De minimis or acceptable risk

Reduce/not reduce risk

Stringency of reduction

Economics

Priority of concern

Legislative mandates

Legal issues

Risk perception

Evaluate

Decide

Implement

Comparative Risk

Fields of Analysis

Science

Effects Compared

- Health to health
- Ecological to ecological
- Short-term to short-term
- Long-term to long-term
- Voluntary to voluntary
- Involuntary to involuntary

Ranking

- Actuarial *vs.* actuarial
- Forecasted *vs.* forecasted
- Quantitatively crude when estimated

Public Policy

Effects Compared

- Health to ecological
- Short- to long-term
- Voluntary to involuntary
- Global to national
- Economics & trade
- Public perception
- Public values

Ranking

- Actuarial *vs.* forecasted
- Policy based, qualitative

Summary --

- Clean Air means a healthier environment
- Research/Assessment/Management must move towards a "***One Atmosphere***" Approach
- Good communication requires clear messages regarding science, policy and its interface

Priority Research Directions in applying risk assessment to airborn environmental risks

- Variation in susceptibility
- Low dose risk estimation
- Effect of less than lifetime exposure
- Effects of multipathway/multichemical exposure
- Improved exposure assessment to reflect activity patterns, temporal variation, etc.

Uncertainties in applying risk assessment to airborn environmental risks

- Variation in susceptibility
- Low dose risk estimation
- Effect of less than lifetlme exposure
- Effects of multipathway/multichemical exposure
- Improved exposure assessment to reflect activity patterns, temporal variation, etc.

Historical Perspective and Future Outlook

N.D. van Egmond

Substantial progress on atmospheric pollution research in recent decades has brought us to our current understanding of the broad aspects of the problems ranging from the street canyon to globalscale atmospheric chemistry and physics, and from fine particles to greenhouse gas radiative forcing and the interactions between them all. Figure 1 summarizes the various issues, the respective dominant precursors and their mutual interactions.

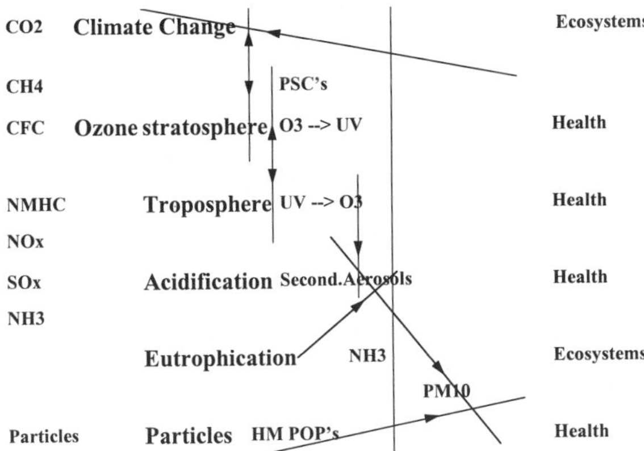

Figure 1. Major air pollution issues, and their interactions.

Current research is to a substantial extent focused on the interactions mentioned, such as:

- the formation of fine particulates / secondary aerosols as a result of tropospheric oxidant formation and oxidation of acidifying compounds,
- the role of these secondary aerosols in both climate change, and effects on health,
- the interactions between stratospheric temperature (climate) changes and ozone depletion by chlorine, bromine and fluorine compounds, including the role of polar stratospheric clouds and,
- the production of ozone in the troposphere in relation to radiative (climate) forcing.

1. CLIMATE CHANGE

Uncertainties in the climate change issue will remain, as illustrated by forthcoming new and plausible theories on the observed global warming. Recently, the theory of Danish researchers on cloud formation, being dependent on cosmic radiation, which in turn is modulated by the radiative activity of the sun, at least explains why the solar sunspot cycle appears so dominant in long-term temperature records. Nevertheless the greenhouse gas radiative forcing theory has gained enough robustness to allow some falsification of the solar activity theory in global warming. According this new theory we can expect stratospheric warming. This, however, contradicts the stratospheric cooling, expected in the greenhouse gas theory, and indeed actually observed / measured in the stratosphere.
All of this illustrates the current position of the state-of-art on climate change. The trends of many observed changes in temperature, precipitation, circulation patterns etc. are in agreement with the theoretical expectations for increased radiative forcing by greenhouse gases.
Simulation studies with climate change models and estimated maximum allowable effects on absolute temperature change, rate of change (to allow migration of ecosystems) and sealevel rise suggest 'safe landing corridors'. Here the global emissions have to stay under 14 GtC for the coming two decades and achieve a 2% per year decrease thereafter. Early action allows more emissions in the long term. Given the expected increase of

CO_2-emissions in developing countries / economies, the industrialized countries at least have to stabilize their current emissions in the coming decades.

2. OZONE LAYER

In situ stratosphere measurements of chlorine and ozone, and intermediate compounds, have confirmed the theory of catalytic ozone depletion by halogenic radicals. Recent research further indicates / confirms:

- substantial to large ozone depletion over both the Antarctic and the Arctic,
- observed downward trends in stratospheric ozone over Europe and the USA,
- indicated and estimated effect from increased UV on health and ecosystems.

Successful international policymaking (Montreal-Copenhagen protocols) has resulted in a large decrease in CFC emissions. From model simulations it can be deduced that the CFC concentrations in the stratosphere have passed their maxima; maximum ozone depletion is expected before the year 2000. Nevertheless, complete ozonelayer recovery will not be reached before 2045.

Remaining issues for the stratospheric ozone issue are:

- the increased production of replacing chemicals like HCFC's. As these compounds are more easily scavenged by OH-radicals in the troposphere, their ozone-depleting potential is reduced but has not yet reached zero; moreover, these compounds contribute to radiative forcing (greenhouse gases).
- CFC's are still produced on an illegal basis,
- methyl bromide is still widely used, especially by developing countries; a phase-out would require an additional ten years.

3. TROPOSPHERIC OZONE

Ozone formation in the troposphere from NOx and Non Methane Volatile Organic Compounds (NMVOC) is well-understood. The Nox concentration is critical for much of the ozone in rural areas, whereas the ozone level in densely populated regions is mainly determined by the NMVOC levels. In many European countries the ozone threshold values are frequently exceeded. European (ECE) policies have resulted in a 3 % NMVOC decrease, and a 10 % Nox decrease between 1990 and 1994.

However, the targets for the 5th European Action Programme and UN-ECE will probably not be met. No consistent downward ozone trend has been detected to date. Further emission cuts, especially for traffic emissions (both NMVOC an Nox), are needed to reduce ozone to acceptable levels. Additional European regulations (Euro3 and 4) are under way. Further Nox reductions will be adopted in the new EU ozone directive and the second (ECE) Nox protocol. This protocol will be based on an integrated approach benefiting from:

- effect-based concepts (e.g. health and critical levels)
- cost estimates of emission reductions and,
- consideration of several pollutants simultaneously.

4. ACIDIFICATION AND EUTROPHICATION

The full process of acidification is now understood to a reasonable extent, both with respect to the atmospheric chemical transformation of primary pollutants like Sox, Nox and NHx into secondary aerosols, the deposition processes under various (micro-meteorological) conditions, and the subsequent soil-chemistry processes (leading to decreased ecosystem vitality). Current research on soil chemistry has indicated that (at least for The Netherlands) the critical levels for the total(potential) acidifying deposition could be relaxed from 1400 moles to about 2000 moles per hectare (average levels for various soil types).

In contrast with this total potential acid deposition, the eutrophying nitrogen flux / deposition into the soil is seen as critical; current (critical) loads cannot be relaxed.

European policies (ECE) have resulted in substantial decreases in SO2 emissions. Nox emissions are decreasing at a much slower rate (see above) and NH3 emissions remain high in certain areas of intense cattle breeding.

Current research is directed both to the role of (acidic) secondary aerosols in the radiative balance in relation to the climate change issue and to the renewed interest in the effects on health of the fine particulate fraction of aerosols.

5. FINE PARTICULATES

Relatively little progress has been made on the issue of fine particulates. After decades of research the problem still is seen as complex and only partly understood. Monitoring of fine particles is under way, but not yet fully operational. The fine particles (< 10 microns or < 2.5 microns, PM10 and PM2.5) appear as divers conglomerates of sulphates, nitrates, polycyclic hydrocarbons, glycerides, dioxins and oxidized hydrocarbons. Epidemiological research has shown for these complexes associations with potential, serious health effects. Cost benefit calculations show that benefits of concentration reductions exceed costs up to a factor of 200. Legislation of PM10 at the European level is under way.

Remaining (scientific) issues:

- the quantification of non-anthropogenic sources; current emission inventories and model simulations underestimate the measured ambient concentrations. Apparently, unidentified sources are missing from the inventory.
- the emission inventory of primary emitted (anthropogenic) particles needs to be improved.
- knowledge of the photo-chemical carbonaceous material in the conglomerates, is unsatisfactory at the moment.
- associations with effects on health remain empirical; the causality and / or plausibility of such associations are still undecided. Given the high variability in the composition of the conglomerates, this is considered to be a serious weakness in the fine particle issue. Given the additional variability in other, highly correlated atmospheric pollutants (in time and space), the statistical relationship, and subsequently deduced evidence of effects on health are weak ('multi-collinearity').

6. THE SCIENTIFIC PERSPECTIVE

Summing up, scientific research, as discussed during the various US-Dutch workshops and symposia over the last decades, has brought insight into the whole spectrum of air pollution-related issues ranging from fine particles to global change (see figure 2). For the issues on the intermediate level such as eutrophication, acidification and ozone in both troposphere and stratosphere, the scientific knowledge achieved is adequate for supporting and legitimating sound policies. However, at the outer ends of the spectrum the extremes meet; complexity and uncertainty remain high for the fine particulate and climate change issues. Both issues have in common that the number of correlated variables within the real ambient world experiment are very high, compared to the variability, which can be observed anyhow. For the fine particulates, the large spatial scale of the concentration (and composition) pattern limits the potential of empirical / statistical analyses, in addition to which the mechanism of human-health impacts (inhalation toxicology) also lack insight in causal mechanisms. This perspective is seen in the convergence of epidemiological (statistical) and toxicological approaches, for instance, by means of exposure via particle concentrators. In this, the test objects (animal or human) are exposed to the real (complex) conglomerates, and not just to simplified single laboratory exposures.

The uncertainty on the global-scale issue given the uniqueness of the global experiment in time, rather than in space, remains high. Here, the strict rules of science apply where new measurement results never prove or confirm the greenhouse gas-theory but, at best, make it more likely. On the other hand, new results (if contradictory to the phenomena as expected) may falsify the theory, paving the way for a new one. Still, the current position, also formally confirmed by the IPCC scientific consensus, is to improve likelihood from current research results.

Both for the global change and the fine particulate issue, policy makers have to learn to live with the principal and inherent uncertainties. Our further scientific research only can make the associations between causes and effects more likely, not proven.

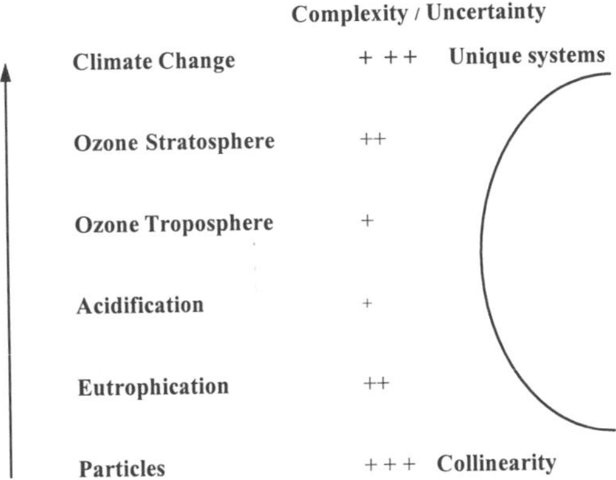

Figure 2. Complexity / Uncertainty for the air pollutant issues

7. THE POLICY / SCIENCE PERSPECTIVE

The focus of research for direct policy support centers, on the following aspects:

- Integrated Assessments and linkage of the various issues as described above, linking these environmental, pollutant-oriented issues to the social / economic actors. Policy measures on CO2, for example, will also bring down the levels of all the other 'fossil fuel' pollutants. Economic sectors will also interact: examples are industrial processes and required transport intensity.
- Burden sharing, partly based on these integrated assessments. As convincingly demonstrated by the UN-ECE process, cost-effective and substantial measures can be taken so that costs can be allocated to the various economies, proportionally and/or justifiably.
- Regionalization, by adopting simultaneous top-down (global scale) and bottom-up (regional) analyses, of both environmental / physical issues, as well as actors in economic sphere.
- Decreasing resource intensity: in the end the remaining - hard to solve - air pollutant problems which are left all depend on the current use of fossil fuels.

More than marginal improvements have to come from alternative options like solar energy, biofuels, and changing lifestyles.

8. CONCLUSIONS

What is the increment of the results presented during the few days of this symposium compared to the preceding decades of scientific research and policy development? The results of the various sessions are summarized, referring partly to the 'Historical Perspective and Future Outlook' presented earlier this week.

Climate change

- Observations made in the stratosphere, troposphere and oceans are in agreement with theoretical (model) expectations.
- The US and Europe differ in their position with respect to the climate change approach. The US aims at long-term budget approaches (tradable emission permits), whereas Europe tends to targets for the short term, i.e. 15 % reduction of greenhouse gas-emissions in 2010. With respect to Joint Implementation, Europe takes the position that this would only refer to industrialized countries.
- The monetary impact of Climate Change is estimated at 1 - 2 % GDP; the former Soviet Union and China are expected to end with net positive global change effects; developing countries will probably be the net losers.
- The 'safe landing emission corridor' for industrialized countries is very narrow, given the minimum restrictions on absolute and relative temperature change and sealevel rise. Stabilization in the coming decades is the upper limit of a development to meet these restrictions.

- Many options for reduction of HFC's (US) and methane (Europe) are in discussion. The reduction of non-CO2 greenhouse gases is complicated by interactions between gases and techniques.

Ozone concentrations

- From both toxicological and epidemiological evidence, the causal relationship between ozone exposure and health effects explicitly has been confirmed. There is a growing concern for chronic health effects.

  The critical ozone levels for health and ecosystems have the same magnitude.
- Background levels of ozone are still increasing on a global scale; this will be a persistent problem in the next century.
- The ozone issue will be linked to pollutants like PM10, according to the idea of an integrated 'one-atmosphere approach'. This will enable a source oriented, cost-effective optimization.

Persistent Organic Pollutants

- Modelling POPs (long range) is complicated and still unsatisfactory; pesticides have been modelled on the European scale; deposition and resuspension (volatizing) processes complicate the modelling efforts.
- For pesticides, interest is shifting from bio-accumulation, via soil contamination to (long-range) atmospheric transport.
- (Very) Long-range transport is a US-Dutch (or CLRTAP) problem; for example Toxaphene is transported from the US Cotton Belt to Europe.

- In European Risk Assessment there is a need for simplified procedures: from allowable concentrations - to allowable emissions, avoiding the complex process of atmospheric transport and transformations.
- The (political) discussion on Tolerable Daily Intake for Dioxins proceeds to lower levels.

  The observed daily intake (mother's milk) decreases as the apparent result of policy measures.

Endocrine Disrupters

- The study on endocrine disrupters is triggered by scientific information on decreasing male fertility / sperma counts.
- There is an evident increase of breast and testicle cancer in the US and Europe; the geographical variation is significant; however the relationship with persistent pollutant exposure remains unclear upto now.
- The relationship of endocrine disrupters with effects on ecosystems is found to be plausible.
- The issue of possible effects of endocrine disrupters should be prioritized with respect to other health issues; the relevance of this issue has not yet been confirmed.

Particulate Matter

- Particulate Matter is definitely (statistically) linked to significant health effects, notwithstanding principal complications with respect to the statistical process (collinearity). The causal relationship has not been confirmed!
- However, it would be prudent to assume that the relationship is causal; this plausibility is supported by toxicological results.

- Research should be directed to source-effect (in addition to dose-effect) relationships; through this way of source apportionment, the variability in the composition of the PM conglomerates can be handled more effectively. This may also be a 'way-out' to the problem of 'mass concentration' versus 'number concentration'.

- Dispersion models (indeed) only partly explain PM10 (and PM 2.5) observations.

Industry

In Europe the policy for control of industrial emissions is focused on the formulation of Best Available Technologies (BAT). This approach is fairly 'demand-driven' and has the consequent advantage of adopting emerging non-BAT technologies into the system.

The US-approach is more supply-driven and based on Verification and Certification of existing, proven technology (Environmental Technical Verification EVT).

Mobile sources

- Technological improvements are not adequate to offset growth trends. This is mainly due to the growth in freight traffic (lorries), which offsets the improvements which have been made with respect to passenger cars.
- The CO2 emissions from traffic form a persistent element of the global change issue.
- The lifecycle analysis for electrical cars questions the environmental effectivity; the energy efficiency indeed increases, but so do the emissions e.g. Lead (Pb) may increase.

Agriculture

- Novel foods are expected to increase the environmental efficiency of food production by a factor of 5 to 20 in terms of energy conversion, waste etc.
- The effects of ozone on agricultural crops is again under study; the production losses are estimated between 10% and 40%.
- Priority should be given to the rural development of the poorest Third World countries. This should be achieved by introducing fertilizers, genetic manipulation (leading to, for example, drought-resistant species) and intensified educational programs. These measures should essentially counteract the ongoing, dramatic urbanization trend.

Energy

- Air pollution problems are almost all related to fossil fuel use. Fossil fuels will become cheap in the future. There will be a shift from oil to gas and from gas to coal.
- Given the remaining uncertainties, CO2 reduction policies should be flexible. More research is needed on solar energy and biomass.

Environment and Economy

- There is general need to base further policies increasingly on full cost pricing, i.e. internalization of environmental costs. This should stimulate a market-driven shift to more environment-effective technologies and production-consumption patterns.
- More general economic / fiscal instruments are seen as being indispensable for further policy development.
- Tradable Emission Permits are seen as a potentially valuable instrument.
- From current cost-benefit analyses, it can be concluded that the benefits are often higher than the costs (long-term / short-term problem).

SESSION A

PARTICULATE MATTER

Human health risks of airborne particles: Historical perspective [1]

Morton Lippmann

Nelson Institute of Environ. Medicine, New York University Medical Center, Tuxedo, NY 10987, USA

1. INTRODUCTION

The historical overview is limited to the past 125 years on the basis that this is the period whose experience with particulate matter (PM) pollution remains relevant to contemporary concerns. A landmark event was the London smog of 9-11 December 1873, which produced a significant excess of human mortality, as well as mortality and pathological changes in show cattle that were on exhibit at that time. A similar smog episode of 5-9 December 1952 caused similar responses, and led to remedial action in the UK during the 1950's and 1960's that greatly reduced both black smoke concentrations and their public health impacts.

In the post coal-smoke era, epidemiological studies in the U.S. showed close associations between the sulfate (SO4=) concentrations in PM and annual mortality, hospital admissions for respiratory and cardiovascular diseases, lost-work time, and respiratory symptoms. With the introduction of monitoring networks for thoracic particulate matter (PM10) in the mid 1980's, many investigators began to show significant correlations between PM10 and daily mortality for a large number of urban areas that differed greatly in climate and in their proportions of other air pollutants. In the relatively few studies

[1] Supported by a National Institute of Environmental Health Sciences Center Program ES-00260.

that also had access to data on the mass concentrations of particles smaller than 2.5 μm in aerodynamic diameter (PM2.5), the associations with health effects were generally comparable to the associations with SO4=, and stronger than those with PM10, leading the EPA Administrator to propose new National Ambient Air Quality Standards (NAAQS) for PM2.5. This paper reviews these historical developments.

2. HIGHLIGHTS OF THE COAL-SMOKE ERA

Quantitative information on adverse health effects associated with particulate matter dates back to the London episode of 1873. A summation of bronchitis mortality during and following the 9-11 December 1873 fog episode was tabulated in the Ministry of Health (1954) report of the 5-9 December 1952 episode. As shown in Table 1, various 19th Century fog episodes produced excesses in bronchitis deaths that were comparable to that reported for the more famous 1952 episode. Also, it is important to note the higher baseline bronchitis mortality for London in the late 19th Century, when the population was below 3 million (compared to about 8 million in 1952), and at a time when that cigarette smoking could not have been a contributory cause.

Table 1

Excess Bronchitis Deaths Associated with Historic London Fogs

Dates of Fog*	Av. Weekly Bronchitis Mortality in Previous 10 Years*	Excess Bronchitis Deaths in Week of Fog and During Succeeding Three Weeks*				Total 4 Week Excess in Bronchitis Deaths
9-11 Dec. 1873		7-13 Dec.	14-20 Dec.	21-27 Dec.	28 Dec.-3 Jan.	
	228	133	424	129	102	788
26-29 Jan. 1880		25-31 Jan.	1-7 Feb.	8-14 Feb.	15-21 Feb.	
	294	258	939	453	167	1817
2-7 Feb. 1882		29 Jan.-4 Feb.	5-11 Feb.	12-18 Feb.	19-25 Feb.	
	357	14	324	186	31	555
21-24 Dec. 1891		20-26 Dec.	27 Dec.-2 Jan.	3-9 Jan.	10-16 Jan.	
	375	35	583	333	437	1388
28-30 Dec. 1892		25-31 Dec.	1-7 Jan.	8-14 Jan.	15-21 Jan.	
	451	-55	208	154	2	309
26 Nov.- 1 Dec. 1948		21-27 Nov.	28 Nov.-4 Dec.	5-11 Dec.	12-18 Dec.	
	65	14	84	33	20	151
5-9 Dec. 1952		1-6 Dec.	7-13 Dec.	14-20 Dec.	21-27 Dec.	
	86	-3	621	308	92	1018

* Source: Ministry of Health (1954). Report # 95 on Public Health and Medical Subjects. MORTALITY AND MORBIDITY DURING THE LONDON FOG OF DECEMBER 1952. London, H.M. Stationery Office.

The first scientific literature citation on the health effects of London smog was also related to the December 1873 episode. Some excerpts from this paper follow:

Excerpts from: The Veterinarian XLVII (JAN. 1874)

THE EFFECTS OF THE LONDON FOG ON CATTLE IN LONDON

Our readers will have heard a good deal already about the terrible disturbance which was caused at the last show of the Smithfield Club by the sudden occurrence of a dense fog during a sharp frost....The atmosphere became dense and pungent on the second day of the Show of 1873....

Before the fog had continued for many hours some of the cattle in the Agricultural Hall evinced palpable signs of distress.

On Tuesday, the first day of the fog, as early as eleven o'clock in the morning several animals were marked as affected with difficult breathing. No abatement of the fog took place during the day; on the contrary, towards evening it became rather worse, and the majority of the cattle in the Hall showed evidence of suffering from its influence.

Sheep and pigs did not however experience any ill effect from the state of the atmosphere either then or during the remaining time of the Show.

During Wednesday night ninety-one cattle were removed from the Hall for slaughter. On Thursday the atmospheric conditions were improved, and no fresh attacks were recorded. On Friday, the air was comparatively clear, and all the animals which remained in the Hall were in good sanitary condition.

The post-mortem appearance were indicative of bronchitis; the mucous membrane of the smaller bronchial tubes was inflamed, and there was also present the lobular congestion and emphysema which belong to that disease.

While we have no air concentration data for the late 19th Century episodes, we do have visual impressions of air quality of that era. Figure 1 is a Gustave Dore woodblock print of 1872 of a downtown London street showing a plume from a coal-fired locomotive, as well as horse drawn traffic and smoke from domestic furnaces. Figures 2 and 3 shows Claude Monet paintings of the Thames River at London in 1871. Figure 4 is an 1884 watercolor by James McNeil Whistler of Piccadilly. Monet was so entranced by the varying colors of the coal smoke that he encountered during his residence in London in the early 1870's that he returned for a 3-year visit at the turn of the Century and produced about 100 oil paintings of London and vicinity. Figure 5 is one of these paintings of the Houses of Parliament.

Figure 1. Ludgate Hill - A Block in the Street. Gustave Dore - 1872.

Figure 2. 1871 painting by Claude Monet entitled 'Boats on the Thames, London'. Private Collection.

Figure 3. 1871 painting by Claude Monet entitled 'Westminster Bridge'. The National Gallery, London.

Figure 4. 1884 watercolor by James McNeil Whister entitled 'Nocturne in Grey and Gold-Picadilly'. National Gallery of Ireland, Dublin.

Figure 5. 1904 painting by Claude Monet entitled 'Houses of Parliament, London, Sun Breaking through the Fog'. Musee d'Orsay, Paris.

As shown in Figure 6, the daily death rate rose rapidly with the onset of the fog on December 5, 1952, and peaked one day after the peak of pollution, as it was indexed by the measured pollutants, i.e., black smoke (BS) and sulfur dioxide (SO2). There was also a rise in hospital emergency bed admissions, which peaked two days after the pollutant peaks. Both the deaths and hospital bed admissions remained elevated for several weeks after the fog lifted (see Tables 1-3). Note also that hospital admissions exhibited declines on Sundays, a finding consistent with the known practices for hospital admissions.

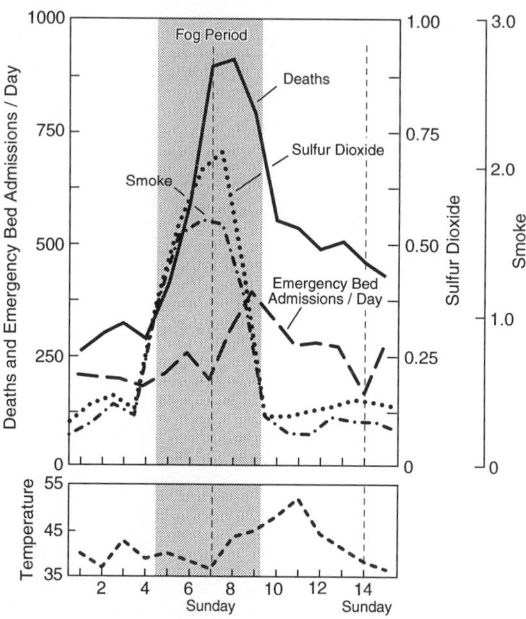

Figure 6. Metropolitan London total mortality and emergency bed admissions during the 1952 pollution episode in relation to black smoke, expressed as mg/m3, and sulfur dioxide, expressed as parts per million by volume.

The Ministry of Health (1954) report attributed an excess of ~ 4,000 deaths from all causes to the exposures during the 1952 episode, and Table 2 shows the deaths, by week, divided according to cause. Deaths peaked in the first full week, and were still above baseline levels two weeks after that. The specific cause with the greatest number of excess deaths over the four weeks was bronchitis (1,156 excess deaths) and it had the greatest relative risk (RR=6.67). The next greatest increase, for heart disease (737 excess deaths), had an RR of only 1.82. The all cause relative risk was somewhat higher (1.96). Table 3 shows that most of the excess deaths occurred in individuals over 55 years of age (2,616 excess deaths over the four weeks), but there was an excess for all age groups beyond 4 weeks of age. Overall, the excess mortality was concentrated among the elderly with pre-existing disease.

It is of particular interest to current concerns that while recent daily mortality studies show much lower absolute risk levels from the much lower peaks in PM pollution, the elevated relative risks among the very young and oldest cohorts and the risk rankings among causes of death are quite similar today to those of December 1952.

Table 2
Greater London Deaths Divided According to Cause - Nov. and Dec. 1952

Cause	Av. Number of deaths in weeks ending 8th, 15th, 22nd, 29th Nov.	Number of deaths registered in week ending 6th Dec.	13th Dec.	20th Dec.	27th Dec.	For weeks ending 6th, 13th, 20th, 27th Dec. Av. No. of deaths	RR
Pulmonary Tuberculosis	17	14	77	37	21	37.25	2.19
Lung Cancer	34	45	69	32	36	45.50	1.34
Heart Disease	226	273	707	389	272	410.25	1.82
High Blood Pressure	14	19	47	36	21	30.75	2.20
Other Circulatory	22	26	46	31	32	33.75	1.53
Influenza	2	2	24	9	6	10.25	5.13
Pneumonia	31	45	168	125	91	107.25	3.46
Bronchitis	51	76	704	396	184	340.00	6.67
Other Respiratory	6	9	52	21	13	23.75	3.96
Ill-defined Causes	20	25	79	35	37	36.50	1.83
All Other Causes	340	411	511	412	316	412.50	1.21
All causes	763	945	2,484	1,523	1,029	1,495	1.96

From: Comm. on Air Pollution: Interim Report, Cmd 9011, London. H.M. Stationery Office (Dec. 1953).

Table 3

Greater London Deaths Divided According to Age - Nov. and Dec. 1952

Dec. Age	Av. Number of deaths in weeks ending 8th, 15th, 22nd, 29th Nov.	Number of deaths registered in week ending 6th Dec.	13th Dec.	20th Dec.	27th Dec.	For weeks ending 6th, 13th, 20th, 27th Av. No. of deaths	RR
Weeks:							
0-4	20	16	28	19	12	18.75	0.94
4-52	8	12	26	15	11	16.00	2.00
Years:							
1-4	7	6	7	13	7	8.25	1.18
5-14	4	4	6	6	2	4.50	1.18
15-24	7	9	7	14	7	9.25	1.32
25-34	11	16	28	17	11	18.00	1.64
35-44	26	36	64	29	34	40.75	1.57
45-54	70	80	204	96	83	115.75	1.65
55-64	133	157	448	251	167	255.75	1.92
65-74	211	254	717	444	258	418.25	1.98
75 and over	266	355	949	619	437	590.00	2.22
All ages	763	945	2,484	1,523	1,029	1,495.00	1.96

From: Comm. on Air Pollution: Interim Report, Cmd 9011, London, H.M. Stationery Office (Dec. 1953).

The Ministry of Health (1954) report also noted that there was a clear association between chronic air pollution and the incidence of bronchitis and other respiratory diseases. The death rate from bronchitis in England and Wales (where coal smoke pollution was very high) was much higher than in other northern European countries (with much lower levels of coal smoke pollution). The very high chronic coal smoke exposure in the U.K., associated with a high prevalence of chronic bronchitis, appears to have created a large pool of individuals susceptible to "harvesting" by an acute pollution episode.

The December 1962 London fog episode was the last to produce a clearly evident acute harvest of excess deaths, albeit a much smaller one than that of December 1952. Commins and Waller (1963) developed a technique to measure H2SO4 in urban air, and made daily measurements of H2SO4 at St. Bartholomew's Hospital in Central London during the 1962 episode. As shown in Figure 7, the airborne H2SO4 rose rapidly during the 1962 episode, with a greater relative increase than that for black smoke (BS).

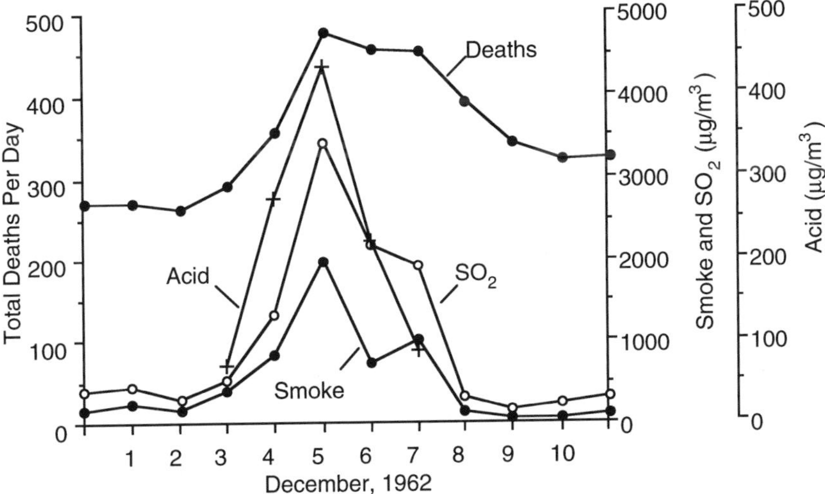

Figure 7. December 1962, London pollution episode.

The U.K. Clean Air Act of 1954 had led to the mandated use of smokeless fuels and, as shown in Figure 8, annual mean smoke levels had declined by 1962, to about one-half of the 1958 level. The annual average SO2 concentrations had not declined by 1962, but dropped off markedly thereafter, along with a further marked decline in BS levels. For the period between 1964 and 1972, the measured levels of H2SO4 followed a similar pattern of decline.

During the later part of the coal smoke era in the U.K., researchers begin to study the associations between long-term daily records of mortality and morbidity and ambient air pollution. In the first major time-series analysis of daily London mortality for the winter of 1958-1959, Martin and Bradley (1960) and Lawther (1963) used the readily available BS and SO2 data. They estimated that both pollutants were associated with excess daily mortality when their concentrations exceeded about 750 μg/m3. However, additional analyses of this data set led to different conclusions. For example, Ware et al. (1981) concluded that there was no demonstrable lower threshold for excess mortality down to the lowest range of observation (BS ≈ 150 μg/m3), as illustrated in Figure 9. Although 150 μg/m3 is now near the upper end of observed concentrations rather than at the lower end, time-series analyses still indicate an increasing slope as concentrations decrease.

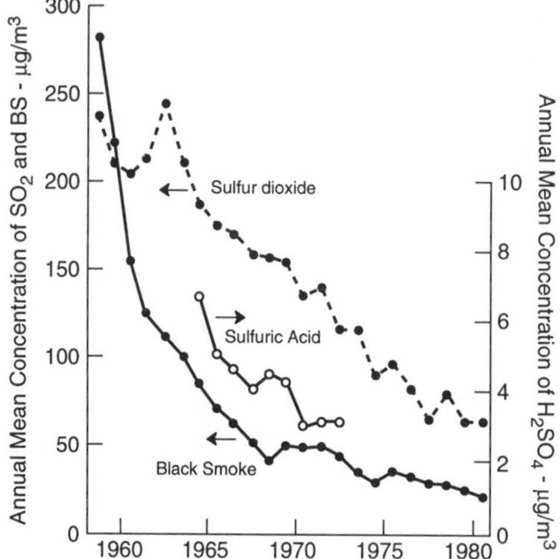

Figure 8. Long-term trends in annual mean atmospheric concentrations of black smoke (BS) and sulfur dioxide (SO2) at seven stations in Greater London, and annual mean concentration of sulfuric acid (H2SO4) at St. Bartholomew's Hospital in Central London.

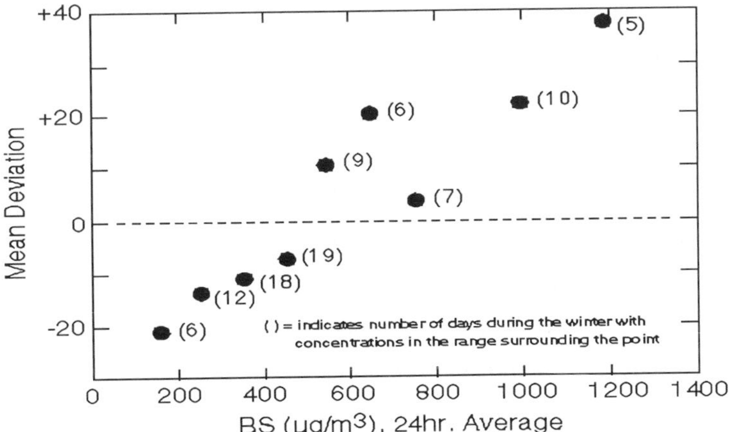

Figure 9. Martin and Bradley (1960) data for winter of 1958-9 in London as summarized by Ware et al. (1981), showing average deviations of daily mortality from 15-day moving average by concentration of black smoke (BS).

In terms of time-series analyses of morbidity, a study by Lawther (1970) reported the daily symptom scores of a panel of patients with chronic bronchitis in relation to the daily concentrations of BS and SO2. As shown in Figure 10, there was a close correspondence between symptom scores and both pollutant indices.

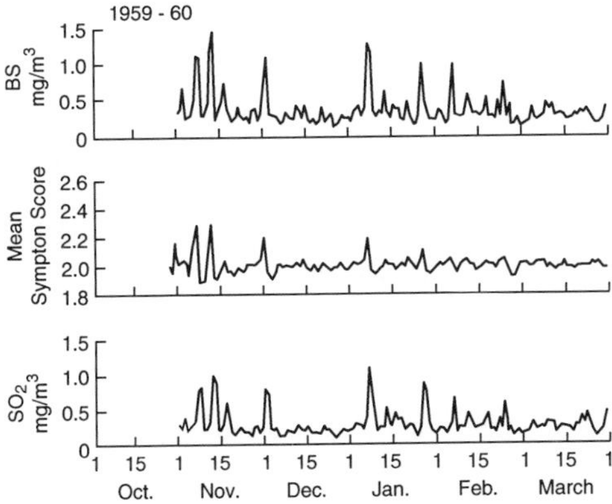

Figure 10. Results from 'diary' studies in London, winter 1959-60, showing day-to-day variations in the illness score for bronchitic subjects together with mean daily concentrations of black smoke (BS) and sulfur dioxide (SO2).

Chronic coal smoke exposure also affected baseline lung function. Holland and Reid (1965) analyzed spirometric data collected on British postal workers in 1965. By that time, pollution levels were well below their peaks, but the postal workers had been exposed out-of-doors for many years when pollution levels were higher. As shown in Figure 11, the London postal workers had lower forced expiratory volumes in one second (FEV1) and peak expiratory flow rates (PEFR) than their country town counterparts. As indicated in Figure 11, the deleterious effects of smoking were accounted for in these analyses. Within each smoking category, the differences between the London and country town means were attributed to pollution on the basis that pollution levels were, on average, twice as high in London as in the country towns.

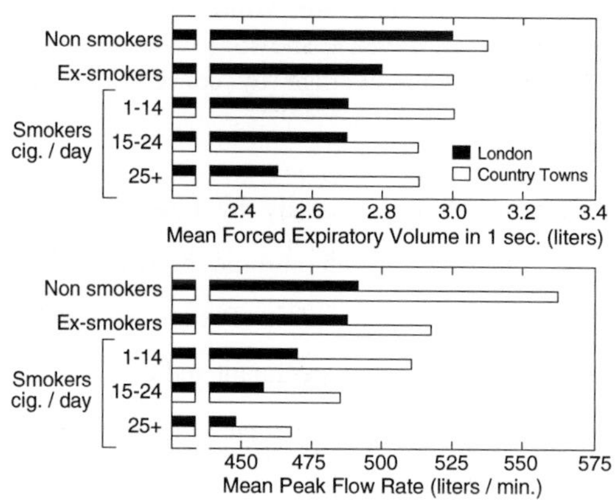

Figure 11. Cross-sectional study of lung function in British Postal Workers in 1965 standardized to age 40. Adapted from: Holland, W.W. and Reid, D.D. (Lancet 1:445-448, 1965) SO2 and smoke levels in the country towns were about half those in London.

The marked reduction in U.K. smoke pollution levels during the 1960's was shown to be associated with a marked reduction in annual mortality in County Boroughs by Chinn et al. (1981). As shown in Table 4, mortality rates in middle-aged and elderly men and women for the 1969-1973 period were no longer associated with an index of smoke pollution. By contrast, for both the 1948-1954 and 1958-1964 periods, the index of smoke exposure correlated strongly with annual mortality rates for both chronic bronchitis and respiratory tract cancers. On the basis of such evidence of improved health status, our U.K. colleagues considered air pollution to be a problem solved, and essentially halted further investigations for the next several decades.

Table 4

Standardized Annual Mortality Rate Regression Coefficients on Smoke* for 64 UK County Boroughs (From: Chinn, S. et al., J. Epid. Comm. Health 35: 174-179, 1981)

Sex	Ages	Mortality in	Cancer of Trachea, Bronchus & Lung	Chronic Bronchitis
Males	45-64	1969-1973	0.07	0.02
		1958-1964	0.53++	0.32+
		1948-1954	0.71+++	0.48+++
	65-74	1969-1973	0.15	-0.06
		1958-1964	0.68+++	0.31
		1948-1954	0.87+++	0.37+
Females	45-64	1969-1973	-0.02	-0.02
		1958-1964	-0.64++	0.33+
		1948-1954	0.49+	0.49++
	65-74	1969-1973	0.07	0.03
		1958-1964	0.25	0.40+
		1948-1954	0.61++	0.31

* Based on index of black smoke pollution 20 years before death of Daly (Br. J. Prev. Soc. Med. 13: 14-27, 1959).
\+ $p < 0.05$
++ $p < 0.01$
+++ $p < 0.001$

Coal smoke pollution affected acute mortality and morbidity in the U.S. and in other countries as well as in the U.K., but there was much less documentation and quantitative analyses prior to the mid 1960's. Among the notable non-U.K. reports are those of Firket (1936) on the December 1930 fog episode in the Meuse Valley in Belgium, and the reports on the October 1948 Donora, PA fog episode.

The December 1930 fog in the Meuse Valley was associated with 60 deaths from a population of ~ 6,000, but the pollutant concentrations were not measured.

In Donora, a valley town of about 10,000 people at a bend in the Monongahela River south of Pittsburgh, there were steel mills, wire mills, zinc works and a sulfuric acid plant along the river bank for the entire length of the town. As reported by Schrenk et al. (1949), a persistent valley fog was associated with 20 excess deaths as well as acute morbidity among 43% of the population. About 10% were reported to have severe effects requiring medical attention. In a ten-year follow-up of the affected population, Ciocco and Thompson (1961) reported greater mortality rates and incidences of heart disease and chronic bronchitis among the residents who had reported acute illness in 1948 in comparison to residents who did not report such illness.

3. EXPERIENCE WITH SULFATE (SO4=) AS AN INDEX OF PM EXPOSURE AND RISK

With the phasing out of bituminous coal as a fuel for domestic heating, the use of the optical density of smoke samples as an index of the health risk associated with ambient particulate matter became increasingly problematic. It has also become clear that total suspended particulate matter (TSP), the standard index of PM pollution in the U.S. prior to 1987 was also far from ideal. Under high wind conditions, gravimetric TSP concentrations are dominated by PM too large to penetrate into the human thorax, even during oral inhalation. Some U.S. investigators chose the sulfate (SO4=) content of TSP samples as an alternate index of PM associated health risk. Because of the nature of its sources, essentially all of the SO4= in the ambient air is on fine particles below 2.5 μm in aerodynamic diameter (PM2.5).

As discussed by Lippmann and Thurston (1996). Sulfate is often a relatively large fraction of PM2.5, it is non-volatile, it is stable on filters used for air sampling, it can be easily extracted from the filters, and it can be accurately analyzed with relatively simple and inexpensive procedures. Furthermore, it generally correlates better with mortality and indices of morbidity in populations than do other frequently measured PM indices, such as TSP, BS, CoH, and PM10. An early example of a health effect closely associated with ambient SO4= concentration, i.e., incidence of protracted respiratory disease among female workers is illustrated in Figure 12.

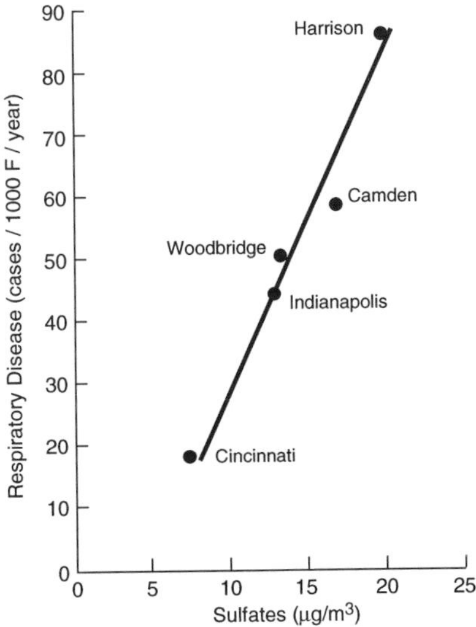

Figure 12. Incidence of respiratory disease lasting more than seven days in women (mean of three years) versus concentration of suspended particulate sulfates in the city air at test sites.
From: Dohan, F.C. and Taylor, E.W. Am. J. Med. Sc., 337-339 (1960).

Lippmann (1989) proposed that SO4= is the best surrogate for H+ exposure, the latter being the most likely causal factor for the observed associations between PM and chronic mortality. The hypothesis, illustrated in Figure 13, was based on the Ozkaynak and Thurston (1987) annual mortality analysis. Their actual data, for SO4= and mortality, are shown in Figure 14. There has been a reluctance on the part of many to accept the SO4=-mortality association as likely to be causal on the basis of ecological analyses such as those of Ozkaynak and Thurston (1987), or the earlier analyses of Lave and Seskin (1977). Many skeptics felt that the results could have been due to confounding by differences among the communities in smoking, occupations, ethnicity, etc. However, later prospective cohort studies, to be discussed in the next section, have reported similar associations. Other examples of associations between human health effects and ambient SO4= concentrations are illustrated in Figures 15-17.

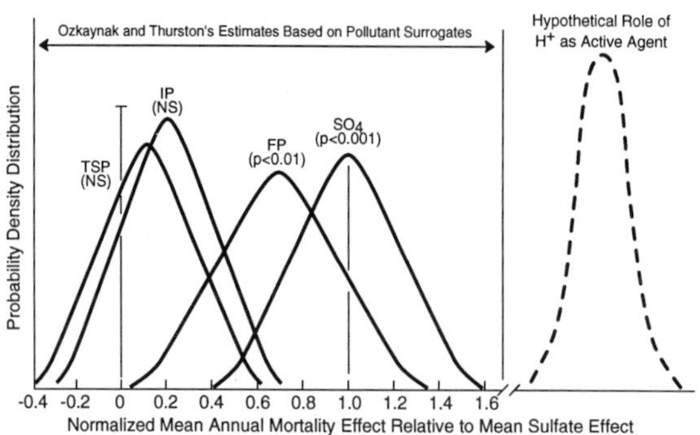

Figure 13. Hypothetical role of H+ in relation to analysis of annual mortality associations in 98 U.S. SMAS by Ozkaynak and Thurston, Risk Anal. 7:449-461 (1987).

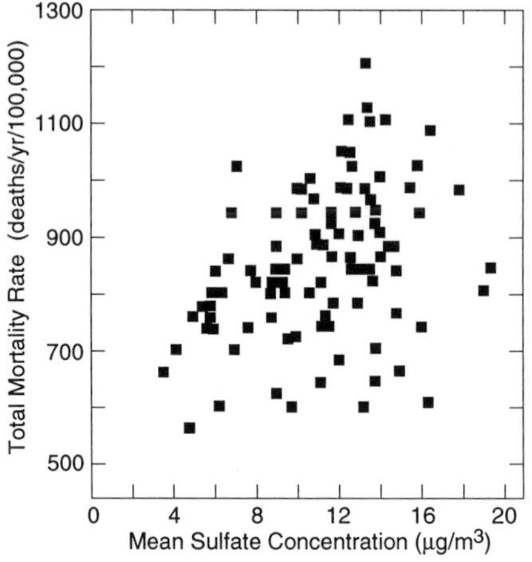

Figure 14. Plot of total mortality rate vs. annual mean SO4= concentration in 98 U.S. SMSA's in 1980. From: Ozkaynak and Thurston, Risk Anal. 7:449-461 (1987).

An important recent study that addressed morbidity in a large number of individuals also adjusted for individual risk factors. Ostro (1990) examined lost-time due to respiratory causes vs. ambient particulate matter. It was based on interview data on a random sample of U.S. households in 25 communities in the national Health Interview Survey (HIS) of 1979-1981. The lost-time was most closely related to SO4=. As shown in Figure 16, the associations, in terms of exposure-response slopes and scatter, were similar in nature to those seen for mortality in Figures 14 and 15.

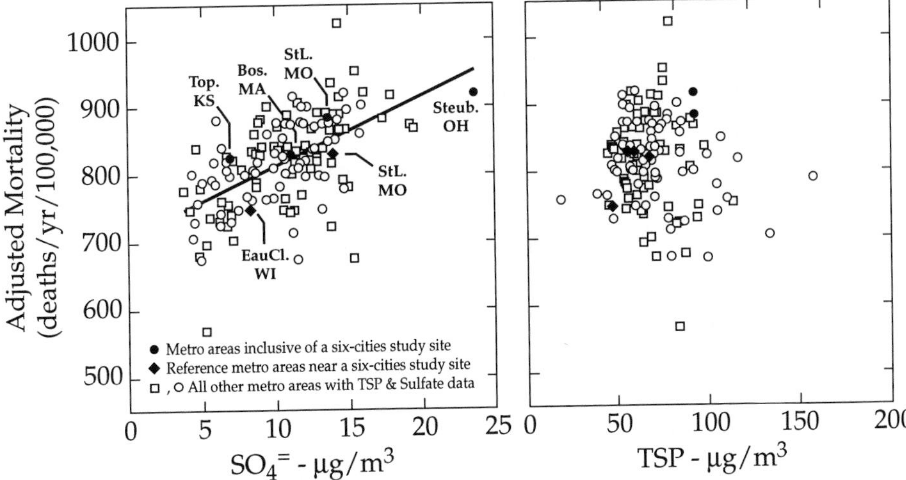

Figure 15. Age-sex-race adjusted mortality rates from Pope et al. (1995), including six communities studied by Dockery et al. (1993). The results are consistent, yet could lead to different interpretations concerning the utility of TSP as a useful measure of risk. Adapted from Figure V-6 of PM Staff Paper (EPA, 1996).

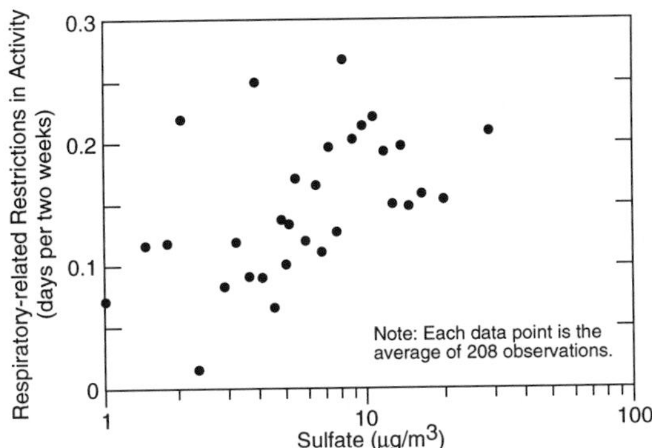

Figure 16. Association between respiratory morbidity and sulfates, controlling for covariates. From: Ostro, B.D., Risk Anal. 10:421-427 (1990).

In terms of hospital admissions, SO4= was significantly associated with hospital admissions in all of the studies for which is was available. The results of one such study, by Burnett et al. (1994), are illustrated in Figure 17.

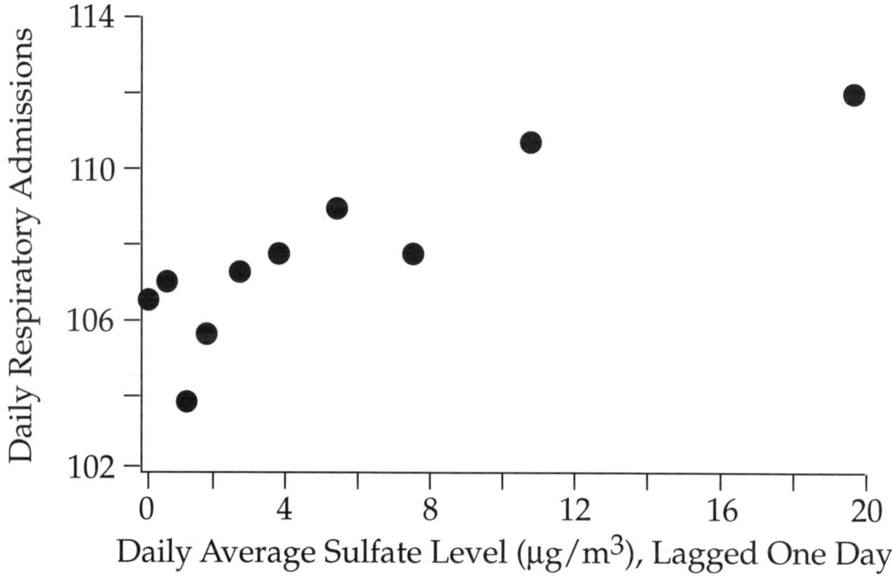

Figure 17. Hospital admissions in Southern Ontario in relation to daily average SO4= concentration in ambient air.

4. RECENT STUDIES BASED ON THORACIC (PM10) AND FINE PARTICLE (PM2.5) MASS CONCENTRATIONS

In the past seven years there has been a great increase in the number of time-series studies of the associations between daily ambient air pollutant concentrations and daily rates of mortality and hospital admissions for respiratory diseases. Also, there is now important information from two prospective cohort studies of annual mortality rates. In terms of morbidity, there has been a rapid growth of the literature showing associations between airborne particle concentrations and exacerbation of asthma, increased symptom rates, decreased respiratory function and restricted activities.

Much of the recent literature was summarized by Pope et al. (1995a). They converted historically measured values for CoH and TSP to estimated levels of PM10, and remarked that very similar coefficients of response were determined in all locations. Table 5 shows Thurston's (1995) independent analysis of acute mortality studies in nine communities with measured PM10 concentrations, including four of the ten studies cited by Pope et al. (1995a). As indicated in this table, the coefficients of response tend to be higher when the PM10 is expressed as a multiple-day average concentration, and lower when other air pollutants are included in multiple-regression analyses. In any case, the results in each city (except for the very small city of Kingston, TN) indicate a statistically significant association.

It is also clear from recent research that the associations between PM10 and daily mortality are not seriously confounded by weather variables or the presence of other criteria pollutants. Figure 18 shows that the calculated relative risks for PM10 are relatively insensitive to the concentrations of SO2, NO2, CO, and O3. The results are also coherent as described by Bates (1992). Figure 19 shows that the relative risks (RR's) for respiratory mortality are greater than for total mortality, and the RR's for the less serious symptoms are higher than those for mortality and hospital admissions.

Table 5

Comparison of Time-Series Study Estimates Total Mortality Relative Risk (RR) for a 100 μg/m3 PM10 Increase

Study Area (Reference)	Measured PM10 Concentrations Mean (μg/m3)	Maximum (μg/m3)	RR for 100 μg/m3	95% CI for 100 μg/m3
1. Utah Valley, UT (Pope et al., 1992)	47	297	1.16*††	(1.10-1.22)
2. St. Louis, MO (Dockery et al., 1992)	28	97	1.16*†	(1.01-1.33)
3. Kingston, TN (Dockery et al., 1992)	30	67	1.17*†	(0.88-1.57)
4. Birmingham, AL (Schwartz, 1993)	48	163	1.11*††	(1.02-1.20)
5. Athens, Greece (Touloumi et al, 1994)	78	306	1.07*†	(1.05-1.09)
			1.03**†	(1.00-1.06)
6. Toronto, Can. (Özkaynak et al., 1994)	40	96	1.07*†	(1.05-1.09)
			1.05**†	(1.03-1.07)
7. Los Angeles, CA (Kinney et al., 1995)	58	177	1.05*†	(1.00-1.11)
			1.04**†	(0.98-1.09)
8. Chicago, IL (Ito et al., 1995)	38	128	1.05**†	(1.01-1.10)
9. Santiago, Chile (Ostro et al., 1995)	115	367	1.08*†	(1.06-1.12)
			1.15*††	(1.08-1.22)

* Single pollutant model (i.e. PM10).
** Multiple pollutant model (i.e. PM10 and other pollutants simultaneously).
† One-day mean PM10 concentration employed.
†† Multiple-day mean PM10 concentration employed.

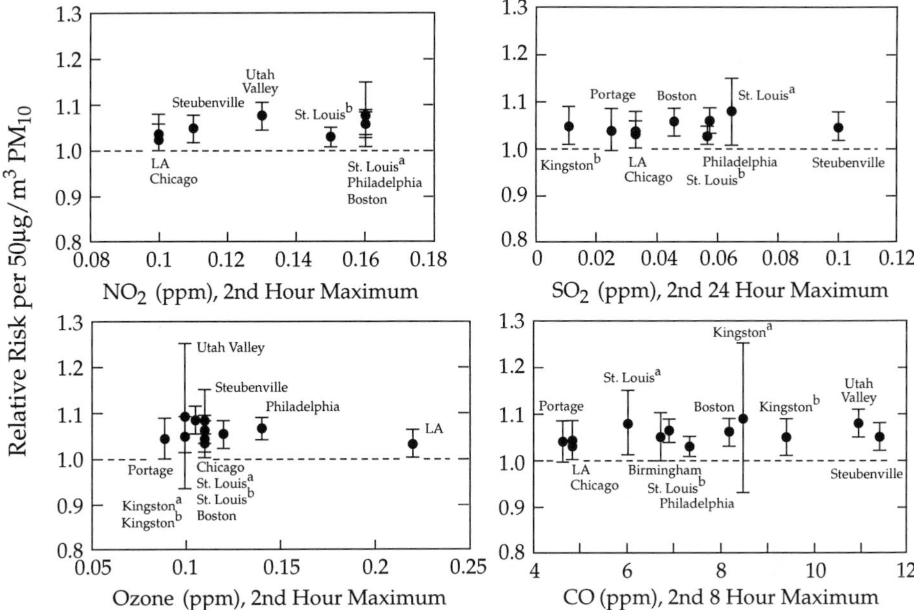

Figure 18. Relationship between RR associated with PM10 and peak daily levels of other criteria pollutants. Adapted from Figure V-3a of PM Staff Paper (EPA, 1996).

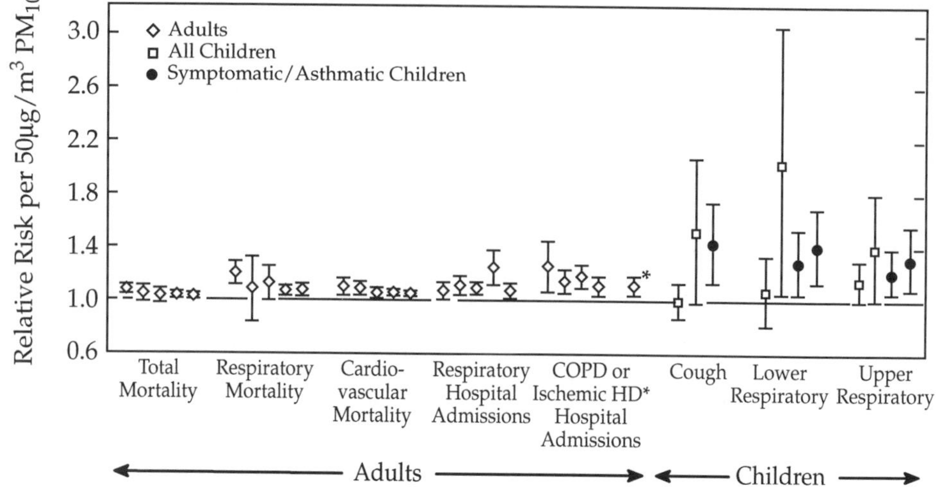

Total, Respiratory, Cardiovascular Mortality
1. Pope et al. (1992) 2. Schwartz (1993) 3. Styer et al. (1995) 4. Ostro et al. (1995a)
5. Ito and Thurston (1996)

Respiratory Hostipal Admissions
1. Schwartz (1995) New Haven, CT 2. Schwartz (1995) Tacoma,WA 3. Schwartz (1996) Spokane, WA
4. Ito and Thurston (1994) Toronto, Canada

COPD or Ischemic HD* Hostipal Admissions
1. Schwartz (1994f) Minneapolis, MN 2. Schwartz (1994c) Birmingham, AL
3. Schwartz (1996) Spokane, WA 4. Schwartz (1994d), Detroit, MI
*5. Schwartz & Morris (1995), Detroit, MI, Ischemic HD

Cough, Lower Respiratory, Upper Respiratory
1. Hoek and Brunekreef (1993) 2. Styer et al. (1994) 3. Pope & Dockery (1992), symptomatic children

Figure 19. Relationships between relative risks per 50 μg/m3 PM10 and health effects. Adapted from Figure V-2 in PM Staff Paper (EPA, 1996).

While there is mounting evidence that excess daily mortality is associated with short-term peaks in PM10 pollution, the public health implications of this evidence are not yet clear. Key questions remain, including:

- which specific components of the fine particle fraction (PM2.5) and coarse particle fraction of PM10 are most influential in producing the responses?
- do the effects of the PM10 depend on co-exposure to irritant vapors, such as ozone, sulfur dioxide, or nitrogen oxides?
- what influences do multiple day pollution episode exposures have on daily

responses and response lags?

- does long-term chronic exposure predispose sensitive individuals being 'harvested' on peak pollution days?
- how much of the excess daily mortality is associated with life-shortening measured in days or weeks vs. months, years, or decades?

The first four questions above are complex, and difficult to answer at this time on the basis of current knowledge. The Discussion section will examine them in greater detail.

The last question above is a critical one in terms of the public health impact of excess daily mortality. If, in fact, the bulk of the excess daily mortality were due to 'harvesting' of terminally ill people who would have died within a few days, then the public health impact would be much less than if it led to prompt mortality among acutely ill persons who, if they did not die then, would have recovered and lived productive lives for years or decades longer. An indirect answer to this question is provided by the results of two recent prospective cohort studies of annual mortality rates in relation to long-term pollutant exposures.

Dockery et al. (1993) reported on a 14-to-16 year mortality follow-up of 8,111 adults in six U.S. cities in relation to average ambient air concentrations of TSP PM2.5, fine particle SO4=, O3, SO2 and NO2. Concentration data for most of these pollutant variables were available for 14-16 years. The mortality rates were adjusted for cigarette smoking, education, body mass index and other influential factors not associated with pollution. The two pollutant variables that best correlated with total mortality (which was mostly attributable to cardiopulmonary mortality) were PM2.5 and SO4=. The overall mortality rate ratios were expressed in terms of the range of air pollutant concentrations in the six cities. The rate-ratios for both PM2.5 and SO4= were 1.26 (1.08-1.47) overall, and 1.37 (1.11-1.68) for cardiopulmonary. The mean life-shortening was in the range of 2-3 years.

Pope et al. (1995b) linked SO4= data from 151 U.S. metropolitan areas in 1980 with individual risk factor on 552,138 adults who resided in these areas when enrolled in a prospective study in 1982, as well as PM2.5 data for 295,223 adults in 50 communities. Deaths were ascertained through December, 1989. The relationships of air pollution to all-cause, lung cancer, and cardiopulmonary mortality was examined using multivariate analysis which controlled for smoking, education, and other risk factors. Particulate air pollution was associated with cardiopulmonary and lung cancer mortality, but not with mortality due to other causes. Adjusted relative risk ratios (and 95% confidence intervals) of all-cause mortality for the most polluted areas compared with the least polluted equaled 1.15 (1.09 to 1.22) and 1.17 (1.09 to 1.26) when using SO4= and PM2.5 respectively. The mean life-shortening in this study was between 1.5 and 2 years. Figure 15 shows the range of values for the adjusted mortality rates in the various communities versus annual average SO4= and TSP concentrations. The results appear, both by inspection and analysis, to be quite similar to those found in the previous studies of Ozkaynak and Thurston (1987) (See Figure 14) and Lave and Seskin (1970). Thus, the results of these earlier studies provide some confirmatory support for the findings of Pope et al. (1995b), while the Pope et al. (1995b) results indicate that the concerns about the credibility of the earlier results, due to their inability to control for potentially confounding factors such as smoking and socioeconomic variables, can be eased.

The Dockery et al. (1993) study had the added strength of data on multiple PM metrics. As shown in Figure 20, the association becomes stronger as the PM metric shifts from TSP to PM10. Within the thoracic fraction (PM10) the association is much stronger to the fine particle component (PM2.5) than for the coarse component. Within the PM2.5 fraction, both the SO4= and non-SO4= fractions correlate very strongly with annual mortality, suggesting a non-specific response to fine particles.

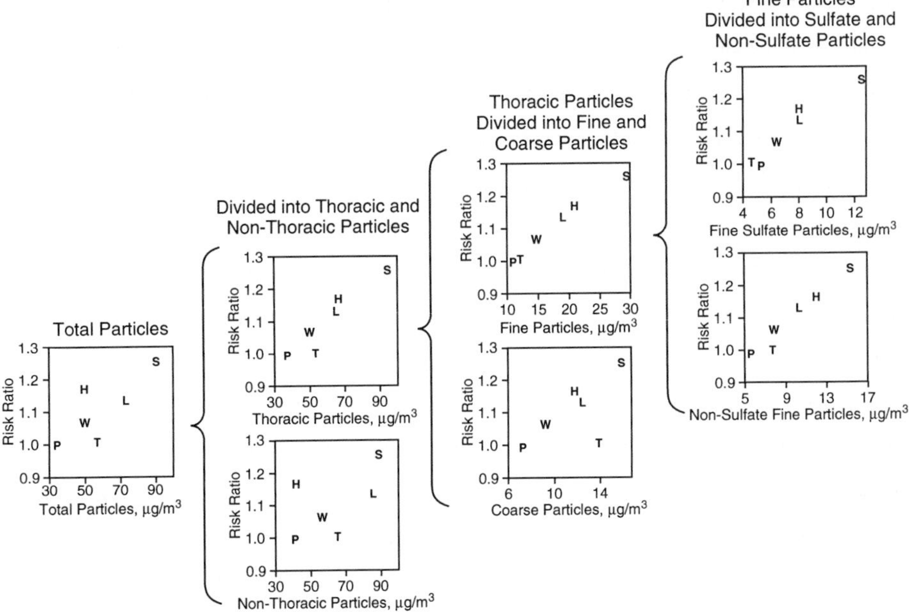

Figure 20. Adjusted relative risks for annual mortality are plotted against each of seven long term average particle indices in the Six City Study, from largest size range (total suspended particulate matter (lower left), through sulfate and nonsulfate fine particle concentrations (upper right). Note that a relatively strong linear relationship is seen for fine particles, and for its sulfate and non-sulfate components. Topeka, which has a substantial coarse particle component of thoracic particle mass, stands apart from the linear relationship between relative risk and thoracic particle concentration. Adapted from Figure V-5 of PM Staff Paper (EPA, 1996).

The importance of the fine particles as a risk factor for subnormal vital capacity in children is illustrated in Figure 21, which shows data collected in the Harvard-Health Canada cross-sectional study of 22 U.S. and Canadian communities (Raizenne et al., 1996). There was a significant association between the percentage of children with FVC < 85% of predicted and fine particle mass concentration, but no apparent association with the coarse component of PM10. Actually, the strongest association observed in this comparison was for the H+ component of the fine particles. Most of the recent epidemiological studies have not had the advantage of available PM2.5, SO4= or H+ data, and have had to rely on PM10 data. A summary of such PM10 epidemiology, in terms of relative risks and 95% confidence intervals, is shown in Figure 19. There is coherence in the data, as defined by Bates (1992), in terms of the relative risk ratings, with mortality risks increasing from total to cardiovascular to respiratory, and with cough and respiratory conditions being more frequent than mortality.

In the absence of any generally accepted mechanistic basis to account for the epidemiological associations between ambient fine particles on the one hand, and mortality, morbidity and functional effects on the other, the causal role of PM remains questionable. However, essentially all attempts to discredit the associations on the basis of the effects being due to other environmental variables that may co-vary with PM have been unsuccessful. As shown in Figure 18, the relative risk for daily mortality in relation to PM10 is remarkably consistent across communities that vary considerably in their peak concentrations of other criteria air pollutants. The possible confounding influence of adjustments to models to account for weather variables has also been found to be minimal (Samet et al., 1997, Pope and Kalkstein, 1996).

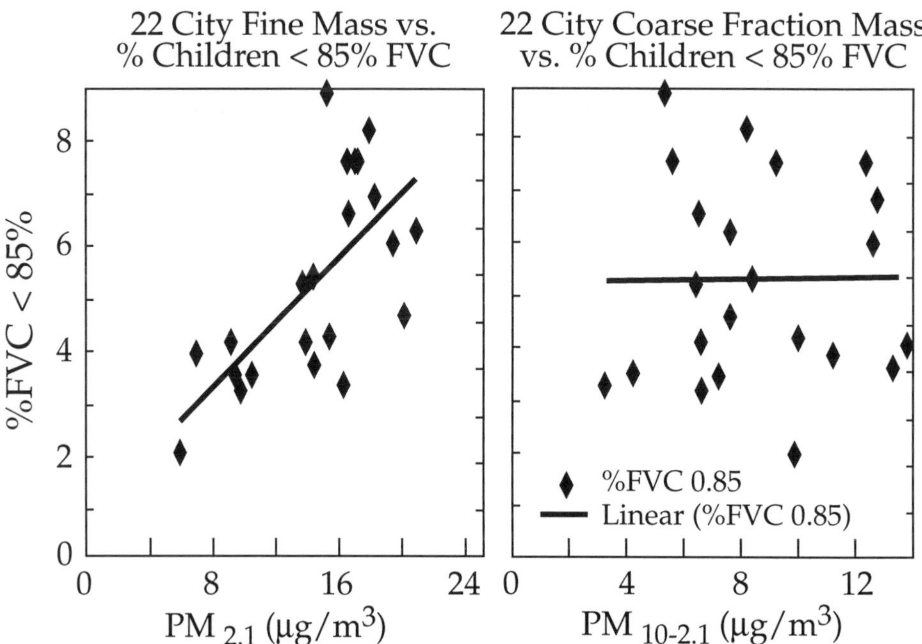

Figure 21. Plot appearing in PM Staff Paper (EPA, 1996). Based on data reported by Raizenne et al. (1996).

The findings of Dockery et al. (1993) and Pope et al. (1995b), in carefully controlled prospective cohort studies, indicating that mean lifespan shortening is of the order of two years implies that many individuals in the population have lives shortened by many years, and that there is excess mortality associated with fine particle exposure greater than that implied by the cumulative results of the time-series studies of daily mortality.

5. DISCUSSION AND CURRENT KNOWLEDGE ON THE HEALTH EFFECTS OF PM

The results of studies in recent years, summarized above, have made it possible to frame the remaining issues in a more coherent and focussed manner.

One key issue is the role of SO4=, and why it consistently correlates with mortality and morbidity as well as, or better than, other metrics of PM pollution. It is extremely unlikely that SO4=, per se, is a causal factor. If it is not, then it must be acting as a surrogate index for one or more other components in the PM mixture.

One possibility is that the effects are really due to the PM2.5 mass, irrespective of particle composition, and that SO4= is a more stable measurement of airborne PM2.5 than is the reported PM2.5 itself. The ambient PM2.5 includes nitrates (primarily ammonium nitrate) and organics formed by photochemical reactions in the atmosphere. There can be considerable volatilization of these species on sampling filters, resulting in negative mass artifacts whose magnitude varies with source strengths and ambient temperature.

Another possibility is that SO4= is serving as a surrogate for H+, a more likely active agent on the basis of the results of controlled exposure studies in humans and animals. The support for this hypothesis is summarized in Table 6. The utility of SO4= as a surrogate for H+, especially for time-series studies in a given region without complex topography, is illustrated in Figure 22, which demonstrates that both H+ and SO4= concentrations are almost the same at two sites sixty miles apart, and that the concentration of both ions tend to rise and fall together.

Table 6

Components of Ambient Air Particulate Matter (PM) that may Account for Some or all of the Effects Associated with PM Exposures

Component	Evidence for Role in Effects	Doubts
Strong Acid (H+)	▸ Statistical associations with health effects in most recent studies for which ambient H+ concentrations were measured ▸ Coherent responses for some health endpoints in human and animal inhalation and in vitro studies at environmentally relevant doses	▸ Similar PM-associated effects observed in locations with low ambient H+ levels ▸ Very limited data base on ambient concentrations
Ultrafine Particles ($D \leq 0.2\ \mu m$)	▸ Much greater potency per unit mass in animal inhalation studies (H+, Teflon, and TiO2 aerosols) than for same materials in larger diameter fine particle aerosols ▸ Concept of 'irritation signalling' in terms of number of particles per unit airway surface	▸ Only one positive study on response in humans ▸ Absence of relevant data base on ambient concentrations
Soluble Transition Metals	▸ Recent animal study evidence of capability to induce lung inflammation	▸ Absence of relevant data on responses in humans ▸ Absence of relevant data on ambient concentrations
Peroxides	▸ Close association in ambient air with SO4= ▸ Strong oxidizing properties	▸ Absence of relevant data on responses in humans or animals ▸ Very limited data base on ambient concentrations

A third possibility is that the causal factor is the number concentration of irritating particles, which would be dominated by the particles in the ultrafine mode (diameters below 50 nm) (Oberdörster et al., 1995). Epidemiologic support for this hypothesis has been provided by Peters et al. (1997), who reported closer associations between peak expiratory flow rates and symptoms in adult asthmatics with particle number concentration than with fine particle mass concentration in Erfurt, Germany.

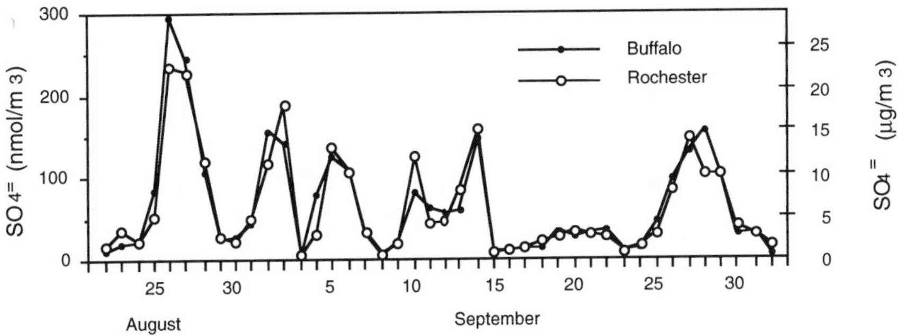

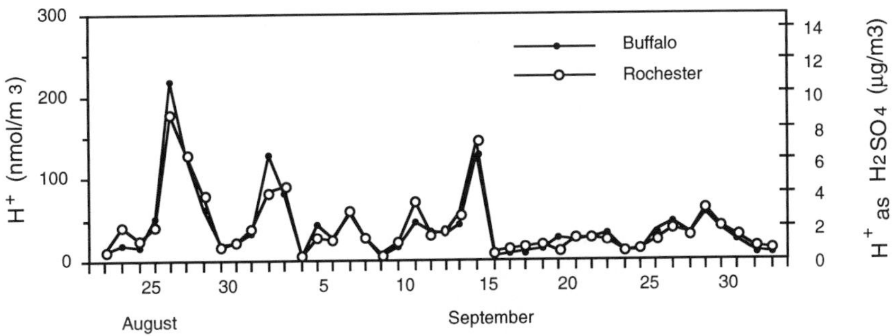

Figure 22. Intercomparison of Rochester, NY and Buffalo, NY sulfate and daily acid aerosol concentrations (August 22 - October 2, 1990).

A fourth possibility is that soluble transition metals in the ambient PM generate sufficient amounts of reactive oxygen species in the respiratory tract airways to cause inflammatory responses and chronic lung damage (Pritchard et al., 1996).

A fifth possibility has been proposed by Friedlander and Yeh (1996), i.e., that reactive chemical species, such as peroxides, are responsible for the health effects associated with fine particles, and that SO4=, being a product of chemical reactions involving hydrogen peroxide, is serving as a surrogate measure of the airborne peroxides.

It is also possible that effects are related to a hybrid of H+ and ultrafines, i.e., acid-coated ultrafine particles. As shown in Figure 23, sulfuric acid coatings on ultrafine zinc oxide particles produce about the same responses as pure sulfuric acid for a given number of equivalent sized particles, yet the coated particles only had one-tenth of the acid content per unit volume of air. Thus, the response may be related to the number of acidic particles that deposit on the lung surfaces rather than the amount of acid deposited. In other words, the total concentration of H+ may be a better surrogate of the active agent than SO4= or PM2.5, but it still is a crude index for the number concentration of irritant particles. Amdur and Chen (1989) suggested that number concentration was important for sulfuric acid aerosol, and Hattis et al. (1987, 1990) gave the concept a name, i.e., 'irritation signalling'. Research of Chen et al. (1995) indicate that acid-coated particles much smaller than those discussed by Hattis et al. (1987, 1990) were capable of producing lung responses.

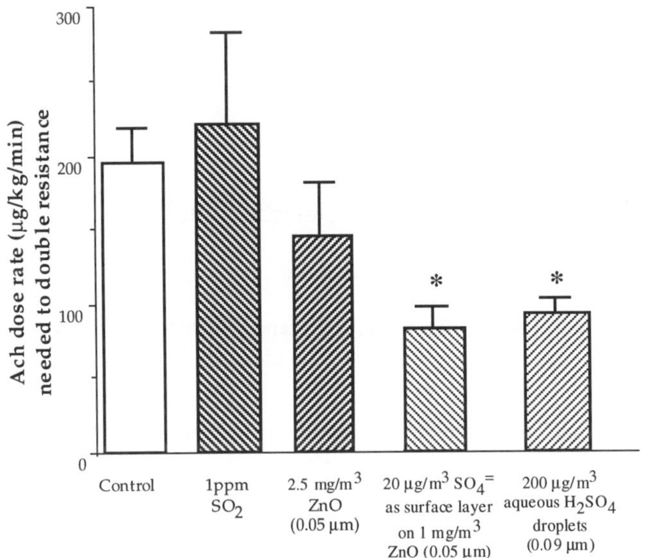

Figure 23. Dose of intravenously infused acetylcholine required to double airway resistance in guinea pigs from baseline levels 2 hours after a 1 hour inhalation exposure. Values are mean ± S.E. The asterisks indicate reductions significant at $p < 0.05$. For the aerosols, the median particle diameters are indicated in parentheses.

If the number concentration of acid-coated particles is the most relevant index of the active agent in ambient PM, then new sampling techniques will be needed to characterize ambient air concentrations and personal exposures.

Other components of the ambient ultrafine aerosol have not been well characterized either, and they may also be important health stressors. One class is the volatile trace metals (such as As, Cd, Cu, Pb, Zn) which condense as ultrafine particles in the effluent airstream of fossil fuel combusters (Amdur et al., 1986) and are inefficiently captured by air cleaners for fly ash collection. Another class is the ultrafine organics from atmospheric photochemical reaction sequences.

Any remaining inconsistency between the epidemiological findings and the results of the controlled exposure studies may be explicable on the basis that the relatively rare individuals who respond in the epidemiological populations are an especially responsive subset of the overall population, and the low probability that such sensitive individuals would be included in the controlled exposure studies in the laboratory. An alternative hypothesis is that the controlled exposure atmospheres have not contained the highly toxic components or ultrafine particle sizes that may be present in ambient atmospheres.

In summary, excess daily mortality and morbidity have been related to ambient pollution at current levels in many communities in the U.S. and around the world using available pollutant concentration data. However, it is not at all clear whether any of the pollutant indices used are causally related to the health effects or, if none of them are, which is the best index or surrogate measure of the causal factor(s). This gap can best be addressed by analyses of pollutant associations with mortality and morbidity in locations where a number of different pollutant metrics are available simultaneously, using analytic methods not dependent on arbitrary model assumptions.

6. POLICY IMPLICATIONS

While more research is needed on causal factors for the excess mortality and morbidity associated with PM in ambient air, and on the characterization of susceptibility factors, responsible public health authorities cannot wait for the completion and peer review of this research. It is already clear that the evidence for adverse health effects attributable to PM challenges the conventional paradigm used for setting ambient air standards and guidelines, i.e., that a threshold for adversity can be identified, and a margin of safety can be applied. Excess mortality is clearly an adverse effect, and the epidemiological evidence is consistent with a linear non-threshold response for the population as a whole. A revision of the Air Quality Guidelines of the World Health Organization-Europe (WHO-EURO) is currently nearing completion. The Working Group of WHO-EURO on PM, at meetings in October 1994 and October 1996 in Bilthoven, The Netherlands, determined that it could not recommend a PM Guideline. Instead, it prepared a tabular presentation of the estimated changes in daily average PM concentrations needed to produce specific percentage changes in: 1) daily mortality; 2) hospital admissions for respiratory conditions; 3) bronchodilator use among asthmatics; 4) symptom exacerbation among asthmatics; and 5) peak expiratory flow. The concentrations needed to produce these changes were expressed in PM10 for all five response categories. For mortality and hospital admissions, they were also expressed in terms of PM2.5 and SO4=. Using this guidance, each national or local authority setting air quality standards can decide how much adversity is acceptable for its population. Making such a choice is indeed a challenge.

In the U.S., the EPA Administrator proposed revised PM NAAQS on November 26, 1996 in recognition of the inadequate public health protection provided by enforcement of the 1987 NAAQS for PM10. For PM10, the 50 μg/m3 annual average would be retained without change, and the 24-hr PM10 of 150 μg/m3 would be relaxed by applying it only to the 98th% value (8th highest in each year) rather than to the 4th highest over 3 yrs. These PM10 standards would be supplemented by the creation of new PM2.5 standards. The annual average PM2.5 would be 15 μg/m3, and the 24 hour PM2.5 of 50 μg/m3 would apply to the 98th% value. It is only by implementing the new PM2.5 NAAQS that the degree of public health protection for ambient air PM would be substantially advanced, and then only in the eastern U.S. and in some large cities in the west where fine particles are major %'s of PM10. In these locations the greatest reductions in adverse health effects are feasible and most cost-effective.

In my view, the proposed PM NAAQS are not too strict. In terms of its introduction of a more relevant index of exposure and a modest degree of greater public health protection, it represents a prudent judgment call by the Administrator. These NAAQS may not be strict enough to fully protect public health, but there remain significant knowledge gaps on both exposures and the nature and extent of the effects that make the need for more restrictive NAAQS difficult to justify at this time. It is essential that adequate research resources be committed to filling these gaps before the next round of NAAQS revisions early in the next decade. The costs of the research, while substantial (on the order of \$50x106 per year), are quite small in comparison to the health benefits resulting from exposure reductions resulting from the controls.

REFERENCES

1. Amdur, M.O. and Chen, L.C., Environ. Health Perspect. 79 (1989) 147.
2. Amdur, M.O., Sarofim, A.F., Neville, M., Quann, R.J., McCarthy, J.F., Elliott, J.F., Lam, H.F., Rogers, A.E., and Conner, M.W., Environ. Sci. Technol. 20 (1986) 138.
3. Bates, D.V., Environ. Res. 59 (1992) 336.
4. Burnett, R.T., Dales, R.E., Raizenne, M.E., Krewski, D., Summers, P.W., Roberts, G.R., Raad-Young, M., Dann, T., and Brook, J., Environ. Res. 65 (1994) 172.
5. Chen, L.C., Wu, C.Y., Qu, Q.S., and Schlesinger, R.B. Inhal. Toxicol. 7 (1995) 577.
6. Chinn, S., Florey, C. duV., Baldwin, I.G., and Gorgol, M., J. Epidemiol. Commun. Health 35 (1981) 174.
7. Ciocco, A. and Thompson, D.J., Am. J. Public Health 51 (1961) 155.
8. Commins, B.T. and Waller, R.E., Analyst 88 (1963) 364.
9. Dockery, D.W., Pope, III, C.A., Xu, X., Spengler, J.D., Ware, J.H., Fay, M.E., Ferris, Jr., B.G., and Speizer, F.E., N. Engl. J. Med. 329 (1993) 1753.
10. Firket, J., Trans. Faraday Soc. 32 (1936) 1191.
11. Friedlander, S.K. and Yeh, E.K. (1996), The submicron atmospheric aerosol as a carrier of reactive chemical species: Case of peroxides. pp. 4-122 to 4-135 in: Proceedings of 2nd Colloquium on Particulate Air Pollution and Human Health, Lee J. and Phalen, R. (eds.). Univ. of CA - Air Pollution Health Effects Laboratory (Dec. 1996).
12. Hattis, D., Wasson, J.M., Page, G.S., Stern, B., and Franklin, C.A., J. Air Pollut. Control Assoc. 37 (1987) 1060.
13. Hattis, D.S., Abdollahzadeh, S., and Franklin, C.A., J. Air Waste Manage. Assoc.

40 (1990) 322.

14. Holland, W.W. and Reid, D.D., Lancet 1 (1965) 445.
15. Lave, L.B. and Seskin, E.P., Science 169 (1970) 723.
16. Lawther, P.J., J. Inst. Fuel 36 (1963) 341.
17. Lawther, P.J., Waller, R.E., and Henderson, M., Thorax 25 (1970) 525.
18. Lippmann, M. and Thurston, G.D., J. Expos. Anal. Environ. Epidemiol. 6 (1996) 123.
19. Lippmann, M., Environ. Health Perspect. 79 (1989) 3.
20. Martin, A.E. and Bradley, W.H., Mon. Bull. Minist. Health-Public Health Lab. Service 19 (1960) 57.
21. Ministry of Health (1954), 'Mortality and Morbidity During the London Fog of December 1952', Her Majesty's Stationary Office, London.
22. Oberdörster, G., Gelein, R.M., Ferin, J., and Weiss, B., Inhal. Toxicol. 7 (1995) 111.
23. Ostro, B.D., Risk Anal. 10 (1990) 421.
24. Ozkaynak, H. and Thurston, G.D., Risk Anal. 7 (1987) 449.
25. Peters, A., Wichmann, E., Tuch, T., Heinrich, J., and Heyder, J., Am. J. Respir. Crit. Care Med. (In press, 1997).
26. Pope, C.A. III and Kalkstein, L.S., Environ. Health Perspect. 104 (1996) 414.
27. Pope, C.A., III, Dockery, D.W., and Schwartz, J., Inhal. Toxicol. 7 (1995a), 1.
28. Pope, C.A., III, Thun, M.J., Namboodiri, M.M., Dockery, D.W., Evans, J.S., Speizer, F.E., and Heath, Jr., C.W., Am. J. Respir. Crit. Care Med. 151 (1995b) 669.
29. Pritchard, R.J., Ghio, A.J., Lehmann, J.R., Winsett, D.W., Tepper, J.S., Park, P., Gilmour, M.I., Dreher, K.L., and Costa, D.L., Inhal. Toxicol. 8 (1996) 457.
30. Raizenne, M., Neas, L., Damokosh, A.I., Dockery, D.W., Spengler, J.D., Koutrakis, P., Ware, J.H., and Speizer, F.E., Environ. Health Perspect. 104 (1996) 506.
31. Samet, J.M., Zeger, S.L., Kelsall, J.E., Xu, J., and Kalkstein, L.S. (1997), Air pollution, weather, and mortality in Philadelphia 1973-1988. Report on Phase 1.B of the Particle Epidemiology Project, Health Effects Institute, Cambridge, MA 02139.
32. Schrenk, H.H., Heimann, H., Clayton, G.D., and Gafater, W.M. (1949), 'Air Pollution in Donora, Pennsylvania', Public Health Bull. No. 306, U.S. Government Printing Office, Washington, DC.

33. Thurston, G.D. (1995), Personal communication of table prepared for draft EPA Criteria Document on Particulate Matter.

34. Ware, J.H., Thibodeau, L.A., Speizer, F.E., Colome, S., and Ferris, B.G., Jr. Environ. Health Perspect. 41 (1981) 255.

Fine and coarse particles: chemical and physical properties important for the standard-setting process

Wm. E. Wilson

National Center for Environmental Assessment, U.S. Environmental Protection Agency, MD-52, Research Triangle Park, NC 27711, USA

1. INTRODUCTION

Recent epidemiologic studies report statistical associations between a variety of health outcomes and indicators of particulate matter (PM). These findings lead to concerns about atmospheric aerosols (PM suspended in air) that are likely to continue well into the 21st century. The U.S. Environmental Protection Agency (EPA) has recommended augmenting the current PM_{10} (thoracic particle) standards with new fine particle standards (24-hour and annual). This recommendation largely resulted from the growing epidemiologic evidence for an association of health effects with indicators of the smaller size fraction of the atmospheric aerosol. Two possible indicators of the smaller size fraction for both research and standard setting are (1) fine-mode particles and (2) respirable particles. EPA scientists have preferred to use measurements of fine-mode particles in preference to measurements of respirable particles, primarily because of the significant distinctions between fine-mode and coarse-mode particles in regard to their sources, composition, and properties (physical, chemical, and biological).

As a result of the increasing concern with fine-mode particles, and with specific components of PM, new techniques for the measurement and analysis of aerosol mass and components are needed. Indicators that measure only one component of PM need to be augmented with measurements of the mass of all components. Sampling techniques are needed that capture and measure the semivolatile PM components, such as ammonium nitrate and organic components of particles in wood smoke, that partially are lost by most techniques in use today. Samplers are also needed that cleanly separate PM into fine-mode and coarse-mode fractions. Improved measurement techniques are needed to characterize the chemical and physical properties of aerosol, to determine compliance with standards, to determine sources of particles, to evaluate models, to assess exposure for acute and chronic epidemiology, to assess risk from PM exposure for setting standards, and to guide development of control strategies.

2. HISTORICAL TRENDS

Over the years, the PM components of interest and the PM measurement techniques used have changed; therefore, it is difficult to obtain long-term trends for PM indicators. EPA's National Air Quality and Emissions Trends Report[1] gives the annual average PM_{10} concentrations for the United States from 1988 to 1995, based on a consistent set of monitors from 955 sites. As shown in Figure 1, there is a small downward trend; however, it would be useful to know the relative changes in fine and coarse components of PM. Figure 2 shows annual average trends in Philadelphia for total suspended particles (TSP), PM_{10}, and $PM_{2.5}$.[2,3,4] TSP decreased significantly between 1973 and 1981 but increased slightly since 1981. There does not appear to have been a significant change in PM_{10} between 1987 and 1994. There appears to have been some improvement in $PM_{2.5}$ concentrations between 1980 and 1994; however, such a conclusion is uncertain because of the changes in $PM_{2.5}$ monitors and monitoring sites.

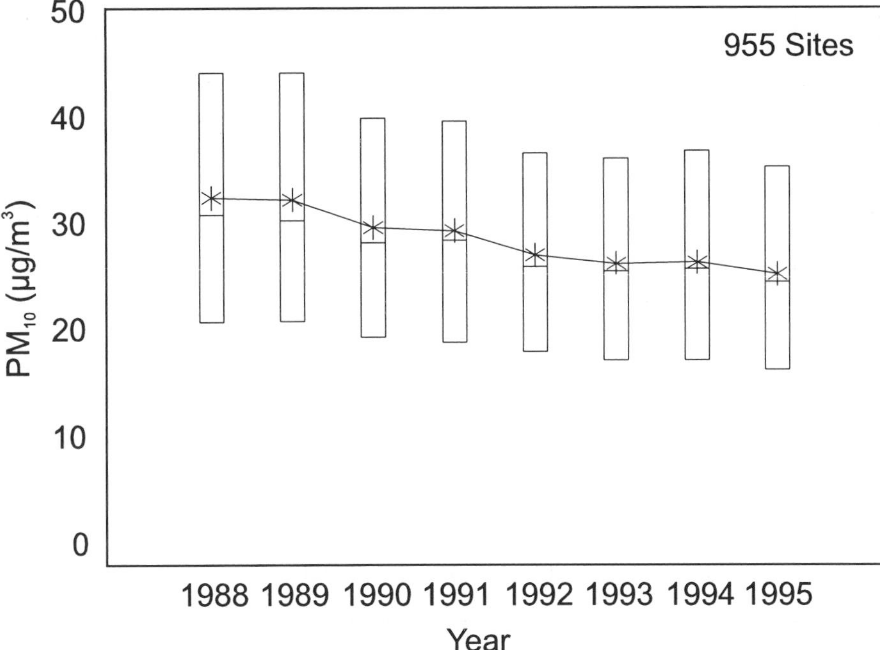

Figure 1. PM_{10} concentration trend from 1988 to 1995, based on yearly averages at 955 monitoring sites in the United States; bars show 90th percentile (top) and 10th percentile (bottom), the asterisk is the mean, and line is the median.[1]

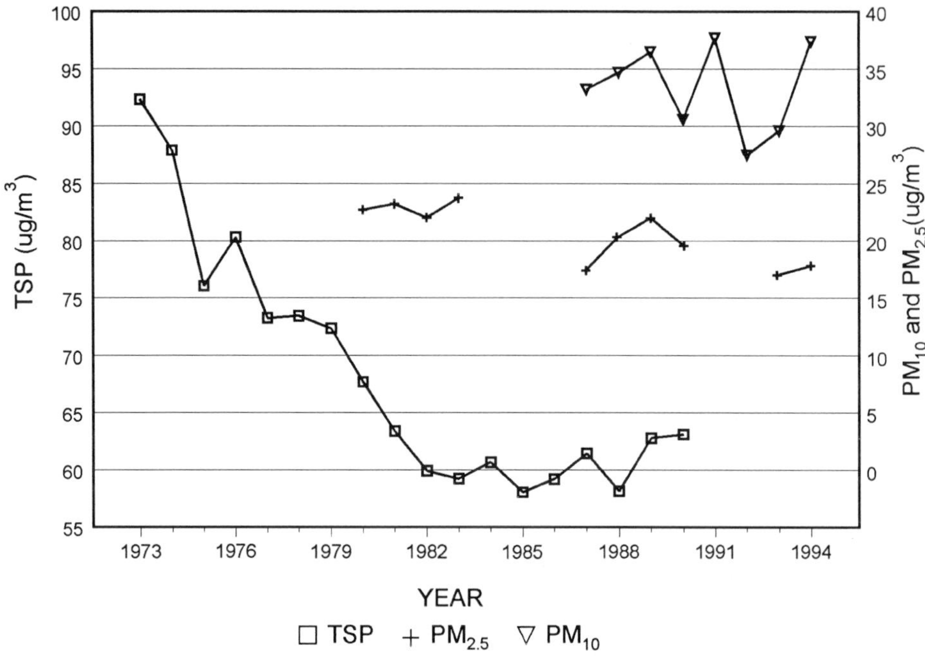

Figure 2. PM trends in Philadelphia, PA, 1973-1994; TSP, PM_{10}, and $PM_{2.5}$ (1987-1990) measurements from the Aerometric Information Retrieval System (AIRS);[2] $PM_{2.5}$ (1992-1994) from Harvard Philadelphia Study;[3] and $PM_{2.5}$ (1979-1983) from IPN.[4]

One source of information on long-term trends in fine particle pollution is the visibility distance data collected routinely at airport weather stations.[5] Visibility distance data, corrected for snow, rain, and relative humidity effects and transformed into the etinction coefficient, provide an indicator (known as 'haze') that increases as fine particle concentrations increase. This indicator, plotted as the 75th percentile, is shown in Figure 3 for several regions of the United States for the period 1948-1992.

Haze trends differ by season and region. In the winter, haze has declined in the Upper Midwest and Northeast but has increased in the Southeast and remained about the same in the Mid-Atlantic region. In the summer, however, there has been a dramatic increase in haze in all regions. A major increase in haze began in the 1950's; this increase accelerated in the 1960's but began to level off in the 1970's.

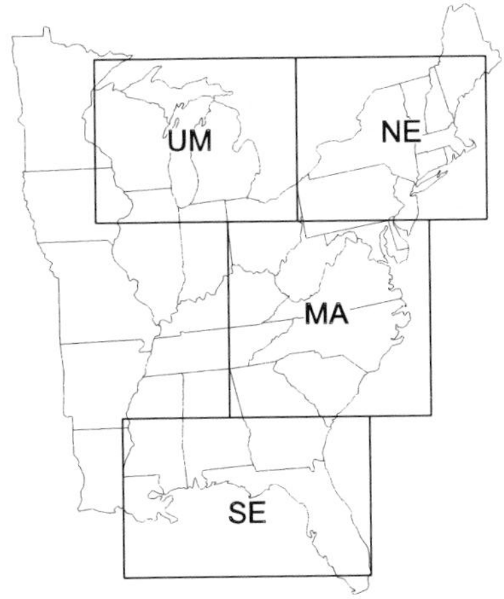

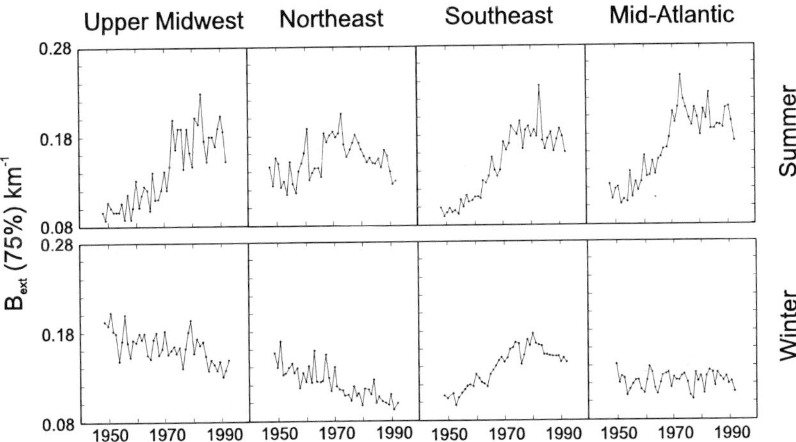

Figure 3. Haze trends in four regions of the United States from 1948 to 1992, shown as the 75th percentile of the extinction coefficient, corrected for relative humidity effects and with days with snow or rain removed.[5]

Because of the decrease in winter and the remarkable increase in the summer, the haze peak in the United States has shifted from winter in the upper Midwest to summer in the Mid-Atlantic region. This may be attributed to an overall decrease in industrial emissions and an increase in emissions of sulfur dioxide (SO_2) and nitrogen oxides (NO_x) in the summer resulting from increased demand for electricity for air conditioning.

The increase in haze since 1950 indicates that the potential exists for significant reductions in fine particles in the United States.

Developed countries, such as the United States, Canada, Japan, and those in Western Europe, will need to continue to be concerned with particle pollution. Megacities in developing countries will experience much greater concentrations of air pollution as industry and traffic expand in the 21st century. Regional pollution from developing countries will continue to impact developed countries.[6]

3. SIZE FRACTIONS

3.1. Fine-mode and coarse-mode particles

Before proceeding further, it will be useful to define the various size fractions that are used in discussions of PM. PM is not a single pollutant; rather it is composed of many different components. PM also may be divided into a variety of size fractions. Many different parameters have been used as indicators of the various size fractions and components of PM. One useful distinction has been to consider PM as two separate classes of pollutants, (1) fine-mode particles and (2) coarse-mode particles. In addition to size, fine-mode and coarse-mode particles differ in sources, formation mechanisms, composition, atmospheric lifetimes, spatial distribution, indoor/outdoor ratios, and temporal variability.[7] Current research is developing evidence that fine-mode and coarse-mode particles also differ in biological effects.[8-11]

The definition of fine-mode and coarse-mode particles is an operational one. It is based on measurements of size distribution, beginning in the early 1970's, that yielded bimodal distributions of particle mass with a minimum between 0.7 and 3.0 μm diameter.[12,13] An idealized distribution, showing the normally observed division of ambient aerosols into fine-mode and coarse-mode particles, is presented in Figure 4. *Aerosol* refers to a suspension of solid or liquid particles in air; however, aerosol is sometimes used to refer to the particles only. Fine-mode and coarse-mode particles may overlap in the intermodal region between 1 and 3 μm.[14]

3.2. Inhalable, thoracic, and respirable particles

It is also possible to define size fractions in terms of their entrance into the various compartments of the respiratory system. This convention classifies particles into inhalable, thoracic, and respirable particles. In a general sense, *inhalable particles* refer to particles with the potential of entering the respiratory tract, including the head airways region. *Thoracic particles* refer to particles with the potential of entering the tracheobronchial region (i.e., lower respiratory tract, including the trachea, bronchi, and the gas-exchange region of the lung). *Respirable particles* are particles with the potential of entering the gas-exchange region. Depending on the particle size, only a portion of the inhalable, thoracic, or respirable particles will enter the target compartment, and portion will be removed higher in the respiratory tract. Also only a portion of each size fraction will be deposited in the target compartment, and a portion will be exhaled. In the past, exact definitions of these terms have varied among organizations.[15] As of 1993 a unified set of definitions was adopted by the American Conference of Governmental Industrial Hygienists (ACGIH),[16] the International Standards Organization (ISO),[17] and the European Standardization Committee (CEN).[18] A historical account of the development of size-selective samplers for collecting size fractions that reach various portions of the respiratory system is given by Lippmann.[19] The particle separation curves for inhalable, thoracic, and respirable particles are shown in Figure 5. Also shown are the separation curves for PM_{10} and for the new Federal Reference Method (FRM) for $PM_{2.5}$ currently being developed.

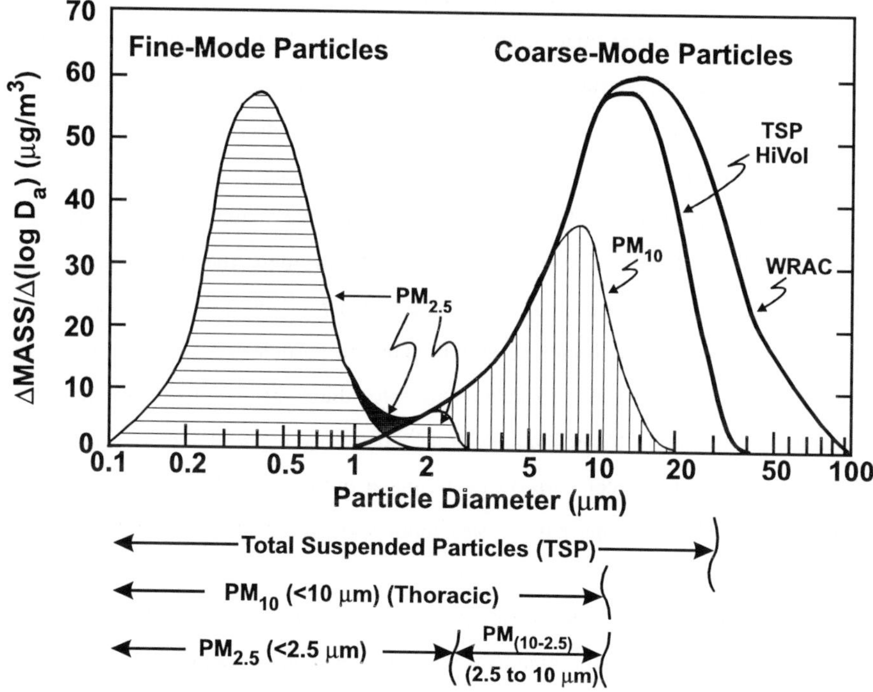

Figure 4. An idealized size distribution of ambient particulate matter showing fine- and coarse-modes and the portions collected by samplers with various size cut points.[7]

3.3. Particle samplers

Samplers have been designed to collect specific portions of the particle size-distribution. For example, TSP is defined by the design of the High Volume Sampler (HiVol), which collects all of the fine particles but only part of the coarse particles. The upper cut-off size of the HiVol is undefined, except by the design of the sampler, and varies with wind speed and direction. Extraordinary measures, such as were undertaken with the Wide Range Aerosol Classifier (WRAC), are required to collect the entire coarse mode.[20] Samplers also have been designed to collect the inhalable, thoracic, and respirable fractions.

The sample collected with a $PM_{2.5}$ cut is sometimes referred to as fine. In this paper, samples specified by relatively sharp cuts will be described numerically (i.e., PM_{10} or

$PM_{2.5}$). That portion of PM between $PM_{2.5}$ and PM_{10}, as shown in Figure 4, will be called $PM_{(10-2.5)}$). Because fine and coarse have lost the precise meaning intended by Whitby,[12] the addition of "mode" is used to emphasize reference to the fine- or coarse-mode particles shown in the distributions in Figure 4. Thus, $PM_{2.5}$ may be considered an indicator of fine-mode particle mass; PM_{10}, an indicator of the thoracic component (the fraction of PM that enters the thorax); and $PM_{(10-2.5)}$, an indicator of thoracic coarse-mode particles (i.e., that portion of the coarse-mode particle mass that reaches the thoracic compartment). In the proposal for new fine particle standards, EPA recommended maintaining the PM_{10} standards and sampling network, with PM_{10} becoming a marker for thoracic coarse-mode particles. However, research studies likely will determine $PM_{(10-2.5)}$ either directly or as the difference between PM_{10} and $PM_{2.5}$. Because of the overlap of fine- and coarse-mode particles in the intermodal region (1-3 μm), $PM_{2.5}$ is only an approximation of fine-mode particles, and $PM_{(10-2.5)}$ is only an approximation of thoracic coarse-mode particles.

In this paper, fine particles and coarse particles will refer to fine-mode particles and thoracic coarse-mode particles, respectively, not to the approximations given by $PM_{2.5}$ or $PM_{(10-2.5)}$. Also, diameter normally will refer to the aerodynamic diameter.

PM regulation began in the United States in 1971 with a TSP standard defined by the HiVol.[22] To focus regulatory concern on those particles small enough to penetrate and deposit in the lower respiratory tract (thoracic region), the indicator for the National Ambient Air Quality Standard for PM was changed in 1987 from TSP, as measured by the HiVol, to PM_{10}.[23] PM_{10} samplers collect all of the fine particles and part of the coarse particles. The upper cut point is defined as having a 50% collection efficiency at 10 ± 0.5 μm diameter. The slope of the collection efficiency curve also is defined.[24] Samplers with upper cut points of 3.5, 2.5, 2.1, and 1.0 μm are also in use. Dichotomous samplers split the particles into smaller and larger fractions that may be collected on separate filters. Detailed information on the design and use of particle samplers may be found in the review by Chow.[25]

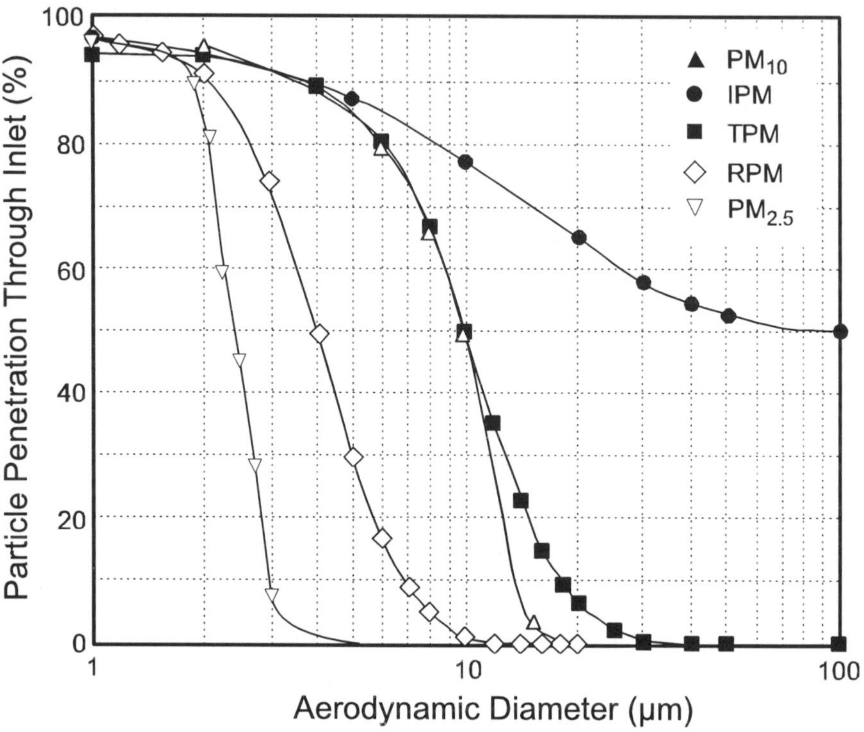

Figure 5. Particle separation curves for inhalable (IPM), thoracic (TPM), and respirable (RPM) particles and for PM_{10}[14] and $PM_{2.5}$.[21]

3.4. $PM_{2.5}$

Interest in fine and coarse particles, as distinct components of the atmospheric aerosol, began in the early 1970's, largely because of size distribution studies by Whitby and co-workers.[12,13] In 1979, EPA scientists endorsed the need to measure fine and coarse particles separately.[26] Based on limitations of existing technology, 2.5 μm, rather than 1.0 μm, was chosen for the cut point between fine and coarse particles. The $PM_{2.5}$ sample contains all of the fine particles but, especially in dry areas or during dry conditions, may collect a small but significant fraction of the coarse particles.[14] In 1979, EPA initiated the Inhalable Particle Network (IPN),[4] which included measurements of fine and coarse particles using dichotomous samplers with upper cut points of 15 and 2.5 μm. $PM_{2.5}$, as well as indicators of coarse particles measured in the Regional Air

Pollution Study in St. Louis, MO,[27] and the Harvard Six Cities Study.[28] Epidemiologic results from the Harvard Six Cities Study[28,29] that implicated $PM_{2.5}$ as being of special concern for health effects intensified interest in a fine particle standard. This interest was further intensified in 1994, when the American Lung Association brought a suit in U.S. Federal Court that forced EPA to begin a review of the PM standard.[30]

The first step in the review process is the preparation, by EPA's Office of Research and Development, of the Air Quality Criteria for Particulate Matter,[31] a critical review of scientific knowledge of PM. In this document, EPA recommended that '... fine and coarse particles should be considered as separate subclasses of pollutants. Consideration of formation, composition, behavior, exposure relationships, and sources argue for monitoring fine and coarse particles separately. Because fine and coarse particles are derived from different sources, it is also necessary to quantify ambient levels of fine and coarse particles separately in order to plan effective control strategies.'[32]

The second step in the review process is the preparation, by EPA's Office of Air Quality Planning and Standards (OAQPS), of the Review of the National Ambient Air Quality Standards for Particulate Matter: Policy Assessment of Scientific and Technical Information (the so-called OAQPS Staff Paper).[33] In this document, OAQPS recommended additional standards for fine particles and that 2.5 μm be used as the cut point between fine and coarse particles. The Clean Air Science Advisory Committee (CASAC), in their review of the staff paper, endorsed $PM_{2.5}$ as the indicator for fine particles.[34] EPA's National Exposure Research Laboratory is preparing the FRM that will include a design specification of the impactor, giving a sharp 50% cut at 2.5 μm aerodynamic diameter.[21]

3.5. EPA chooses $PM_{2.5}$ as the indicator for fine particles

During the review of the PM standard, there was some discussion of the possibility of choosing respirable particles for the new PM standards in order to harmonize U.S. and European particle measurements. However, EPA's decision to recommend a sharp, upper 50% cut point of 2.5 μm when it proposed new standards for fine particles[35] was based on different considerations than those of organizations (such as ACGIH, ISO, and CEN) that have recommended inhalable, thoracic, and respirable particle size fractions (Figure

5). Particle size fractions based on entrance into various compartments of the respiratory tract were developed originally for industrial hygiene purposes. In industrial applications, workers may be exposed predominantly to a single type of particle with a single size distribution. For these situations, inhalable, thoracic, and respirable size fractions may well be appropriate. However, in more complex situations, with more than one type of particle or size distribution, size fractions based on particle source or particle composition may be more appropriate. For example, coal miners are exposed to diesel soot, in the lower range of fine particles, and coal dust, in the coarse particle size range. Coal mining environments in the United States are monitored with a dichotomous sampler with a cut point between fine and coarse particles of 0.8 μm.[36] The ambient atmosphere is an even more complex situation with a wide range of particle sizes and sources. However, a separation into fine and coarse particles still gives useful information on particle sources that would be lost with a respirable sample.

The CASAC Technical Subcommittee for Fine Particle Monitoring[37] has encouraged EPA to 'give careful consideration to harmonizing the definition of PM_{10} with internationally accepted definitions of thoracic particulate matter'. The PM_{10} separation curve, although sharper than the inhalable particle curve, nevertheless gives a reasonable representation of inhalable particles. However, EPA, in its proposal for additional particle standards, choose fine particles instead of respirable particles. The preference for fine particles is based on the differences between fine and coarse particles, in terms of sources, composition, and properties. A respirable sample will combine both fine and coarse particles. However, a $PM_{2.5}$ sample, together with a PM_{10} or thoracic sample, allows a separate determination of fine and coarse particles. The difference between these two size-fractionation schemes is particularly important if fine and coarse particles have different sources and different biological properties. A thoracic and a respirable sample would not give information on which particle component was most prevalent. If one of these samples exceeded a standard, and a control program were required, this sampling scheme would provide little information as to which types of sources to control to most effectively reduce PM levels. A combination of PM_{10} and $PM_{2.5}$, however, would direct controls toward either fine or coarse particles. If fine and coarse particles have different biological properties, one city, whose high concentrations result from the more biologically active particle component, would have a different public health problem than would a city whose particle mix was dominates also easier to identify distinct source categories by factor analysis or chemical element balance if particles are first separated into fine and coarse. For example, potassium in fine particles is a tracer for wood burning, whereas potassium in coarse particles comes from soil.

3.6. Respiratory tract deposition

The selection of a fine particle fraction instead of a respirable particle fraction, either for research or standard setting, might be influenced by the pattern of deposition, as a function of particle size, into two important compartments of the lung, (1) the alveolar (the unciliated, gas exchange) compartment and (2) the tracheobronchial (the ciliated airways) compartment. The respiratory tract deposition fractions for the alveolar (A) and tracheobronchial (TB) compartments, as functions of particle size for a healthy adult, are shown in Figure 6.[38] Thoracic particles, given by a broad particle separation curve with a 50% upper cut point at 10 μm diameter (Figure 5), provide an indicator of those particles that enter the thorax and then can deposit in either the A or TB compartments. Respirable particles, given by 50% upper cut point at 4.0 μm and a broad separation curve (Figure 5), will be dominated by particles that deposit in the A compartment but also will contain a small fraction of particles that deposit in the TB compartment.

A particle sample with a cut at 2.5 μm will have a smaller, although possibly significant, fraction of particles that deposit in the TB compartment. A cut at 1.0 μm could give a better separation of PM into coarse-mode particles, above 1.0 μm, which deposit in both the A and TB compartments, and fine-mode particles, below 1.0 μm, which deposit primarily in the A compartment. Thus the patterns of deposition in the respiratory tract provide no basis for choosing a respirable fraction rather than a fine particle fraction or choosing 2.5 μm rather than 1.0 μm as the cut point for the fine particle indicator.

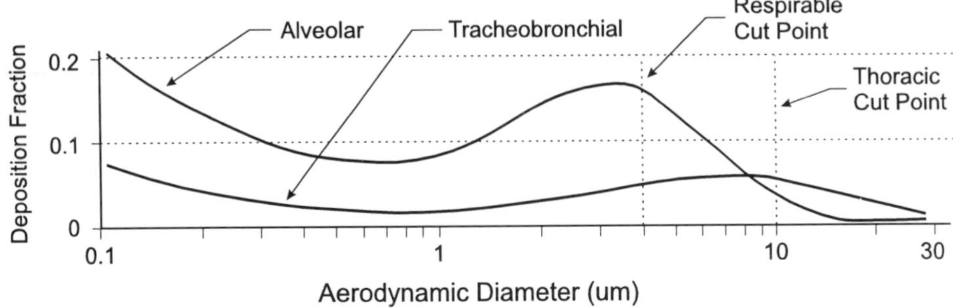

Figure 6. Calculated deposition fractions to the tracheobronchial and alveolar compartments of the respiratory tract for monodisperse aerosols as a function of aerodynamic diameter for quiet breathing (tidal volume = 750 ml, breathing frequency = 15 min^{-1}).[38]

Figure 6 shows the deposition fractions for a healthy adult breathing at a normal rate. The deposition fractions and patterns will change for different levels of exercise and for differences in lung size (e.g., deposition patterns for children will differ from those of adults).[39] Experimental studies also indicate differences in deposition fractions for various degrees of lung impairment.[39] Thus, the deposition fraction for 1-μm diameter particles was found to increase in going from normal subjects (0.14) to smokers (0.16) to smokers with small airway disease (0.21) to asthmatics (0.22) to subjects with chronic obstructive airway disease (0.28).[40] Major variations in deposition can occur even with healthy subjects, leading to local or regional enhancement of deposition.[41] Thus the variability in deposition fractions and patterns, especially for the entire population as opposed to workers, argues that fine particles, rather than respirable particles, are the most appropriate indicator for the protection of public health from the smaller size fraction of thoracic particles.

4. PM COMPOSITION

The major components of $PM_{2.5}$ mass are sulfate, nitrate, and ammonium ions; elemental and organic carbon; and mineral components. Also of potential importance are hydrogen ions or acidity and trace components, including toxic metals, transition metals, and polynuclear aromatic compounds. $PM_{(10-2.5)}$ mass is primarily mineral with a significant unknown fraction that may include biogenic materials. The composition of PM varies geographically. Mass apportionments for $PM_{2.5}$ and PM $_{(10-2.5)}$ in the eastern and the western United States are shown in Figure 7.[42] For $PM_{2.5}$, sulfate is greater in the eastern United States. Minerals, nitrates, and elemental and organic carbon are greater in the western United States. However, due to loss by volatilization, substances that exist in equilibrium with gas phase components, such as ammonium nitrate (NH_4NO_3), organic components of particles in wood smoke, and certain other organic compounds, may be underestimated. For $PM_{(10-2.5)}$, the mineral component is greater in the western United States. Because of the high concentrations of coarse particles in the western United States, it is of concern to measure fine particles with minimum intrusion of coarse particles.

5. DIFFERENCES BETWEEN FINE AND COARSE PARTICLES

The EPA proposal to retain the PM_{10} standard for protection from exposure to coarse particles, and to set new particle standards based on fine particles instead of on respirable particles for protection from smaller particles, is based largely on the desire to consider fine and coarse PM as different pollutants. Fine and coarse particles appear to differ in health effects, as indicated by epidemiology and toxicology; they also differ in sources, composition, and properties (physical, chemical, and biological). Exposure differences, in terms of temporal variability, regional uniformity, and infiltration factors (the ratio of the concentration of ambient particles that have penetrated indoors and remain suspended to the concentration of ambient particles outdoors), make it desirable to measure fine and coarse particles separately for exposure assessment for epidemiologic studies.[7] Some of the various differences between fine and coarse particles are summarized in Table 1.[7]

6. PM INDICATORS

6.1. Single-component measurements

Some PM indicators that have been used for research and epidemiologic studies measure only a single component of the atmospheric aerosol. British Smoke (BS) is a technique frequently used in European studies.[43] It has an upper 50% cut of approximately 4.5 μm diameter, but it can collect some particles up to 10 μm in diameter.

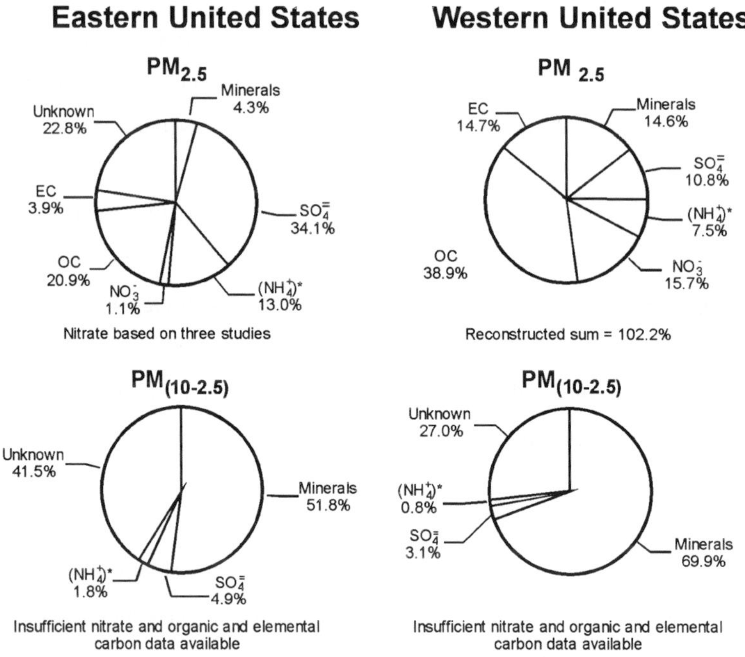

Figure 7. Major chemical components by mass for $PM_{2.5}$ and $PM_{(10-2.5)}$ particles in the western and the eastern United States.[42] (NH_4^+) represents the concentration of NH_4^+ that would be required if all $SO_4^=$ were present as $(NH_4)_2SO_4$ and all NO_3^- as NH_4NO_3; therefore, $(NH_4^+)^*$ represents an upper limit to the true concentration of NH_4^+.

However, because BS is based on an optical measurement of blackness, it responds primarily to elemental carbon (EC). In Britain in the 1950's and 1960's, most EC came from the burning of coal for producing electric power and for home heating. This is still the major source of EC in some eastern European countries. In western Europe, however, EC now comes mostly from diesel exhaust. As a result, the relationship between BS and PM mass will vary regionally, locally, and seasonally. BS measurements can be credibly expressed in micrograms per cubic meter only if site specific calibrations between the BS reflectance readings and collocated gravimetric measurements are available for locations and time periods with relatively similar mixes of PM components. The relationship between BS and PM mass also will vary diurnally. Therefore, BS may not be used as an indicator of hourly PM mass.

Sulfate also may be used as an indicator for fine PM mass. It is a larger component of PM than is EC, but, because of the variation in the nitrate and organic components, the relationship of sulfate to fine PM mass is also variable. Indicators that depend on the measurement of one component of PM are generally not good indicators of PM mass unless the relationship is determined experimentally on a routine and frequent basis. Certain components, such as hydrogen ions, transition metals, polyaromatic hydrocarbons, number of ultrafine particles, etc., may be measured because of their potential health impact even though they are not significant in terms of mass. Measurements of the individual components of PM are useful; however, for most studies, it also will be important to measure a parameter that is closely related to total particle mass.

6.2. Loss of semivolatile components

Several major components of fine PM exist in the atmosphere in equilibrium with gaseous species. These are known as semivolatile materials (SVM). Examples are NH_4NO_3, organic components of particles in wood smoke, and certain other organic compounds. Changes in temperature or atmospheric composition during sampling can lead to losses of these semivolatile components of PM. Some components of fine PM are hygroscopic or deliquescent and will contain liquid water (particle-bound water [PBW]). The amount of PBW increases as relative humidity increases, but, for regulatory purposes, water is not considered a component of PM mass. The difficulty arises in removing the PBW without also removing the other semivolatile PM components.

Table 1

Comparison of ambient fine- and coarse-mode particles

	Fine-Mode Particles	Coarse-Mode Particles
Formed from:	Gases	Large solids/droplets
Formed by:	Chemical reaction or vaporization Nucleation, condensation on nuclei, and coagulation Evaporation of fog and cloud droplets in which gases have dissolved and reacted	Mechanical disruption (crushing, grinding, abrasion of surfaces, etc.) Evaporation of sprays Suspension of dusts
Composed of:	Sulfate, nitrate, ammonium, and hydrogen ions Elemental carbon Organic compounds (e.g., polyaromatic hydrocarbons) Metals (e.g., lead, cadmium, vanadium, nickel, copper, zinc, manganese, iron) Particle-bound water	Resuspended dust (soil dust and street dust) Coal and oil fly ash Oxides of crustal elements, (silicon, aluminum, titanium, and iron) $CaCO_3$, NaCl, and sea salt Pollen, mold, and fungal spores Plant/animal fragments Tire wear debris
Solubility:	Largely soluble, hygroscopic, and deliquescent	Largely insoluble and nonhygroscopic
Sources:	Combustion of coal, oil, gasoline, diesel fuel, and wood Atmospheric transformation products of NO_x, SO_2, and organic compounds, including biogenic organic species (e.g., terpenes) High-temperature processes, smelters, steel mills, etc.	Resuspension of industrial dust and soil tracked onto roads and streets Suspension from disturbed soil (e.g., farming, mining, unpaved roads) Biological sources Construction and demolition Coal and oil combustion Ocean spray
Atmospheric half-life:	Days to weeks	Minutes to hours
Travel distance:	100s to 1000s of km	<1 to 10s of km

The most frequently used indicator for fine particles in the United States is $PM_{2.5}$. Its definition includes, in addition to the 50% cut point diameter and the shape of the particle separation curve, specifications for sampling and equilibration. Those PM components with significant vapor pressure that survive the sampling and handling process also can be lost during the equilibration process. The FRM for PM_{10} calls for an equilibration period of 24 hour at a temperature between 10 and 25°C and a relative humidity between 15 and 45%.[44] There are no specifications on the temperature during the collection and handling of the PM_{10} sample. It is possible that the FRM for the proposed $PM_{2.5}$ standard will impose limitations on the allowable temperature increase above ambient during sampling and handling and will specify tighter limits on temperature and relative humidity during equilibration. This should increase the precision of the method but will still allow significant loss of SVM.

Nitrate loss has received considerable theoretical and experimental study.[45,46,47] Examples of the observed losses are shown in Figure 8.[48] Nitrate loss can be avoided by a sampling system in which gas-phase nitric acid is removed by a denuder and nitrate particles are collected on a nylon filter, which will absorb nitric acid as it vaporizes from NH_4NO_3 particles.[49] Similar results can be obtained with a Teflon® filter, followed either by a nylon or glass fiber filter impregnated with a basic substance such as potassium carbonate.[50,51]

Organic losses, both in terms of specific, toxic SVM and total organic mass, are also of concern.[45] A variety of after filters or absorbents have been used to collect organic species that are distributed between the particle and gas phases.[45,52] Quartz filters are used when an analysis of elemental and organic carbon by combustion is required. A positive artifact, resulting from the absorption of organic vapors by the quartz filter media, has received much attention.[53] Techniques have been proposed for correcting for this artifact by use of a back-up filter.[54] However, organic material may evaporate from particles, either as cleaner air passes through collected particles on the filter or as the filter and collected particles are equilibrated in an atmosphere that has less organic material in the gas phase than was present during sampling. This evaporation of semivolatile organic material causes a negative artifact.

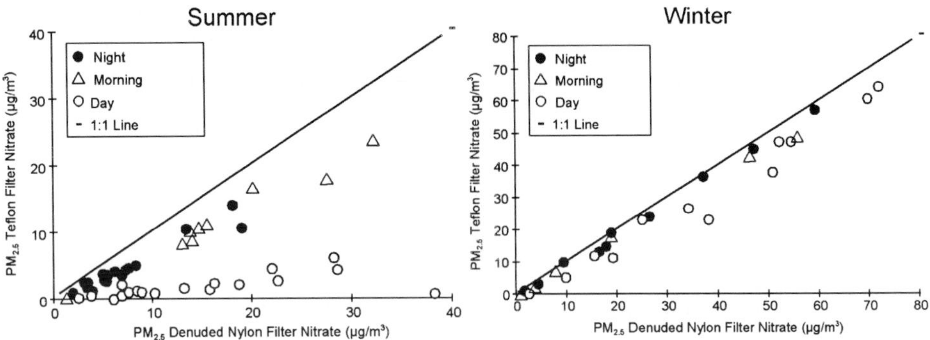

Figure 8. Comparison of nitrate measurements in Los Angeles; $PM_{2.5}$ samples collected on Teflon® filters versus samples collected on nylon filters after removal of nitric acid by denuders.[48]

This is especially a problem with a Teflon® filter, but also may be a problem with quartz filters.[55,56,57] Particles in wood smoke apparently contain a variety of semivolatile organic compounds.[55,58] Techniques for determining the mass and composition the organic semivolatile phase have been developed by Eatough and co-workers.[59,60,61]

Several continuous PM measurement techniques have been developed that deal with SVM in different ways. The Tapered Element Oscillating Microbalance (TEOM®)[62] measures the change in the mass of material collected continuously on a filter that is normally changed only after collecting material for a week or more. However, the collected material is kept at 50°C, a temperature high enough to prevent significant collection of water by hygroscopic or deliquescent materials. Otherwise, mass changes resulting from changes in PBW, as relative humidity (RH) changed, would dominate the mass changes caused by collection of new PM. The TEOM®, therefore, measures that component of the atmospheric aerosol that is nonvolatile at 50°C but excludes NH_4NO_3 and other SVM, including a significant fraction of organic particles in wood smoke.[56]

Two other techniques for continuous PM measurement remove PBW by dehumidification with a diffusion drier before collection. Koutrakis et al.[63] developed a technique based on the change in pressure drop across a filter as particles are collected. The filter surface is changed every 20 min, so that losses resulting from changes in temperature or atmospheric composition will be minimized. Obeidi et al.[64] developed a system that also uses a diffusion drier to remove PBW, but that depends on a carbon-impregnated filter to collect both nonvolatile and semivolatile PM. A denuder must be used before the filter to remove SVM in the gas phase. Then, any material evaporating from semivolatile particles will be collected on the carbon-impregnated filter and correctly measured as PM mass. To reduce the demands on the denuder, Obeidi et al.[64] concentrate the particles with a 0.1-μm virtual impactor before dehumidifying, denuding, and collecting them.

6.3. Cut point for separating fine particles from coarse particles

During the preparation of the PM criteria document, EPA considered whether 2.5 μm, or some smaller diameter such as 1.0 μm, would be the most appropriate size for separating fine and coarse particles.[14] Two considerations led to a preference for 2.5 μm. First, most of the epidemiologic studies of fine particles used $PM_{2.5}$ as the fine particle indicator; no epidemiologic data and little, if any, concentration data were available with a 1.0-μm cut. Secondly, during very high RH conditions, such as in fog or clouds, particle size distributions, measured by impactor-type classifiers, indicated that some fine-mode nitrate and sulfate particles were greater than 1 μm in diameter, presumably because of growth with increasing RH. PM concentrations were high during such periods, and a lower cut point size might result in missing some fine-mode mass. The existence of such large particles outside of fog and clouds has not been confirmed by techniques that measure single particles. The development of techniques to dehumidify aerosols makes it possible to remove some or all of the PBW and to use a lower cut point to more cleanly separate fine and coarse particles. However, it has not yet been demonstrated to what extent PBW can be removed without also removing some SVM.

7. FINE PARTICLE RESEARCH IN THE 21ST CENTURY

EPA has proposed, as a fine particle indicator for determining compliance with the new standards, a $PM_{2.5}$ cut and equilibration to remove PBW.[35] However, a new approach is need for research measurements of fine particles in the 21st century.[65] It is hoped that new techniques can be developed that will remove PBW but not lose SVM (such as ammonium nitrate and some organic component particles in wood smoke) and that will cleanly separate fine particles from coarse particles. Such techniques will be needed for a variety of research applications, including measurement of exposure for epidemiologic studies. Time-series epidemiologic studies of acute health outcomes will be needed for 24-hour and continuous (or hourly average) measurements. Chronic epidemiologic studies will need long-term measurements that integrate samples for a week, a month, or possibly longer periods.

Under the Clean Air Act, EPA is required to review the PM standards every 5 years.[66] Research with improved particle monitoring must begin soon so that new and useful information will be available for the next review. Hopefully, the 21st century will bring a new, improved indicator for the next U.S. fine particle standard. Countries that have not set a fine particle standard should consider carefully what fine particle indicator to use in research and for subsequent standard setting.

Disclaimer

The views expressed in this paper are those of the author and do not necessarily reflect the views or policies of the U.S. Environmental Protection Agency. The U.S. Government has the right to retain a nonexclusive, royalty-free license in and to any copyright covering this article.

REFERENCES

1. U.S. Environmental Protection Agency. (1996) National air quality and emissions trends report, 1995. Research Triangle Park, NC: Office of Air Quality Planning and Standards; report no. EPA 454/R-96-005.
2. AIRS, Aerometric Information Retrieval System [database]. (1995) [PM_{10} and $PM_{2.5}$ data]. Research Triangle Park, NC: U.S. Environmental Protection Agency, Office of Air Quality Planning and Standards.
3. Koutrakis, P. (1995) [$PM_{2.5}$ and PM_{10} data for Philadelphia, 1992-1995]. Boston,

MA: Harvard School of Public Health.

4. Rodes, C. E.; Evans, E. G. (1985) Preliminary assessment of 10 μm particulate sampling at eight locations in the United States. Atmos. Environ. 19: 293-303.
5. Husar, R. B.; Wilson, W. E. (1993) Haze and sulfur emission trends in the eastern United States. Environ. Sci. Technol. 27: 12-16.
6. Mage, D.; Ozolins, G.; Peterson, P.; Webster, A.; Orthofer, R.; Vandeweerd, V.; Gwynne, M. (1996) Urban air pollution in megacities of the world. Atmos. Environ. 30: 681-686.
7. Wilson, W. E.; Suh, H. H. (1996) Fine and coarse particles: concentration relationships relevant to epidemiological studies. J. Air Waste Manage. Assoc.: accepted.
8. Beck, B. D.; Brain, J. D. (1982) Prediction of the pulmonary toxicity of respirable combustion products from residential wood and coal stoves. In: Proceedings: residential wood & coal combustion specialty conference; March; Louisville, KY. Pittsburgh, PA: Air Pollution Control Association; pp. 264-280.
9. Dreher, K.; Jaskot, R.; Richards, J.; Lehmann, J.; Winsett, D.; Hoffman, A.; Costa, D. (1996) Acute pulmonary toxicity of size-fractionated ambient air particulate matter. Am. J. Respir. Crit. Care Med. 153: A15.
10. Godleski, J. J.; Sioutas, C.; Katler, M.; Koutrakis, P. (1996) Death from inhalation of concentrated ambient air particles in animal models of pulmonary disease. Am. J. Respir. Crit. Care Med. 153: A15.
11. Gwynn, C.; Burnett, R. T.; Thurston, G. D. (1996) Acidic particulate matter air pollution and daily mortality and morbidity in the Buffalo, NY region. Am. J. Respir. Crit. Care Med. 153: A16.
12. Whitby, K. T. (1978) The physical characteristics of sulfur aerosols. Atmos. Environ. 12: 135-159.
13. Whitby, K. T.; Charlson, R. E.; Wilson, W. E.; Stevens, R. K. (1974) The size of suspended particle matter in air. Science (Washington, DC) 183: 1098-1099.
14. U.S. Environmental Protection Agency. (1996) Air quality criteria for particulate matter. Research Triangle Park, NC: National Center for Environmental Assessment-RTP Office; report nos. EPA/600/P-95/001aF. Chapter 3. Available from: NTIS, Springfield, VA; PB96-168224.
15. Soderholm, S. C. (1989) Proposed international conventions for particle size-selective sampling. Ann. Occup. Hyg. 33: 301-320.
16. American Conference of Governmental Industrial Hygienists (ACGIH). (1994)

Appendix D: particle size-selective sampling criteria for airborne particulate matter. In: 1994-1995 threshold limit values for chemical substances and physical agents and biological exposure indices. Cincinnati, OH: American Conference of Governmental Industrial Hygienists; pp. 43-46.

17. International Organization for Standardization (ISO). (1991) Air quality—particle size fraction definitions for health-related sampling. Geneva, Switzerland: ISO. Approved for publication as CD 7708.

18. European Standardization Committee (CEN) (1991) Size fraction definitions for measurement of airborne particles in the workplace. Brussels: CEN. Approved for publication as prEN 481.

19. Lippmann, M. (1995) Size-selective health hazard sampling. In: Cohen, B. S.; Hering, S. V., eds. Air sampling instruments for evaluation of atmospheric contaminants. 8th ed. Cincinnati, OH: American Conference of Governmental Industrial Hygienists; pp. 81-119.

20. Lundgren, D. A.; Burton, R. M. (1995) Effect of particle size distribution on the cut point between fine and coarse ambient mass fractions. In: Phalen, R. F.; Bates, D. V., eds. Proceedings of the colloquium on particulate air pollution and human mortality and morbidity; January 1994; Irvine, CA. Inhalation Toxicol. 7: 131-148.

21. Vanderpool, R. W. (1997) [Memorandum to William E. Wilson, U.S. EPA, regarding the $PM_{2.5}$ curve]. Research Triangle Park, NC: Research Triangle Institute; June 13.

22. National Air Pollution Control Administration. (1969) Air quality criteria for particulate matter. Washington, DC: U.S. Department of Health, Education, and Welfare, Public Health Service; NAPCA publication no. AP-49. Available from: NTIS, Springfield, VA; PB-190251/BA.

23. Federal Register. (1987) Revisions to the national ambient air quality standards for particulate matter. F. R. (July 1) 52: 24634-24669.

24. Code of Federal Regulations. (1996) Test procedures. C. F. R. 40: §53.43.

25. Chow, J. C. (1995) Measurement methods to determine compliance with ambient air quality standards for suspended particles. J. Air Waste Manage. Assoc. 45: 320-382.

26. Miller, F. J.; Gardner, D. E.; Graham, J. A.; Lee, R. E., Jr.; Wilson, W. E.; Bachmann, J. D. (1979) Size considerations for establishing a standard for inhalable particles. J. Air Pollut. Control Assoc. 29: 610-615.

27. Altshuller, A. P. (1985) Relationships involving fine particle mass, fine particle sulfur and ozone during episodic periods at sites in and around St. Louis, MO. Atmos. Environ. 19: 265-276.

28. Spengler, J. D.; Dockery, D. W.; Turner, W. A.; Wolfson, J. M.; Ferris, B. G., Jr. (1981) Long-term measurements of respirable sulfates and particles inside and outside homes. Atmos. Environ. 15: 23-30.

29. Dockery, D. W.; Pope, C. A., III; Xu, X.; Spengler, J. D.; Ware, J. H.; Fay, M. E.; Ferris, B. G., Jr.; Speizer, F. E. (1993) An association between air pollution and mortality in six U.S. cities. N. Engl. J. Med. 329: 1753-1759.

30. American Lung Association v. Browner. (1994) CIV-93-643-TUC-ACM CD. Ariz., October 6, 1994.

31. U.S. Environmental Protection Agency. (1996) Air quality criteria for particulate matter. Research Triangle Park, NC: National Center for Environmental Assessment-RTP Office; report nos. EPA/600/P-95/001aF-cF. Available from: NTIS, Springfield, VA; PB96-168224.

32. U.S. Environmental Protection Agency. (1996) Air quality criteria for particulate matter. Research Triangle Park, NC: National Center for Environmental Assessment-RTP Office; report nos. EPA/600/P-95/001aF. Chapter 1. Available from: NTIS, Springfield, VA; PB96-168224.

33. Bachmann, J. D.; Caldwell, J. C.; Damberg, R. J.; Edwards, C.; Koman, T.; Martin, K.; Polkowsky, B.; Richmond, H. M.; Smith, E.; Woodruff, T. (1996) Review of the national ambient air quality standards for particulate matter: policy assessment of scientific and technical information. OAQPS staff paper. Research Triangle Park, NC: U.S. Environmental Protection Agency, Office of Air Quality Planning and Standards; report no. EPA/452/R-96-013. Available from: NTIS, Springfield, VA; PB97-115406REB.

34. Wolff, G. T. (1996) Closure by the Clean Air Scientific Advisory Committee (CASAC) on the staff paper for particulate matter [letter to Carol M. Browner, Administrator, U.S. EPA]. Washington, DC: U.S. Environmental Protection Agency, Clean Air Scientific Advisory Committee; EPA-SAB-CASAC-LTR-96-008; June 13.

35. Federal Register. (1996) National ambient air quality standards for particulate matter: proposed decision. F. R. (December 13) 52: 65,638-65,713.

36. Marple, V. A.; Rubow, K. L.; Olson, B. A. (1995) Diesel exhaust/mine dust virtual impactor personal aerosol sampler: design, calibration and field evaluation.

Aerosol Sci. Technol. 22: 140-150.

37. Wolff, G. T. (1996) Report of the Clean Air Scientific Advisory Committee (CASAC) Technical Subcommittee for Fine Particle Monitoring [letter to Carol M. Browner, Administrator, U.S. EPA]. Washington, DC: U.S. Environmental Protection Agency, Clean Air Scientific Advisory Committee; August 7.

38. Yeh, H. C.; Zhuang, Y.; Chang, I. Y. (1993) Mathematical model of particle deposition from inhaled polydisperse aerosols. In: Nikula, K. J.; Belinsky, S. A.; Bradley, P. L., eds. Inhalation Toxicology Research Institute annual report 1992-1993. Albuquerque, NM: U. S. Department of Energy, Lovelace Biomedical and Environmental Research Institute; pp. 127-129; report no. ITRI-140. Available from: NTIS, Springfield, VA; AD-A277 924/7/XAB.

39. U.S. Environmental Protection Agency. (1996) Air quality criteria for particulate matter. Research Triangle Park, NC: National Center for Environmental Assessment-RTP Office; report nos. EPA/600/P-95/001bF. Chapter 10. Available from: NTIS, Springfield, VA; PB96-168224.

40. Kim, C. S.; Kang, T. C. (1997) Comparative measurement of lung deposition of inhaled fine particles in normal subjects and patients with obstructive airway disease. Am. J. Respir. Crit. Care Med. 155: 899-905.

41. Kim, C. S.; Hu, S. C.; DeWitt, P.; Gerrity, T. R. (1996) Assessment of regional deposition of inhaled particles in human lungs by serial bolus delivery method. J. Appl. Physiol. 81: 2203-2213.

42. U.S. Environmental Protection Agency. (1996) Air quality criteria for particulate matter. Research Triangle Park, NC: National Center for Environmental Assessment-RTP Office; report nos. EPA/600/P-95/001aF. Chapter 6. Available from: NTIS, Springfield, VA; PB96-168224.

43. Katsouyanni, K.; Zmirou, D.; Spix, C.; Sunyer, J.; Schouten, J. P.; Ponka, A.; Anderson, H. R.; Le Moullec, Y.; Wojtyniak, B. Vigotti, M. A.; Bacharova, L. (1995) Short-term effects of air pollution on health: a European approach using epidemiological time-series data. The APHEA project: background, objectives, design. Eur. Respir. J. 8: 1030-1038.

44. Code of Federal Regulations. (1991) Appendix J to Part 50—reference method for the determination of particulate matter as PM_{10} in the atmosphere. C. F. R. 40: §50.

45. U.S. Environmental Protection Agency. (1996) Air quality criteria for particulate matter. Research Triangle Park, NC: National Center for Environmental

Assessment-RTP Office; report nos. EPA/600/P-95/001aF. Chapter 4. Available from: NTIS, Springfield, VA; PB96-168224.

46. Hering, S.; Eldering, A.; Seinfeld, J. H. (1997) Bimodal character of accumulation mode aerosol mass distributions in southern California. Atmos. Environ. 31: 1-11.

47. Zhang, X.; McMurry, P. H. (1992) Evaporative losses of fine particulate nitrates during sampling. Atmos. Environ. Part A 26: 3305-3312.

48. Chow, J. C.; Fujita, E. M.; Watson, J. G.; Lu, Z.; Lawson, D. R.; Ashbaugh, L. L. (1994) Evaluation of filter-based aerosol measurements during the 1987 Southern California Air Quality Study. Environ. Monit. Assess. 30: 49-80.

49. Malm, W. C.; Sisler, J. F.; Huffman, D.; Eldred, R. A.; Cahill, T. A. (1994) Spatial and seasonal trends in particle concentration and optical extinction in the United States. J. Geophys. Res. [Atmos.] 99: 1347-1370.

50. Koutrakis, P.; Wolfson, J. M.; Slater, J. L.; Brauer, M.; Spengler, J. D.; Stevens, R. K.; Stone, C. L. (1988) Evaluation of an annular denuder/filter pack system to collect acidic aerosols and gases. Environ. Sci. Technol. 22: 1463-1468.

51. U.S. Environmental Protection Agency. (1992) Determination of the strong acidity of atmospheric fine-particles (<2.5 μm) using annular denuder technology. Washington, DC: Atmospheric Research and Exposure Assessment Laboratory; EPA report no. EPA/600/R-93/037.

52. Gundel, L. A.; Lee, V. C.; Mahanama, K. R. R.; Stevens, R. K.; Daisey, J. M. (1995) Direct determination of the phase distributions of semi-volatile polycyclic aromatic hydrocarbons using annular denuders. Atmos. Environ. 29: 1719-1733.

53. McDow, S. R.; Huntzicker, J. J. (1990) Vapor adsorption artifact in the sampling of organic aerosol: face velocity effects. Atmos. Environ. Part A 24: 2563-2571.

54. Turpin, B. J.; Huntzicker, J. J.; Hering, S. V. (1994) Investigation of organic aerosol sampling artifacts in the Los Angeles basin. Atmos. Environ. 28: 3061-3071.

55. Kamens, R. M.; Perry, J. M.; Saucy, D. A.; Bell, D. A.; Newton, D. L.; Brand, B. (1985) Factors which influence polycyclic aromatic hydrocarbon decomposition on wood smoke particles. Environ. Int. 11: 131-136.

56. Meyer, M. B.; Lijek, J.; Ono, D. (1992) Continuous PM_{10} measurements in a woodsmoke environment. In: Chow, J. C.; Ono, D. M., eds. PM_{10} standards and nontraditional particulate source controls, an A&WMA/EPA international specialty conference, v. I; January; Scottsdale, AZ. Pittsburgh, PA: Air & Waste Management Association; pp. 24-38. (A&WMA transactions series no. 22).

57. Tang, H.; Lewis, E. A.; Eatough, D. J.; Burton, R. M.; Farber, R. J. (1994) Determination of the particle size distribution and chemical composition of semi-volatile organic compounds in atmospheric fine particles with a diffusion denuder sampling system. Atmos. Environ. 28: 939-947.

58. Cupitt, L. T.; Glen, W. G.; Lewtas, J. (1994) Exposure and risk from ambient particle-bound pollution in an airshed dominated by residential wood combustion and mobile sources. In: Symposium of risk assessment of urban air: emissions, exposure, risk identification, and risk quantitation; May-June 1992; Stockholm, Sweden. Environ. Health Perspect. 102(suppl. 4):75-84.

59. Eatough, D. J.; Wadsworth, A.; Eatough, D. A.; Crawford, J. W.; Hansen, L. D.; Lewis, E. A. (1993) A multiple-system, multi-channel diffusion denuder sampler for the determination of fine-particulate organic material in the atmosphere. Atmos. Environ. Part A 27: 1213-1219.

60. Eatough, D. J.; Tang, H.; Cui, W.; Machir, J. (1995) Determination of the size distribution and chemical composition of fine particulate semi-volatile organic material in urban environments using diffusion denuder technology. In: Phalen, R. F.; Bates, D. V., eds. Proceedings of the colloquium on particulate air pollution and human mortality and morbidity, part II; January 1994; Irvine, CA. Inhalation Toxicol. 7: 691-710.

61. Eatough, D. J.; Cui, W. (1995) Semi-volatile particulate organic compounds in urban fine particles. In: Particulate matter: health and regulatory issues, proceedings of an international specialty conference; April; Pittsburgh, PA. Pittsburgh, PA: Air and Waste Management Association; pp. 320-327. (A&WMA publication VIP-49).

62. Patashnick, H.; Rupprecht, E. G. (1991) Continuous PM-10 measurements using the tapered element oscillating microbalance. J. Air Waste Manage. Assoc. 41: 1079-1083.

63. Wang, E. (1997) Particle mass measurements by flow obstruction [dissertation]. Boston, MA: Harvard School of Public Health.

64. Obeidi, F.; Eatough, D. J.; Meyer, M. B.; Wilson, W. E. (1997) Development of a real-time monitor for volatile fine particulate matter. Presented at: 90th Annual meeting and exhibition of the Air and Waste Management Association; June; Toronto, ON, Canada.

65. U.S. Environmental Protection Agency. (1996) Particulate matter research needs for human health risk assessment [external review draft]. Research Triangle Park,

NC: National Center for Environmental Assessment-RTP Office; report no. NCEA-R-0973; October 25.

66. U.S. Code. (1990) Clean Air Act, as amended by PL 101-549, November 15, 1990. U. S. C. 42: sect. 7401-7626.

Ambient particulate matter: Is there a toxic role for constitutive transition metals?

D. L. Costa

Pulmonary Toxicology Branch, MD-82; Experimental Toxicology Division, National Health and Environmental Effects Laboratory, Research Triangle Park, NC, USA 27711

1. INTRODUCTION

A core of epidemiology studies have demonstrated consistent statistical associations between daily concentration profiles of ambient particulate matter (PM) and mortality / morbidity rates in exposed human populations [reviewed in 1]. That the impact of ambient PM on human health can be observed at low, heretofore thought to be "safe" levels has attracted considerable attention within the environmental health and political communities, and has raised the question of whether the observations are "biologically plausible" [2]. In other words, is there an adverse biologic mechanism(s) of clinical significance that could account for the impact of such low levels of PM on human health? Indeed, if the observations can be substantiated, the question arises as to what subpopulations are most at risk and what are their risk factors?

Until recently, there has been little formal coordination between the epidemiological and toxicological assessments of PM-related health issues. The epidemiologists had, for many years, been without sufficiently sensitive statistical tools to clearly dissect the impact of pollutants like PM from other covariates, while the toxicologists had, for their part, largely focused attention on the effects on healthy laboratory animals of protracted exposures to single acidic aerosols occasionally co-mingled with ozone. However, with the advent and relatively recent application of time-series data analyses to the air pollution issue [3], unperceived, yet significant, associations between PM and human health have been unveiled. The result has been a plea for relevant laboratory studies that might elucidate a "biologically plausible" explanation for the recent epidemiologic observations. This impetus has prompted most current empirical PM research to become more "hypothesis-driven", conceptually integrating aerosol monitoring and chemistry data, animal and clinical toxicology, with existent epidemiology.

Currently, several hypotheses proposed by laboratory investigators to explain the epidemiological findings are receiving particular attention. These hypotheses include: the acidic properties of PM [4], the ultrafine fraction (<0.01 (m) of the PM distribution [5], altered intrapulmonary distribution of the lung PM dose [6], the organic fraction of PM [7, focusing on cancer], the presence of biological materials in PM [8], and the transition metal content of PM [9,10]. Each hypothesis has a sufficient database to merit support and consideration in the PM research arena, but it is the "transition metal" hypothesis which has captured particular attention in our laboratory at EPA. This is not to imply that other explanatory theories have been dismissed or are being ignored off-hand; rather we feel that the "metal hypothesis" merits special attention because the concept of PM-associated metals provides reasonable answers to the following generic questions pertinent to any such theory.

- Are there environmental sources for exposure to the putative toxicant(s)?
- Is there evidence of personal exposure to the toxicant(s)?
- Does the putative toxicant(s) possess sufficient toxicity?
- Are the suspected mechanisms of toxicity able to be extrapolated to the human?
- Is there evidence of an exposure-response relationship with the toxicant(s), especially at low concentrations?
- How well does the theory generalize from one PM sample or locale to another?

2. THE "METAL HYPOTHESIS"

In brief, it is our opinion that PM-associated transition metals can reasonably satisfy each of the criteria as stated. Firstly, metals are ubiquitous in ambient PM, though they vary widely in concentration and in type. Since the early 1970's, it has been appreciated that the fine respirable PM fraction is most enriched with metal [11], which would agree with the apparent stronger associations of health effects with the fine PM (<2.5 (m) that deposits deep in the lung [1]. It is in the deep lung of urban dwellers that metals are retained and readily detected, particularly among inhabitants of air-sheds surrounding heavy industry [12]. Finally, studies in laboratory animals have shown that PM-associated metals can be toxic to the lung, inducing inflammation, altering normal physiologic function, and impairing host defenses [9,10,13].

The bioavailability of the metal (mostly as water or acid soluble cations) appears to govern its acute proinflammatory toxicity [14]. Recently, we demonstrated that a metal-rich emission source PM (a fugitive residual oil fly ash - ROFA) and ambient PM from various urban environments (St. Louis, MO; Washington, DC; Ottawa, Canada; and Düsseldorf, Germany) having only 10% the metal of the ROFA, induced qualitatively and quantitatively similar toxicologic effects when the samples were delivered to the lung in terms of bioavailable metal content, irrespective of PM mass. Likewise, extractable metal from PM samples in the industrial regions of Mexico City were found to be of equal potency as the ROFA when assayed *in vitro* for cytotoxicity, while the samples from the lesser polluted regions of the city were substantially less toxic (J. Bonner and K. Dreher, personal communication). In related studies, Ghio and coworkers [15] have shown that metal associated with humus-like organic material derived from fossil fuel combustion products, as well as ambient PM from several regions in the U.S., induced inflammation in the lungs of rodents in apparent proportion to the amount of extractable metal. Thus, bioavailable metal regardless of the carrier appears to be requisite for the PM toxicity [16]. Other studies suggest that some metals in a complex mixture like ROFA can interact, indeed in an antagonistic fashion, but that total bioavailable metal can be a reasonable metric for PM dose [10,14]. Not only does the theory relate to ambient PM, recent preliminary data from our laboratory indicates the toxicity of domestic indoor air PM in rodents is similarly defined by its bioavailable metal content (unpublished data).

It is our view that transition metals provide the foundation for "biologic plausibility" and the theory appears to be garnering more and more attention [17], however, we remain far from accounting the epidemiologic observations. What is critical at this point is acquiring evidence demonstrating that metals associated with ambient PM can have effects at the low concentrations likely to be encountered in the ambient environment, and whether / how such PM-associated metal can induce effects in humans. *In vitro* studies with human airway epithelium show clearly that metals, such as vanadium, can induce toxicity by perhaps at least two mechanisms [18], but as with most *in vitro* studies, concerns remain regarding dose and extrapolation to the whole organism. Moreover, of incidental, but no less importance is the potential for cumulative metal toxicity, a mechanism yet to be explored.

3. ANIMAL MODEL OF PRE-EXISTENT CARDIOPULMONARY DISEASE

The epidemiology strongly suggests that the aged, particularly those with underlying cardiopulmonary diseases (e.g., COPD, infection) and children with asthma are more susceptible to the reputed PM-associated mortality and morbidity, respectively [1]. The difficulty in experimentally studying such persons has passed the gauntlet to toxicologists to attempt to study relevant animal models bearing analogous cardiopulmonary impairments. A number of models are currently being characterized and pursued in an effort to address this question, but one model recently emerged as having particular sensitivity to certain PM and may be working via metal-mediated mechanism.

When the alkaloid, monocrotaline (MCT), is injected into rodents, a progressive vasculitis appears after several days which results in pulmonary inflammation and vascular remodeling by ~10-12 days such that the animals go on to develop pulmonary hypertension, right cardiomegaly, and ultimately *cor pulmonale* ending in death after several weeks [19]. This injury has been exploited widely as a model to study the pathogenesis of pulmonary hypertension. This model has been shown to be sensitive to

ROFA when administered by intratracheal instillation [14] and more recently by inhalation [20]. The cause of death in the MCT-ROFA animals in the intratracheal studies appears to have been related to altered cardiac function [21], although it is unclear whether death was associated with direct cardiac injury or was secondary to pulmonary failure. Recent electrocardiographic studies in humans [22] suggest that heart rate changes and loss of beat variability with PM is consistent with the current clinical perception that loss of variability is a risk factor for cardiac arrhythmias and heart attack. The MCT model has not as yet been utilized with ambient PM in this laboratory, although Godleski and coworkers [17] have reported mortality in a similar model with inhalation of concentrated ambient Boston air (~350 mg/m^3 for 6 hr/d, 3 days). The significance of these findings demands that these studies be replicated both in Boston and elsewhere.

It is uncertain what the potential underlying mechanisms are determinant of the apparent susceptibility of the MCT treated animals. The induction of inflammation and associated interstitial edema reduces pulmonary diffusing capacity (unpublished data) could lead to ventilation-perfusion mismatches within the lung leading to secondary cardiac changes to an already damaged or stressed heart as suggested by the arrhythmias reported in the work of Watkinson et al. [21] and perhaps implicit in the recent human field study findings [22]. Our recent studies have suggested that the enhancement of the toxicity of intratracheally instilled ROFA in the MCT-treated rat may be mediated by augmentation of metal-catalyzed oxidant injury via the classic Fenton pathway due to the substrate rich milieu of the lung lining fluid during inflammation and the weakened state of the blood-air barrier (preliminary data). *In vitro* assessment of the feasibility of this reaction with ROFA is supportive [9], but direct evidence of this reaction in the *in vivo* model with ROFA, let alone ambient PM, is yet to be verified. More work is needed to assess this hypothesis in light of other pathophysiologic frailties of the impaired animal.

4. CONCLUSIONS

Recent epidemiologic studies indicate a significant association between ambient PM exposure and adverse health consequences, most apparent in susceptible subpopulations. The lack of a clear mechanism implying "biologic plausibility" has stimulated considerable speculation and research in the toxicologic community. Among the several hypotheses suggested, that involving bioavailable transition metals appears, thus far, to be consistent with the epidemiologic observations. However, results to date involve relatively high concentrations of metal in model emission and ambient PM samples. As results accumulate at much lower, and preferably inhalation exposure, concentrations the veracity of this hypothesis will be ascertained.

REFERENCES

1. U.S. Environmental Protection Agency, Air Quality Criteria for Particulate Matter, EPA/600/P-95/001, Vol.2, Washington: Office of Research and Development (1996).
2. S. Moolgavkar and E.G. Luebeck, *Epidem.*, 7 (1996) 420.
3. J. Schwartz and A. Marcus, *Am. J. Epidem.*, 131 (1990) 185.
4. R.B. Schlesinger, *Proced. 2nd Colloquium on Particulate Air Pollution and Human Health*, May 1-3, Eds. J. Lee and R. Phalen, Univ. Utah and Univ. Calif. Press, p.3-72.
5. G. Oberdorster, R.M. Galein, J. Ferin, and B. Weiss, *Inhal. Toxicol.*, 7 (1994) 111.
6. W.D. Bennett, K.L. Zelman, C.S. Minn, and J. Mascarella, *Proced. 2nd Colloquium on Particulate Air Pollution and Human Health*, May 1-3, Eds. J. Lee and R. Phalen, Univ. Utah and Univ. Calif. Press, p. 3-45.
7. J. Lewtas, *Environ. Health Perspect.*, 100 (1993) 211.
8. P.V. Targonski, V.W. Persky, and V. Ramekrishan, *J. Allergy Clin. Immunol.*, 95 (1995) 955.
9. R.J. Pritchard, A.J. Ghio, J.R. Lehmann, D.W. Winsett, J.S. Tepper, P. Park, M.I. Gilmour, K.L. Dreher, and D.L. Costa, *Inhal. Toxicol.*, 8 (1996) 457.

10. K.L. Dreher, R. Jaskot, J. Richards, J.R. Lehmann, A. Hoffman, and D.L. Costa, *J. Toxicol. Environ. Health,* 50 (1997) 285.
11. D.F.S. Natusch, J.R. Wallace, and C.A. Evans, *Science* 183 (1974) 202.
12. Fortoul, L.S. Osorio, A.T. Tovar, D. Salazar, M.E. Castilla, and G. Olaiz-Fernandez, *Environ. Health Perspect.*, 104 (1996) 630.
13. G.E. Hatch, E. Boykin, J.A. Graham, J. Lewtas, F. Pott, K. Loud, and J.L. Mumford, *Environ Res.,* 36 (1985) 67.
14. D.L. Costa and D.L. Dreher, *Environ. Health. Perspect.,* (1997) (in press).
15. A.J. Ghio, J. Stonehuerner, R.J. Pritchard, C.A. Piantadosi, D.R. Quigley, K.L. Dreher, and D.L. Costa, *Inhal.Toxicol.*, 8 (1996) 479.
16. A.J. Ghio, Z.H. Meng, G.E. Hatch, and D.L. Costa, *Inhal. Toxicol.*, 9 (1997) 255.
17. J.J. Godleski, C. Sioutas, M. Katler, M. Catalano, and P.O. Koutrakis. *Proced. 2nd Colloquium on Particulate Air Pollution and Human Health*, May 1-3, Eds. J. Lee and R. Phalen, Univ. Utah and Univ. Calif. Press, p. 4-136.
18. J.M. Samet, W. Reed, A.J. Ghio, R.B. Devlin, J.D. Carter, L.A. Dailey, P.A. Bromberg, and M.C. Madden. *Toxicol. Appl. Pharmacol.* 141 (1997) 159.
19. White and Roth
20. C.R. Killingsworth, F. Alessandrini, G.G.K. Murthy, P.J. Catalano, D. Paulauskis, and J.J. Godleski, *Inhal. Toxicol.*, 9 (1997) 567.
21. Watkinson, W.P., Campen, M.J., and Costa, D.L. Cardiac Arrhythmia Induction after Exposure to Residual Oil Fly Ash Particles in the Pulmonary Hypertensive Rat. *Fund. Appl. Toxicol.*, 1997 (accepted).
22. C.M. Shy, J. Creason, R. Williams, D. Liao, M. Hazucha, R. Devlin, and R. Zweidinger, Presentation at *The 13th Annual Meeting of the Health Effects Institute*, Annapolis, MD May 4-6, 1997.

U.S. status of particulates - monitoring, modeling, background, trends, and standards.

Terence Fitz-Simons, Michelle Wayland, and David Mintz

Air Quality Trends Analysis Group
Emissions, Monitoring, and Analysis Division
Office of Air Quality Planning and Standards, U.S. Environmental Protection Agency
MD-14, Research Triangle Park, NC 27711, USA.

1. INTRODUCTION

Particulate matter (PM) has been measured in the United States as Total Suspended Particulates (TSP) and as particles of diameter 10 microns or less (PM_{10}). The first national ambient air quality standards (NAAQS) for PM were established in 1971 and were based on health data and air quality data assembled by the Department of Health Education and Welfare in 1969 [1]. The original standards were based on TSP as measured by the high volume sampler (or hi-vol) which had a sampling efficiency cut-point of 25 to 45 μm. The standards were set at that time to 1 exceedence of 260μg/m^3 and an annual average limit of 75 μg/m^3. However, a review of the PM NAAQS, initiated in 1979 and completed in 1987 changed the basic indicator of particulate matter in the U.S. to PM_{10} and based the new standards on this indicator. The samplers for this indicator were hi-vols with a sampling efficiency cut-point set at 10μm. The form and the level of the standards were changed at that time to 1 expected exceedence of 150μg/m^3 for daily values and an expected annual average limit of 50μg/m^3. In the ensuing years, the TSP monitoring network grew smaller while the PM_{10} network grew larger. At present, a NAAQS review is underway that will most likely add a new indicator, $PM_{2.5}$ (particles of diameter 2.5 microns or less), to the list of PM standards.

2. MONITORING

The monitoring method for PM in the U.S. is basically gravimetric, as opposed to nephelometer or beta gauge instruments. The reason for this is the ability to analyze a subset of filters for speciated compounds comprising the particulate mix. Such analyses are essential for control strategies and provide a better understanding of the dynamics of a regional or local particulate situation. Unfortunately, gravimetric methods incur high costs per sample and are manual methods. This has resulted in less than everyday sampling for the majority of the U.S. particulate networks (usually 1 in 6 day sampling).

The national U.S. network is composed of monitors that meet certain siting and method requirements and have been classified as part of the state and local air monitoring system

(SLAMS). A subset of these monitors meet more specific siting, purpose, and reporting requirements and are classified as part of the national air monitoring system (NAMS). Other monitors may meet one or more of these requirements and are classified as "special purpose" and "other" monitors. In 1981 the U.S. had a TSP hi-vol network of about 4000 sites. By 1995 the U.S. had a national network of 1737 PM_{10} monitors while the TSP monitors had shrunk in number to about 600 monitors. These trends are shown in Figure 1. The data from these networks are used to develop the trends analyses presented later in this paper. Due to the proposed new standards, a $PM_{2.5}$ network will be deployed nationwide, while some reduction will occur in the number of PM_{10} monitors. However, this reduction will not be as dramatic as the reduction in TSP monitors since the PM_{10} standards will also be retained [5].

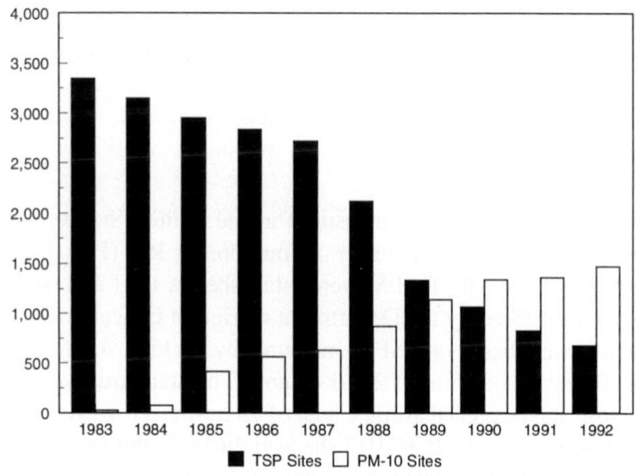

Figure 1. Trends in TSP and PM_{10} Sites in the U.S. from 1983 to 1992.

3. BACKGROUND LEVELS

Natural sources contribute to all size fractions of particulate matter. Since the determination of background usually affects what controls need to be made and where standard levels should be set, defining background concentrations is usually a non-trivial exercise that is complicated by different agendas brought to the scientific table. Add to that the knowledge that background levels can vary widely depending on what area is being discussed and on how background levels are defined. The Environmental Protection Agency (EPA) considers background to be those levels that would result in the absence of anthropogenic emissions of particulates and their precursors. A more complete treatment of background levels can be found in NAPAP, 1991 [2] and Trijonis, 1982 [3]. A less rigorous treatment can be found in Malm *et al.* , 1994 [4].

The actual background levels are shown in Table 1. The lower end of the range is from NAPAP (1991) and Trijonis (1982) while the upper end is from Malm *et al.* (1994). In general, the more industrialized and populated East has higher background levels than the West.

Table 1.
Background Levels of PM_{10} and $PM_{2.5}$

	Western U.S. (μg/m³)	Eastern U.S. (μg/m³)
PM_{10} - annual average	4-8	5-11
$PM_{2.5}$ - annual average	1-4	2-5

4. MODELING ACTIVITIES

Particulate modeling has, in general, been confined to local efforts using gaussian plume model techniques. Urban scale modeling has been performed with the urban airshed model (UAM). The Regulatory Modeling System for Aerosol and Deposition (REMSAD) is more sophisticated and accounts for secondary particle formation. More recent activities revolve around the regional acid deposition model (RADM) and the regional particulate model (RPM). These models are even more sophisticated and incorporate complex chemistry to model secondary particulates. However, the most sophisticated particulate model is to be part of a larger modeling system, Models 3, which will model ozone, carbon monoxide, and particulates on a regional scale. A beta test version is hoped to be available by the Summer of 1997 while the production version may be available in the Summer of 1998 [6].

5. TRENDS

Particulate trends are characterized by data from the national network meeting longevity and data capture requirements. Specifically, for a 10-year period, a site must have enough data to complete 8 valid years before the data from this site is included in the trends analysis. With those requirements and remembering that the U.S. switched to PM_{10} in 1987, the latest analyses

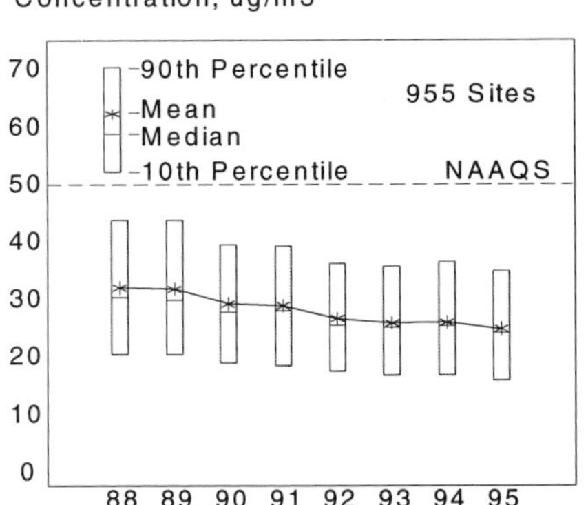

Figure 2. Trends in PM_{10} annual Mean

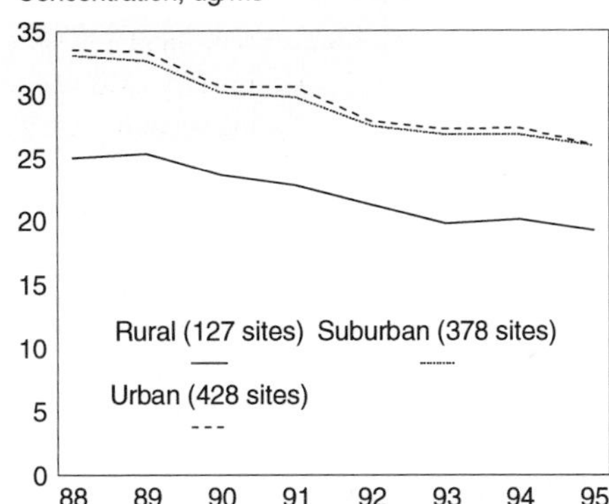

Figure 3. PM_{10} trends by Location.

comprise the 8 year period for 1988 to 1995 during which a monitor had to have 7 valid years of data. In the U.S. trends are reported as national composite statistics from sites meeting the criteria. The specific statistic used is related to the NAAQS for the pollutant being reported. Therefore, for PM_{10}, the composite statistic is based upon the annual mean. Examining these trends, PM_{10} decreased 22% during this period and 4% for the short term period 1994 to 1995. As can be seen in Figure 3, composite averages for urban and suburban sites are almost identical, while the composite average for rural sites is similar but substantially lower.

Looking at trends in traditionally estimated or inventoried emissions (Figure 4), the long-term decrease in emissions was 17% for the period 1988 to 1995 and 6% for the short-term period 1994-1995. These emissions are from fuel combustion, industrial processes, and transportation. For the first time in recent years, emissions from industrial processes were estimated to be higher than fuel combustion emissions. Industrial process emissions increased 1% from 1994 to 1995 while fuel combustion emissions decreased 12% over the same time period. However, the traditionally estimated emissions make up only 6% of total PM_{10} emissions nationwide. The other categories are natural sources such as wind erosion, miscellaneous sources such as agriculture and forestry, fires (both controlled and wildfires), and fugitive dust from paved and unpaved roads. Fugitive dust emissions are the largest component to total PM_{10} emissions, making up 68%. It should be noted that the sharp decrease seen between 1989 and 1990 is an artifact of changes made in data gathering and calculations of emissions for particulates in 1990.

In 1995 there were 24 million people living in 22 counties that had concentrations above either the daily or the annual PM_{10} NAAQS levels [2].

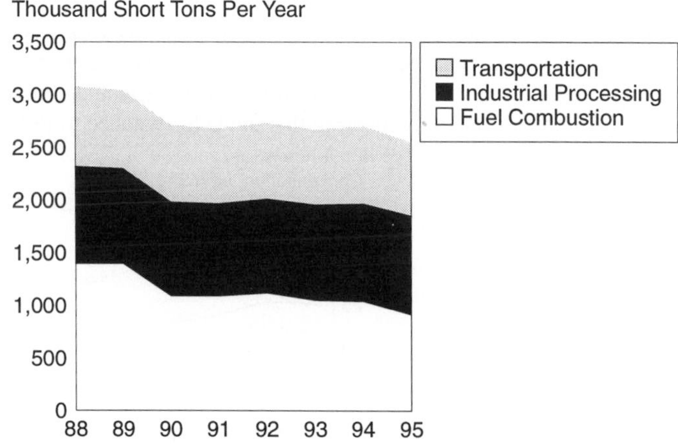

Figure 4. Trends in PM_{10} emissions for 1988 to 1995.

6. NEW STANDARDS

Changes to the particulate matter NAAQS have been proposed in the U.S. Recent health studies have indicated that significant portions of the populations are inadequately protected by the present standards. Several epidemiology studies have suggested that fine particulates might be causing severe health effects such as premature death and morbidity (for example [7, 8, 9]). In response to this new information, the EPA has proposed both daily and annual ambient air quality standards for $PM_{2.5}$. PM_{10} standards will also be retained. Due to variable sampling schedules and uncertain data capture for PM monitors, the daily standard for both PM_{10} and $PM_{2.5}$ has been proposed as a 3-year average of the 98th percentile of yearly data at a site. This approach moves away from the so-called exceedence form used for the present PM_{10} daily standard and towards a concentration based standard recommended by many advisors to the agency. The annual standard remains the same for PM_{10}, the 3-year annual average concentration must be less than $50 \mu g/m^3$. However, for $PM_{2.5}$ the annual average of several monitors will be allowed to produce a spatially averaged annual mean. This is a significant new step in setting air pollution standards for the agency and has generated much discussion.

REFERENCES

1. U.S. Department of Health Education and Welfare (DHEW),(1969). Air Quality Criteria for Particulate Matter, U.S. Government Printing Office, Washington, D.C.
2. National Air Quality and Emissions trends Report, 1995. EPA-454/R-96-005, U.S. Environmental Protection Agency, Office of Air Quality Planning and Standards, Research Triangle Park, N.C., October 1996.
3. National Acid Precipitation Assessment Program (NAPAP),(1991). Office of the Director, Acid deposition: State of Science and Technology, Report 24, Visibility: Existing and Historical Conditions - Causes and Effects. Washington, D.C.
4. Trijonis, J. (1982) Existing and natural background levels of visibility and fine particles in the rural East. Atmospheric Environment. 16:2431-2445
5. Malm, W.C.; Sisler, J.F.; Huffman, D.; Eldred, R.; Cahill, T.A. (1994) Spatial and seasonal trends in particle concentration and optical extinction in the United States. Journal of Geophysical Research 29:1347-1370.
6. Personal communication with Thomas N. Braverman, Air Quality Modeling Group, Emissions, Monitoring and Analysis Division, Office of Air Quality Planning and Standards, Research Triangle Park, N.C.
7. Schwartz, J. (1994). Air Pollution and hospital admissions for the elderly in Detroit, Michigan. Am. J. Respir. Crit. Care Med. 150:648-655.
8. Schwartz, J. And Dockery, D.W. (1996). Is daily Mortality Associated Specifically with Fine Particles? J. Air Waste Manag. Assoc. : Accepted.
9. Spengler, J.D.; Koutrakis, P.; Dockery, D.W.; Raizienne, M.; Speizer, F.E. (1996) Health effects of acid aerosols on North American children: air pollution exposures. Environ. Health Perspect.: in press.

European status - Air quality: Trends, monitoring, background modelling

H.C. Eerens, J.A. van Jaarsveld and J. Peters

1. INTRODUCTION

Over the last decades, aerosols have been subject of many studies which have considerably augmented the understanding of emission, atmospheric fate and deposition of aerosols (e.g. Lee et.al., 1985). The formation and growth of secondary aerosols from the oxidation of sulphur dioxide, nitrogen oxides, ammonium and VOC's, as a result of photo-chemical reactions in the atmosphere, is reasonably well understood and have been shown to be predominantly man made origin. The distribution of these secondary aerosols across Europe is also reasonably well known. Less information is available about the contribution of primary aerosols from anthropogenic and especially natural sources to the mass concentration of aerosols. Studies suggest, however, that their contribution may be significant (e.g. RIVM, 1996).

In most European countries, industrialization and high volumes of traffic mean that anthropogenic sources predominate, especially in urban areas. The most significant of these are traffic, power plants, combustion sources (industrial and residential) and agricultural activities. The main natural sources of airborne particulates in Europe are sea spray and soil resuspension by the wind. In addition, in the Mediterranean basin, Saharan dust and volcano emissions can also be important natural sources of particulates. Although the problems around particles has a long history, emission inventory activities are being undertaken only recently. (*Berdowski et. al., 1997a, 1997b).*

2. HISTORY AND TREND OF AIR POLLUTION BY PARTICULATE MATTER IN EUROPE

Particulate matter has, as already mentioned, a long history in Europe (Brimblecombe, 1987). One of the first quotations stems from the Roman philosopher Seneca who, already in 61 A.D., reported over the conditions in Rome as follows [Stern et. al., 1973]:

> "As soon as I had gotten out of the heavy air of Rome and from the stink of the smoky chimneys thereof, which, being stirred, poured forth whatever pestilential vapors and soot they had enclosed in them, I felt an alteration of my disposition."

Sterns continue to report that air pollution by particulates, caused by burning wood in Tutbury Castle in Nottingham, was considered 'unendurable' by Eleanor of Aquitaine, the wife of King Henry II of England and caused her to move in the year 1257. Somewhat later coal burning was prohibited in London. Looking to the more recent history can the period 1925-1950 be characterized as the period that present day air pollution problems and solutions merged. The Meuse Valley (Belgium) episode, one of the first occasions that excessive deaths were related to a period of elevated air pollution (smog), occurred in 1930. Although air pollution research got a start with first large-scale surveys (Leicester, England, 1939) no significant air pollution legislation or regulations were adopted. Only after the major air pollution disaster hit London in 1952 regulation started in Europe. Although the air quality in Europe has improved since then, the problem of particulates in the air has not been solved. In fact the second US-Dutch International Symposium in 1985 already about Aerosols (Lee et.al., 1986).

3. TRENDS

As a consequence of the long history of the problem many different measurement methods, describing different parts of the particulates in the air, exists now a day in Europe. New regulation (EU/DGXI, 1997) and harmonization activities (cf. CEN/TC 264/WG6) are momentarily in progress to improve this situation. See table 1 for the measurements methods in use (EU/DGXI, 1997) at national networks of EU countries.

Long term trends for PM_{10} are not available for Europe. Historically mostly TSP and black smoke has been measured. Long records are available for e.g. Paris, London, Birmingham, Manchester, Luxembourg, Brussels and Amsterdam (see figure 1). In all cases a strong downward trend in the 60's up to the beginning of the 80's can be observed. The reduced use of coal is the main reason for this downward trend. The increase of diesel traffic, especially visible in the black smoke measurements, flattens out the decrease since the late 80's.

In Europe, ambient concentrations of PM_{10} have been monitored in some urban networks since 1990, but there is currently no coherent overall European PM_{10} data set, mainly because PM_{10} has been systematically monitored only in a few member states (EU/DGXI, 1997). In addition, there is, momentarily no standardized method for monitoring PM_{10} across Europe, although standardization of PM_{10} measurement methods is under development (cf. CEN/TC 264/WG6) and expected to be adopted this year.

Previous studies and the data collected by the Working Group on fine particulates from EU-Member States (see table 2) indicates a reasonably consistent pattern of lower concentrations in the far north of Europe and higher concentrations, possibly due to naturally occurring particles, in the southern countries.

Measurements of $PM_{2.5}$ are still very scarce. Some primary measurements in the UK and The Netherlands suggest a ratio of $PM_{2.5}/PM_{10}$ of approx. 0.6. In the forthcoming particulate EU directive PM_{10} and $PM_{2.5}$ measurements will be mandatory in all EU member states.

An overview of the existing air quality networks and measurement data related to particulates can be found at the European Topic Center for Air Quality of the European Environmental Agency.

4. MODELLING OF FINE PARTICULATES

To describe the PM_{10} concentration a sub-division in 5 groups has been made:

- Primary emitted particulates (calculated by the EUTREND-model)
- Secondary inorganics (NO_3, SO_4, NH_4; calculated by the EUTREND-model)

- Secondary organics (no method available, rough calculates gives an average contribution of 1-2 ug/m^3 for Europe)
- Sea Salt (calculated, using scavenging ratio's, from rain water data)
- Soil resuspension (no calculation method available)

EUTREND

The long range transport model EUTREND is used for the calculation of the annual average mass concentrations of primary (PM_{10}, $PM_{2.5}$) and secondary aerosols (SO_4, NO_3, NH_4) over Europe on the basis of the earlier mentioned emissions inventory. The EUTREND model is an European version of the Operational Priority Substances (OPS) (van Jaarsveld, 1995). This family of models can be characterized as Lagrangian models in which the transport equations are solved analytically. Contributions of the various sources are calculated independent of each other using backward trajectories, local dispersion is introduced via a Gaussian plume formulation. Average concentrations are not determined from sequential (e.g. hourly) calculations but from concentrations calculated for a limited number of meteorological situations (classes) using a representative meteorology for each of the classes. Meteorological data is taken partly from the Numerical Weather Prediction model of the European Center for Medium Range Weather Forecasts (ECMWF) in Reading (UK) and partly from observations at meteorological stations all over Europe.

Dry deposition, wet deposition and chemical transformation are incorporated as first order processes and independent of concentrations of other species (Van Jaarsveld, 1995; Asman and Van Jaarsveld, 1992).

In the EUTREND model, five particle size classes are used, each characterized by a (monodisperse) particle size with corresponding properties calculated by the semi-empirical model of Sehmel and Hodgson (1980) which gives similar results as the more theoretical model of Slinn (1983). Concentrations and depositions are calculated for each of these classes and weighted with the percentage of the total particle mass appointed to the individual classes. Such an approach is especially useful for the modelling of primary-emitted particles because they usually cover a broad range of particle sizes, often including a significant fraction of large particles. Particle growth is not incorporated in the present model but is implicitly assumed to take place in the lowest size-class (d < 1 (m). In support of the European Commission the EUTREND model has been used to calculated the primary and secondary contribution of these sources to the PM_{10} and $PM_{2.5}$ concentration in Europe.

Result of model calculations for rural background locations (using EMEP grids) for secondary inorganics and primary PM_{10} concentrations (using 1990 meteorology and emissions) are given in figure 2 and 3 respectively. For secondary aerosol an average contribution of 8.5 ug/m^3 (rural background between 0.6 and 24.9 ug/m^3) for the European Union has been calculated, with the highest background level in the eastern part of Germany (24.9 ug/m^3). For the primary emitted particulates an average contribution of 2.2 ug/m^3 has been calculated for PM_{10} (rural background between 0.2 and 15.9 ug/m^3) and 2 ug/m^3 for $PM_{2.5}$. The interpretation of these results should happen with some caution because the emission inventory used as input to the model is one of the first in its kind and needs still a lot of improvement and validation.

5. MODELLING OF SEA SALT

As part of its research on acidification RIVM has developed a method to determine the concentration of base-cations in ambient air from rain water measurements (Draaijers et al 1996). With some slight modifications this method has been applied to calculate the concentration of sea salt aerosols in ambient air.

Earlier studies have found (e.g.Woolf et al., 1987) that the typical diameter of sea salt aerosol, that is produced at the ocean surface by the bursting of air bubbles (a minimum wind speed of 3-4 m.s$^-$1 is required), is 1-2 (m (although extending to sizes greater than 10(m). Therefore long-range transport of sea-salt aerosol can be expected. Using scavenging ratios, the air concentrations of these generated sea salt particles can be calculated from precipitation concentrations. Ambient air concentrations derived this way will reflect the large scale background situation.

The scavenging ratios were derived (Draaijers et al 1996) from simultaneous measurements of base-cations concentrations in precipitation and surface-level. This approach is based on the premise that cloud droplets and precipitation efficiently scavenge particles resulting in a strong correlation between concentrations in precipitation and the surface-level air (Eder and Dennis, 1990). Scavenging ratios have been found reasonably consistent when averaged over one year or longer (Galloway et al, 1993). For this reason annual mean precipitation concentration has been used to infer annual mean air concentrations. The scavenging ratio (SR) is defined as:

$$SR= [C]_{rain} * Rho/[C]_{air} \qquad [1]$$

Where $[C]_{rain}$ denotes the concentration in precipitation (mg/l, ~mg/kg), $[C]_{air}$ the concentration in ambient air (in ug/m^3) and Rho the density of air, taken as 1200 g/m^3. For the typical size range of sea salt particles the following relationship between the scavenging ratio and mass median diameter (MMD, in um) can be derived from data of Kane et al. (1994):

$$SR= 188 * e^{(0.227*MMD)} \qquad [2]$$

Rearranging equation [1] and [2] gives a simple empirical model describing the relationship between air concentration at one hand and precipitation concentration and MMD at the other hand (Draaijers et al 1996):

$$[C]_{air}= ([C]_{rain}*1200)/(188 * e^{(0.227*MMD)}) \qquad [3]$$

Precipitation concentrations will reflect atmospheric concentrations of the entire atmospheric column from cloud top to surface level and thus will reflect the large scale 'background' situation. A strong correlation with surface level air concentrations will only be present in well-mixed conditions at sufficient distance from sources. Close to sources surface level air concentrations usually will be considerably higher. As a consequence, near the coast the contribution of sea salt to the total suspended matter *in the surface level* air will be underestimated by using the method described above. On the other hand rain occurs mainly during western wind circulation's, leading to an *overestimation* of the yearly average concentration.

Composition of sea salt

Sea water contains sea salt to about 3.5% by weight, of which 85% is NaCl and it can safely be assumed that the sea-salt content of film and jet drops is similar. The composition of sea salt aerosols will change in relation to the distance to the coast due to the impact of continental air masses. For example, HNO_3 and H_2SO_4 present in continental air masses may volatilize Cl from sea salt and convert NaCl into $NaNO_3$ and Na_2SO_4, respectively, simultaneously releasing HCl(g) to the atmosphere (Mamane and Gottlieb, 1992). If Cl loss is suspected, one can compare the Na/Cl ratio, which is a good indicator of the Cl depletion, since Na is a conservative element. The Na/Cl ratio for the bulk deposition fluxes measured in The Netherlands is very close to the sea-water ratio (0.86), with a small gradient (0.86 to 0.91) over The Netherlands thus leading to the conclusion that Cl loss to the gas phase is insignificant for The Netherlands, See also Draaijers and Hulskotte (1997).

Based on the above described model and parameters the sea salt contribution, based on yearly averaged Na (sea salt $=3.25*[C]_{Na}$) concentrations in rainwater, has been calculated. The precipitation data were taken from Leeuwen et al. (1995). They compiled measurement results from approximately 600 sites scattered over Europe. To calculate the PM_{10} contribution, an average PM_{10}/TSP ratio of 0.7 has been used. The results are presented in figure 4. For the European Union countries the average, calculated, sea salt contribution to PM_{10} is 3.1 ug/m^3, with the highest contribution calculated for the west-coast of Ireland, 9.9 ug/m^3.

6. RESULTS FOR PM_{10}

In figure 5 the sum of the secondary inorganics, primary PM_{10} and the PM_{10} sea-salt contribution is presented. Comparing the levels presented in figure 5 with the measurements of PM_{10} levels in Europe, given in table 2, it can be concluded that there is still a large gap between predictions and measurements. This is especially large in South-Europe. The latter is probably due to natural sources as soil resuspension and Sahara sand who are not yet included in the model.

Also it is clear from the figure that the primary PM_{10} contribution is systematic smaller then the contribution from secondary aerosols. The average PM_{10} background concentration (defined as the sum of the calculated concentrations of primary PM_{10}, secondary inorganics and sea salt) in the Europe Union is, according to these calculation, 15 ug/m3, with a maximum in the eastern part of Germany of 42 ug/m3.

From the calcium concentration (aerosols) map (figure 6, calculated from rain water measurements, corrected for sea-salt contribution), that can be used as a rough indicator for the distribution of resuspended soil aerosols, a clear north-south gradient can be observed. Using the average abundance of Calcium in crustal material (2.1%) an average contribution of resuspended soil of 20 ug/m^3 can be calculated, with the highest contribution in Southern Europe (47 ug/m^3).

7. CONCLUSIONS

- Despite the enormous amount of research that has been carried out over the years our knowledge is still incomplete and more work has to be done, especially concerning the contribution of natural sources to PM_{10} and $PM_{2.5}$.
- Primary emitted particulate matter emission databases are now available, although improvement is needed.
- Primary results indicates that resuspended dust has an important contribution to PM_{10} concentrations in (South)-Europe.
- The contribution of secondary inorganics can be quantified adequately.
- The contribution of secondary organics can't be quantified yet, but results from US research projects indicates that the contribution will be probably small (1-2 ug/m^3)
- A first estimate of the contribution of sea-salt to PM_{10} aerosol in Europe has been presented.

REFERENCES

1. Annema, J.A., Booij, H., Hesse, J.M., Meulen, A. van der, Slooff, W, Integrated Criteria Document Fine Particulate Matter, National Institute of Public Health and the Environment, Bilthoven, The Netherlands, Report No. 601014015, 1996.
2. Asman, W.A.H. en Van Jaarsveld, J.A. (1992) A variable resolution transport model applied for NHx in Europe. *Atmospheric Environment* 26A, 445-464.
3. Berdowski, J.J.M., W. MulderIng C. Veldt, A.J.H. Visschedijk, P.Y.J. Zandveld. "Particulate matter emissions (PM_{10} - $PM_{2.5}$ - $PM_{0.1}$) in Europe in 1990 and 1993", Apeldoorn, The Netherlands, TNO/MEP R 96/472, 1997a
4. Berdowski, J.J.M. et. al., this proceeding, 1997b
5. Brimblecombe, P., The Big Smoke: A history of air pollution in London since medievaltimes, University Press, Manchester, 1987, ISBN 0-416-90080-1
6. Draaijers, G.P.J, Leeuwen, E.P, Jong, P.G.H. de, Erisman, J.W., Deposition of base-cations in Europe and its role in acid neutralization and forest nutrition, Bilthoven, RIVM, reportnr. 722108017, 1996
7. Draaijers, G.P.J. and Hulskotte, J.H.J. (1997), A literature study on some anthropogenic and natural sources of particulate matter in the atmosphere. TNO report no. R96/508.
8. Eder, B.K. and Dennis, R.L., On the use of scavenging ratios for the inference of surface-level concentrations and subsequent dry deposition of Ca^{+}, Mg^{2+}, Na^{+} and K^{+}, Water, Air and Soil Pollution, 52, 197-215.
9. EU/DGXI ,Ambient Air Pollution by Particulate Matter; position paper, concept report April 1997.
10. Galloway, J.N., Savoie, D.L., Keene, W.C., Prospero, J.M., The temporal and spatial variability of scavenging ratios for nss sulfate, nitrate, methanesulfonate and sodium in the atmosphere over the North Atlantic Ocean, Atm. Environ., 25A, 2665-2670
11. Kane, M.M., Rendell, A.R., Jickells, T.D. (1994), Atmospheric scavenging processes over the North Sea. *Atmospheric Environment*, 28, 2523-2530.
12. Lee, S.D., Schneider, T., Grant, L.D., Verkerk, P.J. (eds.) Aerosols, Research, Risk Assessment and Control Strategies, Proceedings Second U.S.-Dutch International Symposium on Aerosols, Wiliamsburg, 1985, ISBN 0-87371-051-7, 1986.
13. Leeuwen, E.P. van, Potma, C., Draaijers, G.P.J., Erisman, J.W., Pul, W.A.J. van

(1995), European wet deposition maps based on measurements. RIVM report no. 722108006, Bilthoven, The Netherlands

14. Mamame, Y. and Gottlieb, J. (1992), Nitrate formation on sea salt and mineral particles - a single particle approach. *Atmospheric Environment*, 26A, 1763-1778.
15. Sehmel G.A. and Hodgson W.H. (1980) A model for predicting dry deposition of particles and gases to environmental surfaces. *AIChE Symposium Series* 86, 218-230.
16. Slinn W.G.N (1983) Predictions for particle deposition to vegetative surfaces. *Atmospheric Environment* 16, 1785-1794.
17. Stern, A.C., Wohlers, H.C., Boubel, R.W., Lowry, W.P.; Fundamentals of air pollution, Academic Press, London, 1973.
18. Van Jaarsveld, J.A. van (1995), Modelling the long-term atmospheric behaviour of pollutants on various spatial scales. Ph.D. thesis, University of Utrecht, The Netherlands.
19. Woolf, D.K., Bowyer, P.A., and Monahan, E.C. (1987), Discriminating between the film drops and jet drops produced by a simulated whitecap. *Journal of Geophysical Research*, 92, 5142-5150.

Table 1. Particle Monitoring in the European Union (situation 94-96)

Type of monitoring	
PM_{10}:	9 countries[1]
Black Smoke:	9 countries
TSP:	8 countries
$PM_{2.5}$:	2 countries

[1] B-gauge (30 min-2h): 6 countries; TEOM (30 min-24 h): 6 countries;Gravimetry (24h): 2 countries

Source: EU/DGXI, 1997.

Table 2. PM_{10} measurements in Europe (1990-1994)

Country	Urban background		Urban Traffic		Urban Industrial	
	N	av. (ug/m3)	N	av. (ug/m3)	N	av. (ug/m3)
Sweden	5	12-16	1	35		
Finland	5	13-44[1]				
Denmark						
Ireland						
United Kingdom	13	20-34				
Netherlands	4	37-41	4	39-43		
Belgium						
Luxembourg	1	30	1	32		
Germany	2	36-65	1	50-58		
France	3	41-67	2	51-54	9	43- 78
Austria	1	40-42				
Portugal	1	72-75				
Spain	6	39-90			4	52-123
Italy						
Greece						

[1] One station 145 ug/m3

Source: EU/DGXI, 1997.

Figure 1. Trend particulate concentrations North-West Europe 1956-1996

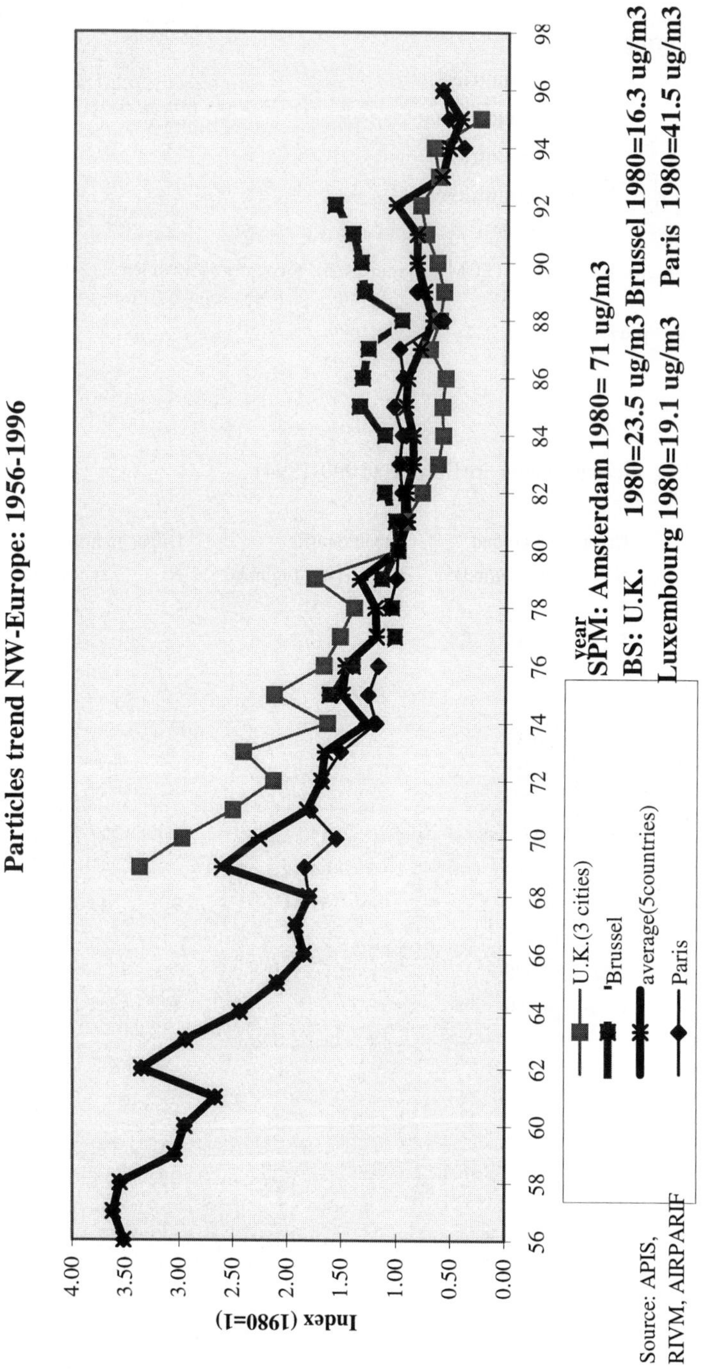

Figure 2. Primary PM_{10} contribution to PM_{10} concentrations of 1990

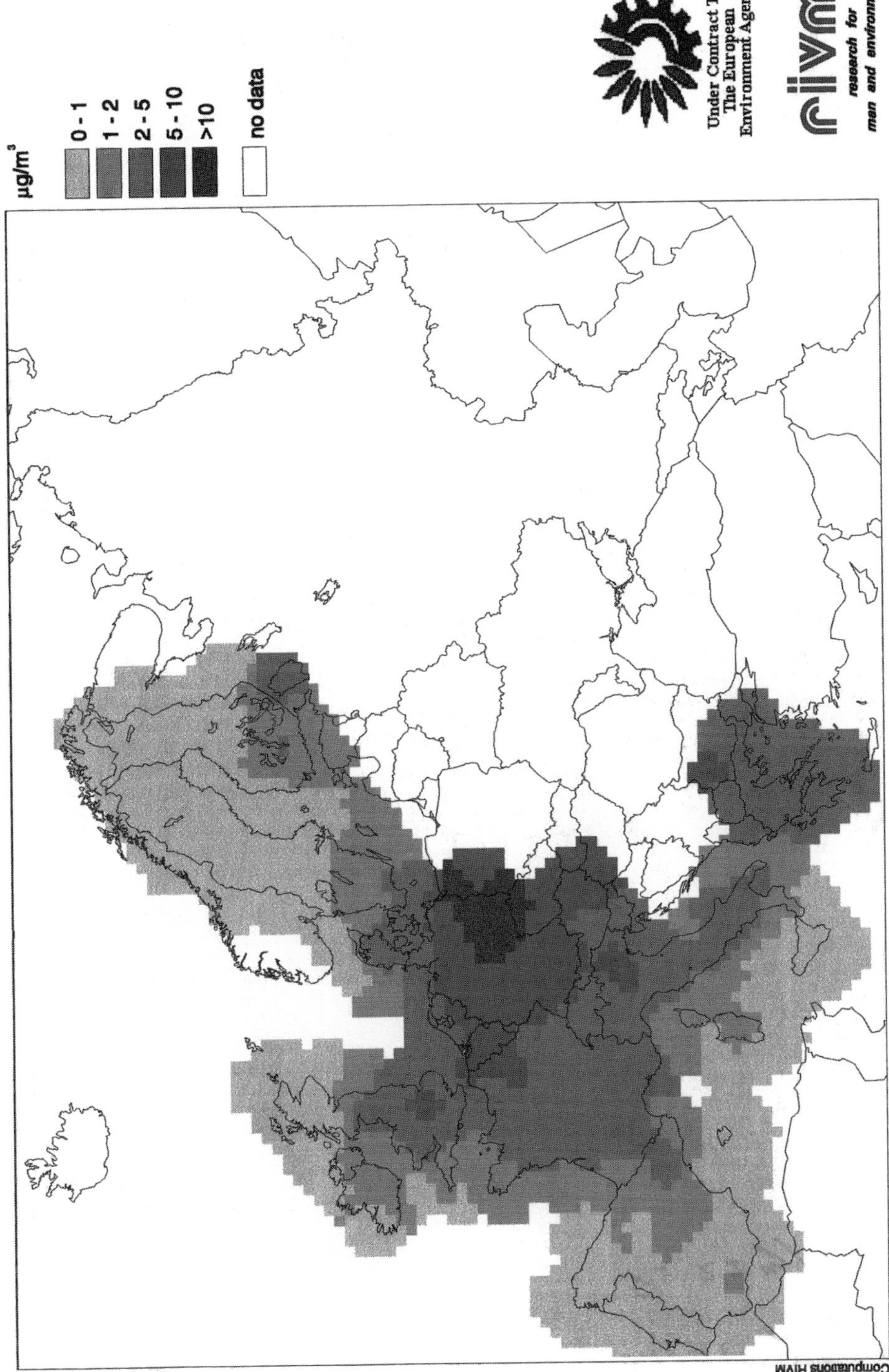

Figure 3. Contribution of secondary inorganics to the 1990 PM_{10} concentrations

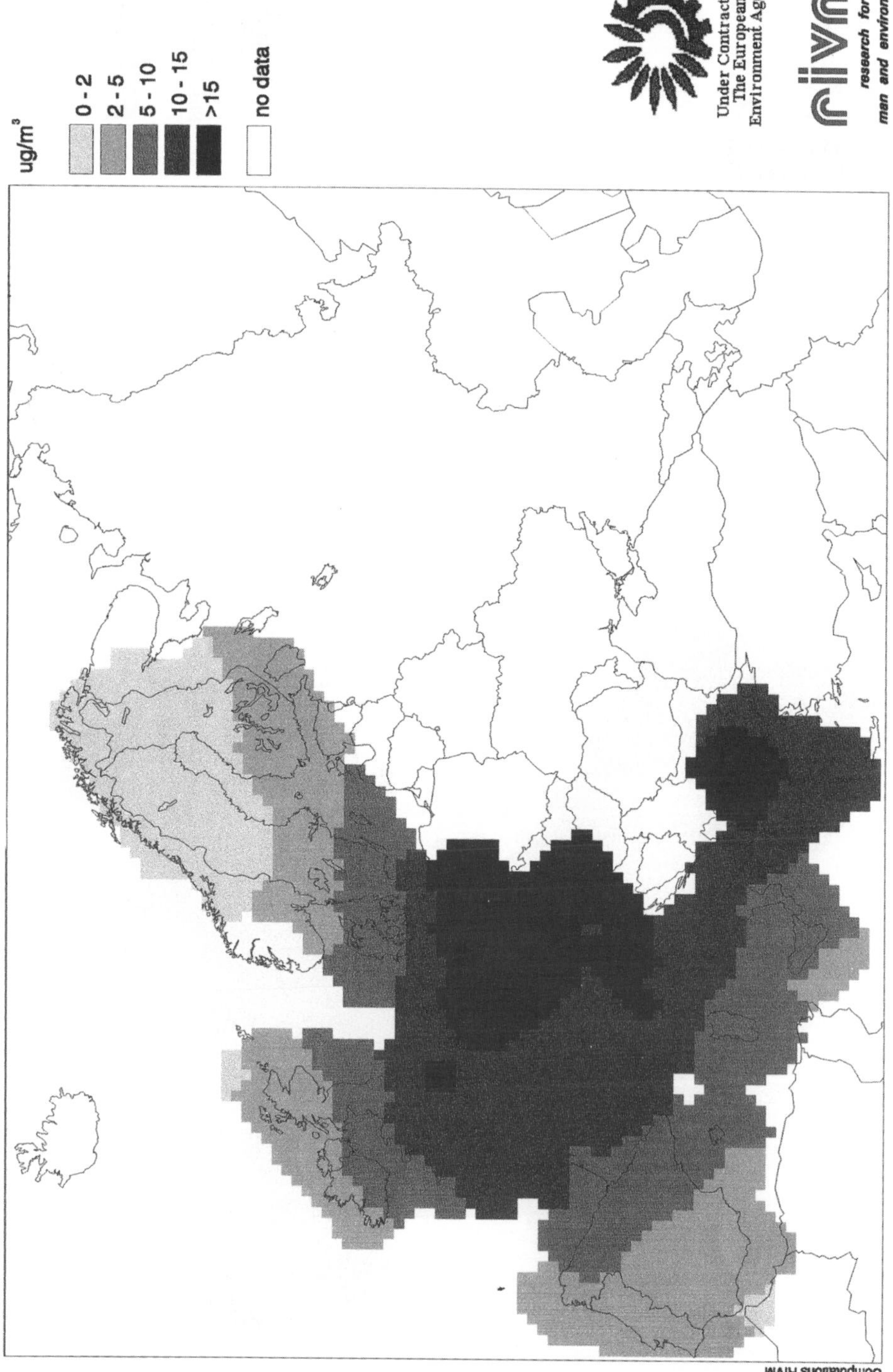

Figure 4. Contribution of sea salt particulates to the 1990 PM_{10} concentrations

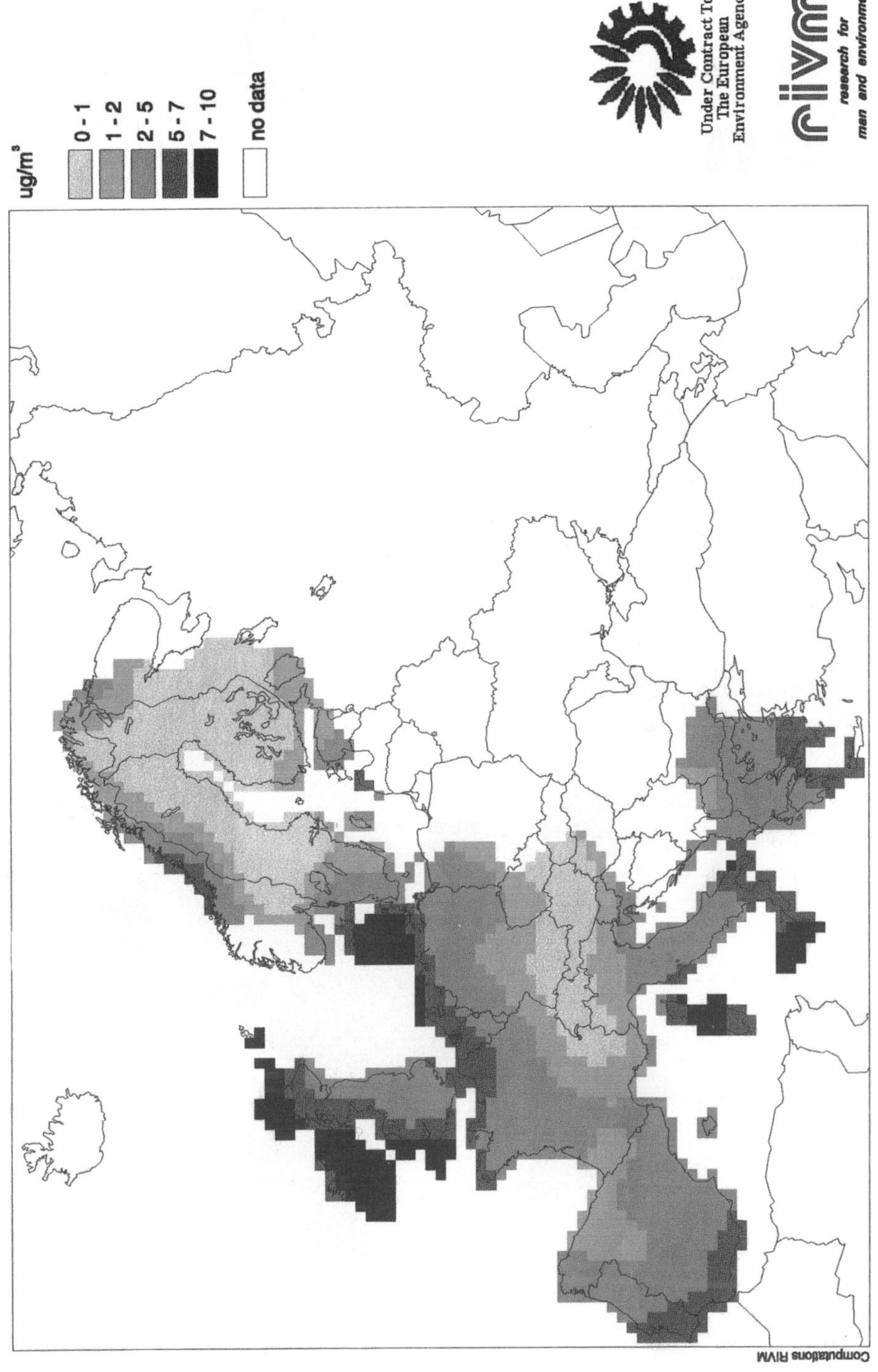

Figure 5. 1990 PM_{10} concentrations

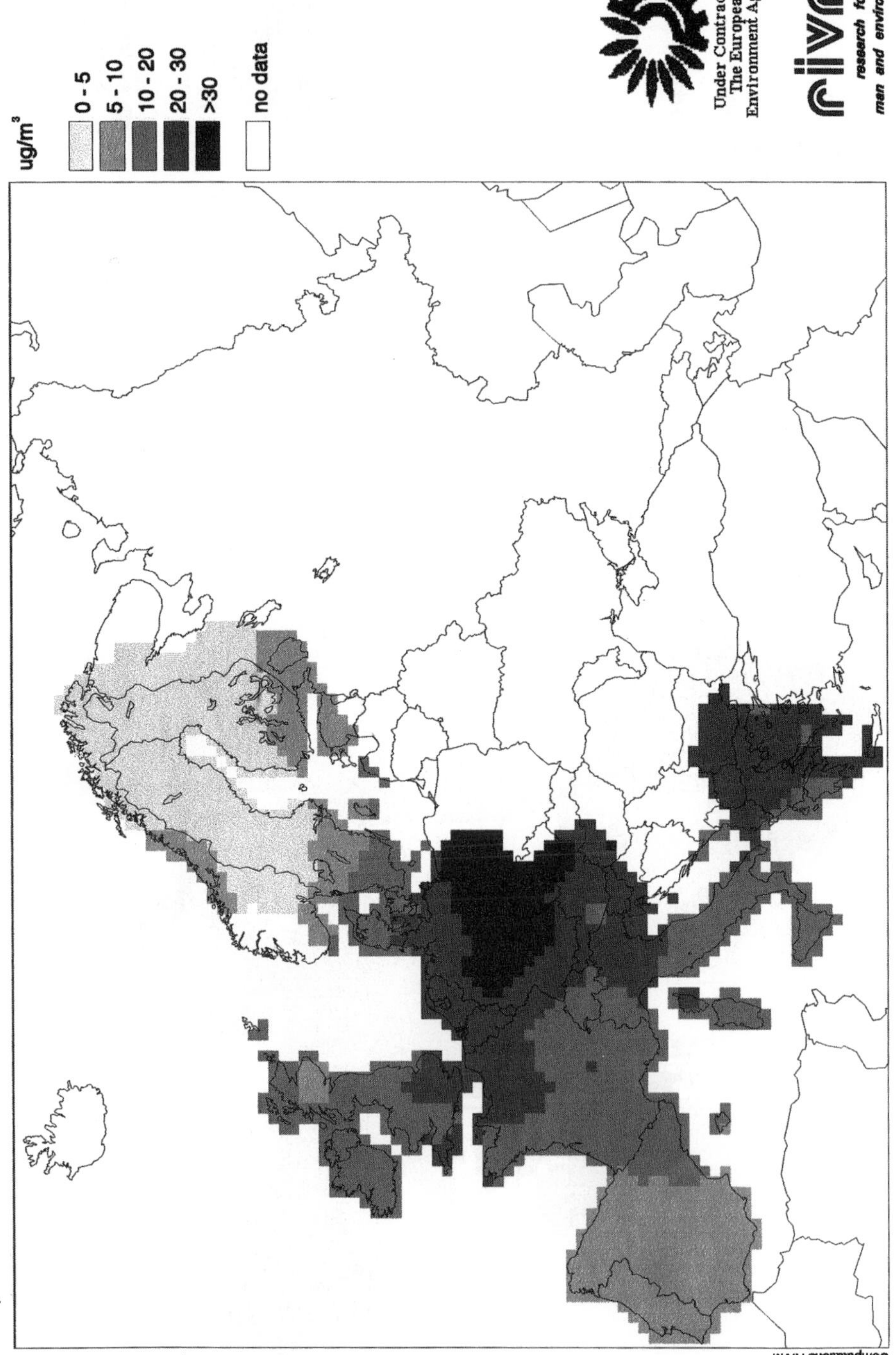

Figure 6. Calcium contribution (*47) to the 1990 PM_{10} concentration

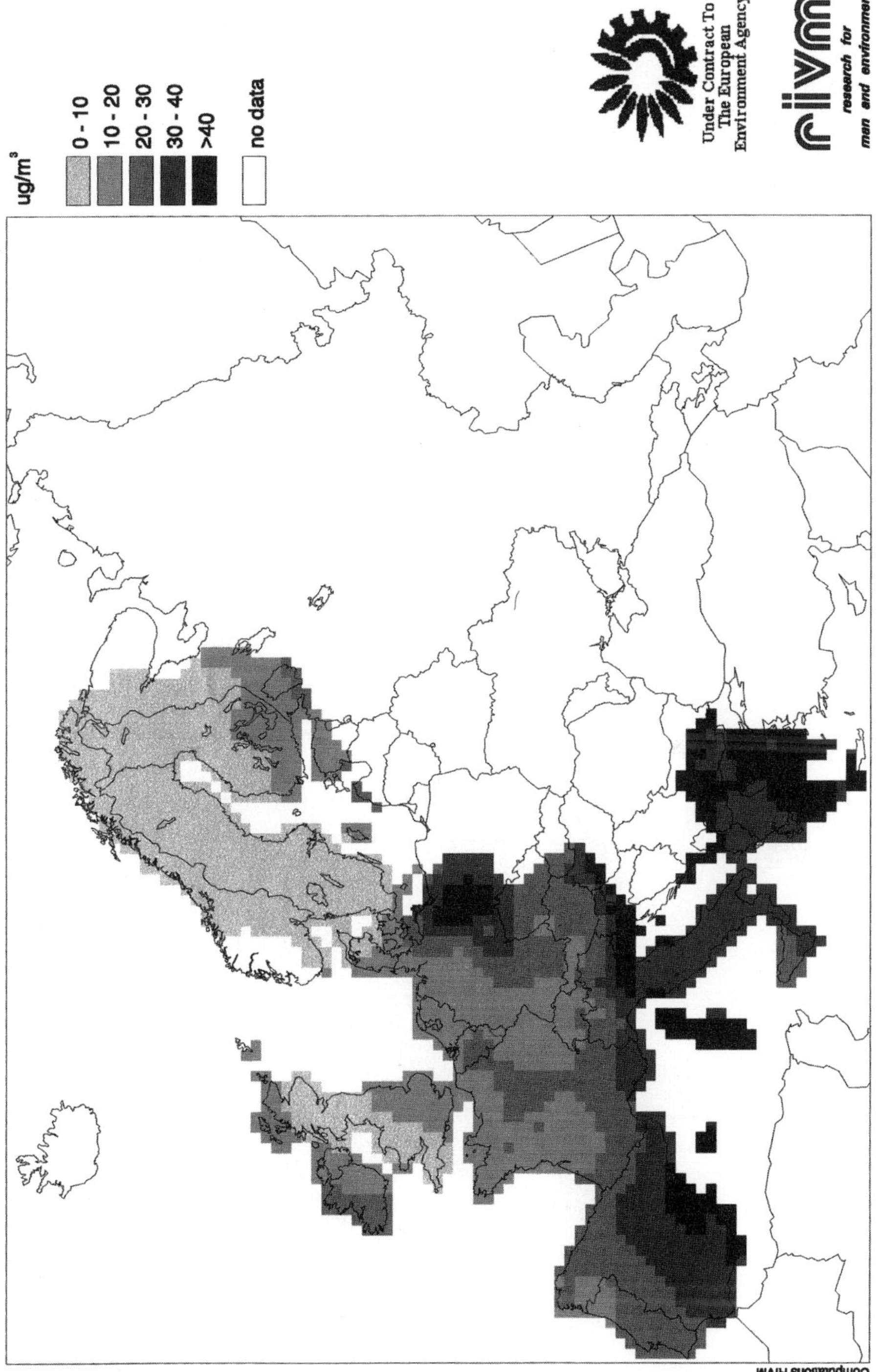

Particulate matter sources, emissions, and control options—USA

W. Gene Tucker

National Risk Management Research Laboratory, Air Pollution Prevention and Control Division, U.S.Environmental Protection Agency, Research Triangle Park,NC 27711,USA

ABSTRACT

A new national standard for particulate matter (PM) in the ambient air is being developed in the United States (U.S.). Whereas the current standard applies to PM less than 10 micrometers (m) in aerodynamic diameter (PM_{10}), the standard being developed would also address particles smaller than 2.5 m ($PM_{2.5}$). To meet either the current standard or a new one, sources of both primary and secondary particles need to be inventoried and controlled. (Control of 'primary' particles, which are emitted directly into the air, involves emissions prevention or collection at the source. Control of 'secondary' particles, which are formed in the atmosphere, requires reducing emissions of precursor constituents such as sulfur oxides, nitrogen oxides, and ammonia.) This paper summarizes the current knowledge on sources of primary particles in the U.S., and options for control of their emissions. Research needs and plans are briefly addressed.

1. BACKGROUND

Because epidemiological associations have been found at particle concentrations below the existing ambient air quality standard, a new national standard is being developed in the U.S. (USEPA, 1996d). Whereas the current standard applies to PM less than 10 m in aerodynamic diameter (PM_{10}), the standard being developed would also address particles smaller than 2.5 m ($PM_{2.5}$). The current PM_{10} standard considers only size and mass concentration; chemical composition and toxicity of the particles are not addressed. The proposed $PM_{2.5}$ standard is also based solely on size and mass concentration. Promulgation of the new standard is scheduled for June 1997. To meet either the current standard or a new one, sources of both primary and secondary particles need to be controlled. Control of 'primary' particles, which are emitted directly into the air, involves emissions prevention or collection at the source. Control of 'secondary' particles, which are formed in the atmosphere, requires reducing emissions of precursor constituents such as condensible organic compounds, sulfur oxides, nitrogen oxides, and ammonia.

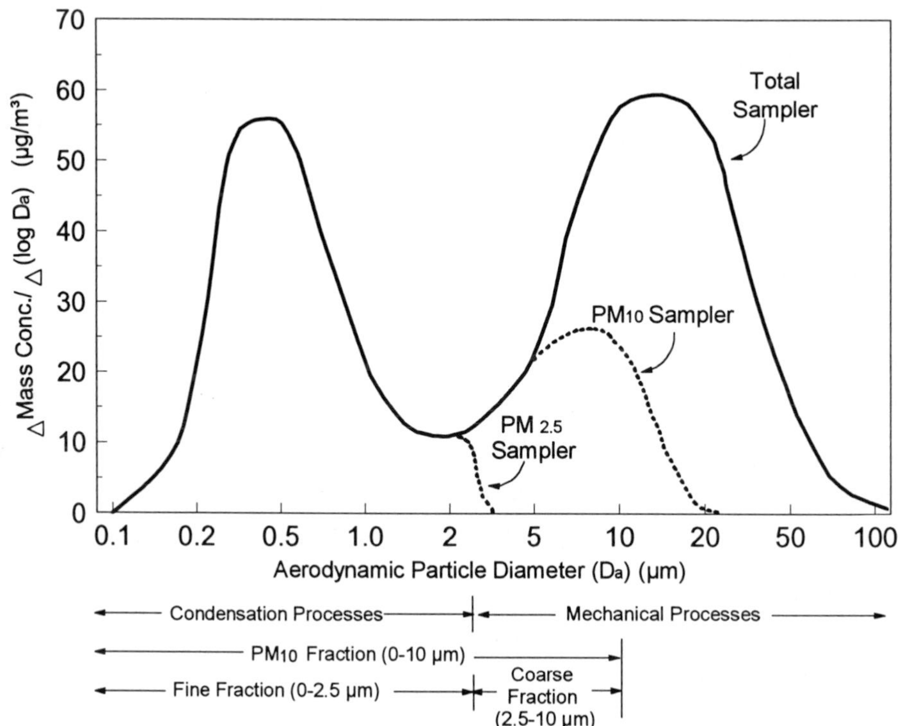

Figure 1. PM_{10} and $PM_{2.5}$ related to a typical size distribution for ambient particles. (Adapted from USEPA, 1996b)

Figure 1 shows a typical size distribution of particles in the ambient air, and how PM_{10} and $PM_{2.5}$ relate to the total distribution. This figure indicates that particles are generally distributed bimodally by size in the atmosphere, with the minimum of the distribution between 1 and 3 μm aerodynamic particle diameter. Particles in the fine mode are created as primary particles either by combustion or other high-temperature processes, or as secondary particles in the atmosphere. Most of the mass of particles in the coarse mode comes from materials that have been ground down by mechanical processes, although fine-mode particles can also become attached (USEPA, 1996b).

Ambient PM is a complex mixture of sizes and types of particles that originate from many sources. The size, chemical composition, and source of particles may all play a role in human exposures, and health effects resulting from those exposures. Little detailed information is available on the specific chemical makeup of ambient particles, especially the metal speciation and semivolatile organic components of fine particles. Chemical characterization of PM from a broad range of locations with a variety of source types is needed to better determine the composition, range of transport, and variability of airborne particles.

2. SOURCES OF FINE PM

The types of sources that contribute to the two modes in the distribution shown in Figure 1 are generally known. Particles in the finer mode include primary particles from high-temperature metallurgical and combustion processes, secondary particles from atmospheric reactions, and an unknown (but theoretically small) amount of fine particles that have been deposited and resuspended by wind or human activities. Particles in the coarser mode include coarse windblown and road dust, pollens and spores, and some industrial particles. PM_{10} samples are generally dominated, on a mass basis, by coarse-mode particles. $PM_{2.5}$ samples can also have substantial amounts, on a mass basis, of coarse-mode particles from the left-hand tail of the coarse-mode distribution; this may be especially likely in areas with significant sources of resuspended dust.

Since the epidemiological studies of PM health effects are based on geographically dispersed (but mostly urban) locations with numerous and varied emission sources, the major sources of the particles are difficult to ascertain. Researchers most familiar with these studies have speculated that two types of sources may be particularly important: primary emissions from combustion sources, and secondary particles formed in the atmosphere. Further hypotheses have pointed toward acidic particles, particles containing sulfates, and particles containing transition metals; all such particles would implicate combustion sources. See Vedal, 1997 and Wilson and Spengler, 1996 for recent reviews.

There is a lot of uncertainty with respect to 'fugitive' particles. Their emission rates are poorly quantified, but potentially large compared to emission rates of combustion and other industrial sources. Table 1 summarizes the U.S. national inventory for particle emissions from general industrial, combustion, and fugitive sources. Note that approximately 70% of the total primary $PM_{2.5}$ is from fugitive sources. This is presumably because the national total PM_{10} emissions are so large, and the mass in the left-hand tail of the coarse mode (see Figure 1) is large compared to the mass of particles being created in the fine mode. Even though the estimates for $PM_{2.5}$ fugitive emissions are being re- evaluated and may be revised downward somewhat, they will still constitute a large percentage of the national total.

Emissions data summarized in Table 1 illustrate the large potential contribution of sulfur oxides [SO_x – principally sulfur dioxide (SO_2)] and nitrogen oxides [NO_x – principally nitrogen dioxide (NO_2)] gaseous precursors to ambient PM concentrations. If their combined 40 million tons per year were completely converted to sulfate and nitrate particles, their emissions would be equivalent to about 60 million tons per year of primary $PM_{2.5}$. However, ambient $PM_{2.5}$ in the U.S. is typically 30 to 50% sulfate and much less nitrate, except in southern California where nitrate may be as much as 45% in some winter periods (USEPA, 1996b). This implies a conversion to $PM_{2.5}$ of somewhat less than 10 million tons per year, or somewhat less than 20% of the gaseous precursor emissions.

Figure 2 shows geographically the areas that do not meet the current PM_{10} standard of 150 $\mu g/m^3$ (24-h average) and 50 &g/m^3 (annual average). It also indicates the source types that are thought to be the major causes of non-attainment with the standard. The number of additional areas of the country that will not meet a new $PM_{2.5}$ standard, and the additional sources needing control, will depend largely on the concentration values and averaging periods that are selected (Paisie, et al., 1997).

There is further uncertainty about exposures to airborne particles inside buildings. Considering the high fraction of time that people (especially those who may be most health-susceptible) spend indoors, there is a need to quantify the three major sources of indoor exposure: particles from outdoors, direct emissions from indoor sources, and resuspended particles from indoor activities (a type of fugitive source). Table 2 summarizes the various types of sources of particles found in buildings. Cigarette smoking, cooking, and penetration of outdoor particles through the building envelope have all been reported to make significant contributions to indoor concentrations of fine particles, but a substantial portion is due to unexplained indoor sources (Wallace, 1996).

Table 1
Summary of U.S. national inventory of particle and precursor emissions

		National Emissions (10^3 Mg/year)			
		$PM_{2.5}$	PM_{10}	SO_x	NO_x
Industrial Processing		600	800	200	1000
Combustion Sources					
industrial & commercial		300	500	19000	11000
residential		600	600	10	70
vehicular		500	600	800	10000
open burning		1100	1300	7	200
Fugitive Sources					
roads		3000	18000		
agricultural production		2000	11000		
construction		2000	8000		
Totals		10000	40000	20000	20000

Adapted from USEPA, 1996a.

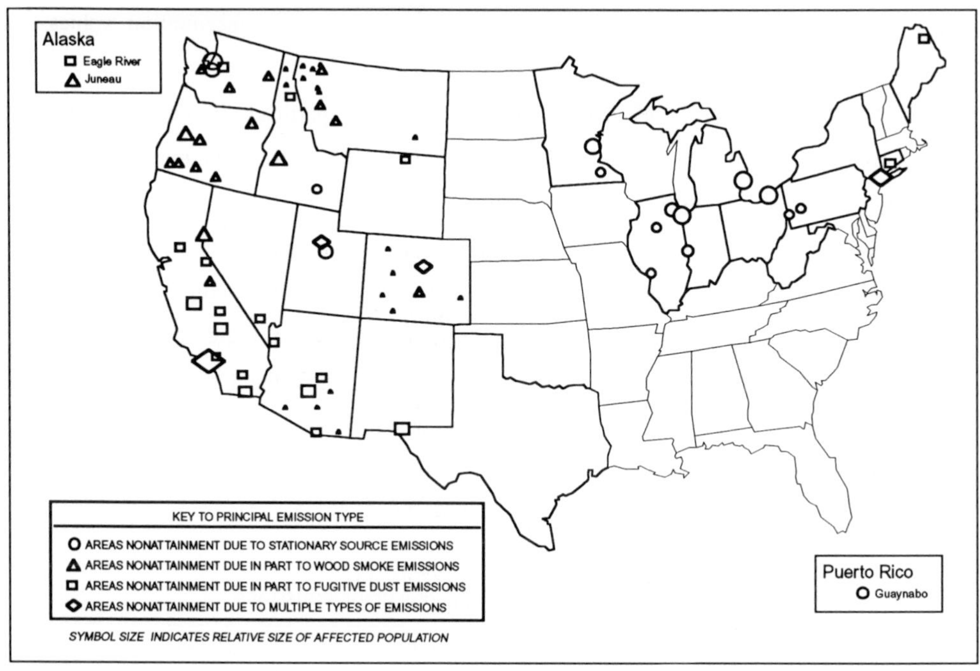

Figure 2. Areas designated non-attainment for PM_{10} particles, by emission type. (Adapted from USEPA, 1996c)

Table 2

Sources of exposure to indoor particles

Outdoor Origin	Indoor Origin
Outdoor air (through infiltration, ventilation)	Smoking, Cooking
Tracked-in soil	Space heating (esp. kerosene, wood)
Carried-in dusts	Ventilation systems, Humidifiers
- from industrial workplaces	Office machines
- from other places	Dust mites, Pets
	Diseased people (bacteria, viruses)
	Personal activities (e.g., hobbies)
	Maintenance/renovation activities
Resuspension of Particles of Either Outdoor or Indoor Origin	
Cleaning activities	
Maintenance/renovation activities	
People movements	

3. EMISSION RATES AND COMPOSITIONS OF PRIMARY PARTICLES

Table 3 presents an overview of emissions of primary particles in the U.S. Note that the four largest source categories are 'fugitive' types. As noted previously, the field data that are the basis of these estimates are being reevaluated. Cascade impactors used in these field studies may have experienced significant 'bounce' of larger particles onto impaction plates designed to collect smaller particles. The $PM_{2.5}$ fraction of PM_{10} may therefore be overstated somewhat. Even after adjustment of the data, however, fugitive sources are likely to remain significant, at least on a national basis, for $PM_{2.5}$.

As noted in Table 1, combustion sources comprise the next largest grouping. However, the constituents of combustion PM may be more toxic on average than the constituents of fugitive PM. Furthermore, combustion sources are generally closer to the population, and may lead to higher exposures per unit of emission.

Table 3
Overview of current knowledge on U. S. emissions of fine particles. (Numbers in regular type are typical values, selected from the referenced literature; *entries in italics are estimates or judgements by the author*)

Source Type [refs. for emissions data]	Constituents of Concern	Total U.S. Emission Rate (10^3 tons/yr)		Approx. U.S. Population in Close Proximity (millions)
		$PM_{2.5}$	PM_{10}	
Roads [a,b,c]	Fine silica and other crustal elements plus re-entrained carbon, asbestos, and metal compounds	3300*	18,000	*250*
Agricultural Production (incl. erosion) [a,b,c]	Fine silica and other crustal elements	2000*	11,100	*Mostly in rural areas*
Construction Activities [a,b,c]	Fine silica and other crustal elements plus industrial reentrainment of carbon, asbestos and metal compounds	1700*	8,500	*Mostly in urban areas*
Open Burning (incl. wildfires, agric. burning, etc.) [a,b,c]	Products of uncontrolled combustion	1130*	1320	*Mostly in rural areas*
Residential Wood Combustion [a,b,c]	Polycyclic Organic Matter (POM)	550	~600	*Mostly in rural and suburban areas*
Diesel Engine Combustion [a,b,c]	Products of incomplete combustion, PM precursor (NO_x)	450	500	*Mostly in urban areas*
Mineral Products Production [a,b,c]	Fine silica and other crustal elements	100	200	*Near urban areas*
Pulverized Coal Boilers [a,d]	Ar, Cr, Hg, Mn, Ni, Pb, Sb, Se, V, Cl, and PM precursors (SO_x, NO_x)	*Unknown*	160	*Utility: mostly rural* *Industrial: mostly near urban areas*

Heavy Fuel Oil Combustion [a,e]	Cr, Fe, Ni, Pb, V, POM, Cl, PM precursors (SO_x, NO_x)	*~30*	30	*Mostly in urban areas*
Residential Fuel Oil Combustion [a]	POM, PM precursor (NO_x)	*~20*	20	*Mostly in urban areas*
Waste Incineration [a,f,g,h]	As, Be, Cr, Cd, Hg, Ni, Pb, PCDD/F, PCB's	*Unknown*	~45	*Mostly near urban areas*
Metal Smelting and Refining [a,i]	Cd, Cr, Pb, Zn, SO_x	*Unknown*	400	*Mostly in rural areas*
Automobiles [j]	V, carbon, organics, PM precursor (NO_x)	*~20*	*~20*	*Mostly near urban areas*
Outdoor Air Introduced into the Indoor Environment	Fine and coarse particles	Unknown	Unknown	*250*
Tracked-in Dust	Pb, other heavy metals, pesticides	Unknown	Unknown	*250*
Indoor Activities (that generate or resuspend particles)	Metals, pesticides, combustion aerosol organics	Unknown	Unknown	*250*

* Estimates of fine particle emissions from these "fugitive" sources, although large compared to other sources in this table, are very uncertain and need to be confirmed.

References for Table 3

a. USEPA (1996a).
b. Cowherd, et al. (1988).
c. Dunkins and Cowherd (1992).
d. Davison, et al. (1974), Kaakinen, et al. Klein, et al. (1975), White, et al. (1984), Markowski and Filby (1985), Kauppinen and Pakkanen (1990), Andren, et al. (1975), Billings and Watson (1972).
e. Bulewicz, et al (1974), Haynes, et al. (1978), Feldman (1982), Chung and Lai (1992)
f. Mumford, et al. (1986), Trichon and Feldman (1989), Trichon and Feldman (1991).
g. Lisk (1988), Greenberg, et al. (1978).
h. Shen (1979).
i. Harrison and Williams (1983).
j. Average of 10 mg/mile; 2×10^{12} vehicle miles traveled.

4. CONTROL OPTIONS

Managing the health risks of exposures to fine particles requires knowledge of the sources and types of particles that are most likely to cause health risks, and knowledge of the performance and costs of risk reduction technologies. Data on the chemical, physical, and toxicological characteristics of the emissions are needed to determine controllability and risk reduction potential.

Since many of the estimates of $PM_{2.5}$ emissions are estimated from data on PM_{10}, there is great uncertainty in the fine particle emissions inventory. In addition, there is a general lack of data on the chemical composition of fine particle emissions. The need for emission characterization is greatest for those sources with constituents (such as metals and acidic components) that are candidates for causal mechanism studies of respiratory health effects, and for those sources having the largest mass contributions to $PM_{2.5}$ in the environment.

As Table 4 implies, there are few data on the effectiveness and costs of emissions prevention, emissions reduction, or exposure reduction technologies for fine particles (i.e., $PM_{2.5}$). Reliable data on emission prevention for either industrial or indoor sources are nearly non-existent. Most of the available data on cost-effectiveness of emission controls are for industrial sources of total PM (and at best, for PM_{10}). Although limited data are available on the <u>efficiency</u> and cost of air cleaning to remove particles from indoor air, there are virtually no data on the <u>effectiveness</u> of air cleaning in reducing exposures to fine particles. Since indoor concentrations of particles approach outdoor concentrations when outdoor concentrations are high, or are about twice outdoor concentrations when outdoor concentrations are low (e.g., Spengler et al., 1981; Sheldon et al., 1989), and since people spend roughly an order of magnitude more time indoors than outdoors, it will be important to develop controls for indoor exposures.

Table 4

Overview of current knowledge on control of fine particles. (Values for efficiencies and costs are estimates or judgements by the author)

Source Type	Primary Control Options, Efficiencies for PM_{10}	Approximate Costs of PM_{10} Controls
Roads	Vacuum sweeping (0-50%) Water flushing & sweeping (0-96%) Paving and roadside improvements Covering trucks Speed and traffic reduction	Dependent on type of control, time of event, frequency of event/yr, and volume of traffic. Very limited published data.
Agricultural Production (incl. erosion)	Low tillage, punch planting, crop strips, vegetative cover, windbreaks Chemical stabilizers, irrigation	Dependent on crop type and regional weather conditions. Little data.
Construction Activities	Wet suppression of unpaved areas, material storage, handling and transfer operations Wind fences for windblown dust	Dependent on type of control, time of event, land area of event, and activity level of equipment. Very limited published data.
Open Burning (incl. wildfires, agric. burning)	Low wind speed and appropriate wind direction	Unknown.
Residential Wood Combustion	Replace with cleaner burning stoves or furnaces	~$1000 per replaced stove or furnace.
Diesel Engine Combustion	Combustion modification Improved fuel characteristics Particle traps	Very limited published data.
Mineral Products Production	Enclosing crushing, transfer areas Water spray suppression Chemical stabilization of unpaved traffic areas	Dependent on type of control and activity level of equipment. Little data.

Pulverized Coal Boilers	ESPs, Fabric Filters	Capital cost \$50-100/kW Annual cost 2-5 mills/kWh Total installed cost \$25-50 per m^3/h
Heavy Fuel Oil Combustion	Cyclones, ESPs	Unkown.
Residential Fuel Oil Combustion	Proper maintenance, modern furnaces	Unknown.
Waste Incineration	Fabric filters, ESP's, venturi scrubbers	Total installed cost \$15-30 per m^3/h.
Metal Smelting and Refining	ESPs, cyclones	Total installed cost \$15-30 per m^3/h.
Outdoor Air Introduced into the Indoor Environment	Air cleaners for ventilation air (30 - 98%)* Whole-building air cleaners (30 - 98%)* In-room air cleaners (30 - 98%)*	Capital cost \$3-10 per m^3/hr of outdoor air treated. Capital cost \$1-10 per m^3/hr of indoor air treated. \$200 to 800 per room.
Tracked-in Dust	Cleaning (e.g., vacuuming) Whole-building air cleaners (30 - 98%)* In-room air cleaners (30 - 98%)*	No published analyses. Capital cost \$1-10 per m^3/hr of indoor air treated. \$200 to 800 per room.
Indoor Activities (that generate or resuspend particles)	Source control, including maintenance Whole-building air cleaners (30 - 98%)* In-room air cleaners (30 - 98%)*	Highly variable; no published analyses. Capital cost \$1-10 per m^3/h of indoor air treated. \$200 to 800 per room.

* Range of single-pass efficiency for removing particles. The effectiveness of air cleaners in reducing exposures to indoor particles is very dependent on installation and operating conditions, and is generally less than the single-pass efficiency.

5. SUMMARY

Growing concerns about the health risks of fine particles have led to the proposal of a revised ambient air quality standard for PM smaller than 2.5 μm in aerodynamic diameter ($PM_{2.5}$) in the U.S. The existing data base on national emissions shows that sources of both primary and secondary particles make significant contributions. On a nationwide basis, the greatest mass of primary and precursor emissions come from combustion, fugitive, and industrial sources, in that order. In specific localities, any of these source categories can be dominant.

It is clear that the new emphasis on fine particles will focus risk management research on large sources and combustion sources of various types. The role of fugitive sources in creating exposures to fine particles is less clear, and field studies to obtain better emissions data are needed. The importance of exposures inside buildings is also unclear; the penetration of outdoor particles into buildings of various types and the role of indoor sources of particles need further study. Current knowledge of the costs and effectiveness of control options for PM_{10} is limited and poorly documented. Carefully documented data on control options for $PM_{2.5}$ are nearly nonexistent.

A risk management research program is now in the early stages of developing new information to address these questions:

- What sources contribute to the fine particle exposures that are of greatest concern to the health research and regulatory community?
- What are the emission rates and physical and chemical characteristics of particles from these sources?
- What are the most cost-effective prevention and control options for these sources (e.g., process changes, upgrades of existing controls, application of new technology)?

Research results will enable the U.S. Environmental Protection Agency to assist state and local regulatory agencies in the development of cost-effective local prevention and control strategies for reducing exposures to fine particles. These strategies are likely to be based on a combination of industrial process changes, improved operation of existing particle control devices, installation of new control equipment on selected sources, increased control of gaseous precursors to ambient fine particles, and control of particles in buildings.

REFERENCES

1. Andren, A.W., Klein, D.H. and Talmi, Y. (1975). *Environ. Sci. Technol.*, **9(9)**, 856-858.
2. Bachmann, J.D., Damberg, R.J., Caldwell, J.C., Edwards, C. and Korian, P.D., (1996c). 'Review of the National Ambient Air Quality Standards for PM: Policy Assessment of Scientific and Technical Information' (OAQPS Staff Paper), EPA-452/R-96-013 (NTIS PB97-115406).
3. Billings, C.E. and Matson, W.R. (1972). *Science*, **176**, 1232-1233.
4. Bulewicz, E.M., Evans, D.G. and Padley, P.J. (1974). *15th Comb. (Int.) Symp.*, 1461-1470, Comb. Inst., Pittsburgh.
5. Chung, S.L. and Lai, N.L. (1992). *J. Air Waste Manage. Assoc.*, **42(8)**, 1082-1088.
6. Cowherd, C., Moleski, G.E. and Kursey, J.S., (1988). 'Control of Open Fugitive Dust Sources'. EPA 450/3-88-008 (NTIS PB89-103691).
7. Davison, R.L., Natusch, D.F.S., Wallace, J.R. and Evans, C.A. Jr. (1974). *Environ. Sci. Technol.*, **8(13)**, 1107-1113.
8. Dunkins, R. and Cowherd, C., (1992). 'Fugitive Dust Background Document and Technical Information Document for Best Available Control Measures'. EPA 450/2-92-004 ((NTIS PB93-122273).
9. Feldman, N. (1982). *19th Comb. (Int.) Symp.*, 1387-1393, Comb. Inst., Pittsburgh.
10. Greenberg, R.R., Zoller, W.H. and Gordon, G.E. (1978). *Environ. Sci. Technol.*, **12(5)**, 566-573.
11. Harrison, R.M. and Williams, C.R. (1983). *Sci. of Total Environ.*, **31**, 129-140.
12. Haynes, B.S., Jander, H. and Wagner, H.G. (1978). *17th Comb. (Int.) Symp.*, 1365-1381, Comb. Inst., Pittsburgh.
13. Kaakinen, J.W., Jorden, R.M., Lawasani, M.H. and West, R.E. (1975). *Environ. Sci. Technol.*, **9(9)**, 862-869.

14. Kauppinen, E.I. and Pakkanen, T.A. (1990). *Environ. Sci. Technol.*, **24(12)**, 1811-1818.
15. Klein, D.H., Andren, A.W., Carter, J.A., Emery, J.F., Feldman, C., Fulkerson, W., Lyon, W.S., Ogle, J.C., Talmi, Y., VanHook, R.I. and Bolton, N. (1975). *Environ. Sci. Technol.*, **9(10)**, 973-979.
16. Lisk, D.J. (1988). *Sci. of Total Environ.*, **74**, 39-66.
17. Markowski, G.R. and Filby, R. (1985). *Environ. Sci. Technol.*, **19(9)**, 796-804.
18. Mumford, J.L., Hatch, G.E., Hall, R.E., Jackson, M.A., Merrill, R.G. and Lewtas, J. (1986). *Fund. and Applied Toxicol.*, **7**, 49-57.
19. Paisie, J. et al. (1997). U.S. PM NAAQS rationale and risk reduction/control policy, 5th U.S.-Dutch International Symposium on Air Pollution in the 21st Century: Priority Issues and Policy Trends, Noordwijk, The Netherlands, April 1997.
20. Sheldon, L. S.; Hartwell, T. D.; Cox, B. G.; Sickles, J. E., II; Pelizzari, E. D.; Smith, M. L.; Perritt, R. L.; Jones, S. M. (1989). An investigation of infiltration and indoor air quality: final report. NYS ERDA contract no. 736-CON-BCS-85. Albany, NY: New York State Energy Research and Development Authority.
21. Shen, T.T. (1979). *J. Environ. Engineering Div.*, 61-74, February 1979.
22. Spengler, J. D., Dockery, D. W., Turner, W. A., Wolfson, J. M., and Ferris, B. G., Jr. (1981). Long-term measurements of respirable sulfates and particles inside and outside homes. *Atmos. Environ.* **15**:23-30.
23. Trichon, M. and Feldman, J. (1989). Chemical kinetic considerations of trace toxic metals in incinerators, *1989 Incineration Conference*, 9.1.1-9.1.18, Knoxville, TN.
24. Trichon, M. and Feldman, J. (1991). Problems associated with the detection and measurement of arsenic in incinerator emissions, *1991 Incineration Conference*, 571-579, Knoxville, TN.
25. USEPA. (1996a). 'The National Particulates Inventory - Phase II Emission Estimates', EPA Office of Air Quality Planning and Standards, revised January 1996.
26. USEPA. (1996b). 'Air Quality Criteria for PM', EPA/600/P-95/001BF (NIIS PB96-168240).
27. USEPA. (1996d). 'National Ambient Air Quality Standards for PM: Proposed Decision', 40 Code of Federal Regulations (CFR) Part 50.
28. Vedal, S. (1997). Ambient particles and health: the lines that divide. *J. Air & Waste Manage. Assoc.*, **47**: 551-581.

29. Wallace, L. (1996). Indoor particles: a review. *J. Air & Waste Manage. Assoc.*, **46**: 98-126.

30. White, D.M., Edwards, L.O., Eklund, A.G., DuBose, D.A., Skinner, F.D., Richmann, D.L. and Dickerman, J.C. (1984). Correlation of coal properties with environmental control technology needs for sulfur and trace elements, EPA-600/7-84-066 (NTIS PB84-200666), Environ. Prot. Agency, Research Triangle Park, NC.

31. Wilson, R. and J. D. Spengler, eds. (1996). Particles in our air: concentrations and health effects, Harvard University Press.

Estimating the Benefits and Costs for the Revised Particulate Matter and Ozone Standards in the United States*

Michele McKeever**

Economist, U.S. Environmental Protection Agency, Mail Drop 15, Research Triangle Park, NC, 27701.

This paper provides an overview of some of the issues, procedures, and decisions associated with preparing a national benefit-cost analysis for revised air quality standards for particulate matter and ozone in the United States. Additionally, the paper presents the results of the analyses in terms of potential monetized benefits and costs that could be estimated for the revised standards in the year 2010.

1. INTRODUCTION

The purpose of this paper is to provide an overview of the methodological issues, procedures, and decisions that are associated with preparing benefit-cost analyses for the revised primary Ozone and Particulate Matter (PM) air quality standards recently promulgated (July 1997) in the United States by the U.S. Environmental Protection

---

*The majority of the information included in this paper was presented in April 1997. However, some details, such as the form and level of the final revised ozone and PM standards, have been updated to reflect the most recently available information.

** This paper summarizes work conducted by a team of analysts working for the United States Environmental Protection Agency. For a more detailed description of the analysis, please refer to the July 1997 RIA for the PM and Ozone NAAQS and Proposed Regional Haze Rule.

Agency (EPA). In December 1996, two separate Regulatory Impact Analyses (RIA's) were prepared for the proposal of revised standards for PM and ozone. The revised PM and ozone standards were promulgated in July 1997 and a revised set of benefit-cost analyses was published in a single document, titled "Regulatory Impact Analyses for the Particulate Matter and Ozone National Ambient Air Quality Standards (NAAQS) and Proposed Regional Haze Rule", at the same time.

Currently, the primary and secondary ozone standards are each set at a level of 0.12 parts per million (ppm), with a 1-hour averaging time and 1 expected exceedance form, such that the standards are attained when the expected number of days per calendar year with maximum hourly average concentrations above 0.12 ppm is equal to or less than 1, averaged over 3 years. The Clean Air Act defines the purpose of the primary standard as one of protecting public health with an "adequate margin of safety'. The purpose of the secondary standard is to "protect the public welfare from any known or anticipated adverse effects". (42 U.S.C. 7409) Examples of welfare effects include, but are not limited to, effects on crops and vegetation, animals and wildlife, visibility and climate, and effects on economic values or personal comfort and well-being. (42 U.S.C. 7602) The revised ozone standards (once again, with the primary and secondary standards set equal to each other) will be met at an ambient air quality monitoring site when the 3-year average of the annual fourth-highest daily maximum 8-hour average ozone concentration is less than or equal to 0.08 ppm. For convenience, the revised ozone standards will be expressed in this paper as 0.08 ppm, 8-hour, 4th highest concentration. (U.S. EPA, 1997a)

The current PM standard specifies the indicator for PM as PM_{10}. PM_{10} refers to particles with an aerodynamic diameter less than or equal to a nominal 10 micrometers. Identical primary and secondary PM_{10} standards were set for two averaging times: 50 $\mu g/m^3$, expected annual arithmetic mean, averaged over 3 years, and 150 $\mu g/m^3$, 24-hour average, with no more than 1 expected exceedance per year. The revised PM standards specify standards for several size-specific classes of particles. The current annual PM_{10} standard is retained at the level of 50 $\mu g/m^3$, which is met when the 3-year average of the annual arithmetic mean PM_{10} concentrations at each monitor within an area is less than or equal to 50 $\mu g/m^3$, with fractional parts of 0.5 or greater rounding up. The form of the current 24-hour PM_{10} standard is revised to be based on the 3-year average of the 99th percentile of 24-hour PM_{10} concentrations at each monitor within an area. Additionally, an annual and a 24-hour standard is specified for particles with an aerodynamic diameter less than or equal to a nominal 2.5 micrometers ($PM_{2.5}$). The annual $PM_{2.5}$ standard is met when the 3-year average of the annual arithmetic mean $PM_{2.5}$ concentrations, from single or multiple community-oriented monitors, is less than or equal to 15 $\mu g/m^3$, with fractional parts of 0.05 or greater rounding up. The 24-hour $PM_{2.5}$ standard is met when the 3-year average of the 98th percentile of 24-hour $PM_{2.5}$ concentrations at each population -oriented monitor within an area is less than or equal to 65 $\mu g/m^3$, with fractional parts of 0.5 or greater rounding up. The revised secondary PM standards are set identical to the suite of revised primary PM standards. (U.S. EPA, 1997b)

This paper focuses on some of the methodological issues and decisions associated with attempting to conduct a national benefit-cost analysis for ubiquitous pollutants that are formed from a variety of anthropogenic and natural activities. This paper provides general descriptions of the analytical issues and may not be applicable to all circumstances within the analyses. Although the quantitative results presented at the end of this paper represent the monetized benefits and costs associated with the revised PM and ozone standards, the procedures and issues presented in this paper equally apply to other alternative standards that were examined in the RIA's.

2. BACKGROUND

Reviewers of the benefit-cost analyses should be aware of the purpose of the analyses within the context of the NAAQS rulemakings. As the introduction explains, the Clean Air Act (CAA) directs the EPA to identify and set national standards for pollutants that may reasonably be anticipated to cause adverse effects to public health and the environment. In setting the primary air quality standards, the EPA's first responsibility under the law is to select standards that protect public health. As interpreted by the EPA and the courts, this decision is a health-based decision that specifically precludes cost or other economic considerations. By contrast, the EPA believes that consideration of cost is an essential decision-making tool for the cost-effective *implementation* of these standards. Given the guidance of the Clean Air Act, it is clear that the appropriate place for cost and efficiency considerations is during the development of implementation strategies, strategies that will allow communities to meet the health-based standards in a cost-effective manner over time.

In addition to complying with the CAA, the EPA must also comply with other Federal requirements for analyses. For example, Executive Order 12866, signed by the President of the United States, directs all Federal agencies to "assess all costs and benefits of available regulatory alternatives... Costs and benefits shall be understood to include both quantifiable measures... and qualitative measures of costs and benefits that are difficult to quantify, but nevertheless essential to consider... Further, in choosing among alternative regulatory approaches, agencies should select those approaches that maximize net benefits..., unless a statute requires another regulatory approach". Another example of a Federal requirement is the Unfunded Mandates Reform Act of 1995 (UMRA). The UMRA directs agencies to provide a qualitative and quantitative assessment of the anticipated costs and benefits of a Federal mandate resulting in annual; expenditures of $100 million or more, including the costs and benefits to State, local, and tribal governments, or the private sector. (Chapter 2 of the 1997 RIA provides additional information on these and other Federal requirements, including one that is concerned with the potential economic impacts affecting small entities).

These examples of the types of Federal requirements under which the EPA operates provides the regulatory framework under which benefit-cost analysis is used. Since the CAA precludes consideration of costs or technological feasibility in determining the ambient standards, the results of the RIA's are not taken into account by the Administrator in her decision regarding the appropriate level of the NAAQS. Within this regulatory setting however, benefit-cost analysis can provide a valuable framework for organizing and evaluating information on the effects of environmental programs. When used properly, benefit-cost analysis helps illuminate important potential effects of changes in policy and helps set priorities for closing information gaps and reducing uncertainty. In this context, the objectives of the RIA's for PM and ozone were to: (1)

assess the potential costs and economic impacts of regulatory alternatives, including economic incentives and alternatives to command and control technologies; (2) assess the potential environmental and public health benefits; and (3) assess potential impacts on industries and small businesses.

In order to accomplish the above goals, a series of methodological steps were followed to develop the benefit-cost analysis. These steps are presented in Figure 1. The rest of this presentation will focus on issues associated with each step and finally, the results of the analyses will be presented at the end of this presentation. The diagram presented in Figure 1 represents broad subject categories included in the RIA but does not provide information on the details of the analysis.

3. ANALYTICAL YEAR AND SCOPE OF ANALYSIS

Two immediate decisions affecting the direction of the benefit-cost analysis are the temporal and geographical scope of the analysis. One of the first steps in preparing the RIA was to decide on an analytical year that would serve as the basis of the analysis. The year of the analysis must be a future year because the year must represent a time period during which the standards would be implemented. A constraint on the RIA was the length of time over which the analysis could span. Resource and related constraints such as data management and modeling capabilities made a multi-year analysis infeasible. Therefore, the choice of the analysis year was limited to a single calendar year. This temporal constraint makes it important to choose a "representative" year during which the standards will be implemented. This year is described as the year during which we expect most (but not all) programs to be in place for implementation of these standards. For the proposal of these standards in 1996, the analytical year of the analysis was 2007. In 1997, during the promulgation of these standards, the year 2010 was chosen as a more representative year. The RIA refers to the analysis year as a "snapshot in time" approach. The decision on the geographical scope of the analysis is based on the geographic scope of the standards. All areas of the U.S. are required to comply with the national PM and ozone standards.

Figure 1

Analytical Steps in Developing a Benefit-Cost Analysis

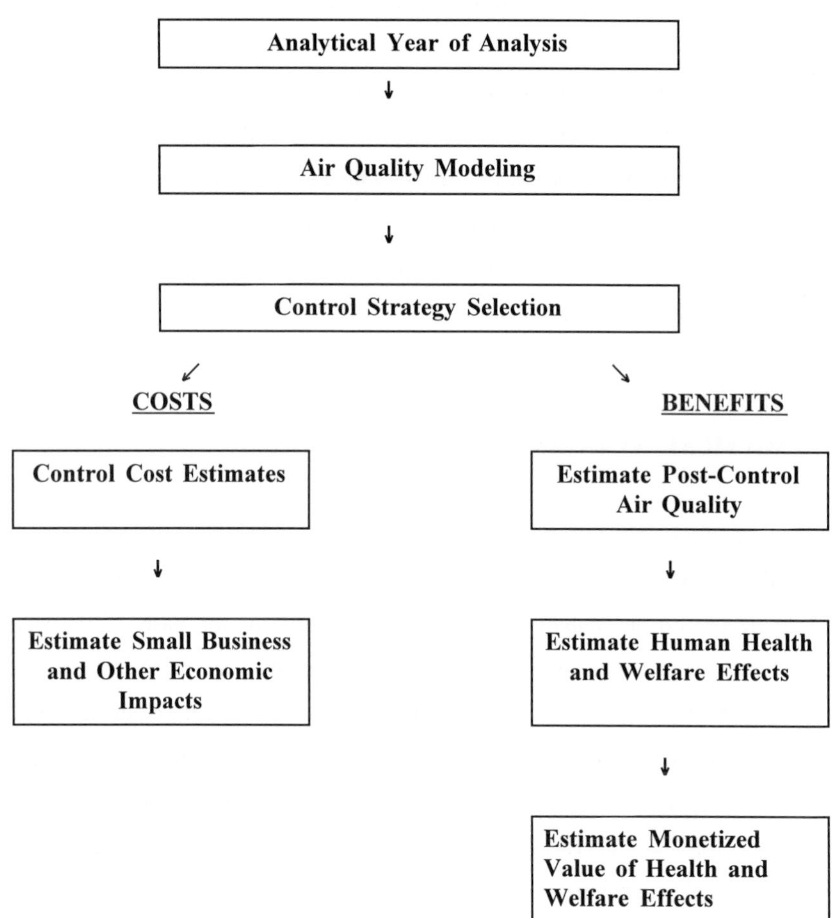

It was therefore decided that the scope of the benefit-cost analyses include all areas within the continental U.S. (Modeling limitations prevented other U.S. areas such as Alaska and Hawaii from being included in the analysis.) Such broad geographical coverage and forecasting horizon place significant limitations on the details that can be included in the analysis.

4. AIR QUALITY MODELING

In the RIA's prepared for the proposal of these standards in 1996, two separate benefit-cost documents, one for ozone and one for PM, were prepared. The analyses were prepared under a constraint of using separate air quality models for estimating ambient concentrations of PM and ozone. The models provided projections of ambient concentrations of PM and ozone for a "baseline" scenario in the year 2007. A baseline scenario represents a scenario in the absence of implementation of the revised air quality standards. In order to establish the baseline projection, the analysis used present-day conditions and then attempted to account for factors such as population growth, economic growth, and additional levels of pollution control mandated by the Clean Air Act during the intervening years. These projections established an emissions inventory, air quality, and population baseline for the analysis year.

Once the analytical baseline was established, it was then possible to evaluate the baseline PM and ozone air quality data against alternative PM and ozone standards. This evaluation allowed the analysis to identify potential areas that would violate a particular standard. The analysis labels these areas as potential nonattainment areas. For example, given the baseline air quality data, which areas in the continental United States would potentially violate the .08 ppm, 8-hour, 4th highest concentration ozone standard?

Having established the degree of potential violation in each nonattainment area, the analysis relied on information from a variety of prognostic and empirical models to determine the amount of emission reductions that would be required in each nonattainment area to enable the area to attain the standard. For ozone, the pollutants of concern are volatile organic compounds (VOC's) and nitrogen oxides (NOx). In addition to VOC's and NOx, the pollutants of concern for PM are sulfur oxides (SOx), primary particles (PM), and ammonium. The use of the various models in combination allowed the analysis to discern a relationship between emissions and ambient concentrations of PM and ozone. [*Note*: This explanation oversimplifies the analyses for ozone and PM. The analyses used separate models to estimate emission reductions and air quality changes. For example, the ozone analysis used separate models to sequentially estimate necessary the emission reductions and resulting air quality changes while the PM analysis used one model that was capable of manipulating both types of information simultaneously. The oversimplified explanation is provided only for the purpose of describing the type of information that is necessary for estimating potential costs and benefits.]

As mentioned earlier, the 1996 PM and ozone analyses were prepared separately. Although the analyses shared a common baseline, the estimation of potential emission reductions and resulting air quality ("post-control" air quality) were conducted independently for PM and ozone. These estimates were calculated separately for PM and ozone because a model integrating both pollutants was not available. Despite this model constraint, the 1997 RIA improved on the previous analyses by partially "integrating" the analyses. (An adjustment was also made to adjust the analytical year or the analysis to the year 2010.) Although the RIA continued to use the same models, the analyses were "integrated" by using the air quality models jointly rather than separately. The previous 1996 analyses evaluated only ozone air quality changes within the ozone analysis and only PM air quality changes within the PM analysis. However, air quality modeling shows that VOC and NOx emissions affect ambient concentrations of both ozone and PM. The joint use of the air quality models allowed the ozone analysis to provide information on the co-benefits for reducing ambient PM concentrations due to the application of controls strategies specifically employed to reduce ozone. A new baseline of lower PM concentrations would thus be established for the PM analysis.

5. CONTROL STRATEGY SELECTION

The air quality and emission reduction information developed during the air quality modeling phase of the analysis are used in conjunction with a database of control measures in order to develop cost estimates. In general, the database consists of known "command and control" technologies (add-on control devices installed downstream from an air pollution source) because quantitative information is most often available for these types of controls. To a smaller extent, less conventional approaches such as pollution prevention measures, trading programs, and educational and advisory measures are also included in the 1997 RIA. Recognizing the analysis is a future year analysis, the 1997 RIA also includes a discussion of technological advancements that might be viable within the next decade. The control measure database does not include these "emerging technologies" due to an inability to reliably quantify the effects of these measures.

Additional information within the database include the types of emission sources (e.g., utilities versus automobiles) to which each control measure can be applied, the types of pollutants that the measure can control, and the average annual incremental cost per ton of pollutant reduction associated with each potential application of the control measure. For the purposes of the analysis, the average annual incremental cost per ton is defined as the difference in the annual cost of a control measure and the annual cost of the baseline control (if any), divided by the difference in the annual mass of pollutant emissions removed by the control measure and the emissions removed by the baseline control.

The database allows control measures to be ranked by the average annual incremental cost information associated with each measure. Upon determining the amount of emission reductions that are required, the analyst can use the database to choose the lowest cost per ton technology first, the second-lowest cost per ton technology second, etc. Within each potential nonattainment area, the analyst applied additional levels of control until either: (1) the required emission reductions were met and the area was classified as being able to attain the standard or (2) the control measure database was exhausted for the nonattainment area before all emission reduction goals were met. In the second case, the analysis was labeled a "partial attainment" analysis since some areas were projected to encounter difficulty in attaining the standard by 2010 using currently well-documented technologies. (Each alternative standard analyzed in the RIA included some areas that could only partially attain the standard.) [*Note:* Although the RIA imposes controls on specific emission sources, the decision on which sources to control will ultimately be made by the individual States.]

6. COST AND ECONOMIC IMPACT ESTIMATION

The purpose of the cost and economic analyses is to estimate potential private and social costs due to implementation of the revised air quality standards. Information within the control measure database allows the estimation of both emission reductions and costs. As explained in the precious section, the database allows the control measures to be matched to specific emission source categories in specific areas. Once the control measures are selected, it is possible to: (1) estimate national control costs associated with a specific air quality scenario and (2) examine potential distributional effects. Cost information is expressed in dollar terms (dollar per ton of pollutant removed) and includes costs for typical factors such as new equipment and additional operation and maintenance costs. A national cost estimate is calculated by summing the cost per ton data for all tons of pollutants removed across all emission source categories.

It is important to recognize that although the benefit-cost analysis uses a set of assumptions to develop cost estimates, the actual determination of how areas will meet the air quality standards will be determined by the individual states within the United States. The states must develop State Implementation Plans based on more detailed area-specific models using more complete information that is available to the EPA for the development of its national analysis. For this reason, while the national benefit-cost analysis may provide a good approximation of the national costs and benefits of the revised standards, this analysis cannot accurately predict what will occur in individual areas.

In addition to facilitating a national cost estimate, the availability of cost data on an industry-specific basis allows the analyst to focus additional economic analyses on industry sectors or areas that might be significantly affected. The estimation of economic impacts (e.g., price, quantity, or employment changes) requires the cost data to be used in conjunction with economic data such as sales revenue, market conditions, and employment data. The impracticality of modeling all market sectors in the U.S. forces the analyst to screen for sectors that might be disproportionately affected. A simple calculation of cost-to-sales revenue may provide useful information for screening purposes. However, it is important to recognize that many uncertainties are associated with the economic impact analysis. For example, it is not possible to know the specific establishments or firms that will be required to implement control strategies. Both cost and revenue (or sales) data only represent national average statistics. A simple calculation of an average cost-to-sales percentage does not provide economic impact information for specific establishments.

7. BENEFITS ESTIMATION

The purpose of the benefits analyses is to estimate potential human health and welfare (defined here as all benefits categories except human health) effects that may be associated with changing the nation's air quality. Ideally, the goal is to assign a monetary value to each benefits category or endpoint so that benefits and costs can be expressed in the same unit. In reality, benefits analysis often falls short of quantifying all effects categories due to information deficiencies.

Assessing the benefits of the PM and ozone regulatory actions is dependent upon the ability to model air quality. Upon the selection of control measures, the analyst can tally all expected emission reductions specific to each geographic (nonattainment) area. The air quality models translate the emission reductions into reduced ambient concentrations of ozone or particulate matter (PM). The ambient ozone and PM concentrations resulting from the application of control measures is referred to as "post-control" air quality. The change in air quality from baseline conditions to post-control is a key element in the estimation of benefits.

A second key factor in benefits estimation is the estimation of concentration-response functions. A concentration-response function specifies a relationship between the presence of a pollutant (e.g., the measured concentration of PM or ozone in a specified environment) and a physical response (e.g., coughing, reduced plant yields). Each concentration-response function is used in conjunction with the estimated air quality changes to quantify the physical effects believed to be associated with the air quality changes. For example, completion of this step allows the analyst to estimate the number of fewer coughs a person might experience due to improved air quality. This quantified effects information provides useful information for evaluating potential air quality changes. However, it is not possible to quantify all effects categories. For example, we may suspect an association between chronic exposure to air pollution and adverse health effects but data is not yet available for the quantification of these effects. These effects are referred to as unquantified benefits and are discussed in a qualitative manner in the analysis.

A third key factor in benefits estimation is to assign an appropriate monetary value to the physical effects that have been valued. As explained earlier, it is important to complete this step so that benefits and costs may be compared in the same terms. In order to accomplish this objective, the analyst must determine the value individuals place on avoiding the adverse physical effects associated with exposure to air pollution. For example, what is the value of avoiding a cough or hospital visit for a respiratory illness? In the case of welfare benefits, what is the value of increasing yields of the nation's wheat crops? The social benefits associated with a change in the environment is calculated as the sum of each individual's willingness to pay for the change.

In the case of valuing human health benefits, the analyses use a concept called willingness-to-pay (WTP). A WTP value is estimated for the majority (but not all, due to data limitations) of physical effects endpoints. The values attempt to incorporate factors such as the direct cost of becoming ill (e.g., hospitalization costs), the change in quality of life due to discomfort, pain and suffering, or limitations on physical activities. The method for valuing agricultural benefits is through the use of computer-simulated supply and demand models that provide information on changes to producer and consumer surplus due to a change in crop yields.

Lack of information often prevents analysts from identifying, quantifying, and monetizing all potential benefits categories that may result from environmental regulation. As a contrasting example, a cost analysis is expected to provide a more comprehensive estimate of the cost of an environmental regulation because technical information is available for identifying the technologies that would be necessary to achieve the desired pollution reduction. In addition, market or economic information is available for the many components of a cost analysis (e.g., energy prices, pollution control equipment, etc.). A similar situation typically does not exist for estimating the benefits of environmental regulation. The nature of this problem is due to the non-market characteristic of many benefits categories. Since many pollution effects (e.g., adverse health or agricultural effects) traditionally have not been traded as market commodities, economists and analysts cannot look to changes in market prices and quantities to estimate the value of these effects. The inability to quantify all benefits categories as well as the possible omission of relevant environmental benefits categories leads to a likely underestimation of the monetized benefits. Although it is not possible to estimate the magnitude of the potential underestimation, it is important to highlight potential benefits categories for which quantitative information is not available at this time. Table 1 lists the anticipated health and welfare benefits categories that are reasonably associated with reducing PM and ozone in the atmosphere, specifying those for which sufficient quantitative information exists to permit benefits estimation.

8. COMPARISON OF MONETIZED BENEFITS AND COSTS

The completion of the benefits and cost analyses makes possible one method for quantitatively evaluating the revised standards (since an effort has been made to express as many factors in the same units (dollars) as possible). The monetized benefits, costs, net benefits, and, for the promulgated PM and ozone standards are presented in Table 2. Note that the estimates represent a "partial attainment" scenario for the promulgated standards. As previously explained, in some projected nonattainment areas, the control measure database was not able to the level of emission reductions that would have been necessary to model the area as an attainment area. The analysis cannot model additional control measures that are not currently in the database and therefore, does not arbitrarily assume all areas will attain the standards in the year 2010. This partial attainment scenario may not be an unrealistic snapshot of the year 2010 since many areas will be given more time to attain the revised standards. Therefore, the year 2010 reflects progress towards attainment but not complete attainment.

Table 1: Particulate Matter and Ozone Benefits Categories

	Unquantified Benefit Categories	Quantified Benefit Categories (incidences reduced and/or dollars)
Health Categories (reduced incidences of ...)	Morphological changes Altered host defense mechanisms Cancer Other chronic respiratory disease Infant mortality Mercury emission reductions Airway responsiveness Pulmonary inflammation and respiratory cell damage Chronic respiratory damage/premature aging of lungs	Premature Mortality Hospital admissions for: all respiratory illnesses congestive heart failure ischemic heart disease pneumonia chronic obstructive pulmonary disease Acute and chronic bronchitis Lower, upper, and acute respiratory symptoms Respiratory activity days Moderate or worse asthma Self-reported asthma attacks Cancer from air toxics Shortness of breath Work loss days Coughs Pain upon deep inhalation Changes in lung function
Welfare Categories (increased value or reduced damage to...)	Materials damage (other than consumer cleaning cost savings) Damage to ecosystems (e.g., acid sulfate deposition) Ecosystem and vegetation effects in Class I areas (e.g., national parks) Damage to urban ornamentals (e.g., grass, flowers, shrubs and trees in urban areas) Reduced yields of tree seedlings and forests Nitrates in drinking water Brown clouds	Commodity crops Fruit and vegetable crops Consumer cleaning cost savings Visibility Nitrogen deposition in estuarine and coastal waters Worker productivity

Table 2: Estimated Benefits and Costs of the PM and Ozone NAAQS
(1990 $; year = 2010; partial attainment)

PM and Ozone NAAQS Combined		
Monetized Benefits	=	\$19.4 billion to \$106 billion
Costs	=	\$9.7 billion
Net Benefits	=	\$9.7 billion to \$96 billion
PM NAAQS		
Monetized Benefits	=	\$19 billion to \$104 billion
Costs	=	\$8.6 billion
Net Benefits	=	\$10 billion to \$95 billion
Ozone NAAQS		
Monetized Benefits	=	\$0.4 billion to \$2.1 billion
Costs	=	\$1.1 billion
Net Benefits	=	\$(0.7) billion to \$1.0 billion

Non-Monetized Benefits Categories:

-Reduced chronic respiratory damage/premature aging of lungs
-Reduced susceptibility to respiratory infection
-Reduced cancer and other air toxics health effects
-Reduced incidence of significant decreases in pulmonary function
-Reduced acute inflammation and cell damage in respiratory systems
-Reduced infant mortality
-Protection of national parks, forests and ecosystems (e.g., from acid sulfate deposition)
-Increased yields of tree seedlings
-Reduced damage to urban ornamentals (e.g., grass, flowers, shrubs, and trees in urban areas)
-Reduced nitrates in drinking water

Notes: PM NAAQS = annual $PM_{2.5}$ standard of 15 $\mu g/m^3$
24-hour $PM_{2.5}$ standard of 65 $\mu g/m^3$
Ozone NAAQS = .08 ppm, 8-hour, 4th highest concentration

( ) denotes negative values

REFERENCES

1. Presidential Document. Executive Order 12866 of October 1993. "Regulatory Planning and Review," 58 Federal Register 51735.
2. U.S. Congress. Clean Air Act Amendments of 1990. 42 U.S.C. 7401-7626, Public Law 159. Washington D.C., U.S. Government Printing Office.
3. U.S. Environmental Protection Agency. (December 1996a). Regulatory Impact Analysis for Proposed Ozone National Ambient Air Quality Standard. Research Triangle Park, NC: Office of Air Quality Planning and Standards.
4. U.S. Environmental Protection Agency. (December 1996b). Regulatory Impact Analysis for Proposed Particulate Matter National Ambient Air Quality Standard. Research Triangle Park, NC: Office of Air Quality Planning and Standards.
5. U.S. Environmental Protection Agency. (July 1997a). National Ambient Air Quality Standards for Ozone, Final Rule. 62 Federal Register 38856-38896. Washington D.C., U.S. Government Printing Office.
6. U.S. Environmental Protection Agency. (July 1997b). National Ambient Air Quality Standards for Particulate Matter, Final Rule. 62 Federal Register 38652-38760. Washington D.C., U.S. Government Printing Office.
7. U.S. Environmental Protection Agency. (July 1997c). Regulatory Impact Analyses for Particulate Matter and Ozone National Ambient Air Quality Standards and Proposed Regional Haze Rule. Research Triangle Park, NC: Office of Air Quality Planning and Standards.

SESSION B

CLIMATE CHANGE

Improving Integrated Assessments for Applications to Decision Making

J.C. Bernabo

Science and Policy Associates, Inc.
Suite 400 West Tower, 1333 H Street NW, Washington, DC, 20005, U.S.A.

1. INTRODUCTION

Most integrated assessments conducted on complex environmental issues, such as climate change, failed to fully meet the needs of the decision makers they are intended to serve. This observation has been recognized by many practitioners over the last two decades (1) and also has been documented by social learning research (2).

When researchers apply assessment models as mechanistic predictive tools, they risk producing results that are not sufficiently relevant and prone to misuse in decision making. Policy-relevant assessment is more of an organic process involving researchers and the issue's stakeholders in an interactive exercise that builds understanding rather than defines "truth." Some valuable general lessons can be drawn from examining past experience in environmental assessments. Based on these lessons, improved solutions can be defined to address better the challenges common to similar endeavors.

This paper generically examines some of the shortcomings of past assessment efforts, identifies possible solutions, and provides examples of recent work on climate change and air quality planning where new approaches have been applied successfully. The emphasis is on providing *practical information* in an easily usable form for those involved in environmental assessments for decision makers.*

2. PROBLEM DEFINITION

The underlying difficulty in conducting scientific assessments to assist decision making is rooted in the divergent purposes and different professional cultures of researchers and managers (3). Scientists ultimately seek to understand an issue but the decision makers' job is to decide. Decisions usually have deadlines and must be based on whatever level of scientific understanding exists, taking account of all the other societal factors that apply to the decision. Indeed, the technical aspects are usually given less weight in the decision process than social and political factors, much to the dismay of researchers pursuing the "truth" about a problem as the scientists define it (4).

Given these deep-seated causes of misunderstanding between the producers of assessments (researchers) and the intended users of the information (decision makers) certain recurring problems are evident in most assessment efforts (3).

---

* Decision making is used broadly and includes policy, management and regulation as specific types. However, because I am focusing on generalities that can apply to all these forms of decision making, these terms are used here as synonymous to avoid monotony.

There are three general shortcomings for many past environmental assessments,

1. Lack of focus on decision makers' specific information needs
2. Inadequate stakeholder involvement and communication at all phases
3. Insufficient integration of quantitative and qualitative factors relevant to the decision makers

2.1 Technical Soundness and Policy Credibility

These problems can be particularly debilitating because assessments not only must be *technically sound* but also *policy credible* to be most useful to decision makers. Researchers are well versed on what is required to ensure scientific credibility, such as using well-documented and peer-reviewed information. But they often find the needs of decision makers require stretching assumptions beyond what they believe "good science" allows, forcing them to make expert judgements and informed guesses or else forfeit contributing effectively to the process. The other crucial assessment attribute, policy credibility, is less well understood by researchers. The assessment's credibility depends on how transparent, inclusive and unbiased the whole process is perceived to be by stakeholders, as much as the credibility of justice depends on the legal process followed. The scientific and decision making participants in the overall assessment process need a greater mutual understanding of each others' arenas because inevitably they must interact in the conduct of policy-relevant assessments. In the past, the interaction typically has been ad hoc, sporadic and often an afterthought rather than a carefully planned process.

Many of the weaknesses of policy-relevant assessments are exacerbated by the lack of adequate dialogue between the producers and users of the assessment, not only before but also during and after it is conducted. There has been a tendency to "ready-fire-aim" rather than systematically ensuring a more productive sequence of: a) ready -- with dialogue; b) aim -- with stakeholder involvement in the design; and c) fire -- with continued dialogue and involvement throughout the implementation and evaluation.

The flawed sequence often results from researchers first discovering environmental issues and formulating them in their terms, before the issue becomes an active policy problem when the decision makers reframe it for their own context. Assessments are often designed based on the researchers' understanding of the scientific issues and their educated guesses of what decision makers would like to know rather than eliciting the decision makers' needs at the onset.

With an approach, that fails to fully engage the assessment's target audience in the initial design, it is unlikely that even the assessment's questions will be focused on those matters most relevant to the decision making. What usually results is a state-of-the-science assessment with minimal usefulness to decision making. In practice such results unintentionally provide fodder for opposing advocates on the issue to use selectively for supporting their predetermined positions in the policy debate.

By applying multi-stakeholder approaches, assessment processes not only focus on the science but also inform the decision makers in a manner that facilitates sustainable outcomes that are technically and politically viable for society. Some of the key factors for integrated assessments as well as the associated benefits and challenges are outlined in table 1.

Table 1
Integrated Assessments: Examples of General Factors, Benefits and Challenges

General Factors in Designing and Conducting Assessments

- Target audience
- Technical methods
- Communication strategy
- Outputs and deliverables
- Bounds: spatial, temporal and issue
- Management approach
- Review and evaluation
- Quality assurance and control

Potential Benefits of Assessment Process

- Identification of key technical and decision making issues
- Development of a problem analysis framework to structure the issue
- Integration of information from multiple disciplines and sources
- Explicit definition of major assumptions and uncertainties
- Establishment of priorities for future policy-relevant research
- Development of broader understanding among stakeholders
- Creation of a basis for more informed decision making

Common Challenges for Assessments

Technical issues

- Poorly understood natural processes
- Extrapolating in space and time
- Data and observation limitations
- Quantifying cumulative uncertainties

Social issues

- Articulating processes
- Politicizing science
- Defining the roles of producers versus users
- Diverse use of language and varied expectations
- Difference between scientific and decision making criteria
- Identifying and recruiting constructive participation of issue stakeholders

3. LESSONS AND SOLUTIONS

Based on the experiences of dozens of environmental assessments on a range of issues in several countries, some significant lessons emerge from past efforts that provide insights for improving the usefulness of assessments for decision makers. The Social Learning Project of Clark and his colleagues at Harvard University is documenting how decision makers have responded to various assessments (2). Five general lessons about assessments that have been judged most effective *by decision makers* (5) are as follows.

The assessment process was inclusive and well designed. Communication can be more important than model integration in determining success. Complex model wiring diagrams suggesting all the pathways to be studied are of limited value unless there is a carefully planned process of interactive flow of information among researchers, synthesizers, stakeholders, reviewers and end users.

The assessment results in expanding options for decision makers. The assessment process can provide a vehicle for finding win-win solutions. Merely providing detailed analysis of specific policies rarely advances the decisions as usefully as when new options are generated in an interactive process between producers and users of the information.

The assessment focuses on real uses at the regional level. Broad generalizations on the national or global scale have limited value at the more local scales where most impacts of a decision are experienced. Averages integrating over large areas can obscure the more localized texture that political representation is based on in national and global fora.

Multiple partial assessments are performed on key components. The end results of large integrations of many factors tend to suffer from the limitations of their weakest links. These limitations and their complexity make them harder to use and interpret by decision makers than sets of comparable, separate analyses.

Assessments are conducted repeatedly by a core group. Single-time grand integration efforts are less likely to benefit from the learning required to be most useful for decision makers. Assessments ultimately need iteration to yield successively better results with a stable core group of assessors who incorporate the learning as they proceed.

These overall lessons indicate the primary importance of the *process* not just the report or other tangible outputs of an assessment effort. In fact, the reports may often be the least valuable output when one examines assessments as representing interactive learning processes involving researchers, assessors, decision makers, stakeholders and the public.

3.1 Defining Policy-Relevant Assessments

One of the difficulties in discussing assessments is that the concept is used in a variety of ways and has no widely accepted definition. While assessments generally apply to some type of focused analysis of a topic, the word is used to mean everything from an informed opinion to complex models to risk analyses. I offer a broad definition here, not as the sole way assessment can be defined, but rather as an explicit definition that covers the kind of policy-relevant and science-based endeavors we are examining.

Policy-relevant assessment can be defined as:

An iterative multi-stakeholder process for systematically analyzing data and synthesizing information into a form that facilitates use in decision making.

These assessments are both a scientific exercise and a social process. While they link the science to the needs of decision makers, they do not usurp the role of the applicable decision making process they support. Such assessments are a form of decision support tool that can be applied along with other tools and factors in the development of actions and plans.

3.2 Attributes of Successful Assessments

Based on the lessons of past efforts there are a number of factors that enhance the effectiveness of policy-relevant assessments.

- Responding to clearly defined needs relevant to decision making
- Involving decision makers in framing the questions and designing the process
- Maintaining interaction with end users during planning and implementation
- Developing and implementing a multi-stakeholder communication strategy
- Establishing credibility with periodic open reviews of the science by research peers and the assessment's process and assumptions by stakeholders
- Using the full spectrum of research approaches and tools to address the issues of importance to decision makers that can be quantitatively modeled and those that only can be qualitatively described, explicitly estimating uncertainties
- Managing the process for successful assessment outputs rather than as research
- Using assessment specialists to integrate the outputs of expert teams

None of these elements alone is sufficient to ensure success. For effectiveness they should be applied as a carefully designed ensemble adapted for the requirements of a specific assessment issue. These are illustrated in Sections 4 to 6 of this paper by their application in recent integrated assessment efforts for climate change and air quality, two issues of widespread and active endeavor where the author has extensive experience.

3.3 Process-Guided Assessment Paradigm

When taken together these attributes suggest a new model for conducting assessments that is less mechanistic and more organic than previous efforts. Assessment can be practiced as a learning-centered dialogue rather than the traditional exercise in defining "truth" or predicting probable future outcomes. The science can and should define what is known and the level of uncertainty, but it is useful to remain humbled by the reality that research is more efficient at converting ignorance into new uncertainties, than it is at turning uncertainties into "facts" (6). Table 2 outlines new emphases for policy-relevant assessments.

Table 2
New Emphases for Policy-Relevant Assessments

Previous Style	Evolving Style
Mechanistic models	Organic models
Reductionist	Holistic
Science focus	Decision context focus
Seeks truth	Provides learning
Generating predictions	Generating options
Defining outcomes	Establishing processes
Producing answers	Framing questions

When science interacts with decision making a new trans-scientific activity results (7), different from unfettered discovery and embedded in the human value judgements that dominate decision making. Because of the challenges this presents, especially to the hyper-rationalistic nature of modern science, a systematic dialogue among the producers and users of an assessment is a crucial element that is often overlooked. Such a dialogue implies communications in both directions aimed at understanding the different perspectives of research and decision making (3).

Ideally before an assessment is even designed, a structured dialogue process should be implemented. With this ongoing dialogue as a foundation, a design can begin with the involvement of not only the assessors and their clients but also the full range of stakeholders who will be impacted by the issue. The assessment process should be designed to ensure the appropriate participation by researcher, assessors, decision makers, stakeholders and the public. Figure 1 illustrates the assessment cycle involving the producers and users of policy-relevant technical information.

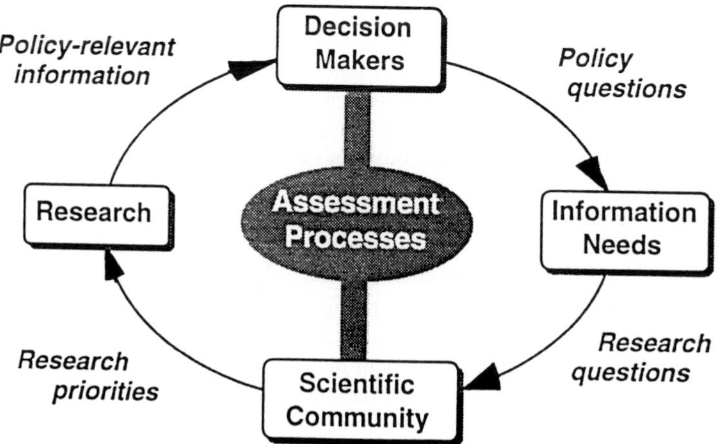

Figure 1. Assessment as an iterative process linking science and decision making.

Process-guided assessments require three phases to achieve the interactive learning effort of the evolving new paradigm,

Phase 1 - Facilitated Dialogue. An ongoing, planned effort that is structured for maximum interaction among participants initially to identify users' needs, determine scientific capabilities, elucidate stakeholders concerns and build mutual understanding for the assessment.

Phase 2 - Interactive Design. An interactive approach is used to clarify goals, define roles, identify the constituency for the assessment and initiate outreach activities. This includes reviewing the proposed roles, methods and outputs to ensure credibility.

Phase 3 - Multi-stakeholder Implementation. The final phase involves conducting the formal assessment, reviewing interim and final results, explaining the significance to users, and maintaining internal and external communications based on a well-defined plan followed by evaluation and then iteration.

In the remaining sections of this paper, examples of actual projects that have been conducted are provided to explain the challenges and benefits of some improved approaches to policy-relevant environmental assessments.

4. IMPORTANCE OF DIALOGUE

A productive dialogue among diverse researchers, decision makers and impacted stakeholders should be structured to facilitate effective communication. Often those seeking to promote dialogue merely get amorphous communication because they fail to develop a viable process for focused interaction. A similar problem would be familiar to researchers if an amateur collected data but neither with a plan for how and where it was to be collected nor adequate documentation to make the findings usable to science.

Dialogues are greatly improved when professional experts in group processes and facilitation are involved. A rich set of theory, information, approaches and experience exists on the subjects of conducting decision making, enhancing dialogue, applying knowledge theory and using elicitation techniques such as focus groups (8).

When a dialogue is well designed and executed it can be the source of useful learning for all the participants in assessments. Participants can learn the context in which they each view the issue, define the range of assumptions they hold and develop the questions of greatest value to the decision making. The involvement of the users and producers of the assessment in this process provides the basis for more informed assessment questions and better planning of the impending analyses. This approach lays the ground work for the ongoing interactive dialogue that can continue throughout the phases and iterations of the assessment and decision-making processes.

4.1 National Climate Change Dialogue

An example from the area of climate change illustrate this process. The "Joint Climate Project to Address Decision Makers' Uncertainties" was begun in the United States in 1989 as a government-industry collaboration with sponsorship by seven federal agencies and the electric power utilities (9). The purpose of the dialogue was to gain a better understanding of what the information needs of future decision makers might be so that more relevant climate change research and assessments could be conducted. The multi-stakeholder dialogue involving researchers and decision makers revealed that most of the $1.6 billion spent annually on climate research was focused on only one of the three areas crucial to decision makers. More than 90% of the planned work was on earth system predictions with practically nothing on climate impacts or response options -- two areas that were of greatest importance to decision making.

The project's interviews and workshops indicated a wide gulf between what researchers considered important to study and what decision makers wanted to know. The most valuable conclusion of the dialogue was the pressing need for more effective means of communication between the science and decision-making communities. The learning the participants gained from their involvement in the dialogue is a major benefit of such a process because it sets the stage for more effective assessment as well as decision making by providing understanding among the actors that otherwise would not have occurred.

4.2 International Dialogue on Climate

During the early 1990's the Dutch National Research Programme on Global Air Pollution and Climate Change initiated a national multi-stakeholder dialogue on climate change as part of their assessment program (10). The multi-stakeholder dialogues underway in both the United States and the Netherlands recognized the fundamentally international aspects to decision making on global climate change. As a result, the Dutch Climate Programme and the U.S. EPA co-funded a pilot international dialogue examining decision makers' information needs over the next decade on the climate issue (11). This cross-cultural dialogue was intended to complement the pioneering efforts under the United Nations to develop an international climate treaty with input from a scientific body it convened, the Intergovernmental Panel on Climate Change (IPCC).

The bilaterally sponsored dialogue project entitled, "Enhancing the Effectiveness of Research to Assist International Climate Change Policy Development" began in 1994 and involved decision makers and researchers from six key nations: China, India, Brazil, Poland, United States, and Netherlands. A structured and professionally facilitated dialogue was conducted to enhance understanding and build a stronger foundation for future assessments, policy-relevant research and decision making.

The project began with a series of individual interviews with decision makers from governments, industries and non-governmental organizations in each of the six nations. From these off-the-record meetings the project team developed an analysis of the range of perspectives on the climate issue in each nation. These interview-based synopses of the *issue landscape* in each nation were shared with a workshop of about 40 decision makers representing key sectors from all six nations. Details of the underlying theory, methods and results are presented in the project report (11).

After reviewing the different national perspectives as seen by diverse stakeholders, the decision makers at the workshop defined a *range of possible policy paths* about which research could provide information on to assist future decision making. A group facilitation process we dubbed the "three P's" was applied by successively defining *probable* (status quo), *preferable* (idealized) and *possible* (achievable) future scenarios for policy. The decision makers developed a series of three P's futures using the approach shown in Figure 2. There was no attempt to agree on one consensus outcome; instead the minimum number of policy paths that reflected all points of view was sought.

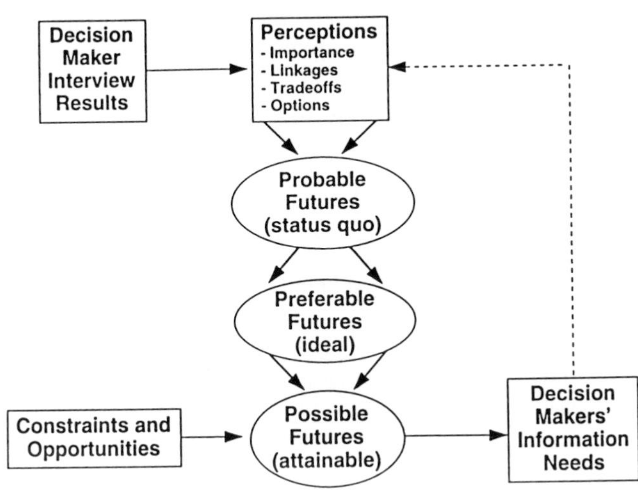

Figure 2. Example of a dialogue process for defining decision makers' information needs.

The international decision makers first established the range of possible outcomes for policy if things continue as they seem now with the *status quo*. The group of decision makers created several policy paths that covered all likely results given current conditions with descriptive titles including: too little/too late; realistic extrapolated trends; strained sustainability; sustainability moderately exceeded; sustainability significantly exceeded; and depletion and exhaustion. As the titles of the probable scenarios suggests, nobody was very hopeful about the status quo outcomes and this built enthusiasm for seeing if better options could be defined.

Then after developing the range of probable outcomes, the workshop participants divided into groups examining what ideal *preferable* future policy scenarios might be, without the constraints of current situations. From these sessions the decision makers agreed on five preferable futures encompassing every bodies aspirations including: globally stabilized emissions; industrial countries emission reductions; transition to renewable energy; high-energy sustainable; and low-energy sustainable futures.

Finally, the group took the five preferable (*most desired*) future scenarios they outlined and applied constrains to come up with the a range for possible (attainable) futures. They then identified the key information needs that research and assessment could address to help develop, evaluate, select and implement these options. Researchers then examined the decision makers' information needs and identified what science could do in the next decade to address the needs of the future. Decision makers from the previous workshop participated in the exercise to ensure continued communication.

The outcome was a set of research objectives that is ahead of the current policy needs. To understand what other options exist within each societies' stakeholders, the dialogue went beyond the upcoming series of negotiations to identify a range of future information needs. Determining the intra-national diversity of views among stakeholders provides insight beyond the current national positions in the negotiations to identify other options that might later appear as the politics change. Even in non-democratic China, a wide range of policy views existed that could emerge in the future. The project's future-looking approach helps to avoid the common situation where by the time research is completed the policy questions have changed.

Table 3
Common Features of Negotiations and Dialogues on Controversial Issues

Negotiations -- often necessary for solutions but not always sufficient because:

- many perspectives are reduced to limited number of official positions
- motivations and range of options for actors are not explicit
- each party seeks maximum advantage
- convergence is toward single least objectionable outcome

Dialogues -- can facilitate more productive negotiations by:

- expanding the range of positions addressed
- increasing understanding by examining underlying motives
- promoting consensus on a range of acceptable options
- establishing a possible basis for win-win solutions

This overview of the climate dialogues is presented to give some examples of how facilitated processes can yield valuable insights not possible with the business-as-usual approaches to linking science and decision making. In all three cited dialogues, participants reported gaining understanding about other sectors that otherwise would have been impossible, which provided a stronger foundation for better assessments and improved decision making.

5. VALUE OF INTERACTIVE DESIGN

The ongoing effort by the United Kingdom to develop an integrated climate impacts assessment program provides a recent example of the effectiveness of interactive design (12). The UK Department of Environment and the Environment Agency wanted to design an impacts assessment program in partnership with other government agencies, regional governments and the private sector. The goal was to establish a long-term program that was highly credible technically and would be accepted and co-funded by a range of stakeholders.

They used an independent third party to conduct and broker a dialogue among the potential sponsors and stakeholders. A series of interviews was conducted with UK stakeholders to determine the range of perspective, types of information needs and prospects for joining a collaborative government-private sector climate impacts assessment effort. Based on the interview results a set of common needs and divergent concerns was compiled and presented at a multi-stakeholder workshop. A dialogue ensued to determine how much support existed for the collaborative effort as well as what elements and process needed to be in place for it to work and garner broad support. Using the consensus criteria developed in the workshop, a group of leading assessment experts (mainly from the UK) formulated an approach to meet the needs of the decision makers and attract stakeholder participation.

The stakeholders desired a staged development rather than single monolithic assessment. The expert team defined three major dimensions for examining impacts over time: 1) natural systems; 2) socio-economic sectors; and, 3) geographic regions. These dimensions (Figure 3) covered 400 million possible interactions for the UK with 10 recognized types of natural systems, 10 geographical regions and 40 defined economic sectors. Obviously, the assessment would have to begin by integrating among subsets of all the possible elements in order to be manageable and technically tractable.

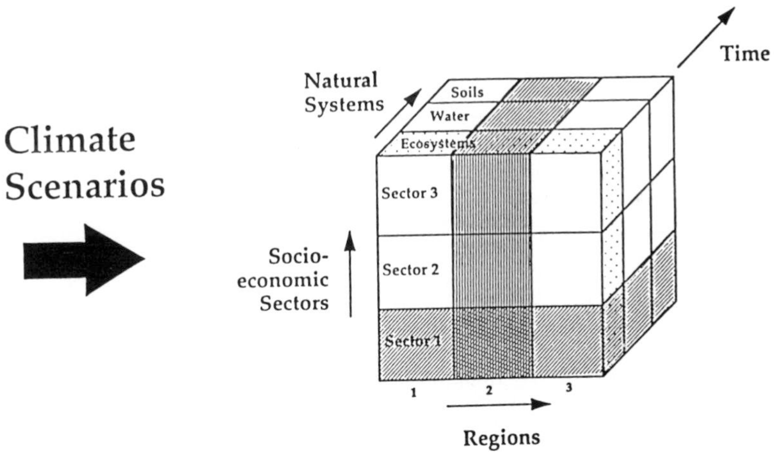

Figure 3. Dimensions of integration for climate impacts assessments

Using pilot assessments integrating across each of the three dimensions can produce useful analyses that will contribute to the overall assessment goals of informing decision makers. The initial assessment activities are proposed to use standardized climate scenarios to examine: 1) how changes in water resources will impact natural and human systems; 2) how impacts of climate on key industries at risk will effect insurance losses; and 3) how the natural and human system in the uplands region will be impacted.

The interactive design project also identified an organizational structure and processes for a collaborative assessment program based on stakeholders input and the desire to ensure policy relevance as well as technical soundness. The design includes a core assessment staff, a government-private sector steering committee of sponsors, and both scientific and decision maker advisory bodies to review the relevancy and soundness of the program (Figure 4). The modular nature of the activities allows flexibility and use of the best expertise for each portion of the assessment as well as continuity of the core group of assessors. This multi-stakeholder program will help build consensus on impacts and response options while reducing the potential for the wasteful dueling-scientists approach that characterizes many controversial issues where cooperative assessments are lacking.

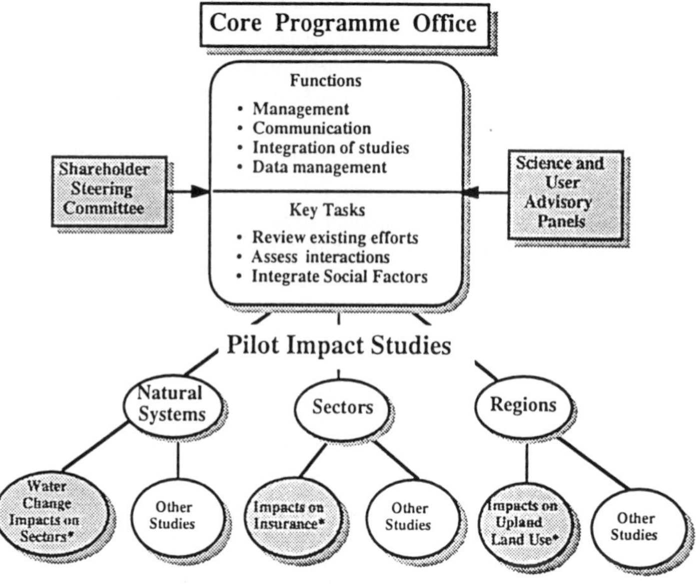

Figure 4. Proposed structure of the UK Integrated Climate Impacts Programme

6. SUCCESSFUL MULTI-STAKEHOLDER IMPLEMENTATION

The success of a policy-relevant assessment is measured best by its direct and indirect contribution to the decision making it was designed to assist. When the newspaper headlines proclaimed that the seven western U.S. states had reached consensus on an air quality plan for the next 40 years (13), it was a testament to the effectiveness of the Assessment of Alternative Emission Management Scenarios conducted for the Grand Canyon Visibility and Transport Commission (14). The integrated assessment they designed and sponsored is an example of a true multi-stakeholder process embodying all of the elements crucial for policy-relevant assessments in Section 3.2.

The Commission was established by the Clean Air Act Amendments of 1990 to involve the seven western governors in developing a strategy for protecting scenic vistas and air quality while accommodating rapid economic growth. They embarked on a six year effort that culminated in unprecedented long-term agreement on emissions planning across a diverse region with different political and economic interests.

The process began with an extended dialogue among the various states, industries, federal agencies, Indian nations, environmental groups and the public on what and how to address the challenge of maintaining and improving regional air quality while enabling economic growth. Facilitated workshops around the region involved the full range of stakeholders in framing the issues and developing mutual understanding. After more than three years of dialogue and planning, the Commission undertook a technical assessment of emissions management alternatives to inform the governors decision making.

A core group of knowledgeable stakeholders from government and the private sector worked together in committees to design the assessment with review by researchers and decision makers. The assessment was overseen by the Western Governors Association, funded by U.S. EPA and conducted by an independent contractor team of leading experts managed by Science and Policy Associates, Inc.

The resulting policy-relevant assessment was successful in large part because it addressed both the quantitative and qualitative aspects of these complex issues. For the factors where quantifiable relationships and adequate models existed, such as for emissions sources, air quality changes and the economics of control, the assessment provided model outputs to illustrate the consequences of five scenarios agreed to by decision makers. The scenarios for emissions management ranged from doing nothing more than current law to doing everything technologically feasible to reduce emissions with three variations in between.

The *primary assessment* provided decision makers with computer-based capabilities to examine the consequences of the various options, all using the same set of scenarios and analyses in order to aid decision making rather than fight over arcane technical difference. At the outset of the project key stakeholders jointly agreed to an approach, data and methods so that their focus was on decision options not debating contradictory results of competing analyses.

Whereas the primary portion of the assessment dealt with those factors that could be quantitatively modeled, the *secondary assessment* tackled the other key factors for decision makers that could not be modeled, such as: most ecological impacts, administrative ease, and equity among stakeholders. These largely qualitative factors where examined for each alternative emission scenario by using approaches such as expert judgement and surveys to avoid them being neglected when considering the quantitative outputs. This two-tiered approach avoided the black-box syndrome where model results lead to conclusions that omit key factors for decision makers simply because they are not yet adequately quantified.

The role of integration in the assessment went beyond linking modules of a model to include packaging the quantitative and qualitative aspects of the analyzes for each option in a manner that facilitated consensus building among the governors. The assessment's overall process provided not only useful information geared to decision makers' explicit needs but a structure for the stakeholders to interact that fostered the level of understanding and cooperation necessary to reach agreement on a technically sound and policy credible outcome.

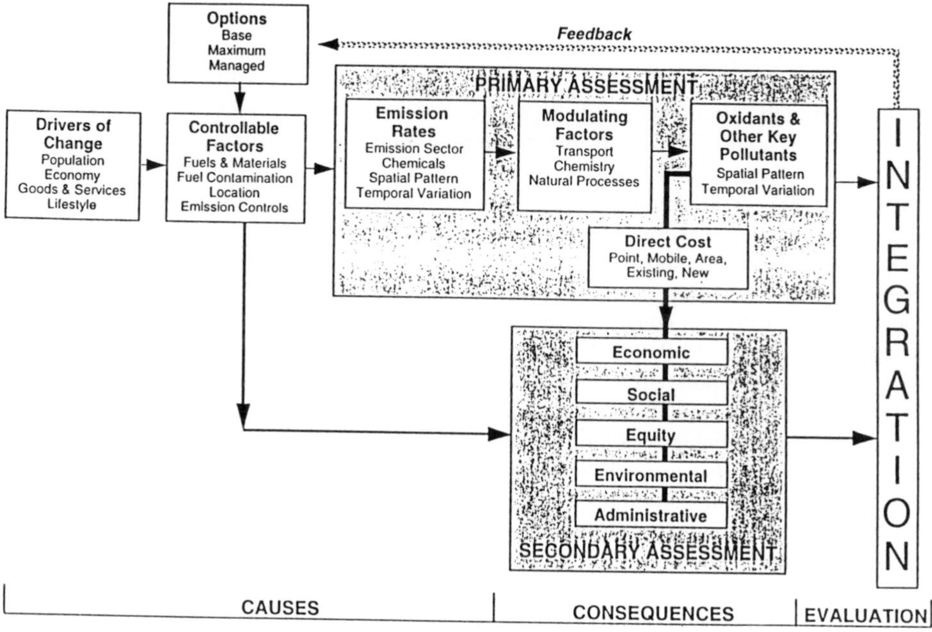

Figure 5. Western US Air Quality Assessment of Emissions Management Scenarios for the Grand Canyon Visibility and Transport Commission (14).

7. CONCLUSIONS

Improving integrated assessments to meet the needs of decision makers requires new approaches that involve stakeholders in interactive learning processes. Facilitated dialogues between the producers and users of assessments build understanding, enhance the relevancy and increase the credibility of the results.

The value of moving beyond traditional mechanistic approaches is demonstrated by recent examples of successful assessments applying structured dialogues, interactive design phases and multi-stakeholder implementation. Policy-relevant integrated assessments are *iterative learning processes* rather than truth defining end-points. To be of greatest value they must include both the quantitative and qualitative factors for analyzing the decision making options.

REFERENCES

1. Bernabo, J. C. (1986). Science and Policy: Notes from a Former Congressional Fellow. EOS, Transactions of the American Geophysical Union, 67, p. 82.
2. Clark, W.C. et al., (1997). A Critical Evaluation of Global Environmental Assessments: The Climate Experience. Harvard University Committee on the Environment.
3. Bernabo, J. C. (1995). Communication Among Scientists, Decision Makers and Society: Developing Policy-Relevant Global Climate Change Research. In Climate Change Research: Evaluation and Policy Implications (S. Zwerver et al., eds.), p. 103-117.
4. Brooks, H. (1984). The Resolution of Technically Intensive Public Policy Disputes. In Science and Human Values, p. 9.
5. Clark, W. C. (1996) Personal Communication and Clark, W. C. and Dickson, N. (1998) Learning to Manage Global Environmental Risks: A Comparative History of Social Responses to Climate Change, Ozone Depletion, and Acid Rain. Cambridge: MIT Press.
6. Brooks, H. (1975). Expertise and Politics: Problems and Tensions. In Proceedings of the American Philosophical Society, 119, p. 257.
7. Weinberg, A. M. (1972). Science and Trans-Science. Minerva 10, p. 207.
8. Herrick, C. and Bernabo, J. C. (1991) Focus Groups Improve Environmental Communication. National Association of Professional Environmental Communicators (NAPEC) Quarterly 2:3, p. 13.
9. Bernabo, J. C. and Eglinton, P. (eds.) (1992) Final Report of the Joint Climate Project to Address Decision Makers' Uncertainties, Electric Power Research Institute, Technical Document No. TR-100772.
10. Klabbers, J. Vellinga, P. et al. (1994). Policy Options Addressing the Greenhouse Effect. National Research Programme on Global Air Pollution and Climate Change, Bilthoven, Netherlands.
11. Bernabo, J. C. et al. (1995). Enhancing the Effectiveness of Research to Assist International Climate Change Policy Development. Dutch National Research Programme on Global Air Pollution and Climate Change, Bilthoven, Netherlands Report # 410-100-090.
12. Bernabo, J. C. et al. (1996). Proposal for a United Kingdom Integrated Climate Change Impacts Assessment. United Kingdom Department of Environment, London, UK. 28 p.
13. Washington Post, 20 June 1996. p. A2.
14. Middleton, P. et al. (1995). Options for Improving Western Vistas, Volumes I-IV. Science and Policy Associates, Inc. Report prepared for the Grand Canyon Visibility and Transport Commission. Western Governors Association, Denver, Colorado.

Estimating socio-economic impacts of climate change

Richard S.J. Tol

Institute for Environmental Studies, Vrije Universiteit, De Boelelaan 1115,
1081 HV Amsterdam, The Netherlands

1. INTRODUCTION

The potential impacts of a possible climate change are many and diverse (cf. also Watson *et al.*, 1996). In this talk, I will address the issue from an economic point of view, in fact, from a decision-analytic point of view. Decision analysis helps a careful, systematic and consistent balancing of the pros and cons of policy strategies towards a problem, climate change and greenhouse gas emission reduction in this case. To that end, different types of impact need to be made comparable, by expressing them in a single indicator. Money is a suitable metric, because of its central importance in almost all exchanges and the long experience of economists in monetary valuation. I will thus try and express the impact of climate change in money. This allows us to interpret the seriousness of climate change (and its regional differences), compared to other problems and compared to the costs of emission reduction. I will even go a step further to present the marginal damages of carbon dioxide, that is, the additional impact that would accrue should we emit one additional ton of carbon. Straightforward mathematics shows that it justified to reduce emissions only up to the point that marginal costs and benefits are equal. Justified, that is, if we want to maximize welfare.

The paper is structured as follows. Section 2 reviews the economic assessments of climate change impacts, following the Second Assessment Report (SAR) of the

Intergovernmental Panel on Climate Change (IPCC). Section 3 discusses more in-depth the contentious issue of valuing non-market impact. Section 4 presents estimates of the marginal damage of carbon dioxide emissions. Section 5 concludes.

2. CLIMATE CHANGE DAMAGE COSTS

The scientific research on global warming impacts has focused predominantly on the $2 \times CO_2$ benchmark, that is, a scenario with an atmospheric CO_2 equivalent concentration of twice the preindustrial level. Despite the fact that this scenario is arbitrary, counterfactual and of limited policy relevance, a large part of the social cost chapter in IPCC SAR (Pearce *et al.*, 1996) is consequently also devoted to "$2 \times CO_2$ damage". A climate change associated with a doubling of the atmospheric concentration of carbon dioxide equivalents is expected to occur in the second half of the next century if no substantial emission reductions are realized.

Information on the impacts of global warming is available for several regions and countries. The best studied regions are developed countries, in particular the United States, where climate change impacts have been analyzed in a series of studies, following initial work by Smith and Tirpak (1989). The most prominent US studies are made comparable by Smith (1996). The most recent, comprehensive US study is by Mendelsohn and Neumann (forthcoming). Other OECD regional studies include CRU/ERL (1992) for the European Union (updated by Plambeck and Hope, 1996); Parry and Duncan (1995) for the United Kingdom); and Nishioka *et al.* (1993) for Japan. In the context of an Asian Development Bank (ADB, 1992) project on climate change in Asia, global warming impacts have also been analyzed for a number of Asian countries. Stzrepek and Smith (1995) contains case studies for Africa, Latin America and Asia. Under the various country study programs brought about by the Framework Convention on Climate Change, most developing countries and countries with economies in transition are now being studied. If these studies go well, the most significant white spot on the climate impact knowledge map will be continental Western Europe.

Studies usually deal with only a subset of damages, and are often restricted to a description of impacts in physical terms. Estimates generally combine, but do not neatly separate the costs of adaptation (such as sea level rise protection) and the costs of residual damages (such as the inundation of unprotected areas).

By far the best studied impact categories are agricultural impacts (e.g., Rosenzweig and Parry, 1994; Adams *et al.*, 1994; Darwin *et al.*, 1995; Schimmelpfennig *et al.*, 1996; Reilly *et al.*, 1996) and the costs of sea level rise (e.g., Fankhauser, 1995b; Turner *et al.*, 1995; Yohe *et al.*, 1995, 1996; Bijlsma *et al.*, 1996). Several types of impacts have largely been ignored so far, because they could not be sufficiently quantified. Other damages were estimated on the back of an envelope.

Attempts at a comprehensive monetary quantification of all impacts are relatively rare, and usually restricted to the United States (Cline, 1992a; Titus, 1992; Mendelsohn and Neumann, forthcoming; Nordhaus, 1991). Preliminary estimates of monetary damage in different world regions are provided by Fankhauser (1995a), Tol (1995), and, subsequent to the finalization of the SAR, Mendelsohn *et al.* (forthcoming).

The impacts of climate change fall apart in two broad categories: impact on marketed and non-marketed goods and services. For market goods, deriving an income change is relatively straightforward. For non-market goods, valuation techniques have to be employed. Four major techniques are available to estimate the 'price' of goods and services which are not traded on markets: the travel cost method, hedonic pricing, household production, or contingent valuation.

In the travel cost method, one analyzes the effort (in time and money) people are willing to spend to visit a particular site with desirable attributes. This effort, expressed in money, is a measure for the monetary worth of the desirable attributes. Advantages of the travel cost method are that it is simple in concept and application, and does not require strong assumptions on the real world. A major disadvantage is that only direct use values can be measured with the travel cost method. Using the travel cost method, one could, for instance, estimate in monetary terms the pleasure people derive from visiting a forest or a beach. People spend time and money to go there, which can be measured and used to derive a demand curve. The value to the average visitor times the number of visitors is an indication of the worth of that forest or beach.

In hedonic pricing, one analyzes price differentials of traded goods and services which have different bundles of non-tradeable attributes. These price differentials are measures for the monetary values of the non-tradeables. Advantages of hedonic pricing are that it is conceptually simple and that it can be used to value a wide range of non-traded goods and services. Disadvantages are that hedonic pricing may involve strong assumptions on information and rationality of economic actors, that it may require complicated statistical techniques, and that only direct use values can be measured. Using the hedonic pricing method, one could, for instance, estimate in monetary terms the pleasure people derive from living in a particular place. A house at a beautiful place is more expensive than the same house at an unattractive location. Measuring the difference -- and controlling for other factors -- one can estimate a demand curve, and thereby obtain an indication of the value of the surroundings of that place.

The household production function approach is similar to, yet more ad than hedonic pricing. In lieu of price differentials in trade goods and services, one analyzes differences in consumption bundles as to the influence of non-tradeables. Demand curves are then derived through an assumed preference structure of the economic agent.

In contingent valuation, one interviews people so as to simulate a hypothetical market on which in reality non-tradeable goods and markets are traded for money. The main advantage of contingent valuation is that it can be used on any good or service, and that it can measure direct and indirect use values as well as option, existence and bequest values. Disadvantages are that contingent valuation assumes that people have well expressed preferences, and that contingent valuation is expensive and elaborate. Using the contingent valuation method, one could, for instance, interview people whether they would be willing to pay amount X to preserve forest Y. Based on the interview results, a demand curve can be derived.

The comprehensive studies mentioned above are based on a mix of these techniques and various approximations. The Fankhauser and Tol figures, which were at the core of the IPCC assessment (Pearce *et al.*, 1996), are reproduced in Table 1. Fankhauser and Tol (1997) have recalculated the initial set of estimates consistently correcting for purchasing power parity and using the same benefit transfer methodology throughout. These results are reproduced in Table 1.

Figures vary between 0 and 7 percent of real (purchasing power parity corrected) GDP.

Table 1 highlights the substantial differences between regions. For the former Soviet Union, for example, damage could be as low as 0.4 percent of GDP, or even negative (climate change is potentially beneficial). Asia and Africa, on the other hand, could face extremely high damages, mainly due to the severe life/morbidity impacts. Developing countries generally tend to be more vulnerable (in relative terms) to climate change than developed countries, because of the greater importance of agriculture, lower health standards and the stricter financial, institutional, and knowledge constraints on adaptation.

Pearce *et al.* (1996) stress the preliminary and incomplete character of these estimates. The estimates do not fully reflect the current state of knowledge, since it takes many years for new insights in climatology and agronomy to trickle down to quantative economic estimates. It should be noted that the above figures are *best guess* estimates. The range does not reflect a confidence interval, but the variation of estimates found in the literature. There is a considerable range of error which has not been quantified. Pearce *et al.* also note that figures on developing countries in particular are largely based on approximation and extrapolation, and are clearly less reliable than those for developed regions. Further, as best-guess estimates, the figures neglect the possibility of impact surprises (such social and political unrest), and of low probability/high impact events (such as a shut down of the ocean conveyor belt). To avoid long-term predictions, damage figures measure the impact of $2\times CO_2$ on a society with today's structure. Vulnerability is likely to change as regions develop and population grows.

Despite these shortcomings, available figures give a rough indication of the possible order of magnitude of $2\times CO_2$ damages and the relative vulnerability of various regions.

Table 1

Annual monetized $2 \times CO_2$ damage in different world regions

	Fankhauser		Tol	
	bn$	%rGDP[a]	bn$	%rGDP[a]
· European Union	63.6	1.4		
· United States	61.0	1.3		
· Other OECD	55.9	1.2		
· OECD America			74.5	1.5
· OECD Europe			57.4	1.6
· OECD Pacific			60.7	3.8
Total OECD	*180.5*	*1.3*	*192.7*	*1.9*
· E. Europe / Former USSR	29.8[b]	0.4[b]	-14.8	-0.4
· Centrally Planned Asia	50.7[c]	2.9[c]	-4.0	-0.1
· South and South East Asia			92.2	5.3
· Africa			46.4	6.9
· Latin America			40.3	3.1
· Middle East			11.5	5.5
Total Non-OECD	*141.6*	*0.9*	*172.8*	*1.7*
World	*322.0*	*1.1*	*364.4*	*1.8*

[a] purchasing power parity corrected GDP; note that the GDP base may differ between the studies.

[b] Former Soviet Union only

[c] China only

Source: Fankhauser and Tol (1997), based on Fankhauser (1995a) and Tol (1995).

3. VALUATION ISSUES

As the previous section makes clear, greenhouse damage estimates still have a number of limitations. A number of important issues concern valuation. A first one is the choice between the two concepts of willingness to pay (WTP) and willingness to accept compensation (WTA), which is essentially an issue of property rights. A second issue is the question of benefit transfer, which asks how estimates for one region or one problem area can be extrapolated to another. A third issue concerns the incorporation of equity issues into comparison and aggregation of estimates. This section deals with each of these in turn.

3.1. Willingness to pay vs willingness to accept

It is a well known empirical fact that economic values derived under a WTP framework tend to differ from estimates that measure the same damage using WTA. The latter can be several times higher. Bateman and Turner (1992), for example, report ratios of WTA over WTP ranging from 1.6 up to 6.5. Climate change damages, too, can therefore be expected to be sensitive to the choice of valuation concept. For practical reasons however, Arrow *et al.* (1993) have recommended the use of WTP for contingent valuation studies, since they tend to produce more reliable results.

Various reasons for this discrepancy between WTP and WTA have been advanced. Firstly, some authors have suggested that the valuation experiments showing the discrepancy have failed to replicate near-market contexts. When respondents are asked to repeat bids, for example, WTP and WTA eventually converge (Coursey *et al.*, 1987). This suggests that lack of time and familiarity with the good might explain the divergence, and that as goods and the context of valuation become more familiar, WTP and WTA values converge. Secondly, respondents may be rejecting the implied property right. WTA implies that the sufferer does not 'own' the environment, or that the sufferer is bereft from an environmental property. If the respondent feels that is incorrect or immoral in some sense, large WTA values may result, including high 'protest' values. Thirdly, psychological prospect theory suggests that respondents are anchored to a reference point, generally defined by the prevailing bundle of goods and assets in their possession. The context of taking some of these goods away is then treated very differently to the context of adding to the set of goods. Increments therefore attract lower values than decrements. This concavity is reinforced if respondents see the good in question as part of their 'identity', i.e., integral to their lifestyle or psychological make-up. The resulting value function is concave or even kinked, contrary to the normal assumption of demand theory (Kahneman and Tversky, 1979). Fourthly, Hanemann (1991) has argued that the discrepancy between WTP and WTA is least where the substitution possibilities for goods are highest. Unique assets, often the context of contingent valuation studies, will tend to have high WTP/WTA discrepancies.

The choice between WTP and WTA constitutes an implicit statement about prevailing property rights, and this is sometimes used as a guideline for the choice of concept. By using WTP -- i.e. asking people how much they would pay to avoid adverse impacts -- a changing climate is implicitly chosen as the reference scenario. People do not have a 'right' to the climate currently observed, but have to pay to obtain it. Conversely, by using WTA, the assumption is that people are entitled to the preindustrial climate or the current climate. They have to be compensated for any damage arising from alterations to it.

However, the appropriate allocation of property rights (and thus the choice between WTP and WTA) in the case of climate change is unclear. On the one hand, the right of future generations to a functioning environment seems hard to question, and is at the core of such notions as sustainable development. This would point toward the use of WTA (assuming, as current evidence suggests, that climate change is predominantly harmful - see Budyko, 1996, and Mendelsohn and Neumann, forthcoming, for a different position): future generations are to be compensated for a climate change induced deterioration in living standards. On the other hand, an equally strong case can be made for the right of developing countries to increase their standard of living, which would imply at least a certain degree of baseline warming, and hence the use of WTP.

In practice, WTP and WTA are often mixed up in actual valuation. This is particularly the case for climate change damages, where the limited number of original studies make it necessary to use whatever information is available, often resorting to benefit transfer. Consequently, estimates tend to be a blend of various approaches, although WTP is perhaps more frequently used. Most estimates in the literature were derived with the benefits of emission reductions in mind, and consequently took business-as-usual climate change as the starting point, asking people about their WTP to obtain a deviation. WTA has also been used, however, as have been a number of second-best measures that were used as approximations in the absence of primary studies. Estimates for the value of a statistical life (VOSL), for example, were derived from a series of predominantly WTA studies (wage-risk studies).

At least as far as the current generation of damage estimates is concerned, the distinction between WTP and WTA is thus blurred. Nevertheless the issue is important conceptually, and it is likely to become increasingly relevant as refinements in damage analysis take place and additional studies are undertaken.

3.2. Benefit transfer and scaling by income

Since primary WTP/WTA data for climate change impacts are still scarce, the damage cost literature relies heavily on what is called benefit transfer. The term refers to an often used short-cut in valuation, in which WTP/WTA results ('environmental benefits') obtained in one study are transferred to a new problem and another site. For the assessment of climate change-induced mortality risk, for example, per-unit values were 'transferred' from a wide range of VOSL studies in various developed countries, most of them using wage differentials (a WTA procedure called the hedonic approach, see e.g. Viscusi, 1993).

Benefit transfer is not without problems. Estimates are often site or problem specific and hence difficult to transfer. In the case of climate change-induced mortality, for example, the cause of the mortality risk in the underlying studies (occupational hazard) is different from that in the new application (climate change related deaths). If people have a different WTP/WTA depending on the type of hazard, the transferred per-unit value would be biased. For example, it is conceivable that the WTP/WTA of the elderly, who will be most at risk from climate-change related heat stress, is different from that of workers in the prime of their life. For other climate change related risks (e.g., malaria, or extreme events), the WTP/WTA may be different again.

To take such effects into account, the values from the underlying study should ideally be corrected for differences in site and socio-economic conditions. Even so, the accuracy of benefit transfer remains open to question. In a direct test of the method, Bergland *et al.* (1995), for example, found a statistically significant difference between estimates based on benefit transfer and the results from a primary study. Alberini *et al.* (1995), on the other hand, secure a consistent set of results for contingent valuation and income-adjusted benefit transfer in a study on morbidity in Taiwan.

Bearing these problems in mind, transferred estimates can provide useful ballpark figures, and the method is often used in situations where the accuracy sought does not justify the costs of a primary study. As for climate change, the use of benefit transfer is primarily necessitated by the absence of primary studies for a large number of countries and damage categories.

In most climate change damage studies, per-unit value estimates were more or less directly transferred from the study site. The only major adjustment made concerned income, which is one of the main explanatory variables for both WTP and WTA. A standard assumption is that WTP/WTA is an increasing function of income.[***] A rich person would normally be willing (and able) to make a higher payment, in absolute terms, than a poor person. By the same token, a compensation of, say $1,000 will appear less attractive to a rich person than to a poor individual. The damage studies reviewed in Pearce *et al.* (1996) therefore usually scale per-unit values according to income, i.e., they use lower values in low income countries. A possible benefit transfer function is

$$V_j = V_i \left[\frac{Y_j}{Y_i} \right]^{\beta} \qquad (1)$$

where subscript j denotes the new application where the value is 'transferred' to, and i the original study site. V denotes the WTP/WTA estimate, Y is per capita income, and β is a scaling parameter, e.g., the income elasticity of marginal utility.

Although there is clear empirical evidence for an income effect, little is known about its magnitude, and scaling is correspondingly controversial. Most of the studies surveyed in

[***] Although the theoretical work of Flores and Carson (1995) does not exclude the possibility of a negative correlation.

Pearce *et al.* (1996) assume an income elasticity of WTP/WTA of 1, or slightly higher, following an early study by Pearce (1980). That is, WTP/WTA as proportions of income are identical across individuals. If a rich person is willing to pay, say, 5% of his income for an environmental good, a poor person would equally be willing to spend 5% of his.

Recent results cast doubt on the assumption of a unitary income elasticity of WTP/WTA, suggesting an income elasticity of less than one (Flores and Carson, 1995; Kristrom and Riera, 1996; see also Krupnick *et al.*, 1996). Given the logic of scaling, a lower income elasticity would imply that damages in developing countries were underestimated initially. (The estimates for developed countries are not affected, since these are the subject of the original study). The evidence is not yet conclusive, though. The few available studies that directly estimate the VOSL in developing countries all came up with substantially lower values than would be obtained through benefit transfer. Thus, Parikh *et al.* (1994) found VOSLs in Bombay of $25,000 using the human capital approach; $20,000 using the wage differential approach and $15,000 based on the Indian Workman's Compensation Act. Da Motta *et al.* (1993) find a VOSL of $15,000 using the human capital approach in Brazil. For comparison, the lowest VOSL assumed in the IPCC social cost chapter is $150,000. The IPCC value is based on an elasticity of WTP/WTA of about one. The Brazilian and Indian evidence would thus imply an elasticity much greater than one. It is worth noting, though, that the human capital approach is largely discredited as an approach to estimate VOSLs. It is also well known that WTP/WTA is likely to exceed the expected value of forgone earnings. It is therefore likely that the Indian and Brazilian studies have underestimated the true VOSL.

Better information is clearly needed in this area. At the same time, it should be recalled that the question of the income elasticity of WTP/WTA has arisen only due to the absence of original damage research which necessitated the use of benefit transfer. Although benefit transfer is likely to continue to be important, primary studies directly concerned with the valuation of climate change damages are therefore at least as important as refinements in benefit transfer. With an increasing number of such studies, issues of benefit transfer will automatically become less relevant. In the meantime, and until clear empirical evidence becomes available, it will be important for subsequent damage assessments to explore the sensitivity of estimates to crucial parameters such as the income elasticity of WTP/WTA.

3.3. Equity weighing

One of the key features of WTP/WTA estimates is that they are a mixture of descriptive and prescriptive concepts. The value of goods is set according to people's own appreciation of them (description) and it is assumed that people's preferences should count (prescription). At the same time, the socio-economic situation from which people make their assessment is taken as given. This can lead to problems if the currently observed situation (say, the distribution of income) is considered to be unfair. WTP/WTA estimates, because they are a function of socio-economic characteristics, will automatically reflect this unfairness. The issue is well known, though, and has a long history in cost-benefit analysis (see, e.g., Pearce, 1986). The solution offered by welfare economics is not to use uniform per-unit values (as some have called for -- Meyer and Cooper, 1995; Hohmeyer and Gaertner, 1992; Ayers and Walter, 1991), but to weight individual estimates by a corrective factor that adjusts values for inequalities in the income distribution. These 'equity weights' are usually derived from a social welfare function. Consequently they strongly depend on the analyst's or policy makers' value judgment and on the welfare function they endorse.

None of the estimates in Pearce *et al.* (1996) had been corrected initially, although the chapter did show how such equity weighing could be carried out. (In addition, equity was the subject of two separate chapters in the SAR: Banuri *et al.*, 1996, and Arrow *et al.*, 1996). Fankhauser *et al.* (1997) fill this gap and calculate equity weights and the corresponding damage figures for a variety of possible welfare and utility functions. In that paper, we show that total, worldwide damage can be expressed as

$$D^{world} = \left[\frac{W_1 \cdot u_Y^1}{W_M} \right] D^1 + \ldots + \left[\frac{W_n \cdot u_Y^n}{W_M} \right] D^n \tag{2}$$

where the terms in brackets denote the equity weights. Equity weights consist of three parts. Firstly, u_Y is a measure how much regional welfare changes as a consequence of damage D. Secondly, W_i is a measure how much global welfare changes as a consequence of a change in regional welfare. Thirdly, W_M is a scaling factor which ensures that equity weights are zero if the underlying income distribution or the distribution of damages is considered just.

Three debatable assumptions underlie (2). Firstly, meaningful welfare functions do exist. Secondly, economic and environmental goods and services are substitutable, at least within the stress imposed by climate change. Thirdly, climate change impacts are small enough to allow for linearization.

A selection of results is reproduced in Table 2. A conventional iso-elastic utility function:

$$u = \frac{a}{(1-e)} \cdot Y^{(1-e)} \tag{3}$$

is used. Different values for parameter *e* (the income elasticity of marginal utility, or risk aversion) will be used below.

The specification for the welfare function is

$$W = \frac{\sum_{i=1}^{n} u^i(\cdot)^{(1-\gamma)}}{1-\gamma} \tag{4}$$

where γ is a parameter of inequality aversion. The larger is γ, the larger is the concern about equality. For $\gamma = 0$, equation (4) reduces to a utilitarian welfare function, letting γ approach 1 gives a Bernoulli-Nash function, $\gamma \to \infty$ represents the maximin (Rawlsian) case, and $\gamma \to -\infty$ is the maximax (Nietzschean) welfare function.

As Table 2 makes clear, estimates are highly sensitive to the assumed welfare concept. World damages are generally (but not necessarily) higher than originally reported in Pearce *et al.* (1996), particularly for high values of risk or inequality aversion. This exercise makes also clear that the original studies (with implicit equity weights of one), implicitly assumed either a just distribution of welfare, or a linear, utilitarian welfare function.

4. MARGINAL DAMAGE ESTIMATES

The analysis so far was confined to comparative statics. All figures in Table 1 are estimates of the impact of one specific change of the climate ($2\times CO_2$) on the current economy. This is clearly insufficient. Not only will we, for the larger part of the future, be confronted with climate change substantially different from $2\times CO_2$, but socio-economic vulnerability to climate change will also shift as a consequence of economic development.

Table 2
Aggregate damages corrected for inequality (in bn$)

	Fankhauser (1995a)	Tol (1995)
Uncorrected damages[a]	322.0	364.4
Utilitarian Welfare Function		
e = 0.0[b]	322.0	364.4
e = 0.5	315.6	411.4
e = 1.0	405.2	614.3
e = 1.5	621.9	1057.6
e = 2.0	1041.7	1930.0
Bernoulli-Nash Welfare Function[c]	405.2	614.3
Maximin Welfare Function		
e = 0.0	50.7	46.4
e = 0.5	95.8	89.4
e = 1.0	181.0	172.2
e = 1.5	342.7	331.8
e = 2.0	646.5	639.3

[a] as in Table 1

[b] e denotes the income elasticity of marginal utility (parameter of the utility function)

[c] Bernoulli-Nash weights are independent of e, and correspond to the case e = 1 of the utilitarian welfare function.

Source: Fankhauser *et al.* (1997), based on indicated sources.

What would be relevant to know from a policy point of view are marginal figures, ie, estimates of the extra damage done by one extra ton of carbon emitted. Unfortunately, the requirements for marginal damage calculations go far beyond the information available from $2xCO_2$ studies. Greenhouse gases are stock pollutants. That is, a ton of gas emitted will affect climate over several decades, as fractions of the gas remain long in the atmosphere. Calculating marginal costs therefore requires the comparison of two present value terms: The discounted sum of future damages associated with a certain emission scenario is compared to the sum of damages in an alternative scenario with marginally different emissions in the base period (in estimates based on optimal control models, the marginal cost is calculated as the shadow price of carbon, i.e., the carbon tax necessary to keep emissions on the socially optimal trajectory; Nordhaus, 1994; Peck and Teisberg, 1993).

The current generation of models deals with this challenge in a rather ad hoc manner, using very simplistic representations of the complex dynamic processes involved. In older studies damage costs were typically specified as a power (usually linear to cubic) function of global mean temperature, calibrated around the $2\times CO_2$ estimates. Damage is usually fully reversible and typically assumed to grow with GNP. Only recently, studies have started to emerge which explicitly incorporate regionally diversified temperatures and sea levels, model individual damage categories (e.g., agriculture) separately, or at least distinguish between damages related to absolute temperature level and those related to the rate of change (Dowlatabadi and Morgan, 1993; Hope *et al.*, 1993; Tol, 1995).

Table 3 provides a list of estimates of the marginal damages obtained from polynomial damage models. Estimates range from about \$5 to \$125 per ton of carbon, with most estimates at the lower end of this range. The wide range reflects variations in model assumptions, as well as the high sensitivity of figures to the choice of the discount rate (Pearce *et al.*, 1996). Estimates are expected to rise over time as a consequence of economic growth and increasing concentration levels.

Using a model called DICE, Nordhaus (1994) finds that the shadow price begins at only about \$5 per ton of carbon in 1995, rises to about \$10 by 2025, and reaches \$21 by 2095 (at 1990 prices). Peck and Teisberg (1993) find values of a similar order of magnitude. Tol's (1994) alternative specification of DICE yields shadow prices of \$13 for 1995, rising to $89 for 2095. These model runs all assume that parameter values are known with certainty. In the case of DICE, expected shadow prices more than double, once uncertainty is added to the model. This result arises because of the skewedness in the damage distribution, which allows for low probability - high impact events (Nordhaus, 1994); risk aversion and concave damage functions further enhance this effect (Tol, 1995). All three authors assume a pure rate of time preference (or utility discount rate) of 3% (for a discussion on discounting see Arrow *et al.*, 1996; Nordhaus, 1994). In contrast, Cline (1993, 1992b) finds significantly higher shadow prices by using a zero utility discount rate. His reproduction of the DICE model generates a path of shadow prices beginning at about $45 per ton, reaching about $243 by 2100. Other parameter specifications provide even higher values.

Fankhauser (1995a) identifies a lower and flatter trajectory for the shadow price of carbon, rising from $20 per ton by 1991-2000 to $28 per ton by 2021-2030, with confidence intervals of $6-45 and $9-64, respectively. Fankhauser uses a probabilistic approach to the range of discount rates, in which low and high discount rates are given different weights. His sensitivity analysis with the discount rate suggest that moving from high (3%) to low (0%) discounting could increase marginal costs by about a factor 9, from $5.5 to $49 per ton of carbon emitted now.

5. CONCLUSIONS

Economic impact estimates of climate change are still in a very early stage of scientific analysis. Some insights can be gained, however. According to current knowledge, the impact of climate change associated with a doubling of atmospheric concentrations of greenhouse gases would be around 1.5% of world income per year, with equally strong arguments that this would be too high as well as too low. Although 1.5% sounds low, it is quite a considerable amount of money. In rich economies, the impact would be a little lower, in poor economies, substantially higher than the world average.

Table 3
The marginal social costs of CO_2 emissions (current value (1990)$/tC)

Study	Type	1991-2000	2001-2010	2011-2020	2021-2030
Nordhaus	MC		7.3		
			(0.3-65.9)		
Ayres and Walter	MC		30 - 35		
Nordhaus, DICE	CBA				
- best guess		5.3	6.8	8.6	10.0
- expected value		12.0	18.0	26.5	n.a.
Cline	CBA	5.8 - 124	7.6 - 154	9.8 - 186	11.8 - 221
Peck and Teisberg	CBA	10 - 12	12 - 14	14 - 18	18 - 22
Fankhauser	MC	20.3	22.8	25.3	27.8
		(6.2-45.2)	(7.4-52.9)	(8.3-58.4)	(9.2-64.2)
Maddison	CBA -MC	5.9-6.1	8.1 - 8.4	11.1-11.5	14.7-15.2

MC = marginal social cost study, CBA = shadow value in a cost-benefit study.
Figures in brackets denote 90% confidence intervals.

Sources: Pearce *et al.* (1996); see also Ayres and Walter (1991), Nordhaus (1994), Cline (1992b, 1993), Peck and Teisberg (1993), Fankhauser (1995a) and Maddison (1994).

These estimates reflect the world as it is, not as we would like to see it. This distinction may be blurred if it comes to valuation of non-market goods and services and to aggregation of impact estimates over countries with widely different living standards. Impact estimates are shown to be quite sensitive to such issues.

Estimates of the marginal impact of carbon dioxide emissions show that there is a clear case to go beyond no-regret emission reduction. How far is open to debate. The minimally justified carbon tax is $5 per ton of carbon, while the best guess is $20/tC.

REFERENCES

1. Adams, R.M., McCarl, B.A., Segerson, K., Rosenzweig, C., Bryant, K.J., Dixon, B.L., Conner, R., Evenson, R.E. and Ojima, D. (1994), *The Economic Effects of Climate Change on US Agriculture*, Report prepared for the Electric Power Research Institute, Palo Alto, CA.
2. Alberini, A., Cropper, M., Fu, T.M., Krupnick, A., Liu, J.T., Shaw, D. and Harrington, W. (1995), *Valuing Health Effects of Air Pollution in Developing Economies: The Case of Taiwan*, mimeo, Resources for the Future, Washington, DC.
3. Arrow, K., Solow, R., Leamer, E., Portney, P., Radner, R. and Schuman, H. (1993), *Report of the NOAA Panel on Contingent Valuation*, Washington DC: National Oceanic and Atmospheric Administration.
4. Arrow, K.J., Cline, W.R., Maeler, K.-G., Munasinghe, M., Squitieri, R. and Stiglitz, J.E. (1996), 'Intertemporal Equity, Discounting, and Economic Efficiency', in Bruce, J.P., Lee, H. and Haites, E.F. (eds.) *Climate Change 1995: Economic and Social Dimensions -- Contribution of Working Group III to the Second Assessment Report of the Intergovernmental Panel on Climate Change*, Cambridge University Press, Cambridge.
5. Asian Development Bank (1992), *Climate Change in Asia*, 8 Country Studies, Manila: ADB.
6. Banuri, T., Maeler, K.-G., Grubb, M., Jacobson, H.K. and Yamin, F. (1996), 'Equity and Social Considerations', in Bruce, J.P., Lee, H. and Haites, E.F. (eds.) *Climate Change 1995: Economic and Social Dimensions -- Contribution of Working Group III to the Second Assessment Report of the Intergovernmental Panel on*

Climate Change, Cambridge University Press, Cambridge.

6. Bateman, I.J. and Turner, R.K. (1992), *Evaluation of the Environment: The Contingent Valuation Method,* **GEC WP 92-18,** Centre for Social and Economic Research on the Global Environment, Norwich and London.
7. Bergland O., Magnussen, K. and Navrud, S. (1995), *Benefit Transfer: Testing for Accuracy and Reliability*, Discussion Paper **D-03/1995**, Department of Economics ans Social Sciences, Agricultural University of Norway, As.
8. Bijlsma, L., Ehler, C.N., Klein, R.J.T., Kulshrestha, S.M., McLean, R.F., Mimura, N., Nicholls, R.J., Nurse, L.A., Perez Nieto, H., Stakhiv, E.Z., Turner, R.K. and Warrick, R.A. (1996), 'Coastal Zones and Small Islands', in Watson, R.T., Zinyowera, M.C. and Moss, R.H. (eds.) *Climate Change 1995: Impacts, Adaptations and Mitigation of Climate Change: Scientific-Technical Analyses -- Contribution of Working Group II to the Second Assessment Report of the Intergovernmental Panel on Climate Change*, Cambridge University Press, Cambridge.
9. Budyko, M.I. (1996), 'Past Changes in Climate and Societal Adaptations', in Smith, J.B., Bhatti, N., Menzhulin, G., Benioff, R., Budyko, M.I., Campos, M., Jallow, B. and Rijsberman, F. (eds.) *Adapting to Climate Change: Assessments and Issues*, Springer, Berlin.
10. Climate Research Unit and Environmental Resources Limited (1992), *Development of a Framework for the Evaluation of Policy Options to Deal with The Greenhouse Effect: Economic Evaluation of Impacts and Adaptive Measures in the European Community*, A Report for the Commission of European Communities, University of East Anglia, Norwich.
11. Cline, W.R. (1992a), *The Economics of Global Warming*, Institute for International Economics, Washington, DC.
12. Cline, W.R. (1992b), *Greenhouse Policy after Rio: Economics, Science and Politics,* September 1992, Laxenburg.
13. Cline, W.R. (1993), *Costs and Benefits of Greenhouse Abatement: A Guide to Policy Analysis,* Paris.
14. Coursey, D.L., Hovis, J.J. and Schulze, W.D. (1987), 'The Disparity between Willingness to Accept and Willingness to Pay Measures of Value', *Quarterly Journal of Economics*, **102**, 679-90.
15. Da Motta, R., Mendes, A., Mendes, F. and Young, C. (1993), *Environmental Damages and Services due to Household Water Use,* Discussion Paper **258**,

Instituto de Pesquisa Economica Aplicada (IPEA), Rio de Janeiro.

16. Darwin, R., Tsigas, M., Lewandrowski, J. and Raneses, A. (1995), *World Agriculture and Climate Change: Economic Adaptations*, Agricultural Economic Report **703**, US Department of Agriculture, Washington, DC.
17. Dowlatabadi, H. and Morgan, G. (1993), 'A Model Framework for Integrated Studies of the Climate Problem', *Energy Policy*, 209-221.
18. Ekins, P. (1995), 'Rethinking the Costs Related to Global Warming. A Survey of the Issues', *Environmental and Resource Economics*, **5**, 1-47.
19. Fankhauser, S. (1995a), *Valuing Climate Change. The Economics of the Greenhouse*, Earthscan, London.
20. Fankhauser, S. (1995b), 'Protection vs. Retreat. The Economic Costs of Sea Level Rise', *Environment and Planning A*, **27**, 299-319.
21. Fankhauser, S. and Tol, R.S.J. (1997), 'The Social Costs of Climate Change: The IPCC Second Assessment Report and Beyond', *Mitigation and Adaptation Strategies for Global Change* (forthcoming).
22. Fankhauser, S., Tol, R.S.J. and Pearce, D.W. (1997), 'The Aggregation of Climate Change Damages: A Welfare-Theoretic Approach', *Environmental and Resource Economics* (forthcoming).
23. Flores, N.E. and Carson, R.T. (1995), *The Relationship between the Income Elasticities of Demand and Willingness to Pay*, Mimeo, Department of Economics, University of California, San Diego.
24. Hahnemann, W.M. (1991),'Willingness to Pay and Willingness to Accept: How Much can they Differ?', *American Economic Review*, **81** (3), 635-647.
25. Hohmeyer, O. and Gärtner, M. (1992), *The Costs of Climate Change*. Report to the Commission of the European Communities, Fraunhofer Institut für Systemtechnik und Innovationsforschung, Karlsruhe (Germany).
26. Hope, C.W., Anderson, J. and Wenman, P. (1993), 'Policy Analysis of the Greenhouse Effect — An Application of the PAGE Model', *Energy Policy*, **15**, 328-338.
27. Kahneman, D. and Tversky, A. (1979), 'Prospect Theory: An Analysis of Decisions uner Risk', *Econometrica*, **47**, 263-291.
28. Kriström, B. and Riera, P. (1996), 'Is the Income Elasticity of Environmental Improvements Less Than One?', *Environmental and Resource Economics*, 7: 45-55.
29. Krupnick, A., Harrison, K., Nickell, E. and Toman, M. (1996), 'The Value of Health Benefits from Ambient Air Quality Improvements in Central and Eastern

Europe: An Exercise in Benefits Transfer', *Environmental and Resource Economics*, **7**, 307-332.

30. Maddison, D.J. (1994), *A Cost Benefit Analysis of Slowing Climate Change,* **WP GEC 94-28,** CSERGE, London.
31. Mendelsohn, R. and Neuman, J., *The Impact of Climate Change on the US Economy* (forthcoming).
32. Mendelsohn, R., Morrison, W., Schlesinger, M.E. and Andronova, N.G., *A Global Impact Model for Climate Change*, draft.
33. Meyer, A. and Cooper, T. (1995), 'A Recalculation of the Social Costs of Climate Change', Working Paper, *The Ecologist.*
34. Nishioka, S., Harasawa, H., Hashimoto, H., Ookita, T., Masuda, K. and Morita, T. (1993), *The Potential Effects of Climate Change in Japan*, Centre for Global Environmental Research, and National Institute for Environmental Studies, Tsukuba (Japan).
35. Nordhaus, W.D. (1991), 'To Slow or not to Slow: The Economics of the Greenhouse Effect', *Economic Journal*, **101** (407), 920-937.
36. Nordhaus, W.D. (1994), *Managing the Global Commons: The Economics of Climate Change*, The MIT Press, Cambridge.
37. Parikh, K., Parikh, J., Muralidharan, T. and Hadker, N. (1994), *Valuing Air Pollution in Bombay*, mimeo, Indira Gandhi Institute of Development, Bombay.
38. Parry, M.L. and Duncan, R. (1995), *The Economic Implications of Climate Change in Britain*, Earthscan, London.
39. Pearce, D.W. (1980), 'The Social Incidence of Environmental Costs and Benefits', in O'Riordan, T. and Turner, R.K. (eds.), *Progress in Resource Management and Environmental Planning*, Vol. 2, Wiley, Chichester.
40. Pearce, D.W. (1986), *Cost Benefit Analysis*, 2nd edition, Basingstoke: Macmillan.
41. Pearce, D.W., Cline, W.R., Achanta, A.N., Fankhauser, S., Pachauri, R.K., Tol, R.S.J. and Vellinga, P. (1996), 'The Social Costs of Climate Change: Greenhouse Damage and the Benefits of Control', in Bruce, J.P., Lee, H. and Haites, E.F. (eds.) *Climate Change 1995: Economic and Social Dimensions -- Contribution of Working Group III to the Second Assessment Report of the Intergovernmental Panel on Climate Change*, Cambridge University Press, Cambridge.
42. Peck, S.C. and Teisberg, T.J. (1993), 'CO2 Emissions Control — Comparing Policy Instruments', *Energy Policy*, 222-230.
43. Plambeck, E.L. and Hope, C.W. (1996), 'PAGE95 - An Updated Valuation of the

Impacts of Global Warming', *Energy Policy*, **24** (9), 783-793.

44. Reilly, J., Baethgen, W., Chege, F.E., van de Geijn, S.C., Lin, E., Iglesias, A., Kenny, G., Patterson, D., Rogasik, J., Roetter, R., Rosenzweig, C., Sombroek, W. and Westbrook, J. (1996), 'Agriculture in a Changing Climate: Impacts and Adaptation', in Watson, R.T., Zinyowera, M.C. and Moss, R.H. (eds.) *Climate Change 1995: Impacts, Adaptations and Mitigation of Climate Change: Scientific-Technical Analyses -- Contribution of Working Group II to the Second Assessment Report of the Intergovernmental Panel on Climate Change*, Cambridge University Press, Cambridge.
45. Rosenzweig, C. and Parry, M.L. (1994), 'Potential Impact of Climate Change on World Food Supply', *Nature*, **367**, 133-138.
46. Schimmelpfenning, D., Lewandrowski, J., Reilly, J.M., Tsigas, M. and Parry, I.W.H. (1996), *Agricultural Adaptation to Climate Change. Issues of Longrun Sustainability*, Agricultural Economic Report **740**, US Department of Agriculture, Washington, DC.
47. Smith, J.B. (1996), 'Standardized Estimates of Climate Change Damages for the United States', *Climatic Change*, **32** (3), 313-326.
48. Smith, J.B. and Tirpak, D.A. (1989), *The Potential Effects of Global Climate Change on the United States*, US Environmental Protection Agency, Washington.
49. Strzepek, K.M. and Smith, J.B. (eds.) (1995), *As Climate Change - International Impacts and Implications*, Cambridge University Press, Cambridge.
50. Titus, J.G. (1992), 'The Cost of Climate Change to the United States', in S.K. Majumdar, L.S. Kalkstein, B. Yarnal, E.W. Miller, and L.M. Rosenfeld (eds.), *Global Climate Change: Implications Challenges and Mitigation Measures*, Pennsylvania Academy of Science, Pennsylvania.
51. Tol, R.S.J. (1994), 'The Damage Costs of Climate Change — A Note on Tangibles and Intangibles, Applied to DICE', *Energy Policy*, **22** (5), 436-438.
52. Tol, R.S.J. (1995), 'The Damage Costs of Climate Change - Toward More Comprehensive Calculations', *Environmental and Resource Economics*, **5**, 353-374.
53. Turner, R.K., Adger, W.N. and Doktor, P. (1995), 'Assessing the Economic Costs of Sea Level Rise', *Environment and Planning A*, **27** (11), 1777-96.
54. Viscusi, W.K. (1993), 'The Value of Risks to Life and Health', *Journal of Economic Literature*, **XXI** (December), 1912-1946.
55. Watson, R.T., Zinyowera, M.C., and Moss, R.H. (eds.) (1996), *Climate Change 1995: Impacts, Adaptation, and Mitigation of Climate Change --Scientific-Technical*

Analysis-- Contribution of Working Group II to the Second Assessment Report of the Intergovernmental Panel on Climate Change, Cambridge University Press, Cambridge.

56. Yohe, G.W., Neumann, J. and Ameden, H. (1995), Assessing the Economic Cost of Greenhouse-Induced Sea Level Rise: Methods and Applications in Support of a National Survey', *Journal of Environmental Economics and Management*, **29**, S-78-S-97.
57. Yohe, G.W., Neumann, J., Marshall, P. and Ameden, H. (1996), 'The Economics Costs of Sea Level Rise on US Coastal Properties', *Climatic Change*, **32**, 387-410.

Greenhouse Gases: Interrelationship with stratospheric ozone depletion

Guus J.M. Velders

Air Research Laboratory, National Institute of Public Health and the Environment, PO Box 1, 3720 BA Bilthoven, The Netherlands

ABSTRACT

Emissions of greenhouse gases can affect the depletion of the ozone layer through atmospheric interaction. In our investigation the increase in emissions of chlorine- and bromine-containing compounds, largely responsible for the change in stratospheric ozone at mid-latitudes, was found to be -5.8% per decade from 1980 to 1990. The increase in CH_4 emissions in the same period changes this ozone trend by +1.4% per decade to -4.4% per decade, which is close to TOMS and Dobson measurements. The increase in N_2O emissions hardly affects this depletion. The decrease in stratospheric temperatures due to increased CO_2 emissions also diminishes the ozone depletion. The effect of these interactions in coming decades is to accelerate the recovery of the ozone layer. The trend in CH_4 emissions described in the business-as-usual scenario IS92a may yield 1980 ozone column levels in 2060 compared with 2080 with CH_4 emissions fixed at 1990 levels. The temperature decrease in the stratosphere may initially also accelerate the recovery of the ozone layer by several years, ignoring a possible large extra ozone depletion by the extra formation of polar stratospheric clouds over large areas of the world.

1. INTRODUCTION

Depletion of stratospheric ozone and the enhanced greenhouse effect are two major global environmental issues receiving a lot of attention from the scientific community, the public and policy-makers. Both phenomena are usually considered independently, initially a valid approximation. Destruction of ozone in the stratosphere is caused mainly by elevated levels of active chlorine and bromine compounds in the stratosphere, which arise mainly from anthropogenic emissions of chlorofluorocarbons (CFC's) and halons. The enhanced greenhouse effect arises from a changing radiation balance in the atmosphere caused by anthropogenic emissions from carbon dioxide (CO_2), methane (CH_4), nitrous oxide (N_2O) and chlorofluorocarbons (CFC's). The stratosphere interacts with the greenhouse effect, or climate change processes, in several ways. Some anthropogenic emissions cause chemical changes in both the troposphere and stratosphere, thereby affecting the concentration of greenhouse gases in the troposphere and also influencing the amount of active chlorine and bromine compounds in the stratosphere. The enhanced greenhouse effect causes changes in temperature of both the troposphere and stratosphere, which may affect the chemical composition of the atmosphere as well as its dynamics. It may also influence the exchange processes between the troposphere and stratosphere and thus the chemical composition. Changes in stratospheric ozone also cause changes in the amount of UV radiation reaching the troposphere, which affects the photolysis rates of several chemical reactions and thus changes the chemical composition of the atmosphere.

The enhanced greenhouse effect, with its social and economic impacts, are under study in the international framework of the Intergovernmental Panel on Climate Change (IPCC, 1994; 1995). This forms an important basis for international negotiations on climate change. The scientific knowledge of the ozone layer is written down in WMO/UNEP reports, for example WMO (1991; 1994), and form the basis for discussions in the framework of the Vienna Convention and the Montreal Protocol. The atmospheric section of IPCC reports focuses on radiative forcing by greenhouse gases, while the WMO/UNEP reports mainly describe processes in the stratosphere. Both types of reports acknowledge interactions between the enhanced greenhouse effect and stratospheric ozone depletion, but do not address them extensively. Here we will quantify the effects greenhouse gas emissions have on stratospheric ozone depletion and on the recovery of the ozone layer.

2. MODEL AND EMISSIONS

Chemical changes in the atmosphere are calculated with the RIVM version of the 2-dimensional model of the stratosphere (Velders, 1995a, 1995b) as originally developed by Harwood and Pyle (1980); see also Law and Pyle (1993a). Details of the simulations and emissions are given in Velders (1997). The basis for the future scenarios (1990-2100) were the IPCC IS92 scenarios (Pepper *et al.*, 1992). The CH_4 and N_2O emissions had to be changed slightly to be in agreement with the current concentrations. The future scenario used for the CFC's in all simulations was IPCC case 6 which closely resembles the Copenhagen amendments to the Montreal Protocol.

3. RESULTS OF THE SIMULATION

The concentration of CO_2, taken from IPCC (1994), increases considerably in the 21st century with possibly large consequences for the temperature in the troposphere and stratosphere. The CH_4 concentration increases in most scenarios. This may affect stratospheric ozone depletion through the formation of water vapor in the stratosphere and enhancement of the production of the reservoir compound HCl. N_2O also increases in concentration but relatively less than for CO_2 and CH_4. The CFC's, methyl chloroform and carbon tetrachloride, decreases in concentration in all future scenarios. Measurements (Montzka *et al.*, 1996; ALE/GAGE network, private communications) already show a decrease in concentration in methyl chloroform and carbon tetrachloride since a few years, while the concentration of CFC-11 has stabilized since approximately 1994. The concentration of CFC-12 still shows a small increase.

Table 1.

Yearly averaged ozone column (DU) in 1990, ozone column trend from 1980 to 1990 (% per decade) at 47°N with trends for CH_4, N_2O and temperature varying.

Simulation	Characterization scenario				Ozone[2]	Relative[3]	Ozone
	CFC_4 halon	CH_4	N_2O	temp	column DU	to run [4] (%)	trend (%dec)
1 CH_4,N_2O fixed	Tr	1900	1900		379.2	-3.0	-5.8
2 N_2O trend	Tr	1900	Tr		379.4	-3.0	-5.6
3 CH_4 trend	Tr	Tr	1900		391.1	0.0	-4.4
4 CH_4,N_2O trend[3]	Tr	Tr	Tr		391.0	reference[3]	-4.3
5 CH_4 trend	1900	Tr	1900		435.5	11.4	0.1
6 N_2O trend	1900	1900	Tr		430.1	10.0	-0.2
7 CH_4,N_2O trend	1900	Tr	Tr		433.3	10.8	0.0
8 Temp. fixed	Tr	Tr	Tr	1900	381.1	-2.5	-4.2
9 Temp. trend	Tr	Tr	Tr	Tr5	390.9	0.0	-3.5

1) Tr = trend in emissions of CFC's, CH_4 and N_2O from 1900 to 1990. For CO and NO_x a trend in surface concentrations is applied; 1900 = fixed 1900 values are used; a blank cell means that fixed 1990 values are used.
2) Estimated total anthropogenic change in ozone column at 47°N in 1980 is -4.5% and in 1990 -8.6% relative to 1900.
3) Simulation 4 is considered as a reference scenario; the others are compared with this one.
4) CFC stands for all CFC's, HCFC's, methyl chloroform, carbon tetrachloride, halons, methyl chloride and methyl bromide.
5) Dynamical feedback and heterogeneous reactions on PSC's are not taken into account.

3.1. Historical changes in ozone: CH_4 trend

Table 1 shows the ozone column in 1990 and the trend in ozone column over the period 1980 to 1990 at 47°N for various time-dependent scenario simulations from 1900 to 1990. The emission trends, based on gradual continuous changes, do not take into account observed fluctuations in the concentration of some of the compounds in the last decade. In most of the simulations summarized in Table 1 the emissions of CFC's and related chlorine and bromine compounds show a trend from 1900 to 1990; the effect of

CH_4, N_2O and temperature changes are studied in addition to this chlorine trend.

A trend in the emissions of CFC's and related compounds alone (simulation 1) gives an ozone column of 379.2 DU and a trend in ozone column of -5.8% per decade from 1980 to 1990. Applying a trend in CH_4 emissions in addition to this (simulation 3) yields a 3% thicker ozone column in 1990 and a smaller ozone trend of -4.4% per decade. That is, the trend in CH_4 emissions this century has reduced the destruction of the ozone layer from 1980 to 1990 by approximately 1.4% per decade. This positive trend comes clearly from the interaction of CH_4 with chlorine in the stratosphere. Figure 1 shows the increase in CH_4 of 80% in the whole atmosphere caused by the trend in CH_4 emissions from 1900 to 1990. The reaction $CH_4 + Cl \Longrightarrow HCl$ results in an increase in HCl of approximately 18% in the lower and 24% in the middle stratosphere. (These percentages are yearly averaged values over the designated areas.) The increase in HCl means that the photostationary equilibrium of active chlorine compounds (ClO_x) shifts more to the reservoir HCl, reducing the concentrations of Cl and ClO, which are the important compounds for catalytic destruction of stratospheric ozone. Cl and ClO are reduced by approximately 20% to 25% in the whole stratosphere (Fig. 1) resulting in an increase of 2% to 4% in ozone in the lower stratosphere. In the upper stratosphere the increase reaches 6% to 8% but this area hardly contributes to the ozone column. The increase in CH_4 in this century causes an increase in H_2O of 1% in the lower, 8% in the middle and 15% in the upper stratosphere. OH, formed in the stratosphere by the reaction $O(^1D)$ + H_2O, increases by approximately 7% in the lower and middle stratosphere. This causes extra ozone destruction by the catalytic HO_x cycles, bus is only a small effect.

Simulation: CH_4 emissions trend

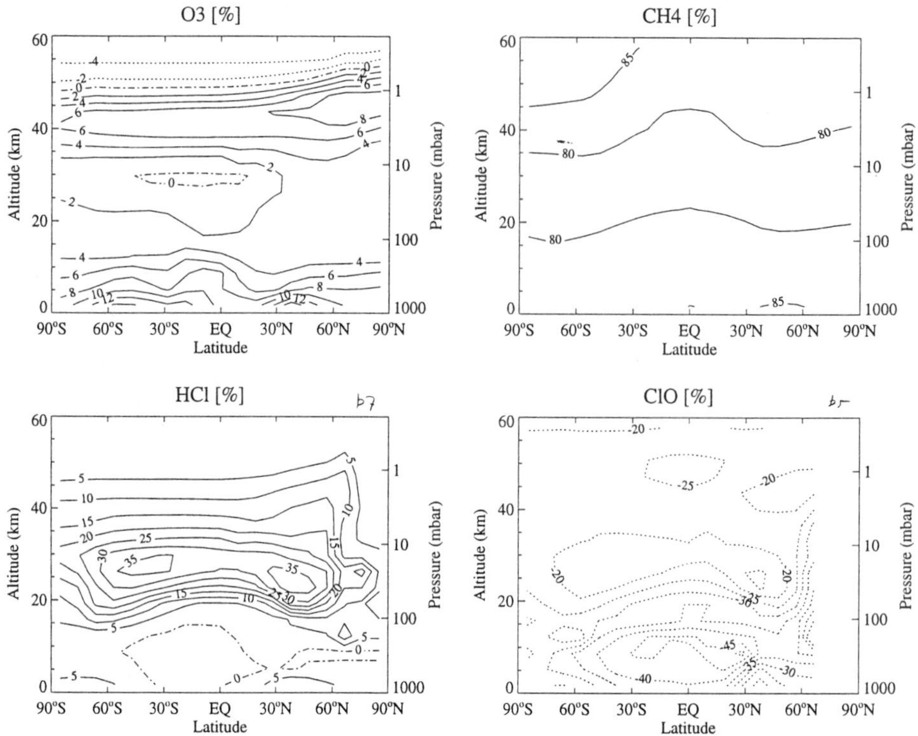

Figure 1. Relative contribution (in %) to the concentrations of O_3, CH_4, HCl and Clo in January 1990 caused by the trend in CH_4 emissions from 1900 to 1990. Fixed 1990 surface concentration of CO and NO_x and a trend in emissions from 1900 to 1990 for N_2O and all chlorine and bromine compounds. Simulation 4 minus 3 of Table 1.

The increase in CH_4 concentration hardly affects the bromine compounds since the bromine reservoir HBr is not produced by the reaction Br + CH_4 but by Br + HO_2. Furthermore, HBr is a less stable compound than HCl and therefore a less effective reservoir. The increase in CH_4 emissions in this century changed the Br concentration in the stratosphere by -4% and BrO by +0.1%.

The increase in methane emissions this century results in less ozone destruction by chlorine compounds (+1.5% per decade) and slightly more destruction by increased H_2O (-0.1% per decade). The importance of the interaction between chlorine and CH_4 for ozone is also clear from simulation 5, in which a trend in CH_4 alone and not in CFC emissions is applied. The CH_4 emission result is a small positive trend in the ozone column of 0.1% per decade, while the much lower chlorine levels in the lower stratosphere, coming only from natural CH_3Cl emissions, yield an 11% thicker ozone column in 1990.

Simulation: N_2O emissions trend

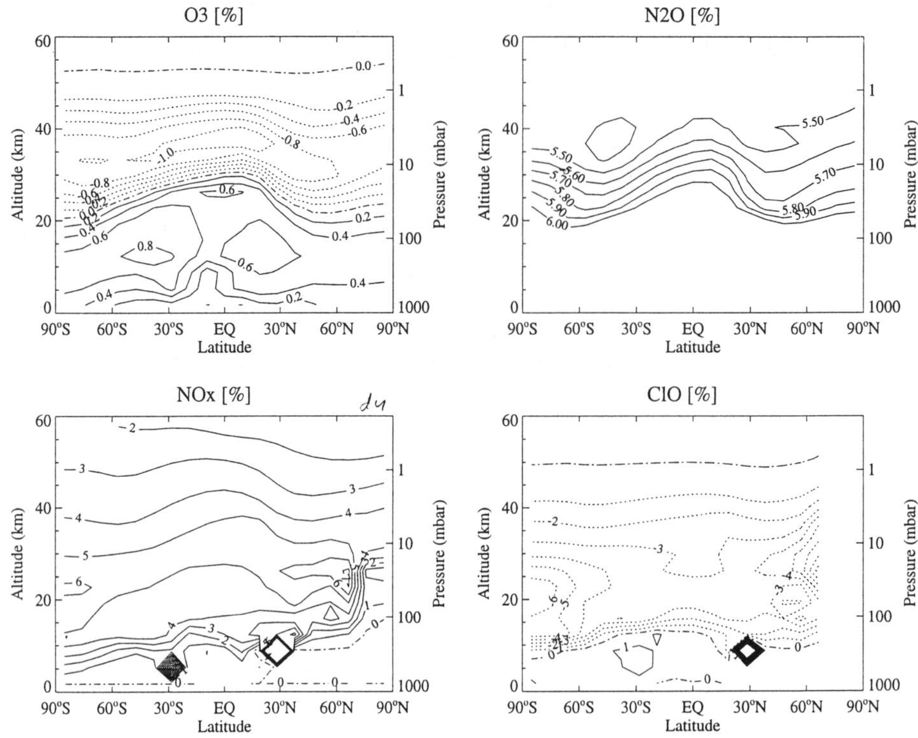

Figure 2. Relative contribution (in %) to the concentrations of O_3, N_2O, NO_x and Clo in January 1990 caused by the trend in N_2O emissions from 1900 to 1990. Fixed 1990 surface concentration of CO and NO_x and a trend in emissions from 1900 to 1990 for CH_4 and all chlorine and bromine compounds. Simulation 4 minus 2 of Table 1.

3.2. Historical changes in ozone: N_2O trend

The trend in N_2O emissions from 1900 to 1990 has a small effect on the ozone column and the ozone column trend. Comparing simulations 1 and 2 or 3 and 4 shows that the N_2O trend yields a trend in ozone of +0.2% per decade or +0.1% per decade, respectively. This positive contribution of N_2O comes from the interaction of nitrogen and chlorine compounds in the stratosphere. N_2O emissions on their own cause an increase in NO_x in the stratosphere which destroys ozone catalytically. Simulation 6 shows a trend in ozone of -0.2% per decade from the trend in N_2O emissions alone (no CFC trend applied). But if CFC emissions also increase the interaction between NO_x and ClO in the stratosphere yields the chlorine reservoir $ClONO_2$, causing a shift in the photochemical equilibrium of chlorine compounds away from Cl and ClO yielding a smaller ozone destruction. This diminishing effect on ozone destruction by $ClONO_2$ formation is larger that the increasing effect by NO_x. Figure 2 shows that an increase in N_2O emissions this century results in a 6% larger N_2O concentration, an increase of NO_x of 5% and of $ClONO_2$ of 1% and a decrease of ClO of 4%, all in the lower stratosphere. There is no large interaction between CH_4 and N_2O emissions which affects ozone formation as can be seen from the simulations 4 and 7 and when comparing them with the others.

3.3. Historical changes in ozone: temperature trend

Radiative-forcing global-mean temperature changes in the atmosphere are calculated from changes in concentrations using the OGI-1D radiative convective model from MacKay and Khalil (1991). The calculated temperature change for 1990 is 0.67 K for the troposphere, which is somewhat higher than the current analyzed global mean-surface temperature increase of 0.3 to 0.6 K (IPCC, 1994). The current radiative forcing calculated with the OGI-1D model agree reasonably well with the values reported by IPCC (1994). The calculated tropospheric temperature changes in 2100 vary between 1.65 and 3.68 K relative to 1900 corresponding with 0.98 to 3.01 K relative to 1990. These values are very close to the best estimates from IPCC for the temperature change in 2100, ranging from 1 to 3.5 K relative to 1990. Larger temperature changes occur in the stratosphere. Our calculations reveal a large amount of cooling in the stratosphere from CO_2 and a slightly smaller amount from ozone. Randel and Cobb (1994) inferred a temperature trend from satellite observations. They found a decrease in temperature in the lower stratosphere ranging from -0.5 to -1.5 K from 1979 to 1990. Radio sonde measurements (WMO, 1994) of the lower stratospheric temperature show a decrease of 0.25 to 0.4 K per decade over the last three decades. These observations are in agreement with our calculations. Our calculations give a change in temperature in the lower stratosphere of approximately 0.45 K from 1980 to 1990. The future scenarios show large decreases in stratospheric temperatures in the 21st century. Relative to 1990, the temperature changes range from -3.26 K for IS92c to -9.75 K for IS92e. The scenarios IS92a and IS92b can be considered as mid-range values, IS92e and IS92f are on the high end of the spectrum and IS92c and IS92d on the low end. Since the radiative forcing from stratospheric ozone depletion decreases in the 21st century, the stratospheric temperature decrease originates mainly from changes in CO_2 emissions.

Simulations 8 and 9 show the effect of the trend in temperature on the ozone column as calculated with a radiative model. The 3K temperature decrease in the stratosphere from 1900 to 1990 yields an increase in the ozone column of 2.5%, while the trend in the ozone column from 1980 to 1990 increases from -4.2% per decade to -3.5% per decade. The higher ozone levels with lower temperatures probably comes from the temperature dependence of the odd oxygen cycle, i.e. the Chapman cycle. Stratospheric ozone levels are, according to the Chapman cycle, proportional to the production of ozone by the reaction $O + O_2$, divided by the ozone destruction by the reaction $O + O_3$. The former reaction rate increases with decreasing temperature, while the latter decreases. So the increase in ozone production and decrease in ozone destruction with decreasing temperature yields higher ozone levels in the stratosphere. The increase in ozone is further assisted by a decrease in NO_x (-2%), OH (-3%) and ClO (-1%) in the lower stratosphere in 1990 relative to 1900, reducing catalytic destruction of ozone by these compounds. The decrease in temperature in the stratosphere comes largely from CO_2 and the effect of a temperature trend can therefore also be viewed as resulting from the increase in CO_2 emissions in this century. The temperature change from stratospheric ozone depletion is therefore estimated to be responsible for a +0.3% per decade ozone trend. It is clear that the effect of ozone destruction on the ozone trend is hard to estimate with a model that does not treat the temperature feedback interactively. The estimated +0.3% per decade contribution of ozone itself is therefore only an indication of the size of the feedback effect and will probably be smaller, since such feedbacks usually diminish the imposed changes.

3.4. Future changes in ozone: 1990-2100

The effects of a trend in emissions in CH_4, N_2O and CO_2 in future scenarios (1990 to 2100) are shown in Table 2. In simulation 10 to 17 the IS92a scenario is applied, a mid-range scenario of the IPCC (Pepper *et al.*, 1992) scenarios. In the original IS92a scenario the CFC emissions are much higher than can be expected regarding the development of these emissions in the last few years (Montzka *et al.*, 1996) and the emissions also do not agree with the latest international agreements as the Copenhagen amendments to the Montreal Protocol. For emissions of the CFC's and related compounds IPCC (Pepper *et al.*, 1992) case 6 scenario is used for all simulations mentioned in Table 2. This scenario, originally already part of the IS92d and IS92e scenarios resembles the Copenhagen amendments. In the IS92a scenario the emissions, and consequently concentration, of CH_4, N_2O and CO_2 increase continuously. Since after 2000 the ozone layer is expected to start to recover, which will increase stratospheric temperatures, and CO_2 concentration is projected to increase according to the IS92 scenarios, the decrease in temperature in the stratosphere can be attributed to CO_2. Measurements of the ozone column do not show a depletion before the end of the 1970's. We will therefore use the ozone column in 1980 as a measure of the recovery of the ozone layer in the 21st century (WMO, 1994). The calculated total anthropogenic change in the ozone column since 1900 was -8.6% in 1990. The column of Table 2 named 'Recovery ozone layer' gives the year in which the ozone column reaches 1980 levels. Until approximately the year 2000 the thickness of the ozone layer will continue to decline because of the high chlorine levels in the atmosphere (Montzka *et al.*, 1996; WMO, 1994).

Table 2. Change in yearly average ozone column (%) at 47°N

Simulation	Characterization scenario:				Relative to 1990 (%)[1]			Recovery ozone layer
	CFC_3 halon	CH_4	N_2O	temp	2020	2050	2100[2]	1980 level
IS92a[3]: temp. fixed								
10 CH_4,N_2O fixed	Tr	1990	1990	1990	-1.9	1.9	5.4	2080
11 N_2O trend	Tr	1990	Tr	1990	-1.9	1.7	4.6	2095
12 CH_4 trend	Tr	Tr	1990	1990	-1.1	3.4	6.9	2060
13 CH_4,N_2O trend	Tr	Tr	Tr	1990	-1.1	3.2	6.1	2065
IS92a[3]: Temp.trend								
14 CH_4,N_2O fixed	Tr	1990	1990	Tr	-0.4	5.4	12.5	2045
15 N_2O trend	Tr	1990	Tr	Tr	-0.4	5.2	11.7	2045
16 CH_4 trend	Tr	Tr	1990	Tr	0.4	7.0	14.2	2040
17 CH_4, N_2O trend	Tr	Tr	Tr	Tr	0.4	6.8	13.4	2040

1) Yearly average ozone column in 1990 is 391.0 DU (simulation number 4 in Table 1)
2) Estimated total anthropogenic change in ozone column at 47°N in 1980 is -4.5% and in 1990 is -8.6% relative to 1900.
3) For the CFC's and related compounds not the original scenario is used but IPCC (Pepper *et al.*, 1992) case 6 scenario, which resembles the Copenhagen amendments to the Montreal Protocol.
4) The change in temperature is caused mainly by increases in CO2 emissions. Dynamical feedback and heterogeneous reactions on PSC's are not taken into account.

According to our calculations (Table 2) the ozone column at mid-latitude will be 1.9% below 1990 levels in 2020 (simulation 10). From approximately 2040 onwards, ozone will reach 1990 levels (see also Velders, 1995b). This is with fixed 1990 emissions for CH_4 and N_2O. In approximately 2080 the ozone column will reach 1980 levels and in 2100 the ozone column will be 5.4% thicker than in 1990. An increase in N_2O emissions in addition to this causes the ozone layer thickness to increase less because of an increase of NO_x in the stratosphere. The lower chlorine levels in the next century reduce the interaction between the NO_x and ClO_x compounds to form the reservoir $ClONO_2$. As shown before, an increase in CH_4 emissions reduces the ozone destruction by ClO_x. The CH_4 increase causes a 1.5% thicker ozone layer in 2100 (comparing simulations 10 and 12). The recovery of the ozone layer will be reduced by 20 years from 2080 to 2060. The stronger interaction between CH_4 and chlorine, compared with NO_x and chlorine, is responsible for the continuous positive influence (thicker ozone layer) with increasing CH_4 emissions. According to simulation 10 the ozone levels are expected to drop below 1980 levels in 2080, which is later than the year 2045 reported by Daniel *et al.* (1995) and WMO (1994). The difference is most likely caused by differences in the emission data in the phase-out period (1990-2010) of CFC's and related compounds, since the Montreal protocol and its amendments are based on production limitations and not on emissions.

A decrease in temperature in the stratosphere of 7 K in 2100 relative to 1990, corresponding with scenario IS92a, has a large impact on the ozone layer in the next century. Comparing simulations 14 to 17 with 10 to 13, the effect of the temperature decrease since 1990 is more than a doubling of the ozone column change in 2100. The inclusion of the temperature trend in the simulations causes a faster recovery of the ozone layer: from approximately 6% to 12% in 2100 relative to 1990. The increase in ozone column thickness in 2100 with the incorporation of the temperature trend yield is larger than the total calculated anthropogenic change in ozone column (-8.6% in 1990). As mentioned before, the effects of temperature changes are difficult to model because of the nature of interactions between ozone and temperature itself. It can best be done using a model which calculates the temperature and ozone changes interactively instead of off-line with a separate model, as has been used here. The changes in ozone originating from temperature (CO_2) changes will be diminished with an interactive model. Large temperature decreases in the stratosphere might also cause extra formation of polar stratospheric clouds (PSC's) in winter in polar areas and possibly also at mid-latitudes and in other seasons. This might cause a large depletion of stratospheric ozone through heterogeneous reactions on the surfaces of the PSC's. If this will happen is speculative, and if so it will probably only occur in winter at high latitudes and in the first half of the next century, when chlorine levels are still high. With a temperature trend in addition to a CH_4 emission trend the ozone column could reach 1990 levels just before 2020 and approximate 1980 values in 2040. The increase in CO_2 emissions, almost completely responsible for the temperature trend in the stratosphere after 2050, causes the ozone layer to reach pre-industrial levels by approximately 2070. A trend in N_2O emissions slows this recovery while a CH_4 trend accelerates it.

4. CONCLUSIONS

Emissions of greenhouse gases affect stratospheric ozone depletion

CFC's and other chlorine- and bromine-containing compounds are responsible for the main destruction of stratospheric ozone; however greenhouse gas emissions can affect the ozone column as well:

Table 3. Contributions to the trend in ozone column from 1980 to 1990 at 47°N and the recovery of the ozone layer as indicated by the year when 1980 ozone levels are reached

Compounds emitted	Ozone column trend (% per decade)	Ozone recovery 1980 levels[2]	Mechanism	Remarks
CFCs, HCFCs, CCl_4, CH_3CCl_3, Halons, CH_3Br	-5.8	2080		
N_2O	+0.1[1]	+15[1]	NO_x, Formation of reservoir	
CH_4	+1.4	-20	Formation of reservoir	
CO_2	+0.4	-25	Temperature	Feedback and PSC's ignored
$CH_4 + N_2O$	+1.5	-15		
$CH_4 + N_2O + CO_2$	+2.0	-40		

1) Contribution of the emissions of the compounds on top of the CFC, HCFC, CCl_4, CH_3CCl_3, halon and CH_3Br contribution (row 1).

2) Ozone layer recovered: the year the ozone layer reaches a level corresponding to the ozone layer in 1980.

3) n.a. = 'not assessed'.

4) Only the effect of the injection of aerosols from volcanic eruptions is calculated, not that of tropospheric aerosols.

Greenhouse gas interactions improve agreement with measurements

Taking into account the effects of a trend in CFCs, CH_4, N_2O and CO_2 emissions results in a good agreement between the calculated (-3.8% per decade) and measured (-4.0% per decade from TOMS and Dobson measurements; WMO, 1994) trend in ozone column at mid-latitude in the Northern Hemisphere from 1980 to 1990.

CH_4 emissions accelerate the recovery of the ozone layer

The interactions between greenhouse gases and stratospheric ozone also have an effect on the recovery of the ozone layer in the 21st century. The international measures (*i.e.* Montreal) are likely to reduce the chlorine levels in the stratosphere in the coming decades. The reduction in emissions of CFC and related compounds alone will result in 2050 in an ozone column 1.9% thicker than in 1990. Anticipating a business-as-usual scenario (IS92a) for N_2O and CH_4 in addition to the chlorine reduction yields 0.2% smaller and 1.5% larger increases, respectively. With fixed CH_4, N_2O and CO_2 emissions an ozone column corresponding with 1980 can be reached in approximately 2080. A methane trend as in scenario IS92a reduces this to approximately 2060, while an additional CO_2 trend (temperature decrease) reduces it to approximately 2040.

CO_2 increases can initially increase the ozone column

The CO_2 emission is largely responsible for a large decrease in temperature in the stratosphere in the 21st century: -7 K in 2100 relative to 1990. This results in an extra increase in ozone column of approximately 3.5% in 2050 relative to 1990. The temperature effect can be diminished if dynamical interactions in the atmosphere are taken into account interactively in a model. The decrease in temperature in the stratosphere might increase the occurrence of PSC's in winter, in polar regions but possibly also at mid-latitudes and in other seasons. If this happens a strong decrease in stratospheric ozone might occur from the temperature decrease. This effect is not considered here. A temperature decrease in the stratosphere, from an increase in CO_2 emissions will therefore, at least initially, yield a thicker ozone layer, as shown above.

REFERENCES

1. Daniel J.S., S. Solomon and D.L. Albritton, On the evaluation of halocarbon radiative forcing and global warming potentials, *J. Geophys. Res.* D **100**, 1271-1285, 1995.

2. Harwood, R.S. and J.A. Pyle, The dynamical behaviour of a two-dimensional model of the stratosphere, *Quart. J. R. Met. Soc.*, **106**, 395-420, 1980.
3. IPCC 1994, Climate change 1994: Radiative forcing of climate, WMO/UNEP, Cambridge university press, Cambridge, 1995.
4. IPCC 1995, Climate change 1995: The science of climate change, Contribution of WGI to the second assessment report of the IPCC, WMO/UNEP, Cambridge university press, Cambridge, 1996.
5. Law, K.S. and J.A. Pyle, Modeling trace gas budgets in the troposphere, 1. Ozone and odd nitrogen, *J. Geophys. Res.*, D **98**, 18377-18400, 1993a.
6. MacKay, R.M. and M.A.K. Khalil, Theory and development of a one dimensional time dependent radiative convective climate model, *Chemosphere*, **22**, 383-417, 1991.
7. Montzka, S.A., J.H. Butler, R.C. Myers, T.M. Thompson, T.H. Swanson, A.D. Clarke, L.T. Lock and J.W. Elkins, *Science* **272**,1318-1322, 1996.
8. Pepper, W., J. Leggett, R. Swart, J. Watson and J. Edmonds, I. Mintzer, Emissions scenarios for the IPCC, An update; Detailed report in support of chapter A3 of "Climate change 1992: the supplementary report to the IPCC scientific assessment", eds. J.T. Houghton, B.A. Callander, S.K. Varney. Cambridge, Cambridge University Press, 1992.
9. Randel, W.J. and J.B. Cobb, Coherent variations of monthly mean total ozone and lower stratospheric temperature, *J. Geophys. Res.*, D **99**, 5433-5447, 1994.
10. Velders, G.J.M., Description of the RIVM 2-dimensional stratosphere model, RIVM report 722201002, The Netherlands, 1995a.
11. Velders, G.J.M., Scenario study of the effects of CFC, HCFC and HFC emissions on stratospheric ozone, RIVM report 722201006, The Netherlands, 1995b.
12. Velders, G.J.M., Effect of greenhouse gas emissions on stratospheric ozone depletion, RIVM report 722201011, The Netherlands, 1997.
13. WMO 1991, Scientific assessment of ozone depletion: 1991, World Meteorological Organization Global ozone research and monitoring project, report no. 25, WMO, Geneva, Switzerland, 1992.
14. WMO 1994, Scientific assessment of ozone depletion: 1994, World Meteorological Organization Global ozone research and monitoring project, report no. 37, WMO, Geneva, Switzerland, 1995.

Fluorocarbons and Sulfur Hexafluoride Emissions Reduction Strategy in the United States

E. A. Dutrow and R. Forte

United States Environmental Protection Agency, 401 M Street, SW (6202J), Washington, DC, USA

1. ABSTRACT

Hydrofluorocarbons (HFCs), perfluorocompounds (PFCs), and sulfur hexafluoride (SF_6), represent possible solutions to the environmental impact caused by chlorofluorocarbons, halons, and other ozone depleting substances; however, most of the PFCs, SF_6, and some HFCs are very potent greenhouse gases. As such, emissions of these gases are under examination by the United States Environmental Protection Agency (USEPA). USEPA is actively tracking emissions of these gases from various sources and has developed programs to control releases of the gases.

2. INTRODUCTION

To protect the earth's ozone layer, substitutes for ozone-depleting chlorofluorocarbons (CFCs) have been developed. At times, these substitute chemicals include gases with extremely high global warming potentials (GWP) and long atmospheric lifetimes. While not posing a major threat to the ozone layer, such substitutes, when emitted, do increase the atmospheric burden of greenhouse gases. Concurrently, emissions of these same high GWP gases may occur aside from applications requiring alternatives to CFCs. Industrial releases may emanate from manufacturing processes or be created as byproducts.

Control of these greenhouse gas emissions is governed internationally by the United Nations Framework Convention on Climate Change (FCCC). The Framework is designed to "achieve ... stabilization of greenhouse gas concentrations in the atmosphere at a level that would prevent dangerous anthropogenic interference with the climate system". Under the FCCC, parties to the convention are compelled to develop national anthropogenic emissions inventories "of all greenhouse gases not controlled by the Montreal Protocol" by sources and to formulate national programs to mitigate climate change.* The United States is a party to the FCCC.

HFCs, PFCs, and SF_6 are among the most potent greenhouse gases. In comparison to carbon dioxide (CO_2), the predominant, anthropogenically emitted greenhouse gas, several HFCs, PFCs, and SF_6 possess high GWPs and extremely long atmospheric lifetimes. Table 1 illustrates these values.

---

*United Nations Framework Convention on Climate Change (1992)

Table 1
Global Warming Potentials and Atmospheric Lifetimes for Selected Greenhouse Gases

Greenhouse Gas	Atmospheric Lifetime (year)	Global Warming Potential (100 year time horizon)
Sulfur hexafluoride (SF_6)	3,200	23,900
Hexafluoroethane (C_2F_6)	10,000	9,200
Tetrafluoromethane (CF_4)	50,000	6,500
Trifluoromethane (HFC-23)	250	11,700
HFC-134a	14.6	1,300
Methane (CH_4)	12 ± 3	21
Carbon Dioxide (CO_2)	250	1

Data from: Climate Change 1995, The Science of Climate Change. Technical Summary. Cambridge University Press. New York, New York. 1996. p. 22.

Concern over emissions of these gases is twofold. First, a discrete emission of a high GWP gas is more potent than the equivalent amount of carbon dioxide. Second, the long atmospheric lifetimes of the gases span to thousands of years. With limited opportunity for removal from the atmosphere by most mechanisms, these gases will persist for a long time. Small emissions have and will continue to contribute to a cumulative atmospheric burden. Environmental effects of these gases will remain for generations of mankind.

3. UNITED STATES STRATEGY TO CONTROL HFC, PFC, AND SF_6 EMISSIONS

In 1993, the United States developed a Climate Change Action Plan (CCAP). Its goal is to "reduce emissions of greenhouse gases to their 1990 levels in 2000".* The plan relies on the cooperation of government and industry to accomplish reductions in greenhouse gas emissions by the year 2000 and is comprehensive in approach including all greenhouse gases as governed by the FCCC. The CCAP includes actions to control emissions of HFCs, PFCs, and SF_6. These actions are classified in two categories: regulatory mechanisms and partnership efforts.

*United States Climate Change Action Plan (1993)

3.1. Regulatory Control of HFC, PFC, and SF_6 Emissions

Regulatory mechanisms restrict the use and emissions of chemicals intended for CFC replacement. Section 612 of the Clean Air Act authorizes the control of uses for high GWP gases if other alternatives to ozone-depleting chemicals exist and pose less risk to human health and the environment. Regulatory actions are developed as part of the USEPA's Significant New Alternatives Policy (SNAP).

SNAP regulation governs new chemicals, new uses of existing chemicals, new product substitutes, and alternative manufacturing processes. Under the SNAP requirements, manufacturers and users of CFC substitutes must submit proposals for use of new alternatives to the USEPA ninety (90) days prior to beginning sales of the product. Manufacturers may not use substitutes deemed unacceptable by USEPA review. USEPA evaluates the ozone depletion potential, GWP, flammability, toxicity, and ecological effects of a substitute in its review. The SNAP has proven effective in reducing the impact on the climate from the phaseout of CFCs by limiting use of high GWP chemicals as alternatives.

3.2. Partnership Efforts

The United States CCAP incorporates partnership programs in the address of greenhouse gas emissions. These efforts are characterized by the development of voluntary programs with emitting industries for reduction of emissions. For HFCs, PFCs, and SF_6, agreements are developed between emitting sources and the USEPA. The agreement, typically called a memorandum of understanding (MOU), specifies that the emitting source track its emissions and agree to implement changes to reduce emissions of the high GWP gases.

Agreements exist between the USEPA and several industrial sources. The first MOU completed controls emissions from aluminum smelting. CF_4 and C_2F_6 are emitted as byproducts. Manufacturers individually have joined with USEPA to make changes in the manufacturing process to reduce the amount of these gases created. By the year 2000, it is expected that 2.7 million metric tons of carbon equivalent (MMTCE) reductions will be achieved.

A second program addresses emissions of the byproduct HFC-23 from HCFC-22 manufacturing. The U.S. producers have committed to reduce their emissions by 5 MMTCE by the year 2000. Opportunities for reduction include process optimization, capture and conversion, and abatement.

The most recent MOU established with U.S. industry covers semiconductor manufacturing. Nitrogen trifluoride (NF_3), CF_4, C_2F_6, SF_6, C_3F_8, and HFC-23 are emitted from the production of microchips. These gases are released due to incomplete use and from reformation following disruption in plasma processes. The U.S. semiconductor industry has implemented a comprehensive research program to identify environmentally protective and cost-effective opportunities for controlling emissions of the PFCs. During 1998, USEPA will work with the current eighteen (18) industry partners to identify actual reduction amounts.

USEPA is continuing to develop voluntary partnerships for the PFC, HFC, and SF_6 reductions. In 1997, programs will be established for the control of SF_6 emissions from electrical power equipment, magnesium production and diecasting, and other uses.

4. CONCLUSION

Opportunities exist for expansion of the voluntary partnerships to countries with similar emitting industries. For instance, the semiconductor industry manufactures microchips worldwide. Individual companies, while formed in one country, operate fabrication facilities in nations around the globe. Extension of voluntary partnerships to the countries of operation would prove beneficial for a global address of PFC control from this industry as well as minimize the barriers that limit a company's ability to compete and implement worldwide climate protection goals.

The U.S. model for controlling of PFC emissions from semiconductor manufacturing has been adopted by Japanese industry and government. In April 1997, the Electronic Industries Association of Japan (EIAJ) announced its agreement with the Japan Ministry of International Trade and Industry (MITI) to implement a program of research and emissions tracking for PFCs in semiconductor manufacturing. The goal of the agreement is to reduce these emissions. It is desirable for such commitments to be made in more countries with significant high technology industry.

Protection of the global climate system is an important activity for government and industry. USEPA and its industry partners are striving to reduce the impact of PFCs, HFCs, and SF_6 on the climate.

REFERENCES

1. United Nations Framework Convention on Climate Change, 1992.
2. Climate Change 1995, The Science of Climate Change. Technical Summary. Cambridge University Press, New York, New York, 1996.
3. United States Climate Change Action Plan, 1993.

A Strategy for Reducing Methane Emissions

Judith Bates, AEA Technology plc
156 Harwell, Didcot, Oxfordshire, OX11 ORA, UK.

ABSTRACT

Methane is an important greenhouse gas whose concentration in the atmosphere has more than doubled since pre-industrial times. It is a more potent greenhouse gas than carbon dioxide, but due to its shorter atmospheric lifetime (of 12 years) it is estimated that global emissions would only need to be reduced by about 8% from current levels to stabilize methane concentrations at today's levels. This is a much smaller percentage reduction than those required to stabilize atmospheric concentrations of the other major greenhouse gases, CO_2 and N_2O.

The main source of methane emissions within the EU is the agricultural sector, where emissions arise mainly from enteric fermentation in ruminant livestock, but also from livestock manure. The other major source is landfills, while coal mining and gas production and distribution are smaller, but still significant contributions. There are a range of possible measures for the reduction of emissions from each of these sectors, varying from technological options such as the collection and combustion of landfill gas, or the recovery and use of methane from animal waste, through to more general measures, often of a longer term nature, such as a reduction in the amount of organic waste going to landfill, or a reduction in livestock numbers. For some sources there are still significant uncertainties in emission factors, which make the development and assessment of abatement options difficult. In addition, there is a lack of data on the cost-effectiveness of many actions and measures. Any strategy for reducing emissions is thus likely to need to combine measures to encourage the deployment of proven techniques, and to encourage research into the cost-effectiveness of options, and to improve knowledge of emissions factors and processes for some sources.

This paper discusses the main options for the reduction of methane emissions and briefly summarizes the strategy paper recently prepared on this subject by the European Commission.

1. INTRODUCTION

Methane is an important greenhouse gas whose concentration in the atmosphere has more than doubled since pre-industrial times, rising from a pre-industrial concentration of about 700 ppbv to a concentration in 1994 of 1720 ppbv [1]. Over the last 20 years there has been a decline in the growth rate of this methane concentration. In the 1970's, the concentration was growing at a rate of about 20 ppbv/yr in the late 1970's, but this fell to a growth rate of 9 to 13 ppbv/yr in the 1980's. Around the middle of 1992, methane concentrations briefly stabilized, but since 1993, the global growth rate has returned to about 8 ppbv/yr.

Methane is a more potent greenhouse gas than carbon dioxide, with a global warming potential (over a 100 year time horizon) 21 times greater than carbon dioxide. However, due to its shorter atmospheric lifetime (of 12 years) it is estimated that global emissions would only need to be reduced by about 8% from current levels to stabilize methane concentrations at today's levels. This is a much smaller percentage reduction than those required to stabilize atmospheric concentrations of the other major greenhouse gases, CO_2 and N_2O.

Total anthropogenic methane emissions in the EU in 1990 were estimated at 23 Mt [2]. This is 5 to 8% of global anthropogenic methane emissions which are estimated at between 300 to 450 Mt [1]. While methane emissions in the EU are much lower than CO_2 emissions (Table 1), the higher global warming potential of the gas means that in global warming terms its emissions are significant, particularly if a short time horizon is considered.

Table 1. Anthropogenic Emissions of CO_2, CH_4 and N_2O in the EU in 1990

	Emissions	20 year time horizon		100 year time horizon	
	Mt	GWP	Mt of CO_2 Equivalent	GWP	Mt of CO_2 Equivalent
CO_2	3,361	1	3,361	1	3,361
CH_4	23	56	1,303	21	488
N_2O	1	280	266	310	294

Source: [2]

The Framework Convention on Climate Change requires (Annex 1) signatories to seek to stabilize emissions of greenhouse gases at 1990 levels by 2000. Negotiations have already begun over emissions targets for post-2000, and the second Conference of the Parties which met in Geneva in July 1996 agreed that developed countries should negotiate '*quantified legally-binding objectives for emissions limitations and significant overall reductions within specified time frames, such as 2005, 2010, 2020*'. It is planned to agree these limits at the next Conference of the Parties in Kyoto, Japan in December 1997.

In considering their initial commitment under the Convention, attention, both within the EU and internationally, has generally been focused on policy options for reducing emissions of carbon dioxide. However, increasing attention is now being paid to potential options for reducing methane emissions, partly because uncertainty over the importance of the various sources of methane has decreased. Interest is likely to continue to grow, if the 'basket' approach, whereby reductions in emissions of the direct greenhouse gases are weighted by their global warming potential, is adopted for future emission targets.

2. METHANE EMISSIONS IN THE EU

Estimated anthropogenic emissions of methane by sector in the EU in 1990 are shown in Table 2. The main source is the agricultural sector, which was responsible for 45% of anthropogenic emissions. These emissions arise principally from enteric fermentation in the digestive tract of ruminant livestock (cattle and sheep), but also from the anaerobic decomposition of livestock manure. The other major source is landfills, where the anaerobic decomposition of organic waste in the landfill leads to the release of landfill gas, which is a mix of CO_2 and CH_4. Coal mining and gas production and distribution are smaller, but still significant contributions. There is significant uncertainty attached to some of the emissions estimates, particularly those arising from landfill sites.

The estimation of methane emissions from natural sources is subject to high uncertainty, but nonetheless, natural emissions are thought to be significant. Globally, they are considered to be at least 20% of total emissions, and in the EU, natural emissions, mainly from wetlands and peat based soils, were estimated to be 9 Mt in 1990 [2]. This is almost 30% of total EU emissions.

Table 2. Main Sources of Anthropogenic Methane Emissions in the EU in 1990

Sector	Emissions	% of total	Of which:	
Agriculture	10.2 Mt	45%	Enteric fermentation	30%
			Livestock manure	15%
Waste	7.3 Mt	32%	Landfills	31%
			Waste water treatment	1%
Energy sector	5.3 Mt	23%	Coal mining, transport and storage	11%
			Gas production and distribution	9%
			Stationary combustion	2%
			Transport	1%

Source: [2]

More recent data on methane emissions from the CORINAIR database for 1994, which will allow a comparison of 1990 and 1994 methane emissions for the EU as a whole, was not available at the time of writing this paper****. 1994 emissions data is already

1. Preliminary CORINAIR data for 1994 has subsequently become available and indicates that total anthropogenic emissions in 1994 were still 23 Mt.

available for several Member States, and as an example, data for 1990-1994 for the UK is shown in Figure 1. This shows that UK anthropogenic emissions are estimated to have fallen by 13% between 1990 and 1994 [3]. This is due mainly to a reduction of almost 60% (435 kt) in mining emissions as a result of a decline in coal production. There has also been a significant reduction of 100 kt (5%) in emissions from landfill sites, due to the increased deployment and use of methane recovery systems. Smaller reductions occurred in emissions from ruminants (13 kt), due to a fall in the number of dairy cattle, and in emissions from stubble burning (12 kt), as this is now banned. Some of these trends e.g. a decline in coal production and hence in methane emissions, and a reduction in dairy cattle numbers, are expected to be replicated in other Member States.

Figure 1. Anthropogenic Methane Emissions in the UK (1990-1994)

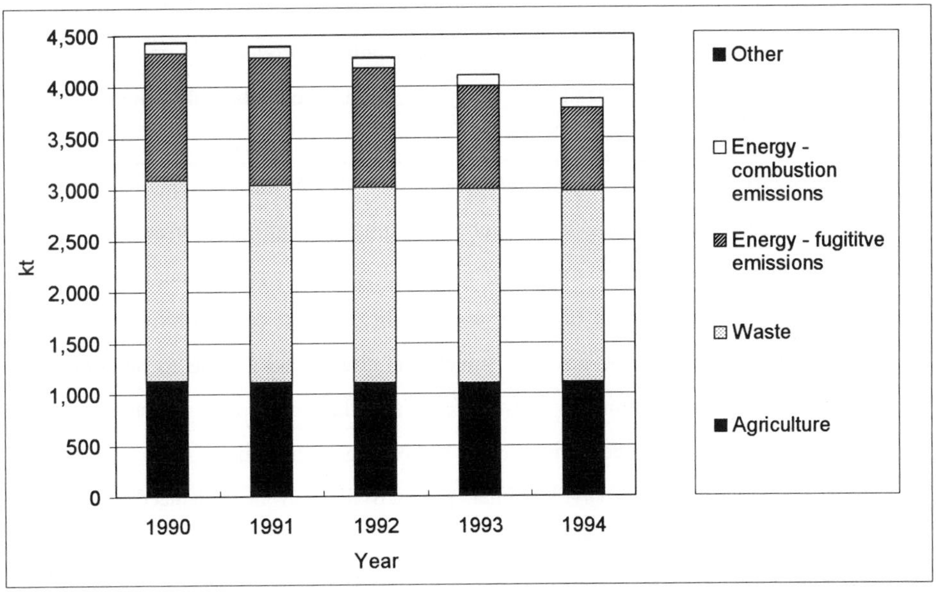

Source: [3]

3. OPTIONS FOR REDUCING EMISSIONS

3.1. Agriculture

3.1.1. Reducing Emissions from Enteric Fermentation

In ruminant animals (e.g. cattle, sheep), microbial fermentation in the large stomach or rumen, breaks down complex vegetable matter into products which can be utilized by the animal. The fermentation process also produces methane, which is then eructed or exhaled by the animal. There are three broad options for reducing emissions from this source:

- *Reduction of livestock numbers*: this is perhaps the most obvious, and while it will produce 'real' savings, if the reduction in livestock numbers corrects overproduction of meat or milk, there is a danger that if restrictions lead to a reduction in useful output, then there may simply be a transfer of production to other countries. If this is in countries where animal productivity is not as high, and/or slurry treatments mean that slurry emissions are higher, then there may be a net increase in emissions.
- *Improving the efficiency with which feed is converted*: the conversion of feed to methane rather than to products which the animal can utilize represents a loss in feed efficiency. Ensuring that animals have a diet which contains sufficient nutrients and which has a high digestibility, and, can minimize methane production. Within Europe however, much has already been done to maximize the efficiency with which animals convert feed, and there may thus be little potential in the short term to reduce emissions by improving animal diet.
- *Increase animal productivity*: Animal productivity can be enhanced by adding production enhancing agents to feed or by injecting animals with steroids. However there is some opposition to this, and it has been banned in some EU countries (e.g. The Netherlands, Denmark and Germany). Improvements in breeding and perhaps, in the future, genetic technologies may lead to increases in productivity.

3.1.2. Reducing Emissions from Manure

In animal wastes which are kept in anaerobic conditions, the fermentation of the organic substrates in the waste by methanogenic bacteria produces methane. Methane emissions from this source are thus highly dependent on how the animal wastes are handled. When manure is stored or treated as a liquid in a lagoon, pond or tank, it tends to decompose anaerobically and produce a significant amount of methane. When manure is handled as a solid or when it is deposited on pastures, it tends to decompose aerobically and little or no methane is produced.

Options for reducing emissions from liquid slurry systems include:

- *Covered lagoons:* In intensive livestock facilities, wastes are commonly washed from the animal shed into large lagoon ponds. By covering the lagoon with an impermeable cover (e.g. a plastic cover) methane generated in the lagoon may be trapped, recovered, and used for heat or power generation on the farm.
- *Anaerobic digestion with flaring or utilization of methane:* Digesters are reactors specially designed to enhance the anaerobic decomposition of waste and thus maximize the production of methane. The methane drawn off from the digester may be either flared, or used for heat or power generation. Digesters for farm waste can range from small scale digesters, suitable for a single farm to large scale digesters, often heated and operating on a continuous rather than batch process, taking waste from several farms. Such large scale digesters can also receive organic waste from other sources such as industry. The solid waste residue can be used as a soil conditioner, or further processed into a peat-substitute compost suitable for horticultural applications, thus providing income.

3.1.3. Cost and Applicability

Several *covered lagoons* are operating successfully in the US at large scale cattle or pig farms. Data from these suggests that if the recovered methane is utilized for heat or power generation on the farm, then abatement costs are in the range of -120 to 130 ECU/t methane, i.e. in some cases the plant could provide cost savings for the farm. [4].

In the case of *anaerobic digestion*, some EU Member States have begun to promote this technology, often as part of programs to encourage and promote renewable energy sources. Denmark has run an Action programme for Centralized Biogas Plants (1988 -1995), and a number of anaerobic digestion plant are now installed. The UK also has some anaerobic digestion plant, partly encouraged by the NFFO scheme, which provides a premium price for electricity generated from renewable energy sources. UK estimates of the cost [5] of abating emissions using a large scale anaerobic digesters are in the range of -90 to 90 ECU/t CH_4 (at a 5% discount rate) depending on cost assumptions, cost of transport of waste, and income for electricity generated and residue produced. A Dutch study also found that such schemes could generates a net income, estimating that the cost of abatement using such plant was about -50 ECU/t CH_4. The same study estimated that costs for farmscale digestion (at 400 to 450 ECU/t CH_4) were however much higher [6].

3.2. Landfills

When anaerobic conditions exist in a landfill, the organic component of waste (food wastes, waste from animal, garden waste, paper and cardboard etc.) is broken down by methanogenic bacteria in a complex biological process which releases CH_4, CO_2 and a number of other trace gases. At a national level emissions from this source are thus influenced by:

- The importance of landfill as a waste disposal option, and more significantly the amount of organic waste going to landfill. This varies significantly between countries in the EU.
- Landfill design. Air can move into shallow or unmanaged sites relatively easily, which inhibits anaerobic degradation. Engineered landfills designed for environmental protection usually exclude air because they are deep and have low permeability liners; hence methane generation is encouraged.
- The proportion of landfill sites at which methane is collected and either flared or utilized.

Finally the *rate* at which methane is produced varies significantly across the EU due to the wide variability in climatic conditions across the EU.

Some of the possible options for reducing landfill emissions are discussed below.

3.2.1. Reducing the mass of degradable organic waste which is landfilled. A number of waste management practices are available to reduce the mass of degradable waste landfilled, and hence reduce methane emissions. These practices are prioritized in the EU's waste hierarchy [7].

1. Waste avoidance and minimization;
2. Re-use;
3. Recovery (of materials and energy)

A number of 'recovery' options are feasible, ranging from composting and anaerobic digestion of organic waste, to materials recycling e.g. of paper and card, glass, to incineration with energy recovery. The choice and appropriateness of the various re-use and recovery options is likely to be affected by a number of factors which are country and region specific, and may be different for different parts of the waste stream. Any national or regional waste management strategy is thus likely to include a number of these options.

Reducing the landfilling of degradable waste is not a short term option, as it will require major changes to waste management practices in many EU countries. In addition, it will take several years for its impact to be fully realized, as the degradation of waste which has already been landfilled will continue for many years.

3.2.2 Collect and burn landfill gas, with energy recovery where economically beneficial

The technology for landfill gas collection and combustion, either for energy recovery or in flares is well understood and demonstrated in many Member States. In the past, gas has typically been collected from sites because of local safety risks (e.g. from explosion or fire caused by a build up of landfill methane in buildings near landfill sites) and /or to reduce local odor nuisance, but there is an increasing appreciation of the need to control emissions because of their significance as a greenhouse gas.

Gas is typically collected from a site using a series of 'wells' consisting of perforated popes which are installed round the site and are connected to 'ring main' gas collection pipes via well heads. The gas is then piped to a central blower/fan or gas compressor unit [8]. Once collected the gas may be flared, or may be used:

- for energy recovery through:
  - direct use in boilers or process heating
  - electricity generation (using spark ignition or dual fuel engines, or gas turbines)
  - upgrading to substitute natural gas
  - vehicle fuelling
  - fuel cells
- as a chemical feedstock

Energy recovery for direct use and electricity generation is already common in some EU member states, e.g. Sweden has 55 energy recovery schemes, mostly supplying heat for district heating schemes, the UK, has 105 projects, the majority of which are for electricity generation, and upgrading landfill gas has been demonstrated in The Netherlands, France and Germany. In the case of vehicle fuelling, prototype plants with a limited number of vehicles running have been demonstrated in the UK and France, but in the case of fuel cells, there are currently no commercial systems in operation, although there is a demonstration study in the US. While it is technically possible to use landfill gas as substitute chemical feedstock, there is no commercial application of this process and its profitability remains to be proven [8].

While the combustion of the methane, either through flaring or in the energy recovery options, converts the methane to carbon dioxide, the higher GWP of methane means that there is still a significant reduction in GWP. There are several additional environmental benefits of energy recovery over flaring, including:

- encouragement of increased efficiency of landfill gas collection, because the gas is seen as an asset;
- indirect reduction of greenhouse gas emissions through the substitution of a non-fossil fossil fuel for a fossil fuel;
- reduced depletion of fossil fuel reserves.

3.2.3. Optimize biological methane oxidation in the cover soils of landfills

In modern landfills, the final operational step in landfilling is the capping and final cover of the site. High rates of methane oxidation take place in landfill cover soils exposed to a methane flux [9, 10], as the methane is oxidized to CO_2 and water by a group of naturally occurring soil bacteria. Research has shown that oxidation is enhanced in well drained soils with a sandy, open structure (which allows easy movement of gases), and by the addition of nutrients [11]. Ensuring that sites have well engineered cover systems on closure and that these are well maintained, thus offers a significant opportunity to reduce residual emissions not captured by a gas collection scheme. However this option has yet to be commercially demonstrated.

3.2.4. Cost and Applicability

A number of countries have already instigated policies to discourage the landfilling of organic waste. Indeed Germany and France have either banned or will soon ban landfilling of biodegradable waste, and in the UK, a landfill tax (at a standard rate of £7/t of waste and a reduced rate of £2/t of inert waste) has been introduced to reflect the environmental impact of landfilling and encourage the use of options towards the top of the waste hierarchy. Several countries have also instigated policies to encourage the recovery and flaring or utilization of landfill gas at new and in some cases existing landfill sites. Indeed, within the EU12 (i.e. excluding Austria, Finland, and Sweden) electricity generation from landfill gas was over six times greater in 1994 (1.4 TWh) than in 1989 (0.2 TWh) [12].

The costs of the various options described above have been estimated [13] and are summarized in Figure 2. It should be noted that many of these costs are subject to considerable uncertainty and are likely to vary through the EU. It was assumed that organic waste which is diverted from landfill is incinerated as this is the most common option in the EU for pre-treating most biodegradable waste prior to landfilling. It can be seen that this is significantly more costly than other options, and it is not known how some of the other waste disposal options available for treating organic waste would compare.

The reduction in landfill emissions which each of the measures might bring about in the future has also been assessed [13]. It is estimated that existing policies which Members States have, or are proposing, to implement will lead to a 45% reduction in emissions from 1990 levels by 2010. Increased recovery and use of landfill gas could reduce emissions by about another 20%, and if policies to divert waste from landfill and to optimize biological oxidation in the landfill cover were also implemented, then emissions in 2010 could be reduced by a further 20%. Thus, implementing all three policies could bring down emissions from landfills to 15% of 1990 levels by 2010.

Figure 2. Estimated Costs of Mitigating Methane Emissions from Landfills

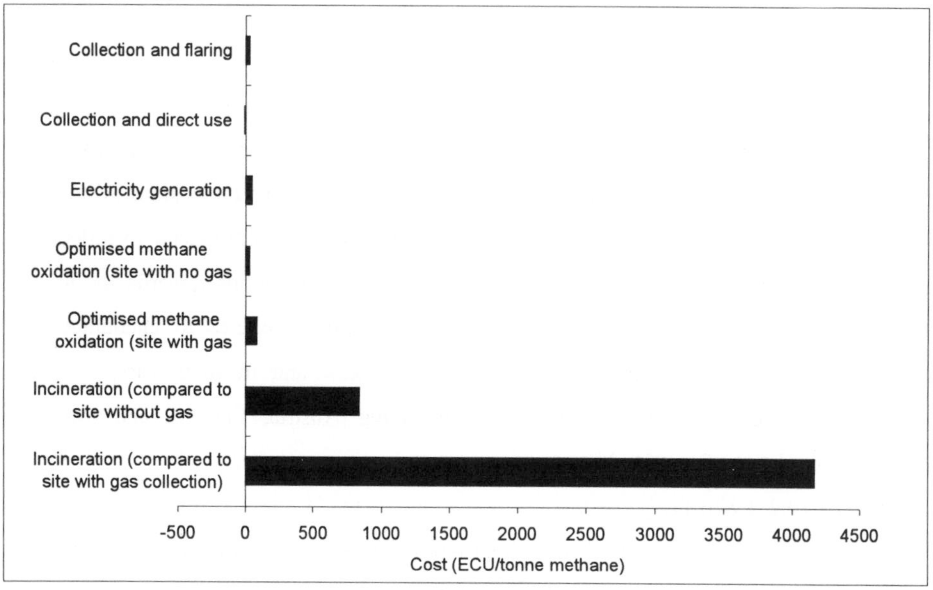

Source: [13]

3.3. Coal mining

Methane and coal are formed together during coalification, the process in which swamp vegetation is converted by geological and biological forces into coal. The amount of methane stored is influenced *inter alia* by the rank (carbon content) and depth of the coal. Higher rank coals generally have a higher associated methane content, and for a given geological setting, deep coal seams have a higher methane content than shallow ones. Much of the methane, which is trapped within the coal seam and the rock strata surrounding the seam, is released when the coal seam is fractured. It is then either (eventually) emitted into the atmosphere or seeps back into the mine workings as the coal is mined. Methane is explosive in concentrations of between 5% and 15% in atmospheric air, so if methane seeps into the mine workings, sufficient ventilation air must be supplied to dilute the methane to a safe level. Typical statutory limits are for a maximum methane content of 1 to 1.25%.

3.3.1. Recovery and Utilization Options

Mine air containing methane is removed from the mine due to the risk of explosion. It is generally vented directly into the atmosphere, but, as with landfill gas, can be flared or used for energy production in a range of ways (e.g. directly in boilers, dryers, kilns and ovens, for electricity generation, for feeding into natural gas pipelines), or used as a chemical feedstock. The potential end use depends partly upon the methane content of the mine gas (and hence the extent to which it has been diluted) and partly on the proximity of potential end users and utility infrastructure.

Removing methane from the mine before it is mixed with air and extracted through the ventilation system has advantages; the gas recovered has a higher calorific value (allowing it to be used as a natural gas feed as well as for power generation) and ventilation requirements for the mine are reduced.

One option, which is widely used in two coal basins in the USA, is to drill boreholes from the surface and drain off some methane well in advance of mining. The gas obtained by this means is virtually pure methane and can be fed into a natural gas transmission system with only minimal treatment. However the technique requires a very permeable coal, otherwise a high number of holes is required per unit area, rendering recovery of the gas uneconomic. European coal is far less permeable than that at the sites in the USA, and while some trials have been carried out, results have been mixed.

Methane may also be drained off before it can enter the ventilation stream in working mines. One method is to drill inclined holes into the strata above the seam and local to the working face, to connect them via a piping system and apply a small suction pressure to bring the gas to the surface. Air leakage through the coal strata, and via the many joints in the piping system, mean that the gas many contain less than 50% methane when it reaches the surface. Nevertheless it's calorific value is still such that it may be utilized in a number of ways, e.g. in boilers to supply space heating or hot water for mine buildings, or for on-site power generation in engines or gas turbines.

Ventilation air vented from mines must have a low methane content for safety reasons (regulations in most countries require that it does not exceed 1%), but may contain more than 50% of the total methane emissions from the mine [14]. Upgrading the air to produce a higher quality gas is expensive and is thus not practiced. It is possible to use the ventilation air as part of the fuel air mixture in boilers or gas turbines which are located on site; if the ventilation air contains 1% methane then it would reduce the fuel requirement by up to 15%. A novel approach under investigation involves passing ventilation air through a fluidised gravel-bed to combust it. In the future, lean burn technology may offer the potential to utilize this very low quality gas.

3.3.2. Applicability of Options

The there are only four EU countries which still mine coal in commercial quantities: Germany, UK, France and Spain. In the two most significant producers, Germany and the UK coal production has fallen significantly from 1990 levels - by 28% by 1993 in Germany [15], and by 48% by 1994 in the UK [16] - leading to a reduction in associated emissions from mining of 12% and 60% respectively [17]. In the longer term, emissions are likely to continue to fall as European coal production continues to decline; projections made by the European Commission's Energy Directorate (DGXVII) indicate that coal production in Europe will have halved by 2010 (from 1990) [18].

3.4. Gas Production and Distribution

Emissions can arise from every stage of the gas production and distribution chain - from exploration through production and processing to transport and distribution. Within Europe, the majority of emissions are thought to be fugitive emissions from natural gas transport and distribution, with emissions from process vents and flares being a smaller, but still significant source [19]. Options for reducing fugitive emissions include:

- improved inspection and maintenance of components such as seals, valves and connections;

- improved pipeline inspection and leakage detection and control;
- replacement of ageing pipelines with corrosion resistant pipelines (e.g. coated steels, PVC, polyethylene).

These measures can reduce fugitive emissions by 50 to 95%. While some component inspection can prove cost effective (i.e. cost are more than recouped by gas savings), a Dutch study [6] found that the cost of improved pipeline leakage control, and pipeline replacement were significant at about 1,200 and 1,800 ECU/t CH_4 respectively. Countries already undertaking pipeline replacement include France, where 'Gaz de France' are planning to replace 1000 km/yr of pipeline between 1993 and 2000, and the UK. In the latter case, British Gas TransCo has a leakage control strategy which aims to reduce emissions from the transmission network by 20% from 1992 levels by 2000, and fugitive emissions from storage installations by 15% [20].

Emissions from process vents and flares arise for a number of reasons and from a number of sources. In oil production, gas associated with the oil must be separated off, and if there is no demand for the gas (e.g. for power generation on the platform) either vented or flared, or reinjected into the field. While venting is strictly controlled there is still some potential to flare or reinject gas rather than vent it. Other sources of emissions include offgases from processes, gas used to purge vent and flare systems, gas used for blanketing storage vessels, and 'leaks' from worn or fouled valves which do not close completely [6]. Many of these emissions may be reduced, often very cost-effectively, by improving control of processes, or small adjustments of systems, although the overall potential for reduction and overall cost is likely to vary significantly from site to site and country to country.

4. THE EU STRATEGY PAPER

In February 1993, the EU in its Fifth Action Programme for the Environment 'Towards Sustainability', defined a series of actions for greenhouse gases: which included the aim of possibly reducing methane emissions. In December 1994, the Environment Council asked the Commission to submit a strategy to reduce emissions of greenhouse gases other than CO_2, in particular methane and nitrous oxide. A strategy paper for methane was produced by the Commission in November 1996, and submitted to the Council and to the European Parliament [21]. It sets out a number of potential actions in the agricultural, waste and energy sector which could be incorporated into a Community emissions mitigation strategy. These are summarized in Table 3.

Table 3. Summary of Actions Suggested in the European Commissions Strategy Paper

Source	Suggested Action	Level for Action
Enteric fermentation	R&D	EU and National
Recover methane from animal waste	1) Demonstration	All
	2) Obligation	EU
Reduce landfilled organic waste	Promotion of measures	All
Landfill gas recovery	Legislation	EU
Energy production from landfill gas	Incentives	EU and National
Mining emissions	Encourage use of best available technologies	EU and National
Gas pipeline leakage	Set standard	EU
	Increase control frequency	National

In the agricultural sector suggested measures include:

- Promoting research and incentives (at an EU and national level) to develop viable policies and measures to reduce emissions from enteric fermentation.
- Promoting the use of anaerobic digesters and covered lagoons, with collected methane preferably being used for energy production, and if not flared. It is suggested this is achieved through a demonstration programme to raise awareness and gain acceptance of the technology, followed at a later date by an EU level obligation to install such systems at larger livestock units.

In the waste sector:

- Promoting (at an EU, national, regional and local level) measures aimed at reducing the amount of organic waste going to landfill. This includes measures aimed at minimizing the generation of organic waste, encouraging separate collection of organic waste, recovery options such as composting and energy recovery. It is also suggested that recycled products could be promoted through economic incentives at an EU and national level.

- Introducing EU legislation requiring that new anaerobic landfills include systems to recover and utilize landfill gas.
- Introducing EU legislation to require the retrofitting of methane recovery systems to existing landfills wherever possible. Encouraging the recovery and use of landfill gas for energy production wherever possible through economic incentives at the EU and national level.

  The Commission has already produced a proposal for a Directive on the landfill of waste [22], which suggests that by 2010, biodegradable municipal waste going to landfill must be reduced to 25% of the total amount of biodegradable waste produced in 1993. The proposal also suggests that landfill gas must be collected from all landfills and either utilized, or if this is not possible, flared.

In the energy sector:

- Encouraging Member States to promote the use of best available technologies for recovering methane from mines which are expected to continue operating for a significant period of time. It is recognized that coal production (and hence methane emission from this source) is expected to continue to decline in the future, so that it may be difficult to justify additional expenditure on methane recovery in many cases.
- Setting an EU standard for allowable levels of leakage from gas transmission and distribution levels.
- At a national level increasing the frequency with which pipelines are inspected.

The measures suggested in the strategy paper reflect a number of parameters: the uncertainty in emissions arising from some sources, the state of knowledge on factors affecting emissions, whether abatement technologies are commercially available and accepted in the market place and whether options have been effectively demonstrated. They thus range from encouraging R&D, through demonstration programs, to legislation based on what is believed to be achievable abatement levels. They also include a mix of measures which could be effective in the short term (such as the retrofitting of landfill gas recovery systems to existing sites) to measures which will may take longer to realize (such as reducing organic waste production and encouraging its diversion from landfill). It should be noted that a number of options also have additional environmental benefits: for example, the use of recovered methane from waste for energy production can reduce CO_2 emissions from the combustion of fossil fuels, and the anaerobic digestion of animal wastes can help to reduce the water pollution which can occur from inappropriately stored wastes.

It has been estimated that measures such as those identified in the strategy might lead to a reduction from 1990 emissions levels of up to 40% [23], although this value is subject to considerable uncertainty.

5. DISCUSSION

Much further work is needed to establish more firmly the reductions which might be achieved in various sectors from different measures. There is also a need for a thorough analysis of the cost-effectiveness of the measures. The indicative costs presented in this paper show that costs for the various options can vary significantly, and that there are a number of options which might be regarded as 'no regrets' policies in that there is no net cost attached to the measure.

Table 4 attempts to broadly summarize the potential reductions which might be achieved in various sectors and gives a relative indication of their expected cost. It should be remembered that the data in the table is very preliminary; nonetheless it indicates that substantial savings might be achieved and that a number of these might be achieved at a relatively low cost. If half of the savings in Table 4 were to be achieved than this would be a reduction equivalent to 20% of total methane emissions in 1990.

Table 4. Indicative Costs and Potential Reductions for Methane Emissions in the EU

Measure	Potential Reduction	Saving (on 1990)	Cost per ton
Enteric fermentation	possibly 10%?	0.7 Mt?	unknown at present
Livestock waste	40%	1.3 Mt	low to medium
Landfill - existing policies	45% (by 2010)	3.4 Mt	expected to be low
+ increased recovery and use of LF gas	additional 15%	1.2 Mt	low
+ diversion and optimization of oxidation	additional 15%	1.1 Mt	low for oxidation; could be high for diversion policies
Gas - process vents and flares	40%	0.01 Mt	low to medium
Gas - pipeline emissions	45%	0.6 Mt	high for majority of savings
Decline in coal production	50% (by 2010)	1.3 Mt	

Acknowledgements - The author wishes to express her gratitude to several colleagues (Martin Meadows, Ian Marlowe, Alison Moore, Katie King, Allan Goode, Jacquie Berry) for their contributions and comments on this paper.

REFERENCES

1. Climate Change 1995 the Science of Climate Change, Cambridge University Press 1996.
2. CORINAIR 90, European Environment Agency.
3. UK Greenhouse Gas Emission Inventory, 1990 to 1994, Annual Report for Submission under the Framework Convention on Climate Change (1996), A.G. Salway, AEA Technology/UK Department of Environment.
4. Inventory of Technologies, Methods, and Practices for Reducing Emissions of Greenhouse Gases, A Report prepared for Working group II of the IPCC, Third External Review Draft, September 1995.
5. Personal Communication, Alison Moore, ETSU.

6. Cost Effectiveness of Emission-Reducing Measures for Methane in The Netherlands, D. de Jaeger and K. Blok, Energy Conversion and Management, vol. 37, Nos.6-8 pp. 1181-1186, 1996.
7. Directive 75/442/EEC, amended by Directive 91/156/EEC, Articles 1 to 12.
8. Methane Emissions from Land Disposal of Solid Wastes Draft Final Report produces for the International Energy Agency Greenhouse Gas R&D Programma, ETSU, 1996.
9. Whalen, S.C., Reeurgh, W.S. and Sandbeck, K.A. (1990), Rapid Methane Oxidation in Landfill Cover Soil. Applied and Environmental Microbiology 56, pp.3405-3411.
10. Knightley, D., Nedwell, D.B. and Cooper, M. (1995). Capacity of methane oxidation in landfill cover soils measured in laboratory-scale microcosms. applied and Environmental Microbiology 61, pp.592-601.
11. Knightley, D. and Nedwell, D.B. (1995). Methane oxidation in landfill cover soils: can bacteria solve the emission problem? Environmental Managers Journal, 3, pp. 24-26.
12. EUROSTAT, 1996. Renewable Energy Source Statistics 1989-1994. EUROSTAT, Luxembourg.
13. Work carried out by AEA Technology in support of the UK Department of the Environment.
14. Methane Emissions from Coal Mining, Report Number PH2/5 (1996) IEA Greenhouse Gas R&D Programme.
15. Energy statistics of OECD Countries - 1990-1991 and 1992-1993, IEA/OECD, Paris.
16. Digest of UK Energy Statistics 1994, HMSO, London.
17. National Submissions to Framework convention on Climate Change.
18. European Energy to 2020: A Scenario Approach, Energy in Europe, Special Issue, Spring 1996, DGXVII.
19. Emissions of Methane by the Oil and Gas System, Report Number PH2/7, January 1997, IEA Greenhouse Gas R&D Programme.
20. Climate Change - the UK Programme UK's Second report under the Framework convention on Climate Change Cm 3558 (1997), HMSO, London.
21. Strategy Paper for Reducing Methane Emissions (Communication from the Commission to the Council and the European Parliament) COM(96) 557 final.
22. Proposal for a Council Directive on the Landfill of Waste, COM(97) 105 final.
23. CITEPA (1994). Strategies for Limiting Methane, B. Oudart, Final Report Contract No. B92/4 - 3040/16177.

A Review of Nitrous Oxide Behavior in the Atmosphere, and in Combustion and Industrial Systems

William P. Linak
Air Pollution Technology Branch, MD-65; Air Pollution Prevention and Control Division;
National Risk Management Research Laboratory; U.S. Environmental Protection Agency; Research Triangle Park, North Carolina 27711 USA

John C. Kramlich
Department of Mechanical Engineering, FU-10; University of Washington, Seattle, Washington 98195 USA

ABSTRACT

Tropospheric measurements show that nitrous oxide (N_2O) concentrations are increasing over time. This demonstrates the existence of one or more significant anthropogenic sources, a fact that has generated considerable research interest for several years. The debate has principally focused on (1) the identity of the sources, and (2) the consequences of increased N_2O concentrations. Both questions remain open, to at least some degree.

The environmental concerns stem from the suggestion that diffusion of additional N_2O into the stratosphere can result in increased ozone (O_3) depletion. Within the stratosphere, N_2O undergoes photolysis and reacts with oxygen atoms to yield some nitric oxide (NO). This enters into the well known O_3 destruction cycle. N_2O is also a potent absorber of infrared radiation and can contribute to global warming through the greenhouse effect.

In combustion, the homogeneous reactions leading to N_2O are principally $NCO + NO \rightarrow N_2O + CO$ and $NH + NO \rightarrow N_2O + H$, with the first reaction being the more important in practical combustion systems. During high-temperature combustion, N_2O forms early in the flame if fuel nitrogen is available. The high temperatures, however, ensure that little of this escapes, and emissions from most conventional combustion systems are quite low. The exception is combustion under moderate temperature conditions, where the N_2O is formed from fuel nitrogen, but fails to be destroyed. The two principal examples are combustion in fluidized beds, and in applications of nitrogen oxide (NO_x) control by the downstream injection of nitrogen-containing agents (*e.g.,* selective non-catalytic reduction with urea). There remains considerable debate on the degree to which homogeneous vs. heterogeneous reactions contribute to N_2O formation in fluidized bed combustion. What is clear is that the N_2O yield is inversely correlated with bed temperature, and conversion of fuel nitrogen to N_2O is favored for higher-rank fuels.

Formation of N_2O during NO_x control processes has been confined primarily to selective non-catalytic reduction. Specifically, when the nitrogen-containing agents urea and cyanuric acid are injected, a significant portion (typically > 10%) of the NO that is reduced is converted into N_2O. The use of promoters to reduce the optimum injection temperature appears to increase the fraction of NO converted into N_2O. Other operations, such as air staging and reburning, do not appear to be significant N_2O producers. In selective catalytic reduction, the yield of N_2O depends on both catalyst type and operating condition, although most systems are not large emitters.

Other systems considered include mobile sources, waste incineration, and industrial sources. In waste incineration, the combustion of sewage sludge yields very high N_2O emissions. This appears to be due to the very high nitrogen content of the fuel and the low combustion temperatures. Many industrial systems are largely uncharacterized with respect to N_2O emissions. Adipic acid manufacture is known to produce large amounts of N_2O as a byproduct, and abatement procedures are under development within the industry.

INTRODUCTION

Nitrous oxide (N_2O) was long neglected as a pollutant species in comparison to the attention given to the other nitrogen oxides (NO_x). Unlike other NO_x species, N_2O is known to be extremely inert in the troposphere. This inertness suggested that little environmental consequence was associated with N_2O emissions, so it was relatively easy to consider it a non-pollutant.

In a more practical vein, the difficulty involved in obtaining N_2O measurements probably contributed to its neglect for many years. At times when NO_x were being measured routinely on many sources, N_2O was still a specialty measurement requiring return of batch samples to the laboratory, followed by a difficult gas chromatographic analysis.

Interest in N_2O emissions was largely started by atmospheric chemists, who observed that the tropospheric concentration was increasing with time at a rate of approximately 0.25%/yr. This increase suggested the existence of at least one unknown, substantial anthropogenic source. It also triggered interest in the consequences of this increased tropospheric N_2O burden.

Examination of the global N_2O budget shows that, while its source (natural and anthropogenic) is through ground level emissions into the troposphere, its primary sink occurs through diffusion to the stratosphere. Here, the N_2O is finally destroyed by either photolysis or reaction with singlet oxygen atoms. The result is that a portion of the N_2O is converted into nitric oxide (NO), which enters the ozone (O_3) destruction cycle. Thus, increased tropospheric N_2O concentrations can lead to increased O_3 removal rates. (It is critical to remember that considerable N_2O is made naturally, and this represents a major contribution to *natural* O_3 destruction in the stratosphere. The concern is that increased anthropogenic N_2O will accelerate this natural rate. This differs from the chlorofluorocarbon (CFC) problem where the natural tropospheric concentrations of CFC's are zero.)

In addition to its impact on stratospheric O_3, N_2O contributes to global warming. The N_2O molecule is a strong absorber of infrared radiation at wavelengths where carbon dioxide (CO_2) is transparent. Although the concentration of N_2O is much less than that of CO_2, it is a much stronger absorber on a molecule-by-molecule basis. This suggests that increased N_2O concentrations in the troposphere could lead to more retention of long wavelength radiation emitting from the surface of the Earth.

The search for the anthropogenic sources has concentrated on (1) industrial processes that may emit globally significant quantities of N_2O, and (2) biological processes that may produce N_2O on a widespread basis. Although there has been extensive work in both of these areas, the work has been hampered by measurement difficulties. Nonetheless, a substantially improved picture of the global N_2O budget has emerged. Based on this understanding, steps are being taken to modify the processes that generate anthropogenic N_2O.

Certain features have come to be recognized as contributing to N_2O emissions from combustion systems. First and foremost is the oxidation of fuel nitrogen under relatively low-temperature conditions. This allows N_2O to form, and to avoid subsequent destruction. Thus, any system in which nitrogen in a combined form is oxidized under relatively low temperatures can lead to N_2O emissions. Practical examples include combustion fluidized beds, and NO_x control processes that involve the downstream injection of nitrogen -containing compounds, such as urea. In most combustion systems, however, the flame temperature is sufficiently high that any N_2O formed in the flame zone is destroyed before the gases are emitted. Thus, most combustion systems do not emit much N_2O.

This paper describes atmospheric behavior and the behavior of N_2O in combustion systems including pulverized coal and fluid bed combustion systems, and thermal waste remediation. The paper concludes with a review of information on N_2O behavior during NO_x control activities, and N_2O from both mobile, and other industrial sources. Notably missing is any discussion of N_2O from biological activities. Although this is an important component of the global N_2O budget, it falls outside of the scope of the paper.

1. ATMOSPHERIC CHEMISTRY AND ENVIRONMENTAL CONSEQUENCES

Role of N_2O in Global Warming

With an effective surface temperature of approximately 6000 K, most of the sun's radiation is emitted within a spectral range of 100 to 3000 nm. These wavelengths include the visible and portions of the ultraviolet and infrared spectra. The Earth's atmosphere is transparent to most of this incident radiation and, as this radiation reaches the Earth, it is either reflected back to space or absorbed to heat the surface. To maintain constant temperatures, heat gained by the sun's incident radiation must be balanced by heat losses through re-radiation. With an average temperature of approximately 300 K, the Earth emits most of its radiation at infrared wavelengths above 3000 nm. Unlike incident solar radiation, the Earth's atmosphere is not entirely transparent to outgoing infrared radiation. Atmospheric gases such as water (H_2O), CO_2, methane (CH_4), N_2O, O_3, and more than a dozen synthetic gases such as CFC's and chlorinated solvents, absorb the Earth's radiation. These gases then re-emit this energy. A portion is radiated toward space at cooler atmospheric temperatures, and another portion is radiated back to Earth's surface where it results in additional surface heating. The net result is increased surface temperatures. This 'greenhouse' effect is necessary for the existence of life on Earth, and accounts for a temperature enhancement from 253 K (-4°F), the calculated average surface temperature without a greenhouse effect, to 288 K (59°F), the Earth's current average temperature.[1,2] Without the greenhouse effect, the Earth would be covered with ice.

While H_2O vapor absorbs radiation across the entire spectrum, other predominant greenhouse gases absorb radiation in distinct bands. These absorption bands are for CO_2 (13000 to 17000 nm), CH_4 (7000 to 8000 nm), N_2O (8000 to 8500 nm), and O_3 (9000 to 10000 nm). In the pre-industrial atmosphere, nearly 80% of the radiation emitted by the Earth was in the spectral range of 7000 to 13000 nm. This region was referred to as the 'window' because of its relative transparency to outgoing radiation.[1] However, as the natural balances of these atmospheric gases are changed (*i.e.*, increased) through human activities, and as previously unknown synthetic greenhouse species, such as CFC's, and chlorinated solvents, are introduced into the atmosphere, this balance is upset, resulting in increased absorption in the CO_2, CH_4, N_2O, and O_3 bands, and new absorption by synthetic gases which absorb strongly in the window region. This increased absorption decreases the Earth's heat loss, and increases the greenhouse effect and net global warming. Since surface temperatures also drive climate and hydrological cycles, the excess energy now available powers changes to weather and precipitation patterns. These effects cannot be easily predicted, however, because of the nonlinear interactions involved.[1]

Table 1 summarizes key greenhouse gas concentrations in the atmosphere.[1,2,3] Comparison is made between pre-industrial concentrations, determined through polar ice core analysis, and current ambient concentrations. The data indicate a trend of increasing atmospheric concentrations for these species. Ramanathan[1] suggests that atmospheric increases in CH_4 and carbon monoxide (CO), such as those seen during the past century, may have increased tropospheric O_3 concentrations by 20%. Levine[4] identifies the relative contribution of several atmospheric gases to global warming: CO_2 (49%), CH_4 (18%), CFC's (Refrigerant-11 and -12) (14%), N_2O (6%), and other trace gases (13%).

Role of N_2O in Stratospheric O_3 Chemistry

Approximately 85% of the Earth's atmosphere, including almost all of its H_2O vapor, is associated with the troposphere, which extends from the surface to about 15 km. The remaining 15% is associated with the stratosphere, which extends from approximately 15 to 50 km above the Earth's surface. Over 90% of the atmospheric O_3 resides in the stratosphere. Stratospheric O_3 shields the earth from biologically lethal ultraviolet (UV) radiation (wavelengths below 310 nm). Most importantly, stratospheric O_3 shields the earth from UV-B radiation with incident wavelengths from 280 to 310 nm. These wavelengths are especially harmful because they lie in a regime where the solar spectrum and DNA (biological) susceptibility overlap.[5]

In 1929, Chapman[6] identified a "classical" mechanism to describe the formation and destruction of stratospheric O_3. According to this mechanism, the chemical production of O_3 is initiated by the photodissociation of molecular oxygen by solar radiation with wavelengths of 242.3 nm or less:

$$O_2 + h\nu \rightarrow O + O \qquad l \leq 242.3 \text{ nm} \qquad (1)$$

Once dissociated, atomic oxygen may combine with molecular oxygen (O_2) and a third body, M [usually nitrogen or O_2], to form O_3:

$$O + O_2 + M \rightarrow O_3 + M \qquad (2)$$

O_3 destruction can also occur through photodissociation:

$$O_3 + h\nu \rightarrow O + O_2 \quad (3)$$

or through reaction with atomic oxygen:

$$O_3 + O \rightarrow 2O_2 \quad (4)$$

At the time, these reactions were thought to fully describe the global stratospheric O_3 balance. However, over the past 20 to 30 years three other destruction routes were discovered involving reactions with hydroxyl radical (OH):

$$OH + O_3 \rightarrow HO_2 + O_2 \quad (5)$$

$$HO_2 + O_3 \rightarrow OH + 2O_2 \quad (6)$$

$$HO_2 + O \rightarrow OH + O_2 \quad (7)$$

chlorine (Cl):

$$Cl + O_3 \rightarrow ClO + O_2 \quad (8)$$

$$ClO + O \rightarrow Cl + O_2 \quad (9)$$

and NO:

$$NO + O_3 \rightarrow NO_2 + O_2 \quad (10)$$

$$NO_2 + O \rightarrow NO + O_2 \quad (11)$$

Most importantly, these three mechanisms are catalytic in nature, resulting in the destruction of O_3, without the net destruction of the OH, Cl, or NO reactant. Thus, these species are recycled and remain available for numerous O_3 destruction steps.

<u>N_2O/NO Chemistry</u>

The stratospheric formation of NO is the result the photolysis of N_2O and reaction with excited singlet-D oxygen, $O(^1D)$, via the reaction set:

$$N_2O + h\nu \rightarrow N_2 + O(^1D) \quad (12)$$

$$N_2O + O(^1D) \rightarrow 2NO \quad (13)$$

$$N_2O + O(^1D) \rightarrow N_2 + O_2 \quad (14)$$

Levine[2] identifies the photolysis reaction (12) as being responsible for approximately 90% of the N_2O destruction, while Reactions 13 and 14 each accounts for about 5% of its destruction. With an atmospheric lifetime of approximately 150 years, N_2O is extremely long-lived. N_2O is also very stable in the troposphere. Its destruction takes place only after its diffusion into the stratosphere.[2] Reaction 13 leads to the production of stratospheric NO and to the subsequent chemical destruction of stratospheric O_3

through the reaction set described by Reactions 10 and 11, with the net result:

$$O_3 + O \rightarrow 2O_2 \qquad (4)$$

Note that NO is not destroyed during this mechanism, and is recycled for further reaction with O_3. Levine[2] identifies this catalytic NO cycle as responsible for about 70% of the global chemical destruction of stratospheric O_3. However, a large portion of this contribution is from natural sources of N_2O. With the exception of relatively minor emissions from high flying aircraft, Reaction 13 is the only known source of stratospheric NO. Other NO_x species released into the troposphere as a consequence of combustion and other industrial activities are quickly removed and do not have the atmospheric lifetimes necessary to reach the stratosphere. Table 2 summarizes the atmospheric concentrations of major gases, selected trace gases (including OH, CFC's, and NO), and trace nitrogen species compiled by Levine.[2,7] It should be noted that N_2O is the second most abundant nitrogen species in the atmosphere after molecular nitrogen.

Tropospheric Measurements of N_2O

Since the mid 1970's, systematic tropospheric measurements of N_2O have been made at locations worldwide.[8,9,10,11] These data, summarized by Khalil and Rasmussen[12] and presented in Figure 1, show the atmospheric concentration of N_2O to be currently increasing at an average rate of approximately 0.80 ± 0.02 ppbv per year or approximately 0.27 ± 0.01% per year .[12] Based on these data, current atmospheric concentrations are estimated to be 313.7 ppbv. It is also interesting to note that Weiss[8] has determined that N_2O is unequally distributed between the northern and southern hemispheres, with the northern hemisphere higher by 0.83 ± 0.15 ppbv at the time that those measurements were made (1976-1980). This may be indicative of the larger land mass or larger population centers and industrial activity in the northern hemisphere.

Data by Pearman *et al.*,[13] Khalil and Rasmussen,[12,14] Ethridge *et al.*,[15] and Zardini *et al.*[16] examining N_2O concentrations in air bubbles trapped in polar ice core samples suggest that this temporal increase is a relatively recent phenomenon. In polar regions, where yearly snow falls do not melt, air associated with the snow is trapped in tiny bubbles as subsequent accumulation and pressure convert older snow to ice. Analysis of the air within these bubbles can yield information concerning the composition of gases in the atmosphere hundreds, thousands, or tens of thousands of years ago. These data, summarized for the past 1000 years by Khalil and Rasmussen[12] and presented in Figure 2, include a composite set of measurements from 0 to 1820 A.D.,[14] 1600 to 1966 A.D.,[15] 1600 to 1900 A.D.,[16] and 1800 to 1900 A.D.[12] Understandably, these data are subject to much more uncertainty compared to the precise measurements of the past 15 years. Evident from Figure 2 is the absence of any significant trend between 0 and 1500 A.D. At that time, however, concentrations were seen to suddenly drop and then rise again. Khalil and Rasmussen[12] suggest that this phenomenon was the result of the 'little ice age' which reportedly occurred at this time and which may have resulted in reduced biological activity. For the period 1880 to 1960, the trend shows a steady increase of 0.07 ± 0.01 ppbv per year. For comparison, the most recent atmospheric data (1976-1988) (see Figure 1 and Table 1) have also been included, and show an even more expanded rate of increase (0.08 ± 0.02 ppbv). Khalil and Rasmussen[12] point out that the ice core data are often imprecise, ambiguous, and subject to potential errors. Moreover, the most recent ice core data (late 1800's - early 1900's) are subject to even greater uncertainties due to problems associated with resolving air bubble formation in ice over short time intervals. The use of ice core data to resolve historical trends of greenhouse gases is reviewed by Raynaud *et al.*[17]

Tropospheric N_2O Balance

Major uncertainties exist concerning the identification and apportionment of the global sources of N_2O. It is known, however, that these global sources must balance the global rate of atmospheric destruction plus the rate of atmospheric accumulation. Rate data suggest that Reactions 12, 13, and 14 destroy approximately 10.5 ± 3.0 teragrams of nitrogen (in the form of N_2O) per year (Tg N per year). Also, the atmospheric accumulation described above requires the production of another 3.5 ± 0.5 Tg N per year. As a result, total global production of N_2O must be approximately 14 ± 3.5 Tg N per year to balance these destruction and accumulation terms.[2]

Table 3 presents estimates of the global sources of N_2O published by several groups in recent years.[2,12,18,19,20,21,22,23] The contributions from natural and anthropogenic sources have been grouped separately for comparison. Natural sources include nitrification and denitrification of nitrogen species in soils and oceans. Denitrification involves the chemical transformation of soil nitrate (NO_3^-) to molecular nitrogen (N_2) and N_2O. Almost all of the N_2O produced by denitrification escapes into the atmosphere. Nitrification involves the oxidation of reduced soil nitrogen species [such as ammonium (NH_4^+)] to nitrite (NO_2^-) and NO_3^- with N_2O as an intermediate product. These same processes are responsible for N_2O production in oceans. However, it is uncertain whether denitrification in oxygen-deficient deep waters or nitrification in oxygen-rich surface waters is responsible.[2] Table 3 indicates general agreement among the four recent studies which have presented natural source data. The single exception is the soils source published by Khalil and Rasmussen [12] which is notably higher than the other estimates.

Table 3 suggests that the global estimates of anthropogenic sources are highly variable, with values ranging from 1.0 to 9.7 Tg N per year. In addition, the different studies often do not include the same set of sources in their anthropogenic estimates. For example, Khalil and Rasmussen [12] include a sewage source that the others do not. IPCC[20] also includes estimates from adipic and nitric acid production which are notably significant. Several data sets[12,18,23] also suggest indirect N_2O formation through atmospheric transformations from NO_x precursors, or heterogeneous mechanisms involving atmospheric nitrates, although these sources are not well quantified. Of particular interest is the potential source identified by Khalil and Rasmussen, [12] which identifies climatic feedback and accelerated biogenic activity from CO_2 increases and global warming as being responsible for approximately 0.2 Tg N per year. This estimate was taken from ice core data and N_2O trends seen during the 'little ice age'. Evident from Table 3 is that not all of the global sources of N_2O have been identified, and that those that have been identified are subject to large error as indicated by the large range of estimates presented. However, if we neglect values presented by de Soete[18] who summarized anthropogenic sources only, we see that, while the magnitude of total global sources ranges widely (5.2 to 19.2 Tg N per year), they bracket the sum of the destruction and accumulation terms determined independently (14 ± 3.5 Tg N per year).

Table 3 indicates that N_2O emissions from fossil fuel combustion sources contribute a relatively small portion of the total anthropogenic source. However, this was not always believed to be true. Only recently, fossil fuel combustion, especially coal combustion, was believed to be the major contributor to the measured increases in ambient N_2O concentrations. These increases also seemed to track measured increases in ambient CO_2 concentrations. Previous research[24] presented data indicating direct N_2O emissions from coal combustion exceeding 100 ppm, and an approximate average N_2O-N:NO_x molar ratio of 0.58:1. These data seemed to confirm earlier suggestions[25,26] that combustion of fossil fuels (and coal in particular) represented a dominant factor in the observed increase of N_2O. In addition, emissions factors generated using N_2O stack concentrations of 100 to 200 ppm were adequate to close the global anthropogenic mass balance.

Additional combustion measurements, gathered by a number of research groups, however, did not always confirm the early results. These numerous studies often used various N_2O sampling and analytical methodologies including samples measured on-line and samples extracted into containers for subsequent analysis in a laboratory environment. An explanation for the resulting growing scatter in the data was proposed by Muzio and Kramlich[27] who suggested the presence off a N_2O sampling artifact. They presented evidence that indicated that N_2O was produced in sampling containers awaiting analysis. They further hypothesized a mechanism for this formation involving NO, SO_2, and H_2O. This evidence questioned the validity of all existing data which involved container sampling. Additionally, since secondary reactions converting NO to N_2O in the sample containers were found to occur easily at room temperature,[27,28,29,30,31,32] a new indirect relationship between anthropogenic NO emissions (including combustion) and global N_2O increases was suggested (see Table 3).[12,18] Although not yet characterized or quantified, this may include NO conversion in plumes, in the troposphere, or on surfaces.[33]

Since the discovery of the sampling artifact, new research has sought to characterize N_2O emissions from fossil fuel combustion sources, and determine the effect that modifications used to control NO_x has on N_2O emissions. For conventional stationary combustion sources (including coal combustion), recent measurements indicate average N_2O emissions less than 5 ppm. These values are more than a factor of 20 times less than emissions believed to be produced several years ago. Interestingly, fluidized bed combustors, and several of the thermal DeNO$_x$ and catalytic processes developed for NO_x

control, also seem to contribute to increased levels of N_2O. At present, these technologies are not in widespread application, and the associated increases in N_2O emissions do not add significantly to the global flux. However, as these technologies are further developed, and their use becomes more common, they have the potential of affecting global emissions. Several of these technologies will be discussed further in later sections.

3. N_2O IN GAS-FIRED COMBUSTION

Kinetic data show that N_2O can be formed as an intermediate in the combustion of fuel nitrogen. Once the fuel nitrogen is consumed, however, fast destruction reactions ensure that little escapes the flame. Higher N_2O emissions can occur only if the flame is quenched, or the fuel nitrogen is introduced downstream of the flame zone. In both cases, N_2O destruction reactions are diminished.

Although the qualitative features of the flame behavior of N_2O are understood, quantitative prediction remains uncertain. In particular, recent measurements of the rate of the critical NCO + NO reaction indicate that (1) its rate is lower at high temperatures than previously thought, and (2) potentially, only a portion of the reaction branches to N_2O. In addition, recent measurements also suggest that the destruction reaction N_2O + OH is approximately 10 times slower than the rate used in most modeling studies. Since most kinetic modeling has used both the earlier, higher NCO + NO rate with a branching ratio into N_2O of unity, and the higher rate for N_2O + NO, these findings have the potential to upset current views of N_2O formation from cyano species.

Elementary Reactions of N_2O Relevant to Gas Flames

Reactions thought to be important to the behavior of N_2O in combustion systems include:

$$N_2O + M \rightarrow N_2 + O + M \tag{15}$$

$$N_2O + O \rightarrow N_2 + O_2 \tag{16}$$

$$\rightarrow NO + NO \tag{17}$$

$$N_2O + H \rightarrow N_2 + OH \tag{18}$$

$$N_2O + OH \rightarrow N_2 + HO_2 \quad (19)$$

$$NCO + NO \rightarrow N_2O + CO \quad (20)$$

$$\rightarrow N_2 + CO_2 \quad (21)$$

$$\rightarrow N_2 + CO + O \quad (22)$$

$$NH + NO \rightarrow N_2O + H \quad (23)$$

The reactions governing NCO concentrations are also important to the problem. These include reactions forming NCO, and alternate destruction pathways that compete with Reaction 20 for available NCO. Rates for these reactions must be carefully selected with an awareness that the understanding is rapidly evolving. In contrast, the reactions and rates governing the concentration of the other precursor, NH, are more firmly established.

Reaction of Fuel Nitrogen

As noted above, the addition of fuel nitrogen to laminar flames leads to the appearance of N_2O as an intermediate. Significant emissions occur, however, only for low temperature flames (obtained either through a low adiabatic flame temperature or through high heat extraction at the burner). Also, addition of fuel nitrogen to large turbulent flames yields no more N_2O in the exhaust than if no fixed nitrogen was included with the fuel.[34,35] Modeling efforts suggested that Reaction 18 is sufficiently fast to be capable of removing all the N_2O formed in the flame zones, even if the N_2O formation rate is artificially augmented by unrealistically rapid char production rates. If N_2O is to be emitted from any fossil fuel systems, what could be the source?

One clue is provided by early data on the oxidative pyrolysis of fuel-nitrogen compounds. Researchers have recognized that, although nitrogen is bound into complex organic structures in coal, it appears in devolatilization products almost exclusively as HCN and NH_3.[36,37,38] It is now known that the HCN and NH_3 arise from secondary tar cracking reactions after primary devolatilization, although the exact mechanism remains an area of research. Early studies on the pyrolysis[39] and oxidative pyrolysis[40,41,42,43] of model fuel-nitrogen compounds showed that, under certain oxidizing, moderate-temperature regimes, compounds such as cyanogen, pyridine, and HCN could yield large amounts of N_2O.

Kramlich *et al.*[34] were able to generate large exhaust concentrations of N_2O in a tunnel furnace by the downstream injection of cyano species. The primary natural gas flame of this furnace was designed to generate 600 ppm NO. The post-flame gases were quenched by heat extraction at a rate of 350 K/s. A side-stream injector was used to introduce HCN into the furnace at various locations and temperatures. The results, shown in Figure 3, indicate that at between 1100 and 1500 K a significant fraction of the HCN was converted to N_2O. Similar results were obtained for an acetonitrile spray, with the maximum conversion temperature offset somewhat due to the time required for the evaporation of the spray. NH_3, however, generated very little N_2O under these conditions.

Application of a plug-flow reaction kinetics model (with distributed side-stream addition) to these results reproduced the major features of Figure 3: a peak in N_2O emissions as a function of injection temperature, and the lack of N_2O production from NH_3. Sensitivity analysis showed that the N_2O behavior is governed by:

$$HCN + O \rightarrow NCO + H \qquad (24)$$

$$NCO + NO \rightarrow N_2O + CO \qquad (20)$$

$$NH + NO \rightarrow N_2O + H \qquad (23)$$

$$N_2O + H \rightarrow N_2 + OH \qquad (18)$$

Above the favorable temperature window, N_2O removal via Reaction 18 was rapid, and alternate pathways for the oxidation of HCN and its intermediates were opened. At lower temperatures, HCN failed to react within the time available. In the case of NH_3 injection, competing oxidation reactions within the temperature window prevented Reaction 23 from generating significant N_2O. One feature of the model was the overprediction of the conversion of HCN to N_2O. Use of a recommended reduced branching ratio[44] of 40% was entered into the model, which resulted in much more realistic N_2O predictions. This lends global support to the fractional branching ratios recently reported.[45,46]

For such a mechanism to explain N_2O emissions in pulverized coal flames, a means of transporting volatile HCN to these cooler environments must be proposed. While late devolatilization or turbulent mixing limitations could provide some HCN within the appropriate temperature window, neither of these is likely to act as a major source in practical systems. This is consistent with the low N_2O emissions reported from oil- and coal-fired furnaces, as will be discussed shortly.

Industrial Gas Flame Data

The emission of N_2O from industrial gas flames has always been found to be quite low. Figure 4 provides a compilation of data from large-scale, turbulent gas flames.[24,25,32,34,35,47,48,49,50,51,52] The left-hand half of the plot compares NO_x and N_2O data, while data in the right-hand panel did not have accompanying NO_x data. In general, emissions are so low as to be of little environmental consequence when it is remembered that the atmosphere contains approximately 0.3 ppm N_2O. Two of these data were obtained with NH_3 doping into the fuel,[34,35] but this failed to generate significant N_2O. The highest concentrations noted in Figure 4 are associated with (1) a Swedish home heating furnace,[49] (2) early data by Hao *et al.*,[24] and (3) early data by Pierotti and Rasmussen.[25]

The home heating furnace likely involves lower overall flame temperatures and greater opportunity for quench of the flame. There is ample evidence that the quench of flames containing fuel nitrogen will produce higher levels of N_2O,[53] although it has not been shown that this is true when fixed nitrogen species are absent from the fuel.

The data of Hao *et al.*[24] were obtained by collection in 50 cm^3 syringes that had an opportunity to age before analysis. Since SO_2 was not present in these gas flames to a significant extent, it is unlikely that a large amount of N_2O was generated in the sample.[27,29,30] However, there is evidence that the formation of N_2O within sample containers can proceed to a limited extent, and at a slower rate, in the absence of one of the necessary ingredients: SO_2 or moisture.[28] Although this question cannot be conclusively settled in retrospect, the large quantity of recent data from Austria[50] and Japan[51,54] suggest that most industrial gas equipment does not produce more than 2 ppm N_2O. Reported concentrations in excess of these values must be carefully examined to determine if special combustion conditions exist which give rise to emissions above the anticipated level.

4. N_2O BEHAVIOR IN OIL AND PULVERIZED COAL FLAMES

Pulverized coal flames were among the first industrial combustion sources to be characterized for N_2O emissions. The widespread use of pulverized coal, coupled with the relatively high amounts of fuel nitrogen contained in the fuel, suggested that this class of sources was worthy of attention.

The initial measurements on utility boilers were reported in 1976.[25,26] This was followed by a limited number of studies through 1986-87. It was later found that large amounts of N_2O could be formed within the flasks that were used for sample storage in many studies. Subsequent studies either modified the storage technique to avoid N_2O formation, or used on-line methods where the samples had little time to age. These later studies have shown very low emissions from pulverized-coal- or oil-fired units, generally less than 5 ppm. There have been, however, several studies of time-resolved N_2O behavior in coal flames, which indicate that much higher N_2O concentrations exist early in the flame. These studies are valuable for their mechanistic insight.

Early Coal Studies

Pierotti and Rasmussen[25] reported three measurements (32.7, 32.8, and 37.6 ppm) from a pulverized-coal-fired university power plant. They used 6 liter, electropolished stainless-steel flasks for their samples, and the subsequent analysis was by gas chromatograph/electron capture detection (GC/ECD). Nominally, these samples would have been subject to the sampling artifact, although the concentrations measured were much lower that those reported in later studies. One reason for this might have been the fact that the source and the analytical laboratory were in close proximity, and the time between sampling and analysis may have been short. Alternatively, the stainless-steel surface may have moderated the pH of the material absorbed on the walls and helped to slow the conversion of NO to N_2O.

Weiss and Craig[26] used 2 liter Pyrex sample flasks to collect stack gas from the Mohave coal-fired station in Nevada. These were preconcentrated by freezing in a liquid-nitrogen bath. The concentrated sample was then evaporated and dried. These concentrated gases (containing mainly N_2O and CO_2) were analyzed by an ultrasonic phase-shift detector.

Thus, these samples would have also been subject to in-container N_2O formation. The reported N_2O emission was 25.8 ppm, which is also low compared to subsequent studies where the sampling artifact was active.

Kramlich *et al.*[35] measured N_2O emissions from a small-scale coal-fired tunnel furnace. The goal was to identify whether the use of air staging for NO_x control would lead to enhanced N_2O emissions. These samples were obtained by an on-line preconcentration procedure similar to that of Weiss and Craig (*i.e.,* the samples withdrawn from the reactor were immediately frozen in a liquid nitrogen trap and then evaporated, dried, and analyzed on-site by thermal conductivity gas chromatography). Although the results were insensitive to the stoichiometry of the rich zone, increased reactor cooling increased N_2O emissions, and premixed coal flames yielded more N_2O than diffusion coal flames. Although the elevated N_2O emissions (20-90 ppm) could be due to sample system chemistry, it is possible that the reduced combustion temperatures in this small facility caused more of the volatile HCN to oxidize in the 1200-1500 K window where N_2O emissions can be formed.[34]

Castaldini *et al.*[55] examined a series of sources associated with NO_x abatement operations. Hao *et al.*[24] surveyed nine oil-fired and three coal-fired furnaces in the northeastern United States, as well as measurements on a pilot-scale furnace.[56] For both Castaldini and Hao a correlation was noted in which the apparent N_2O emissions (obtained using sample flasks and remote analysis) were about 25% of the NO_x emissions measured on-line at the site. This apparent proportionality was due to the conversion of NO_x within the sample flasks to N_2O at an approximate 4:1 ratio. Following the identification of this artifact, both measurement methods development and retesting of combustion sources have been priority activities.

Recent Database on Oil and Pulverized Coal Emissions

Extensive recent surveys of N_2O emissions from pulverized-coal- and oil-fired furnaces have been reported. Table 4 summarizes sources of these data.[23,32,47,50,51,54,57,58,59,60,61] It is clear from reviewing this very large database that the majority of the measurements from industrial combustion systems yield very little N_2O. Emissions in excess of 5 ppm are very rare, with the exception of combustion fluidized beds, which are discussed in the next section. One general trend is that higher emissions tend to be associated with oil-fired units, although the emission levels are still quite low. The reason for this is not known.

At these very low emission levels, the care taken in measurement becomes critical. Simple drying or SO_2 scrubbing will prevent the formation of large amounts of N_2O. However, measurement studies show that residual moisture, SO_2, and/or long sample lines will still allow a few ppm of N_2O to form. Thus, the preponderance of the data suggest very low values, and the few outliers warrant close examination to determine if unusual combustion conditions exist, or if sampling procedures were adequate.

5. COMBUSTION FLUIDIZED BEDS

Among fossil fueled combustion systems, combustion fluidized beds have consistently shown the highest N_2O emissions in field measurements. Although this unwelcome finding has become widely recognized only during the last several years, it has spawned a large research effort. The work has focused on (1) formation mechanisms, (2) emissions as a function of combustion parameters, and (3) control strategies.

Field Data

Field measurements on various full-scale fluidized bed systems have been reported. Table 5 summarizes some of the features of these measurements.[47,49,50,59,60,62,63,64,65,66,67,68] It must be borne in mind that the measurements reported at the third international workshop on N_2O emissions[49] were made before the sample container artifact was discovered, although the vast majority of these measurements were made under procedures that would minimize the in-container generation of N_2O (*i.e.,* most of the samples were dried before storage). The field tests consistently indicate emissions substantially above those found from pulverized coal flames or fuel oil flames.

Attempts to draw general conclusions from the field data are very difficult due to the wide variation in combustor configurations, operating parameters, and fuel types. In spite of this, certain trends are clear from the data. First, N_2O emissions uniformly decrease with increasing bed temperature. At the same time, NO_x emissions increase. In general, lower rank fuels deliver lower N_2O emissions. Most other operating parameters (excess air, addition of limestone) appear to have a weak influence on N_2O. Carefully obtained sub-scale data have been the means of advancing our understanding beyond that available from field data.

Summary of Major Trends

The following presents the major observations reported in the reviews by Mann *et al.*[23] and Hayhurst and Lawrence.[69] First, it cannot be overemphasized how difficult it is to extract mechanistic information from fluid bed experiments. In general, a wide variety of experimental designs and scales have been used to study N_2O formation in fluidized beds. In addition, enough information is frequently not provided to fully rationalize these data.[69] In fluidized beds many of the parameters are coupled in actual experiments (*e.g.*, excess air and bed temperature) which makes it difficult to cleanly extract the influence of a single variable. These observations show the highly empirical nature of the fluidized bed database, and suggest that much care must be taken in data analysis to develop general conclusions. In spite of these difficulties, considerable progress has been made in understanding this complex phenomenon.

The most pronounced trend is that of temperature. Reduced bed temperatures almost universally cause higher N_2O emissions, as the measurements presented on Figure 5 show.[66] Figure 5 also shows that NO increases with bed temperature, suggesting that higher operating temperature is not the best means of controlling N_2O.[70] In addition, higher operating temperatures can cause sintering of added limestone sorbents and reduce SO_2 capture.

The influence of fuel type is less significant than that of temperature. In general, lower rank fuels tend to yield lower N_2O emissions. This has been attributed to the tendency of the lower rank fuels to favor NH_3 release over HCN release.[36,37] The HCN is acknowledged to be more efficiently converted to N_2O.[34,69] Alternately, the higher surface area of lower rank fuels may promote more complete heterogeneous N_2O destruction.[70]

Other correlations have been observed, including an inverse relationship between the O/N ratio of the fuel and N_2O,[23,71]and an increase with the carbon content of the coal.[63]

Although the specific mechanisms underlying these correlations are unclear, both the O/N ratio and coal carbon content can, within limits, be indirect indicators of coal rank. Excess air has been an unusually difficult parameter to distinguish from temperature because they are so closely coupled in fluidized bed operation. In experiments where the temperature effect was removed, higher excess air generally increases N_2O emissions. Mann *et al.*[23] examined the coupling between temperature and excess air, and found that, at higher temperatures, excess air has less of an effect on N_2O emissions, as shown in Figure 6.

Clearly, more experimental work and detailed modeling would help identify the controlling mechanisms in practical fluidized bed combustion. Many results and conclusions are contradictory, and subject to more than one interpretation. Nonetheless, fundamental data have identified two viable mechanisms for N_2O formation in fluidized beds:

- Devolatilization of fuel nitrogen as HCN and NH_3, followed by oxidation to N_2O.
- Oxidation of char nitrogen to NO, followed by the reaction of this NO with char nitrogen to yield N_2O.

On the surface, homogeneous chemistry seems to be capable of explaining many of the major trends. The release of volatile nitrogen as HCN under fluidized bed conditions yields a known N_2O precursor into an environment where N_2O formation is favored. Lower rank coals are known to yield more of their fuel nitrogen as NH_3, which does not convert to N_2O as efficiently under fluidized bed conditions. This tentatively explains the lower emissions observed with lower rank fuels. Alternately, low rank coals may experience greater heterogeneous N_2O reduction due to their different ash composition and morphology. Homogeneous chemistry is capable of reproducing the most prominent characteristic of N_2O behavior in fluidized beds, the decrease in emissions with bed temperature. It does so by showing that the key intermediate, NCO, becomes increasingly converted to NO rather than N_2O as bed temperature increases.

Heterogeneous reactions are also capable of generating N_2O. The reduction of NO at a char surface to yield N_2O does not appear to occur if oxygen is not available to expose fresh char nitrogen.[72] In the presence of oxygen, the effective mechanism appears to be the reaction of NO with exposed char nitrogen to yield N_2O rather than the absorption of

NO on the surface, followed by reaction with a second NO.[73,74]

Extrapolation of these results to practical fluidized beds is more difficult. For example, the data of Tullin *et al.*[73] show that heterogeneous N_2O formation increases with NO doping, a feature used to imply that NO reduction at the char surface is the source of the N_2O. Other data for practical beds show no increase in N_2O with NO doping.[75] This implies that strong N_2O destruction reactions in the bed are active and capable of removing any additional N_2O that may be generated by the reaction of char nitrogen with NO. Since the bed does generate N_2O emissions in spite of this strong reduction reaction, the actual source flux for N_2O must be many times that represented by the emission. The key question is the identity of this source. The hypothesis is that, in a realistically loaded bed, a large amount of volatile nitrogen passes through N_2O as an intermediate. The emission is only a fraction of the total amount of N_2O formed. The yield of N_2O from char nitrogen may be masked by this active volatile chemistry in a realistically loaded bed.

An important message from these data is that both the volatile combustion and char oxidation processes involved in N_2O formation are coupled in systems operating under practical conditions. Experimental systems that seek to decouple the process by moving away from practical conditions (*e.g.,* batch processes, light loading of an otherwise inert bed, use of char instead of coal) are expected to generate sound fundamental data. These results cannot, however, be directly extrapolated to describe trends in full-scale units. To do so requires a fully coupled model that correctly integrates all of the fundamental steps.

6. BEHAVIOR OF N_2O DURING NO_x CONTROL PROCESSES

The goal of NO_x control procedures is to convert NO into N_2 by modifying the combustion environment, introducing a selective agent, or combining a selective agent with a catalyst. Since all of these processes involve nitrogenous intermediates, there is an opportunity for a portion of these intermediates to react and form N_2O. This possibility was recognized early, at least with respect to combustion modifications.[35,55]

This section will review the influence of the major NO_x control technologies on N_2O emissions, including:

- Air staging
- Reburning or fuel staging
- Selective non-catalytic reduction (SNCR), including the Thermal $DeNO_x$ process (NH_3 injection), the urea injection process, the $RapreNO_x$ process (cyanuric acid injection), and the use of advanced or promoted agents
- Selective catalytic reduction (SCR)

In addition, we will briefly review the steps that have been taken to specifically control N_2O from combustion fluidized beds.

Staged Combustion

Air staging generally refers to the division of the combustion air into at least two streams such that the fuel is initially processed through a region of reduced oxygen availability. Under these conditions, conversion of fuel nitrogen to N_2 is improved. The second air stream completes fuel burnout. This basic strategy is executed in a number of ways, including fuel biasing and 'burners out of service'. The low-NO_x burners that are now available from most manufacturers are based on providing staged environments within the burner. The division of the air into two or more streams has been practiced in fluidized beds to create a fuel-rich zone for improved NO_x control.

It is well established that reducing conditions leads to lower N_2O emissions from pulverized coal combustion. Thus, one might suspect that the early formation of N_2O would be reduced during staged combustion. This, however, does not always appear to be the case based on results from brown coal.[76] In spite of the overall fuel-rich conditions in the primary zone, free oxygen still persists early in the flame, and with it, early N_2O.

At the point where the secondary air is added, the fixed-nitrogen species are nominally distributed among NO, HCN, and NH_3. Thus, the oxidation of at least the HCN might be expected to form N_2O. However, the data[76] show no N_2O at the staging point. One probable reason for this is the relatively high temperature, above 1000°C. Another is the concurrent burnout of the CO from the primary zone, which will generate H atoms via

$CO + OH \rightarrow CO_2 + H$. Thus, at the staging point, any N_2O which is formed would likely not survive. Earlier work in a smaller furnace[35] failed to find an influence of staging. The emission levels in this facility were elevated (20-90 ppm) which was attributed to the relatively cool combustion temperature. Another study examining air staging in laboratory-scale furnaces[32] found similar N_2O emissions compared to unstaged operation. This study reported N_2O emissions from the staged combustion of natural gas, No. 2 fuel oil, and No. 5 fuel oil all less than 1 ppm. Coal combustion emissions ranged from 1 to 5 ppm.

Air staging can be approached in circulating fluidized bed combustors (CFBC's) by dividing the air injection location. Mann *et al.*[23] performed a brief examination and found no influence on N_2O emissions. Likewise, Hiltunen *et al.*[68] found no direct influence of air staging beyond that attributable to temperature changes. Jahkola *et al.*[77] found a weak increase in N_2O with staging, while Shimizu *et al.*[78] and Bramer and Valk[79] all saw a concurrent decrease in NO_x and N_2O with increased staging. Hayhurst and Lawrence[69] attribute the latter results to both increasing temperature in the flue gas and the creation of a rich zone at the bottom of the bed. It is clear from the varying results that staging appears to have only a weak influence on N_2O, and its effect is difficult to separate from other parameters.

Reburning

Reburning or fuel staging involves the addition of a second fuel stream after the primary fuel burnout is completed. For example, a low-NO_x burner can be used to complete the burnout of the primary fuel. Secondary fuel is added above these burners to create a moderately fuel-rich zone. Within this zone, radicals generated by the secondary fuel decomposition attack NO to produce N_2, HCN, and NH_3. A final air stream is added to burn out the secondary fuel and convert any remaining reduced nitrogen to NO or N_2. Both coal (containing fuel nitrogen) and natural gas (nitrogen-free) have been proposed as reburning fuels. Process descriptions and development history are available in the literature.[80]

For coal reburning, the fuel is introduced under a reduced temperature compared to a normal industrial coal flame. This suggests that higher conversions of volatile nitrogen to

N_2O may occur than would be normally expected. At present, only very limited measurements have been reported in the open literature. In one subscale study, application of coal reburning to a gas-fired primary flame increased N_2O emissions from less than 1 to 11-13 ppm.[34] Although in terms of concentration this increase appears to be small, it represents a 6.5% conversion of fuel nitrogen to N_2O. This is about an order of magnitude greater than the conversion found in the coal-fired primary flame (0.7%). This is a little surprising because, according to the discussion on air staging, one would expect that the N_2O would be destroyed at the final air staging point. Whether this trend for increased N_2O formation extrapolates to large scale is yet to be seen. In contrast, reburning with natural gas over a coal-fired primary yielded a greater than 50% N_2O reduction.[34]

In their kinetic modeling study, Kilpinen *et al.*[81] studied natural gas reburning. They find no N_2O formation in the fuel-rich zone. If, however, the final air addition temperature is reduced below 1200 K then some of the HCN from the rich zone is irreversibly converted to N_2O. Such an air injection temperature is, however, too low for practical boiler operation since CO burnout times would become unacceptably long. They did not attempt to simulate coal reburning, in which such a temperature is far too low to provide adequate char burnout. They do find that the performance is strongly transport-influenced, so it is difficult to extrapolate the findings to coal. It is, however, clear that gas reburning should be a good tool against primary zone N_2O, and that coal reburning may either form or destroy N_2O, depending on the initial primary zone concentration.

An approach similar to reburning has been attempted in fluidized beds, where natural gas was injected into the cyclone of a CFBC.[67] At substantial firing rates (of the order of 10% of the heat input), N_2O reductions of the order of 50% were achieved, compared with kinetic predictions of 90%. The difference was attributed to the effects of imperfect or finite-rate mixing, possibly with a contribution due to heat loss. Alternately, the recent data of Glarborg *et al.*[82] suggest that the rate of the critical N_2O + OH destruction reaction may be much slower than that used in the model. This may account for the discrepancy.

Selective Non-Catalytic Reduction

Selective non-catalytic reduction (SNCR) had its origins in the observation that NH_3 would selectively react with NO under appropriate temperatures to yield N_2.[83] Many years of work have resulted in an excellent understanding of this process, which is summarized by Miller and Bowman.[84] The following reactions are important:

$$NH_3 + OH \rightarrow NH_2 + H_2O \tag{25}$$

$$NH_2 + NO \rightarrow N_2 + H_2O \tag{26}$$

$$NH_2 + NO \rightarrow NNH + OH \tag{27}$$

$$NNH + M \text{ fi } N_2 + H + M \tag{28}$$

$$H + O_2 \rightarrow OH + O \tag{29}$$

This reaction sequence is self-propagating under the correct conditions. That part of the NH_2 that is consumed by Reaction 27 leads to the generation of the OH radicals (via both Reactions 27 and 29) needed to facilitate Reaction 25. The only significant acknowledged means for generating N_2O is the reaction:

$$NH + NO \rightarrow N_2O + H \tag{23}$$

Both modeling and experimental studies indicate that, while some N_2O is formed, it is a very minor product.[34,85,86,87] Faster oxidation reactions effectively compete for the NH under these conditions.

Alternate agents have been proposed to avoid the handling problems associated with NH_3 and to improve NO_x removal performance. The principal competitor for NH_3 is urea, $CO(NH_2)_2$.[88] The urea is injected as either an aqueous solution or a dry powder. Some controversy has surrounded the products of the initial thermal decomposition reaction. Arguments based on consistency between data and modeling suggest that the products are NH_3 + HNCO,[65,85] which has been confirmed by experiment.[86]

Once released into the gas phase, the HNCO reacts primarily according to the following sequence:[89,90]

$$HNCO + OH \rightarrow NCO + H_2O \tag{30}$$

$$NCO + NO \rightarrow N_2O + CO \tag{20}$$

$$HNCO + H \rightarrow NH_2 + CO \tag{31}$$

In the absence of other reactions, the chain branching associated with NH_2 and N_2O consumption must support the decomposition of the HNCO. Other reactions (*e.g.*, concurrent wet CO oxidation) can also provide the radicals needed to drive Reaction 30. The key difference between urea and NH_3 is that urea can generate HNCO, NCO, and N_2O as major products of reaction, NH_3 does not. Thus, as is well-known, urea generates substantially more N_2O emissions when used as a SNCR agent.[85,87] Typically, less than 5% of the NO reduced is converted to N_2O when NH_3 is used. This compares to conversions greater than 10% for urea.

The RapreNO_x process is another SNCR process based on the use of cyanuric acid as an agent.[91,92] The cyanuric acid thermally decomposes to yield HNCO, which reacts according to Reactions 30, 20, and 31 to destroy NO, in the course of which N_2O is formed as a byproduct.[93]

Comparison of urea and cyanuric acid as agents shows that urea generates less N_2O under equivalent conditions. This is expected since only half of the nitrogen contained in the urea becomes associated with HNCO following injection. The other half forms NH_3 which does not yield significant N_2O. In the case of cyanuric acid injection, all of the nitrogen initially becomes HNCO, and thus final N_2O yields are increased.[85]

A major limitation of SNCR is the relatively narrow temperature window over which the agents are active at removing NO_x. Also, in large-scale facilities, the NO removal at the optimum temperature is not complete. Thus, a considerable research effort has been expended to enhance performance. One approach is the co-injection of combustible compounds (*e.g.*, CO or H_2) with the agents. This has the effect of shifting the optimum temperature window to lower temperatures. Depending on the amount of free oxygen present and the amount of combustible, the performance at the new optimum temperature can be either better or worse than the original unpromoted system. The temperature shifts because the oxidation of the combustible generates excess free radicals that are needed to initiate the reaction of the agent. Without the reaction of the combustible, the agent would not react at the lower temperature because it cannot supply sufficient radicals through chain branching.

Kinetic modeling shows that the combustible fuel acts to generate free radicals. These radicals promote agent decomposition: HNCO + OH $\rightarrow$ NCO + H_2O. Next, N_2O forms via NCO + NO. At higher temperatures, the N_2O is destroyed by N_2O + H. At lower temperatures, the H-atom is still available, but the relatively high activation energy prevents the N_2O + H reaction from being effective. Thus, as the temperature window for NO removal is shifted to lower temperatures by the combustible, most of the reactions 'follow' the window due to their weak temperature sensitivity. The N_2O + H reaction is the exception due to its strong temperature dependence, and the N_2O produced from the agent reaction fails to react further.

The conclusion is that the presence of a combustible may or may not widen the SNCR window for urea injection, but that it appears to (1) reduce the temperature at which peak N_2O emissions are observed, and (2) increase these peak emissions. Note the similarity between this situation and that within a FBC. In a FBC, fuel nitrogen in the form of HCN is released by the coal. This reacts in the presence of oxidizing coal volatiles. Thus, the combustibles within the volatiles can be viewed as "promoters," which tend to reduce the optimum temperature for NO reduction and N_2O formation. With sufficient combustibles present, as would be the case in a FBC, this optimum temperature would be reduced below 700°C. Thus, as temperature is increased, N_2O emissions would be expected to be reduced, and NO emission would increase, which is the normal characteristic of FBC's.

Recent experimental work suggests that sodium additives may be one effective means of reducing N_2O formation in SNCR. Chen *et al.*[94] performed tunnel furnace experiments in which a variety of sodium compounds were co-injected with urea. Figure 7 shows NO_x reduction and N_2O emissions as a function of urea injection temperature. The figure shows that urea promoted by monosodium glutamate gave both an increase in NO_x removal performance and a substantial reduction in N_2O emissions. Other sodium compounds, such as Na_2CO_3, also reduced N_2O emissions, although SO_2 tended to reduce effectiveness. This could be due to the formation of a non-reactive sulfate coating on the sodium particles, or through suppression of sodium volatility due to sulfate formation.

Use of NH_3 for NO_x reduction in FBC's has been extensively examined. The results indicate that little N_2O is generated until NH_3 is added in high stoichiometric excess over NO.[65,77,95] Urea injection, however, is strongly correlated with increased N_2O emissions.[65] However, fixed bed studies using quartz, clay, and ash show that, even with urea injection, N_2O yields are sharply depressed.[96] This suggests that appropriate inorganic surfaces can be used to suppress N_2O formation when urea is used as a SNCR agent.

Selective Catalytic Reduction

In the present context, selective catalytic reduction (SCR) refers to the reduction of NO by added NH_3 over a catalyst. This distinguishes it from processes involving other agents, such as CO, H_2, and CH_4. The process is, at present, applied only to stationary sources of NO_x, with mobile sources being dominated by direct catalytic reduction without use of an agent like NH_3. Selective catalytic reduction is presently being applied to industrial systems in Japan and Germany, and is coming to increasing use in other parts of the world.

The formation of N_2O during SCR was noted in the early 1970's (*e.g.,* Otto *et al.*).[97] A very detailed review of the fundamentals of SCR, including the problem of N_2O production, is available.[98] In addition, a general review of the application of catalysts to environmental problems is also available, which includes SCR and other topics.[99]

Most SCR systems are based on either noble metal catalysts or vanadium in combination with other metals and various substrates. Laboratory work suggests that N_2O can be a major product of SCR over noble metal catalysts.[100] The amount of N_2O formed depends on the state of the surface, and also on the nature of the substrate.[98] The formation appears to be due to a reaction through a Langmuir-Hinshelwood mechanism between two adsorbed NO molecules.[97] The nature of the platinum surface seems to have an effect, with single crystals not yielding significant N_2O. We were unable to find published field measurements on noble metal catalysts, although unpublished information suggests that a significant portion of the reduced NO can be converted into N_2O in practical installations on gas turbine sets.[101]

In addition to the general review of Bosch and Janssen,[98] the problem of N_2O formation over vanadium has been specifically reviewed by Odenbrand *et al.*[102] In earlier work, the proposed mechanism involved the reoxidation of the vanadium surface by adsorbed NO to yield reduced N_2O.[98] Recent work suggests that N_2O arises directly from the reaction of NH_3 and NO at low temperatures, and from direct NH_3 oxidation at high temperatures. Because of this mechanism, the minimum for NO_x emissions falls at a lower temperature than the maximum in N_2O emissions.[98,103] The practical consequence is that, below about 300°C, only negligible N_2O is formed, while the formation becomes significant at higher temperatures due to NH_3 oxidation.[102] This is supported by long-term, pilot-scale testing of a wide number of vanadium catalysts.[104] Here, no significant N_2O was noted, at least to 400°C. Since catalyst sintering begins to become a problem above these temperatures,[102] this is unlikely to be a common operating condition. A survey of N_2O emissions from 22 SCR installations in Japan indicated no emissions exceeding 1 ppm.[54]

Review of the data suggests that, within the broad bounds outlined above, the actual yield of N_2O is highly variable. It depends on catalyst type, and on catalyst treatment (*e.g.,* crystal size), contamination, support, and background gas composition. It will likely depend on catalyst age. The results do suggest that vanadium catalysts do not generate significant N_2O under their normal operating conditions, but that noble metal catalysts may.

Summary

In general, NO_x control procedures have led to significantly increased N_2O if they promote the reaction of cyano species under reduced temperature conditions. Thus, coal reburning may, under some conditions, lead to enhanced N_2O due to the release of fuel nitrogen under reduced temperatures. The N_2O yields are reduced somewhat, however, by the concurrent oxidation of the volatiles, which leads to N_2O scavenging and competitive oxidation of the NCO intermediate.

The downstream injection of urea and cyanuric acid both lead to N_2O formation. Preliminary data suggest that concurrent injection of combustible promoters (*e.g.,* CO or H_2) leads to increased N_2O formation as agent injection temperatures are reduced. The application of SCR suggests that N_2O emissions are negligible from vanadium catalysts if they are operated at their nominally low temperatures. Noble metal catalysts, however, can convert significant quantities of NO into N_2O during SCR.

THERMAL WASTE REMEDIATION

To date, a limited number of field measurements have appeared describing emissions from thermal waste remediation activities. Almost all of these have dealt with municipal solid waste (MSW) or dried sewage sludge. The limited number of results available at present support only a qualitative description of the trends. Nonetheless, some significant differences between waste incineration and coal combustion are apparent.

Most of the measurements are on MSW units. The following data have been reported:

- Iwasaki *et al.*:[48] 10 units (8 stokers, 1 fluidized bed, 1 batch)
- Yasuda and Takahasi:[105] 5 units (2 stokers, 3 fluidized beds)
- Hiraki *et al.*:[106] At least one unit
- Watanabe *et al.*:[107] 12 units (5 stokers, 5 fluidized beds, 2 rotary kilns)

As illustrated in Figure 8, the most striking variation is the decrease in emissions with combustion temperature. In spite of the low combustion temperatures, however, the emissions of N_2O rarely exceed 20 ppm. This appears be due at least partly to the relatively low nitrogen content of the fuel; Iwasaki *et al.*[48] report fuel-nitrogen contents of about 0.5% and emission factors of approximately 70 g N_2O/metric ton waste. This corresponds to a fractional conversion of fuel nitrogen to N_2O of approximately 1%. Interestingly, none of the data suggest a consistent influence of combustor configuration (fluidized bed vs. stoker/grate) on emissions.

A smaller number of sludge incinerators were also examined, including:

- Iwasaki *et al.*:[48] 4 units (2 multi-stage and 2 fluidized beds)
- Yasuda and Takahasi:[105] 5 units (4 fluidized beds and 1 rotary grate)
- Hiraki *et al.*:[106] At least one unit

These units showed much higher emissions: in the case of Yasuda and Takahasi[105] up to 600 ppm. This appears to be primarily a response to the high nitrogen content of the sludge. Iwasaki *et al.*[48] report 5-8% fuel nitrogen, with emission factors corresponding to 400 g N_2O/metric ton sludge. This still represents only a 0.5% conversion of fuel nitrogen. Yasuda and Takahasi[105] evaluated a sufficient number of units at various temperatures to suggest that higher temperatures in fluidized beds can reduce the very high N_2O emissions associated with sludge combustion.

Interestingly, the NO_x emissions were low enough not to be influenced in a significant way by the change in temperature. For MSW incineration the NO_x levels were much higher.[48,105] Mixed MSW and sludge incineration appeared to take on the characteristics of 'diluted' sludge incineration (*i.e.*, increased N_2O emissions in proportion to the increased amount of fuel nitrogen).

Very little work has been reported on other waste treatment activities. Emissions from liquid injection incineration of high nitrogen wastes have not been reported. The high temperatures that are typical of these units are not expected to support high N_2O emissions. Fume incinerators, however, may operate at much lower temperatures. Frequently, the fume represents a relatively inert stream containing dilute fuel-nitrogen compounds. Many of these fumes have low heating values that must be supplemented by gas fuel to obtain a stable flame. For economic reasons, the gas usage is minimized, which can yield a low-temperature flame. Such an environment may favor N_2O emissions.

MOBILE SOURCES

While limited, the historical N_2O database for mobile sources appeared not to have been impacted by the sampling artifact. Although mobile source measurements using chassis dynamometers were often made by sample extraction using Tedlar sampling bags and seldom performed on-line, SO_2 concentrations in mobile source vehicle emissions are many times smaller those from stationary coal and heavy oil combustion. Thus, the sampling artifact which dominated measurements from coal-fired utility boilers did not seem to affect measurements from mobile sources. Mobile source emissions levels established in the early and mid 1980's, before the artifact issue was brought to light, compare favorably with measurements made in later years by researchers who were well aware of the potential sampling problems and took care to avoid the sampling artifact. Table 6 presents a comparison of N_2O emission rates for several classes of vehicles.[108,109,110] Of particular note is the good agreement among the values reported by the three groups, and the fact that catalytically equipped vehicles emit up to 20 times more N_2O than comparable non-catalyst equipped vehicles.

Based on the results above, a conservative estimate of 62.1 mg N_2O/km (100 mg N_2O/mile), and the 1982 estimate for the U.S. vehicle fleet size (115 x 10^6) and distance traveled (2.6 x 10^{12} km, 1.6 x 10^{12} miles),[108] the total U.S. production of N_2O from mobile sources is approximately 1.6 x10^5 metric tons/yr (1.6 x 10^{11} g N_2O/yr, 1.0 x 10^{11} g N/yr, or 0.1 Tg N/yr). Assuming that the world fleet size and distance driven per year are three times those of the U.S., then worldwide mobile N_2O emissions are approximately 0.3 Tg N/yr. This value compares favorably with the values given in

Table 3 and constitutes approximately 8.5% of the total anthropogenic flux. However, in addition to the uncertainties regarding fleet size and distance driven used in the analysis above, many research issues remain including the applicability of different driving cycles to actual use, engine/emission control malfunction or non-optimal operation, quantification of the number of vehicles that use catalysts (including type of catalyst), and the influence of ambient conditions.

EMISSIONS FROM INDUSTRIAL SOURCES

Few industrial sources have been identified as potential emitters of significant N_2O. One receiving recent attention is the manufacture of adipic acid, used primarily in the manufacture of nylon. By one report[111] the manufacture of adipic acid accounts for about 10% of the anthropogenic flux to the troposphere, based on adipic acid production. This inventory failed to take into account existing abatement within the industry, and an improved estimate is 5-8% of the anthropogenic flux.[112]

Adipic acid is formed by the reaction of cyclohexanone and cyclohexanol under nitric acid oxidation. The reaction produces an off-gas that contains 1 mole of N_2O byproduct for each mole of adipic acid produced, along with some NO_x. This stream, which can contain 30-50 mole % N_2O, is usually passed through an absorber to recover the NO_x, and then vented to the atmosphere. Some plants incinerate the stream in process boilers to reduce the NO_x, which coincidentally destroys the N_2O. In 1990, approximately 32% of the off-gas streams were abated in this manner.[112]

Since the recognition of adipic acid as a significant atmospheric source of N_2O, the industry has launched several cooperative projects to evaluate abatement options.[112] Some of these options have the goal of simply eliminating the N_2O from the exhaust stream at the lowest cost. Others focus on converting the N_2O into NO_x, which can then be used as a nitric acid feedstock. Approaches currently under evaluation include:

- N_2O decomposition over a catalyst to yield N_2 or NO_x for byproduct recovery.
- High temperature N_2O thermal decomposition to yield NO_x as a recovered byproduct.
- Thermal destruction in boilers.

It is recognized that no one technology is likely to be applicable to all plants because of site specific technical and economic factors.

Other industrial sources that involve the oxidation of nitrogen compounds under moderate temperature conditions are candidates as N_2O emitters. One example mentioned in the literature are catalytic cracker regenerators.[30] These units are used in gasoline manufacture to regenerate the catalyst used to crack feedstock after it has become coated with a nitrogen-rich coke. The coke is burned off the catalyst in a fluidized bed. Temperatures are moderate in the bed, fuel nitrogen levels are high, and the volatile content of the coke is low. Thus, many of the factors that contribute to high N_2O emissions in fluidized bed coal combustion are present. To date, however, no emissions measurements are reported.

CONCLUDING REMARKS

This review has attempted to bring together the widely scattered literature on the relatively new problem of N_2O emissions from energy conversion and industrial equipment, and the influence of these emissions on the environment. Some of these results are summarized in Table 7.[18] This is still an evolving area, where changes in both quantitative and qualitative interpretations are likely. Far from being the last word, this paper will likely act only as a starting point for future work.

Within homogeneous chemistry, the principal issues include the products of the NCO + NO reaction, and the rates and products of the reactions consuming HCN and HNCO that give rise to NCO and other species in moderate temperature combustion. A significant amount of modeling effort has used rates for NCO + NO that are likely too high. Thus, the adequacy of homogeneous chemistry to explain N_2O yields in processes such as combustion fluidized beds and in selective non-catalytic reduction needs to be revisited.

A considerable amount of effort has gone into defining the influence of basic operating parameters on N_2O emissions from fluidized beds. While basic operating trends are now known, a clear mechanistic understanding is not yet complete. The relative roles of homogeneous vs. heterogeneous N_2O production are shrouded by the fact that char behavior is strongly dependent on the degree of devolatilization. Since fluidized beds contain chars of widely varying ages, the overall behavior represents an ensemble average. This has clearly complicated the task of identifying mechanistic information from actual fluidized bed data. Well defined mechanistic experiments are needed, and these have begun to appear. Particularly useful studies include the examination of simulated differential fluidized bed elements. Another approach is to characterize bed response to perturbations in the gas-phase environment (*e.g.*, the addition of N_2O or HCN into the feed air), or in the solid phase (*e.g.*, the spiking of a well characterized char into a combustor).

ACKNOWLEDGEMENTS/DISCLAIMER

Portions of this work were conducted under EPA Purchase Order 2D1449NAEX with J. Kramlich. The research described in this article has been reviewed by the Air Pollution Prevention and Control Division, U.S. Environmental Protection Agency, and approved for publication. The contents of this article should not be construed to represent Agency policy nor does mention of trade names or commercial products constitute endorsement or recommendation for use.

REFERENCES

1. Ramanathan, V., The greenhouse theory of climate change: a test by an inadvertent global experiment, *Science* **24,** 293 (1988).
2. Levine, J. S., The global atmospheric budget of nitrous oxide, and supplement: global change: atmospheric and climatic, *5th International Workshop on Nitrous Oxide Emissions*, Tsukuba, Japan (1992).
3. Houghton, J. T., Jenkins, G. J., and Ephraums, J. J., *Climate Change: The IPCC Scientific Assessment*, Cambridge University Press, Cambridge, UK (1990).
4. Levine, J. S., Nitrous oxide: sources and impact on atmospheric chemistry and global climate, Presented at: *LNETI/EPA/IFP European Workshop on the Emission of Nitrous Oxide*, Lisboa, Portugal (1990).
5. Janetos, A., Overview and update of EPA's stratospheric ozone and global climate program, Presented at: *LNETI/EPA/IFP European Workshop on the Emission of Nitrous Oxide*, Lisboa, Portugal (1990).
6. Chapman, S., A theory of upper atmospheric ozone, *Quart. J. Roy. Meterol. Soc.* **3,**

103 (1930).

7. Levine, J. S., Impacts of N_2O and other trace gases on stratospheric ozone, *EPA/IFP European Workshop on the Emission of Nitrous Oxide from Fossil Fuel Combustion*, Rueil-Malmaison, France, June 1988, EPA-600/9-89-089 (NTIS PB90-126038) (1989).
8. Weiss, R. F., The temporal and spatial distribution of tropospheric nitrous oxide, *J. Geophys. Res.* **86,** 7185 (1981).
9. Rasmussen, R. A., and Khalil, M. A. K., Atmospheric trace gases: trends and distributions over the last decade, *Science* **232,** 1623 (1986).
10. National Oceanic and Atmospheric Administration (NOAA), *Geophysical Monitoring for Climate Change*, Vol. 16, B.A. Bodhaine and R.M. Rosson, eds., pp. 67, U.S. Dept. of Commerce, Washington, D.C. (1988).
11. Prinn, R., Cunnold, D., Rassmussen, R. A., Simmonds, P., Alyea, R., Crawford, A., Fraser, P., and Rosen, R., Atmospheric emissions and trends of nitrous oxide deduced from 10 years of ALE-GAGE data, *J. Geophys. Res.* **95(18),** 369 (1990).
12. Khalil, M. A. K., and Rasmussen, R. A., The global sources of nitrous oxide, *J. Geophys. Res.* **97(D13),** 14,651 (1992).
13. Pearman, G. I., Etheridge, D., de Silva, F., and Fraser, P. J., Evidence of changing concentrations of atmospheric CO_2, N_2O and CH_4 from air bubbles in Antarctic ice, *Nature* **320(20),** 248 (1986).
14. Khalil, M. A. K., and Rasmussen, R. A., Nitrous oxide: trends and global mass balance over the last 3000 years, *Ann. Glaciol.* **10,** 73 (1988).
15. Ethridge, D. M., Pearman, G. I., and de Silva, F., Atmospheric trace gas variations as revealed by air trapped in an ice core from Law Dome, Antarctica, *Ann. Glaciol.* **10,** 28 (1988).
16. Zardini, D., Raynaud, R., Scharffe, D., and Seiler, W., N_2O measurements of air extracted from Antarctic ice cores: implication on atmospheric N_2O back to the last glacial-interglacial transition, *J. Atmos. Chem.* **8,** 198 (1989).
17. Raynaud, D., Jouzel, J., Barnola, J. M., Chappellaz, J., Delmas, R. J., and Lorius, C., The ice record of greenhouse gases, *Science* **259,** 926 (1993).
18. de Soete. G. G., General discussion, conclusions, and need for future work, Presented at: *LNETI/EPA/IFP European Workshop on the Emission of Nitrous Oxide*, Lisboa, Portugal (1990).
19. Intergovernmental Panel on Climate Change (IPCC, 1990), taken from Minami, K., N_2O emissions from soils and agro-environment, *5th International Workshop on*

Nitrous Oxide Emissions, Tsukuba, Japan (1992).

20. Intergovernmental Panel on Climate Change (IPCC, 1992), taken from Minami, K., N_2O emissions from soils and agro-environment, *5th International Workshop on Nitrous Oxide Emissions*, Tsukuba, Japan (1992).
21. Elkins, J. W., State of the research for atmospheric nitrous oxide (N_2O) in 1989, Intergovernmental Panel on Climate Change, National Oceanic and Atmospheric Administration (1989).
22. Elkins, J. W., Current uncertainties in the global atmospheric nitrous oxide budget, Presented at: *LNETI/EPA/IFP European Workshop on the Emission of Nitrous Oxide*, Lisboa, Portugal (1990).
23. Mann, M. D., Collings, M. E., and Botros, P. E., Nitrous oxide emissions in fluidized-bed combustion: fundamental chemistry and combustion testing, *Prog. Energy Combust. Sci.* **18,** 447 (1992).
24. Hao, W. M., Wofsy, S. C., McElroy, M. B., Beér, J. M., and Toqan, M. A., Sources of atmospheric nitrous oxide from combustion, *J. Geophys. Res.* **92(D3),** 3098 (1987).
25. Pierotti, D., and Rasmussen, R. A., Combustion as a source of nitrous oxide in the atmosphere, *Geophys. Res. Lett.* **3,** 265 (1976).
26. Weiss, R. F., and Craig, H., Production of atmospheric nitrous oxide by combustion, *Geophys. Res. Lett.* **3,** 751-753 (1976).
27. Muzio, L. J., and Kramlich, J. C., An artifact in the measurement of N_2O from combustion sources, *Geophys. Res. Lett.* **15,** 1369 (1988).
28. de Soete, G. G., Parametric study of N_2O formation from sulphur oxides and nitric oxide during storage of flue gas samples, Institut Français du Pétrole, Report No. 36 732-40 ex. (1988).
29. Muzio, L. J., Teague, M. E., Kramlich, J. C., Cole, J. A., McCarthy, J. M., and Lyon, R. K., Errors in grab sample measurements of N_2O from combustion sources, *J. Air Pollut. Control Assoc.* **39,** 287 (1989).
30. Lyon, R. K., and Cole, J. A., Kinetic modelling of artifacts in the measurement of N_2O from combustion sources, *Combust. Flame* **77,** 139 (1989).
31. Lyon, R. K., Kramlich, J. C., and Cole, J. A., Nitrous oxide: sources, sampling, and science policy, *Environ. Sci. Technol.* **23,** 392 (1989).
32. Linak, W. P., McSorley, J. A., Hall, R. E., Ryan, J. V., Srivastava, R. K., Wendt, J. O. L., and Mereb, J. B., Nitrous oxide emissions from fossil fuel combustion, *J. Geophys. Res.* **95(D6),** 7533 (1990).

33. Khalil, M. A. K., and Rasmussen, R. A., Nitrous oxide from coal-fired power plants: experiments in the plumes, *J. Geophys. Res.* **97(D13),** 14,645 (1992).
34. Kramlich, J. C., Cole, J. A., McCarthy, J. M., Lanier, W. S., and McSorley, J. A., Mechanisms of nitrous oxide formation in coal flames, *Combust. Flame* **77,** 375 (1989).
35. Kramlich, J. C., Nihart, R. K., Chen, S. L., Pershing, D. W., and Heap, M. P., Behavior of N_2O in staged pulverized coal combustion, *Combust. Flame* **48,** 101 (1982).
36. Chen, S. -L., Heap, M. P., Pershing, D. W., and Martin, G. B., Fate of coal nitrogen during combustion, *Fuel* **61,** 1218 (1982).
37. Chen, S. -L., Heap, M. P., Pershing, D. W., and Martin, G. B., Influence of coal composition on the fate of volatile and char nitrogen during combustion, *19th Symposium (International) on Combustion*, p. 1271, The Combustion Institute, Pittsburgh (1983).
38. Kramlich, J. C., Seeker, W. R., and Samuelsen, G. S., Chemical effects accompanying the thermal decomposition of pulverized coal particles, *Fuel* **67,** 1182 (1988).
39. Axworthy, A. E., Dayan, V. H., and Martin, G. B., Reactions of fuel-nitrogen compounds under conditions of inert pyrolysis, *Fuel* **57,** 29 (1978).
40. Houser, T. J., and Lee, P. K., Oxidation of cyanogen, *Combust. Sci. Technol.* **23,** 177 (1980).
41. Houser, T. J., McCarville, M. E., and Houser, B. D., Kinetics of oxidation of pyridine in a flow system, *Combust. Sci. Technol.* **27,** 183 (1982).
42. Houser, T. J., McCarville, M. E., and Zhuo-Ying, G., Nitric oxide formation from fuel-nitrogen model compound combustion, *Fuel* **67,** 642 (1988).
43. Axworthy, A. E., Kahn, D. R., Dayan, V. H., and Woolery, D. O., Fuel decomposition and flame reactions in conversion of fuel nitrogen to NO_x, EPA-600/7-81-158 (NTIS PB82-108358) (1981).
44. Zahniser, M. S., McCurdy, K., and Kolb, C. E., Data reported in Miller, J. A., and Bowman, C. T., Mechanism and modeling of nitrogen chemistry in combustion, *Prog. Energy Combust. Sci.* **15,** 287 (1989).
45. Cooper, W. F., and Hershberger, J. F., Measurement of product branching ratios of the NCO + NO reaction, *J. Phys. Chem.* **96,** 771 (1992).
46. Becker, K. H., Kurtenbach, R., and Wiesen, P., Investigation of the N_2O formation in the NCO + NO reaction by Fourier-transform infrared spectroscopy, *5th*

International Workshop on Nitrous Oxide Emissions, Tsukuba, Japan (1992).

47. Muzio, L. J., Montgomery, T. A., Samuelsen, G. S., Kramlich, J. C., Lyon, R. K., and Kokkinos, A., Formation and measurement of N_2O in combustion systems, *23rd Symposium (International) on Combustion,* p. 245, The Combustion Institute, Pittsburgh (1990).
48. Iwasaki, Y., Tatsuichi, S., and Ueno, H., N_2O emissions from stationary sources, *5th International Workshop on Nitrous Oxide Emissions;* Tsukuba, Japan (1992).
49. Ryan, J. V., and Srivastava, R. K., *EPA/IFP European Workshop on the Emission of Nitrous Oxide from Fossil Fuel Combustion*, Rueil-Malmaison, France, June 1988, EPA-600/9-89-089 (NTIS PB90-126038) (1989).
50. Vitovec, W., and Hackl, A., Pyrogenic N_2O emissions in Austria -- measurement at 45 combustion sources, *5th International Workshop on Nitrous Oxide Emissions;* Tsukuba, Japan (1992).
51. Yokoyama, T., Nishinomiya, S., and Matsuda, H., N_2O emissions from fossil fuel fired power plants. *5th International Workshop on Nitrous Oxide Emissions;* Tsukuba, Japan (1992).
52. Gaydon, A. G., and Wolfhard, H. G., Spectroscopic studies of low-pressure flames, *3rd Symposium Combustion, Flame and Explosion Phenomena,* p. 504, Williams and Wilkins, Baltimore, MD (1949).
53. Martin, R. J., and Brown, N. J., Nitrous oxide formation and destruction in lean, premixed combustion, *Combust. Flame* **80,** 238 (1990).
54. Yokoyama, T., Nishinomiya, S., and Matsuda, H., N_2O emissions from fossil fuel fired power plants, *Environ. Sci. Technol.* **25,** 347 (1991).
55. Castaldini, C., DeRosier, R., Waterland, L. R., and Mason, H. B., Environmental assessment of industrial process combustion equipment modified for low-NO_x operation, In *Proceedings of the 1982 Joint Symposium on Stationary Combustion NO_x Control,* Volume II, EPA-600/9-85-022b (NTIS PB85-235612) (1985).
56. Hao, W. M., Wofsy, S. C., McElroy, M. B., Farmayan, W. F., Toqan, M. A., Beér, J. M., Zahniser, M. S., Silver, J. A., and Kolb, C. E., Nitrous oxide concentrations in coal, oil and gas furnace flames, *Combust. Sci. Technol.* **55,** 23 (1987).
57. Sloan, S. A., and Laird, C. K., Measurements of nitrous oxide emissions from P.F. fired power plants, *Atmos. Environ.* **24A,** 1199 (1990).
58. Laird, C. K., and Sloan, S. A., Nitrous oxide emissions from U.K. power stations, *Atmos. Environ.* **27A,** 1453 (1993).
59. Persson, K., N_2O emissions from coal fired boilers during thermal NO_x reduction

tests: Experiences from SWEDCO-supported projects, Presented at: *LNETI/EPA/IFP European Workshop on the Emission of Nitrous Oxide,* Lisboa, Portugal (1990).

60. Sage, P. W., Nitrous oxide emissions from coal-fired plant: An update on the Joule collaborative project, *5th International Workshop on Nitrous Oxide Emissions;* Tsukuba, Japan (1992).

61. Soelberg, N. R., Characterization of NO_x in utility flue gases, Test Report for Canadian Electrical Association, Energy and Environmental Research Corp., Irvine, CA (1989).

62. Åmand, L. -E., Leckner, B., Andersson, S., and Gustavsson, L., N_2O from circulating fluidized bed boilers-present status. Presented at *LNETI/EPA/IFP European Workshop on the Emission of Nitrous Oxide,* Lisboa, Portugal (1990).

63. Åmand, L.-E. and Leckner, B., Influence of fuel on the emission of nitrogen oxides (NO and N_2O) from an 8-MW fluidized bed boiler, *Combust. Flame* **84,** 181 (1991).

64. Åmand, L. -E., Leckner, B., and Andersson, S., Formation of N_2O in circulating fluidized bed boilers, *2nd Nordic Conference on Control of SO_2 and NO_x Emissions from Combustion of Solid Fuels,* Lyngby, Denmark (1990).

65. Braun, A., Bu, C., Renz, U., Drischel, J., and Köser, H. J. K., Emission of NO and N_2O from a 4 MW fluidized bed combustor with NO reduction, *Fluidized Bed Combustion: ASME 1991,* 709 (1991).

66. Brown, R. A., and Muzio, L. J., N_2O emissions from fluidized bed combustion, *Fluidized Bed Combustion: ASME 1991,* 719 (1991).

67. Gustavsson, L., and Leckner, B., N_2O reduction with gas injection in circulating fluidized bed boilers, *Fluidized Bed Combustion: ASME 1991,* 677 (1991).

68. Hiltunen, M., Kilpinen, P., Hupa, M., and Lee, Y., N_2O emissions from CFB boilers: experimental results and chemical interpretation, *Fluidized Bed Combustion: ASME 1991,* 687 (1991).

69. Hayhurst, A. N., and Lawrence, A. D., Emissions of nitrous oxide from combustion sources, *Prog. Energy Combust. Sci.* **18,** 529 (1992).

70. Wójtowicz, M. A., Oude Lohuis, J. A., Tromp, P. J. J., and Moulijn, J. A., N_2O formation in fluidized-bed combustion of coal, *Fluidized Bed Combustion: ASME 1991,* 1013 (1991).

71. Aho, M. J., and Rantanen, J. T., Emissions of nitrogen oxides in pulverized peat combustion between 730 and 900°C, *Fuel* **68,** 586 (1989).

72. de Soete, G. G., Heterogeneous N_2O and NO formation from bound nitrogen atoms during coal char combustion, *23rd Symposium (International) on Combustion,* p.

1257, The Combustion Institute, Pittsburgh (1991).

73. Tullin, C. J., Sarofim, A. F., and Beér, J. M., Formation of NO and N_2O in coal combustion: the relative importance of volatile and char nitrogen, *J. Inst. Energy* **66,** 207 (1993).

74. Tullin C. J., Goel, A., Morihara, A., Sarofim, A. F., and Beér, J. M., NO and N_2O formation for coal combustion in a fluidized bed: effect of carbon conversion and bed temperature, *Energy & Fuels* **7**, 796 (1993).

75. Moritomi, H., and Suzuki, Y., N_2O emissions from circulating fluidized bed combustion, *4th China-Japan Fluid Bed Symposium* , Tsukuba, Japan (1991).

76. Hein, K. R. G., N_2O emissions from the combustion of low calorie coals - research and large scale experience, *5th International Workshop on Nitrous Oxide Emissions;* Tsukuba, Japan (1992).

77. Jahkola, A., Lu, Y., and Hippinen, I., The emission and reduction of NO_x and N_2O in PFB-combustion of peat and coal, *Fluidized Bed Combustion: ASME 1991,* 725 (1991).

78. Shimizu, T., Tachiyama, Y., Souma, M., and Inagaki, M., Emission control of NO_x and N_2O of bubbling fluidized bed combustor, *Fluidized Bed Combustion: ASME 1991,* 695 (1991).

79. Bramer, E. A., and Valk, M., Nitrous oxide and nitric oxide emissions by fluidized bed combustion, *Fluidized Bed Combustion: ASME 1991,* 701 (1991).

80. Chen, S. -L., McCarthy, J. M., Clark, W. D., Heap, M. P., Seeker, W. R., and Pershing, D. W., Bench and pilot scale process evaluation of reburning for in-furnace NO_x reduction, *21st Symposium (International) on Combustion*, p. 1159, The Combustion Institute, Pittsburgh (1987).

81. Kilpinen, P., Glarborg, P., and Hupa, M., Reburning chemistry: A kinetic modeling study, *Ind. Eng. Chem. Res.* **31,** 1477 (1992).

82. Glarborg, P., Johnsson, J. E., and Dam-Johansen, K., Kinetics of nitrous oxide decomposition, Combust. Flame, **99,** 523-532 (1994).

83. Lyon, R. K., Method for the reduction of the concentration of NO in combustion effluents using ammonia, *U.S. Patent 3,900,544* (1975).

84. Miller, J. A., and Bowman, C. T., Mechanism and modeling of nitrogen chemistry in combustion, *Prog. Energy Combust. Sci.* **15,** 287 (1989).

85. Muzio, L. J., Martz, T. D., Montomery, T. A., Quartucy, G. C., Cole, J. A., and Kramlich, J. C., N_2O formation in selective non-catalytic NO_x reduction processes, *American Flame Research Committee 1990 Fall International Symposium* (1990).

86. Caton, J. A., and Siebers, D. L., Comparison of nitric oxide removal by cyanuric acid and ammonia, *Combust Sci. Technol.* **65,** 277 (1989).
87. Kilpinen, P., and Hupa, M., Homogeneous N_2O chemistry at fluidized bed combustion conditions: A kinetic modeling study, *Combust. Flame* **85,** 94 (1991).
88. Arand, J. K., Muzio, L. J., and Sotter, J. G., Urea reduction of NO_x in combustion effluents, *U.S. Patent 4,208,386* (1980).
89. Lyon, R. K., and Cole, J. A., A reexamination of the RapreNO_x process, *Combust. Flame* **82,** 435 (1990).
90. Miller, J. A., and Bowman, C. T., Kinetic modeling of the reduction of nitric oxide in combustion products by isocyanic acid, *Int. J. Chem. Kinet.* **23,** 289 (1991).
91. Perry, R. A., and Siebers, D. L., Rapid reduction of nitrogen oxides in exhaust gas streams, *Nature* **324,** 657 (1986).
92. Perry, R. A., NO reduction using sublimation of cyanuric acid, *U.S. Patent 4,731,231* (1988).
93. Siebers, D. L., and Caton, J. A., Removal of nitric oxide from exhaust gas with cyanuric acid, *Combust. Flame* **79,** 31 (1990).
94. Chen, S. -L., Seeker, W. R., Lyon, R. K., and Ho, L., N_2O decomposition catalyzed in the gas phase by sodium, *1993 ACS Meeting*, Denver, CO (1993).
95. Åmand, L. -E., and Andersson, S., Emission of nitrous oxide (N_2O) from fluidized bed boilers, *Proceedings 10th Int. Conf. on FBC*, San Francisco, CA (1989).
96. Wallman, P. H., and Carlsson, R. C. J., NO_x reduction by NH_3: the effects of pressure and mineral surfaces, *Fuel* **72,** 187 (1993).
97. Otto, K., Shelef, M., and Kummer, J. T., Studies of surface reactions of NO by isotope labeling. II. Deuterium kinetic isotope effects in the ammonia-nitric oxide reaction on a supported platinum catalyst, *J. Phys. Chem.* **75,** 875 (1971).
98. Bosch, H., and Janssen, F., Catalytic reduction of nitrogen oxides: A review on the fundamentals and technology, *Catal. Today* **2,** 369 (1988).
99. Armor, J. N., Environmental catalysis: a review, *Appl. Catal. B* **1,** 221 (1992).
100. Meier, H., and Gut, G., Kinetics of the selective catalytic reduction of nitric oxide with ammonia on a platinum catalyst, *Chem. Engr. Sci.* **33,** 123 (1978).
101. Muzio, L. J., Fossil Energy Research Corp., personal communication (1993).
102. Odenbrand, C. U. I., Gabrielsson, P. L. T., Brandin, J. G. M., and Andersson, L. A. H., Effect of water vapor on the selectivity in the reduction of nitric oxide with ammonia over vanadia supported on silica-titania, *Appl. Catal.* **78,** 109 (1991).

103. de Soete, G. G., Nitrous oxide formation and destruction by industrial NO abatement techniques, including SCR, Institut Français du Pétrole, Report No. 37 755-10 ex., Presented at: Spring 1990 AFRC Meeting, Tucson, AZ (1990).

104. McGrath, T., Energy and Environmental Research Corp., personal communication (1993).

105. Yasuda, K., and Takahasi, M., Emissions of nitrous oxide from waste management system, *5th International Workshop on Nitrous Oxide Emissions;* Tsukuba, Japan (1992).

106. Hiraki, T., Shoga, M., Tamaki, M., and Ota, S., N_2O emission factor of stationary combustion sources and N_2O removal effect by ordinary flue gas treatment, *5th International Workshop on Nitrous Oxide Emissions;* Tsukuba, Japan (1992).

107. Watanabe, I., Sato, M., Miyazaki, M., and Tanaka, M., Emission rate of N_2O from municipal solid waste incinerators, *5th International Workshop on Nitrous Oxide Emissions;* Tsukuba, Japan (1992).

108. Sigsby, J. E., N_2O emission rates from mobile sources, Session III, *EPA Workshop on N_2O Emissions from Combustion*, Durham, NC, February 1986, EPA-600/8-86-035 (NTIS PB87-113742) (1986).

109. Prigent, M. F., and de Soete, G. G., Nitrous oxide (N_2O) in engine exhaust gases. A first appraisal of catalyst impact, *1989 SAE International Congress and Exposition (890492)*, Detroit, MI (1989).

110. Dasch, J. M., Nitrous oxide emissions from vehicles, *J. Air Waste Manage. Assoc.* **42(1),** 63 (1992).

111. Thiemens, M. H., and Trogler, W. C., Nylon production: An unknown source of atmospheric nitrous oxide, *Science* **251,** 932 (1991).

112. Reimer, R. A., Parrett, R. A., and Slaten, C. S., Abatement of N_2O emissions produced in adipic acid manufacture. *5th International Workshop on Nitrous Oxide Emissions;* Tsukuba, Japan (1992).

Table 1. Summary of key greenhouse gases (adapted from Ramanathan[1], Levine[2], and Houghton *et al.*[3])[*].

	CO_2	CH_4	Refrigerant-11 CCl_3F	Refrigerant-12 CCl_2F_2	CH_3CCl_3	CCl_4	N_2O
Pre-industrial atmospheric conc. (1750-1800)	275-280 ppm	0.7-0.8 ppm	0	0	0	0	285-288 ppb
Approx. current atmospheric conc. (1985-1990)[†]	345-353 ppm	1.72 ppm	220-280 ppt	380-484 ppt	130 ppt	120 ppt	304-310 ppb
Current rate of annual atmospheric accumulation	1.8 ppm (+0.46-0.5%)	0.015 ppm (+0.9-1.1%)	9.5 ppt (+4.0-10.3%)	17 ppt (+4-10.1%)	- +15.5%	- +2.4%	0.8 ppb (+0.25-0.35%)
Projected atmospheric conc. mid 21st century[‡]	400-600 ppm	2.1-4.0 ppm	700-3000 ppt	2000-4800 ppt	-	-	350-450 ppb
Atmospheric lifetime[§]	50-200 yr	10 yr	65 yr	130 yr	-	-	150 yr

[*]% = percent by volume, ppm = parts per million by volume, ppb = parts per billion (10^9) by volume, ppt = parts per trillion (10^{12}) by volume.

[†]1990 concentrations are based on extrapolation of measurements reported for earlier years.

[‡]Mid 21st century concentrations are based on current annual rate of atmospheric accumulation, and do not consider activities aimed at reducing emissions.

[§]Atmospheric lifetime is the ratio of atmospheric content to the total rate of removal. CO_2 lifetime is a rough indication of the time necessary for CO_2 concentrations to adjust to changes in emissions.

Table 2. Selected trace gases, and trace nitrogen species in the atmosphere (adapted from Levine[2,7]).

Major gases	Concentration*
Nitrogen (N_2)	78.08 %
Oxygen (O_2)	20.95%
Argon (Ar)	0.93%
Selected trace gases	**Concentration**
Water (H_2O)	0 to 2%
Carbon dioxide (CO_2)	353 ppm
Ozone (O_3)	
Tropospheric	0.02 to 0.1 ppm
Stratospheric	0.1 to 10 ppm
Methane (CH_4)	1.72 ppm
Refrigerant-12 (CCl_2F_2)	0.48 ppb
Refrigerant-11 (CCl_3F)	0.28 ppb
Hydroxyl (OH)	
Tropospheric	0.15 ppt
Stratospheric	0.02 to 0.3 ppt
Trace nitrogen species	**Concentration**
Nitrous oxide (N_2O)	310 ppb
Ammonia (NH_3)	0.1 to 1.0 ppb
Nitric oxide (NO)†	
Tropospheric	0 to 1 ppb
Stratospheric	up to 0.02 ppm
Nitric acid (HNO_3)	50 to 1000 ppt
Hydrogen cyanide (HCN)	200 ppt
Nitrogen dioxide (NO_2)	10 to 300 ppt
Nitrogen trioxide (NO_3)‡	100 ppt
Peroxyacetyl nitrate ($CH_3CO_3NO_2$)	50 ppt
Dinitrogen pentoxide (N_2O_5)‡	1 ppt
Pernitric acid (HO_2NO_2)†	0.5 ppt
Nitrous acid (HNO_2)	0.1 ppt
Nitrogen aerosols	
Ammonium nitrate (NH_4NO_3)	10 ppt
Ammonium chloride (NH_4Cl)	0.1 ppt

*% = percent by volume, ppm = parts per million by volume, ppb = parts per billion (10^9) by volume, ppt = parts per trillion (10^{12}) by volume.
†Exhibits strong diurnal variation with maximum concentration during the day.
‡Exhibits strong diurnal variation with maximum concentration during the night.

Table 3. Estimates of global sources of N_2O (Tg N per year).*

	de Soete[18†]	Levine[2]	IPCC[19,20]	Elkins[22] and Mann *et al.*[23‡]	Khalil and Rasmussen[12]
		NATURAL SOURCES			
Soils					7.6
Tropical soils		3.7		3.7	
wet forests			2.2-3.7		
dry savannas			0.5-2.0		
Temperate soils		0.01-1.5		< 0.5	
forests			0.05-2.0		
grasslands			NK[§]		
Oceans		1.4-2.6	1.4-2.6	1.4-2.6	1.9
Total natural sources		5.1-7.8	4.2-10.3	5.6-6.8	9.5
		ANTHROPOGENIC SOURCES			
Biomass burning	0.5 (0.4-0.6)	0.1-1.0	0.2-1.0	0.02-0.29	1.0 (0.1-1.9)
Sewage					1.0 (0.2-2.0)
Agriculture					
Cattle operations					0.3 (0.2-0.6)
Irrigation (ground water release)				0.5-1.1	0.5 (0.5-1.3)
Use of nitrogen fertilizers	1.0 (0.7-1.3)		0.01-2.2	0.015-1.4	0.6 (0.3-1.9)
on agricultural fields		0.01-1.1			
leaching into groundwater		0.5-1.1			
Land use changes (deforestation)		0.8-1.3		0.8-1.3	0.4
Fossil fuel combustion		0.1-0.3		< 0.5	
Stationary sources	0.2 (0.1-0.3)		0.1-0.3		0.0 (0.0-0.1)
Mobile sources	0.4 (0.2-0.6)		0.2-0.6		0.5 (0.1-1.3)
Industrial activities					
Adipic acid (nylon)			0.4-0.6		NK
Nitric acid			0.1-0.3		
Global warming					0.2 (0.0-0.6)
Atmospheric formation (indirect)	NK				NK
Dry and wet deposition				0.13-5.0	
Total anthropogenic sources	2.1 (1.4-2.8)	1.5-4.8	1.0-5.0	2.0-9.6	4.5 (1.8-9.7)
Total	2.1 (1.4-2.8)	6.6-12.6	5.2-15.3	7.6-16.4	14.0 (11.3-19.2)

*1 Tg = 1 x 10^6 tonnes = 1 x 10^{12} g.
†Anthropogenic sources only.
‡Data presented in the review by Mann *et al.*[23] was taken from Elkins.[21]
§NK - not known.

Table 4. Survey of recent N_2O data from combustion systems.*

Andersson *et al.*†	Several European plants
Dahlberg *et al.*†	17 combustion plants
Electric Power Research Institute†	14 utilities
Laird and Sloan[58]	3 corner-fired, 2 opposed-fired (1 low-NO_x) 3 wall-fired coal, 3 wall-fired oil (2 low-NO_x)
Linak *et al.*[32]	6 coal-fired
Muzio *et al.*[47]	2 oil-fired 7 coal-fired
Persson[59]	1 coal-fired
Sage[60]	2 stokers
Sloan and Laird[57]	4 wall-fired, 3 corner-fired, both normal burners and low-NO_x burners
Soelberg[61]	11 plants
Vitovec and Hackl[50]	9 coal-fired plants, including pulverized coal, lignite, stoker-firing, and briquettes 11 oil-fired plants 4 petroleum-coke plants
Yokoyama *et al.*[51,54]	7 coal-fired 21 oil-fired

*Pulverized-coal fired unless otherwise indicated.
†As reported in Mann *et al.*[23]

Table 5. Summary of field data for combustion fluidized beds.

Åmand and Leckner[63] Åmand *et al.*[62,64]	One 8 and one 12 MW CFB. Detailed parametric variations of excess air, lime addition, and char loading.
Braun *et al.*[65]	One 4 MW_T AFBC used for plant heating. Examined influence of bed temperature, fuel-type, and NO_x control strategies.
Brown and Muzio[66]	One AFBC and one CFBC were examined in detail. Variations included bed temperature, excess air, and sorbent feed. Also more limited parametric variations at three cogeneration CFBC sites.
Gustavsson and Leckner[67]	Examined gas injection for N_2O control at a 12 MW_T CFBC. Emissions range from 80 to 250 ppm as a function of temperature. Examined a number of parameters.
Hiltunen *et al.*[68]	Measurements from 8 CFBCs. Results correlated in terms of mean furnace temperature and fuel type. Emissions range from 10 to 140 ppm.
Muzio *et al.*[47]	Three CFBCs with N_2O ranging from 26 to 84 ppm at 3% O_2. One unit at 3 loads: 55, 75, 100% of full load, yielding 126, 93, 84 ppm N_2O, respectively.
Persson[59]	One 15 MW PFBC and one 40 MW CFB. Varied bed temperature in the larger unit and looked at NO_x control agents.
Ryan and Srivastava[49]	Summary of Third International N_2O Workshop; data presented on 4 Swedish units and 6 Finnish units.
Sage[60*]	Two AFBC's: 30, 22-77 ppm; one CFBC: 91 ppm.
Vitovec and Hackl[50*]	Austrian measurements. Four CFBC's, three AFBC's. CFBC results showed strong dependence on fuel type: bituminous coal = 24, lignite = 7.5, bark/sewage sludge/bituminous coal = 3.3, bark/sewage sludge = 0.8 ppm.

[*]Converted from mg/m^3 to ppm.

Table 6. Mobile source N_2O emission rates. (mg N_2O/km).

	Sigsby[108†]	Prigent and de Soete[109]	Dasch[110]
Non-catalyst auto	3.1-3.7 (5-6)*	2.9 (4.8)	1.5-3.0 (2.4-4.8)
Catalyst auto	4.3-85.1 (7-137)	9.3-62.1 (15-100) oxidation or 3-way cat.	1.9-41.0 (3-66) oxidation cat. 16.2-59.0 (26-95) dual bed cat. 8.1-62.8 (13-101) 3-way cat.
Diesel trucks/buses	19.3-91.3 (31-147)		
Gasoline trucks	29.8-60.3 (48-97)		55.3 (89) light duty, 3-way cat.

*Numbers in parentheses have units of mg N_2O/mi.
†Compilation of test data from several sources.

Table 7. N_2O emission from fossil fuel combustion (adapted from de Soete[18]).

Uncontrolled combustion	N_2O emissions
Conventional stationary combustion (coal, oil, gas)	1-5 ppm
Fluidized bed combustion (depends on temperature, oxygen conc., physical/chemical properties of fuel)	20-150 ppm
Diesel engines (value given for small passenger cars; heavy duty engine emissions may be higher)	0.03 g N per km
Gasoline engines	0.01-0.03 g N per km

NO_x control technology	Effect on N_2O emissions
Fuel staging for conventional stationary combustion	Up to 10 ppmv increase over uncontrolled combustion
Thermal DeNO_x controls (SNCR)	
NH_3 injection	3-5% of NO_x reduction converted to N_2O
Urea or cyanuric acid injection	10-15% of NO_x reduction converted to N_2O
Catalytic processes	
Selective catalytic reduction (SCR)	Limited laboratory studies indicate increased emissions from some catalysts. No field data available.
3-way catalysts (gasoline engines)	
New catalysts	3-5 times the uncontrolled emissions
Medium aged catalysts	10-16 times the uncontrolled emissions

SESSION C

PERSISTENT ORGANIC POLLUTANTS

Dioxins: Dutch/European historical perspective and current evaluation of human health risks

Jeanine A.G. van de Wiel

Health Council of the Netherlands, P.O.Box 1236, 2280 CE Rijswijk, The Netherlands

1. WHY WAS IT A PROBLEM AND WHY HAVE CONTROL POLICIES BEEN DEVELOPED?

Dioxin-like substances are a problem since in the sixties it became clear that these compounds have a highly toxic potential and that nearly everybody is chronically exposed to them because of their widespread occurrence and their persistence. In Europe chemical accidents from 1953 till 1976 in West-Germany, France, The Netherlands, Czechoslovakia, the United Kingdom and Italy made clear that dioxin-like substances could be harmful for man, shown by skin lesions (acné), liver insuffication, central nervous system disturbances, elevated lipid and cholesterol levels in blood, hypertension and prediabetes.

In experimental animals dioxin-like substances also caused skin lesions and liver impairment. Besides that the so called 'wasting syndrome' appeared at high doses. Thymusatrophy, tumors and impairment of reproduction were also observed. Experimental animals show a manyfold difference in vulnerability. The LD50 value for 2,3,7,8-TCDD in male guinea pigs (0,6 μg/kg) is a factor 5000 smaller than the LD50 in male hamsters (3000 μg/kg). This made it difficult to decide what animal model and what extrapolation factors should be used to make a human health risk assessment.

2. WHAT PROGRESS HAS BEEN MADE DURING THE LAST 20-30 YEARS?

Polyhalogenated aromatic hydrocarbons are almost always present in the environment in the form of mixtures of isomers and congeners. This phenomenon, combined with the similarity in their working mechanism, has led to a group approach for these substances for the purpose of risk evaluation. The 2,3,7,8-tetrachloride dibenzo-*p*-dioxin (2,3,7,8-TCDD) acts as reference in assessing the toxic effect of those polyhalogenated aromatic hydrocarbons that have a similar effect. It is both the most thoroughly investigated and the most toxic of this class of compounds with a dioxin-like effect. So-called TEF values have been assigned to all polychlorinated dibenzo-*p*-dioxins and polychlorinated dibenzofurans with chlorine atoms in the 2,3,7 and 8 positions. These toxic equivalency factors indicate the toxic effectiveness of dioxin-like compounds compared with that of the reference compound 2,3,7,8-TCDD; the TEF of this latter compound is, by definition, 1.

In 1977 D.L. Grant was the first to apply weighing factors to grade the toxicity of the different congeners. During the 1980's, several countries and national and international institutions have developed separate models for TEF's like the Federal Swiss Government in 1982, the Danish National Agency of Environmental Protection in 1984, the Federal Republic of Germany's Office for the Environment in 1985. The scientific basis differed between the models. The Danish and Swiss models were based essentially on the relative potency for aryl hydrocarbon hydroxylase induction, whereas the German model was based on a weighing of all available quantitative data on toxicity of the different congeners. In 1987 a Nordic expert group developed a new set of weight factors based on the most recent toxicity data. In 1988, the North Atlantic Treaty Organization Committee on the Challenges of Modern Society (NATO/CCMS) used the Nordic model as the basis for the international toxicity equivalent factors (I-TEF's) and recommended to its member countries to adopt this model. It has indeed been adopted by Germany, Italy, The Netherlands, the United Kingdom, the Nordic countries and beyond Europe Canada and the United States. The I-TEF values for PCDD's and PCDF's have been derived from data originating from *in vitro* and *in vivo* research into acute toxicity, subchronic and chronic toxicity, and carcinogenic and teratogenic effects.

During a meeting in 1993 held under the auspices of the World Health Organization (WHO) - and organized within the context of the International Program on Chemical Safety (IPCS) - the question of whether it was also possible to lay down internationally accepted TEF values for the dioxin-like PCB's was considered. The participants in that meeting have proposed interim TEF values. The criteria for including a PCB in the TEF scheme were that the structure of the relevant PCB resembles that of PCDD's and PCDF's, the PCB binds to the Ah receptor, the PCB causes dioxin-like biochemical and toxic effects and the PCB is not rapidly degraded in the environment and accumulates in the food chain. There is another proposal for TEF_{PCB}'s from mr Safe. Both proposals seem tenable in the light of the present state of scientific knowledge. In addition, the uncertainty in the exposure to PCB's is much greater than the difference in the calculated dose when using one or the other TEF_{PCB} list.

With the help of the TEF values, we can express the exposure to mixtures of PCDD's, PCDF's and dioxin-like PCB's in the form of the so-called toxic equivalency (TEQ). The TEQ value is arrived at by multiplying the concentration of each component in the mixture by the corresponding $TEF_{dioxins}$ or TEF_{PCB} value, and then adding together the products obtained.

The TEF concept is a usable and uniform instrument in the estimation of the risks associated with exposure to PCDD's, PCDF's and dioxin-like PCB's.

During the 1980's a number of regulatory bodies throughout the world generated separate assessments of the risks posed by 2,3,7,8-TCDD. While all of these assessments relied on essentially the same experimental data, they differed in the extrapolation methods. In 1982 The Netherlands based its exposure limit on hepatotoxic effects at 1 ng per kg with a safety factor of 250, resulting in a value of 4 pg per kg per day. Comparable approaches in the United Kingdom, Germany and Switzerland resulted in values from 1 to 10 pg per kg per day. No effect levels for carcinogenicity, reprotoxicity and immunotoxicity were established. In 1989 a tolerable weekly intake has been set by the International Program on Chemical Safety (IPCS) on the basis of rodent cancer studies. The proposal from the IPCS was based on the cancer responses at doses 1, 10 and 100 ng per kg per day in the animal lifetime feeding study of Kociba and coworkers.

An uncertainty factor of 200 was applied to the no-observed effect level (considered to be a threshold) of 1 ng per kg per day and thus a tolerable weekly intake for man turned out to be 35 pg per kg body weight per week. This value coincides with the value given by a Nordic Council working group in 1988.

Since then more about the working mechanism of dioxin-like compounds became known. Techniques to assess other endpoints than death and tumor formation in experimental animals were used to perform studies on developmental and reprotoxicity, neurotoxicity and immunotoxicity. Sampling and analytical techniques became standardized in the EU member states. Food and breastmilk concentrations were measured on a European scale. Epidemiological studies on incidentally high and chronically low exposed populations were reported. Scientific and political discussions about health risk evaluation and lowering of exposure took place on the international level of the WHO, the IARC and the EU. WHO-discussions resulted in 1990 in an acceptable daily intake of 10 pg per kilogram bodyweight per day on a life time base. This was based on both human data of Seveso residents and animal experimental data on carcinogenicity and reprotoxicity. The internal concentration of the liver was taken as the basis for deriving an acceptable daily intake because the liver was considered to be a primary target organ. Using a pharmacokinetic model it was calculated that 1 ng per kg per day during 2 years (the NOAEL for rats) equalled an intake of 100 pg per kg per day during 70 years for humans (the calculated NOAEL for humans). Because of uncertainty about interhuman variability and reprotoxicity a safety factor of 10 was applied. This resulted in a tolerated daily intake of 10 pg per kg per day for humans.

3. WHICH WERE THE 'DRIVING FORCES' BEHIND THE PROGRESS OBTAINED?

Depending on what exposure limit was chosen in a country, the exposure of people was under or above this limit. In The Netherlands, by adopting the WHO-limit there was a larger margin of safety then with the first Dutch exposure limit. Also the extreme low American exposure limit value became known. This gave rise to much concern in citizen groups. It pressed governments to pay scientists to investigate the potential effects of these compounds and to perform risk evaluations.

4. CURRENT SITUATION

As scientists went on with their toxicological and epidemiological research they now come to the conclusion that the health based exposure limit should be lowered. This was stated in an advisory report of the Health Council of The Netherlands to the Dutch government in august 1996 as follows.

In experimental animal studies NOAEL's or LOAEL's have been found that are lower than 1 ng per kg per day. The induction of the enzymes CYP1A1 and CYP1A2, and a more obviously adverse effect such as the increase of endometriosis, serve to illustrate that. According to the Committee, the experiments reported by Bowman and Schantz (1989), by Neubert (1992) and Rier (1993) are the most relevant basis for the derivation of a health-based recommended exposure limit because the laboratory animals were primates (and therefore are relatively close to man) and the observed effects are adverse. In the one study a change in cognitive development in baby Rhesus monkeys was noted; the effect arose in the case of a 2,3,7,8-TCDD dose of 0.1 ng per kg per day in the mothers (which themselves developed endometriosis). The other study showed a change in lymphocytes in Marmosets at a dose of 0.13 ng per kg per day. So 0.1 ng per kg per day was taken as a LOAEL in non-human primates. To accomplish the transition to a NOAEL the dose-effect relationship for all kinds of effects within the dosage range between 0.1 and 1 ng per kg per was extrapolated. This resulted in a factor of 2 lower than the corresponding LOAEL's so 0.05 ng per kg per day. Pharmacokinetic characteristics of PCDD's and PCDF's indicate that, in certain respects (e.g. the way in which these substances are distributed between the liver and the fatty tissue), the monkey occupies a position somewhere between the rat and man. Therefore the Committee proposes a interspecies safety factor of 5. The intraspecies variation (the possible difference in sensitivity within a species and, in this case therefore, between humans) is, according to the Committee, adequately taken into account when the standard safety factor of 10 is used. Accordingly, the Committee derives as a health-based recommended exposure limit for man of 0.001 ng per kg per day (or 1 picogram of 2,3,7,8-TCDD per kg per day). Given the similarity of their effects, the Committee considers this recommended exposure limit to be applicable also to the intake of mixtures of diverse dioxin-like compounds expressed in the TEQ.

Among the most important epidemiological studies concerning exposure to PCDD's, PCDF's and dioxin-like PCB's are that of industrial accidents in the sixties and seventies in Europe and that of Vietnam veterans. The Committee attaches much importance to the results of the research conducted among infants in Rotterdam and Groningen, first because the study deals with the consequences of long-term exposure to low doses - the situation to which the health-based recommended exposure limit relates - and, second, on account of the (as is generally assumed) increased sensitivity of infants (and foetuses) to harmful substances. Effects associated with the building up and development of organ systems are among the first that become apparent. The first point to make from the reported studies from Huisman (1995), Koopman-Esseboom (1994, 1995) and others is that no link has been found between serious, clinically relevant abnormalities in the development of new-born children and their degree of exposure to PCDD's PCDF's and PCB's. However, the infants with the highest prenatal exposure over a period of some time did have a lower neurological optimality score, a lower psychomotor score, and an altered immune function parameter. Among the infants with the highest postnatal exposure, the same subtle signs of a suboptimal development were observable, as well as changes in the thyroid hormone status. Although all the differences referred to fall within the range of the clinical reference values, the Committee thinks that their harmlessness to health in the longer term has not been established. In conformity with the definition of adverse effect such effects are in the Committee's opinion undesirable. The median daily dose expressed in the form of TEQ in the case of the mothers of the infants suffering the highest exposure levels amounts to 0.003 ng per kg per day. The fact that, at this dose, effect arise in infants offers in the opinion of the Committee, further support for the health-based recommended exposure limit of 0.001 ng per kg per day. In the mother themselves, physiological effects related to the exposure to dioxin-like PCB's were also observed, i.e. changes in thyroid hormone levels. The significance of this for health is, in the Committee's opinion, still unclear.

The Council concluded that the possibility that the ingestion of dioxin-like compounds causes adverse health effects in the Dutch population cannot be excluded with reasonable certainty.

5. WHICH PROBLEMS REMAIN AND WHY?

The group of dioxin like compounds contains so many congeners it remains difficult to assess their relative potencies. As the half time of these compounds in humans is 7 years, one should like to know more about the relation of body burden and health effect as a base for setting a daily intake limit. The broad spectrum of effects of these substances remains a problem. There is only a small number of studies at chronic low exposure levels in humans, so inter- and intra species extrapolationfactors are uncertain and large. Besides extrapolation from high to low doses is necessary.

The main problem concerning health risk assessment of dioxin-like compounds is the fact that governments more and more a quantitative risk assessment using the bench mark dose approach. Because it is very expensive to mitigate exposure they need to make a cost/benefit analyses. This is in contrast with the past where scientists were asked to set a level of exposure for chemicals that was safe in the first place, a more qualitative approach.

6. WHAT RELATIONSHIP EXISTS WITH OTHER ENVIRONMENTAL EFFECTS AND VARIOUS SOURCES OF POLLUTION/'CAUSING FACTORS'

There are some other compounds were the same problem causes a lot of ongoing discussion like particulate matter and radon. For these agents it is difficult to extrapolate health effects to the level of the actual exposure. At the same time it is clear that measures to bring down these exposures have a big economical impact.

7. WHICH PROMISING SOLUTIONS ARE AVAILABLE? CAN ADDITIONAL BENEFIT BE OBTAINED FROM VARIOUS LEVELS OF INTEGRATION?

In the health risk evaluation field there are some tendencies to cope with this problem of risk quantification. One of them is the design of sophisticated physiology based pharmacokinetic and pharmacodynamic models to perform more accurate extrapolations from high to low dose and for different experimental animals to humans. In practice this could mean tighter safety or uncertainty margins.

To validate these models it is important to get more data of the exposure - effect and the exposure - response curve.

For dioxins there is already a tendency to standardize and harmonize analytical techniques through Europe to get an European wide exposure assessment that should prove that the source reduction measures that have been taken within the framework of the 5th European Environmental Action Plan do work.

Discussions about exposure limits are going on at the WHO and about carcinogenicity at the International Agency of Research on Cancer. Apart from the progress in international cooperation national stimulation programs like that in the nineties in The Netherlands are very fruitful by building interdisciplinary centers of excellence.

REFERENCES

1. Health Council of The Netherlands: Committee on Risk Evaluation of Substances/Radon. Radon. Assessment of an integrated criteria document. The Hague: Health Council of The Netherlands, 1993; publication no. 1993/03.
2. Health Council of The Netherlands: Committee on the Health Implications of Air Pollution. Particulate air pollution. The Hague: Health Council of The Netherlands, 1995; publication no. 1995/14.
3. Health Council of The Netherlands: Committee on Risk Evaluation of Substances/Dioxins. Dioxins. Polychlorinated dibenzo-*p*-dioxins, dibenzofurans and dioxin-like polychlorinated biphenyls. Rijswijk: Health Council of The Netherlands, 1996; publication no. 1996/10E.
4. Health Council of The Netherlands: Committee on Health-based recommended exposure limits. Toxicology-based recommended exposure limits. Rijswijk: Health Council of The Netherlands, 1996; publication no. 1996/12E.

Dioxins: Dutch/Western European control policy, impact on emission and on human exposure

Job A. van Zorge

Ministry of Housing, Spatial Planning and Environment,
P.O.Box 30945, 2500 GX The Hague, The Netherlands

1. INTRODUCTION

Dioxins are highly persistent chlorinated organic compounds, which because of there lipophilicity after exposure accumulate in the fatty tissues of organisms, including humans. Dioxins are micro contaminants from many industrial, combustion and incineration processes. Dioxins are ubiquitous pollutants and environmental levels are of anthropogenic origin. The most persistent dioxins, the 2,3,7,8-chlorinated congeners, are extremely toxic. They build up in the food chain and therefore present a hazard for predatory animals. Humans exposure is mainly (ca. 95%) by food and is at a level of concern. The hazard of the most toxic dioxin has been recognized for about three decades now and a first Tolerable Daily Intake (TDI) was proposed in 1982 by the Dutch National Institute for Public Health and Environmental Protection. After it was realized that humans and the environment are exposed to a mixture of many polychlorinated dibenzo-p-dioxins (PCDD's) and dibenzofurans (PCDD's), usually together indicated as dioxins, an international accepted system was developed to express the toxicity of mixtures of these compounds [1-2]. This system of toxic equivalency compared to the most toxic dioxin 2,3,7,8-TCDD has been of great help in risk assessments and standard setting. Many countries have developed control policies for dioxins.

2. DEVELOPMENTS IN THE EUROPEAN UNION

2.1. Directives

As a result of an industrial accident in 1976 in Italy, in which 2,3,7,8-TCDD was released to the environment and the Seveso area was contaminated, a directive on major accident hazards of certain industrial activities was introduced [3]. This so-called Seveso-directive requires steps to be taken to prevent major accidents and to limit the consequences of those that do occur.

In the European Directive on the combustion of hazardous waste [4], which has been implemented at the start of this year, an emission limit for dioxins of 0.1 ng TEQ/m^3 is set. This means that facilities for destruction of hazardous waste or facilities combusting hazardous waste as a source of energy, e.g. cement kilns, will have to meet this standard.

The Directive on integrated pollution prevention control (IPPC) [5] was adopted 24 September 1996. It imposes the application of best available techniques (BAT) for indicated categories of industrial activities in order to reduce the impact of on the environment.

The existing EC-directives on municipal waste incineration date back to 1989 and do not dictate an emission limit for dioxins. Installations for municipal solid waste incineration are included in the IPPC-list of activities for which BAT must be introduced.

2.2. Dioxin emission

In February 1993 the Council of Ministers of the European Union agreed upon the 5th European Community Environmental Program in which a reduction target for emissions dioxins of 90% was set for the year 2005 as compared to the reference year 1985. Therefore an inventory of all relevant dioxin sources in the member states is required. A special research project was commissioned by DG XI (the directorate-general for the environment) to the LUA NRW (the North Rhine Westphalia State Environment Agency). The aim of this project is to obtain detailed information about the European dioxin emissions. It was started in January 1995 and is planned to comprise two phases.

Phase 1: inventarization of results of dioxin measurements in EU-member states, to be completed in 1996.

Phase 2: measurements of relevant dioxin sources in EU-member states in order to achieve a comparable level of information, to be started in 1997.

Table 1
Dioxin emissions in some European countries (g TEQ/year) [6]

COUNTRY	REFERENCE YEAR	EMISSION g TEQ/year
United Kingdom	1989	4013
	1994	560 - 1058
	future	315
Sweden	1987	218
	1993	12.4 - 93
Netherlands	1990	612
	1996	90
	2000	63
Germany	1987	1651
	1994	355 - 659
	future	79 - 100

The data collected in phase 1 is very different in quality. For several countries no data are available at all, other countries have estimated their dioxin emissions based on statistical information and on emission factors from the literature, some countries only have data from old measurements and only a few countries have complete inventories for which the estimates are based mainly on measurements and extrapolations. Some results are presented in table 1.

These figures indicate a large reduction in dioxin emissions during the last decade and moreover the prognostic figures show a substantial ongoing decrease for future emissions.

2.3. Dioxin concentrations in air

The reduction of dioxin emissions is substantiated by a reduction of air concentrations. Dioxin concentrations in air have been halved in North Rhine Westphalia, a heavy industrialized state of Germany, in the period 1987/88 till 1993/94 [7].

2.4. Dioxin levels in humans

Also a reduction in body burdens of humans is observed. The human milk studies coordinated by WHO/EURO show a decrease of dioxin levels in the period 1988 till 1993 [8]. In fig. 1 this decrease is presented as an annual reduction percentage compared to the levels of 1988. This shows that there is a positive correlation between the magnitude of the decrease and the levels in 1988.

Trends in dioxin levels of human milk and human blood in individual countries confirm these results [9-11].

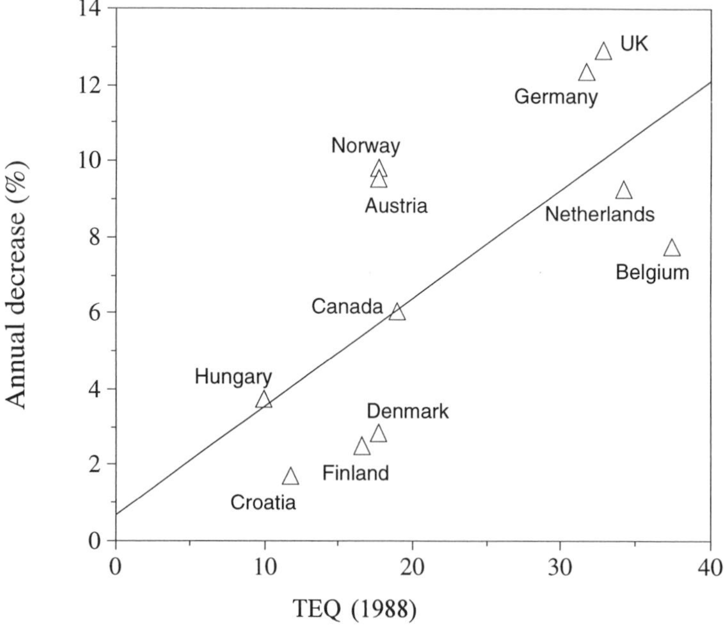

Figure 1. Results of human milk studies WHO/EURO 1988/1993

2.5. Dietary intake

In the framework of the SCOOP program (a scientific cooperation program of the European Union) mid 1997 a project will be started to collect data on dietary exposure of dioxins in the member states. A commission report of these data will be presented to the Scientific Committee which has to advise on the need of standard setting for food.

A more detailed, but less updated review of emission sources, ambient concentrations, dietary exposure and regulations has been presented by Liem and Van Zorge [12].

3. DEVELOPMENTS IN THE NETHERLANDS

The government of The Netherlands has intensively sponsored dioxin research during the past years. The research program was initiated in 1989 by the detection of high levels of PCDD/PCDF in cow's milk in the surroundings of several municipal solid waste incinerators (MSWI's). This program now has been finalized. The main goal of the program was the detection of dioxin sources, quantification of their emissions and use of these data for an emission reduction strategy.

3.1. Emission measures

Because of the milk problem near MSWI's in August 1989 a guideline on incineration of municipal solid waste and related processes, like incineration of hazardous waste, hospital waste and sludges, was published. In this guideline an emission standard for dioxins of 0.1 ng TEQ/m^3 was set for new incinerators. Existing incinerators had to meet this standard the first of December 1993. In February 1993 the guideline was transformed into law and because of problems with the accommodation of the existing incinerators the date of enforcement was delayed to the first of January 1995. In practice it took until the start of 1997 before all the MSWI's complied with the standards. For other dioxin emitting processes no standards have been set until now. License providing authorities however have been instructed to ask for maximum acceptable emission reduction. If necessary for certain sources, emission maxima will be presented in the

National Emission Guideline.

The dioxin emission of MSWI's has to be monitored at least twice a year. As the result of a round robin study an instruction for sampling, sample treatment and analysis was published in 1993, in which also detection limits for the different congeners were dictated [13].

Six old MSW incinerators have been closed, two of them were replaced by new ones and another two new incinerators have been build. Action was taken to reduce the amount of waste which has to be incinerated. Separated collection and composting of the organic fraction of municipal waste now is widely introduced in the country.

3.2. Dioxin emission in The Netherlands

The results of the study on sources and emissions indicated that for The Netherlands MSWI's were by far the most important source (ca. 80%) of dioxin emission into the air. In figure 2 the emissions in 1991 of dioxin sources are presented together with the estimates for the year 2000. Figure 3 presents the proportional contribution of these sources to the total emission in these years. It clearly shows that municipal solid waste incineration changes from a major to a minor source of dioxin emission. In the future wood combustion for house heating, industrial processes and the use of pesticides are expected to be major sources. The latter one is dominated by the extensive former use of pentachlorophenol (PCP) as a wood preserver. The use of PCP, which contains relatively large concentrations of dioxins, was terminated at the start of 1991.

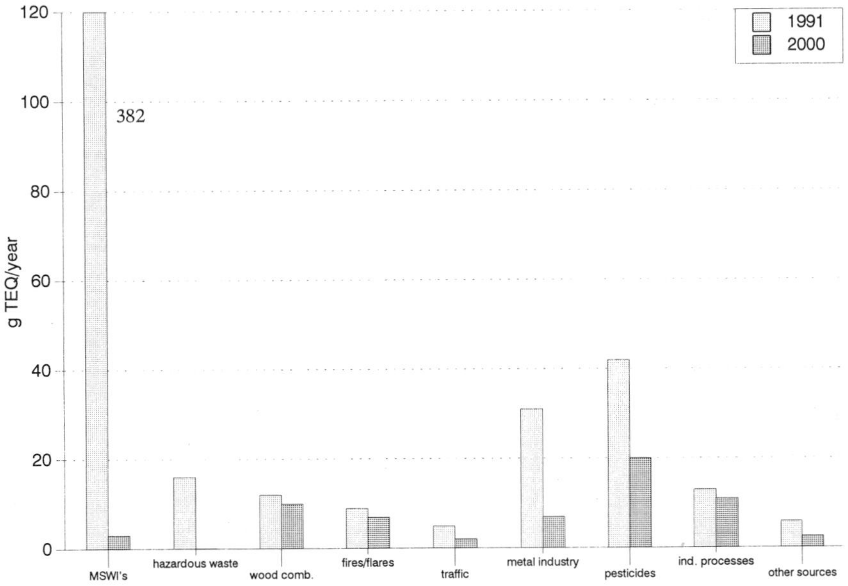

Figure 2. Sources of dioxins in The Netherlands

Table 2 shows the estimated total dioxin emissions and the emissions of MSWI's in The Netherlands from 1985 till 2000 together with the reduction percentage in comparison to the reference year 1989. For MSWI's a more than 99% reduction already has been achieved. For the total emission the reduction percentage compared to the estimated 1985 emission, the European reference year, is given also. It is clear that the European target of 90% reduction in member states in 2005 will be reached much earlier in The Netherlands and in fact already has been achieved.

3.3. Effects of emission measures for MSWI's

As mentioned the detection of high levels of PCDD/PCDF in cow's milk near MSWI's gave the impulse for the emission reduction policy. In order to prevent the consumption of milk with high dioxin levels in 1989 a standard of 6 pg TEQ/g fat was set for dairy products. Milk of dairy farms near MSWI's was monitored and in areas where the standard was exceeded the milk was taken out of the consumption chain. As an illustration of the effectivity of emission reduction measures Figure 4 shows the dioxin levels in milk from reference farms in the Lickebaert area situated near the MSWI AVR, which with a capacity of ca. 1 million tons of waste per year is the biggest MSWI in the country. The retrofitting of this incinerator was completed in 1994 which was clearly reflected in the dioxin levels in milk. Moreover the levels in milk were in good agreement with the calculated values from a chain model that was developed to relate emissions to levels in cow's milk [14].

Table 2
Dioxin emissions in The Netherlands

YEAR	MSWI's g TEQ/year	REDUCTION % *	TOTAL g teq/year	REDUCTION % *
1985			840	
1989	697		899	
1990	410	41	612	32 (27)
1991	382	45	516	43 (39)
1995	13	98	122	86 (85)
1996	5	99.3	90	90 (89)
2000	3	99.6	63	93 (92)

* compared to 1989 emission
(compared to 1985 emission)

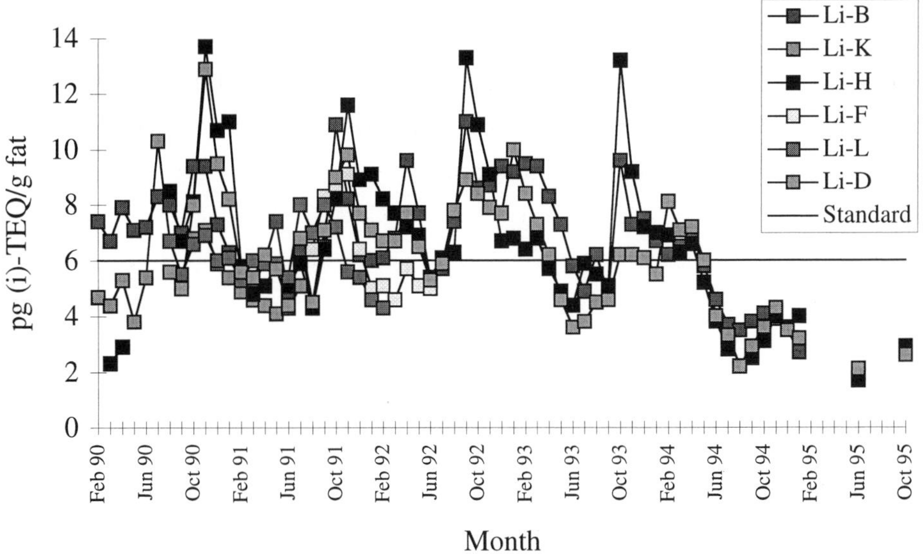

Figure 3. Dioxin levels in cow's milk from the Lickebaert area (1990-1995)

3.4. Dietary intake

Much effort has been put into investigations of dietary exposure. In 1982 a TDI of 4 pg 2,3,7,8-TCDD/kg body weight was introduced [15] and almost unnoticed this was changed into the same value for TEQ's. In 1991 the TDI of 10 pg 2,3,7,8-TCDD/kg b.w. which was proposed in 1990 by WHO [16] was adopted and again this value was interpreted as TEQ's. In spite of this raise of the TDI the policy remained to reduce dioxin levels as far as possible. In 1996 the Health Council of The Netherlands as a result of a reevaluation of dioxin toxicity data advised to reduce the health based exposure level to 1 pg TEQ/kg b.w. [17]. Besides that it was advised to include dioxin like PCB's in risk evaluations and standard setting. Herefore the use of the toxic equivalency factors proposed by WHO/EURO [18] was recommended.

Two tracks were followed in the investigations into dietary exposure:

i analysis of duplicate diets for dioxins and related compounds,

ii analysis of food categories and calculation of exposure by use of the consumption data from the National Food Consumption Inquiry.

The latter approach has the advantage that information is obtained for different age categories and for gender. The results of both investigations were well in line with each other. Figure 4 shows the contribution of food categories to the dietary intake [19]. In this figure dioxin like PCB's are presented separately.

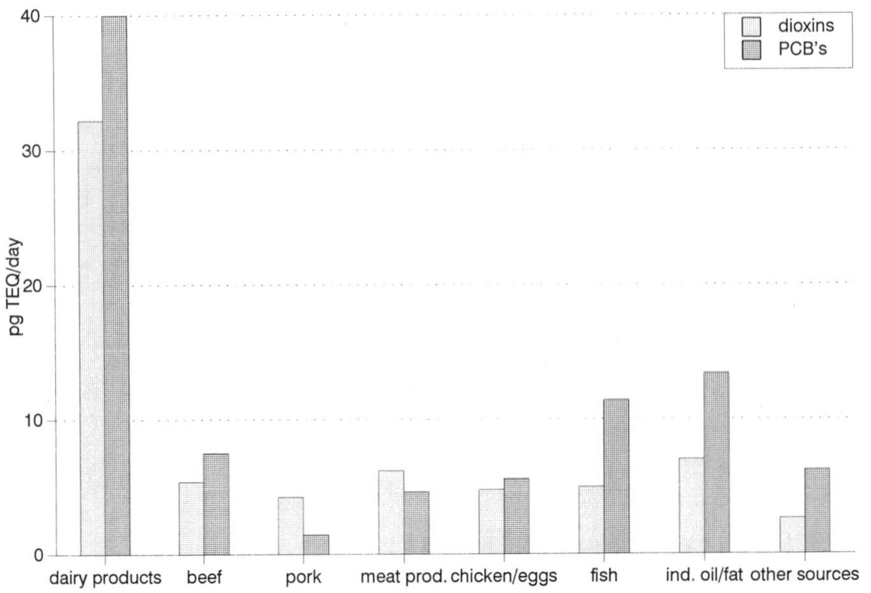

Figure 4. Median contribution of food categories to the dietary intake

The results of duplicate diet studies are given in figure 5. It is clear that there has been a remarkable decrease in the dioxin levels in food from 1978 till 1994. In this study the mean intake of PCDD's, PCDF's and dioxin like PCB's has decreased to 1.5 pg TEQ/kg b.w./day [20]. In view of the fact that after 1994 still important emission repressing measures were introduced it is expected that this downward trend will continue and that a daily intake of 1 pg TEQ/kg b.w. is well within reach as a result of all the measures which have been taken now.

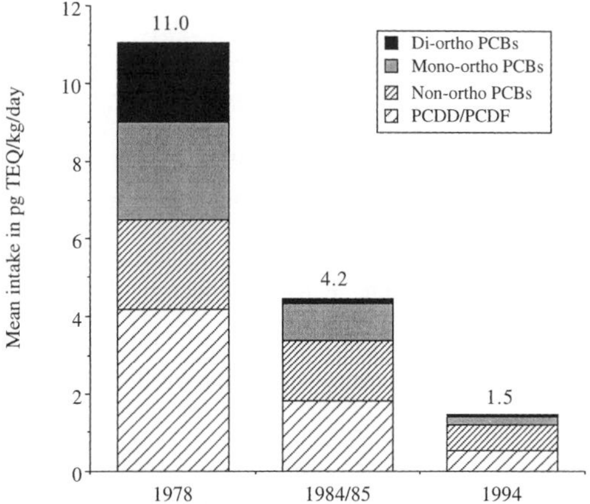

Figure 5. Results of duplicate diet studies in The Netherlands

3.5. Human body burdens

Dioxin body burdens will respond much slower to emission reduction measures. The long halflife in humans causes an accumulation that reaches an equilibrium only after several decades. So past high exposures will work on for a long time. Nevertheless a reduction is observed in human milk which is an indicator of body burden. In The Netherlands the mean levels in human milk decreased from 34.2 to 23.5 pg TEQ/g fat between 1988 and 1993 [21]. An additional study is planned for 1998.

4. CONCLUSION

It is clear that all the efforts and money spend on emission reduction of dioxins have caused a drop in human exposure. It can be expected that human body burdens will decrease below a level that rises concern. A condition for an ongoing downward trend is the maintenance of the measures which have been taken. Because of the transborder transport of dioxin emissions an international approach is necessary. Therefore the EU approach, as laid down in the 5th Environmental Program, is promising.

REFERENCES

1. NATO/CCMS. Pilot Study on International Information Exchange on Dioxins and Related Compounds. International Toxicity Equivalency Factor (I-TEF). Method of Risk Assessment for Complex Mixtures of Dioxins and Related Compounds. North Atlantic Treaty Organization/Committee on the Challenges of Modern Society. Report number 176. August 1988.
2. J.A. Van Zorge, J.H. Van Wijnen, R.M.C. Theelen, K. Olie and M. Van Den Berg, Assessment of the Toxicity of Mixtures of Halogenated Dibenzo-p-dioxins and Dibenzofurans by use of Toxicity Equivalency Factors (TEF), Chemosphere 19 (1989) 1881-1895 .
3. Commission of the European Communities, 1982. Commission Directive 93/21/EC of 5 August 1982 on the Major Accident Hazards of certain Industrial Activities. Off. J. Eur. Communities, L230 (87/216/EC first amendment, 88/610/EC second amendment).
4. Commission of the European Communities, 1994. Commission Directive 94/67/EC of 16 December 1994 on the Combustion of Hazardous Waste. Off. J. EUR. Communities, L 365/34.
5. Commission of the European Communities, 1996. Commission Directive 96/61/EC of 24 september 1996 on Integrated Pollution Prevention Control. Off. J. EUR. Communities, L 257/26.
6. U. Quaß, M. Fermann and G. Bröker, Identification of Relevant Industrial Sources of Dioxins and Furans; Quantification of their Emissions and Evaluation of Abatement technologies, North Rhine Westphalia State Environment Agency (LUA NRW), Interim Report, December 1995.
7. E. Hiester, P. Bruckmann, R. Böhm, A. Gerlach, W. Mülder and H. Ristow, Pronounced decrease of PCDD/PCDF burden in ambient air, Organohalogen Compounds 26 (1995) 147-152.
8. WHO/ECEH, Levels of PCBs, PCDDs and PCDFs in human milk. Second Round of WHO-coordinated exposure study, Environmental Health in Europe, No. 3 (1996), World Health Organization/European Center for Environment and Health, Bilthoven-Nancy-Rome.
9. L. Alder, H. Beck, W. Mathar and R. Palavinskas, PCDDs, PCDFs, PCBs and other organochlorine compounds in human milk. Levels and their dynamics in Germany, Organohalogen Compounds 21 (1994) 39-44.

10. O. Päpke, M. Ball, and A. Liss, PCDD/PCDF in humans, follow-up of background data for Germany, 1994, Organohalogen Compounds 21 (1994) 121-124.

11. A. Schecter, O. Päpke and P. Fürst, Is there a decrease in general population dioxin body burden?, Organohalogen Compounds 30 (1996) 57-60.

12. A.K.D. Liem and J.A. Van Zorge, Dioxins and Related Compounds: Status and Regulatory Aspects, Eviron. Sci. & Pollut. Res. 2 (1995) 46-56.

13. Instructions for the sampling and analysis of 2,3,7,8-substituted polychlorodibenzo-p-dioxins (PCDD) and polychlorodibenzofurans (PCDF) emitted into the air from stationary sources, Publicatiereeks stoffen nr. 1193/8, Ministry of Housing, Spatial Planning and Environment, The Hague, The Netherlands.

14. W. Slob, O. Klepper and J.A. Van Jaarsveld, A chain model for dioxins: from emissions to cow's milk, National Institute of Public Health and Environmental Protection, Report no. 730501039, January 1993.

15. C.A. Van der Heijden, A.G.A.C. Knaap, P.G.N. Kramers and M.J. Van Logten, Evaluation of the carcinogenity and mutagenity of 2,3,7,8-tetrachloro-dibenzo-p-dioxin (TCDD); classification and standard setting, National Institute of Public Health and Environmental Protection, Report no. 627915007, 1982.

16. U.G. Ahlborg, R.D. Kimbrough and E.J. Yrjänheikki (eds.), Tolerable Daily Intake of PCDDs and PCDFs. Toxic Substances Journal 12 (1992) 101-331.

17. Health Council of The Netherlands, Dioxins, Polychlorinated dibenzo-p-dioxins, dibenzofurans and dioxin-like polychlorinated biphenyls, 1996/10E.

18. U.G. Ahlborg, G.C. Becking, L.S. Birnbaum, A. Brouwer, H.J.G.M. Derks, M. Feeley, G. Golor, A. Hanberg, J.C. Larsen, A.K.D. Liem, S,H. Safe, C. Schlatter, F. Waern, M. Younes and E. Yrjänheikki, Toxic Equivalency Factorsfor dioxin-like PCB's. Chemosphere 28 (1994) 1049-1067.

19. A.K.D. Liem, R.M.C. Theelen and R. Hoogerbrugge, Dioxins and PCB's in food. Results of an additional study. National Institute of Public Health and Environmental Protection, Report no. 639102005, March 1996.

20. National Institute of Public Health and Environmental Protection, Report no. 639102020, to be published.

21. A.K.D. Liem, J.M.C. Albers, R.A. Baumann, A.C. van Beuzekom, R.S. Den Hartog, R. Hoogerbrugge, A.P.J.M. De Jong and J.A. Marsman, PCBs, PCDDs, PCDF's and organochlorine pesticides in human milk in The Netherlands. Levels and trends. Organohalogen Compounds 26 (1995) 69-74.

Persistent pesticides: the need for criteria to control atmospheric transport

Marten A. van der Gaag and Janette Worm

Ministry of Housing, Spatial Planning and the Environment
P.O. Box 30945, NL 2500 GX The Hague, The Netherlands

INTRODUCTION

In the fifties persistence was originally seen as a very useful intrinsic property for the efficacy of pesticides. It took almost two decades, however, to realize that adverse effects from the health and environment perspective outweigh the advantages of persistence in most applications.

Another decade was needed to implement adequate measures. Nowadays, persistence is most of the time evaluated at the moment of pesticides registration. Criteria have been developed to assess persistence in relation to biomagnification, accumulation in soil or sediment and leaching to groundwater (geoconcentration). Until recently little or no attention was paid to the evaluation of persistence in elation to long range atmospheric transport. The question is: Have we really learned from the lessons of the past and are we now prepared to anticipate in an early stage to prevent widespread consequences of persistence-related problems.

Accumulating evidence is pointing out that pesticides are found in sites far away from application areas due to medium and long range atmospheric transport [1,2,3]. Concentrations regularly exceed the levels of concern in both water and soil. In order to identify potentially hazardous compounds for atmospheric transport further research is needed and criteria have to be developed to restrict the use or reject registration of such pesticides. These criteria could prevent the introduction to the market of new compounds with hazardous properties, and trigger the need for additional data.

1. A KEY ISSUE IN HEALTH AND ENVIRONMENTAL EFFECTS ASSESSMENT

1.1. Persistence: an intrinsic hazard or a conditional risk?

Persistence as an intrinsic property of a molecule can be quite harmless to the environment if the molecule in question can be recovered and destructed in a controllable way after use. The use of persistent molecules is controllable only in well defined situations, when small volumes are applied. However, crop-protection practice often implies the use of large volumes of an agent in an open situation and diffusion to soil, water, groundwater and air will inevitably happen. In these situations, persistence could lead to bio- or geoconcentration of a molecule or its degradation products and forms a potential environmental threat.

Persistence has been a key issue in environmental effects assessment since the impact on wildlife of very persistent organochlorine pesticides as DDT and drins became evident. Ever since, policy development was triggered by the discovery of new side effects of persistent pesticides on an unexpected target (figure 1), such as potential effects in follow-up crops (bio- magnification), contamination of drinking water sources due leaching into groundwater (several herbicides) and threatening of the ozone layer (methyl bromide). Most of the time the measures taken consisted of restricting the application or eventually banning the use of a specific active substance. Those incidents also gradually led to setting criteria for the registration of pesticides that take into account persistence in direct relation to the target involved.

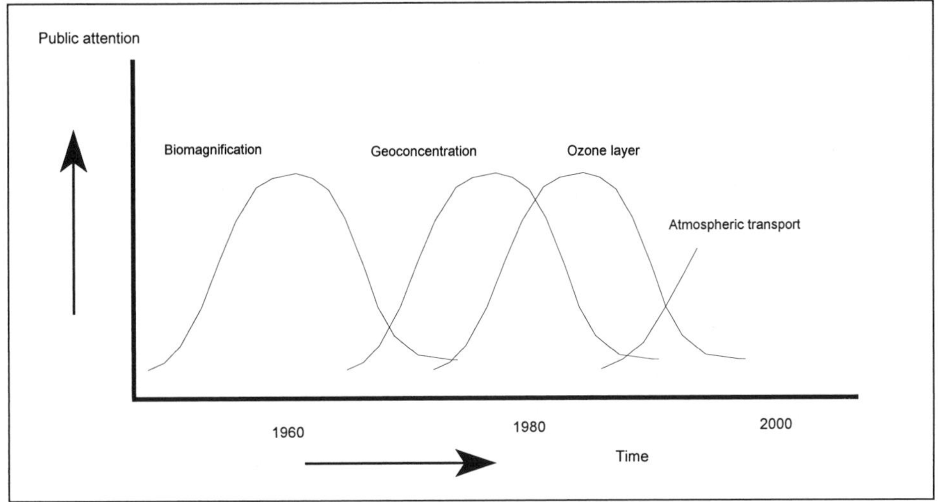

Figure 1: Public attention on the effects of persistent pesticides triggers policy measures. Studies and discussions on persistence of pesticides drew major public attention over the past decades. As the evidence on severe effects on predator populations due to biomagnification, grew to a consensus, the public attention increased, leading to restriction or banning of the use of organochlorine pesticides in many countries. Similar cycles occurred in the eighties in relation to persistence in soil and in ground water and in the early nineties when the effects of methyl bromide on the ozone layer became known. As evidence is building up on the impact of atmospheric transport, it will be necessary to formulate adequate control policies on this issue in the coming years.

1.2. Policies developed to cope with persistence in soil and ground water

Persistence in soil has been a matter of concern for numerous years before legislation was adopted in The Netherlands on this issue [4]. Not only environmental concerns, but also potential effects on follow-up crops of herbicide residues in the soil triggered discussions on persistence and persistence criteria were set in the European plant protection legislation [5]. Active ingredients with a half life time over 90 days have to be screened on the potential effect of their long lasting residues. The residues concentration in a field should be lower than the maximum tolerable risk concentration two years after the last pesticides application. In addition, very persistent agents, with half lives over 180 days will not be registered or will be subjected to very strict use restrictions.

Active ingredients leaching into groundwater at concentrations over 0.1 μg/l (potable water quality standard) are not registered. If leaching of ingredients into the shallow ground water is probable, authorization will only be granted if sufficient degradation will occur in the saturated zone during the travelling time that is necessary to reach the shallowest ground water wells. Only degradation processes that actually happen in these groundwater layers are considered for registration. Ecological effects on deeper groundwater ecosystem are not considered at the moment, although a recent study from the Health Council of The Netherlands [6] indicated that this should be taken into account. The Health Council suggested that the same ecological quality criteria e.g. 0.1 μg/l should apply for both groundwater- and surface water systems.

2. PERSISTENCE AND ATMOSPHERIC TRANSPORT

Persistence in the atmosphere is linked to other time scales than those in the other environmental compartments with a slower transport velocity. Residence times of a few days in the atmosphere may be sufficient to transport a molecule over long distances. The persistence property of a pesticide during transport in the atmosphere, together with its persistence at a potential sink location of geological, biological or tropospheric origin will determine whether dissipation into the atmosphere presents a risk. However, persistence in the atmosphere itself may not be the most critical parameter due to the rapid rate of transport, up to hundreds of kilometers a day [7].

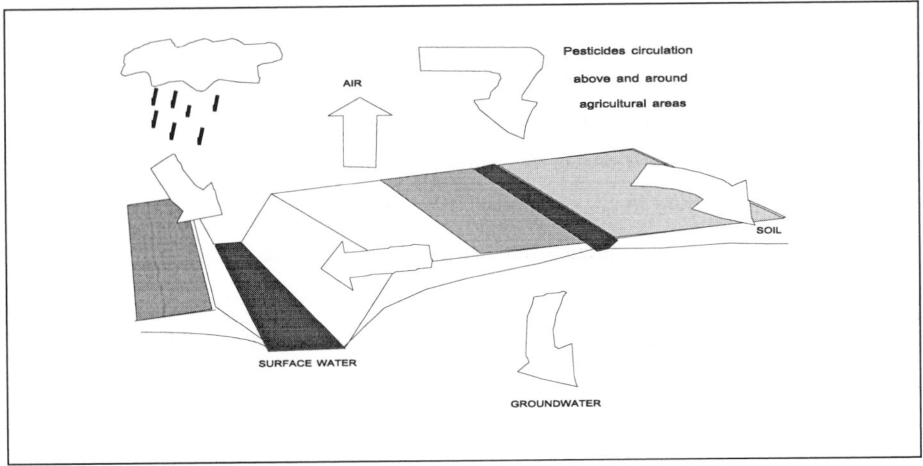

Figure 2: Spray drift is the main cause for atmospheric transport of pesticides over short distances. Medium and long range transport are mainly determined by volatilization, transport of particles and deposition cycles.

2.1. Transport mechanisms of pesticides through the air

During pesticide applications, fine droplets can be carried away on airstreams as spraydrift. Studies of Van Haasteren et al. [8] estimated that about 25 to 40% of the applied dose entered the atmosphere as spraydrift or evaporated from the cropsurface or soil. Due to rapid atmospheric transport, pesticides can have a negative environmental impact on both direct surroundings of application areas as well as further away from their source when they are deposited again (figure 2). Atmospheric transport and deposition of persistent pesticides (or compounds) are determined by molecule characteristics and meteorological circumstances. Deposition and re-entry processes in the atmosphere can be repeated several times until a compound reaches its 'sink'. These processes can lead to migration of persistent compounds from warm climatic regions towards the poles and can be considered as transboundary problems. The residence time in the atmosphere could serve as a valuable screening parameter for possible environmental impact of emitted substances. A second screening parameter could be the persistence after deposition, this is in particular important at the poles where degradation processes are slow due low temperatures.

2.2. A major part of the applied dose of pesticides dissipates into the atmosphere

Recently, the mid-term evaluation of The Netherlands multi-annual crop protection plan estimated that 95% of the total pesticides emission to the environment dissipates into the atmosphere and can be transported over long distances (table 1). This illustrates the importance of persistent pesticides in the atmosphere and the need to control this major pathway for dissipation into the environment.

The presence of pesticides in the air itself can present a potential health risk for workers and for the environment close to the application point. For very toxic and volatile agents as for instance methyl bromide, The Netherlands regulations therefore stipulate a safety zone of 100 m around the object where it is applied. Workers are only allowed to carry out activities for a limited period of time within this area.

Potential threats for locations at medium or long distance from the application areas are more difficult to cope with and no regulations exist. An exception, however, is methyl bromide due to its ozone-depleting potential [9]. The following section outlines different studies on residence time and spatial extent of atmospheric transport of pesticides substances.

Table 1: Estimates of emission of crop-protection agents to water, soil, groundwater and atmosphere [10]. The largest part of the applied dose enters the atmosphere and can be transported rapidly over long distances. The impact of this transport depends on persistence of these agents in the atmosphere and the sinks where they will concentrate.

	Emission in 1995 (in kg/year)	Reduction achieved since 1984/88 (%)	Aimed reduction 1984/88-1995 (%)
Surface water	46.000	72	> 70
Soil/Groundwater	67.000	80	40-45
Air	3.110.000	43	30-35

2.3. Evidence of the impact of atmospheric transport at medium range

Evidence of negative effects of agricultural pesticides at medium range from the application point is growing. Several recent studies in The Netherlands demonstrated contamination of rainwater, leading to concentration levels higher than the ecological target values in natural reserves at significant distance from agricultural areas [11,12,13]. Boom et al [14,15] monitored organostannic compounds in the Flevoland province and studied the impact of intensive pesticides use on atmospheric deposition and concentrations in surface waters. Crop protection agents were detected in the natural reserve of the Oostvaarders Plassen, located in the western area of the province, upwind of the agricultural areas. Pesticides application from planes on potato crops were considered to be the major source for atmospheric transport and environmental threats.

Pruissen et al [15] measured the occurrence of forty pesticides in rainwater at different sites around application areas in the province of Zuid-Holland. High concentrations of fifteen pesticides originated from four main sources: bulb-flowers cultivation (38%), intensive greenhouse farming (31%), regional transport (18%) and long distance transport (13%) of the total detected amount.

The reported levels lead only in incidental cases to exceedance of maximum tolerable risk levels, however, ecotoxicological target values were often exceeded. Non-target areas sometimes located kilometers away from the application area were contaminated with significant amounts of pesticides, leading to concentration that are not acceptable when a high level of environmental protection is required.

2.4. Long range atmospheric transport

Significant accumulation of crop-protection products in remote areas has, until now, only been observed for a number of organochlorinated compounds that were world wide used in large volumes [17]. This fact demonstrates that the phenomenon of long range atmospheric transport can cause problems in regions that are located far away from the pesticide application point.

The process of transport and accumulation happens over a long time scale, and involves deposition that is no longer directly related to a specific use or application point. Due to dilution that occurs, concentrations will remain relatively low, but will steadily build up in areas with relatively vulnerable natural habitats. Possible implications for biodiversity in the long term cannot be excluded, as most ecotoxicologically accepted concentrations were derived for average ecosystems in regions were recovery and recolonization are possible.

3. MEASURES: A NEED FOR AN INTERNATIONAL APPROACH

3.1. Effects of medium and long range transport difficult to prove 'a priori'

In the past decades, regulatory measures were taken after a problem had been clearly identified. However, effects of persistent chemicals in the environment occur after a long period. This in particular is the case with atmospheric diffusion, while during transport dilution occurs and compounds are spread over large areas compared to transportation by rivers.

Once the effects are evident, it will take a long time before they disappear or decrease. The actual authorization practice in most countries is that restriction or total ban of a pesticide has to be based on *a priori* demonstration of no potential negative side-effects. The benefits of pesticide use to a farmer can be made quite clear, however, negative side-effects at close range of the application site can only be estimated with current registration procedures. Since the safety margins between effect-concentrations and trigger values that will dictate restrictions in applications or a ban of a product are small, restrictions or bans will be accepted in most cases. In this way, the effects of spray drift on the neighboring aquatic environment can be taken into account in the registration process.

Medium range transport leads to exposure of non-target areas. This exposure will vary according to meteorological conditions, but will show some consistency over the years. With appropriate models it should be possible to account also for this issue in the authorization procedures.

At long ranges, weighing risks and benefits is even more difficult. Long-range transport involves a long term accumulation process. The lag-phase between use and effects can be long. The built-up of significant concentrations will take a longer time. Persistent chemicals that are only mobile via the atmosphere and that are used in large volumes will potentially accumulate in non-target areas. A direct proof of effects will be difficult. Outside agricultural areas ecological target values are set to protect sensitive environments. These ecological target values are based on a precautionary principle, and the extrapolation factors from effect-concentration will consequently be greater. Discussion over these factors may delay necessary action.

Therefore pesticides that present a serious hazard for long range transport through the atmosphere should preferably be identified in the registration procedure on the basis of intrinsic properties regarding their persistence in the environment and their ability to be transported in the atmosphere. Compounds selected on these criteria should either not be registered or their use should be limited to well controlled applications.

3.2. Global effects require globally harmonized criteria

Because of the global character of atmospheric transport, effective measures can only be taken in an international setting. The UN-ECE working committee on persistent organic pollutants (POP's) is about to reach consensus on the intrinsic properties that determine the risks for long range transport. This consensus was reached over a number of pesticides that are now causing world wide background contamination. There is, however, a need to translate these criteria into triggers to recognize high risk compounds that would be submitted for registration.

An international forum will have to be found to take care of the specific aspects of setting trigger values aimed at preventing unrestricted registration of pesticides presenting a potential risk for long range atmospheric transport.

The current risk assessment procedures of the EU offer a number of possibilities to include atmospheric transport in the hazard and risk assessment for the registration of crop protection products. This includes both the option to include wet and dry atmospheric deposition as one of the inputs into surface waters [18] as well as the possibility to evaluate the risks of concentrations of the active substance or its metabolite in the air [19]. These criteria are not yet operational however.

The EPPO (European and Mediterranean Plant Protection Organization) is preparing a scheme as guidance for decisions on atmospheric transport [20]. Criteria for these decisions have not yet been set in the EU. Because of the often global implications, development of harmonized trigger criteria for atmospheric transport should happen in an international setting familiar with both atmospheric transport issues and pesticide legislation.

REFERENCES

1. T.J. Peterly, Nature 244 (1969), 620.
2. D. Seehars, Report Franuenhofer-Institut für Umweltchemie und Oekotoxicologie, Schmallenberg, Germany (1993), 10.
3. D.J. Gregor and V.J. Gummer (1989), Environmental Science Technology. 23, 5.
4. Decree on environmental criteria for the authorization of pesticides (1994).
5. Official publication of the European Communities. Nr L227, 1.9.1994. Guideline 94/43 (Uniform Principles).
6. National Health Council of The Netherlands (1996). Report GZR 96/11E, Risks of pesticides to groundwater ecosystems.
7. W.A.J. van Pul, F.A.A.M. de Leeuw, J.A. van Jaarsveld, M.A. van der Gaag and C.J. Sliggers (1997, submitted to Chemosphere), The potential long-range transboundary atmospheric transport.
8. J.A. van Haasteren, F. Oeseburg and C. Huygen, Lucht en Omgeving (1987), 136 (in Dutch).
9. United Nations Environment Programme (1995), Report International programme on chemical safety; Environmental health criteria, 166.
10. Anonymous, Final report of the Emission-evaluation project of the Multi-Annual Crop Protection Plan (1996). Phytosanitary department of the Ministry of Agriculture, Wageningen, The Netherlands.
11. D.J. Bakker and K.D. van den Hout (1993). TNO report IMW-RW93/200, Delft, The Netherlands.
11. Bakker D.J., Weststrate J.H. (1996). TNO report IMW-MEP-R96/408, Delft, The Netherlands.
12. L.J.T. van der Pas, J.J.T.I. Boesten, R. Gerritsen and M. Leistra (1995). DLO-Staring Centrum report 387.3, Wageningen, The Netherlands.
13. L. Boom, (1993). Bestrijdingsmiddelen in neerslag en in oppervlaktewater. Report of the Fleverwaard Water Authority, Lelystad, The Netherlands.
14. L. Boom (1995). Ernstige vervuiling van oppervlaktewater en neerslag door bestrijdingsmiddelen in een agrarisch gebied. Report of the Fleverwaard Water Authority, Lelystad, The Netherlands.
15. O.P. Pruissen, M.P.J. Pulles, D.J. Bakker and J. Baas (1995). Report TNO-MW-R 95/103, TNO, Delft, The Netherlands.
16. Province of Zuid Holland, Pesticides in rainfall in South-Holland (1994). Report,

Dienst Water en Milieu, The Hague, The Netherlands.

17. United Nations Economic Commission for Europe (1996). United Nations publication ECE/EB.AIR/50, Convention on long-range transboundary air pollution and its protocols.
18. Official publication of the European Communities. Nr L227, 1.9.1994. Guideline 94/43 (Uniform Principles). § 2.5.1.3.
19. Official publication of the European Communities. Nr L227, 1.9.1994. Guideline 94/43 (Uniform Principles). § 2.5.1.4.
20. European and Mediterranean Plant Protection Organization (1997, in press), Council of Europe. Decision making scheme for the environmental risk assessment of plant protection products. Chapter 12: Air. EPPO-Bulletin).

Endocrine disruption - proof for an environmental factor? European approach

F.X. Rolaf van Leeuwen

WHO European Center for Environment and Health, Bilthoven Division, P.O. Box 10, 3730 AA De Bilt, The Netherlands

1. INTRODUCTION

The threat of impairment of human reproductive function and the impact on the health and reproduction of wildlife as a result of exposure to environmental pollutants have been a topic of growing scientific and public concern. During the last years, numerous scientific papers have been published, reviewing the health impact of endocrine disrupters. From sex changes in fish and alligators to increased incidence of testicular and breast cancer and falling sperm counts, endocrine disrupting chemicals have been accused for causing these effects, but the causal relationship is often not established, and the effects on human reproductive function are sometimes contradictory. These uncertainties and the recognition that, due to long-range transboundary transport of environmental pollutants, this problem is international rather than national, led the European Commission (DG XII), the European Environment Agency, and the European Center for Environment and Health of the World Health Organization (WHO-ECEH) to organize jointly an international workshop on the Impact of Endocrine Disrupters on Human Health and Wildlife. The workshop took place on 2-4 December 1996, Weybridge, England, and was hosted by the UK Department of the Environment. It was supported by OECD, national authorities and agencies of Germany, Sweden, and The

Netherlands and by European industry organizations (CEFIC and ECETOC). The workshop was attended by 78 participants form 11 European countries, the U.S. and Japan, as well as representatives of the Commission, WHO, EEA, OECD, CEFIC, the European Science Foundation and the European Environmental Bureau. It was the first time that this important subject has been discussed at such an international level and where regulators, the scientific community, industry and NGO's have participated.

2. ENDOCRINE DISRUPTERS IN A EUROPEAN PERSPECTIVE

This paper is a reflection of the discussions and the outcome of the European workshop on the Impact of Endocrine Disrupters on Human Health and Wildlife and is based on the Report of Proceedings of the Workshop (1).
The major objective of the workshop was to assess the scope of the problem of endocrine disrupting chemicals in Europe, in order to provide a consolidated basis to define a European strategy for research priorities and guidance for policy making and legislative measures.

Basis for a European approach

To develop such an European approach the workshop evaluated the potential risk with respect to effects on humans and on wildlife, possible relationships with exposure to environmental pollutants, identified gaps in the present knowledge and outstanding epidemiological questions, defined needs for monitoring, screening and testing of chemicals, and identified research priorities. As basis for the discussion a working document was prepared by an editorial group of European scientist, chaired by the Institute of Environment and Health (IEH), Leicester, UK. This document based its review and assessment on relevant reports published so far (e.g. IEH assessment on environmental oestrogens (2), reports from the Danish Environmental Protection Agency (3) and the German Federal Environment Agency (4) on the effects of endocrine active chemicals in the environment, US EPA workshops on research needs (5) and risk strategy (6),and the workshop of the international School of Ethology (7). In order to structure the discussion during the workshop, working groups were established focusing on human epidemiology, wildlife, mechanisms and modelling, exposure, and methodology.

3. MAJOR SCIENTIFIC ITEMS

A central theme in the discussion was the definition of an endocrine disrupter. It was agreed that any definition should focus on adverse effects observed in vivo in intact animals, and should encompass effects both on young or adult organisms and their progeny. The following definition was agreed:

An endocrine disrupter is an exogenous substance that causes adverse health effects in an intact organism, or its progeny, consequent to changes in endocrine function.

It was concluded that substances for which the endocrine disrupting activity is only identified in in vitro systems should be distinguished from the "true" disrupters by the adjective "potential". By using this definition of an endocrine disrupter one should realize that it is crucial that the adverse effects are due to the endocrine disrupting activity per se and are not secondary to the occurrence of overt toxicity to other organs or systems.

3.1. Human epidemiology

Although of varying degree, the incidence of testicular cancer has increased in almost all european countries which have reliable cancer statistics. It can be concluded that these increases are real, and that they are not attributable to improvement in diagnosis, and reporting. For sperm counts the information is less clear and a definite conclusion is hampered by the observed geographical differences. However, the magnitude of the effect is sufficiently large to conclude that "they are unlikely entirely attributable to known confounding variables, such as bias from selection of subjects, differing laboratory methodologies, and the influence of abstinence and frequency of intercourse".

Trends in sperm motility and morphology could not be established with sufficient certainty, but it was recognized that quality control of semen analysis was only recently well recognized. Therefore international coordination (WHO/EC) in this field was recommended. With regard to female endpoints only information on increasing trends in breast cancer are available.

For all the effects established in humans, however, it should be noted that there is no evidence for an association of human health effects and exposure to endocrine disrupting substances in the environment.

To investigate this topic further it was recommended that reproductive health should be investigated in cohorts having different exposure to endocrine disrupting pollutants or industrial chemicals. Also the topic of lifestyle related factors needs further attention. In addition, readily measurable endpoints, such as twining rate and sex ratios, should be included in epidemiological studies.

3.2. Wildlife

Contrary to the information available in the U.S. only few cases exist in the European region where adverse endocrine effects are associated with high environmental levels of endocrine disrupting chemicals. It was stated by the Workshop that the assessment of potential effects on wildlife should concentrate on reproductive effectiveness, because this was considered to be the critical factor in survival of populations, and consequently the maintenance of biodiversity. For field studies a broad testing strategy, including comparison with unimpacted areas was suggested. The need for the development of biomarkers, particularly to predict impact on reproductive effectiveness was underlined, as well as the identification and selection of sentinel species.

3.3. Mechanisms and models

Although there is a vast amount of information on the mechanisms via which hormonally active compounds control basic physiological processes, such as the development of the reproductive system, there is still insufficient information on the effect of exogenous compounds on the endocrine system. Also the link between chemically induced pathophysiological effects and endocrine function is poorly understood. In general there are sufficient animal models to detect adverse reproductive effects with the exception of testicular cancer, but it was recommended that current animal models should be validated with reference to their relevance for human hazard assessment. It was understood that different models are needed for screening of chemicals and for mechanistic studies, but the workshop recommended that priority should be given to studies detecting effects rather than to those unraveling the mechanism of action. It was noted that there are parallels between persistent oestrus in rodents and the phenomenon of polycystic ovaries in humans, but there is an urgent need to establish whether the aetiologies of both effects are comparable. Because persistent oestrus is an estrogen induced effect and readily detectable, it may have considerable utility if both effects are comparable.

3.4. Exposure

It was recognized that information on the presence of endocrine disrupters in environmental compartments, data on sources and release into the environment, and information on dispersion, bioaccumulation, and metabolism is very limited. An integrated strategy on exposure assessment linked with epidemiological studies with humans or field studies with wildlife was recommended. Caution, however, was expressed when undertaking these studies, because there are thousands of chemicals and monitoring should therefore focus on those chemicals which have been shown to exert endocrine alterations in validated in vivo test systems. An Europe-wide strategy for monitoring should be developed in which maximum use should be made of existing databases such as IUCLID. This will facilitate appropriate risk assessment. Where appropriate risk mitigation measures could be taken to reduce exposure of humans and wildlife. For decisions on these measures studies of the cost-effectiveness are essential.

3.5. Methodology

The ability of a compound to disturb endocrine systems could be best determined in a whole organism (in vivo), because the effects observed in such a situation are more relevant than those observed in vitro. The major mechanisms for these effects are receptor interactions and disturbances of hormone synthesis or metabolism. Although the effect of endocrine disruption involving changes in metabolism can be best evaluated in vivo, it is much easier to investigate them in vitro. Therefore, the development of in vitro assays in this field may be a future perspective for research, but one must realize that this type of research only identifies the "potential" endocrine disrupters.

It was recognized that based on the available toxicological data different approaches should be adopted for assign a chemical's potential hazard to disturb endocrine function. During the initial assessment emphasis will be placed on the identification of adverse effects rather than on mechanistic aspects. Based on these considerations a testing strategy was proposed, primarily based on currently available test systems and protocols (1). Besides, it was recommended that whole-organism assays should be developed (and validated) for testing endocrine disrupters in birds and fish. Also the use of structure activity relations (SAR) was recommended in association with the acquisition of new data, in particular for chemicals where, at present, only limited data are available.

4. GENERAL CONCLUSION

As shown above, the workshop concluded on a number of scientific issues related to the potential health threat for humans and wildlife of exposure to endocrine disrupting chemicals. However, also the strategic and policy domain was touched and to this end a number of conclusions was drawn (1). Because of their importance these conclusions are cited literally:

"It was accepted that resource allocation to this area should be balanced against other important public health issues. It was recommended that policy should be based upon scientific principles, following a weight-of-evidence approach and that studies should be performed following rigorous scientific principles and practice. When deemed necessary consideration should be given to measures to reduce exposure to endocrine disrupters in line with the Precautionary Principle, as described in Principle 15 of the 1992 Rio Declaration."

REFERENCES

1. Report of the Proceedings of the European Workshop on the Impact of Endocrine Disruptors on Human Health and Wildlife : 1996. European Commission, DG XII (EUR 17549).
2. Institute for Environment and Health Assessment on environmental oestrogens: consequences to human health and wildlife (Assessment A1): July 1995.
3. Danish Environmental Protection Agency Report: 1996. Male Reproductive Health and Environmental Xenoestrogens. Environ. Health Perspect. 104 (Suppl. 4), 741-803.
4. German Federal Environmental Agency: Endocrinically active chemicals in the environment (Texte 3/96) January 1996.
5. United States Environmental Protection Agency Sponsored Workshop: Research Needs for the Risk Assessment of Health and Environmental Effects of Endocrine Disrupters: 1996. Environ. Health Perspect., 104 (Suppl. 4), 715-740.
6. United States Environmental Protection Agency workshop: Development of a risk strategy for assessing the ecological risk of endocrine disruptors: May 1996.
7. International School of Ethology 11^{the} Workshop, Erice, Sicily: Environmental endocrine disrupting chemicals: neural, endocrine, and behavioral effects: 1996.

Research on Emissions and Mitigation of POP's from Combustion Sources

C. W. Lee, P. M. Lemieux, B. K. Gullett, J. V. Ryan and J. D. Kilgroe

Air Pollution Technology Branch, Air Pollution Prevention and Control Division, National Risk Management Research Laboratory, U.S. Environmental Protection Agency, Research Triangle Park, NC 27711, USA

ABSTRACT

The environmental consequences of persistent organic pollutants (POP's) are of increasing concern due to the serious health effects on animals and humans including reproduction, development and immunological function. Several major classes of POP's, including polycyclic aromatic hydrocarbons (PAH's), chlorobenzenes, chlorinated dioxins and chlorinated furans, have been identified as products of incomplete combustion (PIC's) produced in trace levels in combustion systems. A wide variety of combustion processes, ranging from power plants, industrial boilers, industrial furnaces and incinerators, to home heating devices, are believed to be potential sources of POP's. Full-scale combustion facilities can be significant sources of POP's due to the large mass flow of flue gas released from a plant. Total emissions of POP's from small combustion devices, such as wood stoves and residential oil furnaces, can also be significant due to the large numbers of existing units near high population areas. It becomes increasingly important to understand the formation of POP's from different combustion processes to identify sources of POP's and to develop strategies for their prevention and mitigation.

Research on POP emissions from combustion sources conducted by EPA is largely driven by the need for regulating the emissions of hazardous air pollutants as required by

Title III of the 1990 Clean Air Act Amendments and by the Resource Conservation and Recovery Act. This paper provides a summary of EPA's research on emissions and control of POP's from combustion sources with emphasis on source characterization and measurement, formation and destruction mechanisms, formation prevention and flue gas cleaning. Laboratory experiments conducted to examine the PAH emissions from a wide variety of combustion processes, ranging from pulverized coal utility boilers to wood stoves, have shown that they exhibit widely different emission characteristics. Waste incineration research conducted by the National Risk Management Research Laboratory, Air Pollution Prevention and Control Division (NRMRL/APPCD) has also shown that complex mechanisms, including physical mixing and chemical kinetics, are involved in the formation of chlorinated PIC's.[1] Research has also indicated that the formation of ultra-trace levels of chlorinated-dioxins and -furans in combustion/incineration processes includes the complex interaction of several factors including temperature, chlorine content and catalyst. The beneficial effect of sulfur and sorbents for dioxin formation prevention is demonstrated. This Laboratory's effort to develop and evaluate state-of-the-art technologies for on-line measurements of PAH's, volatile PIC's, dioxins and furans is also discussed. The promising potential of applying artificial-intelligence-based control systems for improving combustion processes operating conditions as a POP prevention approach is demonstrated.

1. INTRODUCTION

The wide distribution of persistent organic pollutants (POP's) in the environment has become a global issue of growing concern. International intentions to control such substances are expressed in the Washington Declaration made at the conclusion of a United Nations Environmental Program conference attended by environment ministers of 108 countries held in Washington, DC, during November 1995 [1]. POP's are long-lived organic compounds which survive long distance migration in the global environment and

[1] NRMRL/APPCD was previously named the Air and Energy Engineering Research Laboratory (AEERL).

become concentrated as they move through the food chain with serious health effects on animals and humans including reproduction, development and immunological function. Much of the concern on POP's involves 12 chemicals or chemical classes, including polychlorinated biphenyls (PCB's), polychlorinated dibenzo-p-dioxins (PCDD's) and -furans (PCDF's), polycyclic aromatic hydrocarbons (PAH's) and pesticides such as DDT and chlordane.

Although the production and use of many of the anthropogenic chemicals which are considered as POP's, such as PCB's and dichlorodiphenyltrichloroethane (DDT), are banned in most developed countries, they are still widely distributed and used in developing countries. However, there are also POP's which are produced unintentionally as byproducts from combustion processes. Several major classes of POP's, including PAH's, chlorobenzenes, PCDD's and PCDF's, have been identified as products of incomplete combustion (PIC's) emitted at trace levels from various combustion systems. The development of control and mitigation strategies for POP's requires a better understanding of the major sources of these pollutants. It is very important to identify the industrial and residential combustion processes which may also be the potential sources of POP's. This paper summarizes EPA's research on POP emissions and control from different combustion sources.

Research on POP emissions from combustion sources conducted by the Air Pollution Prevention and Control Division (APPCD) of EPA's National Risk Management Research Laboratory (NRMRL) is motivated by the need for regulating the emissions of air toxic pollutants as required by Title III of the 1990 Clean Air Act Amendments (CAAA's) [2] and by the Resource Conservation and Recovery Act (RCRA) [3]. Title III of the CAAA's lists 189 compounds and compound classes and application of maximum achievable control technology (MACT) is required by any non-utility source that emits over 9,091 kg/year (10 t/year) of any one air toxic pollutant, or 22,727 kg/year (25 t/year) of total air toxic pollutants. PAH's, chlorobenzenes and PCDD's/PCDF's are among the classes of the 189 regulated air toxic pollutants which are also considered as important POP's. In addition, revised RCRA standards were proposed for hazardous waste combustors (hazardous waste burning incinerators, cement kilns and lightweight aggregate kilns) in April 1996 [4]. These proposed standards would limit the

PCDD/PCDF emissions to 0.2 ng International -Toxic Equivalency (I-TEQ)/dscm. Characterization of air toxic emissions from a wide range of sources and development of their control strategies have become an important part of the Agency's CAAA's implementation efforts. Research on dioxin emission prevention and control from hazardous waste combustion is very important to the Agency's RCRA regulation development.

2. POP EMISSIONS

Characterization of organic air toxic emissions from combustion sources in general is much more difficult compared to that from chemical production facilities which involve predominately fugitive emissions of individual product compounds. Air toxic emissions generated by combustion, including the three major classes of POP's mentioned above, are typically contained in a complex mixture of organic compounds at trace levels. Because of large volumes of flue gas produced during the combustion process, even trace levels (ppm) of air toxic compounds in the flue gas can exceed the limits specified under Title III of the CAAA's. Due to a wide variety of combustion boilers that exist for utility, industrial, commercial and institutional applications across the country and because of the relatively small amount of information available characterizing organic air toxic emissions from these sources, research has been undertaken recently by EPA to provide such information.

2.1. PAH Emissions

PAH's are among the most common PIC's found in combustion flue gases at trace levels. PAH's will likely be produced as a result of localized improper mixing of fuel and combustion air in a large size industrial combustor. A study was conducted by EPA to characterize organic air toxics emissions in the flue gases from the combustion of pulverized coal [5]. A small-scale combustor was operated under different conditions to simulate high excess air firing and nitrogen oxide (NO_x) controls by combustion modifications of a utility boiler. Only a few organic air toxics were found above the detection limits, with naphthalene as the only PAH identified from the tests. Results of the tests also indicate that total air toxics emissions from a large coal-fired utility plant are not likely to increase as a result of installation of combustion modifications for NO_x control. In other research performed recently by EPA to determine the emissions levels of organic air toxics from combustion of various grades of oils, tests were conducted on a commercial fire tube package boiler running on fuel oil [6]. It was found that carbonyls dominated the trace organic emissions and PAH's constituted only minor components of the emissions.

The emissions of PAH's have been identified as the major pollutants emitted from residential combustion units in studies conducted by EPA. Large numbers of such units, including wood stoves, oil combustors and coal-fired furnaces, are operating in urban/suburban regions and their emissions may have serious impacts on indoor and ambient air quality. With collaboration of EPA's National Health and Environmental Effects Research Laboratory as part of the Integrated Air Cancer Project, we have conducted studies to characterize the organic emissions from different residential combustion devices. High levels of PAH's (0.7 g/kg wood burned) have been observed in EPA's wood stove research [7]. Some operating variables, such as wood species, stove type and altitude of the stove, were found to have a strong effect on PAH emissions, while burn rate has very little effect on PAH emissions. Pine was found to produce more PAH's than oak; conventional stoves showed higher PAH emission rates than those from a catalytic stove; and lowering the operating altitude of the stove from 825 to 90 m caused an increase in PAH emission rates. It was also found from this study that emissions from burning oak are less mutagenic than those from pine and emissions from the catalytic stove were more mutagenic than those from conventional stoves.

The organics emitted from residential heating furnaces have been found to be closely related to the chemical structure of the fuel burned in EPA's home heating fuels studies [8]. This is particularly true for wood and coal which contain mainly polymeric chemical structures where thermal cleaving during combustion accounts for a significant portion of the emitted organics. It was found that combustion of wood, which contains largely lignin and cellulose, produces mainly oxyaromatics and naphthalene accounts for almost all the PAH's in its emissions. The burning of coal, which contains fused ring structures, produces emissions with three-, four- and five-ring PAH's, a class that includes benzo(a)pyrene and other known carcinogens. In the case of oil, the organic emissions contain mainly the unburned droplets of the oil itself, with substituted naphthalenes dominating the PAH emissions. Natural gas was found to be a clean-burning home heating fuel. Only a few PAH's with levels at least 100 times less than those emitted from wood stoves were identified in natural gas furnace emissions [9].

PAH's were identified as the most common pollutants produced from open burning of a wide variety of waste materials. Open burning is still widely practiced as a waste management method for several types of wastes, ranging from agricultural plastics to land-clearing debris. Because of concerns of their potential impacts on ambient air quality, EPA has undertaken a series of experimental studies to characterize the emissions from simulated open burning of waste materials. It was found that emissions produced from simulated open burning of agricultural plastic contain a complex mixture predominated by high molecular weight PAH's, only a minor fraction of which can be identified [10]. No mutagenic effects were found for the emissions as a whole; however, organic extracts of the particulate emissions contained mutagenicity comparable to that measured from wood stoves. In response to public concern over health hazards caused by the growing incidence of tire fires, EPA has conducted research to characterize the emissions generated from simulated scrap tire fires. It was found that the emissions contained mainly PAH's, which ranged from 10 to 50 g/kg of tire material burned and alkyl-substituted PAH's were the predominant PAH's identified [11]. EPA has also conducted a study to characterize the emissions produced from open burning of land-clearing debris, in order to assess the environmental risk associated with such land-clearing practice. Substantial emissions of a large number of pollutants including carbon monoxide (CO), fine particulate matter, volatile organic compounds (VOC's) and high molecular weight organics dominated by PAH's were observed from simulated open burning experiments [12]. Only 14 of the PAH's were identified, with a majority of PAH's only tentatively identified through searches of mass spectral libraries.

2.2. Formation of Chlorinated PIC's

We are continuing our research to understand the formation and control mechanisms associated with toxic organic pollutants from waste incineration, with emphasis placed on the emissions of chlorinated PIC's. Chlorocompounds are commonly contained in waste streams, which are difficult to destroy thermally during incineration with the resulting formation of chlorinated PIC's. Many of the chlorinated PIC's, such as chlorobenzenes, are toxic and also listed as a major class of POP's. Our research has focused mainly on PIC formation and control from rotary kiln incinerators. This particular incinerator design is very versatile and is widely used for treating industrial wastes in the form of liquids, sludges, or solids.

The secondary combustion chamber (SCC) is an important piece of control equipment for rotary kiln incinerators [13-14]. The SCC should be capable of destroying any unburned organic material that exits the primary combustion chamber due to rogue droplets, transients, quenching, or incomplete mixing. SCC's are also commonly used to combust liquid wastes that have high heating values. Design criteria in the past have been mostly limited to a time-temperature requirement, such as 2 s at 1000°C (1800°F). Although a time-temperature requirement is not written into the hazardous waste incinerator regulations as defined in the Resource Conservation and Recovery Act (RCRA), it appears to have been adopted as a criterion by regulators and the regulated community alike. A disadvantage of this "apparent" policy is that mixing, known to be of critical importance in incineration systems [15], is largely ignored and no economic incentives exist with which to improve afterburner designs, given that any new design would likely require a certain time-temperature profile before being allowed to be installed, even if such a design could meet required emissions limits with a much more compact configuration.

The emissions that the SCC must deal with generally result from some sort of system failure in the primary chamber, since steady-state operation of the primary chamber generally eliminates the need for an SCC. Liquid injection incinerators, for example, typically don't require an SCC. The failure modes that can cause elevated levels of organic compounds to enter the SCC include mixing failures, such as those caused by: poor microscale mixing intensities or poor macroscale mixing; poor atomization; flow stratification; batch charging and depletion of oxygen in the primary chamber; and reaction quenching, such as that caused by unburned material entering cold regions of the combustion device, or by cold walls. Rotary kilns in particular exhibit high levels of flow stratification [16-18] and typically have some of their waste feed fed in batches and, as such, generally employ an SCC.

Part of the reason that a time-temperature requirement is used as a common SCC design criterion is that the effects of turbulence and complex chemical kinetics are not understood well enough to incorporate their use into the permitting process. It is very important, however, to work toward gaining an understanding of kinetics and mixing in incinerators, since it is possible to have excessive levels of PIC's even after having successfully met the necessary time-temperature requirement. EPA, in cooperation with the New Jersey Institute of Technology (NJIT) and Massachusetts Institute of Technology (MIT), has been performing research on a pilot-scale rotary kiln incinerator simulator (RKIS) to complement laboratory-scale research being performed at both of the previously mentioned academic institutions, with the ultimate goal of furthering the state-of-the-art of SCC design by incorporating gas-phase mixing and kinetic considerations into the design criteria, particularly in regards to chlorocarbon combustion. Initial pilot-scale experiments have consisted of system characterization tests.

In order to incorporate gas-phase mixing and kinetic phenomena into afterburner design, it is necessary to achieve several goals, including:

1. Development of reaction pathways and kinetic data for combustion of the principal organic hazardous constituents (POHC's) present in the waste, along with possible mechanisms of formation of PIC's from POHC decomposition products. Although mechanistic information is not available for complex compounds, mechanisms do exist for C1 and C2 chlorocarbon combustion [19]. Initial tests focused on combustion of carbon tetrachloride (CCl_4) and methylene chloride (CH_2Cl_2), compounds for which kinetic mechanisms exist. Combustion of CH_2Cl_2 results in levels of 1,2 dichlorobenzene and monochlorobenzene much higher than those found from CCl_4 combustion. It may be possible that CH_2Cl_2 can readily form chlorinated intermediate structures that are ring-growth precursors, resulting in direct formation of monochlorobenzene rather than from chlorination of benzene. Combustion of CH_2Cl_2 produced higher quantities of

identified PIC's than combustion of CCl_4, particularly during fuel-lean combustion. Although CCl_4 may be useful as a POHC due to its high thermal stability and provide a useful measure of a system's ability to meet the required 99.99 % destruction and removal efficiency (DRE), it may not challenge an incinerator's ability to produce or destroy PIC's.

2. Development of models that take into account macromixing and micromixing phenomena to aid in the scale-up of results from very small-scale experiments to pilot and full-scale systems. Kinetics and thermodynamics alone cannot account for emissions of PIC's from incinerators. Mixing must eventually be considered [20]. Initial tests have attempted to characterize the macromixing in the EPA's RKIS SCC using sulfur dioxide (SO_2) as a tracer. By measuring residence time distributions (RTD's) in various portions of the SCC, it is possible to determine the unknown reactor volumes in a series of ideal reactors. These reactor volumes can then be modeled numerically using detailed reaction mechanisms [21].
3. Development of techniques to measure trace organic species or surrogates for trace organic species in the field, given that many of the advanced diagnostics available in a laboratory setting cannot easily be transferred to a field application. Semi-continuous measurement of key organic compounds can potentially be used to characterize the overall destruction of all hazardous trace organics of concern. Since POP's are frequently in the semi-volatile range of boiling points (>150°C) and thus more difficult to measure using existing monitoring methodologies, it is useful to find a surrogate compound for the POP's that is more easily measured than the POP of interest.

2.3. Dioxins

On-going research has been investigating the fundamental mechanisms behind formation of PCDD's/PCDF's, collectively termed 'dioxins'. Dioxins are formed in the post-combustion region from components of chlorine (Cl), unburned organics and catalytic surfaces. A complex interaction of surface-induced catalysis, organic ring structure formation and chlorination countered by destruction/dechlorination reactions can result in the formation and emission of dioxins and other chloro-organics The synthesis reactions that lead to the formation of dioxins occur at temperatures ranging from approximately 200 to 600°C [22, 23]. They may be catalyzed by combustor deposits and by entrained and collected fly ash. These deposits can be composed of fly ash or sooty materials containing catalytic metals such as copper [24, 25]. With appropriate fly ash properties and operating temperatures, air pollution control devices such as electrostatic precipitators and fabric filters can act as chemical reactors that form dioxins and other chloro-organics[25].

While commonly associated with waste combustion, dioxin formation can also occur from burning coal, oil and wood. The mechanisms of formation are being studied at EPA to enable source-specific predictions and to develop pollution prevention strategies that discourage or prevent dioxin formation from occurring. Work at EPA has coupled mechanistic aspects related to the effect of different types of metal catalysts [26], the ability of Cl to chlorinate aromatic ring structures [27] and the role of catalysts in formation of the biaryl structure [28]. The effect of these interactive mechanistic parameters, along with combustor operating parameters, has been examined for their relationship with dioxin formation [29-30]. This work has shown on a combustor-specific basis that reduction of hydrogen chloride/chlorine (HCl/Cl_2) concentration, completeness of combustion and/or increases in quench rate can reduce formation of dioxin.

Recent work has focused on the effect of sorbent injection technologies to remove Cl and thereby prevent formation of dioxins [31]. Research programs on two pilot-scale combustors have shown that injection of calcium (Ca)-based sorbents at moderately high temperatures has the effect of preventing formation of dioxins [29, 32]. This is likely to hold for all systems in which formation is Cl limited. This technology is the subject of two U.S. patents [33, 34] and is currently undergoing field demonstration in the U.S.

Another preventive strategy derived from mechanistic studies is the suppression of dioxin formation by the presence of sulfur (S) as SO_2. The effect of SO_2 may relate to its effect on an important, catalytic chlorination reaction and biaryl synthesis [35] and/or its effect on gas-phase reactions with chlorinating compounds [36]. This SO_2 effect has been shown on a large pilot scale and is currently undergoing field demonstration. Results to date [30, 32] suggest significant (90%) suppression of dioxin formation across all congener classes at S/Cl ratios >1/1.

Upcoming work at EPA will undertake a mobile sampling program for dioxin emissions from heavy duty diesel vehicles. This work, part of EPA's Dioxin Reassessment effort, will provide important information in the effort to quantify dioxin sources in the environment. This effort is significant in that U.S. mobile sources, with the exception of a recently completed tunnel study (still in draft form), are entirely uncharacterized as potential dioxin sources. Further, it is possibly the first mobile sampling effort for diesel engine dioxins.

Another recently initiated program will be using isotopically labeled fuels to study the mechanisms and rates of formation. This work is geared toward discerning the critical pathways of dioxin formation and determining rate information for development of a global reaction model. Expected results will discern the importance of incomplete combustion (either gas phase or carbonaceous, solid phase) on supplying organic dioxin precursors and, thus, will be relevant to a wide variety of fuel types and combustors. This program is being designed with the assistance and advice of a team of international researchers and reviewers.

3. ON-LINE POP MEASUREMENT

Part of the importance of determining what POP's are released from combustion systems is the measurement of those POP's. This creates a major technical challenge since many of the compounds of interest are present in very small quantities, plus are normally in the semivolatile boiling point range (>150°C), which partitions the organic compounds of interest between the gas phase and condensed phase (usually bound on particulate matter). This partition makes continuous measurement difficult. A useful approach to circumvent this limitation is to measure another, more easy-to-measure compound, that is a precursor to the POP of interest, or at least is a well-correlated indicator of not only the presence/absence of the POP, but also of the relative concentration of the POP in the stack. This concept of using a surrogate indicator has been used in current regulations of municipal waste combustors (MWC's). CO, for example, has been used as a useful surrogate for PCDD's/PCDF's in MWC systems; however, its usefulness as a surrogate indicator breaks down in a system that is very well operated.

A potentially more useful approach is to measure a volatile organic precursor to the semivolatile POP's of concern. In a recently completed test program [37] at the U.S. EPA Incineration Research Facility (IRF) in Jefferson, Arkansas, several continuous emission monitors (CEM's) for measuring trace quantities of various organic and inorganic pollutants were tested. One of the CEM's tested was a dual-detector on-line gas chromatograph (GC) system developed by EPA/APPCD in-house. This system consisted of a sample delivery system, a sample concentrator and a GC equipped with a flame ionization detector and an electron capture detector [38]. The IRF's rotary kiln incineration system was operated at conditions indicative of normal incinerator operation, while injecting varying concentrations of 10 VOC's that are usually found in incinerator stack gases as PIC's. Target VOC concentrations included low level (1-2 $\mu g/m^3$), medium level (10-20 $\mu g/m^3$) and high level (160-240 $\mu g/m^3$). Tests were performed to compare CEM results to the EPA's standard reference methods using a relative accuracy test audit protocol [39]. The EPA/APPCD on-line GC performed successfully for most compounds at all concentration levels. Instrument sensitivities were sufficient to measure all 10 compounds at levels typically found in well-operated systems.

The high molecular weight PAH's are predominantly adsorbed on soot (carbon aerosols) due to their low vapor pressure at stack temperatures. Currently, PAH's are measured using a sampling train (EPA Modified Method 5, MM5) followed by solvent extraction and analysis using GC and mass spectrometry (MS) [40-42]. The measurement is time consuming and no real-time data can be obtained from such captive sampling techniques. EPA recently conducted tests to evaluate the application of a photoelectric aerosol sensor for on-line measurement of particle-bound PAH's in combustion flue gas. The PAH monitor works on the principle of photoionization of carbon aerosols. After being exposed to the ultraviolet (UV) light of the monitor, the carbon aerosols which have PAH molecules adsorbed on the surface pass through the monitor, then emit electrons and the resulting electric current is proportional to the particulate-bound PAH's. Initial tests of the PAH monitor for measuring emissions from burning tire-derived fuel in EPA's pilot-scale RKIS indicated that the monitor appeared to track transient operation of the combustor well [43]. The PAH monitor was also evaluated during a recently completed test program, which gave excellent relative accuracy for measuring the three selected PAH's (naphthalene, phenanthrene and pyrene) at an intermediate concentration level [37]. Operating problems caused by the high moisture content of the combustion flue gas were experienced and the monitor's manufacturer is modifying the design of the instrument's moisture removal system.

A recent international research program has assisted in the development of Jet-REMPI™ (resonance-enhanced multiphoton ionization) as an analytical technique for monitoring extremely low concentrations of chlorinated dioxins. The DLR Jet-REMPI apparatus was tested in the EPA laboratories for its ability to detect chlorinated dioxins. This work [44] has detected spectra for a dichlorodibenzodioxin isomer and determined a detection limit of less than 30 ng/dscm. Identification work for more highly chlorinated isomers is currently underway. It is likely that suitable correlations between the lower chlorinated dioxin isomers and the more toxic, higher chlorinated isomers can be established, enabling Jet-REMPI to be used as a correlative monitor for dioxin toxic equivalency (TEQ) emissions.

4. APPLICATIONS OF AI FOR COMBUSTION CONTROL

The objective of this research [45] was to apply fuzzy logic artificial intelligence (AI) to control combustion systems, in particular the pilot-scale RKIS and its secondary combustion chamber (SCC) mentioned above. The mechanistic details of system response in a system such as the RKIS are complex and are not usually known *a priori*. However, a relatively small number of generic control rules based on operating experience can be linguistically stated and then translated into a fuzzy logic system for automatic control. The purpose of the control system in this case was to reduce emissions of PIC's from the RKIS, when significant transients were artificially imposed on the system by spraying liquid surrogate wastes into the RKIS based on a time-based algorithm.

Rotary kilns have the advantage of being flexible enough to handle a wide variety of waste streams in a wide variety of forms, including flammable liquids, aqueous streams, sludges and whole drums. When a drum containing volatile material is fed into a rotary kiln, it ruptures and releases its contents into the hot kiln environment in a transient event that occurs over a short period of time. If the instantaneous stoichiometric requirements to combust the volatile waste are greater than the amount of oxygen being added to the kiln through the main burners or auxiliary air sources, then a plug of unburned material, called a transient "puff" leaves the kiln and must be dealt with by the SCC. The transient puffs include large quantities of soot, CO and organic compounds of both a volatile and semivolatile nature. Past experiments have shown high quantities of PAH's contained in transient puffs (46).

One possible control option is to equip the SCC with a system that can inject additional oxidizer in response to a puff leaving the kiln. This option has been shown to dramatically decrease the emissions of PAH's and other PIC's (35). If air is used as the oxidizer, then the downstream equipment, such as baghouses or scrubbers, must be sized to handle the increased volume of flue gases, based on maximum flow rates. If oxygen (O_2) is used as the oxidizer, then the downstream equipment can be sized smaller; however, significant costs are associated with the use of pure O_2. A tradeoff is realized between increased capital costs and increased operating costs. If a means is available to maximize the efficiency of O_2 usage, then O_2 injection will become more economically attractive.

A series of experiments were performed to evaluate the effectiveness of fuzzy logic control schemes by injection of pure O_2 into the SCC's afterburner. A statistically designed test matrix was repeated for each of four control schemes: no control, feedback control (based on stack O_2) and two fuzzy logic control schemes based on CO and total hydrocarbon (THC) emissions at the kiln exit. Results were evaluated using a combined performance indicator that utilizes stoichiometrically weighted integrated emissions of CO, THC and soot at the inlet and outlet of the SCC. The following conclusions were observed:

- The speed of analyzer response is a significant hindrance to the effectiveness of any sort of process control for reduction of transients from incinerators.
- The feedback control tended to open the O_2 valve all the way in order to attempt to restore the stack O_2 to a level near the set point.

- The fuzzy logic control scheme was more effective than feedback in controlling transients of short duration.
- The fuzzy logic control scheme was significantly more efficient in its use of the injected oxygen, particularly where the transients were small.

5. CONCLUSIONS

Major classes of POP's including PAH's, chlorobenzenes and chlorinated dioxins and furans can be emitted from various combustion processes as unintentional and unwanted byproducts. Research has been conducted by EPA to study POP emissions from combustion sources, with emphasis on understanding the fundamental mechanisms which lead to POP emissions and identifying mitigation strategies for such emissions. Our research found that good combustion conditions achieved in well-operated industrial units in general prevent significant POP emissions. POP emissions result from complex interactions of a wide variety of physical and chemical mechanisms occurring in combustion systems. POP emission potentials strongly depend on the design and operating characteristics of the combustion units. The presence of fuel/waste components which contain difficult-to-destroy chemical structures in fuels/wastes also plays an important role for promoting POP emissions. Significant progress has also been made in our research on developing technologies for continuous monitoring and reducing POP emissions from combustion sources.

REFERENCES

1. F. Wania and D. Mackay, Env. Sci. Technol., 30 (9), 190A-196A, 1996.
2. Public Law 101-549, Clean Air Act Amendments of 1990, November 15, 1990.
3. 40 CFR Part 60, "Hazardous Waste Combustors; Revised Standards; Proposed Rule", Part II, U.S. Environmental Protection Agency, April 19, 1996.
4. 40 CFR Part 60, et al., "Hazardous Waste Combustors, Revised Standards, Proposed Rule", Part II, U. S. Environmental Protection Agency, April 10, 1996.

5. C.A. Miller, R.K. Srivastava and J.V. Ryan, Environ. Sci. Technol., 28 (1994) 1150.

6. C.A. Miller, J.V. Ryan and T. Lombardo, J. Air & Waste Manage. Assoc., 46 (1996) 742.

7. R.C. McCrillis, R.R. Watts and S.H. Warren, J. Air & Waste Manage. Assoc. 42 (1992) 691.

8. R.S. Steiber, J. Air & Waste Manage. Assoc., 43 (1993) 859.

9. J.V. Ryan and R.C. McCrillis, "Analysis of Emissions from Residential Natural Gas Furnaces," in Proceedings: 87th Annual Air and Waste Management Association Meeting, Paper No. 94-WA75A.04, Cincinnati, OH, June 1994.

10. W.P. Linak, J.V. Ryan, E. Perry, R.W. Williams and D.M. Demarini, J. Air & Waste Manage. Assoc. 39 (1989) 836.

11. P.M. Lemieux and J.V. Ryan, J. Air & Waste Manage. Assoc, 43 (1993) 1106.

12. C.C. Lutes and P.H. Kariher, "Evaluation of Emissions from the Open Burning of Land-Clearing Debris," Final Report, EPA-600/R-96-128 (NTIS PB97-115356), October 1996.

13. P.M. Lemieux, W.P. Linak, J.A. McSorley, J.O.L. Wendt and J.E. Dunn, Combust. Sci. & Technol., 74 (1990) 311.

14. R.W. Rolke, R.D. Hawthorne, C.R. Garbett, E.R. Slater, T.T. Phillips and G.D. Towell, "Afterburner Systems Study," EPA-R2-72-062 (NTIS PB-212-560), August 1972.

15. J.O.L. Wendt, W.P. Linak and P.M. Lemieux, Hazardous Waste & Hazardous Materials, 7 (1) (1990) 41.

16. V.A. Cundy, T.W. Lester, A.M. Sterling, A.N. Montestruc, J.S. Morse, C.B. Leger and S. Acharya, J. Air Poll. Cont. Assoc., 39 (1989) 944.

17. V.A. Cundy, T.W. Lester, A.M. Sterling, A.N. Montestruc, J.S. Morse, C.B. Leger and S. Acharya, J. Air Poll. Cont. Assoc., 39 (1989) 1073.

18. V.C. Cundy, A.M. Sterling, T.W. Lester, A.L. Jakway, C.B. Leger, C. Lu, A.N. Montestruc and R.B. Conway, Environ. Sci. Technol., 25(2) (1991) 223.

19. W.P. Ho and J.W. Bozelli, "Validation of a Mechanism for Use in Modeling CH_2Cl_2 and/or CH_3Cl Combustion and Pyrolysis," Proceedings of the Twenty Fourth Symposium (International) on Combustion, Sidney, Australia, (1992) 743.

20. J. Brouwer, G. Sacchi, J.P. Longwell and A.F. Sarofim, Combust. Sci. Technol., 101(1-6) (1994) 361.

21. C. Bass, R. Barat, G. Sacchi and P.M. Lemieux, "Fundamental Studies on the

Characterization and Failure Modes of Incinerator Afterburners," Proceedings of the 1995 International Incineration Conference, University of California, Irvine, CA, (1995) 223.

22. R. Addink and K. Olie, Env. Sci. & Technol., 29 (6) (1995) 1425.
23. S.B. Gorishi and E.R. Altwicker, Env. Sci. and Technol., 29 (5) (1995) 1156.
24. C.W. Lee, J.F. Ryan, R.E. Hall, G.D. Kryder and B.R. Springsteen, Combust. Sci and Tech., 116-117 (1996) 455.
25. J.D. Kilgroe, Journal of Hazardous Materials, 47 (1996),163.
26. B.K. Gullett, K.R. Bruce and L.O. Beach, Chemosphere, 20 (1990) 1945.
27. K.R. Bruce, L.O. Beach and B.K. Gullett, Waste Management, 11 (1991) 97.
28. B.K. Gullett, K.R. Bruce and L.O. Beach, Chemosphere, 25 (1992) 1387.
29. B.K. Gullett, P.M. Lemieux and J.E. Dunn, Environ. Sci. Technol., 28 (1994) 107.
30. B.K. Gullett and K. Raghunathan, Chemosphere, 34:5-7 (1997) 1027.
31. B.K. Gullett, W. Jozewicz and L.A. Stefanski, Ind. Eng. Chem., 31(11) (1992) 2437.
32. K. Raghunathan, B.K. Gullett, et al., 'Prevention of PCDD/PCDF Formation by Coal Co-firing', Proceedings of the Fifth Annual North American Waste to Energy Conference, RTP, NC, April 22-25, (1997) 779.
33. B.K. Gullett, Reduction of Chlorinated Organics in the Incineration of Wastes, U.S. Patent No. 5 021 229, (1991).
34. B.K. Gullett, Reduction of Chlorinated Organics in the Incineration of Wastes, U.S. Patent No. 5 185 134, (1993).
35. B.K. Gullett, K.R. Bruce and L.O. Beach, Environ. Sci. Technol., 26 (1992) 1938.
36. K. Raghunathan and B.K. Gullett, Environ. Sci. Technol. 30(6) (1996) 1827.
37. L.R. Waterland, E. Whitworth and M.K. Richards, "Innovative Continuous Emission Monitors: Results of the EPA/DOE Demonstration Test Program," Proceedings of the 1996 International Conference on Incineration and Thermal Treatment Technologies, University of California, Irvine, CA, (1996) 373.
38. P.M. Lemieux, J.V. Ryan, C. Bass and R. Barat, J. Air & Waste Manage. Assoc., Nashville, TN, 46 (1996) 309.
39. 40 CFR Part 266, Appendix IX, Section 2 "Performance Specifications for Continuous Emission Monitoring Systems."
40. U.S.EPA, "Modified Method 5 Sampling Train," Method 0010, in Test Methods for Evaluating Solid Wastes, SW-846 (NTIS PB88-239223), Revision 0,

September 1986.

41. U.S.EPA, "Extraction of Semivolatile Analytes Collected Using Modified Method 5", Method 3542, in Test Methods for Evaluating Solid Wastes, SW-846 (NTIS PB88-239223), Revision 0, January 1995.

42. U.S.EPA, "Semivolatile Organic Compounds By Gas Chromatography/Mass Spectrometry (GC/MS): Capillary Column Technique," Method 8270C, in Test Methods for Evaluating Solid Wastes, SW-846 (NTIS PB88-239223), Revision 3, January 1995.

43. P.M. Lemieux, "Pilot-Scale Evaluation of the Potential for Emissions of Hazardous Air Pollutants from Combustion of Tire-derived Fuel," Final Report, EPA-600/R-94-070 (NTIS PB94-169463), May 1994.

44. H. Oser, R. Thanner, H.-H. Grotheer, B.K. Gullett, N. French and D. Natschke, 'On-line Detection of Dichlorodibenzodioxins by DLR Jet-REMPI', Presented at: Japanese Flame Days '97, Japanese Flame Research Committee, May 16-17, 1997, Osaka, Japan.

45. P.M. Lemieux, C.A. Miller, K.J. Fritsky and P.J. Chappell, "Development of an Artificial-Intelligence-Based System to Control Transient Emissions from Secondary Combustion Chambers of Hazardous Waste Incinerators," Proceedings of the 1995 International Conference on Incineration and Thermal Treatment Technologies, University of California, Irvine, CA, (1995) 527.

46. W.P. Linak, J.A. McSorley, J.O.L. Wendt and J.E. Dunn, J. Air Poll. Cont. Assoc., 37 (8) (1987) 934.

Perspectives on future risk assessment and prevention for control of POP's

A.W. van der Wielen

Ministry of Housing, Spatial Planning and the Environment
P.O.Box 30945, 2500 GX The Hague, The Netherlands

1. RISK ASSESSMENT PROCEDURE IN THE EUROPEAN UNION

The production, distribution, use and disposal of substances lead almost inevitably to their presence in the environment. The management of the risk involved of these substances represents an international challenge. It occupies a prominent place on Agenda 21, Chapter 19. The development and harmonization of risk assessment methods is a necessary prerequisite to successful risk management. Recent developments in the legislation of the European Union (Council Directive 92/32/EEC[1] on new substances; Council Regulation 793/93/EEC[2] on existing substances) have accelerated the need to define the principles and practical considerations of this process. In the scope of new and existing substances the Commission Directives laying down the principles of risk assessment (Commission Directive 93/67/EEC[3] and Commission Regulation 1488/94/ EEC[4]) were implemented in the practice of evaluating substances. Technical guidance documents (TGD's) were developed to provide a helpful tool in performing risk assessment and to facilitate harmonization in the technical process. On the EU level the former separate TGD's for both new and existing substances are consolidated to one integrated TGD[5]. In parallel to the consolidation of the TGD's the attunement of on national level developed computerized systems to support the risk assessment were prepared and upgraded to a European system consistent with the consolidated TGD (EUSES 1.0[6]).

2. GENERAL PRINCIPLES OF THE RISK ASSESSMENT IN THE EU

Since mid eighties scientific methods were used in The Netherlands to predict and assess systematically the potentially hazard effects for environment and indirectly for man related to production and use of notified substances. The essential information for risk assessment were environmental exposure and effect data. To evaluate all possible effects a set of tools, like emission scenarios, predictive models and extrapolation methods were available. These tools were based on the concept that the EU harmonized minimal set of data at base set level was sufficient for risk assessment. The results of the already developed tools are joined together in the risk assessment model for the evaluation of new chemical substances based on causality between emissions and effects (fig. 1).

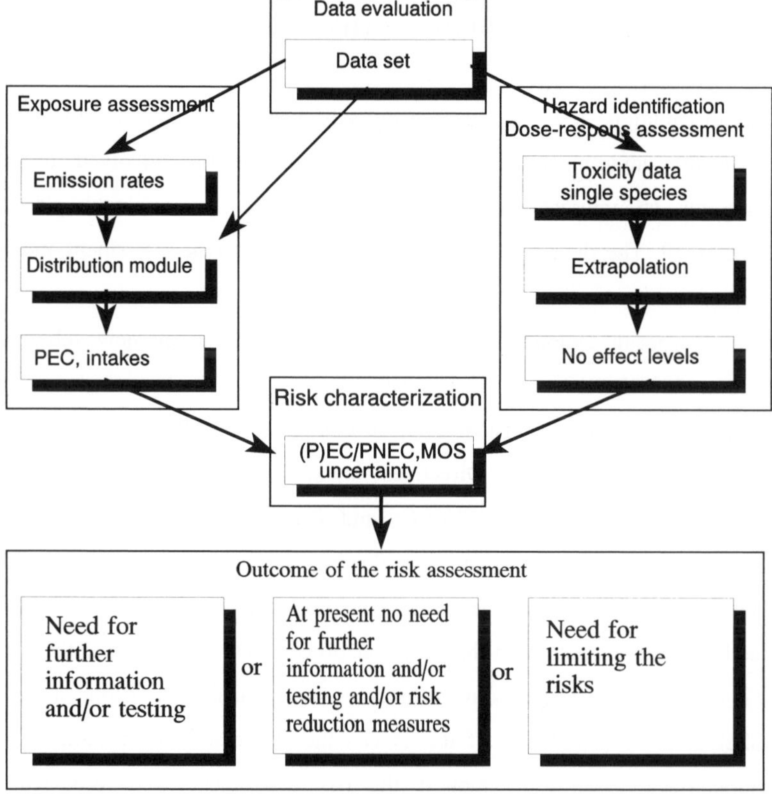

Figure 1: Principle of Risk Assessment

The exposure assessment is formed by models describing the processes as they occur in nature to predict the environmental concentration (PEC). The effect assessment shows the extrapolation of the toxicological test resulting into one or more predicted no effect concentrations for the distinguished biological endpoints (PNEC). Risk assessment for ecosystems takes place by comparing the predicted environmental concentration (PEC) and the predicted no effect concentration for ecosystems (PNEC-eco). The risk assessment for man compares the predicted no-observed-adverse-effect concentration derived from subacute, subchronic or chronic animal tests with the total predicted daily human intake. The ratio showing the margin-of-safety (MOS) is a quantitative result of the risk assessment to predict the risk in a specific situation. In the risk assessment for man no extrapolation factors and/or safety factors are used so far because of differences of opinion between the member states concerning the magnitudes of these factors.

The applied model (see figure 2) is a concise reflection of the major emission and distribution pathways for substances as they move through the environment in the European Union. The potential groups considered to be at risk are:

- bacteria in a waste water treatment plant.
- aquatic organisms;
- human beings on the workplace;
- the general public exposed directly via consumer products;
- the general public exposed indirectly via the environment.

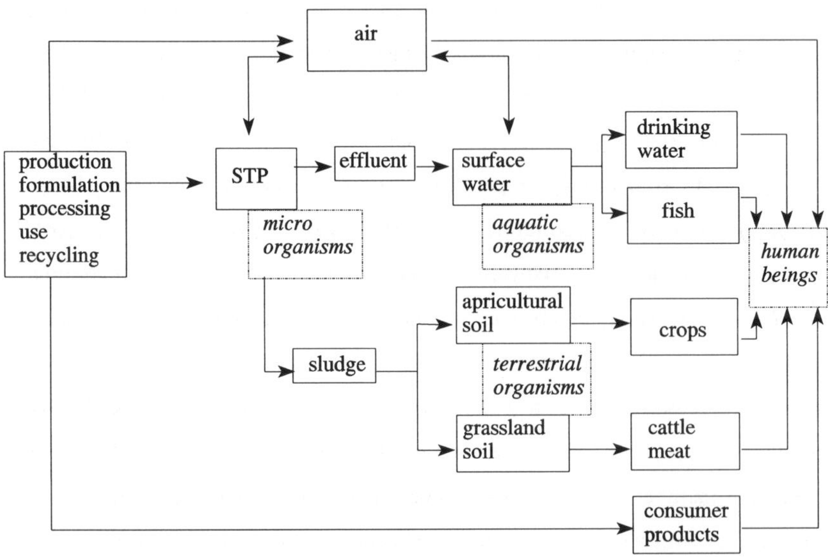

Figure 2: Elements of the emission/exposure assessment

In this model a substance is assumed to enter the surface water only via a sewage treatment plant (STP). The effluent of the STP is diluted in the surface water, where it is available for uptake by aquatic organisms. Each box reflects data resulting from previous processes and are, in its turn, input data for the following processes. The arrows reflect processes modelled using fixed and variable data.

For aquatic organisms, the predicted dissolved concentration of the substance in surface water is compared with the predicted no effect level that is obtained from extrapolation of toxicological tests. The tests, however, give only information about the effects of the substance on single species. For a more realistic risk assessment, the effects was extrapolated from single species to ecosystems. The sludge from the STP is considered to be used as fertilizer on grass- or arable land. Consequently the substance reaches human beings through the crops, the meat of cattle, fish and via drinking water. The total daily dose for human beings is compared with the dose without effect. If inhibition test results are available, a comparison is made between the concentration of the substance in the STP and the extrapolated no inhibition concentration for bacteria.

Potential pathways of the substance entry into the environment and its emission points are presented in the hazard assessment model. Only the emission to the influent of a STP is considered. All direct emissions to the soil are excluded because in most of the EU member states only so called "controlled dumping" on landfills is allowed and ecotoxicity data of terrestrial test organisms are mostly not available. Only after receiving additional ecotoxicity data other environmental compartments will be assessed. Emission to the air is evaluated when it is likely that the air concentrations will exceed the no effect levels. The air concentration will be evaluated in relation to the inhalation exposure of man.

When the risk characterization leads to the conclusion "need for limiting the risks", risk reduction measures will be recommended. The risk reduction measures will be based on a risk-benefit analysis taking into account (a) technical feasibility, (b) economical factors, (c) social/cultural factors, (d) legislative/political factors, and (e) research to reduce the uncertainties in methodologies and measurements. This is the most difficult step in the risk management process, because it is a multifactorial task. On the EU level a guidance concerning the applied methodologies to develop proposals for risk reduction measures is nearly finished. It will become available as chapter 6 to the TGD on risk assessment of new and existing substances.

3. EXPERIENCES WITH THE EU EXISTING CHEMICALS PROGRAM

Assuming that high volume chemicals may lead to high exposure, the OECD developed a list of HPVC chemicals based on the criteria > 1000 tons production per annum in more then one country or > 10,000 tons per annum in one country. Based on the present overall list containing approximately 2500 chemicals working lists in batches of about 50 chemicals for every phase of the program were developed given initially priority to chemicals lacking critical information ("data poor" chemicals) and since 1994 also priority to potentially high risk chemicals. The EU Existing Substances Regulation is the basis for the EU contribution to the OECD high production volume chemicals (HPVC) program. While the HPVC-program was initially focused on completing data of "data poor" HPVC’s, the EU program is mainly focused on potentially high risk HPVC’s.

The EU existing substances regulation comprises of required delivery of available information of chemicals produced/imported > 1000 tons per annum in the first phase (ended in June 1994) and delivery of available information for chemicals produced/imported > 10 tons per annum in the second phase (deadline June 1998). The first phase realized data delivery of some 2500 substances whose datasets vary in completeness. The working lists for preparing risk assessments have been selected from the 2500 HPVC’s given priority to potential high risk for man and the environment. Since 1993 three priority lists have been published, the first list of substances in 1994, the second list of substances in 1995 and recently in 1997 the third list containing substances.

The planning of the EU existing substances program was based on the optimism in the possibility to cover a working program of about 20 selected substances per year itemized over the 15 EU member states in function of rapporteur responsible for preparing the risk assessment. However, the practice of preparing risk assessment reports learned very soon from the start that (a) harmonized scientific quality standards and a harmonized format for reporting the risk assessment were needed before risk assessment reports could be prepared in a harmonized way, (b) a critical review procedure must be set up to get international acceptance, and (c) substantial human and financial resources in each member state must be made available to fulfil these criteria. Preparing the harmonized preconditions before the EU risk assessment procedure of priority substances could start led to^Ia delay of several years. Already during the development of the third priority list it was made clear that because of limited resources and major delay in developing risk assessments of the previous lists many member states were not able to take substances of the third list on board. At this moment the EU existing substances program is still waiting for the first risk assessment reports to reach the level of EU agreement. A more realistic proposition of the EU existing substances program in final risk assessment reports could be 5 substances per year.

The situation is even more critical because the regulation also requires recommendations for risk reduction based on criteria of the still to approve missing chapter 6 of the TGD with respect to cost-benefit analysis and risk reduction methodology. Because the priority were made for potential high risk chemicals and risk reduction recommendations could already be foreseen, the first draft risk assessment reports still need to be completed with recommendations for risk reduction before formal adoption can be reached. So, the outlook on the progress of the existing substances program is none too bright.

We must conclude that the present schemes in developing risk assessment reports for priority substances is not a ideal working method to implement also for future POP's. It is only a question of time before the EU member states call for reconsidering the whole existing substances procedure because of the involved human and financial resources and the lack of effectiveness in the final aim to develop a acceptable risk management for chemicals in general.

4. A MORE EFFECTIVE RISK MANAGEMENT APPROACH: EXAMPLE PAPER CHEMICALS

4.1. Background

In the European Union, Directive 92/32/EEC[1] requires a risk assessment to be carried out on new notified substances. A major part of the new notified substances are dyes, which are predominantly used in the paper industry. Based on the results of the risk assessments there was serious concern for most of the notified substances that the aquatic environment was at risk at processing and recycling of paper. In discussing specific notified substances on EU level the assessment of the emission from production and processing paper was improved by agreement on a harmonized set of emission parameters based on detailed analysis of the emission from the paper industry. Because the production and processing processes in the paper industry were quite uniform, we explored the idea to develop simple aquatic environmental criteria for the assessment of paper chemicals based on harmonized models concerning emission and distribution in the environment[6]. The aquatic environmental criteria pertain to parameters which are important for the estimation of emissions (use category, function of the substance. and specific physical-chemical properties) and the aquatic toxicity of the substance {classified as R50 (L(E)C50 <= 1 mg/l), R51 (10 <= L(E)50 < 1 mg/l) or R52 (10 <

L(E)50 <= 100 mg/l)}. In addition the environmental criteria should follow the accepted risk policy in The Netherlands and the accepted EU risk assessment methods. The environmental criteria should be transparent and usable without the help of computers.

4.2. General overview of the results

A "standard scenario" which includes a fixed quantity level, Henry's law constant, octanol-water partition coefficient and fixed aquatic toxicity were used to calculate potential PEC/PNEC ratios for the function categories coloring agents, reprographic agents and other types of agents (see table 1; only the category coloring agents is presented as an example). Next, aquatic environmental criteria were derived, which describe the critical range of physical-chemical and aquatic toxicity properties for which paper chemicals belong to the classes of "low risk: no immediate concern" (PEC/PNEC <= 1, "immediate risk; concern - risk assessment required" (PEC/PNEC < 1000) or "high risk: high concern - restriction in use required" (PEC/PNEC >= 1000). Taking together the readily and non-readily degradable paper chemicals (see table 2; only coloring agents are presented as an example), the classes following the environmental criteria corresponded for 97% with the PEC/PNEC ratio for the aquatic environment for all life cycle stages of the notified paper chemicals. The results show that univocal and generally applicable aquatic environmental criteria can be derived based on an accepted risk policy and accepted risk assessment methods for chemicals grouped in use categories with a quite similar emission pattern.

Table 1: Estimated emission and potential PEC/PNEC ratios at different aquatic toxicity values for paper chemicals (assuming 100% release to the sewage treatment plant.

Use category/life cycle	Release estimate (fraction)	Daily emission rate (kg/d)	Potential PEC (mg/l)	Pot. PEC/PNEC dependent on L(E)C50 (mg/l)		
				R52 10 - 100 Mean 31.6	R51 1 - 10 Mean 3.16	R50 0.1 - 1 Mean 0.316
Reprographic agents						
Production	0.02	20.0	1	31.6	316	3165
Formulation	0.02	3.33	0.17	5.27	52.7	527
Processing:						
- Printing and allied processes	0.0005	0.50	0.025	0.79	7.91	79.1
- Others	0.815	815	40.75	1290	12896	128960
Recovery						
- cardboard/sanitary	0.01	0.33	0.02	0.53	5.27	52.7
- others	0.14	4.67	0.23	7.38	73.8	738

Table 2: Aquatic environmental criteria for paper chemicals used as reprographic agents

Reprographic agents	**R52** **10 - 100 mg/l**	**R51** **1 - 10 mg/l**	**R50** **0.1 - 1 mg/l**	**R50** **< 0.1 mg/l**
Production	concern + RA required if logH<=2 and logKow<=4, unless logH=2 and logKow=4	concern + RA required	high concern + restrictions in use if logH<=2 and logKow<=4, unless logH=2 and logKow=4	high concern + restrictions in use
Formulation	no immediate concern	concern + RA required	concern + RA required	concern + RA required /high concern + restrictions in use
Processing:				
- printing and allied process	no immediate concern	no immediate concern	concern + RA required	concern + RA required
- others	high concern + restrictions in use if logH<=1 and logKow<=3	high concern + restrictions in use	high concern + restrictions in use	high concern + restrictions in use
Recovery:	no immediate concern	concern + RA required	concern + RA required	concern + RA required /high concern + restrictions in use
- cardboard/sanitary	no immediate concern	no immediate concern	concern and RA required	concern and RA required
- others	no immediate concern	concern and RA required	concern and RA required	concern + RA required /high concern+ restrictions in use

Table 3: Comparison of environmental criteria and risk assessment results for notified paper chemical in The Netherlands.

No	L(E)C50	Class according to the environmental criteria				PEP/PNEC calculations			
	mg/l	pd	f	pc	rc	pd	f	pc	rc
1	1.88	II	II	II	II	458	51	1207	51
2	2.8	II	II	I	II	893	60	8.9	60
3	3.15	II	II	II	II	794	53	1254	53
4	3.24	II	II	I	II	768	51	7.7	51
5	10.7	II	I	II	I	29	1.9	46	1.9
6	11.63	II	I	II	I	69	4.5	409	4.5
7	16.7	II	I	II	I	180	10	237	10
8	25	II	I	I	I	100	6.7	1.0	6.7
9	34.6	II	I	II	I	25	2.8	67	2.8
10	46	II	I	II	I	33	3.6	30	3.6
11	115	I	I	I	I	26	1.5	34	1.4
12	225	I	I	I	I	11	0.7	18	0.7
13	532	I	I	I	I	4.7	0.3	7.4	0.3
14	>1	I	I	I	I	-	-	-	-
15	>1	I	I	I	I	-	-	-	-
16	>1000	I	I	I	I	-	-	-	-
17	>1000	I	I	I	I	-	-	-	-
18	>1000	I	I	I	I	-	-	-	-
19	26.1	II	I	II	n.a.	13	0.7	49	n.a.
20	0.23	III	II	II	II	2527	421	63	842
21	0.33	II	II	II	II	286	48	7.2	95
22	0.73	III	II	I	II	1348	222	34	452
23	2.6	II	II	I	II	152	25	3.8	51
24	10	II	II	I	II	100	17	2.5	33
25	>0.1	I	I	I	I	-	-	-	-
26	>0.1	I	I	I	I	-	-	-	-
27	>100	I	I	I	I	-	-	-	-
28	>1000	I	I	I	I	-	-	-	-

5. CONCLUSION

The idea to develop a more effective risk management based on accepted risk policy and accepted risk assessment methods seems to be successful for paper chemicals. It needs further discussion, but the approach could be explored for other groups of substances, like POP's. Because the main concern of POP's in the environment are predominantly related to persistency and environmental toxicity, and furthermore the distribution in the environment is quite similar for POP's, we would suggest not to follow the way of detailed risk assessment reports for each POP, but to develop derived environmental criteria for POP's based on "standard emission" scenarios. This approach only needs well improved emission scenarios for the relevant use categories of POP's.

REFERENCES

1. Directive 92/32/EEC amending for the seventh time Directive 67/548/EEC on approximation of the laws, regulations and administrative provisions relating to the classification, packaging and labelling of dangerous substances; Official Journal of the European Communities L 154, 1992.
2. Council regulation (EEC) nr. 793/93 on the assessment and risk reduction of existing substances; Official Journal of the European Communities L 84, 1993.
3. Commission directive 93/67/EEC laying down the principles for the assessment of risks to man and the environment of substances notified in accordance with Directive 67/548/EEC; Official Journal of the European Communities L 229, 1993.
4. Council regulation (EC) nr. 1499/94 laying down the principles for the assessment of risks to man and the environment of existing substances in accordance with Regulation (EEC) nr. 793/93; Official Journal of the European Communities L 161, 1994.
5. Technical guidance documents in support of the Commission direction 93/67/EEC on risk assessment for new notified substances, and the Commission regulation (EC) 1499/94 on risk assessment for existing substances; Official Journal of the European Communities L, 1997.
6. G.B. Janssen and E.H. Hulzebos: Aquatic environmental criteria for paper chemicals; National Institute of Public Health and the Environment (RIVM); Report no. 601505001, February 1997.

SESSION D

OZONE/NO_x

Photochemical Oxidant Air Pollution: A Historical Perspective

Arthur Davidson
Davidson and Associates Environmental Consulting
4814 Somerset Drive SE
Bellevue, Washington 98006

Abstract

Photochemical smog first came into prominence in July 1943, in Los Angeles. In 1947, the Los Angeles County Air Pollution Control District was formed to deal with the growing smog problem, which was caused by a combination of poor atmospheric ventilation, strong solar radiation, confining topography, and generally uncontrolled pollutant emissions. In the early to mid 1950s, the basis of photochemical smog theory was set forth by Professor Arie J. Haagen-Smit of the California Institute of Technology, who concluded that ozone and other photochemical oxidants were not emitted directly, but were formed in the atmosphere when precursor emissions of oxides of nitrogen and hydrocarbons reacted with each other in the presence of sunlight. Beginning in 1959, the State of California moved to control vehicle emissions, passing a series of acts which culminated in the formation of the California Air Resources Board in 1967. On the federal level, congress passed the first Clean Air Act in 1963, with subsequent major amendments added in the 1970 Act (which also established the Environmental Protection Agency), and the 1990 Act. Although some controls had been required on motor vehicles in the 1960s, it was not until 1975 that the first catalytic converters, designed to reduce emissions of hydrocarbons and carbon monoxide appeared, with the three way catalyst (to also control oxides of nitrogen) being introduced on 1981 model vehicles. In 1976, recognizing the regional nature of photochemical smog, the four local agencies of the Greater Los Angeles region combined to form the South Coast Air Quality Management District. In the Los Angeles region, ozone concentrations showed a moderate overall decline from the late 1950s to the late 1970s, and have declined substantially from the late 1970s until the present time, even in the face of large increases in population and vehicle miles traveled. Ozone trends for the period 1986-95 are also presented for several major cities in the United States. Also discussed are current trends to incorporate market forces in the control of pollutant emissions, and the future outlook for reducing pollutant emissions still further by a combination of technological and educational means.

1. BACKGROUND: THE EARLY YEARS

1.1 Initial Awareness of Photochemical Smog

Photochemical smog first came into prominence in July 1943, in Los Angeles, California,[1] although, based on visibility statistics, it must have been present at least as early as the 1930s.[2] The area of southern California has a particularly high meteorological air pollution potential for photochemical smog and other types of air pollution, due to the high frequency of strong, low, temperature inversions, light morning wind speeds, and strong solar radiation. The high mountains fringing the area to the north and east also play a role in confining airflow so as to increase the potential for smog formation. The high air pollution potential of the area began to be realized with the sharp increase in the region's industrial growth and resulting pollutant emissions which accompanied World War II. Postwar growth only added to the problem.

1.2 Formation of Los Angeles County Air Pollution Control District in 1947♦

In 1947, the Los Angeles County Air Pollution Control District (LAAPCD) was formed to deal with the growing smog problem,[3] replacing other agencies that had previously dealt with the problem in a less comprehensive manner. By the end of 1947, the LAAPCD had put in place the beginnings of an air pollution control program which required all major industrial sources of air pollution to obtain air pollution permits in order to continue to operate. Some of the early air pollution controls resulted in the reduction of smoke from factories and waste disposal operations, and in particular in 1958, the banning of the use of backyard incinerators. Other stationary source controls included those on gasoline composition, organic solvents, oxides of nitrogen, and sulfur compounds. [3] Ozone concentrations however continued to be high.

1.3 Development of Photochemical Smog theory by Professor Arie. J. Haagen-Smit, of the California Institute of Technology (Caltech)

In the early to mid 1950s the basis of photochemical smog theory was set forth by Professor Arie J. Haagen-Smit of the California Institute of Technology, who concluded that ozone and other photochemical oxidants were not emitted directly, but were formed in the atmosphere when precursor emissions of oxides of nitrogen and hydrocarbons reacted with each other in the presence of sunlight.[4,5] Auto exhaust, which contains both oxides of nitrogen and hydrocarbons, immediately became to be seen by many as one of the major sources of photochemical smog. Other major sources of hydrocarbons included refineries, and activities associated with the use of paints and solvents, while other major sources of oxides of nitrogen included electrical generating facilities.

2. EXPANDED EFFORTS TO ADDRESS PHOTOCHEMICAL SMOG PROBLEM

2.1 Role of Motor Vehicle in Photochemical Smog Problem: Formation of the California Air Resources Board

By the late 1950s in the Greater Los Angeles region, there had been significant reductions in emissions of oxides of nitrogen and hydrocarbons from stationary sources, but the motor vehicle still remained as a major and uncontrolled source of these precursors to ozone. Motor vehicle emissions also constituted a threat to the air quality of other regions of California, including the San Francisco Bay area, and San Diego. Beginning in 1959, the State of California moved to control vehicle emissions, passing a series of acts which

---

♦ Other early air pollution control districts formed in California were those in the counties of Orange (1950), Riverside (1955), and San Bernardino(1956), the San Francisco Bay Area (1956), and San Diego County (1955).

made it the first governmental entity to regulate the emissions of new automobiles, beginning with positive crankcase ventilation (PCV) controls for hydrocarbons on domestic 1963 model year automobiles. Subsequent legislation culminated in the formation of the California Air Resources Board (CARB) in 1967.[6] Among the key responsibilities of CARB were promulgating regulations for motor vehicle emissions, adopting State Ambient Air Quality Standards, and assisting and overseeing the programs of local air pollution control districts, more of which had formed by that time.[6] CARB has continued to play a major role in reducing emissions from motor vehicles by requiring industry to develop the technology to meet ever more stringent vehicle emission standards, and by funding research in a wide variety of areas dealing with mobile source emissions, and their environmental impacts.

2.2 Regional Nature of Photochemical Smog

In an effort to understand more about the formation and transport of air pollution, a massive air sampling survey was carried out in the greater Los Angeles region in 1972 and 1973.[7] ♦In addition to obtaining data from a dense network of surface stations monitoring air quality and meteorological variables, fixed wing light aircraft were employed to measure the vertical and horizontal distribution of pollutant concentrations and meteorological elements over much of southern California, encompassing the four major counties of the Greater Los Angeles region. The most important conclusion of the study was that photochemical smog was a regional problem and, to be effective, air pollution control had to be carried out on a regional basis. Prior to this time, individual air pollution control agencies had tended to take a more parochial view of the problem. In 1976, recognizing the regional nature of photochemical smog, the four local air pollution control districts of the Greater Los Angeles region combined to form the South Coast Air Quality Management District (SCAQMD), and began to refer to the common airshed which they shared, as the South Coast Air Basin (SoCAB).

2.3 Development of Catalytic Converter in 1975

Perhaps no single air pollution control device has had such a favorable impact on air quality as the catalytic converter, which was first introduced on 1975 model year vehicles sold in the United States, and designed to reduce exhaust emissions of hydrocarbons and carbon monoxide. The subsequent development in 1981 model year cars of a 3-way catalytic converter, to also reduce emissions of oxides of nitrogen, benefited air quality still more. In addition to the reduction in emissions of carbon monoxide, oxides of nitrogen, and hydrocarbons, use of the catalytic converters also resulted in a dramatic improvement in lead air quality, as shown in Figure 1, since the catalytic converters required the use of unleaded gasoline. Also, the 3-way catalytic converter with its associated oxygen sensor yielded the best combination of performance , economy, and emission reductions.

♦ Other air monitoring studies carried out about the same time included the Los Angeles Reactive Pollution Project (LARPP) in 1973, and the Regional Air Pollution Study (RAPS) in St. Louis, during the period 1974-77.

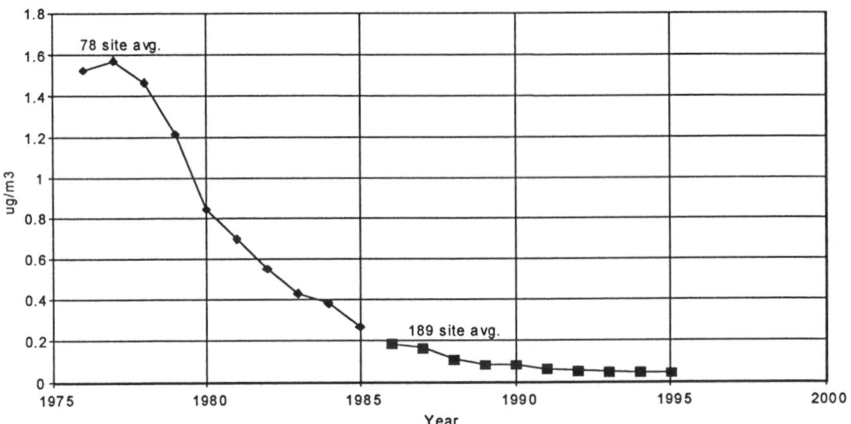

Figure 1. Lead: U.S. trends in maximum quarterly concentration, 1976-95.

2.4 Selected Federal Air Pollution Legislation

1963 Clean Air Act – Under this act, the federal government agreed to use its authority to deal with interstate air pollution problems, provide grants for program development for state and local air pollution authorities, and assume research responsibility for motor vehicle pollution.

1970 Clean Air Act Amendments – The far reaching amendments of this act, established the Environmental Protection Agency, set uniform national ambient air quality standards (NAAQS), required the submission to EPA of State Implementation Plans (SIPs) to achieve the NAAQS, and set strict emission standards for new motor vehicles.

1990 Clean Air Act Amendments – The amendments of this act were also very far reaching, including titles dealing with nonattainment, mobile sources, hazardous air pollutants, acid rain, permits, and even stratospheric ozone. With respect to ground level ozone, the act set ozone compliance deadlines ranging from November 1993 for areas in marginal attainment of the ozone standard, to November 2010 for areas with extreme ozone concentrations, the latter category applying to the Greater Los Angeles region, now referred to as the South Coast Air Basin (SoCAB). The Act also called for the use of reformulated fuels in the worst ozone areas, a reduction in gasoline volatility, and set tighter emission standards for cars and trucks.

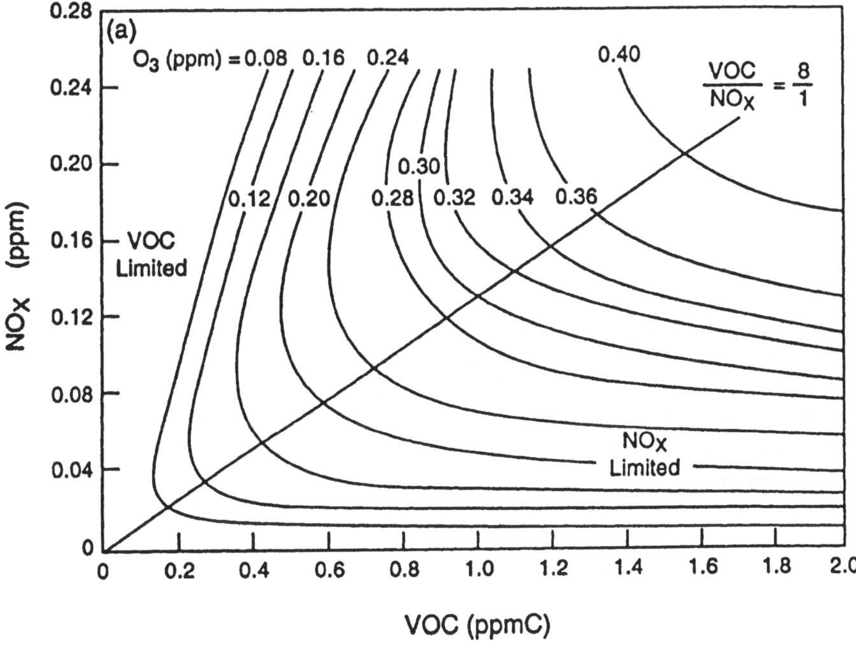

Figure 2. Empirical Kinetic Modeling Approach (EKMA) diagram and associated ozone isopleths.

3. MODELING OF PHOTOCHEMICAL SMOG

Over the years two basic approaches to modeling ozone concentrations have predominated, the empirical kinetic modeling approach (EKMA), and the grid-based photochemical model.

3.1 EKMA - The EPA first proposed the use of EKMA in the mid 1970s. It has since been updated to include more sophisticated chemistry such as the carbon bond IV mechanism, and has also been made city-specific. Figure 2 above shows a typical EKMA diagram for determining the effect of changes in emissions of volatile organic carbons (VOC)[♦], and oxides of nitrogen (NO_x) on ozone air quality. One's place on the chart is at the intersection of a source area's characteristic VOC/NO_x ratio during the morning hours, say 6AM to 9AM, and a characteristic maximum downwind ozone concentration later in the day. EKMA does not predict ozone maxima. Rather, the path chosen to go from the existing ozone maximum isopoleth to the desired ozone maximum isopleth determines the relative percent changes in emissions of VOCs and NO_x that are needed to accomplish the desired change in ozone concentrations. It is apparent that depending on one's position on the chart, ozone control strategies may differ considerably. For example, a position in the NO_x-limited right side of the chart, would indicate that a strategy of reducing NO_x

[♦] Historically, in addition to being referred to as VOCs, hydrocarbon emissions which react to form ozone have also been referred to as non-methane hydrocarbons (NMHC), reactive organic gases (ROG), and non-methane organic compounds (NMOC).

emissions while leaving VOCs alone would lead to the greatest ozone improvement. The opposite situation would obtain if one's position were in the VOC limited, middle and upper portion of the left side of the chart. Indeed, in such a case, a strategy of reducing NO_x emissions alone would actually *increase* ozone concentrations. Historically, this latter situation has often been, and continues to be, a subject of controversy. Although computationally efficient, EKMA has many limitations. It is really only appropriate for regions with a well defined urban core and a relatively simple transport pathway to the location of the ozone maxima downwind.[8]

3.2 Grid-Based (Eulerian) Airshed Models

These complex models seek to represent the effect of emissions, chemistry, meteorology, and topography, on the formation and transport of ozone and its precursors, through the use of a fixed Cartesian reference system, usually consisting of cells several kilometers on a side horizontally and of varying height, depending on how many layers are being employed, and upon the mixing height of the polluted layer that is being modeled. The atmospheric dynamics are carried out within each grid cell in a repeated fashion over small increments in time, thereby simulating the formation and transport of the polluted air mass.

Grid based airshed models require detailed, accurate, spatially and temporally resolved data on emissions, meteorology, and air quality, in addition to data on topography. In recent years, the Urban Airshed Model (UAM) has been applied to a number of areas in the U.S., including Los Angeles, Sacramento, the San Joaquin Valley of California, New York, and Chicago. The Regional Oxidant Model (ROM) has been used to model ozone concentrations in the northeastern U.S., and the CIT Airshed model has been used to model photochemical smog in Mexico City.[9] One of the prime goals of each of these models is to determine the degree to which emissions of VOCs and NO_x must be reduced to achieve the desired air quality objective or standard.

Experience has shown that a more spatially and temporally detailed emission inventory, including speciated VOCs, is necessary for better model performance.[9] Also needed are better data on the meteorology, chemistry, and air quality of the polluted layers aloft, which are known to play a major role in the formation and transport of photochemical smog.[7,9] Some of the needed information is currently being pursued by a variety of special air monitoring studies.[9] In 1994, the EPA began a program to provide for enhanced monitoring of ozone and its precursors, by means of PAMS (Photochemical Assessment Monitoring Stations) for 22 ozone nonattainment areas in the U.S. rated as extreme (1 location), severe (9 locations), or serious (12 locations).[10] As of October 1996, there were 65 PAMS sites in operation, with the total expected to reach 90 sites by 1998. Some of the complexities of grid-based airshed models are shown in Figure 3.

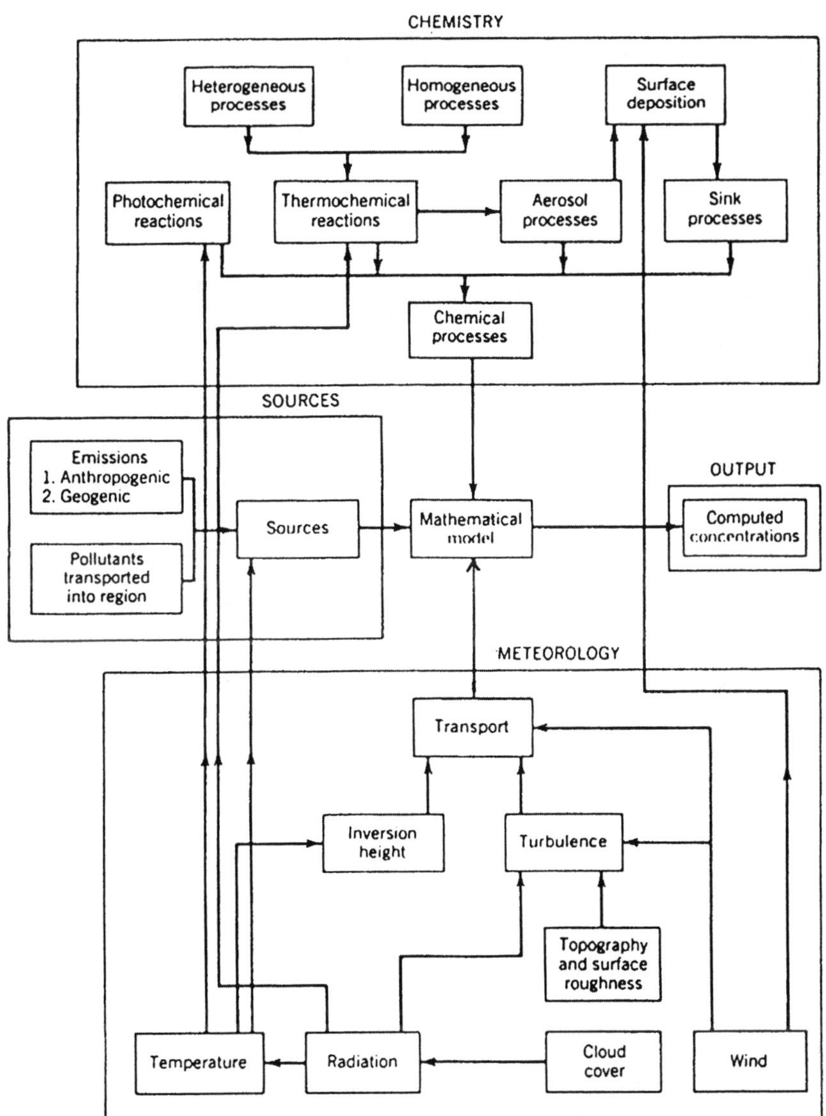

Figure 3. Elements of a typical airshed model (Source: McRae et al.[11]).

4. OZONE AIR QUALITY TRENDS

For 44 major urban areas of the United States considered as a whole, the EPA reports that there was an average decrease of about 10 percent in meteorologically adjusted concentrations over the period 1986-95.[10] This statistic refers to the 99th percentile of ozone concentrations. Figure 4 , based on EPA data,[10] shows ozone trends at four major U.S. cities during the period 1986-95. The data have been presented as three-year moving averages in order to reduce the impact of year to year differences in meteorology on the ozone trends.

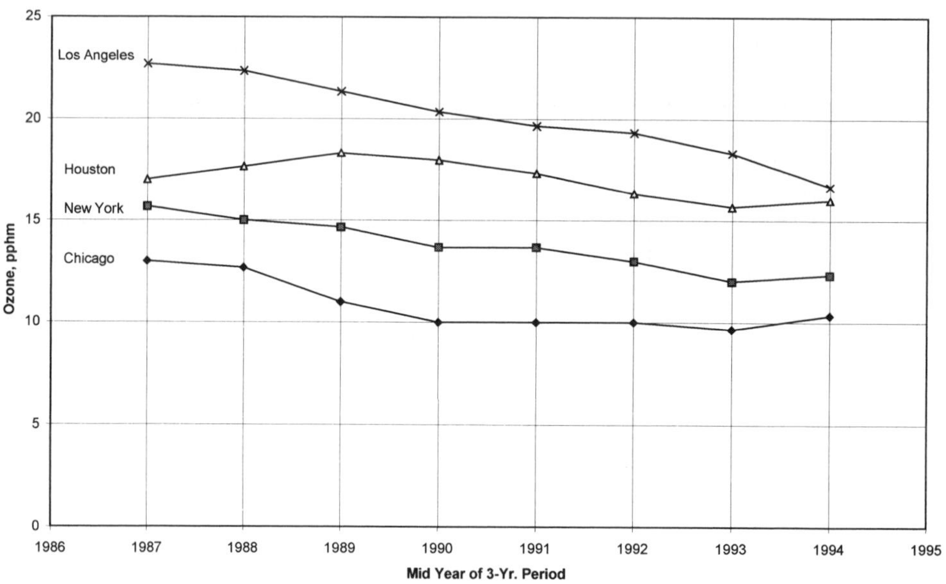

Figure 4. Ozone: 3-year moving average of 2nd highest annual 1-hr. maximum.

For the Greater Los Angeles area, based on 14 trend stations, the 3-year moving average of the 2nd highest daily 1-hour maximum ozone concentration decreased steadily over the entire period. Over the length of the period, ozone maxima decreased by 26.4 percent (from 22.7 pphm in 1986-88 to 16.7 pphm in 1993-95). In the Houston area, based on 10 trend stations, ozone maxima increased from 1986-88 to 1988-90, decreased from 1988-90 to 1992-94, and increased slightly in 1993-95. For the period as a whole ozone maxima decreased by 5.9 percent (from 17.0 pphm in 1986-88 to 16.0 pphm in 1993-95).

In the New York area, based on 4 trend stations, ozone maxima generally decreased from 1986-88 to 1992-94, before increasing slightly in 1993-95. For the period as a whole, ozone maxima decreased by 21.7 percent (from 15.7 pphm in 1986-88 to 12.3 pphm in 1993-95). At Chicago, based on 15 trend stations, ozone maxima decreased from 1986-88 to 1989-91 and showed little change thereafter. For the period as a whole, ozone maxima decreased by 20.8 percent (from 13.0 pphm in 1986-88 to 10.3 pphm in 1993-95). Of the four cities, Chicago clearly has the best ozone air quality. It is interesting to note that Los Angeles ozone maxima are now only about four percent above those of Houston, based on the latest period of 1993-95.

As would be expected, the historical data base of ozone measurements in the greater Los Angeles area is lengthy, reaching back to 1955. Although this area still exhibits the worst ozone air quality in the U.S., it is instructive to see just how much ozone concentrations have decreased since control agencies in the region geared up to respond to the problem. Figure 5 shows 3-year moving averages of the single highest 1-hour ozone concentration of the year, from 1955 through 1996. The extremely high annual ozone maxima of the early years showed little overall change from 1955-57 to 1965-67, but decreased sharply from 1965-67 to 1975-77. There was a small increase in annual ozone maxima from 1975-77 to 1978-80, but from 1978-80 until the present time there has been a long, steady decrease in ozone maxima. For the period as a whole, annual ozone maxima have decreased by 52 percent (from 56.0 pphm in 1955-57 to 26.7 pphm in 1994-96), even in the face of a doubling of population in the South Coast Air Basin.

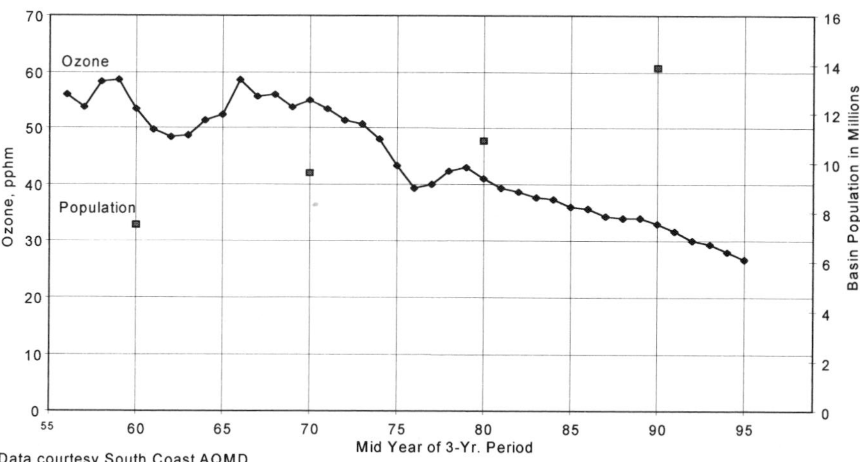

Figure 5. Ozone: 3-year moving average of annual 1-hour South Coast Air Basin maximum vs. population, 1955-96.

Further evidence of the great improvement in ozone air quality in the Los Angeles area is given in Figure 6, which shows that over the period 1976-96, based on 3-year moving averages and 17 trend stations, there has been a 100 percent reduction in the number of Basin-days♦ with ozone at or above 35 pphm, an 86 percent reduction in the number of Basin-days with ozone at or above 20 pphm, and a 48 percent decrease in the number of Basin-days exceeding the current federal standard of 12 pphm. The great improvement in ozone air quality of the South Coast Air Basin and other areas of California was brought about by increasingly strict controls on emissions of oxides of nitrogen and volatile organic carbons from both stationary and mobile sources.[12]

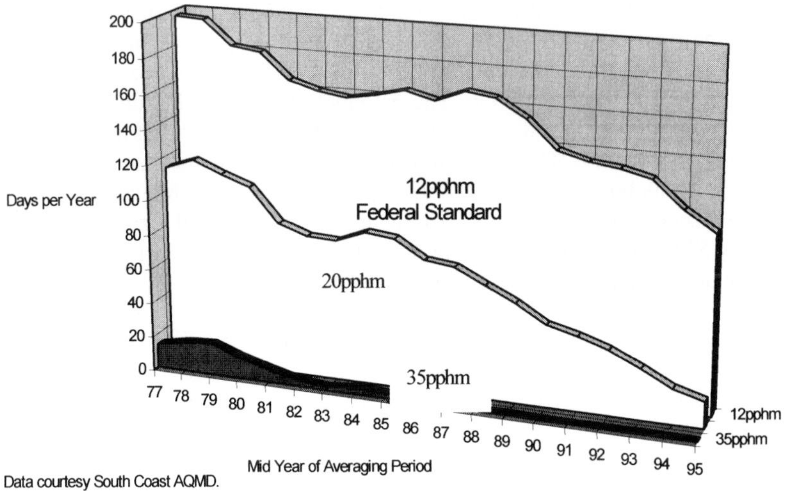

Figure 6. Ozone: 3-year moving average of Basin-days at or above indicated values.

Further improvement in U.S. ozone air quality will depend greatly on the degree to which emissions from motor vehicles can be controlled still further. One of the key areas that must be addressed is the detection and repair of vehicles which are gross emitters, that is, which emit pollutants in amounts several to many times the allowable amount for a given model year. Data from a random roadside survey of more than 11,000 vehicles in California indicate that 10 percent of the vehicles are responsible for 60 percent of fleet exhaust hydrocarbon emissions, and 10 percent of vehicles (although not exactly all the same vehicles) are responsible for 60 percent of fleet carbon monoxide emissions.[13] Another major challenge is to reduce the growth in vehicle miles traveled (VMT). In the U.S. over the period 1970 to 1995, VMT increased at a rate four times that of population (116 percent vs. 28 percent).[10]

♦ A Basin-day is defined as a day on which one or more trend stations equal or exceed a key concentration.

5. RECENT AND FUTURE DEVELOPMENTS

5.1 EPA's Proposed New Ozone Standard

On December 13, 1996, EPA announced[14] that, based on the most recent health effects research, it was proposing a change in the NAAQS for ozone from the current level of 12 pphm: 1-hour average, to 8 pphm: 8-hour average, and solicited comments on the proposal (the announcement also included a proposed change to the NAAQS for particulate matter). To insure more statistical stability in the compliance statistic, EPA also proposed that compliance be determined by taking the 3rd highest 8-hour average ozone each year, averaging the results over a 3-year period, and then comparing it to the standard. For the U.S. as a whole, the proposed standard if adopted would increase the number of ozone nonattainment counties from 106 to 336, based on statistics for the period 1993-95.[15] A final regulation is expected to be issued in July 1997, after which it will be reviewed by Congress.

5.2 Control Approaches

Throughout the history of air pollution control, one approach has dominated, that of "command and control." In this approach, regulatory agencies basically tell those in charge of emission sources the type and quantity of emissions to be reduced, and how to reduce them. This method has often proven to be successful, and for certain regions, and in many situations, may still be the preferred approach. However, when market forces can be brought to bear on the problem, air pollution controls can often be implemented at far less cost to industry, and therefore to society. For example, Title IV of the Clean Air Act Amendments of 1990, allows power plants that reduce their emissions of oxides of sulfur (SO_x) to a point below their allowable emissions, to bank the excess emission reductions, and sell them to other utilities for whom the costs of controls to meet their emission reduction targets may be very large. In this manner, the utility industry as a whole can meet its prescribed emission reduction targets, while at the time minimizing the overall costs of control.

In 1994, another market-based program, known as RECLAIM (Regional Clean Air Incentives Market) was started in the South Coast Air Basin.[16] By the early 1990s, as a result of the many rules in place for SO_x and NO_x in the SoCAB, the cost of further emission reductions had become very high, \$25,000. per ton for NO_x controls at power plants, and \$32,000. per ton for SO_x, for proposed catalytic crackers at refineries.[16] Clearly the time had arrived for a new approach to achieve the further emission reductions needed for these pollutants. Under RECLAIM, a facility-wide emissions limit is initially established, and then gradually reduced over a 10 year period in proportion to the emission reduction needs of the region. The operators of the facility determine how best to meet the emissions limit for the given year. Those facilities that reduce their emissions to a level below their allowable emissions, receive emission reduction credits which can then be sold on the open market. Those facilities who find it more advantageous to meet their emission limits by purchasing emission reduction credits may then do so, and thereby avoid incurring the expense and inconvenience of installing added control equipment. RECLAIM thus allows operators of a facility to choose the path they find most advantageous, while still achieving the needed emission reductions for the region. As of early 1996 there were 353 facilities in RECLAIM[17], all emitting more than the 4 tons per year threshold needed to be included in the program. It has been estimated by the SCAQMD that controlling emissions from facilities in the RECLAIM program achieves a savings of 42 percent in average annual costs of emission reduction compared to the command and control approach.[16]

5.3 Inter-comparison of Regional Oxidant Studies

The North American Research Strategy for Tropospheric Ozone (NARSTO) has begun a project to compare methodologies, findings, and real world experiences associated with regional oxidant studies carried out in recent years.[18] This needed effort gives promise of providing valuable guidance to those charged with making key decisions on ozone air quality. The results will be presented in a paper as part of NARSTO's 1998 state-of-the science assessment.

5.4 Future Measures For Further Improving Ozone Air Quality

In order to continue to improve ozone air quality, along with other accompanying air pollution problems, it will be necessary to improve upon and go beyond what is currently being done. Possible measures, some of which are already in use in certain areas include:

- Use of lower-polluting reformulated fuels for motor vehicles.
- On-road remote sensing devices to identify gross emitting vehicles, paired with programs to require their repair.
- Breakthroughs in battery and fuel cell technology, to power electric vehicles; development of hybrid electric-gasoline powered vehicles.
- Development of paints, stains, coatings and solvents with lower or no reactive hydrocarbon emissions.
- Where appropriate, use of market forces to achieve emission reductions.
- Increased use of telecommuting and teleconferencing to reduce traffic congestion, and thereby reduce emissions.
- Employer subsidies to employees who take public transit or ride share.
- Use of public education programs to encourage a less polluting lifestyle (when millions of citizens each pollute a little less, the net emission reductions are large, and very cost-effective).

6. ACKNOWLEDGMENT

The author gratefully acknowledges the help of Margaret Hoggan and Lawrence Kolczak of the South Coast AQMD, in providing key data used in the preparation of this paper. Valuable information was also readily available from the EPA's "National Air Quality and Emissions Trends Report, 1995," the authors of which should be commended.

7. REFERENCES

[1]Los Angeles Times, Heavy Gas Fumes in Air Blamed for Smarting Eyes , July 10, 1943.
[2]R.W. Keith, "Downtown Los Angeles Noon Visibility Trends, 1933-1969," Los Angeles County Air Pollution Control District, December 1970.
[3]Profile of Air Pollution Control , Los Angeles County Air Pollution Control District, 1971.
[4]A.J. Haagen-Smit, C.E. Bradley, M.M. Fox, "Ozone Formation in Photochemical Oxidation of Organic Substances," Industrial and Engineering Chemistry, **45**:2086 (1953).
[5]A.J. Haagen-Smit, M.M. Fox, "Automobile Exhaust and Ozone Formation," SAE Transactions, **63**:575 (1955).
[6]T.C. Austin, R.H. Cross, P. Heinen, "The California Vehicle Emission Control Program - Past, Present and Future," California Air Resources Board (1982).
[7] D.L. Blumenthal, T.B. Smith, W.H. White, et.al., "Three-Dimensional Pollutant Gradient Study, 1972-73 Program," Meteorology Research, Inc., Submitted to California Air Resources Board (November 1974).
[8]J.H. Seinfeld, "Ozone Air Quality Models: A Critical Review," *J. Air Waste Manage. Assoc.* **38**:616 (1988).
[9] P.A. Solomon, "Regional Photochemical Measurement and Modeling Studies: A Summary of the Air and Waste Management Association International Specialty Conference," San Diego, CA, November 8-12, 1993, *J. Air Waste Manage. Assoc.* **45**:253 (1995).
[10] U.S. Environmental Protection Agency, Office of Air Quality Planning and Standards, "National Air Quality and Emissions Trends Report, 1995," October 1996.
[11] G.J. McRae, W.R. Goodin, J.H. Seinfeld, "Mathematical Modeling of Photochemical Air Pollution," Final Report to the California Air Resources Board, Contract No. A5-046-87 and A7-187-30, April 27, 1982.
[12]B.J. Finlayson-Pitts, J.N. Pitts, Jr., "Atmospheric Chemistry of Tropospheric Ozone Formation: Scientific and Regulatory Implications," *J. Air Waste Manage. Assoc.* **43**:1091 (1993).
[13]D.R. Lawson, "Passing the Test – Human Behavior and California's Smog Check Program," *J. Air Waste Manage.* Assoc. **43**:1567 (1993).
[14]Federal Register: Volume 61, Number 241, December 13, 1996.
[15] Personal communication, Warren Freas, EPA Research Triangle Park, NC, March 20, 1997.
[16]J.M. Lents, P. Leyden, "RECLAIM: Los Angeles' New Market-Based Smog Cleanup Program," *J. Air Waste Manage. Assoc.* **46**:195 (1996).
[17]AQMD Advisor, Volume 3, Number 4, March 1996.
[18]Personal communication, Paul Solomon, Pacific Gas and Electric, April 1, 1997.

Numerical forecasting of ozone at the surface

Øystein Hov[1)] and Frode Flatøy[2)]

[1)] NILU, P.O.Box 100, N-2007 Kjeller, Norway

[2)] Geophysics institute, University of Bergen, N-5007 Bergen, Norway

1. INTRODUCTION

Exposure to elevated concentrations of ozone gives rise to a health risk for the human population and impairs agricultural productivity. National air quality standards, directives or guidelines have been introduced in many countries around the world. In many countries a warning system is established to alert citizens to take precautionary steps if the concentrations exceed certain values (see Dobris, 1995). According to the European Council Directive (92/72/EEC) on air pollution by ozone, thresholds are established for population information and warning. The threshold for warning of the public is 180 ppb as an hourly concentration, while the threshold for information of the public is hourly concentrations above 90 ppb. In the summer of 1996 in Europe, the threshold for warning of the public was exceeded at three sites (two monitoring stations in Athens and one in Firenze, Italy). The threshold for information of the public was exceeded in all member states of the European Union in the summer of 1996, except Ireland. It was calculated that 46% of the population in cities with operational ozone monitors may have been exposed to ozone exceedances in the summer of 1996 (Sluyter and van Zantvoort, 1996).

The forecasting of ozone analogous to numerical weather prediction is discussed in this paper. During the summer months, some countries have introduced routine procedures to predict ozone based on the combination of the weather forecast and the monitoring of ozone, or by running chemical models alongside the numerical weather prediction model. In this paper a coupled chemistry-transport model (CTM) is described and it is shown how an ozone forecast can be made when a modern numerical weather prediction model is run in combination with a state of the art chemistry model.

2. MODEL DESCRIPTION

The general approach used in the model is similar to that used in several other three dimensional models developed in the past decade [Liu et al., 1984; Chang et al., 1987; Carmichael et al., 1991; McKeen et al., 1991, Hass et al., 1993]. The mesoscale chemical transport (MCT) model is closely coupled to a numerical weather prediction (NWP) model with an advanced treatment of cloud physics and precipitation processes. This allows a detailed parameterization of convective transport and the calculation of photolysis rates in cloudy conditions to be made. A circumpolar grid or smaller grid can be applied (Figure 1). If a large circumpolar grid is chosen for the calculations, it is ensured that most of the conversion of anthropogenic and natural emissions of NO_x, VOC and other species over North America, Europe and Asia, take place within the model volume, and the effect of not well known upwind boundary conditions is minimized. Since the advection field has a predominant westerly component the air masses will remain inside the model domain for a substantial part of the integration period. This is further discussed in Flatøy and Hov [1996], while if a smaller grid domain is chosen (e.g. Figure 1), the boundary conditions more rapidly influence the results.

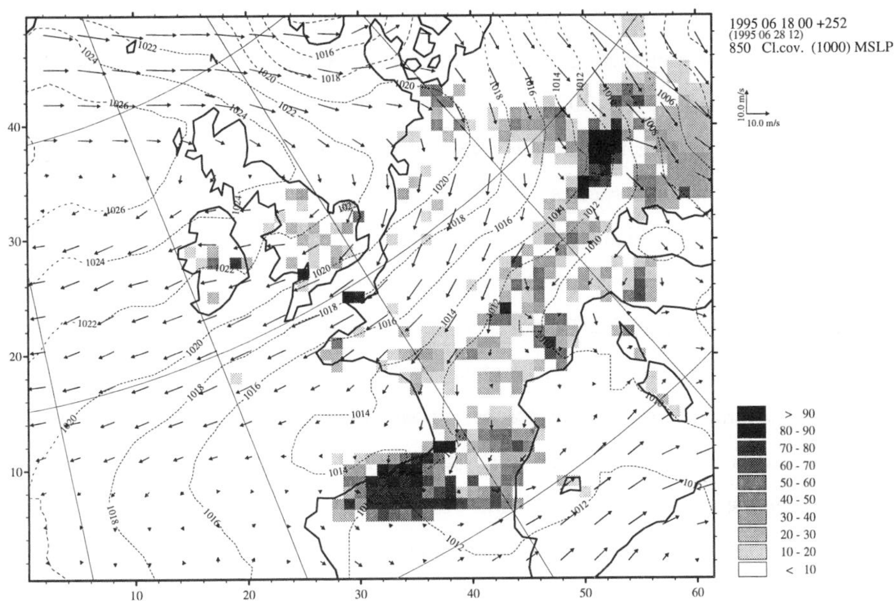

Figure 1. Mean sea level pressure, the 850 hPa cloud cover, and the 850 hPa winds at 1200 UT on June 28

The NWP model is based on the limited area model NORLAM from the Norwegian Meteorological Institute, described in Grønås et al. [1987], Nordeng [1986], and Kvamstø [1992]. Sigma coordinate primitive equations which are integrated on a stereographic map projection are used. The horizontal domain and the size of the grid elements are the same in the NWP and MCT models. Meteorological fields from the NWP model are in the present setup stored once every hour. The horizontal grid resolution of the input and the results is 50 kilometer at 60°N and there are 18 unequally spaced vertical layers extending up to 100 hPa in the version used here. With this resolution no interpolation in time of the meteorological data is thought necessary before use by the MCT model. The NWP model also supplies data for land use, topography, vegetation and ground albedo. Analysis every 6 hours from the ECMWF are used as initial and boundary conditions for a set of 18-hour prognosis with the NWP model. The first 6 hours of each prognosis are considered as spin up and are skipped, while the following 12 hours are used. The meteorological input to the MCT model then consists of a chain of independent 12 hour prognosis segments. In a forecast mode, a prognosis would have to be generated based on the most recent analysis which is available.

The transport part of the MCT model is described by Flatøy et al. [1995], and Flatøy [1993] and the chemistry part by Flatøy et al. [1995]. The chemistry scheme includes more than 40 chemical species and more than 120 chemical reactions for the gas phase formation of photooxidants. The emissions of volatile organic compounds (VOC) are represented by six nonmethane hydrocarbons (NMHC) and several oxygenated VOC's, in addition to isoprene. Nighttime chemistry is assumed to occur through nitrate radical attack on VOC's. Both peroxy radical + NO and peroxy radical + peroxy radical reactions are included to assure validity of the reaction scheme over a wide range of NO concentrations. Aerosol and liquid phase chemistry is parameterized. The chemistry scheme is documented by Strand and Hov [1994], who applied it in a study of global ozone. The choice of initial values and boundary conditions is discussed in Flatøy et al. [1995].

EMEP emissions' inventory has been utilized for NO_x, SO_2 and anthropogenic NMHC emissions. The natural VOC emissions represented by isoprene are derived by an emission-temperature relationship from Lübkert and Schöpp [1989] using the forest cover and the variable surface temperature from the result of the NWP model calculations. The formula gives low isoprene emissions at night, following the diurnal cycle in the surface temperature. The temperature is here also a surrogate for the effect of the variation in the photosynthetically active radiation (PAR), which strongly affects the isoprene emissions. Empirical relationships for the dependence of the isoprene emissions on both temperature and PAR have been reported in the literature [Guenther et al., 1993]. Aircraft NO_x emissions are taken from the global ANCAT/ECAC Emission inventory database group [1995] and the lightning NO_x emissions are from Køhler et al. [1995].

3. FORECAST OF OZONE

The period of calculation to be discussed here is June 24 to 29, 1995 where a high pressure center north west of Scotland gave rise to a westward air flow off the coast of

South England and France. During part of the period a low pressure center was found over the Iberian peninsula which increased and narrowed the air flow and gave a flow situation with a continental plume from Europe flowing into the North Atlantic. We have performed several 18 days' simulations (18 June-5 July 1995) with different grid resolutions to investigate large and small scale features of this situation. The same period was also used in a study of the importance of lightning emissions of NO_x and how results from simulation models are influenced by uncertainties in this source. Lightning emissions are connected with convective activity. A set of simulations were made on a circumpolar grid covering the northern hemisphere from the pole to 35°N with a horizontal resolution of 150 km and 10 unequally spaced layers. The simulations were run with different types and localizations of emissions to study the combined effect of North American and European emissions on the atmospheric chemistry over the North Atlantic. Part of this work is presented in Flatøy et al. [1996]. It was shown how the transport of pollutants in the free troposphere across the Atlantic only takes in certain atmospheric flow situations.

Here the focus will be on a simulation where the horizontal scale was 50 km and with 18 layers in the vertical and covering a small part of the northern hemisphere (see Figure 1). This gives a possibility to study the small scale meteorological and chemical structures which a plume of continental emissions contains. Especially the effect of topography, better resolved convection and land-sea interactions are important for the meteorology and more detailed inventories of surface emissions and improved convective transport are important for the chemistry computations.

Figure 1 shows the mean sea level pressure, the 850 hPa cloud cover, and the 850 hPa winds at 1200 UT on June 28 1995 from the simulation with the 50 km grid. Before reaching the ocean the air masses passed over large sources of oxides of nitrogen, volatile organic compounds and other pollutants in the United Kingdom and in Central Europe. This gave rise to a well defined plume of polluted continental air downwind of the sources. The period when the plume developed and existed was characterized by fair weather and the plume was not disturbed by frontal motions and only to a limited degree by convective activity.

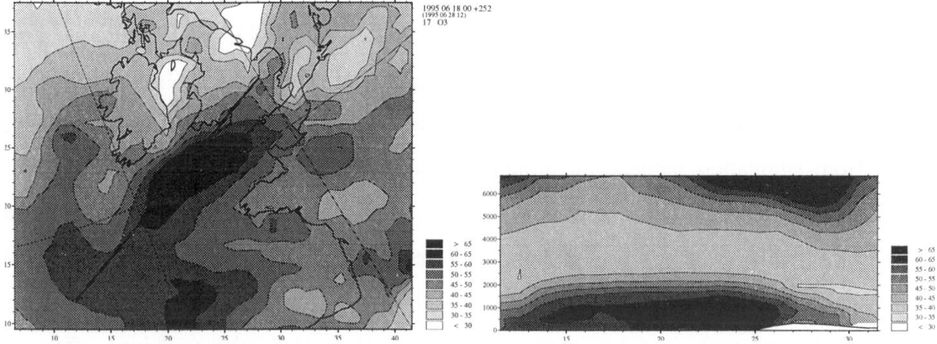

Figure 2. Ozone concentration at 1200 UT on June 28 for model layer 17, ~250 m a.s.l. (left panel) and vertical cross section of ozone concentration (right panel).

Figure 2 shows the O_3 plume at 1200 UT on June 28 1995 for model layer 17 (~250 m a.s.l.). Air with increased ozone concentration is transported with the large scale winds westward off the United Kingdom and the European continent. Ozone production (not shown) takes place close to the source regions, the largest ozone production is not found where the NO_x concentration has its maximum but in areas where NO_x is from 1-3 ppb.

Figure 3 shows scatter plots from June 28 1995 for model layer 17 of the nett ozone production by chemistry between 1200 UT and 1500 UT versus the NO_x concentration (left panel) and versus the NMHC concentration (right panel). The most favorable conditions for ozone production is seen to occur for NO_x concentrations around 2 ppb and NMHC concentrations around 20 ppbC.

Lightning emissions of NO_x can have an important influence on the chemical ozone production in the troposphere. To investigate if this can be properly handled by simulation models we have made a simulation for the above period where NO_x from lightning is distributed according to the convection modelled in the employed numerical weather prediction model, and compared this to the base simulation when a constant lightning inventory is used. The results indicate that when a realistic distribution of the lightning sources of NO_x in time and space is calculated, the upper tropospheric fields calculated for ozone and nitrogen oxides, differ significantly from the results of a standard run where the lightning sources of NO_x are distributed according to some monthly or seasonal climatological average. The differences are larger than e.g. the effect calculated from aircraft emissions of NO_x. This work is further described in Flatøy and Hov [1997].

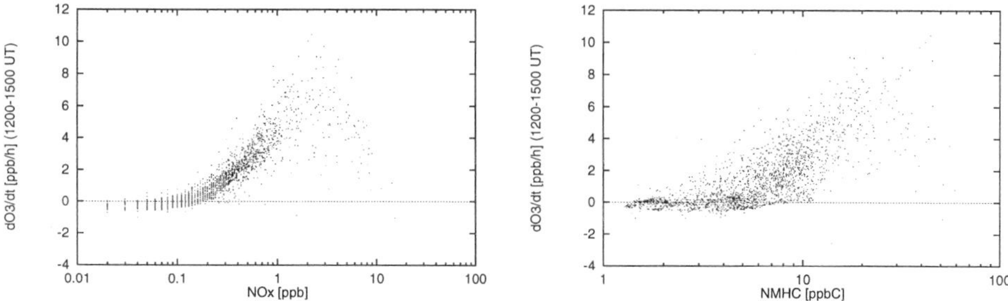

Figure 3. Scatter plots from June 28 for model layer 17 of the ozone production by chemistry between 1200 UT and 1500 UT versus the NO_x concentration (left panel) and versus the NMHC concentration (right panel).

The employed chemistry transport model gives a possibility to make estimates of the oxidation capacity of the atmosphere and the role of VOC and NO_x in the chemical ozone formation. This is of importance when measurement campaigns are planned and also when measured data is analyzed. An other important aspect of the model is the possibility to separate the different physical and chemical processes that modify the concentration distribution and thereby increase the understanding of the mechanisms behind observed concentration, see examples in Flatøy et al., (1995, 1996), Flatøy and Hov (1996, 1997) and Hov and Flatøy (1997).

The coupled NWP and chemistry model can be run with sufficient computational efficiency to be used in a forecast mode, provided that an appropriate weather analysis is available to initialize the numerical weather prediction model. The initial and boundary conditions of the chemical compounds are of particular importance, however, in particular of those species which have a chemical decay time comparable to the transport time through the model domain. Such compounds are ozone, CO, oxides of nitrogen, and a long range of hydrocarbons and oxygenated species. In particular the boundary conditions for ozone aloft and on the lateral boundaries are essential for the absolute levels of ozone calculated in the free troposphere. Run in a forecast mode, i.e. covering a few days only, the ozone distribution in the free troposphere is to a large extent determined by dynamical processes, and consequently the aloft and lateral boundary conditions are important.

4. MODEL VALIDATION AND DISCUSSION

The calculations shown are examples of historic forecasts of ozone on a regional or continental scale, based on meteorological forecast by a numerical weather prediction model and based on precursor emissions estimated from annual values. The model has been compared with surface measurements, ozone sondes and aircraft measurements. In Flatøy et al. [1995] comparison were made of the calculated distribution of ozone over Europe and surface ozone as well as ozone sonde measurements. It was found that the upper boundary condition chosen for ozone (at 100 hPa) is quite important for the absolute level of ozone in the troposphere, while the relative changes with height and with time are caused mainly by dynamical/physical processes when a time period of a few weeks is studied.

The model results have also been compared with aircraft measurements made in international projects like the EU sponsored OCTA and POLINAT in 1993 and 1995 over parts of the North Atlantic, and where also CO, NO_x and NO_y were measured in addition to ozone. More details on the model validation are given in Flatøy et al. [1996] and Flatøy and Hov [1996].

A comprehensive field campaign to measure the transformation of European precursor emissions into secondary products like ozone and PAN is being undertaken in the summers of 1996 and 1997. In the EU-funded project Testing Atmospheric Chemistry in

Anticyclones (TACIA) the British Meteorological Research Flight aircraft C130 is instrumented to measure ozone, NO, NO_x, NO_y, HNO_3, PAN, CO, carbonyls, peroxides, photolysis rate coefficients and perhaps peroxy radicals on a continuous basis and individual nonmethane hydrocarbons on an offline basis, in air masses flowing off the European continent over the North Atlantic in anticyclonic conditions. These measurements will serve as a comprehensive basis for model validation and an assessment of the prognostic capability of the CTM.

ACKNOWLEDGEMENT

The work reported here is supported by the Commission of the European Communities through the project Testing Atmospheric Chemistry in Anticyclones TACIA (ENV4-CT95-0038).

REFERENCES

1. CAT/ECAC Emission inventory database group, Abatement of Nuisances Caused by Air Transport/European Civil Aviation Conference: A global inventory of aircraft fuel and NO_x emissions for AERONOX, First AERONOX day, DLR, Cologne 28 February 1995.
2. Benkovitz, C. M., J. Dignon, J. Pacyna, T. Scholtz, L. Tarrason, E. Voldner and T. E. Graedel. Global Inventories of Anthropogenic Emissions of SO_2 and NO_x , 1995 (in preparation).
3. Brost, R. A., R. B. Chatfield, J. P. Greenberg, P. L. Haagenson, B. G. Heikes, S. Madronich, B. A. Ridley and P. R. Zimmerman, Three-dimensional modeling of transport of chemical species from continents to the Atlantic Ocean. Tellus, 40B, 358-379, 1988.
4. Carmichael, G. R., L. K. Peters, and R. D. Saylor, The STEM-III regional acid deposition and photochemical oxidant model, I, An overview of model development and applications, Atmos. Environ., 25A, 2077-2090, 1991.
5. Chang, J. S., R. A. Brost, I. S. A. Isaksen, S. Madronich, P. Middleton, W. R.

Stockwell, and C. J. Walcek, A three-dimensional Eulerian acid deposition model: Physical concepts and formulation, J. Geophys. Res., 92, 14,681-14,700, 1987.

6. Dobris (1995) Europe's environment. The Dobris Assessment. D. Stanners and P. Bourdeau, eds. European Environment Agency, Copenhagen, 676 p.

7. Flatøy, F. Balanced wind in advanced advection schemes when species with long lifetimes are transported. Atmos. Environ., 27, 1809-1819, 1993.

8. Flatøy, F., Ø. Hov, and H. Smit, Three-dimensional model studies of exchange processes of ozone in the troposphere over Europe, J. Geophys. Res., 100, 11465-11481, 1995.

9. Flatøy, F., and Ø. Hov, 3-D model studies of the effect of NO_x emissions from aircraft on ozone in the upper troposphere over Europe and the North Atlantic, J. Geophys. Res., 101, 1401-1422, 1996.

10. Flatøy, F., Ø. Hov, C. Gerbig, and S. J. Oltmans, Model Studies of the Meteorology and Chemical Composition of the Troposphere over the North Atlantic During August 18-30, 1993, J. Geophys. Res. 101, 29,317-29,334, 1996.

11. Flatøy, F. and Ø. Hov, (1997) NO_x from lightning and the calculated chemical composition of the free troposphere, J. Geophys. Res., accepted.

12. Grønås, S., A. Foss, and M. Lystad, Numerical simulations of polar lows in the Norwegian Sea. Tellus, 39A, 334-353, 1987.

13. Guenther, A. L., P. R. Zimmerman, P. C. Harley, R. K. Monson and R. Fall, Isoprene and monoterpene emission rate variability: model evaluations and sensitivity analyses. J. Geophys. Res., 98, 12,609-12,617, 1993.

14. Hass, H., A. Ebel, H. J. Jakobs, and M. Memmesheimer, Interaction of the dynamics and chemistry in photo-oxidant formation, Transport and Transformation of Pollutants in the Troposphere, Proceedings of EUROTRAC Symposium '94, Garmisch-Partenkirchen, April 11-14 1994, P. Borrell, P. M. Borrell and W. Seiler, editors, SPB Academic Publishing, The Hague, The Netherlands, 65-68, 1993.

15. Hov, Ø. and F. Flatøy, Convective redistribution of ozone and oxides of nitrogen in the troposphere over Europe in summer and fall. Tellus (in press), 1997.

16. Kvamstø, N. G., Implementation of the Sundqvist Scheme in the Norwegian Limited Area Model, Meteorol. Rep. Series, 2-92, Dep. of Geophys., Univ. of Bergen, Bergen, Norway, 1992.

17. Køhler., I., R. Sausen, and L. Gallardo Klenner, NO_x production by lightning, The Impact of NO_x emissions from aircraft upon the atmosphere at flight altitudes

8-15 km (AERONOX), edited by U. Schuman, Draft, Final report to the CEC, DLR, Oberpfaffenhofen, Germany, 343-345, 1995.

18. Liu, M. K., R. E. Morris, and J. P. Killus, Development of a regional oxidant model and application to the northeast United States, Atmos. Environ., 18, 1145-1161, 1984.

19. Lübkert, B., and W. Schöpp, A Model to Calculate Natural VOC Emissions from Forests in Europe, Rep. A-2361, Int. Inst. for Appl. Syst. Anal., Laxenburg, Austria, 1989.

20. McKeen, S. A., E.-Y. Hsie, M. Trainer, R. Tallamraju, and S. C. Liu, A regional model study of the ozone budget in the eastern United States, J. Geophys. Res., 96, 10,809-10,845, 1991.

21. Nordeng, T. E., Parameterization of physical processes in a three-dimensional numerical weather prediction model, Tech. Rep. 65., Norwegian Meteorol. Inst., Oslo, Norway, 1986.

22. Sluyter, R. and van Zantvoort, E. (1996) Information document concerning air pollution by ozone. Overview of the situation in the European Union during the 1996 summer season (April-July). Report to the Commission by the European Environment Agency, European Topic Center on Air Quality, Copenhagen.

23. Strand, A., and Ø. Hov, A two dimensional global study of the tropospheric ozone production, J. Geophys. Res., 99, 22,877-22,895, 1994.

Ecological Effects of Tropospheric Ozone: A U.S. Perspective - Past, Present and Future

W.E. Hogsett and Christian P. Andersen
U.S. EPA National Health and Environmental Effects Research Laboratory
Western Ecology Division
Corvallis, Oregon

ABSTRACT

Understanding the effects of tropospheric ozone on vegetation, as called for in the U.S. Clean Air Act, has involved collection of experimental data at the species level and, in particular, at the level of the individual and populations. Frequently the studies have been regression designs involving single species and single pollutants resulting in quantitative exposure-response functions that characterize the effects on biomass or reproduction (crop yield). Recently, the 1996 EPA Oxidant Criteria Document reviewed the published research on oxidant effects on crops, forests, and ecological resources, and concluded that the current secondary National Ambient Air Quality Standard (NAAQS) for ozone is neither protective or appropriate. The subsequent discussion and decision on what is an appropriate form and level of the secondary NAAQS utilized results from single pollutant and single species studies. To estimate concentrations causing effects composite response-functions for crops and trees from a large number of species were developed, predicting crop yield loss or annual biomass loss in seedlings. This approach is useful based on available quantitative data on biological effects; however it assumes that individual plant response does not change in the presence of other stresses or in natural systems which are considerably more complex (e.g. species' assemblages and competition for resources). For example, the importance of biological complexity is illustrated when we examine the below-ground ecosystem, an often overlooked portion of the ecosystem. We have found that incorporating natural biological complexity into potting soils can result in carbon fluxes opposite those predicted from individual plant studies using artificial media lacking natural soil foodwebs. A future research approach to understanding O_3 effects on ecosystems is required that will develop the necessary linkages to extrapolate experimental data taken at the individual level, often in artificial conditions, to predict changes on individuals or populations in more complex native environments. We will present experimental and modelling activities from our laboratory that show how we are beginning to address the problems of scale, complexity, and multiple stresses in forested ecosystems exposed to ozone stress. Questions frequently posed in ecological risk assessments.

1. INTRODUCTION

The Clean Air Act mandates the protection of ecological resources from adverse effects of air pollutants through setting a secondary National Ambient Air Quality Standard (NAAQS). Ecological resources are part of public welfare and include crops, forests, soils, ecosystems. As part of the standard setting process, EPA critically reviews all the pertinent literature every 5 years, and presents the conclusions in the Ozone and other Photochemical Oxidants Criteria Document (U.S. EPA, 1996). This document is scientifically peer-review and reviewed by the Clean Air Science and Advisory Committee (CASAC). The revised and approved document then provides the scientific basis for EPA's decisions regarding the NAAQS for ozone.

The most recent Oxidant Criteria Document was completed in 1996 (U.S. EPA, 1996), and concluded that the general conclusions of the previous 1986 Criteria Document and the 1992 Supplement were not altered. The general conclusions were (a) current ambient ozone concentrations in many areas of the country were sufficient to impair growth and yield of plants; (b) effects occur with only a few hourly occurrences above 0.08 ppm; (c) several species exhibited growth and yield effects when the mean ozone concentration exceeded 0.05 ppm for 4-6 hr/day for at least 2 weeks; (d) regression analyses of the National Crop Loss Assessment Network (NCLAN) data developing exposure-response functions for yield loss indicated that at least 50% of the crops were predicted to exhibit a 10% yield loss at 7-hr seasonal mean ozone concentrations of 0.05 ppm or less; (e) European crop yield loss studies substantiated the effects observed in this country; and (f) studies of forest tree seedlings substantiated pre-1986 studies indicating the sensitivity of a number of species, at least as seedlings.The experimental studies showed that tree seedling growth is altered at ozone concentrations observed in many areas of the U.S.

The 1996 Oxidant Criteria Document also reported that the conclusions of the 1992 Supplement were still valid: (a) The current 1 hour form of the standard (120 ppb) is not protective nor is it an appropriate air quality indicator for plants; (b) the 7- and 12-hr seasonal mean was also not appropriate because of it treats all concentrations equally, and does not consider exposure duration; and (c) the experimental studies suggest the indicator of air quality for a secondary NAAQS should cumulate all hourly concentrations during the daylight hours and weight the higher concentrations (the 'best' weighting scheme or the relative importance of concentration ranges is not known; however, concentrations above 60 ppb are generally considered important in causing effects)

As a result of the scientific data, EPA has concluded that the current secondary NAAQS is not protective of ecological resources, in particular protecting crops and forests in the U.S. The proposal for an alternative secondary NAAQS consists of 2 options for consideration. (1) Set the secondary NAAQS equal to the newly proposed alternative primary NAAQS (8 hr maximum of 80-90 ppb with 1-5 exceedances). While this proposal is based on air quality analyses, and is not biologically related to plant response, the attainment of the alternative proposed primary NAAQS would result in a reduction in ozone concentrations harmful to plant growth.; or (2) set the secondary as a 3 month, 12 hr SUM06 of 26.4 ppm hr. This air quality indicator is based on current biological response data and would prevent tree seedling biomass losses and crop yield losses greater that 10% in 50% of the species. The decision regarding which proposal to adopt is still pending.

The process of assembling the Criteria Document also reveals the scientific gaps in our knowledge of ozone effects on plants. Primary among these are (a) the problems associated with relating 2-4 year-old tree seedling response to ozone over 1-3 seasons to the response of large, mature trees over a 40-200 year life cycle; (b) the problems associated with extrapolating responses developed on individuals and relating it to the response of a forest stand; (c) how little we know about natural system complexity, wherein the interaction of species, genotypes, and the multitude of past and present environmental influences dictate the eventual response of the species or community in question. We will provide a brief overview of approaches we are taking to address these major areas of uncertainty.

2. PROBLEM: *Predicting Effects of Ozone on Long-lived Species over time, across varied environments, and in assemblage with other species.*

The knowledge gaps pointed out in the 1996 Criteria Document represent the source of much of the uncertainty in any risk characterization for any atmospheric pollutant, i.e. the ability to predict effects over long time periods, across diverse regions, and across scales of biological organizational. Fig. 1 illustrates three important considerations when extrapolating data: (1) the influential role of *size and ontogeny* : seedlings to mature trees; (2)*complexity of the environment* , including belowground processes and trophic groups, species diversity and stand structure: chambers to forests; and (3) the role of *time* : single season to multiple years. The whole tree in Fig. 1, including the aboveground and belowground tissues, is the experimental subject. Experimental studies have developed quantitative exposure-response functions to a single pollutant, for a single species at the individual to population level, of a given age, size and phenotypic stage, over a single season, and growing in a defined environment, either artificial or natural (lower left hand corner of the triangle). The information gained from each of these exposure experiments is then invariably used in risk characterizations for prediction of the effect along one or more of the lines of the triangle, i.e. mature trees, forests, and multiple years. Areas where the experimental data is limited. The prediction of risk to forests over multiple years requires an understanding of all the components of a forest, including the belowground component, species assemblages and age structure, as well as the spatial and temporal factors, in order to make inferences as to how a forest will respond to a given level of pollutant over time. The first concern is the significant limitations in extrapolating effects at the individual level. Changes in physical size as the tree ages, as well as changes in processes that occur with maturation need to be accounted for in making inferences beyond the experimental. For example, forest tree seedlings that range 1-10 ft in height are studied and potential effects are inferred on mature trees that will attain heights of 150 ft. Because of differences in resource utilization in trees of different size and age (Wareing, 1993; Hinckley et al., 1997), we anticipate their being differences in both the

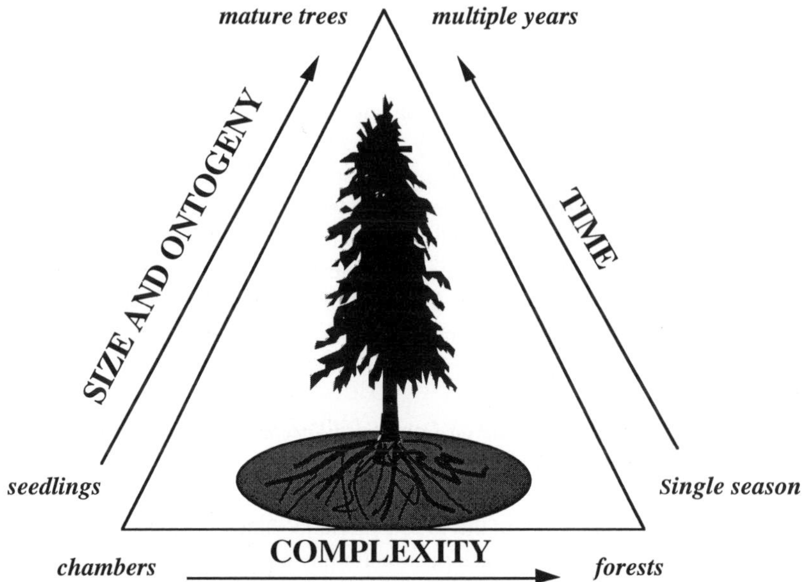

Figure 1. Extrapolation of stressor effects data (e.g. exposure response functions) across: (1) ***SIZE AND ONTOGENY*** of species, e.g.from seedlings to mature trees; (2) ***TIME***, including the duration of the stress exposure, season of exposure, year-to-year variation of exposure, phenology and life span of species; and (3) ***COMPLEXITY***, from experimental chamber studies to 'real-world' conditions, artificial soils to native soils with inclusions of intact soil food webs, or single individuals to communities to landscapes

uptake of ozone, as well as the ultimate effect on growth, to result from age and size differences. A few studies have indicated such differences in response and attributed those differences to uptake of ozone (Grulke et. al., 1989; Samuelson and Edwards, 1993; Fredericksen et al., 1996).

In addition to these age/size limitations, the spatial and temporal distribution of environmental and exposure factors influencing the response to a pollutant need to be considered in the extrapolation. Water and nitrogen availability is of particular importance in understanding response to a pollutant over the species' range. The ozone exposure varies from year to year with weather and emissions. The year-to-year variation of exposure is not accounted for in most experimental data. How to appropriately use available data in risk characterizations is a question for future activity, as well as conducting new research to address questions of the influence biological scale and multiple environmental factors on the effect of ozone.

3. RESEARCH APPROACH - ***How to address the shortcomings in existing data and reduce uncertainty in extrapolations needed for risk characterization***

A multifaceted, interactive research approach is required, including experimental and modelling components, with each informing the other. The objective is a mechanistic understanding of each of the lines of extrapolation illustrated in Fig. 1.

Plants are exposed to a wide range of spatially distributed environmental extremes, including anthropogenic stressors (Taylor et al., 1994). Plants utilize available resources in order to optimize growth and reproductive outputs under constantly changing conditions. Air pollution represents a stress similar to other stresses in that plant response is the result of both avoidance and tolerance mechanisms (Ariens et al., 1976; Tingey and Andersen 1991). Avoidance factors include factors that restrict the pollutant from reaching the target site, e.g., stomatal closure. Tolerance factors include physiological shifts that result when the stress reaches the target site and causes a metabolic change. Despite the nature of tolerance mechanisms, plant response to a stress involves a shift from one metabolic condition to another, and therefore represents a stress response.

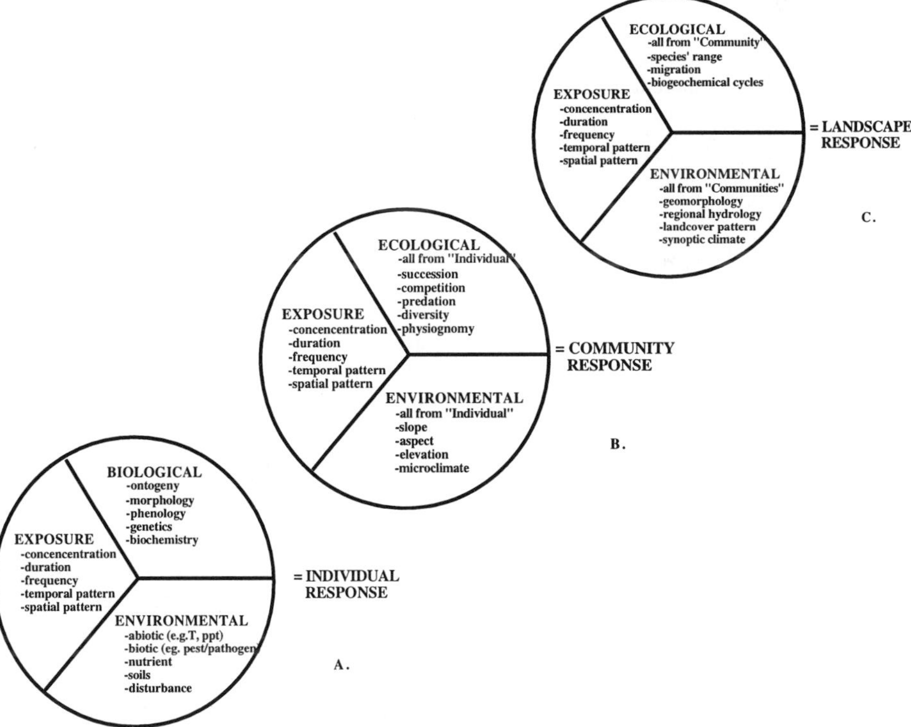

Figure 2. Factors influencing plant response to stress at the Individual/Population level (A) or Community/Stand level (B) or the Landscape level (C)

There are several factors that affect the magnitude of plant response to ozone stress (Hogsett et al., 1988), and Figure 2 shows a list of important factors at each spatial scale of interest. Figure 2a provides a list of factors influencing individual plant response to ozone, including (1) the past and present environmental milieu, including daily and seasonal weather conditions, soil fertility and water holding capacity, soil food web, and disturbances of various kinds (pest infestations, wildfire, floods, land-use, and so on); (2) biological factors, including genetic variability and constraints, seasonal phenology, limits to biochemical responses and (3) components of stressor exposure, including concentration of pollutants, frequency and duration of exposure, and daily, seasonal and annual patterns of exposure. All of these factors need to be considered to extrapolate the individual plant's response across temporal and spatial scales.

Understanding the factors that influence response to ozone stress is considerably more complex at the level of communities (Fig 2B) than at the level of the individual (Fig 2A). Predicting behavior of individuals growing in mixed age, mixed species stands includes all of the factors that influence individual responses, plus such factors as variations in elevation, steepness of slopes, and aspect, and the predictable characteristics of climate near the ground (microclimate). In addition, ecological processes (plant succession from least to most shade tolerant species; competition for sunlight, nutrients, and soil moisture; effects of biodiversity and the physiognomy of vegetation) cannot be derived directly from knowledge of the biological attributes inherent in individual plants. Yet, these processes play a critical role in determining individual plant response to stressors as they form groups of species (populations). Indeed, exposure to stressors must now include a careful review of plant spatial distribution as controlled by the environmental variables included in Figure 2B.

The shift from consideration of relatively uniform stands of vegetation gathered into communities, to highly diverse ecosystems gathered into watersheds and landscape units involves additional processes that modify response to stressors (Fig 2C). Environmental effects of geomorphology (landforms, geology, drainage patterns) and regional hydrology (seasonal movement of moisture in soils, into streams and the atmosphere) combine with synoptic climate variables (air mass frequencies, teleconnected patterns such as el niño) to generate a regional response which will not equal the average response of populations in the region. Land uses create varying sizes and connectivities of wildlands, each of which functions differently. Related ecosystem processes become significant at this dimension. For example, they define the geographic ranges of animal and plant species distribution, and the daily and seasonal movements (animal migrations) or long-term changes in their boundaries (plant migration). Although the exposure variables at the landscape level are mostly those useful at the level of individual plants, the converse is not true: the landscape-level phenomena can be predicted without much consideration of individual plant characteristics or their environmental constraints.

The extrapolation from individual to community to landscape and on to the biome level requires multiple steps (Hinckley et al., 1997). Typically, one works with empirical information generated at one level of organization, using it to model and validate at the next level of organization, and subsequently using that information (model) to predict (implications) at a third level of organization. The complexity characteristic of each level of hierarchy precludes extrapolation beyond 2 or 3 levels of scale. For the purposes of studying ozone effects, it is necessary to bridge scales across each of several levels of complexity: cells-organs-individuals; organs-individuals-stands; individuals-stands-ecosystems; stands-ecosystems-landscapes.

Ideally, the complexities described above for each level of scale could be addressed experimentally using experimental designs that are quantitative and enable prediction of effects over a range of concentrations, including the variation in that response (Hogsett et al. 1988). Due to logistical problems, exposure studies over several years are not easily accomplished on whole forests, or even whole trees. Therefore, extrapolation from smaller scaled exposure studies is necessary. To accomplish the extrapolations, we use a combination of process-based and stand-level model simulations. Models can be parametrized for large trees or stands that simulate the interaction of multiple factors in response to ozone based on the experimental studies of the multiple interactions, such as those illustrated in Figure 2.

Models are essential components of any risk characterization of ecosystems response to natural and anthropogenic stressors (Rastetter, 1996). Driving forces such as temperature, precipitation, nutrient inputs, topography, soil moisture, and soil biota make it impossible to predict future response to ozone based on single-factor experiments alone (Rastetter et al., 1996). Process- based models at the individual or the stand level can provide a self-consistent synthesis of the results of many experiments. The synthesis provided by these models includes the interactions among physiological and ecological processes that give rise to the synergistic responses to multiple environmental factors and stresses (Rastetter et al., 1996).

Coupled with modelling efforts, correlative, observational studies can be carried out in forested areas along pollution gradients where monitoring is concurrent and other environmental gradients (e.g. water availability) are present. These studies are possible in some regions of the U.S. (Miller et al., 1982;1989), and will be useful in helping us understand population-level responses as well as validating model simulations.

As a way to illustrate how we are combining experimental and modelling efforts, several examples of our current research activity in Corvallis are discussed. We have associated each example with a particular level of biological organization, i.e., individual/population, community, and landscape; however, there is overlap among the levels both in the experimental information obtained and its subsequent use.

3.1. Individual/Population

3.1.1. Extrapolation of effects on a few species to reflect the effect on all species

A major portion of the research activity over the past decade has been directed at developing exposure-response functions for both crops and forest tree species (Heagle et al., 1988; Fuhrer et al., 1989; Hogsett et al., 1989; Tingey et al. 1991; Neufeld et al., 1995; Karnosky et al., 1995.). Many of the studies were designed to predict growth and yield effects at the individual and population level at a range of ozone concentrations. The primary means of exposing plants during the last 15 years has included the use of the open top chamber methodology for developing the empirical database of ozone effects on crop and tree species (Heagle et al., 1988; Taylor et al., 1994). Open-top chambers provide a sufficient range of ozone treatments for quantifying effects (i.e. exposure-response functions), and still provide growing conditions that closely match those in the plants' natural growing environment. In addition, the multitude of chamber sites has permitted adequate replication of studies from year to year. Examples of these studies include the National Crop Loss Assessment Network (NCLAN) (Heck et al, 1991), the European Open-Top Chamber studies (Weigel et al., 1987), and several studies of tree species (Hogsett et al.,1989; Matyssek et al., 1993; Neufeld et al., 1995; Karnosky et al., 1995; Rebbeck, 1996). The studies included a wide range of response functions illustrating that ozone response is species dependent. For the first few years of the our project studying forest tree species, we developed exposure response functions for 11 forest tree species during 1-2 year exposure seasons using open-top chambers at locations across the U.S. (Table 1). Sites were selected in different regions of the U.S. where the species of interest were indigenous. A common research protocol was used at each site, and several of the species were examined at more than one site.

Table 1. Tree Species: Locations and exposure durations used to construct composite exposure response function.

Species/family	Site	Exposure duration (days)
Aspen—wild	Oregon	112
	Michigan	98
Douglas fir	Oregon	234
	Oregon	234
Ponderosa pine	Oregon	230
	Oregon	280
Red alder	Oregon	118
	Oregon	112
Black cherry	Smokey Mtn Nat'l Park	76
	Smokey Mtn Nat'l Park	140
Red maple	Smokey Mtn Nat'l Park	55
Tulip popular	Smokey Mtn Nat'l Park Ohio	184 222
Virginia Pine	Smokey Mtn Nat'l Park	98
Loblolly	Alabama	555
Sugar Maple	Michigan Ohio	180 222
E. white pine	Michigan Ohio	180 222

Figure 3 shows examples of exposure-response functions for total biomass (foliage, stem and root) after one growing season. Douglas-fir (*Psuedotsuga menziesi* (Mirb.) Franco) seedlings were relatively insensitive to ozone (Fig. 3a), while ponderosa pine (*Pinus ponderosa* Laws.) (Fig 3b) and quaking aspen (*Populus tremuloides* Michx.) (Fig. 3c) were relatively sensitive to ozone. The response of the other 7 species varied from insensitive (e.g. Virginia pine, *Pinus virginiana* Mill.) to moderately sensitive (e.g. Sugar Maple, *Acer saccharum* Marsh.) to sensitive (e.g. Black cherry, *Prunus serotina* Ehrh.). Given the range in response among species, it is difficult to generalize or extrapolate an effect across all species in order to provide information for decision making regarding risk characterization and standard setting. However, one approach to accomplishing such a task is to develop a composite exposure-response function for tree species and one for crops. Figure 4 shows the composite functions for crops and tree species. The percentile distributions illustrate the range of predicted relative biomass loss as a result of ozone exposure. At 30 ppm-hr ozone, the predicted biomass loss of tree seedlings ranges from 0-20%.

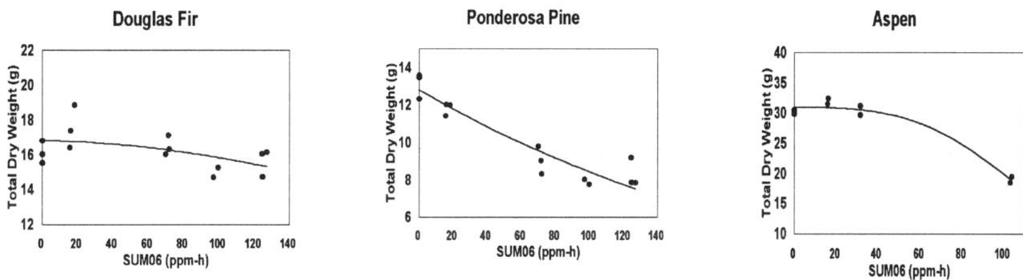

Figure 3. Exposure-response functions for Douglas-fir, Ponderosa Pine and Aspen. Total biomass response after one season of exposure. Exposure is expressed as SUM06 (sum of all hourly values equal to or greater than 60 ppb during the daylight hours (12 hr) over the growing season)

The Weibull or linear exposure-response model was used to relate final harvest biomass and exposure, expressed as the 12-h SUM06 index for 31 NCLAN crop studies (Fig 4A) and 26 tree seedling studies (Fig 4B). Separate regressions were calculated for studies with multiple harvests or cultivars, resulting in a total of 54 individual equations from the 31 NCLAN studies and 56 equations from the 26 seedling studies. For crops, each equation was used to calculate the predicted relative yield loss at 10, 20, 30, 40, 50, and 60 ppm-h, and the distributions of the resulting loss were plotted (Fig 4A).

For seedlings having been exposed to ozone for one or multiple exposure seasons of varying durations, each equation was used to calculate the predicted biomass loss at 10, 20, 30, 40, 50, 60, and 70 ppm-h per 92 days a year, and the distributions of the resulting loss were plotted (Fig 4B).

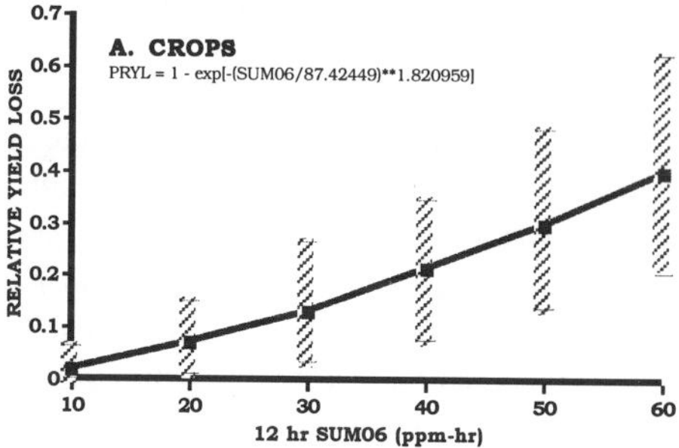

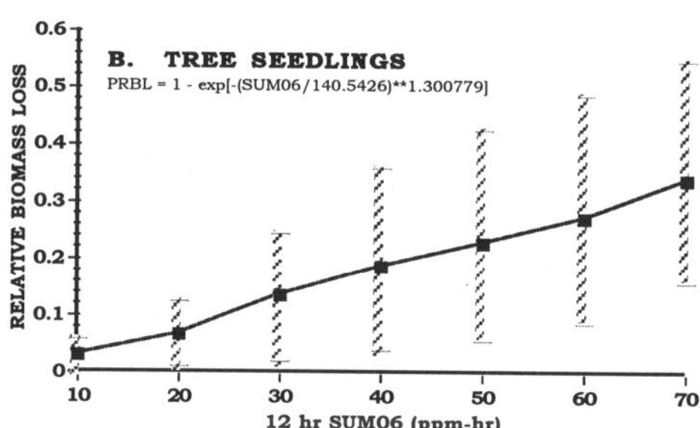

Figure 4. Composite exposure response functions for crops (A) and tree seedlings (B). Median response (50th percentile) is illustrated with solid line and weibull model given in each figure. 25th and 75th percentile shown with shadowed bars at each exposure value.

At each SUM06 value, the median (50th percentile) response was used to characterize the typical plant response to ozone across studies. The composite curve for crops or seedlings is the calculated Weibull model fit to the median response points.

Using these composite functions which composit species response, one can select an ozone concentration for a standard that is protective of trees or crops, depending on the level of protection sought. The line represents the Weibull curve that describes the response in 50% of the seedlings or crops. Using this as an estimation of the mean response in tree species, concentrations can be determined that are protective against 10, 20, or 30% biomass losses (or crop yield losses), in 50% of the species. Although this approach provides some means to estimate the impact of ozone on crop and forest species, it includes large uncertainties because of the species-dependence of the response, and it does not account for the variation in the response resulting from different growing environments, exposure or other biological/ecological considerations (Fig. 2).

3.1.2. Extrapolation of effects along the line of increasing complexity

As controlled experimental studies begin to incorporate the multitude of factors typical of natural systems, (e.g. temporal and seasonal variation in exposure concentrations, variation in availability of water and nitrogen, presence of intact soil food web and rhizosphere biota, presence of competing species, etc), an understanding of both the direction and the magnitude of the modification by these influences on the exposure-response will result. The information can be used to reduce uncertainty in risk assessments, as well as improving process-based models simulating long-term effects of ozone on forest communities. Many earlier studies have examined the influence of these factors, including water stress on crops(e.g.Tingey & Hogsett, 1985; Miller et al. 1989; Temple et al 1988.), relative humidity (McLaughlin & Taylor, 1980), nutrient status (e.g. Tjoelker & Luxmore 1991; Bytnerowicz et al, 1990), exposure components (Walmsley et al., 1980; Musselman et al., 1983;1986; Hogsett et al., 1985; 1988; Hogsett and Tingey, 1991).

Three examples of ongoing studies at our lab looking at this question from the perspective of the individual, community (Sec. 3.2.1), and landscape (Sec. 3.3) (Fig 2) are presented.

Nitrogen Status:

In a study of the influence of nitrogen supply on the response of ponderosa pine to ozone, we have measured both an individual response and a 'system' level response (Andersen and Scagel, 1997). Three levels of nitrogen- full, 2/3 or 1/3 rate was provided the year prior to exposure and continued through the year of exposure. The exposures were continued for 2 years. Figure 5 shows the response in biomass and foliar nitrogen content after one year of ozone exposure as not obviously different with varying nitrogen. The observed response suggests there is not a significant influence of nitrogen status on the growth response to ozone, and thus it may be appropriate to extrapolate exposure-response functions describing biomass at the level of the individual across regions differing in nitrogen availability. One measure of system level response, however, CO2 flux from the soil, did indicate an influence of nitrogen status. Figure 5 shows a more pronounced response to ozone at the low nitrogen level, suggesting a possible interaction. The low nitrogen status may influence the ability to maintain roots and increased root mortality may be contributing to increased metabolic activity of soil biota and thus account for the increased CO_2 flux from soil. These studies are not completed and are not thoroughly analyzed, but do suggest possible implications in our ability to extrapolate exposure response functions spatially as a result of available nitrogen. The measured response, biomass loss or CO2 flux from soil, are both valid and provide critical data for risk characterization. However, from this study two different conclusions might be drawn on the role of nitrogen depending on whether the endpoint for assessment was at the level of the individual, or at the system level.

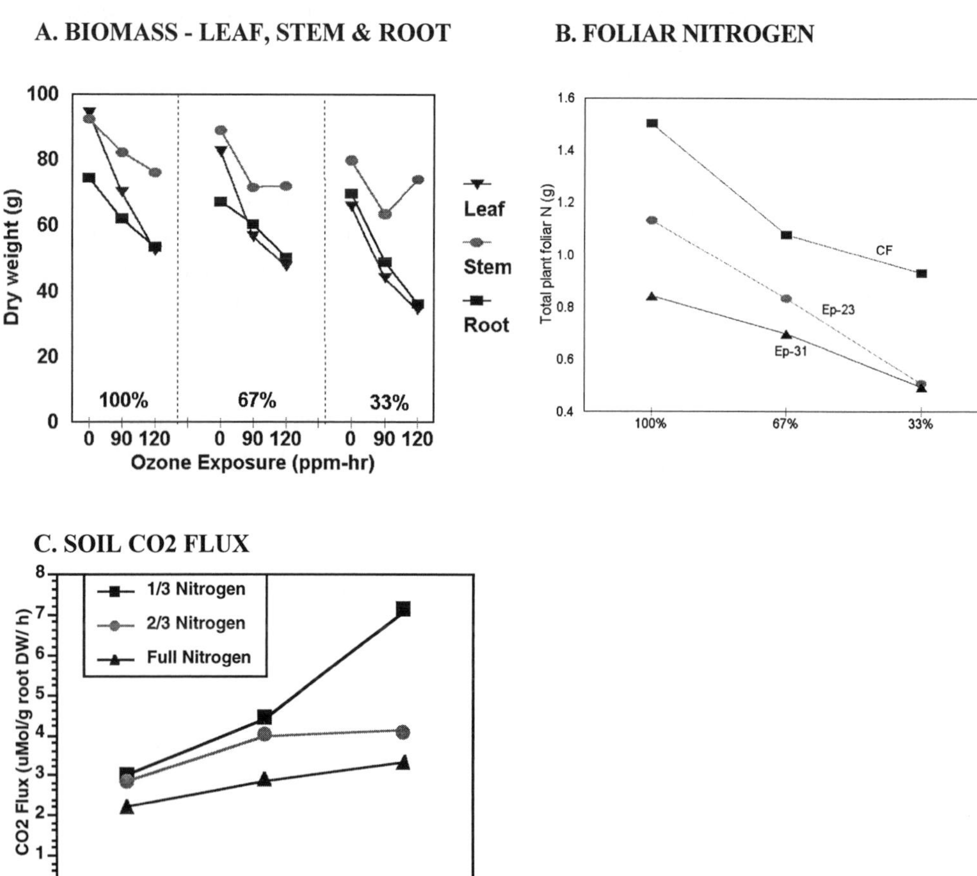

Figure 5. Response of ponderosa pine to 3 levels of ozone (Episodic [Ep] regimes having a SUM0 of 90 & 120 ppm hr and charcoal filtered[CF]) under 3 levels of nitrogen (100%, 67% & 33% of precribed application) in native soils. (A) biomass response after one season. (B) Total foliar nitrogen after one season. (C) Soil CO_2 flux after one season.

3.1.3. Extrapolation of effects from seedlings to mature trees:

Understanding the role of size and age in determining tree response to ozone is very important. As discussed above, much of the data we have today was obtained on various species of trees in the seedling stage, usually 1-5 years of age. As a result, the uncertainty in using this data to extrapolate the effects on large mature trees is very significant for any risk assessment. How maturation (ontogeny) effects tree response to air pollution is not known. It is thought that changes in pollutant uptake may occur with size, as well as differences in carbon sink sizes, the size of respiratory pools, and priorities (such as growth versus maintenance respiration) may affect the way the tree responds to ozone. Very few studies have addressed this question due to the technical difficulties of exposing large trees to ozone (Samuelson and Edwards, 1993; Grulke et al., 1989; Fredericksen et al., 1995).

Studies in Corvallis are planned focusing on resource utilization (C, N, H_2O) by trees (e.g. Ponderosa pine) as a function of size and age. The research will develop a mechanistic basis for scaling (Wareing, 1993), and will provide data to parametrize models that will incorporate age differences in resource utilization (Hinckley et al., 1997) affected by ozone, as well as incorporating gas exchange differences which effect ozone uptake.

3.2. Community

3.2.1. Extrapolation of effects along the lines of complexity:

Presence of other species and the soil food web:

In the second example, studies at Corvallis have indicated the influence of intact soil food webs and the importance of including belowground complexity in when studying tree response to ozone. Mesocosms in open-top chambers were planted with varying densities of blue wild rye (*Elymus glaucus* Buckl.) (a natural competitor of ponderosa pine seedlings) and a single ponderosa pine seedling in soil from a ponderosa pine forest (Andersen, et al. 1997). The response to ozone exposure was measured as changes in carbon dynamics, including allocation in individuals (species) and soil respiration as CO_2 flux, over 3 years. This was not a true competition study design because it lacks the necessary permutations, however, it does indicate the influence of increasing complexity of the environment on the exposure response. The preliminary data would suggest that there is not an interaction between plant density and ozone. Figure 6 shows an increase in CO_2 flux and soil organic matter as grass density and ozone concentration increases. However, what we are learning from this study is the importance of an intact soil food-web, which was included here with the native ponderosa pine forest soil. Earlier exposure studies of carbon allocation belowground, both from our lab and results from other laboratories, showed a decrease in allocation of carbon belowground with ozone exposure (Gorissen and van Veen, 1988; Spence et al. 1990; Andersen and Rygiewicz,1995).

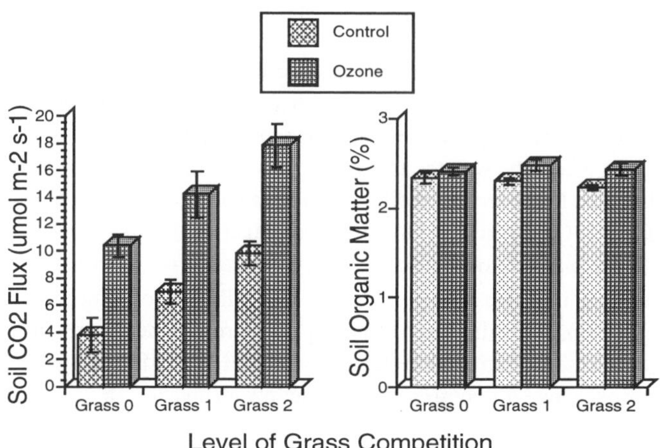

Figure 6. Soil CO2 flux (A) and Soil organic matter (B) following exposure to ozone for 1 year in mesocosms containing pine seedling and blue stem rye grass at 3 increasing densities ("Grass 0,1,2"). Bars indicate 95% confidence interval. Differences were significant at the 5% level

From these individual plant studies we would have predicted that ozone exposure would decrease CO_2 flux from soil due to less allocation of carbon belowground, rather than increase it as shown in Figure 6. The increase in both soil organic matter and soil CO_2 flux may be the result of increased root turnover and subsequent degradation by soil organisms in ozone exposed systems, which could increase total soil CO_2 flux. If this is the case, perhaps with time a decrease in flux will be observed as the seedlings get older and the total root system size decreases. Exactly how this information can be included in extrapolation of results for risk characterization is not known, however it does indicate that when studies are accomplished in simple systems and over short time frames (1-3 years), some degree of uncertainty should be included.

3.3 Landscape level

We have employed an approach for extrapolating tree biomass and crop yield response functions across each species' range to determine the extent and magnitude of possible ozone risk to ecological resources in the U.S. (Hogsett et al., 1997). The approach uses a geographical information system (GIS) to integrate all the necessary data in a spatial context., including (1) estimated exposure over a growing season, (2) characteristics of the growing environment (e.g. precipitation, temperature, soil water, soil nitrogen levels, etc), (3) species' distribution, (4) species' inventory data (e.g. standing biomass, productivity) and (5) landscape features (e.g. elevation) (Figure 7).

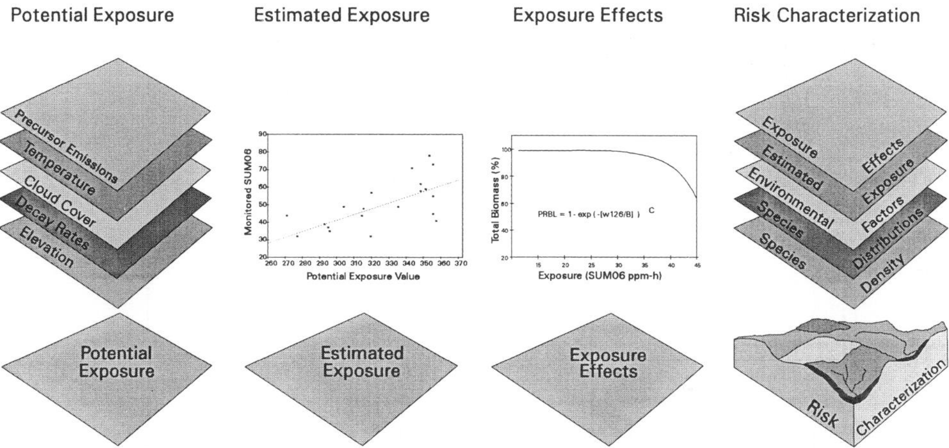

Figure 7. The GIS data layers for characterizing risk of ozone to tree species, including GIS-based estimation of ozone exposure in non-monitored areas for a given year, seedling or model-simulated tree exposure exposure response functions for each species, and the spatial data on environmental factors such as precipitation, and species density. The combination of these layers results in a new map of risk.

The approach is useful for distributing experimentally-derived exposure-response functions from previous crops or tree seedling exposure studies , as well as extrapolating process-based whole-tree model simulations of large trees using TREGRO (Weinstein et al. 1991) over a 3 year time period incorporating the role of environmental influence such as water or nitrogen, or stand-level model simulations of community structure (ZELIG) incorporating very long-term effects (100 years) as a result of ozone, climate, soil, water and nutrient availability. Theses extrapolation are based on the spatial and temporal distribution of ozone exposure and climate variables, e.g., precipitation and temperature that are present across the species' range or the region of the forest-type. The results of spatially distributing seedling exposure-response functions and then area-weighting that response is given in Figure 8. A sensitivity-ranking based on ozone sensitivity of each species and the potential area of the species impacted can be made (Hogsett et al 1997).

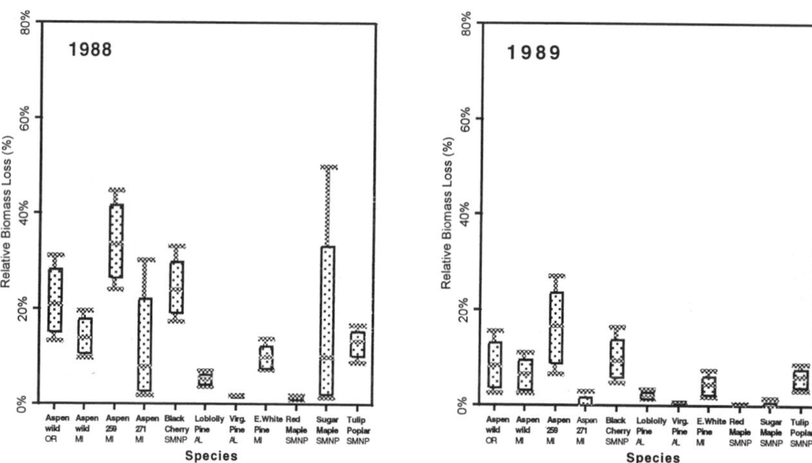

Figure 8. Area-weighted annual biomass loss in seedlings from 1988 and 1989 ozone exposure. Percentile distributions of species' area is shown for 10th, 25th, 50th, 75th and 90th percentile. The response for 50% of the species' area is shown as the midline in the shadowed box. The 25 and 75% of the area are the ends of the shadowed box.

To illustrate the process-based model simulation approach, Figure 9 shows the response in loblolly pine biomass after 3 years of exposure. We compared a hot, dry year having relatively high ozone concentrations (1988) to a cooler, wetter year having relatively low ozone concentrations (1989). TREGRO was parametrized from available literature on

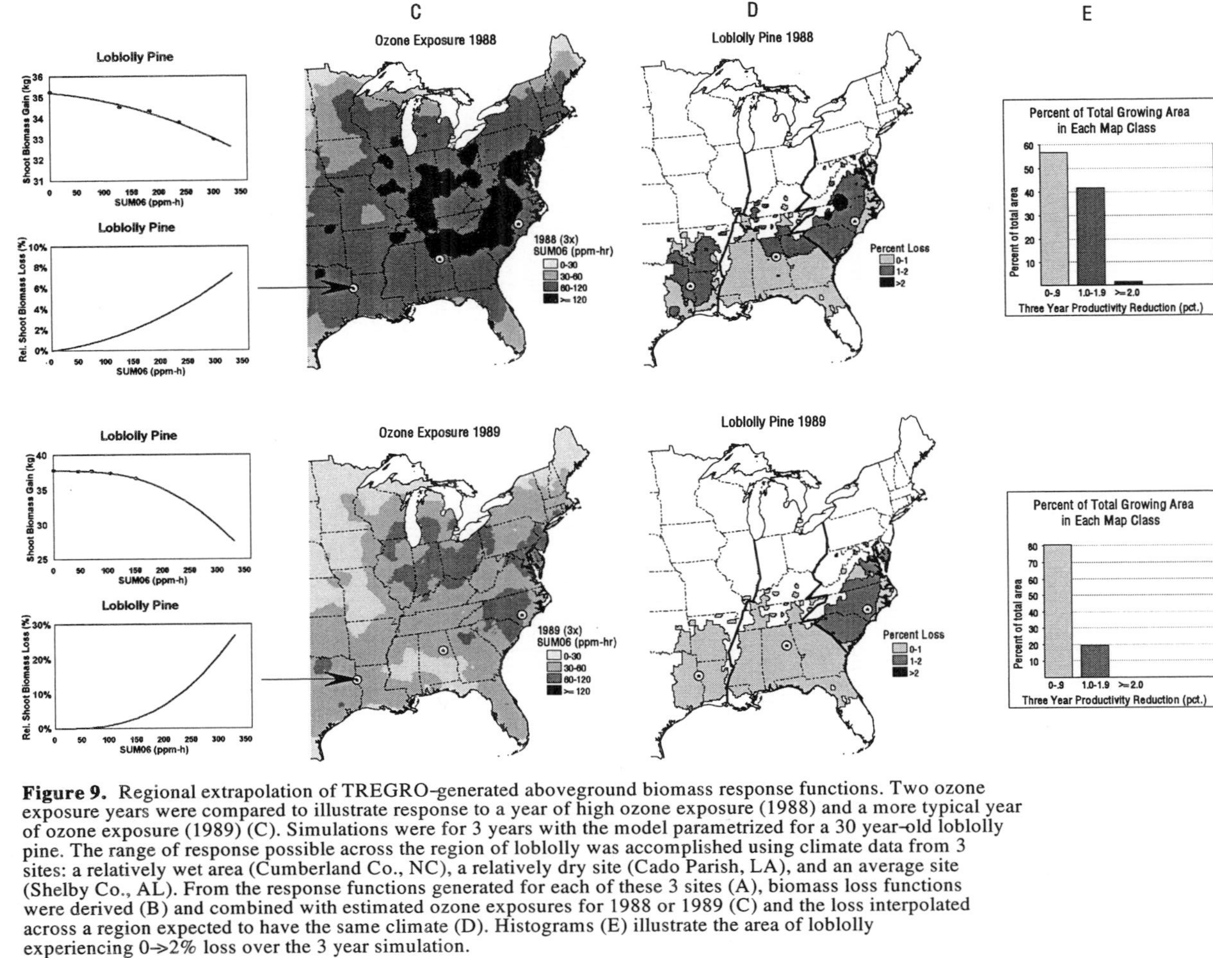

Figure 9. Regional extrapolation of TREGRO–generated aboveground biomass response functions. Two ozone exposure years were compared to illustrate response to a year of high ozone exposure (1988) and a more typical year of ozone exposure (1989) (C). Simulations were for 3 years with the model parametrized for a 30 year–old loblolly pine. The range of response possible across the region of loblolly was accomplished using climate data from 3 sites: a relatively wet area (Cumberland Co., NC), a relatively dry site (Cado Parish, LA), and an average site (Shelby Co., AL). From the response functions generated for each of these 3 sites (A), biomass loss functions were derived (B) and combined with estimated ozone exposures for 1988 or 1989 (C) and the loss interpolated across a region expected to have the same climate (D). Histograms (E) illustrate the area of loblolly experiencing 0->2% loss over the 3 year simulation.

large mature loblolly pines and used meteorological and ozone exposure data from 3 sites across the range of loblolly. The sites represented a relatively drier region (Cado Parish, Louisiana), a relatively wetter region (Cumberland Co. North Carolina) and an average site (Shelby Co., Alabama) based on a 10 year average of all met sites in this region. Using 3 sites for the simulations allows a means to demonstrate the range of response to ozone across the species' range based on the meteorological variables of precipiation and temperature. The biomass response functions resulting from the 3 year runs (1988-1988-1988 and 1989-1989-1989) for each met site are interpolated over the region that is thought to be experiencing the same climate. Cado Parish response functions for the two 3-year scenarios are shown in Figure 9A. The biomass loss function (Fig 9B) is derived from the shoot biomass response function (Fig 9A) and is applied across the appropriate region with the estimated ozone exposure (Fig 9C) and the resulting biomass loss maps are generated (Fig 9D). The histograms (Fig 9E) show the area of loblolly predicted to experience various reductions in growth over the 3 year period. Clear differences are apparent comparing the higher ozone exposures of 1988 to the 1989 scenario of relatively low ozone. Sixty percent of the loblolly area is predicted to experience 1-2% reductions in productivity over the three years of 1988 ozone, compared to 20% of the area when 1989 exposure scenarios are used.

The TREGRO response function generated using the climate data for the relatively wet year of 1989 compared to the function generated using the relatively dry year predicts changes in growth at less ozone concentration (150-200 ppmhr for 3 year exposure) than that from the dry year of 1988 (250-300 ppm-hr for 3 years) (Fig 9A) suggesting the role of precipiation in the response to ozone. In the drier year, the reduced conductivity would result in less ozone being taken up and thus higher exposures are needed to cause effects. To illustrate the effect of precipitation on the biomass loss across the region of loblolly pine, the response function from 1989 is used with the estimated ozone exposures of 1988 and 1989 (Figure 10). With the higher ozone exposure of 1988 and the greater ozone response simulated using the higher precipitation there is a greater loss exhibited than when this same response function is used with the 1989 estimated ozone exposure (Fig 10D). If the same ozone exposure year (1988) is compared using the 3 year 1988 and 3 year 1989 generated response function, there is substantially more area experiencing greaeter than 2% productivity loss with the wet year (1989)-generated response function (Fig 9D and 10D). This simulation is in keeping with experimental data demonstrating the reduced effect of ozone under drought conditions (Tingey and Hogsett 1985). Although there are higher exposures in 1988, because of the higher temperatures, and less precipitation, there is presumably less ozone uptake due to stomatal closure. The model simulated response function demonstrated this. These results also suggest a possible weighting scheme as year to year variation in exposure is taken into consideration for exposure-response functions and developing air quality indices. The result of such an analysis in a risk characterization is that less weight would be given to warm years where precipitation is low compared to cooler, wetter years (even though less ozone exposure would be anticipated).

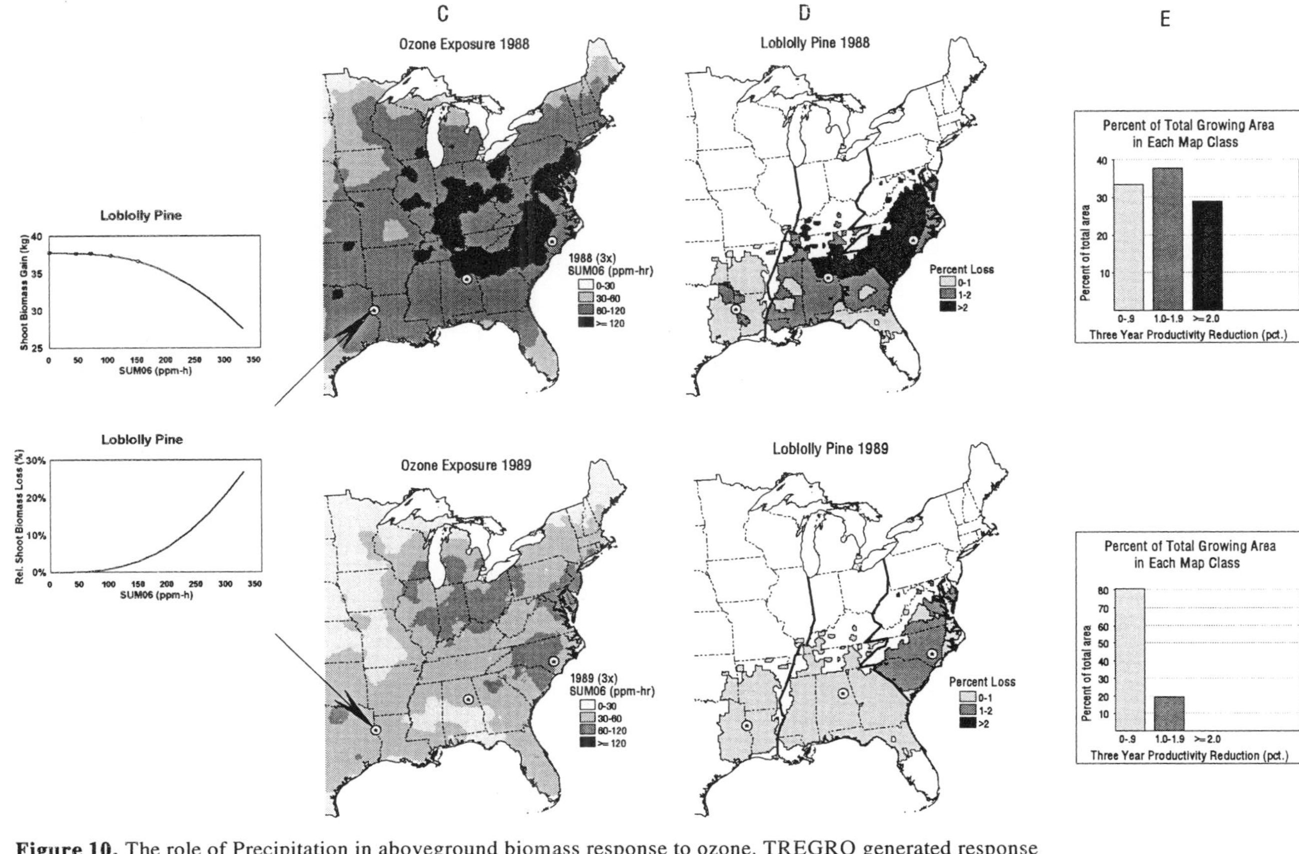

Figure 10. The role of Precipitation in aboveground biomass response to ozone. TREGRO generated response function of shoot biomass response for a relatively wet year of 1989 (A) was used to derive a biomass loss function (B) and that was applied to 1988 and 1989 estimated ozone exposure (C) across the region of loblolly using the GIS. The loss was interpolated across the region expected to have the same climate to produce maps (D) and from the maps, histograms reflect the area of loblolly predicted to experience various reductions in productivity over 3 years (E).

Assessing effects of ambient ozone on crops in The Netherlands with ethylenediurea (EDU): spatial and temporal variation

A.E.G. Tonneijck and C.J. van Dijk

DLO Research Institute for Agrobiology and Soil Fertility (AB-DLO), P.O. Box 14, 6700 AA Wageningen, The Netherlands

ABSTRACT

As part of UN/ECE ICP-Crops program, plants of clover (Trifolium subterraneum cv. Geraldton) and bean (Phaseolus vulgaris cv. Lit) were exposed to ambient air at four rural sites in the growing seasons of 1994 through 1996. The influence of ozone on visible injury, growth and yield was determined by comparing the response of plants treated with ethylene diurea (EDU) to that of untreated plants. EDU is an antioxidant that protects plants from ozone damage. A considerable degree of ozone-induced injury in both crop species has been observed at all sites. Although injury in each species varied between years and sites, the site-dependent pattern was similar for each year. Adverse effects of ozone on pod yield in bean were observed each year but did not vary between sites. Plant responses were not related to the measured ozone levels (AOT40) and environmental factors such as VPD seem to have a strong influence on the ozone sensitivity of plants in terms of foliar injury. Evidence suggests that the short-term critical level to protect plants against ozone injury should be lower than those that have been proposed recently.

1. INTRODUCTION

It has been recognized over the past two decades that levels of air pollutants within Europe are sufficiently high to cause adverse effects on sensitive crops. Thus, policies to reduce these levels and their associated impacts on the environment need to be developed on an international basis. For Europe, these policies are being formulated mainly within the Convention on Long Range Transboundary Air Pollution (LRTAP Convention) of the United Nations/Economic Commission of Europe (UN/ECE). In 1988, the International Cooperative Program on effects of air pollution and other stresses on crops and non-woody plants (ICP-Crops) came into force under this convention. The aim of this program is to provide quantitative information regarding the responses of crops and non-woody plants to air pollution.

Ozone is the pollutant of primary concern to the ICP-Crops which aims to verify the long-term critical level of ozone for yield reduction and the short-term critical level for protection of crops against visible injury (Benton et al., 1995). Critical levels have been defined as the concentrations of pollutants in the atmosphere above which direct adverse effects on receptors, such as plants, ecosystems or materials, may occur according to present knowledge (UN/ECE, 1988). Recently, short-term and long-term critical levels to protect crops against significant effects by ambient ozone have been proposed (Kärenlampi and Skärby, 1996). These critical levels are expressed as cumulative exposures over the threshold concentration of 40 ppb ozone during daylight hours and are referred to as AOT40.

Ozone-induced foliar injury has been frequently observed after episodes with elevated concentrations and growing seasonal mean concentrations of ambient ozone are considered to be sufficiently high to adversely affect yield of sensitive crops in the Netherlands (Tonneijck, 1989). However, data on effects of ozone in ambient air on crop productivity were lacking for our country. Furthermore, information on temporal and spatial variation of plant responses to ambient ozone and data to determine relationships between exposure and plant response were hardly available.

As part of the UN/ECE ICP-Crops program, a three-year project was started in the Netherlands in 1994 to (1) quantitatively assess the degree of ozone injury, (2) to assess growth and yield reductions resulting from chronic exposures, (3) to study the spatial and temporal variations of ozone-induced effects and (4) to develop relationships between exposure and plant response. In this project, the antioxidant ethylenediurea (EDU) was used to investigate the responses of ozone sensitive plant species such as clover (Trifolium subterraneum) and bean (Phaseolus vulgaris) to ambient air at four sites in rural areas. EDU is currently the best known systemic antioxidant and has been used extensively to study the effects of ambient ozone on crops and trees (Manning and Krupa, 1992). Some results concerning plant responses at the various sites during the growing seasons of 1994 through 1996 are presented in this paper and discussed briefly in relation to exposure levels and the proposed critical levels for ozone.

2. EXPERIMENTAL PROCEDURES

Experimental procedures were performed according to the UN/ECE ICP-Crops protocol (UN/ECE, 1994) with minor adjustments. For the growing seasons of 1994 through 1996, potted plants of clover (Trifolium subterraneum cv. Geraldton) and bean (Phaseolus vulgaris cv. Lit) were exposed to ambient air at four sites in rural areas: Westmaas (51°47' N, 04°27' E), Schipluiden (51°59' N, 04°16' E), Zegveld (52°08' N, 04°50' E), and Wageningen (51°58' N, 05°38' E). For experimental details see Table 1. The influence of ambient ozone on visible injury (proportion of leaves injured) was assessed and the effects on growth and yield were determined by comparing the responses of plants treated with EDU to those of untreated plants.

Table 1.
Timetable of events for the experiments with bean and clover in the growing seasons of 1994 through 1996

Year	Experiment	Start of exposure	Harvest 1	Harvest 2
Clover				
1994	1	June 21	July 19	August 16
	2	August 23	September 20	October 18
1995	1	June 20	July 18	August 15
	2	August 15	September 12	October 10
1996	1	June 18	July 16	August 13
	2	August 13	September 10	October 8
Bean				
1994	1	June 28	August 9	September 13
1995	1	June 27	August 8	September 5
1996	1	June 25	August 20	October 2

Two successive eight-week experiments with clover were performed each year. Three-week-old plants (three plants per pot) were transported to the sites and the first application with EDU was carried out. EDU was applied at 2-week intervals as a soil drench using 100 ml per pot (10x10 cm) of a 150 mg l^{-1} solution. Control plants received an equal volume of distilled water. After four weeks (Harvest 1), fully expanded leaves were removed to determine the amount of foliar injury and biomass, and the plants were then left to regrow. The final harvest was performed after another four weeks and the amount of foliar injury and leaf dry weight were assessed (Harvest 2).

One experiment with bean was performed each year. Three-week-old plants (one plant per pot) were transported to the sites and the first application with EDU was carried out. EDU was applied as a soil drench using 200 ml per pot (5 liter) of a 100 mg l^{-1} solution. Additional applications of EDU were given at 2-week intervals with 100, 150, 200 and 250 mg l^{-1} solutions, respectively. To assess the amount of visible injury and dry weight of green pods, an intermediate harvest (Harvest 1) was performed when the green pods were ready for market. A dry harvest (Harvest 2) was carried out when the pods were mature and contained loose seeds, and pod dry weight was determined.

Concentrations of O_3, NO_x and SO_2 were recorded at monitoring stations of the National Air Quality Monitoring Network (RIVM, 1989), that were located within 500 m of the biomonitoring sites. Exposures to ozone are expressed as AOT40 (Accumulated exposure Over a Threshold of 40 ppb). This exposure index is calculated as the sum of the differences between the hourly ozone concentrations in ppb and 40 ppb for each hour when the concentration exceeds 40 ppb. The AOT40 for each site is calculated only for daylight hours with a mean global radiation 50 W m^{-2} (Kärenlampi and Skärby, 1996) at Wageningen.

3. RESULTS AND DISCUSSION

3.1. Foliar injury

The first symptoms in non-EDU-treated beans were observed two (in 1994 and 1995) and five weeks (in 1996) after the start of the exposures. Symptoms characteristic of O_3 stress consisted of dark brown blemishes and occurred mainly on the adaxial leaf surfaces. These ozone-induced symptoms have been recorded on field-grown beans in the Netherlands since the early eighties (Tonneijck, 1983) and in many other European countries (Sanders and Benton, 1995). The degree of injury varied between sites and years (Figure 1). On average, injury amounted to 27% in 1994, to 8% in 1995 and to 1% in 1996. In the last year, symptoms were observed at one site (Schipluiden) only. The greatest amount of injury (35%) was recorded at Wageningen in 1994. Injury differed significantly between the sites each year and was greatest at Schipluiden in two of the three growing seasons.

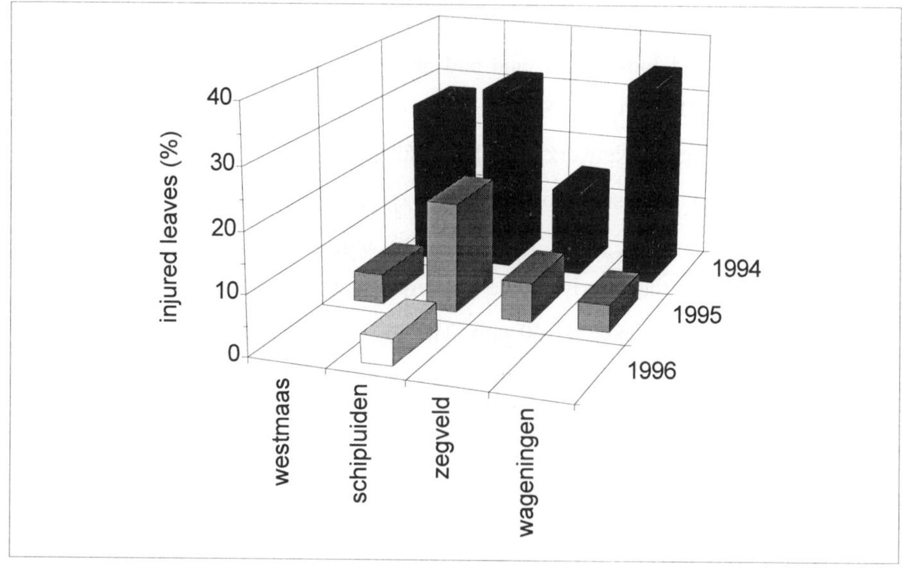

Figure 1. Ozone injury (% leaves injured) in bean after exposure to ambient air at four rural sites in the growing seasons of 1994 through 1996 (Harvest 1).

Subterranean clover was sensitive to air pollution by ozone as it displayed the characteristic symptoms of chlorotic or necrotic lesions until Harvest 1 of Experiment 2 each year. Thus, concentrations of ambient ozone were sufficiently high to cause foliar injury in this species until the end of the summer. As in bean, the degree of injury in clover also varied between sites and years (Figure 2). On average, injury as assessed at Harvest 1 of Experiment 1, amounted to 9% in 1994, to 22% in 1995 and to 1% in 1996. Injury generally differed between the sites each year and was greatest at Schipluiden. The greatest amount of injury (31 %) was recorded at this site after four weeks of exposure (Harvest 1 of Experiment 1) in 1995.

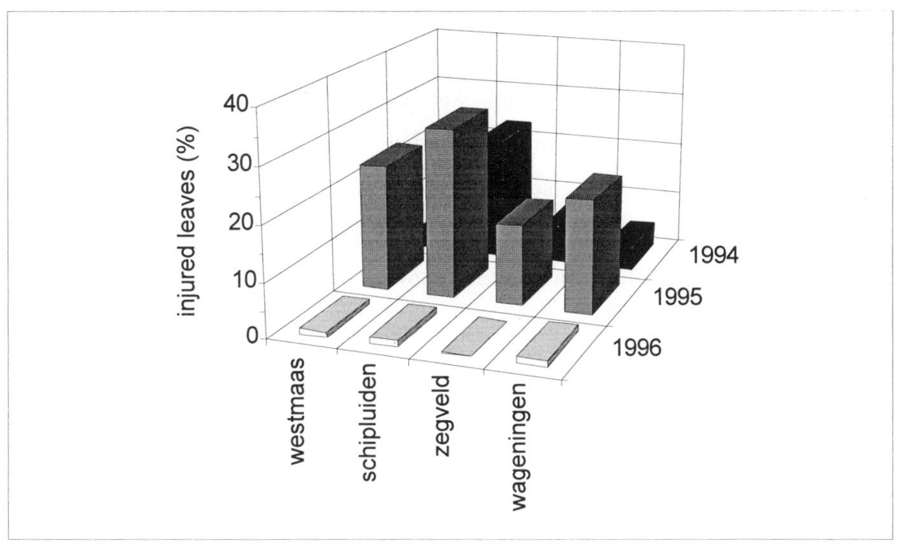

Figure 2. Ozone injury (% leaves injured) in clover after exposure to ambient air at four rural sites in the growing seasons of 1994 through 1996. Data relate to Harvest 1 of Experiment 1.

3.2. Growth and yield

Pod yield in non-EDU-treated plants of bean was reduced each year in comparison to EDU-treated plants (Table 2). Although EDU itself may reduce or increase pod yield in bean (eg Kostka-Rick and Manning, 1993), the observed reduction of pod yield in non-EDU-treated bean plants could be attributed to enhanced concentrations of ambient ozone (Tonneijck and Van Dijk, 1997b) and was observed especially when mature pods were harvested. Yield reduction of mature pods was greatest in 1994 and amounted to 17%, on average. Analyses indicated that the adverse effects of ozone on pod yield did not differ between sites for each year, which contrasted with the site-dependent pattern of ozone injury in this species. Furthermore, interannual variation concerning ozone-induced yield reduction appeared to differ from that concerning visible injury, thereby indicating that the degree of ozone injury is not linked to proportional yield reduction by ozone in bean.

The effect of ambient ozone on leaf biomass production in clover is not yet clear. Leaf biomass in non-EDU-treated clover plants was generally lower than in EDU-treated plants when ozone concentrations were high. Differential responses in leaf biomass between EDU and non-EDU-treated plants were variable when concentrations of ambient ozone were relatively low (Tonneijck and Van Dijk, 1997a). Results showed that proportional reduction in leaf biomass production in non-EDU-treated plants as compared to EDU-treated plants generally did not depend on the site and was smaller than 10 % on average for each year (data not shown).

Table 2
Proportional decrease (%) in yield of green and mature pods of non-EDU-treated plants of bean as compared to EDU-treated plants (adverse effect of ozone), after exposure to ambient air at four rural sites in the growing seasons of 1994 through 1996.

Plant stage	Reduction pod weight (%)		
	1994	1995	1996
Green pods	-0.5	10.7	2.7
Mature pods	16.6	3.0	13.5

3.3. Plant response in relation to ozone exposure

Research is currently being performed to study various aspects of the relationships between ozone exposure and plant response and to verify the short-term and long-term critical levels for ozone as defined in Kärenlampi and Skärby (1996). Analyses of the data of 1994 have shown that observed responses concerning injury, biomass production and yield in bean and subterranean clover were not related to ozone exposures (AOT40) measured at the sites (Tonneijck and Van Dijk, 1997a,b). This may indicate that AOT40 is not a good exposure index to describe the response of bean and clover to ambient ozone under Dutch conditions. Environmental factors such as temperature, relative humidity and the presence of other air pollutants can influence the sensitivity of plants to ozone (Guderian et al., 1985). The lack of correlation between ozone exposure and response may also be due to environmental conditions that are likely to vary between sites and exposure periods. Sulphur dioxide and nitrogen oxides do not seem to be relevant since the measured concentrations appeared to be low and similar among the sites (Tonneijck and Van Dijk, 1997a,b).

To elucidate various aspects of plant response in relation to ambient ozone, some data of our study are presented concerning visible injury in clover. According to Balls et al. (1996), vapor pressure deficit (VPD) has a relatively strong influence on the degree of ozone injury in this species by influencing pollutant uptake of the leaves; the lower the VPD the greater the injury response to ozone. In 1994 (Figure 3), daily mean values of VPD and daily maximum temperatures in Wageningen were lower in the second experiment with clover (August 23-October 18) than in the first experiment (June 21-August 16). Daily mean values of VPD were generally below 1.5 kPa, a limit in the proposed short-term critical levels (Benton et al., 1996), throughout the growing season. Daily concentrations of ambient ozone (AOT40) were relatively high up to and including August 4 and declined strongly after this date. However, the degree of ozone injury after four weeks of exposure in Experiment 2 was about equal to the observed injury in Experiment 1. Thus, clover was more sensitive to ambient ozone in the second experiment than in the first experiment. These results indicate that injury response in clover may not only depend on ozone exposure (AOT40). Furthermore, many observations within the UN/ECE ICP-Crops program have shown that sensitivity of clover to ambient ozone increased with decreasing VPD (Benton et al., 1996).

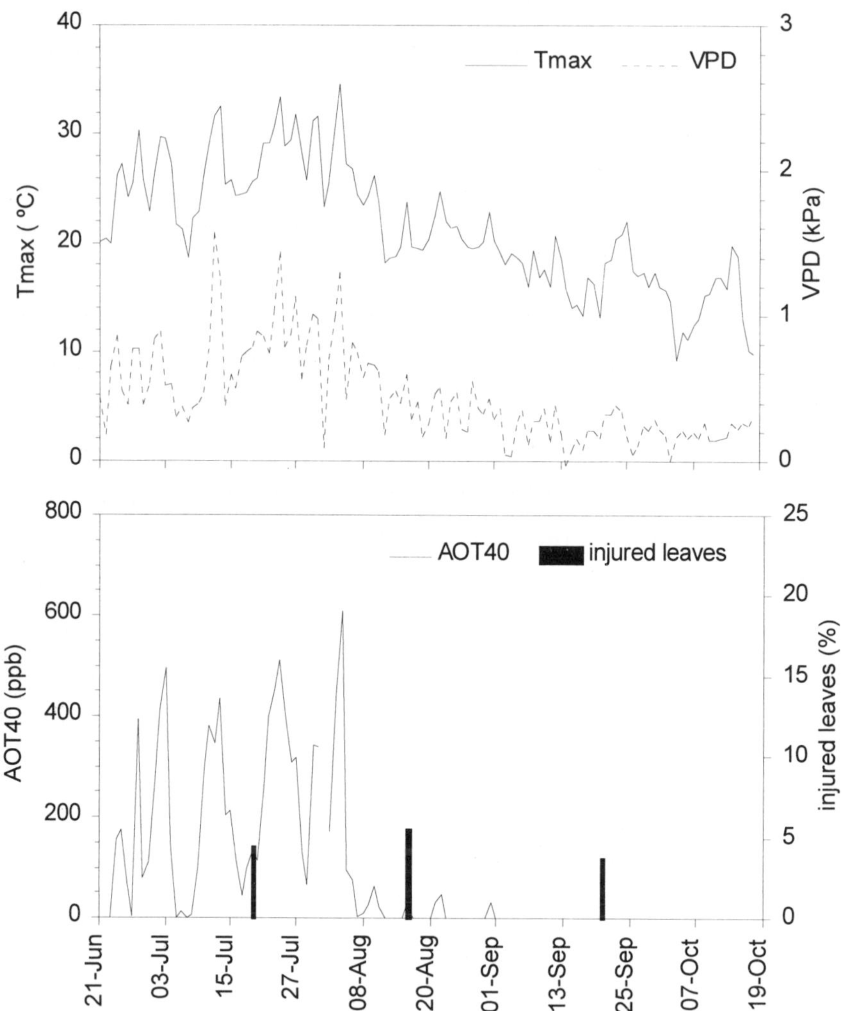

Figure 3. Ozone injury (% leaves injured) in clover after different exposures of four weeks to ambient air and day-by-day variation in maximum temperature (Tmax), vapor pressure deficit (VPD) and concentrations of ambient ozone (AOT40) at Wageningen in 1994.

Recently, Benton et al. (1996) proposed two short-term critical levels (AOT40 accumulated over five consecutive days) for injury development by ozone depending on the level of VPD: an AOT40 of 500 ppb(h when mean VPD (9.30-16.30 h) exceeds 1.5 kPa and an AOT40 of 200 ppb(h when mean VPD is below 1.5 kPa. Calculations for 1994 and 1995 (Table 3) have shown that the AOT40 accumulated over five consecutive days prior to ozone injury in clover generally exceeded the proposed short-term critical levels at all sites each year except for the second experiment in 1994. In this experiment, foliar injury was observed at all sites whereas the maximal five-days AOT40 ranged from 1 to 81 ppb(h. The mean VPD was rather low during the corresponding five-days period at Wageningen and amounted to 0.16 kPa. The occurrence of such low levels of VPD may be an important reason that injury was observed at exposures below the short-term critical level (AOT40) of 200 ppb(h ozone. The results suggest that the short-term critical level to protect plants against ozone injury should be lower than those that have been proposed recently. Since injury in clover was observed once after a five-days exposure (AOT40) to 1 ppb(h ozone, the threshold value of 40 ppb used to calculate the accumulative exposures may be too high and should also be reconsidered.

Table 3

Maximal AOT40 (ppb(h) at different sites accumulated for five consecutive days in the week prior to the first observation of injury in Trifolium subterraneum cv. Geraldton during the growing seasons of 1994 and 1995.

Site	Experiment 1		Experiment 2	
	Before Harvest 1	After Harvest 1	Before Harvest 1	After Harvest 1[a]
1994				
Schipluiden	924	1724	81	0
Zegveld	963	1297	1	2
Wageningen	1102	2033	33	0
Westmaas	947	1251	27	0
1995				
Schipluiden	697	781	781	22
Zegveld	836	670	710	3
Wageningen	1252	704	1011	11
Westmaas	912	1048	1008	26

[a] Since no injury has been observed, data relate to the maximal AOT40 accumulated for five consecutive days during the four weeks exposure period between Harvest 1 and Harvest 2.

4. CONCLUDING REMARKS

The concentrations of ambient ozone in the Netherlands are sufficiently high to cause adverse effects on sensitive crop species such as clover and bean. Foliar injury has been observed at all sites till the middle of September in each year. The degree of injury differed between sites and the site-dependent variation generally was similar for both species each year. A year-to-year variation in the degree of injury was observed for both crop species. Studies with the antioxidant EDU led to the conclusion that ambient ozone can adversely affect pod yield in bean in the Netherlands while the effects of ozone on biomass production in clover are not yet clear. In contrast to injury, effects on growth and yield did not vary between sites.

Knowledge concerning spatial variation of plant responses is necessary when attempting to evaluate the risk from exposures to ambient ozone. Short-term exposures to high concentrations generally result in visible injury while chronic exposures to low concentrations can cause physiological alterations that ultimately result in growth and yield reductions. These physiological alterations are probably not linked to visible symptoms. Our results indicate that the occurrence of spatial variation depends on the nature of plant response (injury versus growth and yield) and, thus, possibly on the type of exposure to ambient ozone (acute versus chronic).

Guidelines or critical levels to protect crops against adverse effects of ozone are generally deduced from artificial fumigation experiments in open-top chambers. This also applies to the AOT40 concept. At present, there is an ongoing debate about the relevance of results from this type of experiments to explain the response of crops to ambient ozone in the field (eg Grünhage and Jäger, 1996). Key issues relate to ozone exposure dynamics such as the importance of ozone peaks versus mid-range hourly values of ozone and to factors influencing pollutant uptake such as atmospheric transport properties and VPD. Our results underline this problem since the observed plant responses could not as yet be related to AOT40 values. Clover appeared to be relatively sensitive to ambient ozone at the end of the growing season under conditions of low VPD. Evidence concerning the occurrence of injury in this species suggests that a threshold value of 40 ppb ozone may be too high to calculate cumulative exposures. To extend our knowledge on the risk of exposures to ambient ozone, more research is needed to study the relation between plant responses and exposures under 'real world' conditions.

ACKNOWLEDGEMENTS

The authors acknowledge the financial support of the authorities of the province of South-Holland for the research in this province and the supply of concentration data by the National Institute of Public Health and Environmental Protection.

REFERENCES

1. Balls, G.R., D. Palmer-Brown and G.E. Sanders, 1996. New Phytologist 132: 271-280.
2. Benton, J., J. Fuhrer, B.S. Gimeno, L. Skärby and G. Sanders, 1995. Water Air and Soil Pollution 85: 1473-1478.
3. Benton, J., J. Fuhrer, B.S. Gimeno, L. Skärby, D. Palmer-Brown, C. Roadknight and G. Sanders-Mills, 1996. In: Kärenlampi, L. and L. Skärby (Eds.), Critical levels for ozone in Europe: Testing and finalizing the concepts. UN-ECE Workshop Report, Department of Ecology and Environmental Science, University of Kuopio, Finland, pp. 44-57.
4. Grünhage, L. and H.-J. Jäger, 1996. In: Kärenlampi, L. and L. Skärby (Eds.), Critical levels for ozone in Europe: Testing and finalizing the concepts. UN-ECE Workshop Report, Department of Ecology and Environmental Science, University of Kuopio, Finland, pp. 151-168.
5. Guderian, R., D.T. Tingey and R. Rabe, 1985. In: Guderian, R. (Ed.), Air pollution by photochemical oxidants. Formation, transport, control and effects on plants. Berlin-Heidelberg: Springer-Verlag, 129-333.
6. Kärenlampi, L. and L. Skärby, 1996. Critical levels for ozone in Europe: Testing and finalizing the concepts. UN-ECE Workshop Report, Department of Ecology and Environmental Science, University of Kuopio, Finland, 363 pp.
7. Kostka-Rick, R. and W.J. Manning, 1993. Environmental Pollution 82: 63-72.
8. Manning, W.J. and S.V. Krupa, 1992. In: Lefohn, A.S. (Ed.) Surface level ozone exposures and their effects on vegetation. Lewis Publishers, Inc., Chelsea, USA, 93-156.
9. RIVM, 1989. National Air Quality Monitoring Network. Technical Description, Report No. 228702017, Bilthoven, National Institute of Public Health and Environmental Protection.

10. Sanders, G.E. and J. Benton, 1995. Ozone pollution and plant responses in Europe - An illustrated guide. Nottingham: The ICP-Crops Coordination Centre, The Nottingham Trent University.
11. Tonneijck, A.E.G., 1983. Netherlands Journal of Plant Pathology 89: 99-104.
12. Tonneijck, A.E.G., 1989. In: Schneider, T., S.D. Lee, G.J.R. Wolters and L.D. Grant (Eds), Atmospheric Ozone Research and its Policy Implications. Elsevier, Amsterdam, 251-260.
13. Tonneijck, A.E.G. and C.J. van Dijk, 1997a. Agriculture, Ecosystems and Environment, accepted.
14. Tonneijck, A.E.G. and C.J. van Dijk, 1997b. New Phytologist, 135: 93-100.
15. UN/ECE, 1988. UN/ECE critical levels workshop report. Bad Harzburg, FRG, March 1988.
16. UN/ECE, 1994. ICP-Crops Experimental Protocol. Nottingham: The ICP-Crops Coordination Centre, The Nottingham Trent University.

OZONE HEALTH EFFECTS:
Repeated Exposure and Sensitive Subjects

Lawrence J. Folinsbee

National Center for Environmental Assessment
U.S. EPA
Research Triangle Park, NC 27711

Ozone has a broad range of health effects in humans. I intend to focus primarily upon two aspects of these health effects: (1) responses to repeated ozone exposures and the potential implication of these studies for understanding chronic effects of ozone and (2) the mechanisms of response to ozone in sensitive subjects and how these may be useful in understanding ozone-associated morbidity. In this regard, the focus will be mainly on human controlled exposure studies.

1. REPEATED EXPOSURES

Reported effects of long-term residence in communities with high ozone levels include a seasonal variation in ozone-induced pulmonary function responses, impediments to lung development in children, and acceleration of the age-related decline in lung function in adults. There is very limited evidence that long-term repeated ozone exposure may be related to an increased incidence of asthma in adults. Finally, several recent studies are suggestive of an association between acute ozone exposure and cardiorespiratory mortality.

1.1 Community Exposures

Studies of effects of ozone and lung function have been reported in the literature for decades. A lower sensitivity to ozone in people resident in communities with high ambient ozone levels is well known [1]. Moreover, studies indicated that this effect has a seasonal fluctuation that is consistent with a depression of response during repeated ozone exposure and the return of a "normal" response with a period of several months of limited ozone exposure [2]. A recent pilot study [3] of incoming freshman at the University of California (Berkeley) indicated that the level of performance on some tests of lung function (especially FEF75% and FEF25-75%), in relation to that expected for age, race and gender, were inversely related to estimated lifetime exposure to ozone (i.e., a higher lifetime ozone exposure resulted in a poorer performance on lung function tests). These observations suggest that chronic ozone exposure in children may be associated with interference with lung development.

1.2 Controlled Exposures

Over the past twenty years, many studies of repeated exposure to ozone in volunteer subjects have been performed. In general, these studies have shown that humans exposed to ozone on successive days gradually develop an attenuated response to ozone which, over the course of 3-5 days, leads to the absence or near absence of pulmonary function or symptom responses to ozone. This has been referred to as "adaptation," although in the usual biological sense it is not an adaptation since, with no further exposure, the responses return to the pre-exposure levels within a week or so. If the initial exposure conditions are severe enough to cause marked pulmonary function and symptom responses, the individual often has an exaggerated response on the second day of exposure [4,5]. Repeated ozone exposure has also been associated with a temporary decline in baseline lung function [5,6]. Studies in which a seasonal attenuation of pulmonary function response is observed [2] show that the seasonal attenuation persists for several months whereas the attenuation observed with repeated laboratory exposures persists for no more than a couple of weeks [7]. The absence of pulmonary function or symptom responses could have suggested the erroneous conclusion that once one became "adapted" to ozone, that there were no further effects.

Acute exposure to ozone causes an inflammatory response characterized by release of various pro-inflammatory mediators (e.g., interleukins IL-6, IL-8, and prostaglandin PGE2), infiltration of neutrophils (PMN), activation of alveolar macrophages, and increased epithelial permeability. Only recently has there been any attempt to examine the inflammatory and cellular correlates of an attenuated function and symptom response in humans. Devlin et al. [9] showed that some

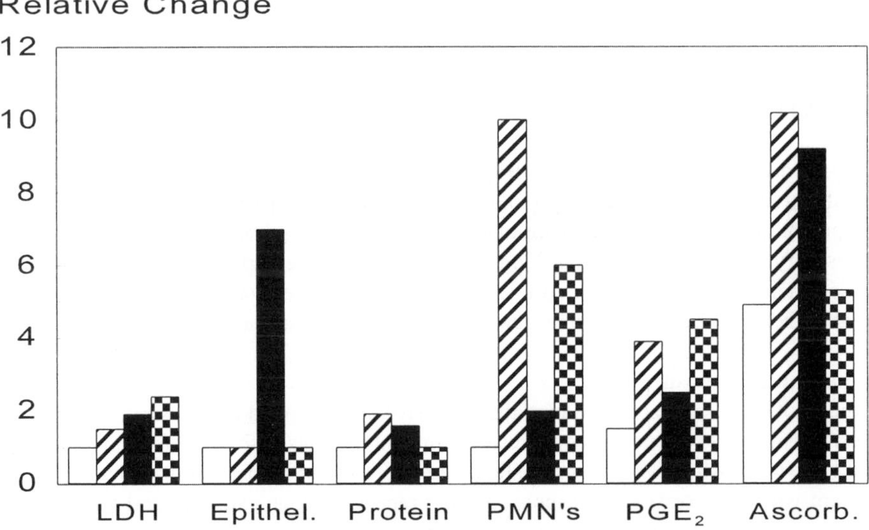

Figure 1. Lactate dehydrogenase (LDH), epithelial cell number, protein, neutrophils (PMN's), prostaglandin E_2 (PGE_2), and ascorbate content of BAL fluid after air (open bars) or ozone exposure (Day 1, hatched; Day 5, solid; return exposure, checkered) ([Devlin et al., [9]).

inflammatory and cellular responses associated with acute ozone exposure were also attenuated after five consecutive days of ozone exposure (Figure 1). Notably absent were increases in airway neutrophils and IL-6 hallmarks of the inflammatory response in healthy individuals acutely exposed to ozone. In addition, prostaglandin E_2 levels, usually markedly elevated after ozone exposure, showed only a slight increase after the fifth exposure. Epithelial cells, not normally seen immediately after an acute exposure, were present in the BAL after the fifth exposure, possibly the result of already damaged cells physically knocked loose during the lavage process or from delayed sloughing of damaged cells. In addition, the bronchoalveolar lavage (BAL) fluid level of the enzyme lactate dehydrogenase (LDH), an indicator of cellular damage, was increased. This latter marker indicates that ozone continued to cause cell damage even in the absence of inflammation, symptoms, and lung function changes. When subjects were reexposed to ozone either 10 or 20 days later, the absence of an increase in BAL protein, epithelial cells, or the antioxidant ascorbate suggested a continued attenuation of these responses, although pulmonary function and symptom responses were similar to "pre-adaptation" levels. The changes in BAL protein suggest that the epithelium which replaces ozone-damaged epithelial cells is less "leaky." It is likely that cellular repair processes leading to a modified epithelium, that is more resistant to attack by ozone (such as observed in laboratory animals repeatedly exposed to ozone) [10], may also alter cellular/inflammatory responses. However, the more rapid return of pulmonary function and symptom responsiveness is consistent with the likely neural mediation of these responses.

Relationships between changes in spirometric lung function and symptoms and biochemical and cellular markers, in response to ozone exposure, have been of interest for some time. In this study [9], spirometry and symptom responses generally showed an initial decline, accentuated on the second day, and gradually diminishing by the fifth exposure. Despite the temporal alignment of the changes in spirometry and symptoms, there was no significant correlation between symptoms and function within the group of subjects at any specific time point. However, individuals exhibited consistent relationships between the magnitude of their individual changes in spirometry and their symptoms across the series of exposures. There was no evidence of a relationship between the biochemical and cellular markers and the changes in spirometry and symptoms. An increase in airway responsiveness to bronchoconstrictor agents has been shown, in other studies [4], to be increased during acute and repeated exposure to ozone (Fig 2). The changes in airway responsiveness were not correlated with the spirometric and symptom responses.

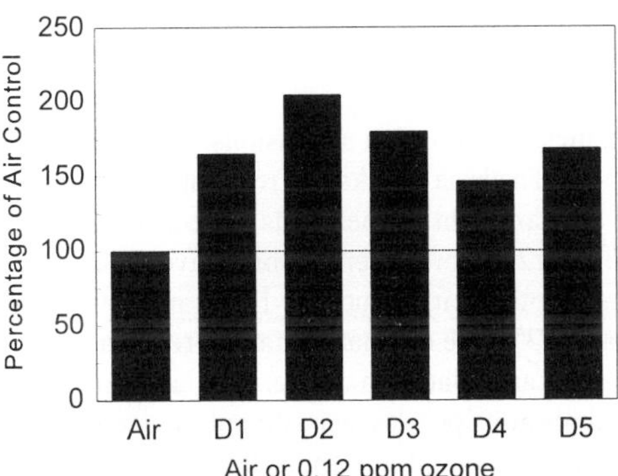

Figure 2. Airway responsiveness following exposure to air or 0.12 ppm ozone for 6.6 h on five consecutive days (Folinsbee et al., [4])

2.0 SENSITIVE SUBJECTS

There is a broad range of responsiveness to ozone within the population. Of particular concern from a public health standpoint is the identification of specific groups within the population who may be more sensitive as a result of specific characteristics, especially those characteristics associated with respiratory disease. Individuals with asthma, a chronic inflammatory airway disease, appear to be more responsive to ozone exposure.

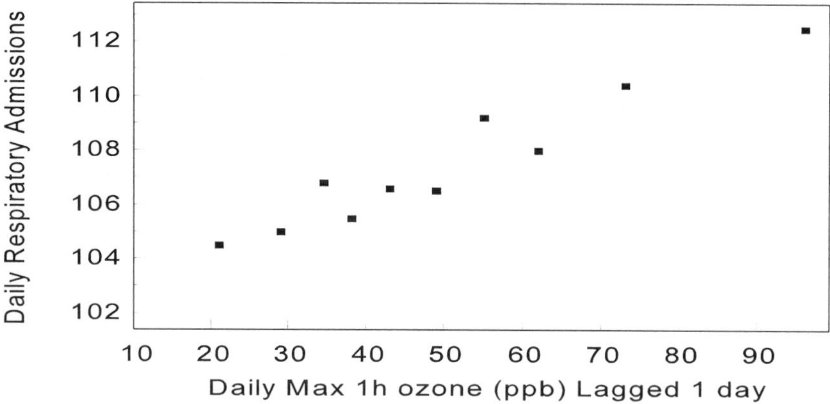

Figure 3. Respiratory admissions for 168 hospitals in Ontario, Canada in relation to daily 1 h maximum ozone concentration, lagged one day (Redrawn from Burnett et al., [11])

2.1 Community Exposures

Ambient exposure to ozone is associated with increased hospital admissions, increased visits to hospital emergency rooms, and increased numbers of asthma attacks and respiratory symptoms associated with asthma. Approximately 30 years ago, ambient ozone levels were reported to be associated with hospital admissions in Los Angeles [12]. More recent reports have detailed the association with respiratory hospital admissions in other cities such as London, New York, Toronto (Figure 3), and Rotterdam. Thurston et al., [13] have estimated that approximately 10-20% of summertime respiratory hospital admissions are related to ozone. This amounts to an estimated 1-3 ozone-related respiratory hospital admissions per day per 100 ppb ozone for each million in the population. In addition to increased asthma symptoms and asthma attacks, ozone accounts for increased numbers of asthmatics who present to emergency rooms (ER) in areas such as Atlanta [14], Los Angeles, Montreal, New Jersey, and Mexico City. It has been estimated that ozone accounts for as much as 15-20% of asthma ER visits during the summer and that asthma ER visits may increase as much as 30-40% on the highest ozone days [13,14]. These observations suggest that asthmatics may be more responsive to ozone than non-asthmatics.

2.2 Controlled Exposures

Exposure of asthmatics, usually with mild disease, to low concentrations of ozone in a controlled exposure facility has been conducted by a number of investigators. Although many of these studies showed little, if any, difference in spirometric responses between asthmatics and non-asthmatics, the interpretation of these studies may have been limited by the low concentration of ozone used [15]. More recent studies [16-21] conducted at higher ozone concentrations or for longer durations show somewhat greater changes in spirometry or airway resistance in asthmatics (Figure 4). The differences between healthy and asthmatic subjects exposed under the same conditions vary considerably among studies and range from no difference to a two-fold greater response in asthmatics. Nevertheless, these differences in response to ozone are considerably smaller than the differences in response to SO_2 [22] or other

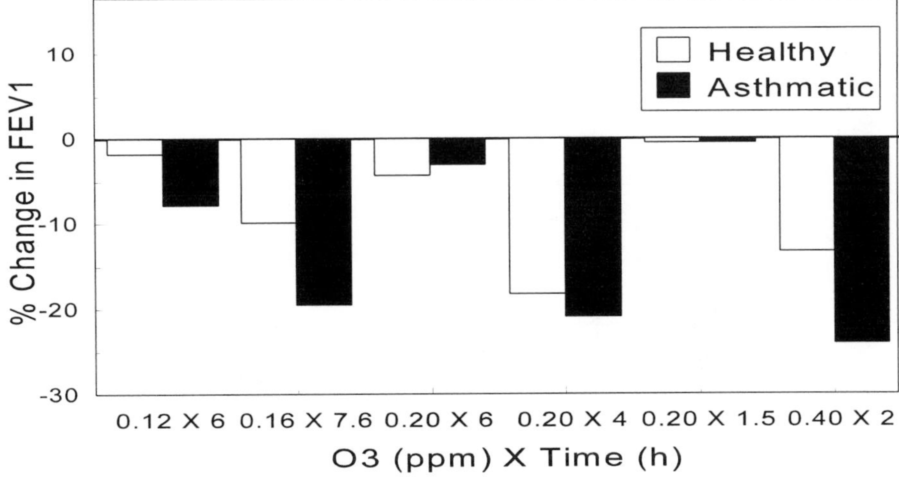

Figure 4. Change in FEV_1 (forced expired volume in one second) in response to ozone exposure of the specified concentration (ppm) and duration (h) from 6 different studies. Data from Linn et al., [17]; Horstman et al., [18]; Basha et al., [19]; Scannell et al., [20]; Linn et al., [21]; Kreit et al., [16]; respectively, left to right.

non-specific broncho-constrictor agents for which differences between asthmatics and non-asthmatics may be as much as 100-fold or more. The presence of inflammation in the airways of asthmatics led a number of investigators [19,20; RB Devlin, U.S. EPA, personal communication, 1997) to hypothesize that the inflammatory response would be increased in asthmatics exposed to ozone. In these studies, asthmatics had a greater number of neutrophils (a key cellular marker of ozone-induced inflammation in non-asthmatics) in their lavage fluid following ozone exposure than did non-asthmatics (Figure 5). In 3 studies conducted under comparable conditions, the mean percentage of neutrophils in BAL after ozone exposure in non-asthmatics ranged from 7-9% and in asthmatics ranged from 8-16%. The study with the smallest difference between asthmatics and non-asthmatics was one which employed only asthmatics with a known allergic response to house dust mite antigen. These individuals had elevated levels of eosinophils (the characteristic inflammatory cell in asthma) in their lavage fluid.

These above observations provide a mechanistic connection with the observation, from epidemiologic studies, that induction of asthma attacks is associated with ozone exposure. Horstman et al. [18] studied a group of moderate asthmatics exposed to 0.16 ppm ozone for 7.6 hours. The FEV_1 response of the asthmatics was about twice as large as that of the non-asthmatics and wheezing was experienced by more than half of the asthmatics. Treatment with an inhaled beta-adrenergic agonist partially reversed the decline in FEV_1 and alleviated symptoms, indicating that a substantial portion of these responses were due to bronchoconstriction (Figure 6). In non-asthmatics, in contrast, beta-agonists do not alleviate the decrease seen in FVC (Forced Vital Capacity) or FEV_1. In non-asthmatics, the FEV_1 response appears to be due primarily to a neurally mediated restriction in maximum inspired volume that can be promptly alleviated by a topical local anaesthetic [23] or by systemic opioid analgesics [24].

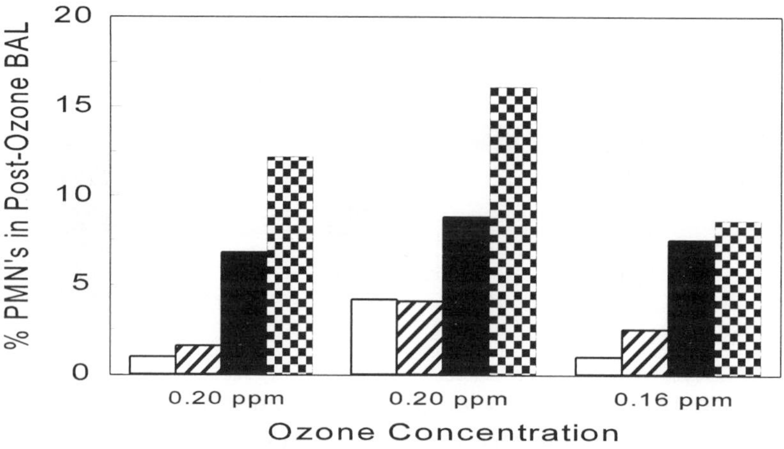

Figure 5. Percentage of neutrophils (PMN) in bronchoalveolar lavage (BAL) performed after air and ozone exposure (ppm) in both healthy and asthmatic individuals. (Clear bars: Healthy-Air; Hatched bars: Asthmatic-Air; Solid bars: Healthy-Ozone; Checkered bars: Asthmatic-Ozone). Data from Basha et al. [19], 6 h @ 25 L/min; Scannell et al. [20], 4 h @ 45 L/min; Unpublished data courtesy of R. Devlin, 7.6 h @ 25 L/min.

Since ozone is known to increase non-specific bronchial responsiveness and induce inflammatory processes, it was clear that ozone had the potential to increase bronchial responsiveness to inhaled specific antigens. An increased response to specific antigen after 3 h exposure 0.25 ppm ozone has been demonstrated by Jörres et al. [25] in both mild allergic asthmatics and allergic rhinitics, although the response was much less pronounced in the individuals with allergic rhinitis. In asthmatics, the dose of allergen which caused a 10% decline in FEV_1 after air exposure caused a 28% decline after ozone exposure. In contrast, the allergic rhinitics experienced no significant change after air exposure and an 8% change after ozone exposure. At lower ozone concentrations, the augmentation of allergen responses is less clear; one study has reported such an effect [26] and another has shown no significant effect under the same conditions [27]. Kehrl et al. [28] also reported an increase in response to house dust mite antigen in mild allergic asthmatics 16-18 h after completion of a 7.6 h exposure to 0.16 ppm ozone. These latter results not only support the findings of Jörres et al. [25] but more importantly indicate that an increased response to antigen can persist beyond the immediate post-exposure period. The association of hospital admissions or ER visits with ozone exposure is often strongest with a one-day lag, suggesting that the ozone-induced increase in reactivity to antigen may trigger a more severe asthma attack than would antigen exposure in the absence of ozone.

3. MORTALITY AND OZONE

A question that may be answered by the 21st Century is whether there is an association between ozone exposure and mortality. Although some studies have suggested such a relationship, there have been many questions raised about methodology and statistical approaches, precluding the possibility of coming to a satisfactory conclusion. In many of these

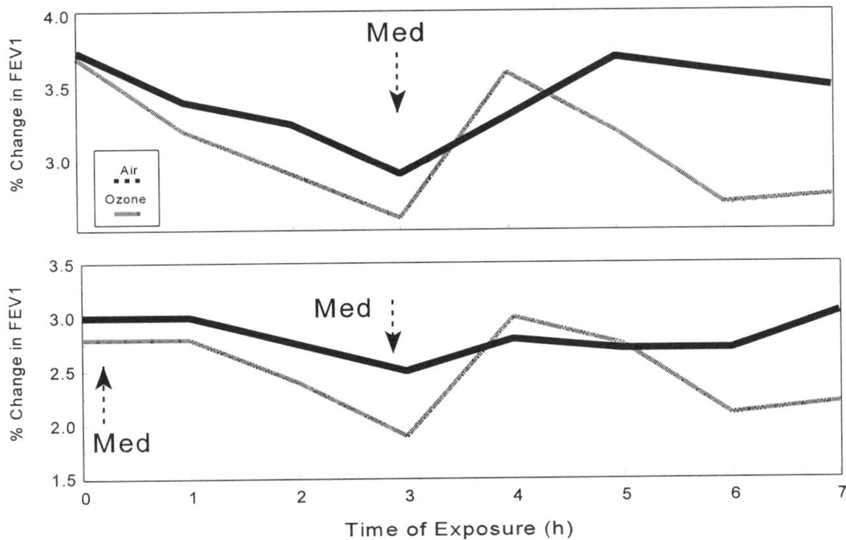

Figure 6. FEV_1 of two asthmatic subjects exposed to 0.16 ppm ozone or air for 7.6 h including mild exercise. Beta-adrenergic agonist medication was self-administered, as indicated by the arrows, in both air and ozone exposures. Note the marked improvement in FEV_1 after medication and the larger overall decline during the ozone exposure. (Redrawn from Horstman et al., [18])

analyses, the effect of particulate matter may have been a confounding factor [29]. A number of new studies have begun to look at this question again, using time-series analyses, and tend to show a positive relationship between ozone exposure and daily mortality. The association of ozone and mortality in these newer studies appears to be separate from the effects of other pollutants. [30-32].

What remains puzzling is that, in two cities with high ozone levels, Mexico City and Los Angeles, the association for ozone and mortality does not remain significant when particulate levels are taken into consideration. [29,33]. Asthma-related mortality cannot account for the apparent increase in mortality associated with ozone. Although an accelerated rate of decline of lung function [34] could contribute to mortality this also seems to be an unlikely explanation.

4. CONCLUSIONS

Clearly, human responses to ozone exposure are modified by repeated exposure. Ongoing lung cell damage was previously demonstrated in laboratory animals repeatedly exposed to ozone. Despite the attenuation of spirometric lung function and symptom responses in repeatedly exposed humans, we now know that lung cell damage continues, as does the repair process. The processes that lead to the changes in ozone responsiveness, either from a series of acute exposures in a controlled exposure chamber or from seasonal exposure in an oxidant polluted community, are not well understood from a human lung cell biology perspective. There is some suggestion that seasonal ambient exposures may have some implications for long term lung health. It is likely that persons with impaired or hypersensitive respiratory systems may be impacted to the greatest extent. The association of asthma morbidity end-points with ozone exposure is clear. Although several ppm of ozone can induce life threatening effects in healthy humans, there is presently no mechanistic explanation of how very low (tens of parts per billion) concentrations of ozone could induce death.

This document has been peer-reviewed in accordance with U.S. Environmental Protection Agency policy and approved for publication. The views expressed are those of the author and do not necessarily reflect the views of the agency. Mention of trade names or commercial products does not constitute endorsement or recommendation for use.

REFERENCES

1. Hackney JD, Linn WS, Buckley RD, and Hislop HJ. Studies on adaptation to ambient oxidant air pollution: effects of ozone exposures in Los Angeles residents vs. new arrivals. Environ Health Perspect 18:141-146 (1976).
2. Linn WS, Avol EL, Shamoo DA, Peng RC, Valencia LM, Little D, and Hackney JD. Repeated laboratory ozone exposures of volunteer Los Angeles residents: an apparent seasonal variation in response. Toxicol Indust Health 4:505-520 (1988).
3. Kunzli N, Lurmann F, Segal M, Ngo L, Balmes J, Tager IB. Association between lifetime ambient ozone exposure and pulmonary function in college freshmen — Results of a pilot study. Environ Res 72: 8-23, 1997.
4. Folinsbee LJ, Horstman DH, Kehrl HR, Harder S, Abdul-Salaam S, Ives PJ. Respiratory responses to repeated prolonged exposure to 0.12 ppm ozone. Am J Respir Crit Care Med 149:98-105, 1994.
5. Folinsbee LJ, Devlin RB, Koren HS. Time course of pulmonary response to repeated ozone exposure in man. Presented at the Annual Meeting of the American Thoracic Society, Boston, MA, May 1994. Am Rev Respir Dis 149 (#4, pt. 2): A151, 1994.
6. Horvath SM, Gliner JA, Folinsbee LJ. Adaptation to ozone: Duration of effect. Am. Rev. Respir. Dis. 123:496-499, 1981.
7. Linn WS, Medway DA, Anzar UT, Valencia LM, Spier CE, Tsao FS, Fischer DA, and Hackney JD. Persistence of adaptation to ozone in volunteers exposed repeatedly for six weeks. Am Rev Respir Dis 125:491-495 (1982).
8. Koren HS, Devlin RB, Graham DE, Mann R, Horstman DH, Kozumbo WJ, Becker S, McDonnell WF, and Bromberg PA. Ozone-induced inflammation in the lower airways of human subjects. Am Rev Respir Dis 139:407-415 (1989).
9. Devlin RB, Folinsbee LJ, Biscardi FH, Hatch G, Becker S, Madden M, Robbins MK, Koren HS. Inflammation and cell damage induced by repeated exposure of humans to ozone. Inhalation Toxicology 9:211-235, 1997.
10. Tepper JS, Costa DL, Lehmann JR, Weber MF, and Hatch GE. Unattenuated structural and biochemical alterations in the rat lung during functional adaptation to ozone. Am Rev Respir Dis 140: 493-501 (1989).
11. Burnett RT, Dales RE, Raizenne ME, Krewski D, Summers PW, Roberts GR, Raad-Young M, Dann T, Brook JR,. Effects of low levels of ambient ozone and sulfates on the frequency of respiratory admissions to Ontario hospitals. Environ Res 65:172-194, 1994.
12. Sterling TD, Pollack SV, Phair JJ. Urban hospital morbidity and air pollution. Arch Environ Health 15:362-374, 1967.
13. Thurston GD, Ito K, Kinney PL, Lippmann, M. A multi-year study of air pollution and respitatory hospital admissions in three New York State metrpolitan areas: results for 1988

and 1989 summers. J Exposure Anal Environ Epidemiol 2:429-450, 1992.

14. White MC, Etzel RA, Wilcox WD, Lloyd C. Exacerbations of childhood asthma and ozone pollution in Atlanta. Environ Res 1994; 65:56-68.
15. Linn WS, Buckley R, Speir C, Blessey R, Jones M, Fischer D, and Hackney JD. Health effects of ozone exposure in asthmatics. Am Rev Respir Dis 117:835-843 (1978).
16. Kreit JW, Gross KB, Moore TB, Lorenzen TJ, D'Arcy J, and Eschenbacher WL. Ozone-induced changes in pulmonary function and bronchial responsiveness in asthmatics. J Appl Physiol 66:217-222 (1989).
17. Linn WS, Shamoo DA, Anderson KR, Peng R-C, Avol EL, Hackney JD. Effects of prolonged repeated exposure to ozone, sulfuric acid, and their combination in healthy and asthmatic volunteers. Am J Respir Crit Care Med 1994; 150:431-440.
18. Horstman DH, Ball BA, Brown J, Gerrity TR, Folinsbee LJ. Comparison of pulmonary responses of asthmatic and nonasthmatic subjects performing light exercise while exposed to a low level of ozone. Toxicol Ind Health 11:369-385, 1995
19. Basha MA, Gross KB, Gwizdala CJ, Haidar AH, Popovich J. Bronchoalveolar lavage neutrophilia in asthmatic and healthy volunteers after controlled exposure to ozone and filtered purified air. Chest 1994; 106:1757-1765.
20. Scannell CH, Chen LL, Aris R, Tager I, Christian D, Ferrando R, Welch B, Kelly T, Balmes JR. Greater ozone-induced inflammatory responses in subjects with asthma. Am J Respir Crit Care Med 1996; 154:24-29.
21. Linn WS, Anderson KR, Shamoo DA, Edwards SA, Webb TL, Hackney JD, Gong H. Controlled exposures of young asthmatics to mixed oxidant gases and acid aerosol. Am J Respir Crit Care Med 1995; 152:885-891.
22. Folinsbee, LJ. Sulfur oxides: Controlled human exposure studies. In: Lee SD, Schneider T. (eds). Comparative Risk Analysis and Priority Setting for Air Pollution Issues, Proceedings of the 4th U.S.-Dutch International Symposium, June 1993, Keystone CO. Pittsburgh: Air & Waste Management Assoc., 1995. pp. 326-334.
23. Hazucha MJ, Bates DV, Bromberg PA. Mechanism of action of ozone on the human lung. J Appl Physiol 67:1535-1541 (1989).
24. Passanante A, Hazucha MJ, Seal E, Folinsbee L, Bromberg PA. The role of analgesia in modulating lung function response to ozone in man (abstr) . Anesth Analg 1995; 80:S371.
25. Jorres R, Nowak D, Magnussen H. Effects of ozone exposure on allergen responsiveness in subjects with asthma or rhinitis. Am J Respir Crit Care Med 1995; 153:56-64.
26. Molfino NA, Wright SC, Katz I, Tarlo S, Silverman F, McClean PA, Szalai JP, Raizenne M, Slutsky AS, Zamel N. Effect of Low Concentrations of Ozone on Inhaled Allergen Responses in Asthmatic Subjects. The Lancet 338:199-203, 1991.
27. Ball BA, Folinsbee LJ, Peden DB, Kehrl HR. Allergen bronchoprovocation of patients with mild allergic asthma after ozone exposure. J Allergy Clin Immunol. 98:563-572, 1996.
28. Kehrl HR, Ball B, Folinsbee L, Peden D, Horstman D. Increased specific airway reactivity of mild allergic asthmatics following 7.6 hr exposures to 0.16 ppm ozone. Presented at the Annual Meeting of the American Thoracic Society, San Francisco, CA, May 1997. Am J Respir Crit Care Med 155 (#4, pt. 2): A731, 1997.
29. Kinney PL, Ito K, Thurston GD. A sensitivity analysis of mortality/PM10 associations in Los Angeles. Inhalation Toxicol 7:59-69, 1995.
30. Samet JM, Zeger SL, Kelsall JE, Xu J. Air Pollution and Mortality in Philadelphia, 1974-1988. Report to the Health Effects Institute. (March, 1997).
31. Sartor F, DeMuth C, Snacken R, Walckiers D. Mortality in the elderly and ambient ozone

concentration during the hot summer, 1994, in Belgium. Environ Res 72: 109-117, 1997.

32. Anderson HR, Ponce de Leon A, Bland JM, Bower JS, Strachan DP. Air pollution and daily mortality in London: 1987-92, Brit Med J 312:665-669, 1996.
33. Loomis DP, Borja-Aburto VH, Bangdiwala SI, Shy CM. Ozone Exposure and Daily Mortality in Mexico City: a Time-series Analysis. Research Report #75, Cambridge, MA: Health Effects Institute, 1996.
34. Detels, R., Tashkin, D.P., Sayre, J.W., Rokaw, S.N., Coulson, A.H., Massey, F.J., and Wegman, D.H. The UCLA population studies of chronic obstructive respiratory disease: 9. Lung function changes associated with chronic exposure to photochemical oxidants; a cohort study among never smokers. Chest 92:594-603 (1987).

Toxicology of ozone as characterized by laboratory animals and extrapolated to humans

J.A. Graham[a], J. Overton[b] and D.L. Costa[b]

[a] National Exposure Research Laboratory, Research Triangle Park, NC 27711, USA

[b] National Health and Environmental Effects Laboratory, U.S. Environmental Protection Agency, Research Triangle Park, NC 27711, USA

ABSTRACT

The adverse effects of ozone have been amply demonstrated in animal toxicology, human clinical, and epidemiology studies. Each of these research approaches has various strengths and weaknesses, but together the coherence is remarkable. This paper interprets the effects of ozone as demonstrated and inferred from laboratory animal toxicology studies. The major classes of ozone effects are: decrements in pulmonary function and symptoms, respiratory morbidity, inflammation, alterations of host defenses, and chronic effects on lung structure and function. All have been observed in laboratory animal studies, except for those calling for 'uniquely' human events (e.g., the reporting of symptoms or admission to hospitals). However, the findings in animals are often more extensive because a wider range of endpoints can be used. This raises the issue of whether such findings can be quantitatively extrapolated to humans. Most experts accept the premises supporting the qualitative extrapolation. That is, if ozone causes the effect in several animal species, it likely could cause the effect in humans, albeit at an unknown exposure. The quantitative extrapolation (i.e., the estimate of exposure that is likely to cause the effects in humans) is more controversial. This paper discusses the principal results of animal inhalation toxicology studies that greatly expand our understanding of the effects of ozone and provides a preliminary estimate of the potential for humans to experience chronic effects. The predominant topics relate to the impacts of several days of exposure and chronic exposure for months to years. Human clinical studies suggest that the pulmonary function of people accommodates to repeated intermittent ozone exposure. A similar pattern of attenuation of pulmonary function is shown in laboratory animals; however, cellular damage continues. This has been confirmed in humans. Chronic exposures of rats and monkeys result in structural remodeling of the lung, especially in the region where the conducting airways and gas exchange region meet. Various changes are involved, which many experts interpret as being consistent with incipient peribronchiolar fibrogenesis within the interstitium of the lung. In both rats and monkeys, such structural changes persisted after exposure ceased. Furthermore, seasonal exposure caused equivalent or sometimes greater effects than continuous exposure in both species. Using an extrapolation model with numerous assumptions and exposure data from a hypothetical child who plays outdoors and an outdoor worker exposed to ozone over a 214-day season in New York city, rat and monkey lung structural studies suggest potential chronic health effects of ozone. The combined use of animal and human data as linked by quantitative extrapolation has not only refined the assessment of the health impact of ambient ozone exposure but has established a viable paradigm to assess the health effects of other potentially harmful air pollutants.

1. INTRODUCTION

Controlled laboratory animal studies were the first to show the potential for ozone to cause a range of health effects. Even today animal studies provide unique insights into health effects because the exposures are controlled and the endpoints invasive. For example, rats have been exposed to known concentrations of ozone for several years and their lungs have been autopsied with advanced scientific methods. Even though laboratory animal studies can provide insights unavailable in human studies, there is considerable uncertainty in extrapolating the results of animal toxicology studies to humans. Humans in the ambient environment are exposed to imprecisely known concentrations of ozone in combination with other pollutants, and there are interspecies differences in the lungs and in the impact of ozone on them. The optimal risk assessment strategy for ozone (and other pollutants) thus involves an integrated interpretation of evidence from animal toxicology, human clinical, and epidemiology studies. This enables the strengths of each discipline to contribute to the overall assessment. The health effects database for ozone is one of the best, facilitating such an integration; analogous effects are observed with all three of these research approaches.

Public health concerns for ozone center on effects on pulmonary function and symptoms, respiratory morbidity, inflammatory responses, effects on host defenses, and chronic effects on the respiratory tract. Animal toxicology studies have directly enabled and expanded understanding of all these classes of effects, except for respiratory morbidity (e.g., increased hospital admissions for respiratory causes in relation to increased ozone concentrations). The goal of this paper is to summarize this contribution of laboratory animal research by focusing on the quantitative extrapolation of such studies to humans. The information to be summarized here is drawn from the ozone criteria document by the U.S. Environmental Protection Agency [1], and readers wanting more detailed information should consult that document containing numerous references. Only a very few critical papers are cited in this paper.

Effects of ozone observed in several animal species are very likely to be observed in humans given the substantial similarities between species and the commonality of the mechanism(s) of action of ozone. This qualitative extrapolation has been recognized and accepted for some time. Although the molecular mechanism(s) of action of ozone is not known fully, molecular targets (e.g., carbon-carbon double bonds, sulfhydryl groups) are identical across species. However, there are significant interspecies differences which cast uncertainty on the quantitative extrapolation. Thus, even those who agree on the qualitative extrapolation argue about what exposures cause effects in humans. Recent scientific advances in quantitative extrapolation, discussed in the ozone criteria document, expand the understanding of several effects, especially chronic effects.

2. ACUTE AND SHORT-TERM EFFECTS

Ozone affects host defenses. Alveolar macrophages, which remove particles and microbes from the alveolar region by phagocytosis, have less ability to engulf microbes after exposures to ozone as low as 0.1 ppm (2 hours) in rabbits [2] and 0.08 ppm (6.6 hours, moderate exercise) in humans [3]. One of the most investigated indicators, susceptibility to lung bacterial infection in mice, is increased after a 3-hour exposure to 0.08 ppm [4]. There is no direct corollary in humans, but given the interspecies similarity of the mechanisms of antibacterial lung host defenses, impacts on human host defenses can be hypothesized. Impaired clearance also has implications for interactions with other pollutants, especially particulate matter. Rats exposed to an urban pattern of ozone for 6 weeks had increased retention of asbestos fibers in the lung tissue 30 days after exposure ceased [5], suggesting an impact of ozone on the biological impact of the known carcinogen, asbestos.

The variety of studies of the lung and systemic immune system show mixed effects, but the database is not yet robust enough for clear conclusions. The effects of ozone on antiviral defenses are uncertain. There is no evidence yet of effects on the early course of viral lung infections, but laboratory animal studies suggest an increase in postinfluenza lung damage after a few months of exposure.

Pulmonary function decrements in humans are one of the hallmarks of ozone exposure. Many chamber, field, and epidemiology studies have shown that ozone decreases forced vital capacity (FVC) and forced expiratory volume in 1 second (FEV_1). These changes primarily result from inhibition of inspiration that is thought to be caused by neurogenic and/or inflammatory mechanisms. This response has been observed in healthy subjects exposed to levels as low as 0.08 ppm for 6.6 hours while undergoing moderate exercise [6]. Somewhat higher concentrations cause an increase in breathing frequency in humans. Several species of animals have similar shifts to rapid, shallow breathing after ozone exposure. Figure 1 demonstrates the commonality of rat and human changes in FVC after ozone exposure at different exercise levels. As can be observed here (and in all other exercise studies), a concomitant increase in exercise increases the degree and harmonization of response in humans and rats.

Reports of attenuation of ozone-induced pulmonary function changes and symptoms with several days of exposure (exposures for 1 to 3 hours/day) began to appear in 1977 [9]. The body of work clearly indicates that the spirometric responses of humans are observed on the first 1 to 4 days of a series of intermittent exposures to ozone. At higher ozone concentrations, greater responses may be seen on Day 2. However, as exposures are repeated on the next few days, responses decreased or were absent. Further research in humans showed that responsiveness returned at about 1 or 2 weeks after the original series of exposures. Symptom changes paralleled the functional effects. Because environmental exposures to elevated levels of ozone often persist for a few days, interpretation of the impacts of these findings on risk was important. Such an interest stimulated animal toxicology studies. Tepper and co-workers [10] found that the pulmonary function of rats had a similar pattern of attenuation to ozone exposure. Even so, lung remodeling progressed, and signs of inflammation (lavageable protein) were sustained over the 5-day exposure period. In humans, some inflammatory markers attenuate and some persist; damage, as indicated by a biochemical marker, persists. Joint interpretation of these studies results in a conclusion that risk persists, and perhaps is even enhanced, even though pulmonary function changes revert towards normal. Perhaps there is a linkage. As breathing returns to normal, more ozone may penetrate to the more distal regions of the lung, which are thought to be more vulnerable.

One of the earliest observations of the effects of ozone was the production of edema in the lungs of animals exposed acutely to high concentrations. Years of follow-up in animal studies showed that ozone affected the barrier function of the lung. This function is central to health because the barrier must allow air exchange, while keeping microbes and other undesirable environmental elements out of the body and keeping essential fluids and cells in the body rather than flooding the air spaces. This barrier function is also linked to inflammation, which for the lung, is commonly characterized by an increase in polymorphonuclear leukocytes (PMN's), protein, and other bioactive compounds in lung lavage fluid.

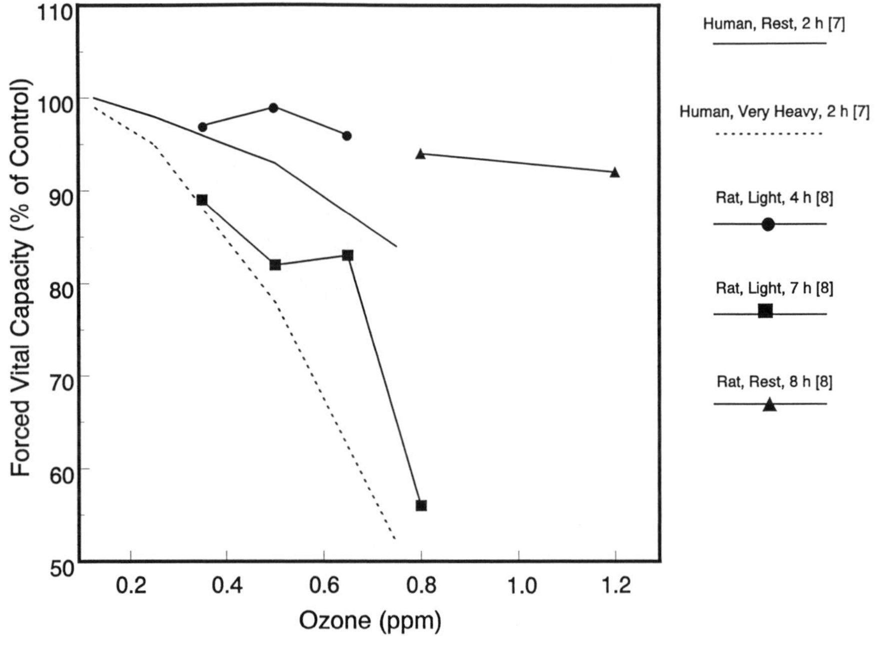

Figure 1. Effects of Ozone on Forced Vital Capacity in Rats and Humans (shaded area is predicted range of changes in humans expected between light exercise [top line] and very heavy exercise [bottom line].) The key also indicates whether the rats were exposed at rest or under light exercise [1].

Numerous animal toxicology studies have been conducted on these endpoints in search of better understanding of the impacts of such changes. Such research stimulated human clinical studies [3] that found that 6.6-hour exposures of healthy humans during moderate exercise increased the PMN's (0.08 ppm), bioactive mediators (0.08 ppm), and protein (0.1 ppm) in lung lavage fluid. Figure 2 compares the protein in bronchoalveolar lavage fluid in three species of animals and in humans as a function of pulmonary tissue dose of ozone. Briefly, Miller and co-workers [11] used dosimetry models to calculate the dose to the pulmonary region (i.e., alveolar region) as a function of concentration and duration of exposure, exercise level, and species. The pulmonary region was chosen because it likely is the site of damage that allows protein to enter the air spaces. They observed a remarkable similarity in the dose-response across species. Additional studies in laboratory animals and humans exposed to ozone have shown that inflammatory responses persist for about a day after exposure ceases. Inflammation is of concern because it is evidence that injury has occurred. Even though there are many uncertainties, a role of repeated inflammation in the causation of chronic lung disease cannot be ruled out.

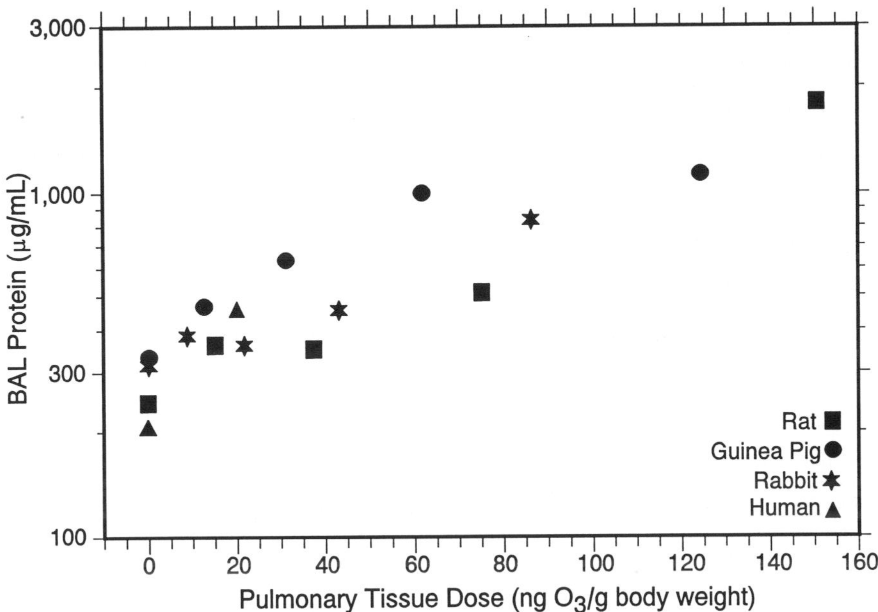

Figure 2. Protein in the bronchoalveolar lavage (BAL) for several laboratory animal species and humans, as related to the estimated pulmonary dose (normalized per gram of body weight) [1,11].

3. CHRONIC EFFECTS OF OZONE

3.1. The Nature of the Effects

The few studies of the immune system or antibacterial host defenses after chronic exposure of laboratory animals either demonstrated no effects or did not increase the magnitude of effects caused by short-term exposure. Thus, lung structural and functional changes are of most concern after months to years of exposure to ozone. At present, most knowledge about the effects of chronic exposures is derived from laboratory animals. Epidemiology studies have not been performed with the ability to discern such chronic effects in humans, assuming they exist. Even so, the epidemiological evidence that exists is qualitatively supportive of lung function decrements in people living in more highly polluted communities.

Research findings in several species of animals, including nonhuman primates, are remarkably similar. Structural changes occur along the whole respiratory tract, but those in the small airways and in the centriacinar regions (where the conducting airways and the gas exchange region join) are of greater health concern because of the importance of these areas to gas exchange, the primary function of the lungs. Figure 3 provides a schematic of the types of changes that occur. Figure 4 contains a description of the time course of the response during and after exposure.

As duration of exposure increases from days to months, a pattern emerges. Inflammatory exudate usually peaks in the first few days of exposure, resulting in increased permeability and the movement of cells into the air spaces. This response falls off in a few days, remaining at a low level as exposure continues and attaining recovery when exposure stops. Epithelial hyperplasia increases over the first few days and then plateaus, dropping off after exposure ceases. In this process, ciliated cells (that move debris and other material up and out of the lung) are sloughed off and replaced with nonciliated cells. Within the centriacinar region, there is hypertrophy of the epithelial cells of the proximal alveolar region of the transition zone of the alveoli and smallest airways while interstitial fibroblasts increase and create an exudate; collagen fibers accumulate; and the interstitium thickens. Even after exposure to ozone ceases, collagen can continue to accumulate. Several of these changes are similar to the earliest lesions found in human respiratory bronchiolitis, some of which may progress to fibrotic lung disease. Functional changes have been sought in several studies, but have been variable. Most pulmonary function tests do not measure changes in small airways and the centriacinar region. Even so, some studies suggest 'stiffer' lungs, which are consistent with fibrotic-like changes.

The influence of different exposure patterns is complex. For example, Tyler et al. (1988) exposed monkeys to a 'daily' regimen (8 hours/day, 7 days/week, 18 months) or a 'seasonal' regimen (same as for daily, but only every other month, for a total of

9 months of exposure over an 18-month period). Both groups of monkeys were affected (e.g., both had respiratory bronchiolitis), and for a few endpoints, the seasonal group had more changes (e.g., increased lung collagen content and increased chest wall compliance), suggesting delayed lung maturation. Qualitatively similar changes have been observed in rats exposed to such 'on-off' regimens. Rats also have been exposed to urban patterns of ozone (0.06 ppm background 7 days/week on which were superimposed 9-hour peaks, 5 days/week, slowly rising to and falling from 0.25 ppm) for up to 78 weeks [12].

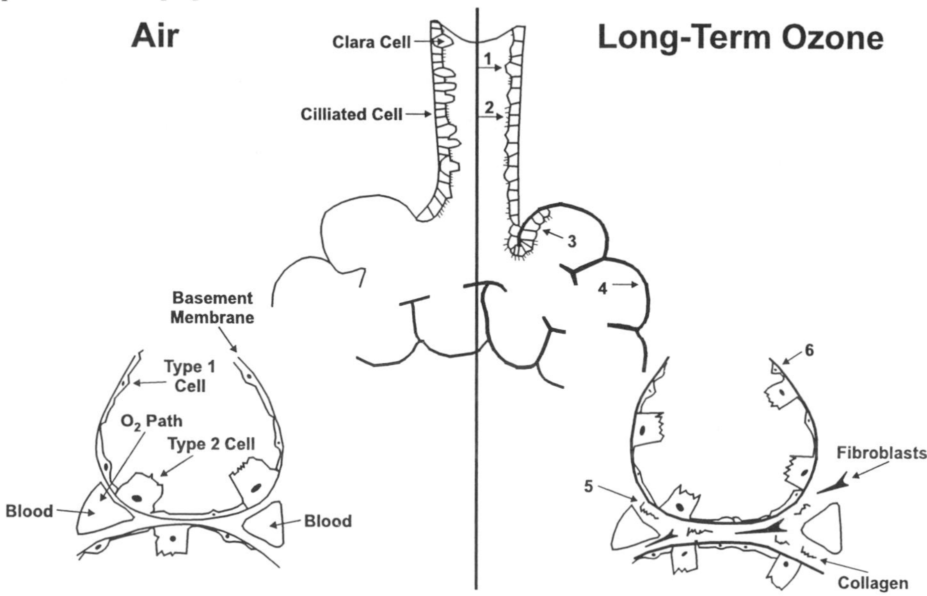

Figure 3. Schematic of zone-induced structural changes in terminal bronchiolus and centriacinar regions of the lung. 1=nonciliated cells which increase and/or have altered shapes, 2=ciliated cells which lose cilia or are sloughed off, 3=bronchiolar epithelium which begins to extend into the alveoli, 4=thickened epithelium, 5=thickened interstitium, 6=thickened basement membrane. Also shown (unnumbered) is an increase in fibroblasts and collagen in the interstitium and a decrease in the number of Type 1 cells accompanied by an increase in the number of Type 2 cells (modified [1]).

Effects observed in the centriacinar region were similar to those described above. Interestingly, many of the changes resolved by 17 weeks after exposure ceased, but the epithelial and endothelial basement membrane were thickened and accompanied by increased collagen fibers at this 17-week postexposure period. These findings had functional correlates indicating that the lung was indeed stiffer perhaps due to the remodelling of the distal airway-alveolar interface [13,14].

3.2. The Quantitative Extrapolation

The impact of ozone on increasing the thickness of the acellular and total interstitial volumes in the centriacinar region was chosen as the effect to be extrapolated because of the importance of this region to the overall function of the lung and the availability of high quality data over a range of concentrations and durations in more than one species of animal. The previous discussion refers to the centriacinar region.

Figure 4. Schematic comparison of the duration-response profiles for the centriacinar region of the lung exposed to a constant low concentrations of ozone [15].

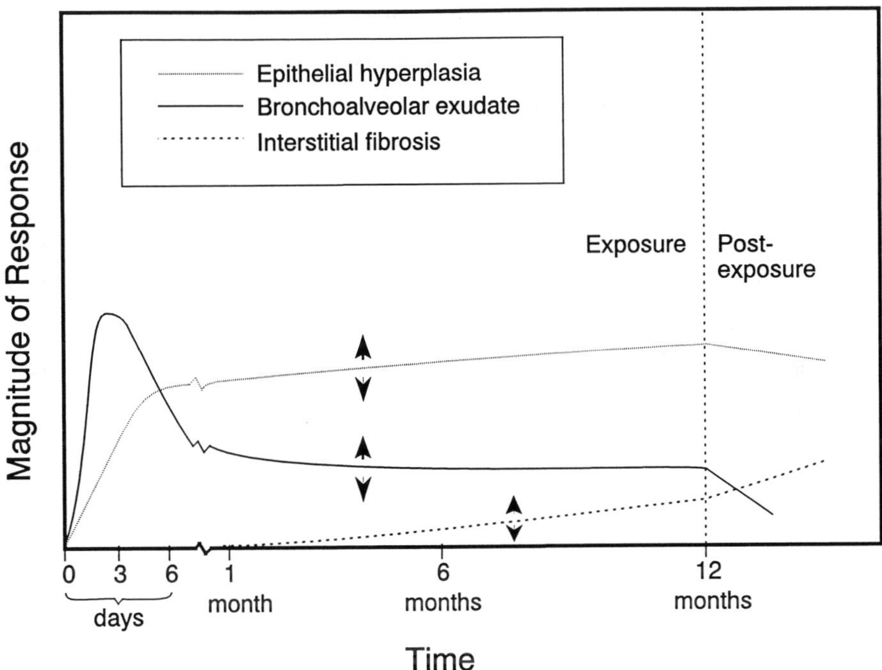

The studies to be used here involved morphometric analyses of a subregion, called the proximal alveolar region (PAR), which for the purposes of this paper can be considered equivalent to the centriacinar region. The general approach taken was to convert the exposure-response information to dose-response curves for these animal studies using dosimetry models and then compare these curves to the doses calculated from exposure models of children and adults. Many assumptions, with varying and sometimes unknown validity, were used. For example, it was assumed that there was a relationship between effect and total cumulative dose to the target tissue. One of the most significant assumptions is that the rate of change of interstitial thickness is related to the rate of ozone uptake. This ignores potential differences in species sensitivity to defensive or reparative mechanisms. Even so, given the likely irreversibility of fibrogenesis and degenerative lung disease, such an assumption is not extreme. More specifics follow (see the ozone criteria document [1], Chapter 8, for details).

The two rat studies selected were reported by Chang and co-workers [12,16]. The initial one used the urban pattern of exposure mentioned above (background of 0.06 ppm on which were superimposed a 9-hour spike of 0.25 ppm, 5 days/week) for up to 78 weeks with periodic measurements; the later study used rats exposed for 5 days/week (6 hours/day) to 0.12, 0.50, or 1.0 ppm for up to 87 weeks. The nonhuman primate studies selected used bonnet monkeys exposed for 90 days (8 hours/day) to 0.15 or 0.3 ppm [17] and cynomolgus monkeys exposed to 0.25 ppm (8 hours/day, 7 days/week), for 18 consecutive months or for every other month for a total of 9 months of exposure over the 18-month period [18].

Figure 5 shows these dose-response curves for the rats and monkeys. The similarity in the dose-response between the two different rat studies and the two different monkey studies is notable, increasing confidence in the representativeness for these species. As can be seen, the monkey is more responsive than the rat. The reason is unknown, but in addition to differences in species sensitivity, it partly may be because of the rats being exposed during the daytime, when they are more quiet (i.e., reduced ventilation and hence ozone dose). The data show an apparent linear dose-response, which is well supported by the strong correlation coefficients. This gives credence to the opinion that interstitial injury is cumulative with exposure.

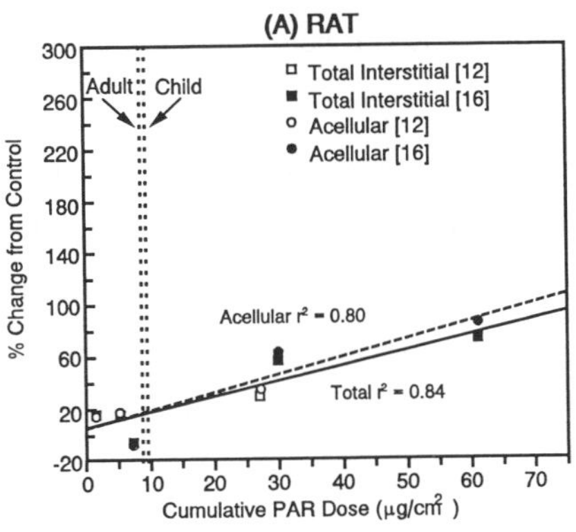

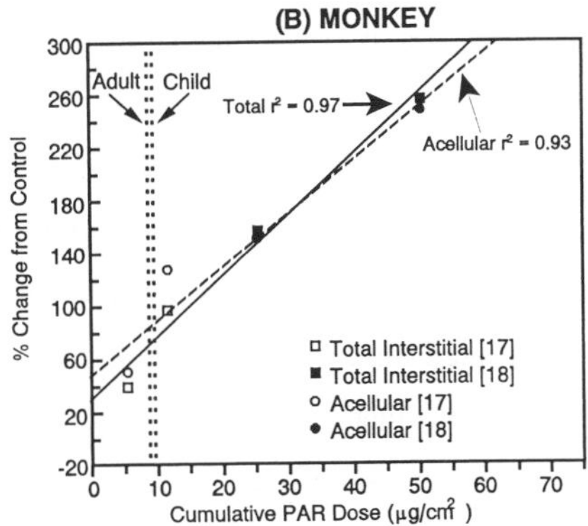

Figure 5. Extrapolation of chronic effects from a laboratory animals to humans. The solid lines represent the linear regression for total interstitial thickness, and the dashed line represents the linear regression for acellular thickness for both studies in each panel of the figure [1].

For the human part of the extrapolation, a hypothetical New York City adult outdoor worker and a 9-year-old New York City child who plays outdoors were selected from the population groups used for the risk assessment prepared by the U.S. Environmental Protection Agency [19]. The activity patterns and ozone exposure information was obtained from the probabilistic NAAQS exposure model [20] using data collected in New York City from April through October 1991. These activity patterns and concentrations were extrapolated to estimate the dose to the PAR of the subjects using a mathematical dosimetry model of Miller and co-workers [21] and Overton and co-workers [21] and assumptions about anatomy, mass transfer coefficients, and ventilation of the selected subjects. The model predicted that the theoretical child would have received a cumulative dose of 9.9 μg/cm^2 to the PAR over the 214 days of the ozone-season; the adult outdoor worker would have received a cumulative dose of 8.6 μg/cm^2. Comparing these doses to those on the rat and monkey dose-response curves suggest that a child might experience a 20 to 75% increase in PAR thickness and an adult might have a 15 to 70% increase over the predicted ozone season. The ranges are bounded by the rat curves (low end) and monkey curves (high end). There is no evidence to determine whether the rat or the monkey better represent the potential human response. However, it is reasonable to assume that the human response lies somewhere in between.

Prolonged exposures cause changes in the distal lung and airways of monkeys and rats that appear consistent with incipient peribronchiolar fibrogenesis within the interstitium and these changes may progress even when exposure ceases. The animal studies have also demonstrated that these structural changes do not reverse between ozone seasons. Thus, the estimate that a child and an adult with a 'real-world' ozone exposure receive a dose that may increase PAR thickness is of concern, especially since only one "season" for the humans was plotted.

4. CONCLUSIONS

The results of short-term animal toxicology, human clinical, and epidemioly studies are strongly correlated. This supports a homology of responses across species, which is strengthened by theoretical equivalencies based on hypothesized mechanisms of action of ozone. Examining the wide array of effects of ozone, the laboratory animals continue to provide more qualitative reasons to be concerned about humans. The quantitative extrapolation by necessity contains numerous assumptions. Nevertheless, it suggests that children and adult workers who spend time outdoors may receive a cumulative dose of ozone during a single season sufficient to increase thickness of the interstitium in the PAR of the lung. In spite of the many uncertainties about the degree of response and its medical interpretation, there is reason for concern that long-term ozone exposure could impart a chronic effect in humans.

REFERENCES

1. U.S. Environmental Protection Agency. (1996) Air quality criteria for ozone and other photochemical oxidants. Research Triangle Park, NC: Office of Health and Environmental Assessment, Environmental Criteria and Assessment Office; EPA report nos. EPA-600/P-93/004aF-cF. 5v. Available from: NTIS, Springfield, VA.
2. Driscoll, K.E.; Vollmuth, T.A.; Schlesinger, R.B. (1987) Acute and subchronic ozone inhalation in the rabbit: response of alveolar macrophages. J. Toxicol. Environ. Health 21: 27-43.
3. Devlin, R.B.; McDonnell, W.F.; Mann, R.; Becker, S.; House, D.E.; Schreinemachers, D.; Koren, H.S. (1991) Exposure of humans to ambient levels of ozone for 6.6 hours causes cellular and biochemical changes in the lung. Am. J. Respir. Cell Mol. Biol. 4: 72-81.
4. Miller, F.J.; Illing, J.W.; Gardner, D.E. (1978) Effect of urban ozone levels on laboratory-induced respiratory infections. Toxicol. Lett. 2: 163-169.
5. Pinkerton, K.E.; Brody, A.R.; Miller, F.J.; Crapo, J.D. (1989) Exposure to low levels of ozone results in enhanced pulmonary retention of inhaled asbestos fibers. Am. Rev. Respir. Dis. 140: 1075-1081.
6. Horstman, D.H.; Folinsbee, L.J.; Ives, P.J.; Abdul-Salaam, S.; McDonnell, W.F. (1990) Ozone concentration and pulmonary response relationships for 6.6-hour exposures with five hours of moderate exercise to 0.08, 0.10, and 0.12 ppm. Am. Rev. Respir. Dis. 142: 1158-1163.
7. Hazucha, M.J. (1987) Relationship between ozone exposure and pulmonary function changes. J. Appl. Physiol. 62: 1671-1680.
8. Costa, D.L.; Stevens, M.S.; Tepper, J.S. (1988) Repeated exposure to ozone (O_3) and chronic lung disease: recent animal data. Presented at: 81st annual meeting of the Air Pollution Control Association; June; Dallas, TX. Pittsburgh, PA: Air Pollution Control Association; paper no. 88-122.3.
9. Hackney, J.D.; Linn, W.S.; Mohler, J.G.; Collier, C.R. (1977) Adaptation to short-term respiratory effects of ozone in men exposed repeatedly. J. Appl. Physiol.: Respir. Environ. Exercise Physiol. 43: 82-85.
10. Tepper, J.S.; Costa, D.L.; Lehmann, J.R.; Weber, M.F.; Hatch, G.E. (1989) Unattenuated structural and biochemical alterations in the rat lung during functional adaptation to ozone. Am. Rev. Respir. Dis. 140: 493-501.
11. Miller, F.J.; Overton, J.H.; Gerrity, T.R.; Graham, R.C. (1988) Interspecies

dosimetry of reactive gases. In: Mohr, U.; Dungworth, D.; McClellan, R.; Kimmerle, G.; Stöber, W.; Lewkowski, J., eds. Inhalation toxicology: the design and interpretation of inhalation studies and their use in risk assessment. New York, NY: Springer-Verlag; pp. 139-155.

12. Chang, L.-Y.; Huang, Y.; Stockstill, B.L.; Graham, J.A.; Grose, E.C.; Ménache, M.G.; Miller, F.J.; Costa, D.L.; Crapo, J.D. (1992) Epithelial injury and interstitial fibrosis in the proximal alveolar regions of rats chronically exposed to a simulated pattern of urban ambient ozone. Toxicol. Appl. Pharmacol. 115: 241-252.

13. Tepper, J.S., Wiester, M.J., Weber, M.F., Fitzgerald, S., and Costa D.L. (1991) Chronic exposure to a simulated urban profile of ozone alters ventilatory responses to CO_2 challenge in rats. Fund. Appl. Toxicol. 17: 52-60.

14. Costa, D.L., Tepper, J.S., Stevens, M.A., Watkinson, W.P., Doerfler, D.L., Gelzleichter, T.R., and Last, J.A. (1995) Restrictive lung disease in rats exposed chronically to an urban profile of ozone. Am. J. Resp. Crit. Care Med. 151: 1512-18.

15. Dungworth, D. L. (1989) Noncarcinogenic responses of the respiratory tract to inhaled toxicants. In: McClellan, R.O.; Henderson, R.F., eds. Concepts in Inhalation Toxicology. New York, NY: pp. 273-298.

16. Chang, L.-Y.; Stockstill, B.L.; Ménache, M.G.; Mercer, R.R.; Crapo, J.D. (1995) Consequences of prolonged inhalation of ozone on F344/N rats: collaborative studies. Part VIII. Morphometric analysis of structural alterations in alveolar regions. Cambridge, MA: Health Effects Institute; pp. 3-39; research report no. 65.

17. Harkema, J.R.; Plopper, C.G.; Hyde, D.M.; St. George, J.A.; Wilson, D.W.; Dungworth, D.L. (1993) Response of macaque bronchiolar epithelium to ambient concentrations of ozone. Am. J. Pathol. 143: 857-866.

18. Tyler, W.S.; Tyler, N.K.; Last, J.A.; Gillespie, M.J.; Barstow, T.J. (1988) Comparison of daily and seasonal exposures of young monkeys to ozone. Toxicology 50: 131-144.

19. U.S. Environmental Protection Agency. (1995) Review of national ambient air quality standards for ozone assessment of scientific and technical information. Research Triangle Park, NC: Office of Air Quality Planning and Standards; OAQPS staff paper.

20. Johnson, T.R. (1994) One-minute ozone exposure sequences for selected New York cohorts [letter to Dr. John H. Overton, U.S. EPA]. Durham, NC: International Technology Corporation; November 22.

21. Miller, F.J.; Overton, J.H., Jr.; Jaskot, R.H.; Menzel, D.B. (1985) A model of the regional uptake of gaseous pollutants in the lung: I. the sensitivity of the uptake of ozone in the human lung to lower respiratory tract secretions and exercise. Toxicol. Appl. Pharmacol. 79: 11-27.

22. Overton, J.H., Jr.; Graham, R.C.; Miller, F.J. (1987) A model of the regional uptake of gaseous pollutants in the lung: II. the sensitivity of ozone uptake in laboratory animal lungs to anatomical and ventilatory parameters. Toxicol. Appl. Pharmacol. 88: 418-432.

Overview of Ozone Human Exposure and Health Risk Analyses Used in the U.S. EPA's Review of the Ozone Air Quality Standard

R.G. Whitfield

Argonne National Laboratory, Decision and Information Sciences Division

H.M. Richmond

U.S. Environmental Protection Agency, Office of Air Quality Planning and Standards

T.R. Johnson

TRJ Environmental, Inc.

ABSTRACT

This paper presents an overview of the ozone human exposure and health risk analyses developed under sponsorship of the U.S. Environmental Protection Agency (EPA). These analyses are being used in the current review of the national ambient air quality standards (NAAQS) for ozone. The analyses consist of three principal steps: (1) estimating short-term ozone exposure for particular populations (exposure model); (2) estimating population response to exposures or concentrations (exposure-response or concentration-response models); and (3) integrating concentrations or exposure with concentration-response or exposure-response models to produce overall risk estimates (risk model). The exposure model, called the probabilistic NAAQS exposure model for ozone (pNEM/O_3), incorporates the following factors: hourly ambient ozone concentrations; spatial distribution of concentrations; ventilation state of individuals at time of exposure; and movement of people through various microenvironments (e.g., outdoors, indoors, inside a vehicle) of varying air quality. Exposure estimates are represented by probability distributions. Exposure-response relationships have been developed for several respiratory symptom and lung function health effects, based on the results of controlled human exposure studies. These relationships also are probabilistic and reflect uncertainties associated with sample size and variability of response among subjects. The analyses also provide estimates of excess hospital admissions in the New York City area based on results from an epidemiology study. Overall risk results for selected health endpoints and recently analyzed air quality scenarios associated with alternative 8-hour NAAQS and the current 1-hour standard for outdoor children are used to illustrate application of the methodology.

INTRODUCTION

As stated in the U.S. Clean Air Act, the U.S. EPA is required to set, review, and revise, as appropriate, the primary NAAQS for criteria pollutants. This review includes a determination on whether the scientific basis for a NAAQS has changed sufficiently to warrant revisions. The primary standards, which are to be set at levels sufficient to protect public health with an adequate margin of safety, are based on scientific evidence reviewed in "criteria documents" (CDs). A CD summarizes and evaluates the scientific literature relevant to setting ambient air quality standards fo a given pollutant or class of pollutants (e.g., particulate matter). With respect to health effects, the ozone CD evaluates the human clinical and field studies and the epidemiological and animal toxicological evidence regarding physiological and adverse health effects that result from exposure to ozone and other photochemical oxidants.

The NAAQS review process consists of three principal components: (1) an assessment of the scientific and technical basis for deciding on the primary (health effects) and secondary (welfare effects) standards; (2) development of regulatory decision packages for standards and implementation; and (3) development of guidance, technical requirements, and attainment schedules for State implementation of new or revised standards. The Agency obtains comments through an extensive review process involving the public, the Clean Air Scientific Advisory Committee (CASAC), and other federal agencies including the Office of Management and Budget (OMB).

NAAQS are set for ubiquitous pollutants to protect the most sensitive population group(s) such as asthmatics and emphysematics who are exposed to the ambient environment in the course of their normal activities. The standards must protect (1) the sensitive population as a whole, but not necessarily the most sensitive individuals and (2) public health with an adequate margin of safety against effects that have not yet been uncovered by research and effects whose medical significance is a matter of disagreement.

Central to the review process is the definition of adverse health effects. Language of the Washington, D.C. Circuit Court discusses a precautionary mandate to protect against reasonable medical concerns. This does not require a medical consensus that effects are clearly harmful. Adverse effects may include aggravation of preexisting conditions such as asthma and emphysema. While medical experts may differ, the law states that determination of adverse health effects is a matter of judgment by the EPA Administrator.

The concept of providing an adequate margin of safety is a result of EPA's rejection of any attempt to set a risk free standard. Instead, EPA has recognized the importance of assessing the relative acceptability of various degrees of uncertainty about the level of protection against adverse effects afforded by limiting exposures to low levels.

EPA considers many factors in the NAAQS review process, including: the nature and severity of health effects; the degree of human exposure; and health risk estimates for sensitive population groups when a standard is just attained. Exposure and risk analyses are tools used to aid the Administrator's judgments about which standard(s) provide an adequate margin of safety.

EPA has adopted several basic principles for conducting NAAQS exposure and risk analyses. These include: explicit treatment of major uncertainties using (where feasible) probabilistic methods; use of experimental data, models, and expert judgment (as appropriate) to develop probabilistic relationships; deliberate avoidance of conservative assumptions; generation of multiple exposure and risk measures to estimate central tendencies and uncertainties; comprehensive presentation of qualitative information on the nature and severity of effects; and discussion of limitations, caveats, and additional uncertainties that were not characterized quantitatively.

As part of its review of the NAAQS for ozone, EPA's Office of Air Quality Planning and Standards (OAQPS) has for several years led the development of tools and methods for assessing the public health risk associated with attaining alternative ozone NAAQS. The purpose of these developments is to characterize, as explicitly as possible, the range and implications of uncertainties in the existing scientific database, while fully using current scientific knowledge, available animal and human experimental and observational data, and scientific expertise. This risk assessment addresses the effects of acute exposures to ozone. It combines exposure-response relationships with exposure estimates to produce overall risk estimates. In addition, hourly air quality data in New York City are used to estimate excess respiratory-related hospital admissions of asthmatics during the ozone season. A summary of the results of acute risk assessment and its role in the ozone NAAQS review can be found in the OAQPS Staff Paper (EPA 1996a).

The acute risk assessment addresses the effects of exposure to ozone for populations engaged in either heavy or moderate exertion. The heavy exertion effects are based on 1- to 3-h controlled human exposure studies by McDonnell et al. (1983), Avol et al. (1984), and Kulle et al. (1985). The moderate exertion effects are based on results from 2-h controlled human studies by Seal et al. (1993) and from 6.6-h controlled human studies by Folinsbee et al. (1988), Horstman et al. (1990), and McDonnell et al. (1991). The hospital admissions estimates are based on a multiyear study of air pollution and respiratory hospital admissions in New York City (Thurston et al. 1992).

Previous risk assessments also studied the acute health effects of ozone (Hayes et al. 1987). Methods developed in these assessments (Hayes et al. 1987, 1989) and earlier assessments for lead (Wallsten and Whitfield 1986; Whitfield and Wallsten 1989) provide a foundation for the current risk assessment.

RISK ASSESSMENT APPROACH

The basic risk assessment approach, illustrated in Figure 1, includes developing an exposure model and a health model.* The exposure model accounts for human contact with a specific criteria pollutant. The contact can be described in terms of a cumulative exposure over a specified time. For this assessment, exposure estimates were generated by pNEM/O_3 for nine urban areas and that portion of the population thought to be potentially at greatest risk to ozone exposure. The at-risk groups are outdoor children and outdoor workers. This determination is based on the discussion of at-risk populations in the ozone CD (EPA 1996b).

Two types of exposure measures have been made: persons and person-occurrences. The persons measure counts the number of individuals exposed one or more times per ozone season to the exposure indicator (ozone level and breathing rate) of interest. The person-occurrences measure first counts the times per ozone season that an individual is exposed to the exposure indicator of interest and then accumulates counts over all individuals. Therefore, the person-occurrences measure confounds persons and occurrences: 1 occurrence for 10 persons is counted the same as 10 occurrences for 1 person. The maximum number of daily maximum hourly exposure occurrences is equal to the population multiplied by the number of days in the ozone season. This paper includes both types of measures.

* Strictly speaking, Figure 1 applies only to headcount risks for acute effects. However, the differences for hospital admissions are subtle. One difference is that hospital admissions estimates are based on a concentration-response relationship involving ambient air quality data observed at a fixed-site monitor rather than on estimates of exposure.

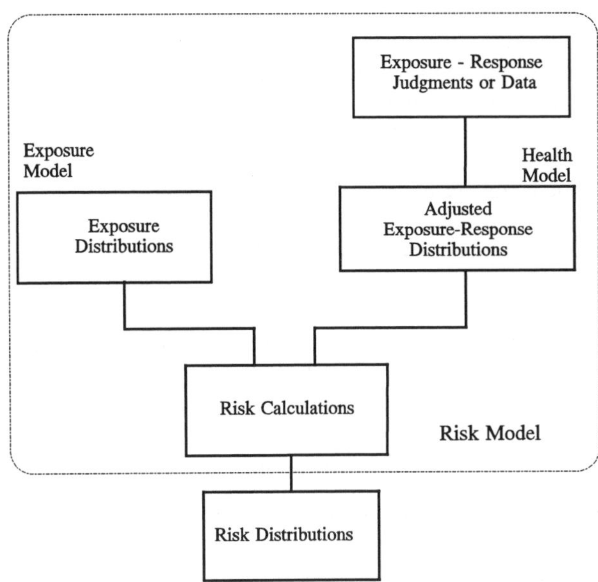

FIGURE 1 Basic Risk Assessment Approach: Risk Model Incorporating Both an Exposure Model and a Health Model

A health model accounts for human reaction to contact with a criteria pollutant. Reactions include symptoms or physiological changes (e.g., reduced pulmonary function or increased respiratory symptoms). A health model can be based on either data, judgment, or a combination of both. One important aspect of a health model is the "most at-risk population" — persons believed to be most at risk because they are either highly reactive (e.g., children whose physiological development may be impaired by exposure to ozone) or more frequently exposed (e.g., outdoor children and outdoor workers). The Clean Air Act requires NAAQS to be set at a level that protects the populations most at risk with an adequate margin of safety.

In this paper, the exposure-response relationships that characterize the effects of ozone exposure on pulmonary function and the respiratory system are based on controlled human exposure data obtained in clinical studies. Controlled human exposure studies, in contrast to epidemiological or field studies, were thought to be most appropriate for specifying the data needed for estimating exposure-response relationships. The concentration-response relationship central to our model for estimating excess hospital admissions is based on epidemiological studies by Thurston et al. (1992).

EXPOSURE MODELING

Evaluating alternative NAAQS proposed for a particular pollutant requires assessing the risks to human health associated with ozone exposures that result while just attaining each of the standards under consideration. Important factors that need to be considered in an ozone exposure assessment are magnitude of ozone concentrations; duration of ozone concentrations; spatial distribution of concentrations; frequency of repeated peak concentrations; ventilation state of the individual at time of exposure; and movement of people through zones of varying air quality, which affects the actual exposure patterns of people living within a defined area. Figure 2 shows how the pNEM/O_3 methodology fits into the risk assessment.

In evaluating alternative NAAQS proposed for a particular pollutant, OAQPS assesses the risks to human health of air quality meeting each of the standards under consideration (Richmond and McCurdy, 1988). This assessment of risk requires estimates of the number of persons exposed at various pollutant concentrations for specified periods of time. The estimates may be specific to an urbanized area such as Los Angeles or apply to the entire nation. These estimates are obtained by simulating the movements of people through zones of varying air quality so as to approximate the actual exposure patterns of people living within a defined area. OAQPS has implemented this approach through an evolving methodology referred to as the NAAQS Exposure Model (NEM). From 1979 to 1988, IT Air Quality Services (formerly PEI Associates, Inc.) assisted OAQPS in developing and applying pollutant-specific versions of NEM to ozone, particulate matter, and CO. These versions of NEM are referred to as "deterministic" versions in that no attempt was made to model random processes within the exposure simulation.

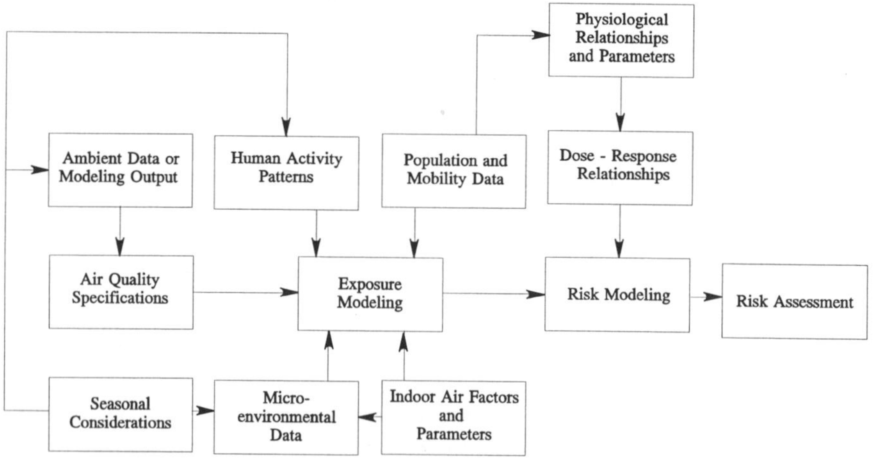

FIGURE 2 Major Components of the NEM Series of Exposure Models and Associated Health Risk Assessment Procedures

In 1988, OAQPS began to incorporate probabilistic elements into the NEM methodology and to apply the resulting model (pNEM) to the criteria pollutants. In 1992, ITAQS developed a special version of pNEM applicable to ozone (pNEM/O_3) that incorporated mass balance techniques. ITAQS subsequently made the following enhancements to this model and its input data bases.

- Use of more recent (1990-91) fixed-site monitoring data for estimating ambient ozone concentrations. The earlier analysis was based on 1981-84 monitoring data.

- An increase in the number of fixed-site monitors used to represent each urban area.

- Use of more recent (1990) census data for estimating cohort populations. The earlier analysis used 1980 census data.

- A new methodology for adjusting ambient ozone data to simulate attainment of one-hour and eight-hour NAAQS.

- Revision of the algorithm used to determine limiting values for equivalent ventilation rate.

- Development of origin/destination tables through the use of a new commuting algorithm.

A report by Johnson et al. (1996) describes these enhancements and summarizes the results of applying the enhanced model to the general population of each of nine U.S cities.

EPA also directed ITAQS to develop a special version of pNEM/O_3 applicable to outdoor workers and to use it to estimate the ozone exposures of outdoor workers residing in each of the nine cities. A summary of this work can be found in a report by Johnson et al. (1996).

In follow-up work for EPA, ITAQS developed a second special version of pNEM/O_3 applicable to children who tend to be active outdoors (hereafter referred to as "outdoor children"). A report by Johnson et al. (1996) summarizes the results of applying this version of pNEM/O_3 to outdoor children residing in the nine cities.

In each of the recent applications of pNEM/O_3, the ozone levels reported by monitors within each study area were adjusted to simulate the attainment of a specified ozone standard. The standard was defined as a limit to the number of daily maximum one-hour or eight-hour ozone concentrations expected to exceed a specified ozone concentration. The attainment status of each study area was determined by a single year of monitoring data, and the adjustment procedure was applied to this year. EPA is currently evaluating a variety of alternative forms of the ozone NAAQS. Each of these alternative NAAQS limits the average value of the n-th ranked daily maximum 8-hour concentration to a specified concentration, based on three years of data. Five of the alternative NAAQS are formulations of 0.08 ppm and 0.09 ppm ozone standards that permit rounding ov values to the nearest 0.01 ppm:

- 8HA2H-0.084: the second highest daily maximum 8-hr concentration averaged over three years shall not exceed 0.084 ppm (abbreviation: 8284)

- 8HA3H-0.084: the third highest daily maximum 8-hr concentration averaged over three years shall not exceed 0.084 ppm (abbreviation: 8384)

- 8HA5H-0.084: the fifth highest daily maximum 8-hr concentration averaged over three years shall not exceed 0.084 ppm (abbreviation: 8584)

- 8HA2H-0.094: the second highest daily maximum 8-hr concentration averaged over three years shall not exceed 0.094 ppm (abbreviation: 8294)

- 8HA3H-0.094: the third highest daily maximum 8-hr concentration averaged over three years shall not exceed 0.094 ppm (abbreviation: 8394)

In determining attainment status with respect to the 0.084 ppm standards, ozone concentrations between 0.080 ppm and 0.084 ppm are rounded down to 0.08 ppm. Consequently, an ozone concentration of 0.084 ppm is not considered to exceed a standard level of 0.08 ppm. In a similar manner, values between 0.090 ppm and 0.094 ppm are rounded down to 0.09 ppm for the 0.094 ppm standards.

Each of the five standards specifies the maximum ozone concentration permitted for the three-year average of a ranked value. Researchers developed an air quality adjustment procedure (AQAP) for each eight-hour standard that could be used to simulate the ozone levels expected to occur during a single year when a city just attained the standard over the three-year period. They also developed a comparable AQAP that was applicable to the current one-hour standard. This standard is defined as follows:

- 1H1EX-0.124: the daily maximum 1-hr concentration expected to be exceeded once per year shall not exceed 0.124 ppm (abbreviation: 1124)

The new AQAPs used in these exposure assessments differ from the AQAPs used in previous applications of pNEM/O_3 to outdoor children. Each of new AQAPs is applied to a three-year period so as to simulate the attainment of the associated standard over the entire period. The adjustment required for the three-year period is applied to a single year (1990, 1991, 1992, or 1993) within the period and the resulting single-year of ozone data is used in the exposure assessment to represent typical attainment conditions. The selected year is the middle year of the three-year period with respect to relative ozone level. This approach contrasts with the previous AQAPs in which attainment was determined by examining a single year (1990 or 1991) of ozone data. The ozone data for the selected year were adjusted to exactly attain the specified standard and then used in the exposure assessment to represent attainment conditions.

The new AQAPs also use a proportional adjustment procedure for all standards in all cities. In the previous AQAPs, a non-proportional adjustment procedure based on the Weibull distribution was used for all standards in six of the nine cities and for a few standards in the remaining three cities (Chicago, Denver, and Miami). A proportional adjustment procedure was used for the other standards in these three cities.

Exposure estimates are based on a set of 10 runs of pNEM/O_3 for a particular combination of city and standard. To reduce the effects of run-to-run variability caused by the probabilistic elements of pNEM/O_3, each set of 10 runs used the same 10 random number generator "seeds." This approach retained the desired variability within each set of 10 runs but removed the set-to-set variability. Consequently, the differences in the mean exposure estimates associated with the seven standards were assumed to reflect primarily the differences in ambient ozone concentrations permitted by the standards. In the previous AQAPs, each set of 10 runs used a different set of random number seeds. The resulting variability in random number sequences used in the pNEM/O_3 runs may have produced some of the differences observed among the mean exposure estimates associated with the various standards under evaluation.

Exposure-response probabilities for each run are computed by dividing the number of children at each ozone concentration by the number of children who reached a specific exertion level in a run. For example, the data in Table 1 (which are for just attaining standard 1124, Philadelphia, children, 8-h exposures at moderate exertion, pNEM/O_3 run 2) show that the number of children for this particular run is about 270 thousand. About 40 thousand children were exposed to ozone at concentrations of 0.041-0.06 ppm. Dividing this number by the total number of children at moderate exertion results in a probability of 0.16. The same computations are performed for the remaining ozone concentrations. The sum of the calculated probabilities does not equal 1 because exposures at ≤ estimated background are not included. The probabilities for this example are presented in Table 1. Such probabilities are needed to estimate risk distributions. Figure 3 is an example of the variation in exposure estimates among 10 pNEM/0_3 runs for Philadelphia children given that the current standard (1 h, 1 expected exceedance, 0.124 ppm) is just attained.

Table 2 is an example of the summary statistics about air quality scenarios that have been developed for each city. Entries in Table 2 are estimates of the 8-hr maximum dosage exposures experienced by outdoor children during which ozone concentrations exceeded 0.08 ppm and EVR ranged from 13-26 L/min/m^2.

The new exposure and air quality estimates developed by Johnson et al. (1997) were later used in a quantitative risk assessment by Whitfield (1997). Some results of that risk assessment are presented in this paper.

TABLE 1 Calculating Exposure Probabilities for Outdoor Children Exposed for 8 Hours at Moderate Exertion from pNEM/O_3 Estimates for Run 2, Philadelphia, Scenario 1124 Just Attained

Ozone Intervals (ppm)	Number of Children: In Interval or Higher[a]	Number of Children: In Interval[b]	Probabilities[c]
0.000	268,923	0	(NR)[d]
0.001 - 0.020	268,923	9,058	(NR)
0.021 - 0.040	259,865	26,012	(NR)
0.041 - 0.060	233,853	43,123	0.1604
0.061 - 0.070	190,730	52,135	0.1939
0.071 - 0.080	138,595	68,048	0.2530
0.081 - 0.090	70,547	44,090	0.1640
0.091 - 0.100	26,457	19,555	0.0727
0.101 - 0.110	6,902	6,902	0.0257
0.111 - 0.120	0	0	0.0000
0.121 - 0.130	0	0	0.0000
0.131 - 0.140	0	0	0.0000
0.141 - 0.150	0	0	0.0000
0.151 - 0.160	0	0	0.0000
0.161 - 0.170	0	0	0.0000
0.171 - 0.180	0	0	0.0000
0.181 - 0.190	0	0	0.0000
0.191 - 0.200	0	0	0.0000
0.201+	0	0	0.0000

[a] A total of 268,923 children reached a moderate exertion level in run 2.

[b] Number in interval i equals number in interval i or higher minus the number in interval $i + 1$ or higher (e.g., 9,058 = 268,923 − 259,865).

[c] Probability of interval i equals the number in interval i divided by 268,923. This probability is also the fraction of children who reached a moderate exertion level while exposed to the ozone concentration for interval i in run 2.

[d] NR means not required. Calculations were not made for these concentrations because they are $\leq$ the estimated background level (0.04 ppm).

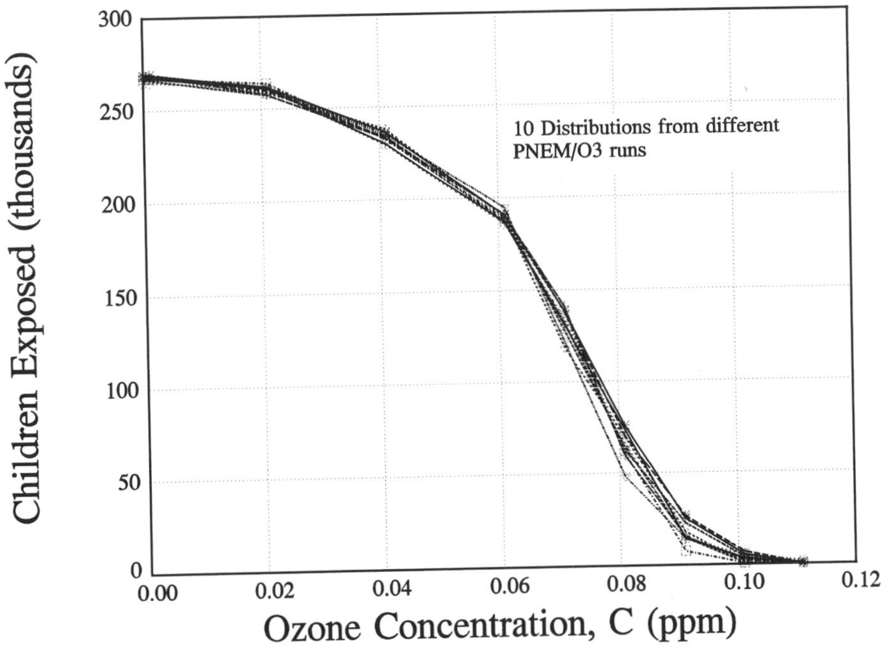

FIGURE 3 Illustration of Variation in Exposure Estimates among 10 pNEM/03 Runs for Philadelphia Children Given that the Current Standard (1 h, 1 expected exceedance, 0.124 ppm) Is Just Attained

TABLE 2 Estimates of 8-H Maximum Dosage Exposures Experienced by Outdoor Children in Philadelphia during which Ozone Concentration Exceeded 0.08 ppm and EVR[a] Ranged from 13-27 $L/min/m^2$

	Regulatory scenario					
Statistic[b]	1H1EX-0.124[c]	8HA3H-0.094	8HA2H-0.094	8HA5H-0.084	8HA3H-0.084	8HA2H-0.084
Mean Estimate of the Number of Outdoor Children	65,153	40,133	24,752	17,147	10,012	4,243
Percent of Total Outdoor Children Population	23.66	14.58	8.99	6.23	3.64	1.54
Range in this percentage for 10 runs						
	17.43-27.64	10.22-19.47	6.08-13.02	2.37-9.61	1.64-7.12	0.10-3.05
Mean Estimate of Person-Occurrences						
Percent of Total Person-Occurrences	78,918	44,645	26,526	17,899	10,169	4,243
Range in this percentage for 10 runs	0.13	0.08	0.05	0.03	0.02	0.01
	0.11-0.17	0.06-0.10	0.03-0.06	0.01-0.05	0.01-0.03	0.00-0.01
Mean Estimate of Occurrences/Person Exposed						
	1.21	1.11	1.07	1.04	1.02	1.00
Percentage exposed for indicated number of days						
1 Day						
2 Days	82.47	90.22	93.30	94.92	97.52	100.00
3 Days	14.11	8.29	5.89	4.81	2.48	0.00
>3 Days	3.05	1.41	0.81	0.27	0.00	0.00
	0.38	0.07	0.00	0.00	0.00	0.00

[a]Equivalent ventilation rate = (ventilation rate)/(body surface area).
[b]Mean or range for 10 runs of pNEM/O_3.
[c]Current NAAQS.

HEALTH MODELING

The health models discussed in this section are described more fully in Whitfield et al. (1996). The basis for the development of health models for lung function and symptoms endpoints is a number of controlled human exposure studies:

- 1-h exposures at heavy exertion* studies by McDonnell et al. (1983), Avol et al. (1984), and Kulle et al. (1985),
- 1-h exposures at moderate exertion† studies by Seal et al. ((1993), and
- 6.6-h exposures at moderate exertion‡ by Folinsbee et al. (1988), Horstman et al. (1990), and McDonnell et al. (1991). Results of these studies were used to develop exposure-response relationships for 8-h exposures.

These studies were selected because they best satisfied the following criteria:

- *Applicability to the population groups potentially at greatest risk*. Studies of persons exposed while engaged in moderate or heavy exertion are of greatest interest because such subjects are thought to be at greater risk than those at rest.
- *Comparability*. The total dose must be compared with the level of exertion and the exercise protocol of particular interest.
- *Number of subjects*. To limit the effects of small sample size, studies with at least 10 subjects per exposure level are desired.
- *Exposure concentrations*. Studies with multiple concentration levels in the range of ambient levels are desired.
- *Individual subject data*. These data are needed to develop exposure-response relationships.

It is important to note that, although the controlled human exposure studies used in the ozone risk assessment included adults aged 18-35, exposure-response relationships derived for both "outdoor children" and "outdoor workers" are used. Recent findings support the use of adult-based results to characterize children. These findings include results from other chamber

---

* Equivalent ventilation rates (EVRs) >30 L/min/m^2

† EVRs between 16-30 L/min/m^2

‡ EVRs between 13-27 L/min/m^2

studies (McDonnell et al. 1985a) and summer camp field studies in at least six different locations in the northeast United States, Canada, and southern California. These locations reported changes in lung function in healthy children similar to those observed in healthy adults exposed to ozone under controlled chamber conditions. As stated in the CD, "although direct comparisons cannot be made because of incompatible differences in experimental design and analytical approach," the range of response in the summer camp studies "is comparable to the range of response seen in chamber studies at low O_3 concentrations" (EPA 1995, pp. 9-7 and 9-8).

One or more of these studies recorded data that make it possible to construct a probabilistic exposure-response relationship for the following health endpoints:

- decrease in forced expiratory volume in 1 s (FEV_1),
- cough,
- chest pain on deep inspiration, and
- lower respiratory symptoms (any of a number of symptoms that included cough).

We considered three levels of FEV_1 decrements (FEV1 decrements ≥10%, ≥15%, and ≥20%) and two levels for cough, chest pain, and lower respiratory symptoms (any, moderate-to-severe). Although all of the studies did not investigate all of these endpoints, it was possible to develop exposure-response relationships for 33 endpoints (i.e., each exposure-response relationship is associated with a specific endpoint and a specific study). Results for only four of these endpoints are presented in this paper: moderate-to-severe cough (for 1-h exposures at heavy exertion, based on McDonnell et al. 1983), moderate-to-severe chest pain on deep inspiration (for 1-h exposures at moderate exertion, based on Seal et al. 1993), and FEV_1 decrements ≥15% and ≥20% (for 8-h exposures at moderate exertion, based on Folinsbee et al. 1988, Horstman et al. 1990, and McDonnell et al. 1991).

Developing exposure-response relationships for acute endpoints is a three-step process. (See Appendix A of Whitfield et al. 1996 for details of this process.) The process starts with data from the laboratory experiments described earlier. Before developing the probabilistic exposure-response relationships, we "corrected the data for exercise in clean air," which means that we attempted to remove any systematic bias in the data that might be attributable to an exercise effect. These data become the "observations" shown in Figure 4 (step 1). In step 2, a function is fit to the data via regression techniques. Specifically, the numbers of subjects responding at 0.08, 0.10, and 0.12 ppm are 5 of 60, 5 of 32, and 6 of 30, respectively, which leads to the response rates shown in Figure 4. This step is necessary to estimate response rates at ozone concentrations that differ from those at which laboratory data are available. Step 3 develops, for example, the 90% credible interval (CI) about the fitted (predicted) median response rate at ozone concentrations needed for the risk assessment calculations (i.e., those used in pNEM/O_3). This last step is accomplished by applying the inverse beta function with parameters X and $N - X$, where X is the predicted median response rate at a particular ozone concentration, and N is the number of subjects associated with the chosen ozone concentration.

The 90% CI is defined by the 0.05 and 0.95 fractiles. For this risk assessment, response rates are calculated for 21 fractiles (for cumulative probabilities from 0.05 to 0.95 in steps of 0.05, plus probabilities of 0.01 and 0.99) at a number of ozone concentrations that depend on the health endpoint. The functions chosen "best fit" the data according to the following principles and rules:

- *Linear functions were favored, especially when the number of observation points (i.e., ozone concentrations at which laboratory data are available) was small.* As few as two usable observation points and as many as six observation points were available for the 33 endpoints mentioned earlier.

- *Functions with high regression r^2 values were more desirable than functions with low r^2 values.* This principle allowed choosing a nonlinear function over a linear function — even if the number of observation points was small — if the r^2 value of the nonlinear function was considerably larger than that for the linear function.

- *All functions for each of the fractiles must be monotonic increasing (i.e., they must never decrease) as ozone concentration increases.* This factor is a logical *rule*, and it came into play when the number of subjects varied considerably at different ozone concentrations. Such a condition made it necessary to use an average number of subjects at all ozone concentrations.

A linear function with slope 2.925 and intercept -0.1462 best fits the data subject to the above principles for FEV_1 decrement $\geq$20% and 8-h exposures. This function defines the fractional* median exposure-response relationship. Because the linear function intercepts the X-axis at 0.05 ppm and response rates cannot be negative, the response rate is 0 at ozone concentrations $\leq$0.05 ppm. Thus, the median exposure-response relationship used in this risk assessment may be thought of as having a "hockey stick" shape with an "inflection point" at 0.05 ppm and zero response for ozone concentrations $\leq$0.05 ppm. In Figure 4, note that the curves for fractiles other than the median are not linear. This is attributable to the construction of credible intervals at the ozone concentrations needed for the subsequent risk calculations; the credible interval widths depend in part on the number of subjects at each ozone concentration, which, for this case, varies. Since 60 subjects were subjected to exposures at 0.08 ppm, the credible interval is smaller than at, for example, 0.10 ppm, at which 32 subjects were exposed.

* To obtain response rate in percent, multiply the fractional result by 100%.

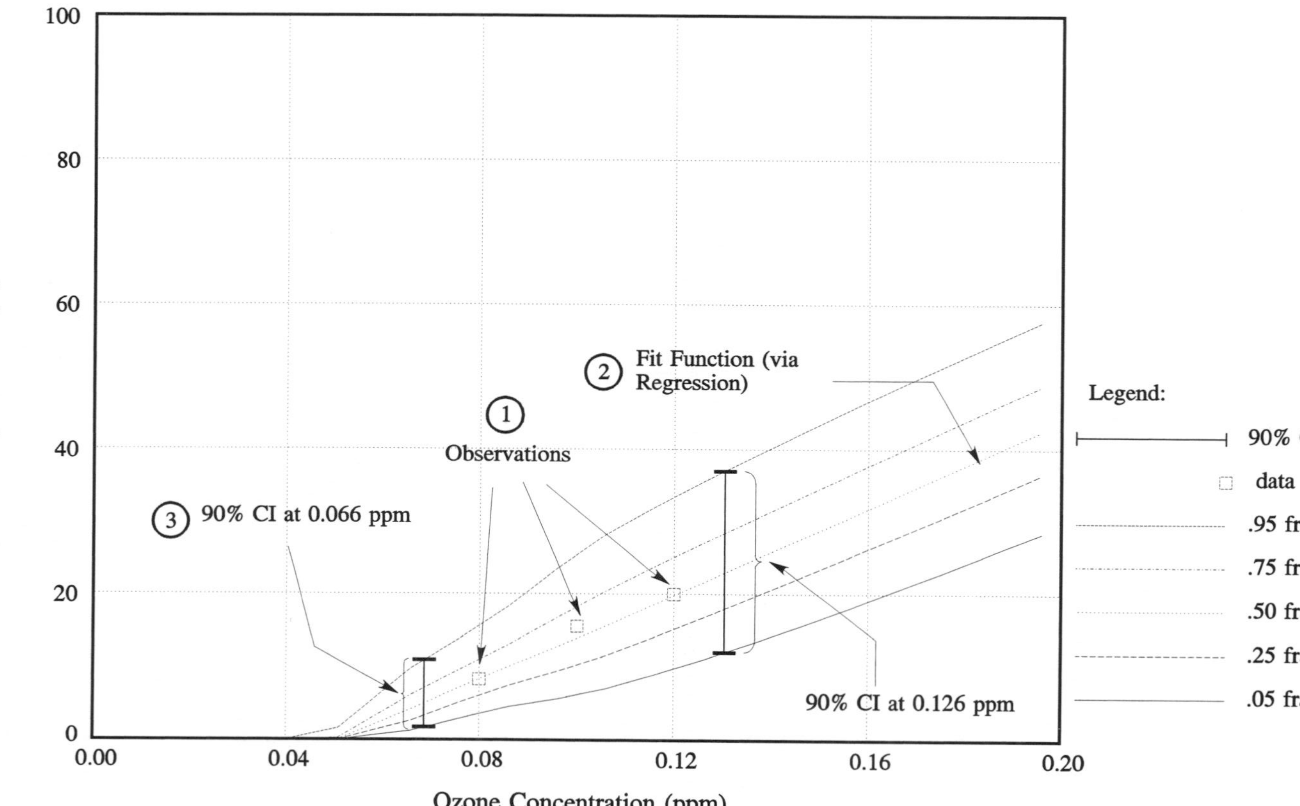

FIGURE 4 Steps Used to Develop Probabilistic Exposure-Response Relationships, FEV_1 Decrement ≥20%, 8-h Exposure, Moderate Exertion (Derived from: Folinsbee et al. 1988; Horstman et al. 1990; and McDonnell et al. 1991)

RISK ESTIMATES FOR 8-HOUR EXPOSURES TO OZONE

The risk computations (which comprise the risk model) are conceptually simple and based on exposure and health models. In general, the risk (which is an expected fractional response rate) for the *k*'th fractile R_k is

$$R_k = \sum_{j=1}^{N} P_j \times [(RR_k|e_j) - (RR_k|\text{background})] \quad , \tag{1}$$

or

$$\begin{aligned} R_k &= \sum_{j=1}^{N} P_j \times RR_k|e_j - \sum_{j=1}^{N} P_j \times (RR_k|\text{background}) \\ &= R_{k,1} + R_{k,2} \quad , \end{aligned} \tag{2}$$

where:

P_j = fraction of the population having personal exposures at an ozone concentration of e_j ppm in a single pNEM/O_3 run;

$RR_k|e_j$ = *k*-fractile response at ozone concentration e_j;

$RR_k|\text{background}$ = *k*-fractile response at background; and

N = number of ozone concentrations.

The *Pj* values used to calculate risk results are based on the assumption that the air quality scenario under study is just attained.

As discussed earlier, it was possible to correct exposure-response relationships for exercise in clean air. The $RR_k|e_j$ values reflect this correction for the FEV_1 decrements ≥20% endpoint for 8-h exposures at moderate exertion (based on study results from Folinsbee et al. 1988, Horstman et al. 1990, and McDonnell et al. 1991). Table 3 shows the risk computations for the 0.5-fractile (the median), for pNEM/O_3 run 2, exposures to outdoor children, based on air quality in Philadelphia that just attains the current standard (1124) for the persons measure. Background ozone is ≤0.04 ppm. The entries in column C are the fractions of the population engaged in moderate exertion who are exposed to the corresponding ozone concentrations in column B. The entries in column D are the expected fractional response rates at the corresponding ozone concentrations for the 0.5 fractile (i.e., median). The result is about 0.06, or 6%. Since $RR_{0.5}$ | background is exactly zero, this result is unchanged after correcting for background. The final step is to multiply the fractional response rate by the number of outdoor children who achieved moderate exertion for 8 h to obtain, in this case, the median estimate of the number of individuals that experience FEV_1 decrements $\geq 20\%$. For pNEM/O_3 run 2, nearly 270 thousand outdoor children achieved the heavy exertion level, so the median estimate is about 16 thousand outdoor children.

To develop a probability distribution over outdoor children that accounts for all of the conditions mentioned earlier, the above computations are repeated for any number of fractiles. Results usually appear "smoother" if a large number of fractiles are used. In this analysis, 21 fractiles are used to calculate risk results.

TABLE 3 Calculating the Median of a Risk Distribution

A	B	C	D	E
Index j	e_j (ppm)	P_j	$RR_{0.5} \mid e_j$	$RR_{0.5} = C \times D$
1	0.051	0.1604	0.000233	0.0000
2	0.066	0.1939	0.041839	0.0081
3	0.076	0.2530	0.071412	0.0181
4	0.086	0.1640	0.100205	0.0164
5	0.096	0.0727	0.128108	0.0093
6	0.106	0.0257	0.156623	0.0040
7	0.116	0	0.186317	0.0000
8	0.126	0	0.216072	0.0000
			Column E Sum:	0.0559

Headcount risk results have been "corrected for background," that is, 0.04 ppm for acute exposures. This correction, as explained earlier, "subtracts" the probability distribution over response at the background concentration from the probability distribution over response for ozone concentrations above background. As with many computations in this analysis, this subtraction assumes perfect correlation and allows corrections to be made by simply subtracting corresponding responses on a fractile-by-fractile basis. If no response occurs at a particular background concentration, the "uncorrected" and "corrected" results are identical.

Table 4 lists risk results for this endpoint when the 1124 standard is just attained in Philadelphia. Ten distributions are provided — 1 for each of the 10 available pNEM/O_3 runs. The first column lists cumulative probabilities for the distributions. Each row in the table lists the number of children having FEV_1 decrements ≥20% for each of 10 pNEM/O_3 runs. For example, the 0.50 cumulative probability (0.5 fractile) estimates range from about 13.5 thousand (run 6) to about 15 thousand (run 10) outdoor children. The mean, standard deviation, and total number of people (TotPop) are listed at the bottom of the table. For run 1, the mean is about 14 thousand, the standard deviation is about 7 thousand, and TotPop is about 270 thousand. Note that TotPop is based on the total number of outdoor children who reached a moderate level of exertion in each pNEM/O_3 run. This number varies from run to run.

TABLE 4 Probability Distributions over Number of Outdoor Children Having FEV_1 Decrements ≥20% During One Ozone Season, Philadelphia, 8-Hour Exposures, Moderate Exertion, Just Attaining Standard 1124[a]

Cumulative Probability	Number of Outdoor Children Having FEV_1 Decrements ≥20% by pNEM/O_3 Run									
	1	2	3	4	5	6	7	8	9	10
0.01	3,794	3,988	3,773	3,667	3,645	3,424	3,609	3,747	3,772	4,028
0.05	5,920	6,229	5,889	5,741	5,730	5,443	5,679	5,876	5,910	6,277
0.10	7,355	7,739	7,316	7,144	7,143	6,822	7,085	7,315	7,354	7,792
0.15	8,451	8,892	8,406	8,216	8,226	7,882	8,163	8,416	8,457	8,949
0.20	9,400	9,889	9,348	9,144	9,164	8,803	9,097	9,369	9,412	9,948
0.25	10,269	10,802	10,212	9,995	10,025	9,650	9,955	10,243	10,288	10,863
0.30	11,094	11,669	11,032	10,803	10,844	10,456	10,771	11,073	11,119	11,731
0.35	11,897	12,512	11,829	11,590	11,641	11,243	11,566	11,880	11,928	12,575
0.40	12,693	13,348	12,621	12,371	12,433	12,025	12,356	12,682	12,731	13,413
0.45	13,497	14,191	13,419	13,159	13,232	12,815	13,154	13,491	13,541	14,257
0.50	14,321	15,055	14,237	13,967	14,052	13,626	13,972	14,320	14,371	15,122
0.55	15,177	15,953	15,088	14,808	14,905	14,471	14,825	15,183	15,235	16,021
0.60	16,083	16,901	15,988	15,697	15,808	15,366	15,728	16,095	16,147	16,972
0.65	17,058	17,921	16,955	16,655	16,781	16,330	16,701	17,077	17,130	17,994
0.70	18,129	19,041	18,019	17,707	17,850	17,391	17,771	18,157	18,209	19,116
0.75	19,336	20,303	19,218	18,894	19,056	18,588	18,981	19,375	19,426	20,381
0.80	20,746	21,775	20,618	20,282	20,465	19,988	20,395	20,798	20,846	21,858
0.85	22,478	23,581	22,338	21,987	22,198	21,710	22,137	22,546	22,591	23,671
0.90	24,794	25,993	24,639	24,270	24,518	24,015	24,473	24,887	24,924	26,094
0.95	28,509	29,854	28,329	27,936	28,243	27,719	28,231	28,644	28,664	29,977
0.99	36,354	37,986	36,123	35,695	36,123	35,558	36,206	36,587	36,562	38,166
Mean	15,393	16,161	15,301	15,031	15,139	14,721	15,077	15,409	15,451	16,235
StdDev[b]	7,040	7,358	6,993	6,920	7,022	6,953	7,039	7,098	7,092	7,383
TotPop[c]	267,352	268,923	265,077	267,324	266,275	268,386	268,292	269,471	269,538	268,874

[a] Based on Folinsbee et al. 1988, Horstman et al. 1990, and McDonnell et al. 1991.

[b] StdDev means standard deviation.

[c] TotPop means total population (i.e., children).

Figure 5 is a plot of the data listed in Table 4 for just attaining standard 1124. It also provides results for scenario 8284. Note that the risks are lower when standard 8284 is attained than when the current standard is attained (indicated by the fact that all distributions for the 8284 standard are closer to the Y axis than all distributions for the 1124 standard).

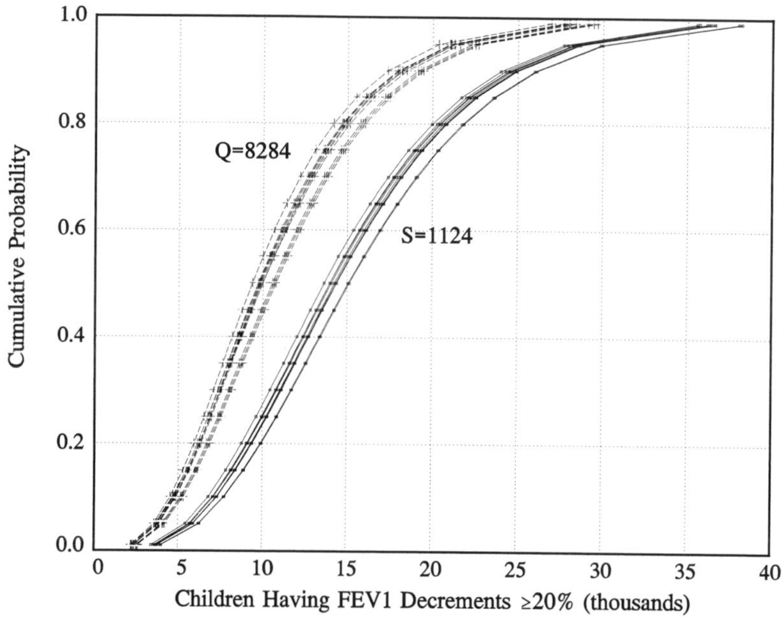

FIGURE 5 Comparison of 2 Sets of 10 Risk Distributions (just attaining scenarios 1124 and 8784, FEV_1 decrements ≥20%, Philadelphia, outdoor children, 8-h exposure, moderate exertion level; based on Folinsbee et al. 1988, Horstman et al. 1990, McDonnell et al. 1991)

On this scale, it is not usually helpful to plot the distributions for the other scenarios because, for many urban areas, many of them are nearly identical and, if they were included, the figure would be quite cluttered. This is especially true for Denver and Miami. Figure 6 provides an indication of the range of results within each set of 10 distributions. Six plots are shown, one for each air quality scenario. Each plot is "representative" of the 10 distributions for a particular scenario. Because only 6 plots are shown rather than 70, it is easier to see patterns. Each of these plots is a valid cumulative probability distribution.* For Philadelphia,

---

* The representative distribution is obtained by computing the average cumulative probability at selected points along the X-axis. This calculation, like the risk calculations described earlier, implicitly assumes that the distributions are perfectly correlated. It may be argued that perfect correlation, while not correct, is more reasonable that perfect independence, and no basis exists for choosing any other degree of correlation between these 2 extremes.

the following insights can be gained about the effects of attaining each standard by examining Figure 6:

- Attainment of any alternative standard provides more protection than that afforded by attainment of the current standard (because the representative risk distribution for 1124 lies to the right of the distributions for all other standards).
- Standard 8284 provides the greatest protection (i.e., results in the lowest risk).
- The level of protection associated with standard 8294 is less than that associated with standard 8584 and greater than that associated with standards 8394 and 1124 (the current standard).

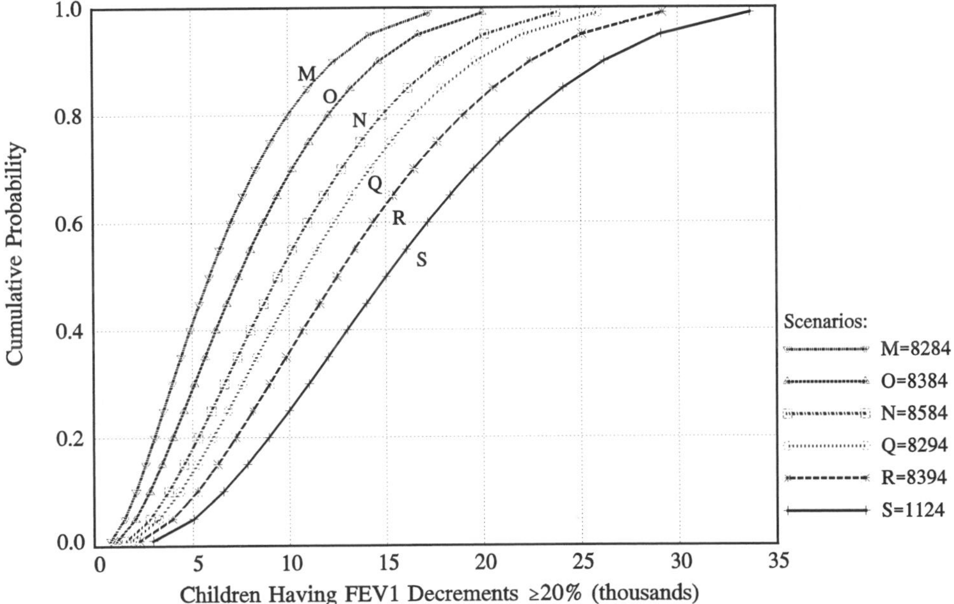

FIGURE 6 Representative Distributions for 6 Sets of 10 Risk Distributions (just attaining 6 standards, FEV_1 decrements ≥20%, Philadelphia, outdoor children, 8-h exposure, moderate exertion; based on Folinsbee et al. 1988, Horstman et al. 1990, and McDonnell et al. 1991)

These observations about standard 8294 are specific to Philadelphia do not apply to all other urban areas. Standard 8294 is less protective than the current standard (1124) for three other urban areas (Chicago, Houston, Los Angeles, and New York City). This finding is illustrated in Figure 7, which shows risk results for 9 urban areas for which exposure estimates have been developed using pNEM/O_3. The figure consists of 54 sets of connected rectangles, which are variations of the Tukey box plot described in Morgan and Henrion (1990), that indicate uncertainty in the risk distributions. For example, there are three rectangles above letter S (the code letter for standard 1124; letter codes for all standards considered here are listed in Table 5) for each urban area. The top rectangle represents the range of the 0.95 fractiles; the middle rectangle represents the range of the medians; and the bottom rectangle represents the range of the 0.05 fractiles. A line connects the bottom of the 0.95-fractile rectangle and the top of the 0.05-fractile rectangle and passes through the 0.5-fractile rectangle.

When the risk distributions for a particular air quality scenario are quite "similar," the rectangles are small. When the variances of a set of risk distributions are small, the rectangles are close together. When the distributions are spaced far enough apart (indicative of widely varying risk estimates for different pNEM/O_3 runs), the rectangles overlap. There are no overlapping cases in Figure 7.

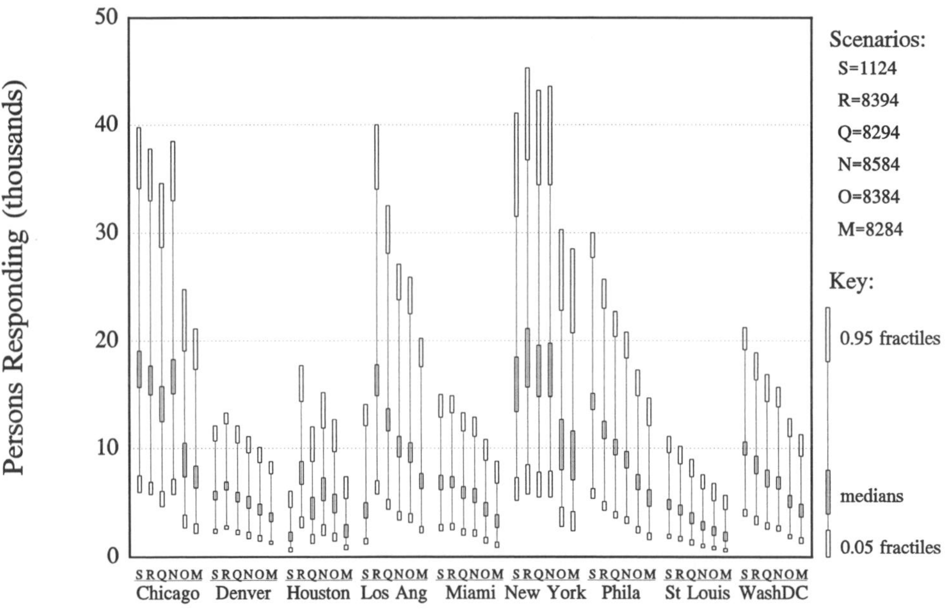

FIGURE 7 Using Box Plots to Represent Variability in 54 Sets of 10 Risk Distributions Attributable to Differences in pNEM/O_3 Exposure Estimates (health endpoint: 8-h exposures at moderate exertion, number of children having FEV_1 decrements ≥20%)

Figure 7 indicates that the current standard (1124) is most protective of children in Houston and Los Angeles; least protective in Philadelphia, St. Louis, and Washington, D.C. Scenario 8284 is most protective in all areas except Houston and Los Angeles. For areas other than Houston and Los Angeles, scenario 8284 results in a reduction of about 50% in the number of children responding compared to the current standard.

Another way to view the same data is shown in Figure 8, in which the measure is percentage of children responding rather than children responding. Under attainment of the current standard, the smallest percentages of children responding result in Houston and Los Angeles, with median response rates less than 1% in both urban areas; for the other areas, the median response rates are about 4-6%. Under scenario 8284, the median response rates are about 1% in Houston and Los Angeles and about 1-4% in the other areas. So, while the current standard is best for Houston and Los Angeles, scenario 8284 results in a level of protection that is more comparable across all nine urban areas. Thus, from a risk equity point of view, it may be argued that scenario 8284 is preferable to the current standard.

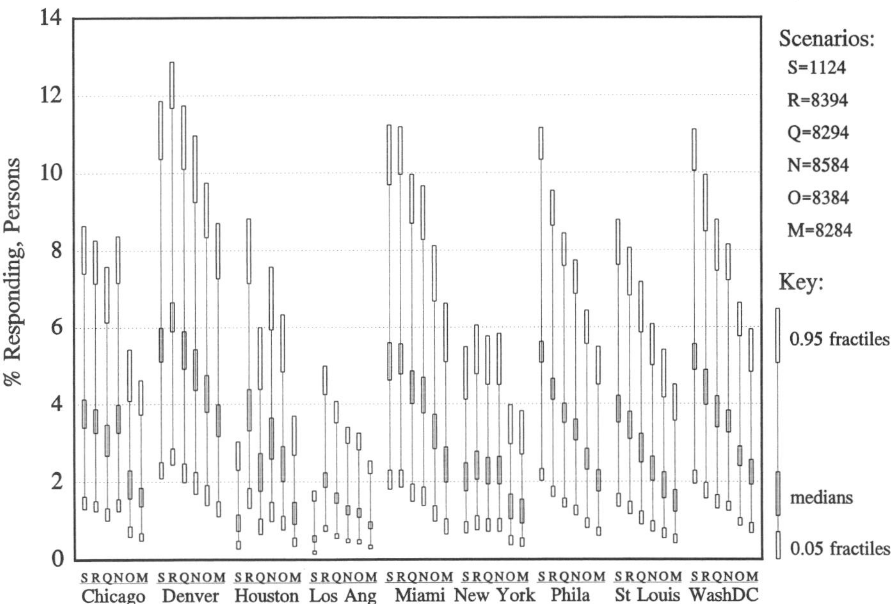

FIGURE 8 Headcount Risk Results for Percentage of Children Responding for FEV_1 Decrement ≥20%, 8-h Exposures, and Moderate Exertion (persons basis; underlying exposure-response relationship based on Folinsbee et al. 1988, Horstman et al. 1990, and McDonnell et al. 1983)

TABLE 5 Summary of Air Quality Scenarios

Letter Code	Abbreviations: Short	Abbreviations: Long	Definition of Scenario
M	8284	8HA2H-0.084	an alternative NAAQS: 8-hr averaging time, 2nd highest average daily maximum of 0.084 ppm ozone
N	8584	8HA5H-0.084	an alternative NAAQS: 8-hr averaging time, 5th highest average daily maximum of 0.084 ppm ozone
O	8384	8HA3H-0.084	an alternative NAAQS: 8-hr averaging time, 3rd highest average daily maximum of 0.084 ppm ozone
Q	8294	8HA2H-0.094	an alternative NAAQS: 8-hr averaging time, 2nd highest average daily maximum of 0.094 ppm ozone
R	8394	8HA3H-0.094	an alternative NAAQS: 8-hr averaging time, 3rd highest average daily maximum of 0.094 ppm ozone
S	1124	1H1EX-0.124	a variation of the current ozone NAAQS: 1-hr averaging time, 1 expected exceedance, 0.124 ppm ozone

EXCESS ADMISSIONS OF ASTHMATICS IN NEW YORK CITY

The hospital admissions model is based on (1) regression coefficients and corresponding standard errors developed by Thurston et al. (1992) and (2) 1-h daily maximum ozone concentrations developed by Johnson et al. (1997). The model applies only to New York City and includes two types of respiratory admissions: asthmatics and members of the general population (including asthmatics) for any of a number of respiratory ailments (i.e., acute bronchitis or bronchiolitis, pneumonia, chronic obstructive pulmonary disease not related to asthma).

Regression coefficients and corresponding standard errors (for asthmatics, the regression coefficient and standard error are 11.7 and 4.7, respectively) define "concentration-response" relationships that include related uncertainties. Figure 9 is a graph of the relationship (which is a set of 21 "curves," one for each of 21 fractiles) for asthmatics. Only the 0.05, 0.50 (median), and 0.95 fractiles are shown to avoid clutter. Although the concentration-response relationship is defined over the range of 0-0.04 ppm ozone, the risk calculations in this section pertain, unless otherwise stated, to ozone levels greater than the estimated background (0.04 ppm). The fractiles at each ozone concentration are obtained from the normal probability distribution.

The main results are shown in Figure 10 and Tables 6-7. Figure 10 shows cumulative probability distributions over excess annual hospital admissions of New York City asthmatics for seven air quality scenarios (based on air quality data from Queens County monitor with background ozone at 0.04 ppm). For example, the 0.05, 0.5 (median), and 0.95 fractiles of the distribution for scenario R (8-h averaging time, 3rd highest daily maximum value, allowed ozone level of 0.094 ppm) are 50, 150, and 250, respectively. In the figure, the distributions for scenarios Q and N appear to be identical. In fact, this is the case as can be seen in Table 6, which lists the data for Figure 10. Each distribution is defined in terms of number of admissions at each of 21 fractiles. Similar results (not shown here) are available for background ozone of 0 ppm. As expected, the numbers of admissions are higher for the 0 ppm case compared to the 0.04 ppm case.

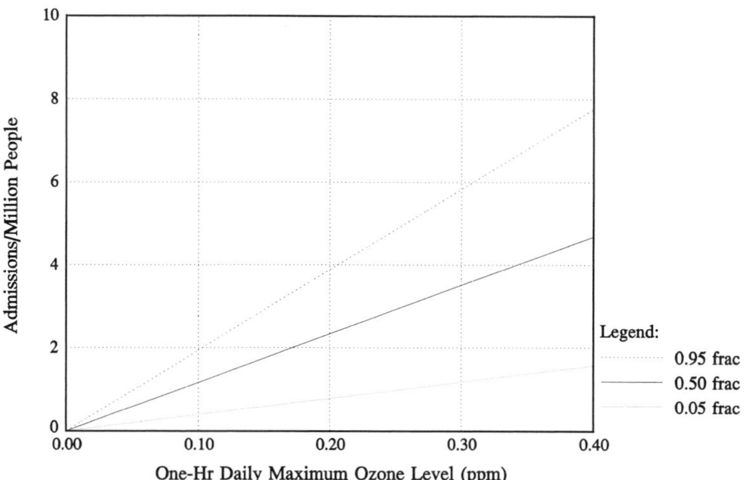

FIGURE 9 Uncertainty about Daily Hospital Admissions of Asthmatics in Relation to Ozone Concentration

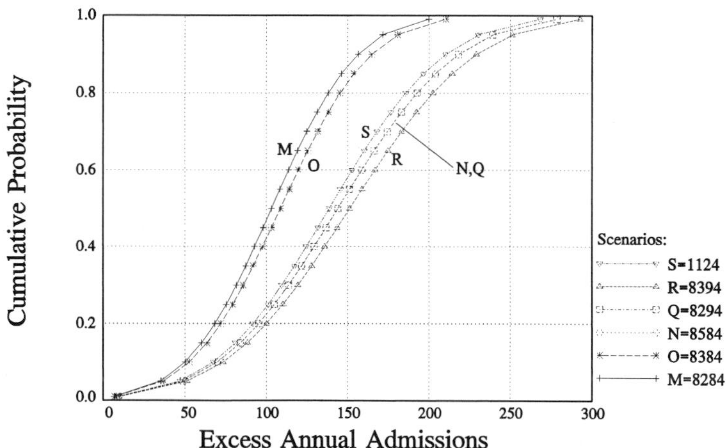

FIGURE 10 Excess Annual Hospitalizations of New York City Asthmatics for Seven Air Quality Scenarios and Monitor 9 Air Quality Data with Background of 0.04 ppm

TABLE 6 Annual Excess Hospital Admissions of Asthmatics Based on the Queens County Monitor Data with a Background Ozone Level of 0.04 ppm (1-h daily maximum ozone levels)

	Air Quality Scenario					
Fractile	S=1124[a]	R=8394[b]	Q=8294	N=8584	O=8384	M=8284
0.01	9	10	9	9	7	7
0.05	47	51	49	49	37	35
0.10	67	73	70	70	53	50
0.15	81	88	84	84	63	60
0.20	92	100	95	95	72	68
0.25	101	110	105	105	79	75
0.30	109	119	114	114	86	82
0.35	117	128	122	122	92	87
0.40	125	136	129	129	98	93
0.45	132	144	137	137	103	98
0.50	139	151	144	144	109	103
0.55	146	159	151	151	114	108
0.60	153	167	159	159	120	114
0.65	160	175	166	166	126	119
0.70	168	183	174	174	132	125
0.75	176	192	183	183	138	131
0.80	186	202	193	193	145	138
0.85	196	214	204	204	154	146
0.90	210	229	218	218	165	156
0.95	230	251	239	239	181	171
0.99	268	293	279	278	210	200
Mean	139	151	144	144	109	103
StdDev	56	61	58	58	44	42

[a] Key to code for scenario S: character 1 specifies averaging time (1 h); character 2 specifies the number of exceedances (1); characters 3 and 4 define the ozone level (0.124 ppm)

[b] Key to code for scenario R and the remaining scenarios: character 1 specifies averaging time; character 2 specifies n of the nth highest 8-h average daily maximum ozone level; and characters 3 and 4 define the allowed ozone level (e.g., 8394 means 8-h averaging time, 3rd highest 8-h average daily maximum ozone level, and 0.094 ppm).

It is important to discuss precision of the estimates listed in Table 6. The estimates are model results that have been rounded to the nearest whole number. This level of precision is presented because further calculations in Table 7 could be distorted if results in Table 6 were rounded to the nearest multiple of 10 or to 2 significant digits. The final calculations in Table 7 involve percentage reductions from the current standard. These estimates have been rounded to either 2 significant whole number digits or 1 significant decimal digit to emphasize the fact that these distributions cannot be known precisely. Besides, a more important indicator of the precision of estimates is the width of the 90% CI associated with each distribution, which in turn relates to variance; without exception, the CIs are large and further emphasize the uncertainty associated with these estimates.

Table 7 compares meeting the current standard (1124) vs. meeting the other standards. The excess admissions estimates come from the hospital admissions model. The estimates for all admissions are based on (1) the 14-16 thousand admissions per ozone season estimates provided by Thurston (1995) and (2) excess admissions attributable to exposures at ozone levels >0.04 ppm. For example, for the current standard (1-h, 1 expected exceedance, 0.124 ppm), the median number of admissions of asthmatics for any respiratory-related reason is 14,800 [which is approximately equal to 15,000 (the median of total admissions in the As-Is scenario) *minus* 388 (the median number of excess admissions in the As-Is scenario attributable to ozone exposure at concentrations greater than the estimated 0.04 ppm background level; from Whitfield et al. 1996) *plus* 139 (the median number of excess admissions associated with attainment of the current standard [scenario 1124] for ozone exposures at concentrations > 0.04 ppm).

As expected, as the base for comparison increases (from excess admissions with background at 0.04 ppm, to excess admissions with background at 0 ppm, to admissions for any respiratory-related reason), the percentage reduction relative to admissions under the 1124 (current standard) scenario decreases substantially. For example, excess admissions of asthmatics attributable to exposures to ozone levels >0.04 ppm decrease from about 140 for scenario 1124 to about 100 for scenario 8384 - a decrease of about 40 admissions or 20%. Considering exposures to any level of ozone, the decrease for the same scenarios is about 50 admissions or 8%. For admissions for any respiratory-related reason, the decrease is about 40 admissions or 0.2%.

One important conclusion that can be drawn from the results in Tables 6-7 is that the scenarios 1124 (the current standard), 8294, and 8584 are quite similar with respect to estimates of excess hospital admissions. In fact, scenarios 8294 and 8584 lead to identical results that are larger than those for the current standard, and estimates for scenario 8394 are slightly higher than those for scenarios 8294 and 8584. Scenarios 8384 and 8284 lead to admissions estimates that are lower than those for the current standard.

TABLE 7 Admissions of New York City Asthmatics — Comparison Relative to Meeting the Current Standard (1-h, 1 expected exceedance, 0.124 ppm)

Case No./Issue	Air Quality Scenario					
	S = 1124 (1 h, 1 ex,[a] 0.124 ppm)	R = 8394 (8 h, A3H, 0.094 ppm)	Q = 8294 (8 h, A2H, 0.094 ppm)	N = 8584 (8 h, A5H, 0.084 ppm)	O = 8384 (8 h, A3H, 0.084 ppm)	M = 8284 (8 h, A2H, 0.084 ppm)
1. Excess admissions[b] (background = 0.04 ppm)	139[c] (47 to 230)[d]	151 (51 to 251)	144 (49 to 239)	144 (49 to 239)	109 (37 to 181)	103 (35 to 171)
Percent reduction from current standard[e]	–	-9	-4	-4	22	26
2. Excess admissions[b] (background = 0 ppm)	697 (236 to 1,157)	720 (236 to 1,157)	707 (236 to 1,157)	707 (236 to 1,157)	643 (236 to 1,157)	631 (236 to 1,157)
Percent reduction from current standard[e]	–	-3	-1	-1	8	9
3. All admissions[f] (thousands)	14.8 (13.9 to 15.6)	14.8 (13.9 to 15.6)	14.8 (13.9 to 15.6)	14.8 (13.9 to 15.6)	14.7 (13.9 to 15.5)	14.7 (13.9 to 15.5)
Percent reduction from current standard[e]	–	-0.08	-0.03	-0.03	0.2	0.2

[a] ex means expected exceedance.
[b] Admissions of asthmatics because of exposure to ozone.
[c] Median estimate.
[d] 90% credible interval (about the median).
[e] Because of the necessary assumption that results across scenarios are highly correlated (i.e., if admissions are high for one scenario, they are high for all scenarios), very little variation occurs in the percentage reduction from the current standard. Decreases are positive and increases are negative numbers.
[f] Admissions of asthmatics for any respiratory-related reason; for scenario *i*, based on estimates of all admissions and excess admissions attributable to ozone levels >0.04 ppm for the As-Is scenario, and estimates of excess admissions attributable to ozone levels >0.04 ppm for scenario *i* (e.g., for scenario 1124: 14,800 ≈ 15,000 − 388 + 139). Fifteen thousand is the number of admissions of New York City asthmatics for any respiratory-related reason during the 1988-1990 ozone seasons (Thurston 1995); 388 is the median number of excess admissions associated with the As-Is scenario (Whitfield et al. 1996); and 139 is the median number of excess admissions for scenario S with a background ozone level of 0.04 ppm. The corresponding 0.05- and 0.95-fractile estimates for excess admissions associated with the As-Is scenario, which are needed to calculate the 0.05- and 0.95-fractile estimates for the third case (all admissions), are 132 and 644, respectively (Whitfield et al. 1996).

LIMITATIONS OF THE ANALYSIS

Reports by Johnson et al. (1996) and Whitfield et al. (1996) present results of earlier exposure and risk assessments, respectively. The earlier ozone risk assessment was similar to the one reported here in that both assessments used results from the pNEM/O_3 exposure model to estimate ozone exposures. Differences between the two risk assessments are primarily due to differences in the exposure assessments: the standards that were evaluated, the procedures used to adjust ozone data, and the treatment of run-to-run variability.

Johnson et al. (1996) provide a comprehensive discussion of the principal limitations of the pNEM/O_3 methodology, which in turn necessarily impact the risk results. These limitations include:

- The availability and level of detail of input data bases (e.g., time and activity diary data);
- The necessarily broad assumptions used in estimating cohort populations; and
- The lack of available data bases integral to the mass balance model used by pNEM (e.g., air exchange rates).

and these limitations also apply to the new pNEM/O_3 exposure estimates used here to obtain risk results.

The new air quality adjustment procedures (AQAPs) for estimating ozone levels and exposures used in these assessments differ from the AQAPs used in the previous application of pNEM/O_3 to outdoor children. Each of the new AQAPs was applied to a three-year period to simulate attainment of the associated standard over the entire period. The adjustment required for the three-year period was applied to a single year (1990, 1991, 1992, or 1993) within the period and the resulting single year of ozone data was used in the exposure assessment to represent typical attainment conditions. The selected year is the middle year of the three-year period with respect to relative ozone level. This approach contrasts with the previous AQAPs in which attainment was determined by examining a single year (1990 or 1991) of ozone data. The ozone data for the selected year were adjusted to exactly attain the specified standard and then used in the exposure assessment to represent attaining conditions.

The new AQAPs use a proportional adjustment procedure for all standards in all cities. In the previous AQAPs, a non-proportional adjustment procedure based on the Weibull distribution was used for all standards in six of the nine urban areas and for a few standards in the remaining urban areas (Chicago, Denver, and Miami). A proportional adjustment procedure was used for the other standards in these three urban areas.

The new AQAPs reduce hourly ozone concentrations in each portion of the distribution by the same proportion. It should be noted that while these "proportional" AQAPs may provide a better representation of the central part of the distribution, they may over-estimate concentration values in the upper tail of the distribution.

Of concern is the effectiveness of any AQAP for estimating ozone concentrations in cities, such as Los Angeles, that have "as-is" ozone levels that are much higher than those in most U.S. cities. Most cities evaluated in this report require "moderate" reductions to attain the various standards under condition. Historical records of high- and low-year ozone level patterns make it possible to reasonably calibrate results. However, because current conditions in Los Angeles must be radically adjusted downward to simulate attainment of all of the standards, and because Los Angeles has never approached any of the specified conditions, there are no empirical data available that could be realistically used to calibrate a proposed AQAP.

Finally, with respect to exposure modeling, is the issue of excessive random influences on exposure estimates. In the current work, Johnson et al. (1997) used a special version of pNEM/O_3 that allows the user to specify the random number generator seed for each model run. Each seed produces a unique, repeatable sequence of random numbers for the probabilistic elements of pNEM/O_3. Thus, Johnson et al. (1997) constrained each set of 10 pNEM/O_3 runs to use the same 10 random number generator seeds. This procedure retained desirable variability within each set of 10 runs and removed undesirable set-to-set variability. As a result, it seems plausible that the differences in the exposure estimates associated with the seven standards are mainly attributable to differences in ambient ozone concentrations permitted by the standards. Consequently, this variability propagates through the risk calculations and leads to a similar conclusion.

With respect to risk results, Whitfield et al. (1996) discussed a number of limitations of exposure-response modeling and risk modeling that remain applicable. These include the following issues:

- *Length of exposure.* Data from controlled human exposure studies were assumed to be appropriate for modeling responses of 1-h exposures at heavy exertion, 1-h exposures at moderate exertion, and 8-h exposures at moderate exertion. Exposure protocols differed among the studies and did not match exactly the conditions of the regulatory standards analyzed here, but it is unlikely that these differences would appreciably affect the results described here.

- *Extrapolation of exposure-response relationships.* The lowest ozone level at which controlled human exposure studies have been conducted is 0.08 ppm. Because large numbers of people are exposed to ozone concentrations between 0.04-0.08 ppm, it was necessary to estimate exposure-response relationships at ozone concentrations much lower than those for which data are available. The accuracy of these estimates is unknown.

- *Reproducibility of ozone-induced responses.* This study assumed that ozone-induced responses for individuals are reproducible. This assumption is supported by the criteria document, which cites several confirming studies.

- *Age and lung function.* None of the studies used to develop the exposure-response used in this study exposed children. However, a number of studies indicate that children aged 8-11 experience FEV_1 changes similar to those of adults aged 18-35 and exposed to ozone concentrations of 0.12 ppm at equivalent ventilation rates of 35 L/min/m^2.

- *Age and symptoms.* None of the studies involving children (primarily summer camp studies) reported symptoms. Therefore, it was necessary to use the same exposure-response relationships involving symptoms for adults and children.

- *Interaction between ozone and other pollutants.* All of the controlled human exposure studies, which formed the basis for the exposure-response relationships used here, controlled only for ozone. Although there is some evidence that the presence of other pollutants might enhance the respiratory effects of ozone (or even cause the same effects), it is not consistent across studies. Therefore, it was assumed that the estimates of ozone-induced health effects would not be affected by the presence of other pollutants such as sulfur dioxide, nitrogen dioxide, carbon monoxide, sulfuric acid, or other aerosols.

- *Smoking status.* All of the subjects in the controlled human exposure studies used here excluded smokers. There is some evidence that smokers may be less responsive to ozone exposures than nonsmokers. To the extent that is true, the risk estimates in this report are overstated.

- *Exposure history.* It is assumed that ozone-induced response at any particular exposure period (1 or 8 h) is not affected by previous exposures. It is possible that ozone-induced responses can be enhanced or attenuated by previous exposures. The absence of data concerning this issue increases uncertainty about results.

- *Naturally occurring ozone.* Risk results do not include exposure to background ozone levels (i.e., ozone concentrations that would be observed in the U.S. in the absence of anthropogenic precursor emissions of volatile organic compounds and nitrous oxides in North America). Responses attributable to exposure to naturally occurring ozone levels were removed from the risk estimates. The criteria document estimates that the summer, 1-h average range for background ozone is 0.03-0.05 ppm. The midpoint, 0.04 ppm, is used in this report.

- *Correction for background ozone.* The procedure for removing responses attributable to background ozone assumed that distributions over total response and response at background are perfectly correlated (i.e., correlation is 1).

- *Output graphs.* The box plots render indistinguishable characteristics of results for individual pNEM/O_3 runs. Therefore, the box plots should be used as a guide for identifying possible trends and developing insights that can by verified only by investigating detailed results.

ACKNOWLEDGMENTS

The authors gratefully acknowledge the efforts of colleagues who were instrumental in preparing this paper: James Capel, Ted Palma, and Jill Mozier of IT Air Quality Services, Inc. This work was funded through Interagency Agreement DW89935085-01-3 between the U.S. Department of Energy and the U.S. Environmental Protection Agency and (IT contract). Any opinions, findings, conclusions, or recommendations are those of the authors and do not necessarily reflect the views of the U.S. Environmental Protection Agency, the U.S. Department of Energy, IT Air Quality Services, Inc., or Argonne National Laboratory.

REFERENCES

Avol, E.L., et al., 1984, "Comparative Respiratory Effects of Ozone and Ambient Oxidant Pollution Exposure during Heavy Exercise," *Journal of the Air Pollution Control Association* 34:804-809.

EPA: see U.S. Environmental Protection Agency.

Folinsbee, L.J., et al., 1988, "Pulmonary Function and Symptom Responses after 6.6-Hour Exposure to 0.12 ppm Ozone with Moderate Exercise," *JAPCA* 38:28-35.

Horstman, D.H., et al., 1990, "Ozone Concentration and Pulmonary Response Relationships for 6.6-Hour Exposures with Five Hours of Moderate Exercise to 0.08, 0.10, and 0.12 ppm," *American Review of Respiratory Disease* 142:1158-1163.

Johnson, T., et al., 1996, *Estimation of Ozone Exposures Experienced by Outdoor Children in Nine Urban Areas Using a Probabilistic Version of NEM*, prepared by IT/Air Quality Services for U.S. Environmental Protection Agency, Office of Air Quality Planning and Standards, Research Triangle Park, N.C. (April). (For copies, contact H.M. Richmond, U.S. Environmental Protection Agency, Office of Air Quality Planning and Standards, MD-15, Research Triangle Park, N.C. 27711; phone 919-541-5271.)

Johnson, T., J. Mozier, and J. Capel, 1997, "Supplement to 'Estimation of Ozone Exposures Experienced by Outdoor Children in Nine Urban Areas Using a Probabilistic Version of NEM'", prepared by IT/Air Quality Services for U.S. Environmental Protection Agency, Office of Air Quality Planning and Standards, Research Triangle Park, N.C. (Jan.). (For copies, contact H.M. Richmond, U.S. Environmental Protection Agency, Office of Air Quality Planning and Standards, MD-15, Research Triangle Park, N.C. 27711; phone 919-541-5271.)

Kulle, T.J., et al., 1985, "Ozone Response Relationships in Healthy Nonsmokers," *American Review of Respiratory Disease* 132:36-41.

McDonnell, W.F., et al., 1983, "Pulmonary Effects of Ozone Exposure during Exercise: Dose-Response Characteristics," *Journal of Applied Physiology: Respiratory Environmental Exercise Physiology* 54:1345-1352.

McDonnell, W.F., et al., 1985, "Respiratory Responses of Vigorously Exercising Children to 0.12 ppm Ozone Exposure," *American Review of Respiratory Disease* 132:875-879.

McDonnell, W.F., et al., 1991, "Respiratory Response of Humans Exposed to Low Levels of Ozone for 6.6 Hours," *Archives of Environmental Health* 46:145-150.

Morgan, M.G., and M. Henrion, 1990, *Uncertainty: A Guide to Dealing with Uncertainty in Quantitative Risk and Policy Analysis,* Cambridge University Press, New York, N.Y.

Seal, E., Jr., et al., 1993, "The Pulmonary Response of White and Black Adults to Six Concentrations of Ozone," *American Review of Respiratory Disease* 147:804-810.

Thurston, G.D., et al., 1992, "A Multi-Year Study of Air Pollution and Respiratory Hospital Admissions in Three New York State Metropolitan Areas: Results for 1988 and 1989 Summers," *Journal of Exposure Analysis and Environmental Epidemiology* 2:429-450.

Thurston, G.D., 1995, New York University Medical Center, personal communication (Dec. 5)

U.S. Environmental Protection Agency, 1996a, *Review of National Ambient Air Quality Standards for Ozone: Assessment of Scientific and Technical Information*, draft Office of Air Quality Planning and Standards (OAQPS) Staff Paper, EPA-450/R-96-007, June. (For copies, contact David McKee, U.S. Environmental Protection Agency, OAQPS, MD-15, Research Triangle Park,

N.C. 27711; phone 919-541-5288.)

U.S. Environmental Protection Agency, 1996b, *Air Quality Criteria for Ozone and Related Photochemical Oxidants (Review Draft)*, EPA/600/P-93/004 af, bf, cf, National Center for Environmental Assessment, Office of Research and Development, Research Triangle Park, N.C.

Whitfield, R.G., et al., 1996, "A Probabilistic Assessment of Health Risks Associated with Short-Term Exposure to Tropospheric Ozone," Argonne National Laboratory Report No. ANL/DIS-3 (June).

Whitfield, R.G., 1997, *A Probabilistic Assessment of Health Risks Associated with Short-Term Exposure to Tropospheric Ozone: A Supplement*, Letter report prepared by Argonne National Laboratory for the U.S. Environmental Protection Agency, Office of Air Quality Planning and Standards covering analysis of seven new air quality scenarios (Jan.). (For copies, contact H.M. Richmond, U.S. Environmental Protection Agency, Office of Air Quality Planning and Standards, MD-15, Research Triangle Park, N.C. 27711; phone 919-541-5271.)

California's Hydrocarbon Reactivity Experience

R. J. Pasek

California Air Resources Board, Research Division, P.O. Box 2815, Sacramento, California 95812-2815, USA

1. INTRODUCTION

Despite the progress made over the past 30 years, California still has some of the dirtiest air in the United States. The enjoyable climate of constant sunshine and warm temperatures ensures that many people choose to live in the state. The population of California has grown at an annual rate about twice that of the other 49 states, and this trend is expected to continue. Additionally, the unique topography of large basins surrounded by mountains and the presence of meteorological conditions conducive to formation of emissions-trapping inversion layers creates ideal conditions for high pollutant concentrations. However, notwithstanding the large increases in mobile and stationary sources, California's air quality has shown steady improvements since the early 1970's due to hydrocarbon (HC) and nitrogen oxides (NO_x) controls. Figure 1 shows the reduction in the annual maximum hour ozone concentrations in the Los Angeles area since 1965.

This success of California's policy of controlling HC and NO_x has been tied to aggressive and innovative control strategies based on standard command and control designs. Recently, a new paradigm has emerged that recognizes that more than one control method may exist to gain the required benefits, and that flexibility in control strategies should be emphasized. Hydrocarbon reactivity, or ozone production potential, is one such concept that has received and is receiving increased attention for the control of ozone. In this paper, reactivity and California's experience in implementing the concept into control strategies, plans, and regulations is explored.

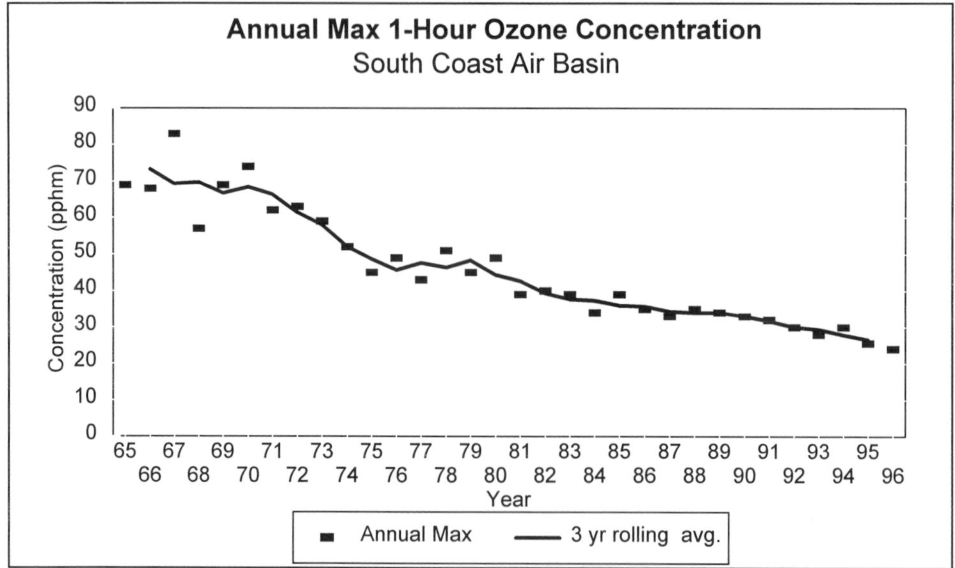

Figure 1. Trend of annual maximum ozone concentrations in South Coast Air Basin.

2. HYDROCARBON REACTIVITY

The amount of ozone formed in the atmosphere due to the reactions of a specific hydrocarbon defines its reactivity or ozone production potential. The earliest studies of photochemical air pollution showed that the amount of ozone formed was dependent on the nature of the organic compounds initially present (Haagen-Smit et al., 1953). The first reactivity scale grew out of these studies, in which organic compounds were ranked by their ozone-forming ability. In the mid-1970's a reactivity scale based on the rate at which a hydrocarbon reacts with the hydroxyl (OH) radical was developed in recognition of the importance that OH radical/hydrocarbon reactions had on ozone formation

(Darnell et al., 1976; and Wu et al., 1976). However, the OH radical reactivity scale is based only on the first of many reactions a hydrocarbon will be involved in before ozone is formed. The subsequent reactions will also depend on the mixture of hydrocarbons in the atmosphere. To take these remaining reactions into account, the reactivity of a hydrocarbon was defined to consist of two parts -- kinetic reactivity and mechanistic reactivity (Carter and Atkinson, 1989; Carter 1994). The kinetic portion is based on the OH radical reaction rate. The mechanistic portion takes into account the subsequent reactions the molecule undergoes with and how much additional ozone is formed because of these reactions. In general a hydrocarbon's reactivity depends on three factors: 1) how rapidly it reacts (kinetic reactivity); 2) the nature of its reaction mechanism (mechanistic reactivity); and 3) the type of environment into which it is emitted (a factor in both kinetic and mechanistic reactivity).

3. HYDROCARBON REACTIVITY AND REGULATIONS

Incorporation of reactivity into California's Low-Emission Vehicles/Clean Fuels (LEV/CF) regulations began in 1987 with the formation of the Advisory Board on Air Quality and Fuels. The Advisory Board's mandate was to define a low-emission vehicle that either was fueled by conventional gasoline and had hydrocarbon exhaust mass emissions equal to half of the existing standard for new cars, or operated on an alternative fuel with an equivalent or lower effect on ozone formation. The staff of the California Air Resources Board recommended that the reactivity of the exhaust emissions be used as the basis for comparing the air quality impacts of the various fuels. Reactivity adjustment factors (RAF's) were developed and adopted as a part of the regulation.

The RAF is defined as the ratio of the reactivity of the low-emission vehicle operated on the alternative fuel to the reactivity of the low-emission vehicle operated on conventional gasoline. Because the reactivity concept was to be adopted into the regulation, it had to withstand scientific and legal scrutiny. The remainder of this paper describes the technical issues that were addressed before the concept of reactivity could be formally adopted, the areas where reactivity may be used in the future, and the research focused on refining the reactivity concept.

4. DEFINITION OF HYDROCARBON REACTIVITY

Perhaps the most fundamental technical question concerning reactivity is how should reactivity be defined. The most relevant definition (from a regulatory perspective) of reactivity would be the change in ozone in the region of concern caused by a change in the emissions of the hydrocarbon being regulated. However, this definition is not very practical because the value calculated is affected by changes in the quantities emitted of the compound being studied. To avoid this dependence, an incremental reactivity value is typically used. Incremental reactivity is defined as the change in the amount of ozone caused by an incremental change in the amount of the compound's emissions divided by the incremental change in the amount of the compound's emissions (i.e., the partial derivative of amount of ozone with respect to the change in amount of the compound of interest). This definition for incremental reactivity takes into account all aspects of the compound's reaction mechanism and the effects of the other compounds present in the region, and was determined to represent a more comprehensive measure of the effect a hydrocarbon would have on ozone formation (Carter and Atkinson, 1989; and Lowi and Carter, 1990).

Ultimately, the maximum incremental reactivity (MIR) was chosen as the value to be used in the LEV/CF regulations because it is defined where hydrocarbon controls would be most effective. Figure 2 shows a plot of incremental reactivity versus the hydrocarbon-to-NO_x ratio.

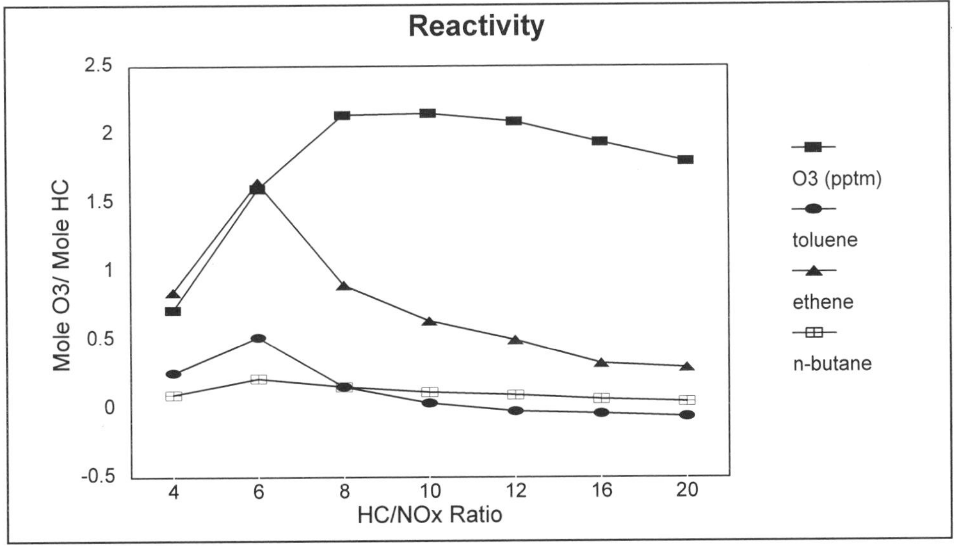

Figure 2. Incremental reactivity of selected hydrocarbons.

The y-axis corresponds to amount of ozone formed per addition of the three compounds plotted (toluene, ethene, and n-butane). As mentioned above, the incremental reactivity depends on the mixture into which the hydrocarbon is emitted. An important parameter of the mixture is the amount of NO_x present relative to the amount of hydrocarbons present. This is typically represented by the ratio of hydrocarbons to NO_x (HC/NO_x). The MIR values occur at HC/NO_x ratios where the incremental reactivities are at their maximum, and, coincidentally, the HC/NO_x ratios where this occurs are very close to the ratios found in typical urban environments in the United States. Controls based on the MIR values correspond to conditions where changes in hydrocarbon emissions have their greatest effect on ozone concentrations, and because the MIR values are defined at HC/NO_x ratios similar to the ambient levels found in major urban areas, the MIR values are particularly relevant to control plans for California cities. California control plans have always favored both hydrocarbon controls, which are more effective in low

HC/NO_x areas (i.e., cities), and NOx controls, which are necessary in high HC/NO_x areas downwind of urban centers, to improve the air quality. Using the MIR scale complements this approach by emphasizing the benefits of hydrocarbon controls in areas especially sensitive to them.

Returning to Figure 2, some other characteristics of hydrocarbon reactivities can be seen. The relative reactivities of the major groups of compounds are illustrated. The olefins (alkenes) are the most reactive, followed by the aromatic compounds (toluene) with the saturated hydrocarbons (alkanes) the least reactive. The peak or maximum incremental reactivity is easily seen at a HC/NOx ratio of around 6, and the difference between the compound reactivities drops off quickly at ratio values greater than 6. The *relative* reactivity of each type of compound stays the same because the reactivities rise and fall together. Toluene is atypical in that its incremental reactivity value drops below that of the alkanes and becomes negative. This is the result of NO_x sinks in the aromatic reaction mechanisms.

Incremental reactivities of compounds cannot be easily measured. Typically, the values are determined from computer models that simulate atmospheric conditions (Carter and Atkinson 1989; Bergin et al., 1995; Carter 1994), and the values are validated by appropriately planned experimental chamber studies (Carter et al. 1995). Computer models have inherent uncertainties because of approximations made in developing the models, and the inputs used in the models. Moreover, uncertainties are introduced from environmental chamber experiments because it is difficult to exactly simulate the ambient environment. Addressing and reducing these uncertainties was necessary before the concept of reactivity could be used in a regulation.

5. REACTIVITY SCALE UNCERTAINTIES

Many factors contribute to the uncertainty of the reactivity value determined for any hydrocarbon, and a reactivity value containing large uncertainties is of no use for developing and implementing a regulation. Uncertainty in the MIR values is introduced by the chamber experiments used to validate chemical mechanisms and estimate reactivity values, the choice of environmental conditions used in the models used to calculate MIR values, and the assumptions used in developing the chemical mechanisms. For the reactivity adjustment factors used in the LEV/CF regulations, the variability of the hydrocarbon exhaust emissions from mobile sources introduces additional uncertainties. These uncertainties and how the uncertainties were addressed are described below.

Environmental chamber experiments are used (1) to estimate the reactivities of the compounds of concern and (2) to validate chemical mechanisms use to model the relevant atmospheric chemistry (Carter and Atkinson, 1989; Carter et al., 1995, Carter and Lurmann, 1991). Chamber-derived reactivity values are not expected to well represent actual values because of chamber effects such as radical generation on the walls, and more important, the choice of an appropriate hydrocarbon mixture for the chamber run. Reactivity values are affected by the nature of the ambient atmosphere, and the choice of the hydrocarbon mixture limits the usefulness of these derived reactivity values to areas where the chamber mixture represents the actual atmosphere.

A more important concern is the appropriateness of the environmental conditions assumed in the modeling runs used to calculate MIR values. Ideally, the reactivity value would be applicable to all environmental conditions found in any region where ozone control plans are needed. An unfortunate limitation of any reactivity definition is that both the absolute and the relative reactivity values change with different environmental conditions (different hydrocarbon mixtures and different hydrocarbon-to-NO_x ratios). Therefore, no reactivity scale can be expected to accurately represent all environmental conditions.

In developing the MIR values incorporated into the LEV/CF regulations, the issue of variable environmental conditions was addressed by using an average of the environmental conditions found in 39 cities in the United States. To further reduce variability, a normalized reactivity value was defined by dividing the hydrocarbon's reactivity value by a total-species-weighted-average reactivity based on all the hydrocarbon species in the specific city and multiplying by the 39-city average of the total species-weighted average reactivity of all the cities. The normalized version of the MIR reduced the variability from about 20% to 12%.

Bergin et al. (1995) compare the MIR values generated by Carter (1994) who used a box model to represent the atmospheric conditions, to the MIR values calculated with a much more sophisticated urban airshed model (the Carnegie/California Institute of Technology (CIT) Model). The urban airshed model incorporates the spatial variations in the hydrocarbon and NO_x concentrations. As mentioned above, the MIR values are very sensitive to the hydrocarbon-to-NO_x ratio, and the use of the more sophisticated airshed model gives an indication of the magnitude of the effect of these spatial variations on the MIR value.

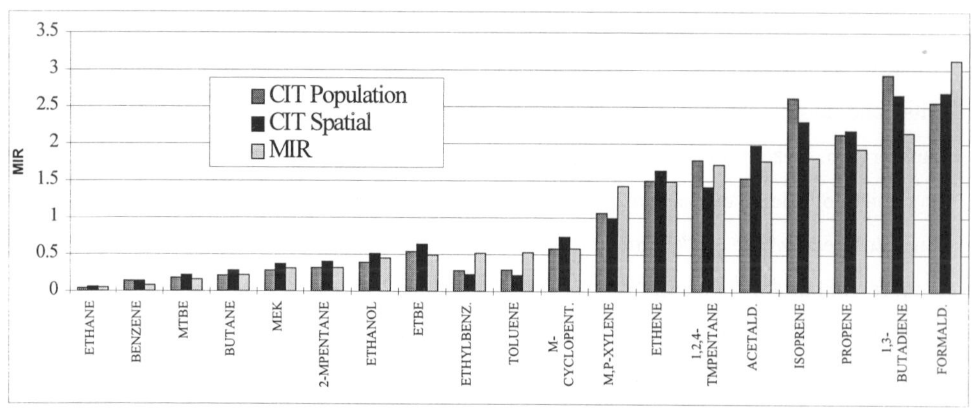

Figure 3. Comparison of Carter's MIR values to those calculated from airshed model.

Figure 3 shows the results of the Bergin study. The "CIT population" and "CIT spatial" labels refer to the urban airshed model results and correspond to ozone exposure measures or metrics that are based on population-weighted exposure or spatially-weighted exposures. The MIR legend refers to the values calculated using the more simplistic box model. The results shown have been normalized by dividing by the reactivity or exposure metric of a mixture of hydrocarbons that represent the composition of automobile exhaust. In this case the correlation between the reactivity values calculated for both the box model and the urban airshed models are very good, indicating that the reactivity values calculated with the simple box model adequately represent the values calculated using a more realistic urban airshed model.

Another important factor that can contribute to the uncertainty of reactivity determinations is the chemical mechanism used to model the atmospheric reactions that result in ozone formation. The chemical mechanism may involve many tens of reaction steps, each defined by an individual rate constant. In the best circumstances, each rate constant has been experimentally measured and its associated experimental uncertainty is known. However, in many cases the rate constant is an estimate based on similar reactions or educated guesses. This adds more uncertainty to the chemical mechanism.

Several investigators have studied the uncertainties associated with the chemical mechanisms and the extent to which the uncertainties affect reactivity value determination. Yang et al. (1995) propagated the uncertainties in rate constants of the more important chemical reactions through a box model run and determined the uncertainties of the MIR value for 26 hydrocarbon species. The uncertainties (i.e., 1 standard deviation) ranged from 27% to 68%, even though the uncertainties of some of the rate parameters in the chemical mechanism were considerably higher (>100%). The analysis also showed that changes in rate parameters of many of the reactions correlated well with the changes in the MIR values of the 26 compounds. This indicates that the uncertainties in the relative reactivity values would be less than those calculated for the absolute reactivities. Bergin et al. (1997) looked at rate parameter uncertainty using the more sophisticated urban airshed model. They found that the relative reactivities are not very sensitive to reaction rate uncertainties. In summary it appears that uncertainty in reaction rates does not contribute too large an uncertainty to restrict the use of reactivity in regulations.

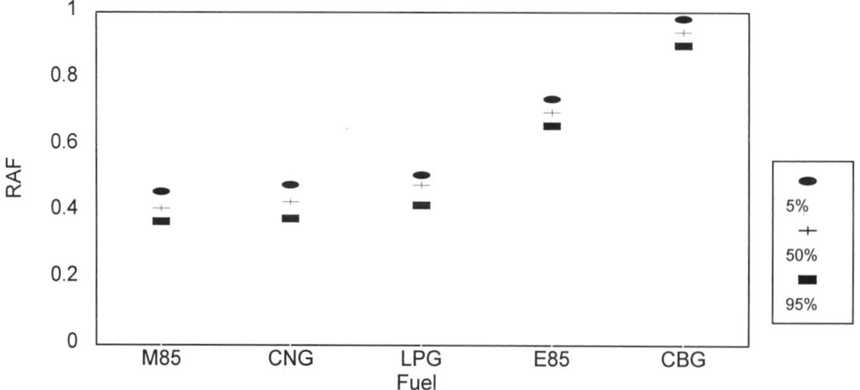

Figure 4. Variation in RAF's due to exhaust composition.

For the LEV/CF program, the reactivity values are used to compare exhaust emission reactivities of vehicles burning different types of fuels. Hydrocarbon species concentrations in exhaust emissions are inherently variable and the variability introduces uncertainty into determination of the RAF described earlier. The effect of the exhaust composition variability was studied by Yang et al. (1996) and showed that the additional uncertainty added by considering the exhaust variability was about 10%. An independent analysis (Russell et al. 1995) determined the effect of exhaust emission variability on the RAF for four alternative fuels and standard gasoline. The results displayed in Figure 4 show that the RAF's are relatively insensitive to the variability of the exhaust composition. The 5%, 50%, and 95% correspond to the 5th, mean, and 95th percentiles of each of the RAF values. The fact that the RAF's are ratios or represent relative reactivities contributes to the insensitivity and reinforces the use of relative or normalized values to minimize the uncertainty.

6. CONSUMER PRODUCTS

With the successful implementation of reactivity into the LEV/CF regulations, other potential categories are now being investigated for incorporation of reactivity into their controls. This includes the consumer products category. Consumer products include many categories such as air fresheners, antiperspirants and deodorants, automotive windshield washer fluids, glass cleaners, hair sprays, floor polishes, and hair styling gels. Consumer products contribute about 15% of the total stationary source hydrocarbon emissions in California, or about 240 tonnes per day. To meet the requirements of California's air quality plan, consumer product hydrocarbon emissions are required to be reduced by 85% by 2010. To achieve this aggressive goal, hydrocarbon reactivity may be used as part of the strategy to "fine-tune" the 85% reduction goal.

As with the LEV/CF regulations, there are many challenges unique to the consumer products industry that need to be addressed before reactivity can adequately be incorporated into a regulation. Uncertainty, as with the LEV/CF regulations, is important. As described above, the RAF's developed for the LEV/CF regulation are defined as ratios of the reactivities of vehicle exhaust composed of similar hydrocarbon mixtures. Because of the similar hydrocarbon mixtures, the uncertainties of the different mixtures are somewhat correlated and by taking the ratio of the reactivities the uncertainties are reduced. Typical reformulations for some consumer product categories will involve changing relatively dissimilar (in that the uncertainties are not expected to be correlated) compounds and the magnitudes of the uncertainties associated with the formulations must be accounted for. For example, how will a regulation incorporate uncertainties to ensure that the reactivity of a new formulation is actually less than that of the old product? A trivial but realistic example would be reformulating a product using a hydrocarbon with a reactivity value of 4+/-1 with a new hydrocarbon with a reactivity value of 3+/-0.5. Does this switch actually reduce the reactivity of the product?

Another challenge for consumer products is that some of the consumer product hydrocarbons have not been studied as thoroughly as the hydrocarbon species found in auto exhaust. About 60% of the hydrocarbons in the consumer product inventory is comprised of ethanol, n-butane, isobutane, propane and isopropanol which have been well studied and their reactivity values are well known. Another 30% of the inventory is composed of petroleum distillates which contain alkanes, alkenes and aromatic compounds which also have better known reactivity values. However, the remaining 10%

of the hydrocarbons in the consumer products inventory have not been adequately studied, in part, due to lack of suitable analytical techniques for measuring these compounds. Therefore, reactivity value uncertainties are higher, and in some instances a useful reactivity value is not available. Resources must be directed toward studying these hydrocarbon species. Additionally, product formulation information may be proprietary, and the additional limitations this may bring about in developing and enforcing such a regulation need to be considered.

However, reactivity considerations (if the uncertainties are manageable) do offer the benefit of ensuring that less reactive compliant consumer products are realized. With existing mass based regulations requiring reduction of overall hydrocarbon mass in consumer products, there is a concern that more reactive hydrocarbon compounds will replace the existing compounds in products as they reduce their hydrocarbon content. In this instance, mass based regulations, operated in the absence of reactivity considerations, would potentially increase air pollution by increasing the overall reactivity of the complying product while reducing its mass of hydrocarbon compounds. Incorporation of reactivity in the regulation would prevent this scenario from occurring, and is an important reason to consider its use in the new regulations.

7. EXPERT ADVICE

Recognizing that there are many uncertainties associated with the use of hydrocarbon reactivity in regulations, the ARB formed two reactivity committees to assist in guiding its reactivity research activities and to provide expert advice on any hydrocarbon-reactivity-related topic. The Reactivity Research Advisory Committee (RRAC) is composed of several experts from the consumer products industry and provides a forum for informed dialogue between the industry and the Air Resources Board (ARB). The Committee's overall goals are to provide a forum for determining the reactivity research needs for the consumer products industry; to enhance reactivity research efforts by coordinating research activities (coordination would avoid duplication of studies and would provide opportunities for co-funding or augmentation of existing and planned studies); to provide information relating to ARB and industry reactivity research activities; and to provide overall guidance on implementing hydrocarbon reactivity into consumer product regulations.

The Reactivity Scientific Advisory Committee (RSAC) is made up of independent respected scientists whose responsibility it is to make recommendations to the ARB on future reactivity-related issues at the request of the Board Chairman. Such recommendations will be advisory only, and will not be binding on the Board. The committee has considered topics such as what is the best reactivity scale to use, how should negligibly reactive compounds be defined, and how should compound trades and substitutions be implemented. Future meetings are planned to address reactivity concepts specific to the consumer product regulations being developed.

8. RESEARCH ACTIVITIES

Many aspects of hydrocarbon reactivity need further study to ensure that the use of the concept in regulations is appropriate. The ARB has supported research on hydrocarbon reactivity for several years as part of its overall program to study regional scale air quality problems. Current studies being funded by the ARB include determining the MIR values for compounds specific to consumer products. These compounds have typically not been studied for a variety of reasons but are now important because of the consumer product control strategies adopted by the ARB. Another study involves determining the uncertainty of calculated MIR's using a combination of mathematical techniques that will result in a more objective estimate of a hydrocarbon's reactivity value. Other studies are being conducted to improve hydrocarbon speciation profiles for aerosol paints, and to improve airshed models. Product studies are being conducted of poorly studied hydrocarbons. All of these research activities will ultimately lead to improved hydrocarbon reactivity estimates.

9. SUMMARY

The concept of reactivity has been and continues to be pursued by the ARB to provide more flexibility to industry to meet the increasingly stringent requirements that are necessary to maintain improvement of California's air quality. Hydrocarbon reactivity was successfully incorporated into the LEV/CF regulation by developing RAF's. The RAF's definition allows for reduction of many of the uncertainties associated with the hydrocarbon reactivity estimates. Other categories of compounds may not allow for similar reductions in uncertainties and it will be more difficult to use their reactivities in developing regulations with the current state of the science. This is the case with the consumer product regulations currently being developed. To aid the ARB in appropriately incorporating hydrocarbon reactivity into the regulatory process, expert panels have been formed and further research into reactivity is underway. The ongoing research and input from the expert panels will further the state of the science and ultimately will lead to more certain treatment of hydrocarbon reactivity.

REFERENCES

1. Bergin, M. S., A. G. Russell, and J. B. Milford (1995) "Quantification of individual VOC reactivity using a chemically detailed, three-dimensional photochemical model", *Environ. Sci. Technol.*, **29:** 3029-3037.
2. Bergin, M.S., A.G. Russel, J.B. Milford (1997) "Effects of Chemical Mechanism Uncertainties on the Reactivity Quantification of Volatile Organic Compounds Using a Three-Dimensional Air Quality Model", in preparation.
3. Carter, W. P. L., and Atkinson, R., (1989) "A computer modeling study of incremental hydrocarbon reactivity", *Environ. Sci. and Technol.*, **23:** 864-880.
4. Carter, W.P.L., and F. W. Lurmann (1991) "Evaluation of a Detailed Gas-Phase Atmospheric Reaction Mechanism using Environmental Chamber Data", *Atmos. Environ.*, **25A**: 2771-2806.
5. Carter, W. P. L. (1994) "Development of ozone reactivity scales for organic gases", *J. Air Waste Manage. Assoc.*, **44:** 881-899.
6. Carter, W.P.L., J.A. Pierce, D. Luo, and I.L. Malkina (1995) "Environmental

chamber studies of maximum incremental reactivities of volatile organic compounds", *Atmos. Environ.*, **29:** 2499-2511.

7. Darnell, K.R., Lloyd, A.C., Winer, A.M., and Pitts, J.N., (1976) "Reactivity Scale for Atmospheric Hydrocarbons Based on Reaction with Hydroxyl Radical", *Environ. Sci. Technol.*, **10**: 692.

8. Haagen-Smit, A.J., Bradely, C.E., and Fox, M.M., (1953) "Ozone Formation in Photochemical Oxidation of Organic Substances", *Ind. Eng. Chem.*, **45**, 2086.

9. Lowi, A., Jr., and W. P. L. Carter (1990) "A Method for Evaluating the Atmospheric Ozone Impact of Actual Vehicle Emissions", SAE Paper No. 900710, presented at the SAE International Congress and Exposition, Detroit, Michigan, February 26 - March 2.

10. Russell, A., J. Milford, M. S. Bergin, S. McBride, L. McNair, Y. Tang, W. R. Stockwell, and B. Croes (1995) "Urban ozone control and atmospheric reactivity of organic gases", *Science*, 269: 491-495, July 28.

11. Wu, C.H., Japar, S.M., Niki, H., (1976) "Relative Reactivities of HO-Hydrocarbon Reactions from Smog Reaction Studies", *J. Environ. Sci. Health, Environ. Sci. Eng.*, **A11**: 191.

12. Yang, Y-J., W. R. Stockwell, and J. B. Milford (1995) "Uncertainties in incremental reactivities of volatile organic compounds", *Environ. Sci. Technol.*, **29:** 1336-1345.

13. Yang, Y-J., and J. B. Milford (1996) "Quantification of uncertainty in reactivity adjustment factors from reformulated gasolines and methanol fuels", *Environ. Sci. Technol.*, **30:** 196-203.

EU/ECE perspective on NO_x/O_3 environmental and control policies

L. Björkbom

Chairman, Working Group on Strategies, Convention on Long-range Transboundary Air Pollution
c/o Swedish Environmental Protection Agency, Blekhomsterrassen 36,
S-106 48 STOCKHOLM, SWEDEN

INTRODUCTION

Within the framework of the Convention on Long-Range, Transboundary Air Pollution (CLRTAP), signed in 1979 and in force since 1983, at present 41 of the 55 Member States of the UNECE plus the Commission of the European Communities are co-operating to curb emissions of pollutants to the atmosphere, that are transported over national boundaries within the UNECE region. The main objectives are to protect the natural environment and human health, but also materials and historical buildings. To date, select numbers of Parties have signed and ratified legally binding Protocols on reductions on sulphur emissions (Reduction of Sulphur Emissions or their Transboundary Fluxes by at least 30 percent, 1985 and Further Reduction of Sulphur Emissions, 1994) on nitrogen oxides (Control of Emissions of Nitrogen Oxides, 1988) and hydrocarbons (Control of Emissions of Volatile Organic Compounds, 1991). At present, negotiations are underway for Protocols on Nitrogen Oxides and related substances, Heavy metals and Persistent Organic Pollutants, hopefully all three to be finalized during 1998/99.

In the EU and its forerunner, air pollution control has been over the two last decades and remains an important element in the overall environmental policy. Although much attention has been given to ambient air quality, a number of directives have also had as objectives to ameliorate environmental and health effects of air pollutants in the whole of the Union. All EU Member States and the Commission are Parties to the CLRTAP, but until recently their have been few instances of coordinated EU inputs to the Protocol negotiations under the CLRTAP. The reason has been that most of the directives, relevant to CLRTAP Protocols - like the LCP directive - have been minim directives. The obvious exception was the ceiling-harmonized directive on car exhausts, which had great impact on the 1988 NO_x Protocol. Recently, a steering body has been established by the Commission and the Executive Body of the CLRTAP to safeguard that relevant decision making within the two bodies should be at least compatible with each other.

1. METHODS OF APPROACH

The methods of approach to achieve the objectives of air pollution control both within the CLRTAP remit and the EC/EU context have developed over time.

In the CLRTAP sphere the initial efforts were to control and/or reduce emissions - pollutant by pollutant (SO_x, NO_x, VOCs)- through legally binding Protocols for each type of pollutant. Reduction targets agreed upon were based on conservative estimates of what could be achieved by applying BAT (not entailing excessive costs!) and the reduction targets were equal to all Parties to the respective Protocols.

In the EC/EU context you could see similar approaches, although the legislative methods tended to be still more piecemeal, where control of a type of pollutant was limited to certain categories of emission sources, without any apparent common strategy behind the legislative pieces.

This has been an approach, which has been considered unsatisfactory by scientists because of the knowledge about chemical interaction in the atmosphere between different types of pollutants and their reactions, when deposited in soils, surface and marine waters and that the sensitivity of the environmental receptors to different pollutants differs widely in the geography. This has been most evident when focusing on acidification, eutrophication and tropospheric oxidant formation, which so far has been the main targets for air pollution policy within the CLRTAP and EC/EU contexts.

The approach was also considered unsatisfactory by policy makers because there were serious doubts as to its cost effectiveness and did not respond to their overriding needs to know, how far you have to reduce emissions of pollutants to achieve politically assessed, environmental and human health targets. Still it was applied, with reference to the precautionary principle, but essentially because neither scientists nor policy makers could, at that time, give satisfactory, quantifiable definitions of what constitutes sound and sustainable environmental quality.

The first breakthrough towards an approach based on environmental quality came when scientists round 1990 could agree on quantifiable measures for sulphur depositions in aquatic and terrestrial ecosystems, beyond which no harmful effects should occur, *i.e.* the so called *critical load.* This concept was used when negotiating the second CLRTAP Protocol on further reductions of Sulphur Emissions, signed in 1994. This achievement was a first step to base international legislation on an effects related approach and also responded to the needs of getting a fair and cost effective division of measures between Parties to the Protocol. However, it did not respond to the wider need of attacking, at the same time, many interacting pollutants, that contribute to the same environmental and human health problems and thus achieving the cost effective potentials of developing and implementing integrated abatement and control programs.

2. THE MULTI-EFFECTS/MULTI-POLLUTANTS APPROACH TO AMELIORATE ENVIRONMENTAL AND HEALTH EFFECTS FROM NO_X AND RELATED SUBSTANCES

The scientific development over the last few years, that has been initiated within the CLRTAP framework and then also applied in the EU context now makes it possible to follow a multi-effects/multi-pollutant approach, using critical loads for acidification and eutrophication and critical levels for ozone relating to forests, crops and human health. Owing to historical developments, the pollutants and the environmental effects attacked are slightly different in the two legislative fora, but these discrepancies should be manageable. In the CLRTAP context, taking into account the existence of the 1994 separate sulphur abatement regime, the nodes are nitrogen compounds, which (together with sulphur compounds) contributes to acidification but also to eutrophication and, in combination with VOC's, to formation of tropospheric ozone. In the EU context the Commission has recently adopted an acidification strategy (SO_x, NO_x and NH_3) and is presently developing an ozone strategy (NO_x and VOC's). In both fora political negotiations based on these effects approaches will soon be initiated.

This is not the place to go into detail in describing the scientific and technical assumptions on which the multi-effects/multi-pollutants approach is based and from which strategies for abatement of the relevant pollutants are derived, which will serve as guidelines for intergovernmental negotiations on emission reductions in the CLRTAP as well as in the EU context. Suffice it to say that the main components are mapping of critical loads for (sulphur), nitrogen compounds and ozone all over Europe, EMEP transport and atmospheric chemistry models and cost curves for abatement measures. Several computer models are used to draw up different cost effective emission scenarios based on integrated assessment modelling. The main instrument is however the RAINS model, developed by IIASA, while other models serve to explore uncertainties in their specialties. In principle, participating governments are responsible to provide input data on emissions and emission projections, critical load data for sensitivity mapping and costs of emission reduction measures. The international co-ordination of all these data is however managed by a number of task forces, mainly manned by government designated experts and scientists involved in the scientific networks that supports the many-faceted R&D needed to solve the issues at hand. The output from the task forces are assessed by working groups of government designated experts on effects, EMEP, abatement techniques and strategies, which are advising the Executive Body of the CLRTAP on the soundness of the scientific and technical state of the art. The Executive Body, when satisfied, mandates the Working Group on Strategies to negotiate a draft protocol for adoption by the EB, which then generally meets at ministerial level.

Through the integrated assessment modelling you can provide negotiators with a number of optional optimized scenarios. It is already evident that the costs, implementation difficulties and thus political readiness to achieve critical loads and levels in one step will not be realistic, although it will be an ultimate target in the Protocol. What you can foresee is a step wise approach in which you, as a first step, will be closing the gaps of present (or from a given historical date, say 1990) depositions of (sulphur), NO_x and NH_3 and concentrations of O_3 to the year 2010. Further steps in narrowing the gaps could then be taken, after thorough review processes in assessing the problems and when and if the Parties through its Executive Body will find it suitable to do so. The whole process is foreseen to be regulated within the framework of the Protocol.

In the Acidification strategy, adopted by the European Commission a fifty percent gap closure has been suggested, but they will refrain from target setting until they will see the implications of the optimized ozone strategy, which is presently under development and also, most likely, until they have seen reactions from the EU Council of Ministers for the Environment. Where the CLRTAP Parties will land on target setting for a first step is still an open question.

A first step with reasonably ambitious emission reduction targets will be expensive to implement for all or most of the Parties. But there are also significant benefits to be reaped, both in monetary and environmental terms. The study that the EC Commission has performed on benefits to be gained, following different reduction scenarios in its Acidification Strategy, indicates, that almost every EU Member State would get benefits well beyond their reduction costs, even when only focussing on effects areas which can be reasonably well assessed in monetary terms, *i.e.* health and materials. Similar studies performed in the CLRTAP context give the same message. The gains in ecosystem protection are still beyond the capacity of economic theory to assess in monetary terms, but are related to natural and cultivated resources of great importance to all national economies in the region.

Although the multi-effects/multi-pollutant strategy has been adopted by the Executive Body as a basis for negotiations, there are still most likely a number of Parties, that will have reservations as to the realism of conducting negotiations on such very complicated and far from well understood simulations of reality, which is the output of the integrated assessment modelers' creative work. Still, everyone seem to be convinced of the potentials for international co-operation in air pollution control policy by adhering to the strategy and therefore prepared to give it a good try to get it transformed into an international, legally binding agreement. My own assessment, as chairman of the negotiations, is that such an agreement will be possible to achieve, provided that

- all elements in the strategy can be made transparent and thus fully understood by the negotiators and their political peers;
- it will be considered to result in a fair distribution of emission reductions among Parties;
- it will reasonably well respond to the different priorities of each of the signing Parties;
- it will be possible to evaluate.

I should also like to add, provided that the very good ambiance in which negotiations have been conducted within the CLRTAP framework over the last decade will prevail and not be disturbed by external problems in the wider sphere of international politics in the UNECE region.

I think it is fair to say, that the strategic work model that has been developed within the CLRTAP framework has been taken over by recent strategic developments in the EU in the field of air pollution policy. The Commission has used the same scientific network and other actors, not least the RAINS model, to provide them with the necessary input for policy making. Baring the different institutional set up in the EU and the different geographical scope, the streamlining between the CLRTAP and EU in air pollution policy making relating to the primary effects areas and pollutants we are here discussing, has come far, and does not need a separate description in this context.

3. CLRTAP FUTURES IN RELATION TO AN EXPANDING EU

This "policy merger" between CLRTAP and EU is, of course, in a way a necessity. After all, the 15 EU Member States, the Commission and the 3 EEA States and the Central and East-European States that have association agreements with the Commission and consider themselves as incumbent Member States in a foreseeable future are together forming a strong majority in the CLRTAP family. There is and will be a demand of a coordinated and coherent air pollution policy in order to safeguard that the countries obligations in the two different international environmental legislative fora will be compatible between themselves, when implementing their obligations in their respective countries.

It is, however, a development, which has certain complications in the CLRTAP set-up, considering the fact that some very important Parties to the Convention are and will not likely become EU Member States, like the US and Canada. The effects-based approach is closely linked to the geographical scope of EMEP, which is, simply put, Europe west of the Urals. The interchange of pollutants like sulphur and NO_x over the Atlantic has so far been deemed to be insignificant. This might not be the case when you are considering the increase of background ozone in Europe and clearly not in relation to *e.g.* persistent organic pollutants.

When The US and Canada joined the CLRTAP in 1979 this was probably more an act of foreign policy than to safeguard national and international environmental policy objectives - as was incidentally the case of most of the initial signatories to the Convention. At least at that time there were no scientific evidence on transboundary fluxes from North America to Europe and still less the opposite way. The US has also had a particular relationship to the CLRTAP. The US is only Party to one of the Protocols, *i.e.* the 1988 NO_x Protocol - although it signed the 1991 VOC Protocol but has so far not ratified its signature. While participating very constructively in the development of the 1994 Sulphur Protocol, the US Government found that they could not sign it because it would run into conflicts with the Clean Air Act. There is a clear risk that this dilemma will occur also when negotiating the multi-effects/multi-pollutants Protocol. Although health concerns relating to NO_2 and O_3 are added to the environmental objectives of this Protocol and thus closer to the main objective of the Clean Air Act than was the case of the 1994 Sulphur Protocol, it is still unclear whether the health concerns will be the limiting factor for reduction targets of the relevant pollutants in Europe.

The Canadian situation is slightly different. Canada was one of the active driving forces behind the 1985 Sulphur Protocol (obviously then motivated by their bilateral difficulties with the US on the Acid Rain issue). It is also Party or Signatory to the other reduction protocols and much involved in the development of the multi-effects/multi-pollutants protocol and, particularly, of a POP's protocol. Although Canada is developing a system of critical loads mapping for acidifying substances and ozone, this, to my understanding, is not integrateable with the European system.

You might, of course, uphold, that the importance in an international air pollution policy lies in the level of reduction targets achieved and not in the methods, from which these targets are derived. If for instance the reduction achievements in North America under the NAFTA Agreement give satisfactory contributions also to the sustainability of European ecosystems and the health of its population, and *vice versa,* if that is at all relevant, so much the better for all practical and political purposes. I think that this should be the pragmatic approach to the issue. But you can not avoid reflecting on the potential risk that the CLRTAP in the long run might fall apart, if the critical loads and levels approach, for very good reasons used on the European scene, will not be the overall method of approach from Vladivostok to Alaska counted westward.[1)]

[1)] A separate point in case is, of course, the Russian Federation and Ukraine, which are not likely to become EU Member States. Part of their combined huge realm is covered by EMEP and the interchange of airborne pollutants between an (expanded) EU and at least the European part of Russia and Ukraine is significant. Russia and Ukraine are no doubt explicit supporters of the critical load's approach, but their countries' capacity to implement abatement policies, but in a distant future, in line with the shares allotted to them in optimized emission reduction scenarios may be questioned. Also this situation may have certain problematic implications for the CLRTAP when EU expands eastward.

The benefits and costs of air pollution control

Richard D. Morgenstern

Visiting Scholar, Resources for the Future; and Associate Assistant Administrator for Policy, U.S. Environmental Protection Agency (on leave). Portions of this paper are adapted from Morgenstern (1997).

The U.S. Environmental Protection Agency estimates that we devote about $150 billion annually to environmental protection. While we can probably afford to spend such a sum on environmental protection, we must spend it wisely. Certainly the magnitude of our environmental expenditures—on the order of 2% of the Gross Domestic Product (GDP)—raises questions about the efficiency and effectiveness of the overall environmental management system. Are we getting good value for our money? Could we do better?

To answer such questions we turn to the field of economics and, particularly, the burgeoning field of environmental economics. Economic analysis can provide an accounting framework for tracking and exploring the implications of environmental decisions. It can serve as a tool for arraying information about the benefits and costs of environmental policies, and as a mechanism for revealing the cost-effectiveness of alternative approaches.

This paper addresses the basic question of how much value we are getting for the resources committed to one particular environmental problem, namely air pollution. The focus is on the aggregate or economy-wide benefits and costs of clean air policies carried out in the U.S. over the past several decades.

1. ECONOMICS AND ENVIRONMENTAL PROTECTION

The great transformation from primitive life to civilized society that occurred over many centuries has altered the environments of entire continents, as humans converted forests into farms and domesticated plants and animals. More recently, our ability to use and adapt the environment has increased dramatically with the development of new polluting technologies, e.g., chlorofluorocarbons. Rising human population and per-capita consumption levels have further contributed to the speed by which we are currently able to alter the environment. The environmental impacts of our newly acquired powers are forcing us to recognize that the environment consists of scarce, even exhaustible resources, e.g., clean air. That is where economics enters, for economics is the science of allocating scarce resources among competing ends.

Economic analysis of environmental programs and policies can serve multiple purposes. It can help allocate our resources more efficiently, encourage transparency in decision making, and provide a framework for consistent data collection and identification of gaps in knowledge. Economic analysis also allows for the aggregation of many dissimilar effects, e.g., improvements in health, visibility, and agricultural output, into one measure of net benefits expressed in a single currency. The challenge for environmental policy is to determine when public intervention in the affairs of firms and individuals is desirable for environmental reasons, and which policies are most appropriate in various circumstances. Economic analysis can help policy makers identify interventions that generate more benefits than costs and assist them in choosing the best intervention from among those that do.

1.2. Environmental benefits and costs

Environmental policies can improve human health, increase output of forests and other natural resources, reduce corrosion or soiling of economic assets, and enhance recreational and other environmental assets. A taxonomy, incorporating broad categories of environmental benefits, is shown in Table 1. Some environmental benefits, e.g., increased output of forests, are measured in commonly used indicators of economic activity like the gross domestic product (GDP). Other benefits are the non-market, welfare-enhancing type that typically are not represented in the GDP, e.g., improved human health or greater biodiversity. It is estimated that more than 90% of the environmental benefits of the Clean Air and Clean Water Acts are of the non-market, welfare-enhancing type not represented in the GDP (Freeman (1982)). Although researchers are trying to develop comprehensive measures of economic activity that capture a broad set of environmental benefits (and costs), e.g., "green GDP", there are, in fact, strong theoretical and practical reasons for excluding such welfare-enhancing benefits from a commonly used measure of economic activity like GDP. Yet, no one doubts that human welfare - rather than GDP - is what societies are ultimately concerned with. The fact that such welfare-enhancing benefits are difficult to value and often involve specialized terminology and measurement techniques does not mean that they are any less valuable than those benefits that are measured by the GDP.

The estimation of costs can be as difficult an undertaking as the estimation of benefits - a fact not often appreciated by even the most knowledgeable practitioners in the field. For example, the most commonly used measure of environmental costs is reported out-of-pocket expenditures for regulatory compliance. However, this is a narrow measure that may either under- or overstate total compliance costs. On the one hand, the omission of items like legal expenses and diverted management focus suggests that reported out-of-pocket expenditures would tend to understate total compliance costs. On the other hand, failure to account for improved worker health or increased innovation tends to overstate total compliance costs. Table 2 contains a taxonomy of environmental compliance costs borne by the private sector, including firms and households, and all levels of government, including the EPA. The reader should note, however, that the cost estimates reported in this paper generally represent reported out-of-pocket expenditures for regulatory compliance. Depending on the importance of the other (often less well measured) cost items listed in Table 2, reported compliance costs may under or overstate total compliance costs.

Table 1
Taxonomy of environmental benefits

To Individuals
Mortality
Morbidity (acute, chronic)
To Production/Consumption
Crops/Forests/Fisheries
Water-Using Industries
Municipal Water Supply
To Economic Assets
Materials (corrosion, soiling)
Property Values
To Environmental Assets
Recreation
Other use values (visibility)
Nonuse (passive use)

2. BENEFITS AND COSTS OF THE CLEAN AIR ACT: AGGREGATE ANALYSIS

Given this brief introduction to environmental benefits and costs, we now may turn to our basic question, namely, what do we know about the overall health, welfare, ecological and other economic benefits of the Clean Air Act and how do these benefits compare with estimated costs? Can we apply an economic framework to determine whether we are getting value for our money?

2.1. Early Research

In a major study Freeman developed the first and - until quite recently - the only comprehensive benefit-cost analysis of the Clean Air Act (Freeman, 1982). His study was controversial upon publication because it involved a great many assumptions - some would say 'leaps of faith' - and attempted to reduce a complex set of issues to a few numbers. One of the key issues in such studies is the question of the baseline: what would ambient air conditions have been in the absence of federal legislation? Ideally, one would compare environmental quality levels with and without the federally mandated controls, holding all other things constant, including the patterns of production, technology, and demand for goods and services which, in turn, determine the generation of pollutants. Such a measure should compare an observed outcome resulting from the policy with a hypothetical or counter-factual position reflecting the same underlying economic conditions and differing only with respect to the impact of the environmental policy. Unfortunately, data and resource limitations prevented Freeman from making such a comparison in his original study. Instead, he measured the benefits of air pollution by examining the actual improvements in air quality observed between 1970 and 1978. As Freeman notes, such measures are likely to underestimate the true benefits because they fail to account for the significant economic growth over the period.

Table 2
Taxonomy of environmental costs

Government Administration of Environmental Statutes and Regulations
Monitoring
Enforcement
Private Sector Compliance Expenditures
Capital
Operating
Other Direct Costs
Legal and Other Transactional
Shifted Management Focus
Disrupted Production
Offsetting Benefits
Resource Inputs
Worker Health
Increased Innovation
Economy-Wide Effects
Product Substitution
Discouraged Investment
Retarded Innovation
Transition Costs
Unemployment
Obsolete Capital
Social Impacts
Loss of Well-Paying Jobs
Economic Security Impacts

Source: Adapted from Jaffee, et al.(1995), page 139.

Table 3 presents Freeman's results for air pollution benefits and costs in 1978 (converted to 1995 dollars). Total benefits of the air pollution program range from $12.3 to $123.3 billion with what Freeman calls a 'most reasonable' point estimate of $49.1 billion. Overall, more than 75% of the benefits are health related. Costs of the air pollution program are estimated to be $35.0 billion. Freeman concludes that, taken as a whole, it is highly likely that air pollution control has been worthwhile on benefit-cost grounds.

Table 3
Air pollution control benefits and costs in 1978 (in billions of 1995 dollars)

Category	Range	Most Reasonable Point Estimate
Benefits		
Health		
Mortality	5.9 - 59.5	29.3
Morbidity	0.6 - 26.2	6.5
Soiling and Cleaning	2.1 - 12.7	6.3
Vegetation	0.2 - 0.8	.6
Materials	1.6 - 5.3	1.5
Property Values	1.9 - 18.8	4.9
Total	12.3 - 123.3	49.1
Costs		35.0

Source: Freeman (1982), page 128.

Drawing on the sizable body of research conducted in the intervening years, a recent EPA study expands on Freeman's original work. Like Freeman's estimates, the new study synthesizes and integrates a large body of information derived from the scientific and economics literatures. The EPA study, which was mandated by Congress as part of the 1990 Air Act Amendments, assesses the benefits and costs of the Clean Air Act, 1970-1990. It was developed by EPA in conjunction with a Congressionally mandated panel of distinguished economists and scientists.

The new EPA study updates the Freeman methodology by developing and comparing two scenarios as a basis to evaluate progress under the Clean Air Act: 'control' vs 'no-control'. The 'no control' scenario essentially freezes federal, state, and local air pollution controls' at levels of stringency and effectiveness which prevailed in 1970 and attributes the benefits and costs of all air pollution controls from 1970-1990 to federal law. In all likelihood, state and local regulation would have required some air pollution controls even in the absence of the Clean Air Act. Certain states, e.g., California, might have required very tight controls. It is also likely that industry would have acted on its own to reduce at least some of its emissions. If one assumed that state and local regulations would have been equivalent to federal regulations then a benefit-cost analysis of the federal clean air act would be a meaningless exercise: the incremental benefits and costs of any federal initiatives would equal zero. On the other hand, any attempt to predict how states' and localities' regulations or voluntary efforts would have differed from the Clean Air Act is extremely speculative. Thus, the freezing of emissions at 1970 emission rates is a reasonable, albeit unrealistic, assumption. Both the 'control' and 'no control' scenarios are evaluated by a sequence of economic, emissions, air quality, physical effect, economic valuation, and uncertainty models to estimate the benefits and costs of the Act. The analytical sequence incorporating these various steps is shown in Figure 1.

The air quality modeling involves a number of key issues worthy of mention. For sulfur dioxide, nitrogen oxides, and carbon monoxide, improvements in air quality under the control scenario are roughly proportional to the estimated reduction in local area emissions. In contrast, differences in estimated ground level ozone concentrations vary significantly from one location to another, because of local differences in the relative proportion of VOC's and NO_x, and weather conditions. Many pollutants contribute to ambient concentrations of particulate matter and, in fact, specific sources vary according to region, urban vs. rural, and other factors. From a human health standpoint, fine particles which can be respired deep into the lungs are the greatest concern. Many of these fine particles are formed in the atmosphere through chemical conversion of gaseous pollutants. They are referred to as secondary particles. The three most important secondary particles are (1) sulfates, which derive primarily from sulfur dioxide emissions, (2) nitrates, which derive primarily from nitrogen oxides emissions, and (3) organic aerosols, which can be directly emitted or can form from volatile organic compound emissions.

Overall, estimated improvements in air quality across the country compared to baseline were substantial: 40% reduction in sulfur dioxide, 30% reduction in nitrogen oxides, 50% reduction in carbon monoxide, 15% reduction in ozone and about 45% reduction in particle concentrations (including both directly emitted and secondary particles). Human health effects or benefits are derived by combining air quality improvements with estimates of dose-response functions derived from the scientific literature. (The EPA study conducted a comprehensive literature review. See study for details.) Table 4 presents selected health benefits of the Clean Air Act 1970-1990, in thousands of cases reduced per year. The mid-range estimates of reduced mortality, for example, show that in 1975 the air pollution controls in place reduced premature deaths attributable to airborne particles (PM_{10}), ozone, sulfur dioxide, and lead by an estimated 20,000 cases. By 1990 the corresponding number of premature deaths avoided stood at 79,000. Similarly, the mid-range estimate for heart attacks avoided rose from 1,000 in 1975 to 18,000 in 1990, largely due to the reduction of lead in the environment.

To develop estimates of economic benefits it is necessary to translate these physical effects into dollar terms. This is often the most contentious aspect of any benefit-cost analysis. Table 5 displays the economic values, drawn from the economics literature, used in the EPA study. Heart attacks and reduced IQ points, for example, are valued at $587,000 per case, and $5500 per point, respectively. In the case of mortality it is not possible to "value" the lives of victims in a benefit-cost sense. One can, however, determine the compensation required for individuals to accept relatively small reductions in mortality risk. Typically, they are inferred from observed behavior, for example, sales of safety devices such as smoke detectors, or wage differentials associated with high risk occupations. For expository purposes this valuation is expressed as "dollars per life saved" even though the actual valuation is really based on small changes in mortality risk. The estimate of $4.8 million per life saved represents an average value from the literature.

The total monetized economic benefit attributable to the CAA was derived by applying the valuation estimates discussed above to the complete stream of physical effects calculated for the 19701990 period. In developing these estimates, steps are taken to avoid the double counting of benefits and costs. EPA reports that the estimated benefits of the Clean Air Act realized during the period from 1970 to 1990 range from $5.6 to $49.4 trillion, with a central estimate of $22.2 trillion.

Figure 1. Summary of Analytical Sequence and Modeled versus Historical Data Basis

Control Scenario		No-Control Scenario
	Compile historical compliance expenditure data	
Develop modeled macroeconomic scenario based on actual historical data		Develop modeled macroeconomic scenario by rerunning control scenario with compliance expenditures added back to the economy
Project emissions by year, pollutant, and sector using control scenario macroeconomic projection as input to sector-specific emissions models		Re-run sector-specific emissions models using no-control scenario macroeconomic projection
Develop statistical profiles of historical air quality for each pollutant based on historical monitoring data (plus extrapolations to cover unmonitored areas)		Derive no-control air quality profiles by adjusting control scenario profiles based on differences in air quality modeling of control scenario and no-control scenario emissions inventories
Estimate physical effects based on application of concentration-response functions to historical air quality profiles		Estimate physical effects based on application of concentration-response functions to no-control scenario air quality profiles
	Calculate differences in physical outcomes between control and no-control scenario	
	Estimate economic value of differences in physical outcomes between the two scenarios*	
	Compare historical, direct compliance costs with estimated economic value of monetized benefits, considering additional benefits which could not be quantified and/or monetized	

* In some cases, economic value is derived directly from physical effects modeling (e.g., agricultural yield loss).

Source: *U.S. EPA, "The Benefits and Costs of the Clean Air Act, 1970 to 1990", Draft Report, April 1997.*

Table 4
Selected health benefits of the CAA, 1970-1990 (in thousands of cases reduced per year, except as noted)

Health Effect		1975	1980	1985	1990
Mortality	High	38	97	124	140
(PM_{10}, O_3, SO_2, Pb)	Mid	20	54	70	79
(thousands)	Low	11	30	40	45
Heart Attacks	High	1	9	19	24
(Pb)	Mid	1	7	14	18
(thousands)	Low	1	5	10	13
Strokes	High	1	5	10	13
(Pb)	Mid	1	4	8	10
(thousands)	Low	1	3	6	7
Respiratory symptoms					
(SO_2)		66	187	165	146
(thousands)					
Respiratory illness					
(NO_2)		1	4	9	15
(millions)					
Hypertension	High	1	6	12	16
(Pb)	Mid	1	5	10	13
(millions)	Low	1	4	8	10

Source: *U.S. EPA, "The Benefits and Costs of the Clean Air Act, 1970 to 1990," Draft Report, April 1997.*

By comparison, the value of direct compliance expenditures over the same period equals approximately $.5 trillion. Comparing central estimates, Americans received roughly $45 of value in reduced risks of death, illness, and other adverse effects for every one dollar spent to control air pollution

A result of this sort stimulates many questions, e.g., what specific regulations account for the most benefits? How do the net benefits of recently promulgated regulations compare to those of older regulations? While the EPA study methodology does not allow one to answer such questions directly, a number of comparisons are possible. Table 6 displays benefits by endpoint category.

Table 5
Central estimates of economic value per unit of avoided effect (in 1995 dollars)

Endpoint	Valuation (mid-estimate)
Mortality	$4,800,000 per case
Heart Attacks	$587,000 per case
Strokes	$587,000 per case
Hospital Admissions	
Respiratory	$7,500 per case
Ischmic Heart Disease	$10,000 per case
Congestive Heart Failure	$8,000 per case
Respiratory Illness and Symptoms	
Upper Respiratory Illness	$18 per case
Lower Respiratory Illness	$10 per case
Acute Bronchitis	$45 per case
Acute Respiratory Symptoms	$17 per case
Work Loss Days	$83 per day
Restricted Activity Days	$38 per day
Asthma Attacks	$32 per case
IQ Changes	
Lost IQ Points	$5,550 per IQ point
Incidence of IQ < 70	$52,700 per case
Hypertension	$682 per year per case
Decreased Worker Productivity	Direct Economic Valuation
Visibility	Direct Economic Valuation
Household Soiling	Direct Economic Valuation
Agriculture (Net Surplus)	Estimated Change in Economic Surplus

Source: *U.S. EPA, "The Benefits and Costs of the Clean Air Act, 1970 to 1990," Draft Report, April, 1997.*

Note that PM mortality alone accounts for three-fourths of the total benefits. Lead mortality benefits are also quite substantial. At first glance this might suggest that control of particulate matter and lead are the most important and, perhaps, the only important actions taken under the Clean Air Act, 1970-1990. Unfortunately, environmental analysis is not that simple. For example, secondary particles, which have been associated with premature mortality, account for a large portion of the total benefits. As noted, these particles are formed in the atmosphere from SO_2, NO_x and VOC emissions which, in turn, are governed by multiple EPA regulations, including those affecting existing as well as new sources. Thus, one cannot say with confidence that the calculated PM benefits derive strictly from regulations explicitly aimed at controlling particles. Some of the calculated PM benefits are clearly associated with these broader control efforts.

A somewhat different approach to estimating benefits and costs of environmental regulations was developed by Hahn (1996). Rather than attempt to integrate all benefits and costs into a single analysis, he examines individual regulations promulgated by EPA between 1990-1995. Using Agency numbers but applying a common discount rate as well as a consistent set of values for reducing health risks, Hahn estimates the present-value benefits and costs of major rules. For all Clean Air Act regulations issued 1990-1995 Hahn calculates benefits at 214.7 billion and costs at $124.7 billion (Table 7).

Table 6
Total monetized benefits by endpoint category for 48 state population for 1970 to 1990 (in billions of 1990 dollars)

Endpoint	Pollutant(s)			Present Value
		5th%	***Mean***	***95th%***
Mortality	PM-10	$2,369	$16,632	$40,597
Mortality	Pb	$121	$1,339	$3,910
Chronic Bronchitis	Pb	$409	$3,313	$10,401
IQ (Lost IQ Pts + Children w/IQ<70	Pb	$271	$399	$551
Hypertension	Pb	$77	$98	$120
Hospital Admissions	PM-10, O_3, Pb & CO	$27	$57	$120
Respiratory-Related Symptoms, Restricted Activity and Decreased Productivity	PM-10, O_3, NO_2 & SO_2	$123	$182	$261
Soiling Damage	PM-10	$6	$74	$192
Visibility	Particulates	$38	$54	$71
Agriculture (Net Surplus)	O_3	$11	$23	$35

Source: *U.S. EPA, "The Benefits and Costs of the Clean Air Act, 1970 to 1990", Draft Report, April, 1997.*

Table 7
Costs and benefits of Clean Air Act regulations 1990-1995 (present value in billions of 1995 dollars)

Number of Regulations	25
Costs	$124.7
Benefits	$214.7
Net Benefits	$90.0

Source: *Hahn, Robert, "Regulatory Reform: What Do the Government's Numbers Tell Us?" in Reviving Regulatory Reform, (Robert Hahn, ed.), American Enterprise Institute, 1996, page 222.*

Although he expresses serious concerns about the credibility of EPA estimates, he concludes that: "if one takes the Agency numbers at face value...there is reason to be gleeful. They basically say that the [EPA] has done more good than harm in promulgating regulations since 1990."[1b]

3. CONCLUSION

While our understanding of the link between the economy and the environment has clearly advanced in recent years there is still controversy over some of the most basic issues. Some of the controversy stems from the imprecision of the questions themselves, e.g., what do 'value' and 'cost' really mean in the context of environmental health or natural resources? Other controversy stems from differences in methodology of various empirical studies used to quantify the sometimes-abstract concepts involved in applying economic analysis to the environment.

Notwithstanding these controversies, this brief foray into the economics of environmental protection suggests a clear conclusion. Taken as a whole, the benefits of the Clean Air Act since enactment in 1970 clearly outweigh the costs. Several analyses, using different methods and different data sources, conclude that aggregate benefits exceed aggregate

[1b]Hahn (1996), page 240.

costs. As the recent EPA analysis notes, " ...even considering the large number of uncertainties permeating each step of the analysis, it is extremely unlikely that the converse could be true."[2c] One study, finds that benefits exceed costs for rules promulgated 1990-1995. Consistently, the studies show that the health benefits are by far the largest part of the monetized benefits.

To paraphrase Winston Churchill, economic analysis may be the worst approach to environmental policy making except for all the others that have been tried. Indeed, economic analysis is critical to the development of sound environmental policies. If economic analysis is not done explicitly, it almost certainly occurs implicitly. In that case, decision making is driven by public fears, special interest lobbying and bureaucratic preferences. Often, the resulting policies do not reflect the best interests of our citizenry. Future work needs to push beyond the aggregate analyses showcased in this paper and focus on more detailed assessments of individual programs and policies. Such assessments can help assure we get the maximum possible environmental protection for the resources committed.

[2c]EPA (1997), page ES-11.

REFERENCES

1. Freeman, A. Myrick III, *Air and Water Pollution Control: A Benefit-Cost Assessment,* John Wiley & Sons, New York, 1982.
2. Jaffee, Adam B., Peterson, Steven R., Portney, Paul R., and Stavins, Robert N., "Environmental Regulation and the Competitiveness of U.S. Manufacturing: What Does the Evidence Tell Us?", *Journal of Economic Literature*, Vol XXXIII (March 1995), pp 132-163.
3. Hahn, Robert W. "Regulatory Reform: What Do the Government's Numbers Tell Us?" in *Risks, Costs and Lives Saved: Getting Better Results From Regulation,* (Robert W. Hahn, editor), Oxford University Press, New York, 1996.
4. Morgenstern, Richard D. "Efficiency Chapter" in *Regulating Pollution: Does the System Work?* by J. Clarence Davies and Jan Mazurek, Resources for the Future, Washington, D.C.,1997 (forthcoming).
5. U.S. Environmental Protection Agency, *"The Benefits and Costs of the Clean Air Act, 1970 to 1990"*, Draft Report, April 1997.

SESSION E

MOBILE SOURCES AND TRANSPORT

Sustainable transport, the challenge ahead

Michael P. Walsh

3105 N. Dinwiddie Street, Arlington, Virginia 22207, USA

1. BACKGROUND AND INTRODUCTION

Four trends continue to dominate the global consideration of motor vehicle pollution control issues:

- the continued growth in the world's population;
- the rising affluence of many rapidly industrializing developing countries, increasing the affordability of motor vehicles;
- the increasing number of health studies showing adverse effects at lower and lower levels of pollution; and
- the response of governments by adopting more and tighter emissions standards for new vehicles or other incentives to stimulate the introduction of pollution controls on vehicles.

Across the entire globe, motor vehicle usage has increased tremendously. As we approach the 21st century, more than 700 million vehicles are on the world's highways - almost 500 million light duty vehicles, about 150 million commercial trucks and buses and another 100 million motorcycles. Over the last thirty five years, on average, the fleet has grown by about 12 million automobiles per year, 3.7 million commercial vehicles and 2.5 million motorcycles per year.[1] While the growth rate has slowed in the highly

[1] "World Motor Vehicle Data, 1996 Edition," American Automobile Manufacturers Association, 1996.

industrialized countries, population growth and increased urbanization and industrialization are accelerating the use of motor vehicles elsewhere.

One result is that most of the major industrialized areas of the world have been experiencing serious motor vehicle pollution problems. To deal with these problems North America, Europe and Japan have developed comprehensive motor vehicle pollution control programs which have led to tremendous advances in light duty vehicle pollution control technologies. At present, similar technologies are under intensive development and commercial introduction for heavy duty diesel trucks and buses, as well as two stroke motorcycles.

Motor vehicle related air pollution problems are not limited to the highly industrialized countries of the Organization for Economic Cooperation and Development (OECD), however. Areas of rapid industrialization are now starting to experience similar air pollution problems to those of the industrialized world. Cities such as Mexico, Delhi, Seoul, Singapore, Hong Kong, Sao Paulo, Manila, Santiago, Bangkok, Taipei and Beijing to cite just a few already experience unacceptable air quality or are projecting that they will in the relatively near future.

The purpose of this report is to survey what is presently known about transportation related air pollution problems, to summarize briefly the adverse impacts which result, to review actions underway or planned to address these problems, and to highlight future problems.

2. HISTORIC PATTERNS OF VEHICLE PRODUCTION AND USE

A. Trends in World Motor Vehicle Production

Overall growth in the production of motor vehicles, especially since the end of World War II, has been quite dramatic, rising from about 5 million motor vehicles per year to almost 50 million. Between 1950 and now, production increased almost linearly from about 10 million vehicles per year to about 50 million per year, approximately 1 million additional vehicles produced each year compared to the year before.

Over the past several decades, motor vehicle production has gradually expanded from one region of the world to another. Initially and through the 1950's, it was dominated by North America. The first wave of competition came from Europe, and by the late 1960's European production had surpassed that of the United States. Over the past two decades the car industry in Asia, led by Japan, has grown rapidly and now rivals both those in the United States and Europe. Both Latin America and Eastern Europe appear poised to grow substantially in future decades.

B. Trends in World Motor Vehicle Registrations

As for worldwide vehicle registrations, the long term trends are sharply upward and are actually accelerating. Europe (including Eastern Europe and the USSR) and North America each have about 35 percent of the world's motor vehicle population. The remainder is divided among Asia, South America, Africa, and Oceania (Australia, New Zealand, and Guam), in that order.

In terms of per capita motor vehicle registration for various regions, the United States, Japan, and Europe also account for the lion's share of the ownership and use of motor vehicles. Indeed, the non OECD countries of Africa, Asia (excluding Japan) and Latin America are home to more than four fifth's of the world's population, yet account for only one fifth of world motor vehicle registrations!

3. FUTURE TRENDS IN MOTOR VEHICLE REGISTRATIONS

Worldwide, the number of motor vehicles is growing far faster than the global population - about 5 percent per year, compared with about 2 percent per year. Analyzing trends in global motor vehicle registrations reveals that the global fleet has been growing linearly since before 1970 and that each year for four decades an additional 18 million motor vehicles have been added to the world fleet. If this linear trend continues, the global vehicle population will reach about 1.06 billion by the year 2010.

Analyzing growth in registrations per capita yields an even higher estimate for the world motor vehicle fleet. Each year worldwide registrations grow by about 2.8 vehicles per thousand persons. If this trend were to continue until 2010, there would be 178 motor vehicles per 1000 persons. If this figure is multiplied by the United Nation's medium variant estimate for global population in 2010 - 7.2 billion -- the motor vehicle fleet will be an estimated 1.2 billion, about 10 percent greater than would result from the strictly linear projection.

In addition to the continued growth in the global population, another factor contributing to the increased vehicle population is the growing affluence in certain rapidly industrializing developing countries, especially in Southeast Asia and Latin America. There is a very good correlation worldwide between the vehicles per capita in a given country and the GNP per capita.

A. Underlying Factors Which Foster Vehicle Growth

The growth in demand for motorized travel in recent decades is well understood. As urban areas populate and expand, land which is generally at the edges of the urban area and previously considered unsuitable for development is developed. The distance of these residential locations from the city center or other sub-centers increases, increasing the need for motorized travel. Motorized travel often in private vehicles supplants traditional modes of travel such as walking, various bicycle forms, water travel, and even mass transit. The need for private vehicles is reinforced as declining population densities with distances from urban centers reduce the economic viability of mass transit.

The evolution of the form of urban areas is influenced by the growth of income and accompanying increases in the acquisitions of private motor vehicles and changes in travel habits. As incomes rise, an increasing proportion of trips shifts, first to motorcycles and, as income increases further to private cars. The trend toward private motorization is not inevitable but is also influenced by public policy towards land use, housing, and transportation infrastructure. While the proportion of middle and upper income households in developing and newly industrialized countries able to afford cars and motorcycles is lower than in industrialized nations, the number of private vehicles still becomes very large as the middle and upper income groups grow in megacities. The number of vehicles and levels of congestion are comparable to or exceed that of major cities in industrialized countries. With the increase in motorized travel and congestion come increases in energy use, emissions and air pollution.

Focusing on Southeast Asia, the region currently experiencing the most rapid growth in road vehicles, as an example, all projections of population trends indicate both rapid increases and increasing urbanization of that population. In short, these trends generally increase the geographical spread of cities, both large and small, increasing the need for motorized transit to carry out an increasing portion of daily activities. Further, when coupled with expanding economies as is increasingly the case in Asian countries, a greater proportion of the urban population can afford personal motorized transportation, starting with motor cycles and progressing as soon as economically feasible to cars.

B. Trends in the Global Motor Vehicle Fleet (By Region)

Weighing the underlying factors influencing vehicle population growth especially population growth and economic development, projections of the future vehicle population have been made. In making these estimates, it was assumed that vehicle saturation, increased congestion and increasing policy interventions by governments would restrain future growth especially in highly industrialized areas.

As noted earlier, the global vehicle fleet has tended to be dominated by the highly industrialized areas of North America and Western Europe. This pattern is gradually changing not because these areas have stopped growing but because growth rates are accelerating in other areas. By early in the next century, based on current trends, the rapidly developing areas of the world (especially Asia, Eastern Europe and Latin America) and the OECD Pacific region will have as many vehicles as North America and Western Europe. Forty years from now, North America and OECD Europe could represent less than half the global fleet.

4. INCREASING HEALTH AND ENVIRONMENTAL CONCERNS

Cars, trucks, motorcycles, scooters and buses emit significant quantities of carbon monoxide, hydrocarbons, nitrogen oxides and fine particles. Where leaded gasoline is used, it is also a significant source of lead in urban air. As a result of the high growth in vehicles and these emissions, many major cities around the world are severely polluted. Reviewing the available evidence, the World Health Organization recently released new air quality guidelines for Europe. In addition to tightening requirements overall, the were unable to conclude that there is any acceptable level of particulate; therefore for this pollutant they could not recommend any acceptable threshold.

Beyond direct adverse health effects, there are other concerns with vehicle emissions. Among these is global warming or the greenhouse effect. Greenhouse warming occurs when certain gases allow sunlight to penetrate to the earth but partially trap the planet's radiated infrared heat in the atmosphere. Some such warming is natural and necessary. If there were no water vapor, carbon dioxide, methane, and other infrared absorbing (greenhouse) gases in the atmosphere trapping the earth's radiant heat, our planet would be about 60 F (33 C) colder, and life as we know it would not be possible.

Over the past century, however, human activities have increased atmospheric concentrations of naturally occurring greenhouse gases and added new and very powerful infrared absorbing gases to the mixture. Even more disturbing, in recent decades the atmosphere has begun to change through human activities at dramatically accelerated rates. According to a growing scientific consensus, if current emissions trends continue, the atmospheric build up of greenhouse gases released by fossil fuel burning, as well as industrial, agricultural, and forestry activities, is likely to turn our benign atmospheric "greenhouse" into a progressively warmer "heat trap", as Norway's former Prime Minister, Ms. Gro Harlem Brundtland, has termed this overheating.

Various human endeavors contribute to climate change. Recent estimates indicate that by far the largest contributor (about 50 percent) is energy consumption, mostly from the burning of fossil fuels. The release of chlorofluorocarbons (CFC's), the second largest contributor to global warming, accounts for another approximately 20 percent. Mostly known for depleting the stratospheric ozone layer, these stable, long lived chemicals are also extremely potent greenhouse gases. Deforestation and agricultural activities (such as rice production, cattle raising, and the use of nitrogen fertilizers) each contribute about 13 to 14 percent to global warming.

Carbon dioxide (CO_2) accounts for about half of the annual increase in global warming. The atmospheric concentration of carbon dioxide, now growing at about 0.5 percent per year, has already increased by about 25 percent since preindustrial times. Half of this increase has occurred over just the past three decades.

Globally, about two-thirds of anthropogenic carbon dioxide emissions arise from fossil fuel burning, the rest primarily from deforestation. In the United States, electric power plants account for about one third of the carbon dioxide emissions, followed by motor vehicles, planes, and ships (31 percent), industrial plants (24 percent), and commercial and residential buildings (11 percent).

The third largest contributor (after the CFC's) is methane (CH_4), accounting for about 13 to 18 percent of the total warming. Sources of this gas include anaerobic decay in bogs, swamps, and other wetlands; rice growing; livestock production; termites; biomass burning; fossil fuel production and use; and landfills. Methane may also be arising from the warming of the frozen Arctic tundra. The atmospheric concentration of methane is growing by about 1 percent annually.

Ozone (O_3) in the lower atmosphere (the troposphere) is the principal ingredient of smog. This gas is created in sunlight driven reactions involving nitrogen oxides, NO_x (as distinct from nitrous oxide, N_2O) given off when either fossil fuels or biomass are burned, and volatile organic compounds from a wide spectrum of anthropogenic and natural sources. In the United States, highway vehicles are the source of about 31 percent of NO_x emissions and about 44 percent of volatile organic compounds. Tropospheric ozone contributes about 8 percent to global warming.

Exactly where nitrous oxide (N_2O) comes from is still uncertain, but prime suspects include the use of agricultural fertilizers and, perhaps, the burning of biomass and coal. A growing source of N_2O is motor vehicles with three way catalysts. Nitrous oxide accounts for about 6 percent of current enhanced warming and also contributes to depletion of the stratospheric ozone layer.

As greenhouse gases accumulate in the atmosphere, they amplify the earth's natural greenhouse effect, profoundly and perhaps irreversibly threatening all humankind and the natural environment. While most scientists agree on the overall features of such warming, considerable uncertainties still surround its timing, magnitude, and regional impacts. Major unanswered questions include whether the additional clouds that are likely to form will have a net cooling or warming effect, how the sources and sinks of greenhouse gases will change, and whether the polar and Greenland ice sheets will grow or retreat. The complexity of the global climate system is daunting and the interactions between the atmosphere and the oceans are still imperfectly understood.

Unless measures are soon taken to reduce the release of greenhouse gases, by as early as 2030 they could reach levels equivalent to twice the carbon dioxide concentrations of pre industrial times. Two recent events have heightened concerns with global warming. In late November 1995, the IPCC Working Group 1 concluded that 'the balance of evidence suggests that there is a discernible human influence on global climate'. (Science 1995). Even more recently, a provisional report issued by the British Meteorological Office and the University of East Anglia concluded that the earth's average surface temperature climbed to a record high in 1995. In spite of commitments by most industrialized countries to stabilize or reduce CO2 emissions, very little progress has occurred in the transportation sector. Strategies such as mandatory increases in fuel economy or substantial increases in fuel taxes have proven elusive in recent years.

5. MAJOR REGULATORY PROGRAMS ARE MAKING PROGRESS IN REDUCING URBAN AIR POLLUTION

A great deal has been accomplished around the world in reducing motor vehicle pollution control. Achievement of this goal generally requires a comprehensive strategy encompassing emissions standards for new vehicles, clean fuels, strategies designed to assure that vehicles are maintained in a manner which minimizes their emissions and traffic and demand management and constraints. As a result of these efforts, there are clear signs of progress. For example, emissions of CO, HC and NOX from passenger cars in the United States and Germany, respectively, are down substantially in recent years in spite of the continued growth in the vehicle populations in these countries.

6. PROGRESS IN ADDRESSING GLOBAL WARMING IS LIMITED

In contrast with the success in reducing CO, HC and NOx from vehicles, there has been very little progress in reducing CO_2 emissions as will be discussed in the next section.

i. The US Experience

In the US, the Corporate Average Fuel Economy (CAFE) standards increased significantly during the 1970's and early 1980's but have remained fairly flat since the mid 1980's, actually declining over the past few years. In spite of the auto improvements, overall transportation energy consumption has continued to go up. This is due to a variety of reasons. For example, as auto fuel efficiency improved, sales of light trucks have increased substantially. Since many of these light trucks tend to be used much like passenger cars and have much lower fuel efficiency than the cars they are replacing, overall light duty vehicle efficiency gains are less than it would appear. In addition, CAFE did not apply to heavy duty vehicles, this category being subject only to a voluntary program. Further, national legislation was modified during the 1980's to allow speed limits to climb to 65 MPG thereby further undercutting energy conservation since fuel consumption increases as highway vehicle speeds climb. During the last decade the auto industry has been moving back toward the horsepower wars of the 60's, substantially increasing power output and reducing 0 to 60 MPH wide open throttle acceleration times over this period.

Simultaneously, people are using private cars and light trucks to drive to work much more than in the past with the result that public transit usage is down. When considering the improvements in new car fuel economy and the return of gasoline prices to pre OPEC levels, the cost of fuel for driving in real terms is much lower in the US today than at any time in the last two decades further encouraging additional driving. Not surprisingly, annual vehicle miles traveled has been increasing across the US by about 50 billion miles per year and because the amount of driving is increasing faster than the improvement in M.P.G., annual fuel consumption continues to increase.

Significantly, U.S. efficiency improvements began with the industrialized world's least-efficient car fleet. Only after the dramatic improvements observed to date are typical U.S. cars as generally efficient as those in the same weight class in other countries.

EPA data suggest that the fuel efficiency wars of the late 1970's gave way to horsepower wars in the 1980's. Throughout the decade, manufacturers substantially increased cars' power output. Unfortunately, in the real world, drivers with more horsepower available tend to accelerate their vehicles faster, thus using more fuel and needlessly increasing on-road emissions of nitrogen oxides, volatile organic compounds, and carbon monoxide.

There is little doubt that the rise in sales of trucks and big cars and the decline in U.S. motor-vehicle fuel efficiency both stem largely from the substantial drop in real fuel prices since the mid-1980's. Gasoline prices (expressed in constant 1989 dollars) were about as low in 1989 as they had been in the previous 39 years. Apparently, fuel prices speak louder than the federal CAFE standards that require manufacturers to produce lighter and more efficient vehicles.

ii. Rest of the World

The trends in U.S. vehicle growth and fuel consumption generally resemble global patterns. While most countries' autos were much more efficient than those in the US, the gap has narrowed substantially. The primary reason, is that despite a great deal of rhetoric, gasoline car fuel efficiency has hardly improved anywhere but in the US.

CO_2 from passenger cars accounts for about half of CO_2 emissions from Transport, and about 12 percent of total CO_2 emissions in the European Union.[2] Under a 'business as usual' scenario, CO_2 emissions from cars are expected to increase by about 20 percent by the year 2000 and by about 36 percent by the year 2010 from 1990 levels. In one year, an average medium size car in the European Union emits some 3 tons of CO_2.[3]

[2] Derived from 'A Community strategy to reduce CO2 emissions from passenger cars and improve fuel economy', COM (95) 689, Communication from the Commission to the Council and the European Parliament, Adopted by the Commission on December 20, 1995.

[3] Assuming 12,600 km per year and an average on road fuel consumption of 9.6 liters per 100 kilometers.

The road transport sector has stood out in recent years as one of the few sectors in the Union experiencing CO_2 emissions growth.

In the UK, a recent government report noted that 'fuel consumption' is rising fastest in the road transport sector, and there has been no improvement in fuel efficiency over the last 20 years. Fuel use for road transport has increased by 90 percent since 1970, accounting for a quarter of total energy consumption. Gasoline prices rose by just 2 percent during that period, compared with an 11 percent rise in household fuel.

In an era of low oil prices (~17$ per barrel) and ready availability of oil, there seems to be very little real interest in addressing energy concerns. This is dramatically illustrated by the continued increase in energy use for personal transportation in many parts of the world.

7. THE DIESEL AS A POTENTIAL SOLUTION TO HIGH CO_2 EMISSIONS

Driven in part by concerns regarding global warming, there is a clear trend toward increased sales of light duty diesel vehicles in many parts of the world. This trend can result in many positive environmental benefits including low fuel consumption, and therefore low levels of CO_2, low levels of exhaust CO and HC (especially during cold start conditions), and very low levels of evaporative hydrocarbons. However, increased diesel sales have a downside, relatively high NOX and particulate emissions. These pollutants continue to receive high priority attention in most areas of the world. As a result, countries around the world are increasingly tightening diesel regulations with the result that technology for reducing emissions continues to advance. However, NOx and PM levels from diesels remain much higher than from gasoline fueled vehicles raising the question whether the improved fuel economy of the diesel is an adequate trade of for the negative impacts on urban pollution.

There is a clear trend toward increased sales of light duty diesel vehicles in many parts of the world. Nowhere is this more true than in Europe, approximately one out of four new cars sold in 1995 was diesel fueled. Further, projections indicate even higher penetration is likely in future years.[4]

Increased sales of light duty diesels are not limited to Europe, however; the diesel is increasingly capturing the Japanese market, as well. In fact, in the United States where light duty diesel penetration has been almost nonexistent in recent years, there is some indication that diesel technology may be emerging as the front runner in the PNGV sweepstakes.[5]

Will the increased sales of diesels result in substantial reductions in carbon dioxide. Unfortunately, it seems unlikely. When considering total life cycle CO_2 equivalent emissions, diesel cars are estimated to have a 12.8% to 13.7% benefit compared to gasoline cars.[6] As direct injection gasoline engines emerge, the greenhouse gap between diesel and gasoline may even narrow.

8. CONCLUSIONS

Continuing air pollution problems from vehicle related pollution have been stimulating innovative pollution control approaches around the world. As these approaches are implemented, steady progress in reducing certain air pollution problems is occurring. An example is the experience in Southern California's Los Angeles Basin, which has had the most aggressive motor vehicle pollution control program in the world over the past forty years.[7] From 1955 to 1993, peak ozone concentrations were cut in half. The number of days on which Federal ozone standards are exceeded fell by 50 percent from the 1976-78

---

[4] 'AID Diesel Car Prospects to 2004', Automotive Industry Data Ltd., 1995

[5] 'Review of the Research Program of the Partnership for a New Generation of Vehicles', Second Report, National Research Council, 1996.

[6] 'Emissions of Greenhouse Gases from the Use of Transportation Fuels and Electricity', M.A. DeLuchi, Argonne National Laboratory, November 1991.

[7] 'The Automobile, Air Pollution Regulation and the Economy of Southern California, 1965-1990', Jane Hall et al, Institute for Economic and Environmental Studies, California State University, April 1995.

time frame to the 1991-1993 interval. Further, the average annual number of days above the Federal carbon monoxide standard fell from 30 to 4.3 during this same period and lead levels are now 98 percent lower than in the early 1970's. Most remarkably, this achievement occurred while the regional economy out-paced the national economy in total job growth, manufacturing job growth, wage levels and average household income. In short, a strong focus on environmental protection is not only not incompatible with strong economic development, they seem to be mutually reenforcing.

Where great progress has occurred, across the entire world, two motor vehicle related pollution problems stand out as needing the development of more creative and effective approaches. The first is the growing concern with urban particulate. As the available health information continues to raise more and more serious concerns regarding diesel particulate, they will need to get more attention in the future. As stated by the UK government in response to the reports from the Committee on the Medical Effects of Air Pollutants and the Expert Panel on Air Quality Standards, "the central element of any strategy will concentrate on technology based measures to secure further long term abatement of vehicle emissions, particularly from diesel vehicles".[8]

Secondly, the increasing problem of CO_2 emissions from the transport sector cries out for more innovative approaches. As recently noted by the European Commission, "Under a 'business as usual' scenario, CO2 emissions from cars are expected to increase by about 20 percent by the year 2000 and by about 36 percent by the year 2010 from 1990 levels. The road transport sector has stood out in recent years as one of the few sectors in the Union experiencing CO_2 emission growth".[9]

While technological improvements in petroleum-powered vehicles are essential to achieving short-term increases in the vehicle fleet's fuel efficiency, they will not be sufficient for the long haul if the global vehicle fleet continues growing. Focusing on

---

[8] Press Release issued on Behalf of Department of Environment, Department of Health and Department of Transport, 'Government Acts On Airborne Particles, Achieving Particle Standard Will Mean Significant Health Benefits', November 8, 1995.

[9] 'A Community Strategy to reduce CO_2 emissions from passenger cars and improve fuel economy', COM (95) 689, December 1995.

increased use of diesel technology seems short sighted in this regard not only because of the increased NOx and PM which diesel technology currently is burdened with but also because the greenhouse gas reduction potential is so marginal. For this reason, longer-term international efforts to develop new transportation energy sources that emit no carbon dioxide will have to be intensified as emissions are reduced. A program of research, development, demonstration, and, ultimately, the introduction of such vehicles should become a matter of high public priority for all the principal vehicle-producing nations. Technologies involving fuel cells and biomass based fuels should play an increasingly important role in solving transportation related environmental problems in the future.

Motor Vehicle Related Air Pollution Issues in the 21st Century

5th US-Dutch International Symposium

Global Trend In Motor Vehicles

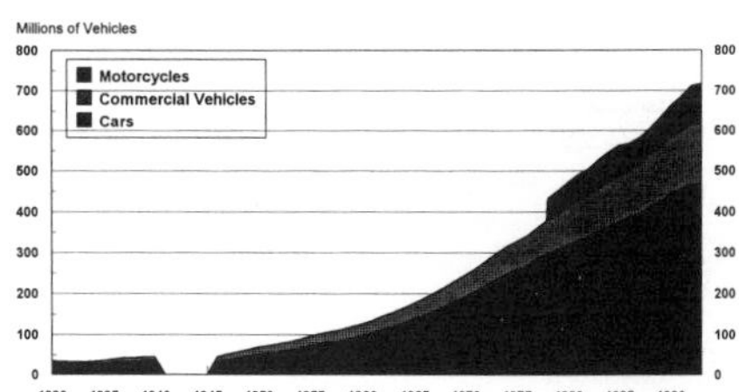

Global Trend In Motor Vehicles

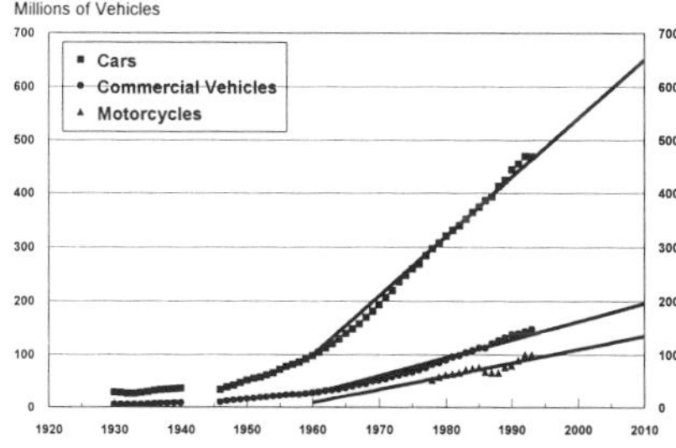

Global Trends In Motor Vehicle Production

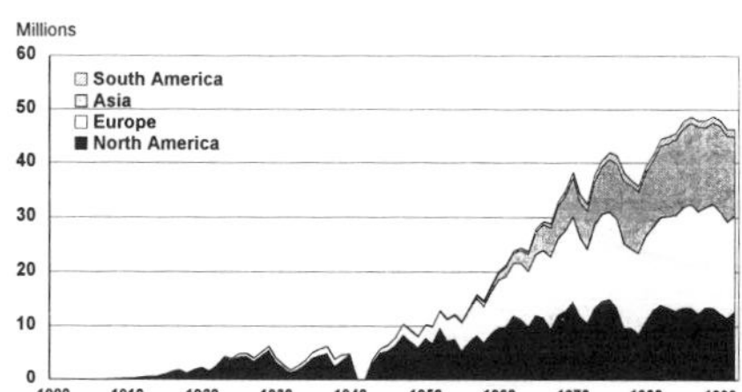

Global Trends In Motor Vehicle Production

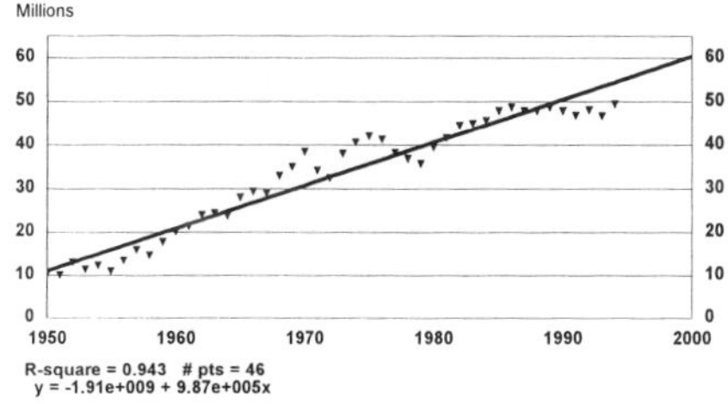

Global Trends In Motor Vehicle Production

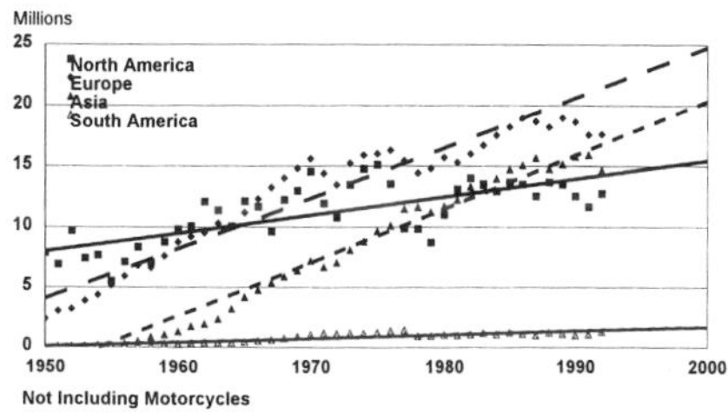

Global Distribution of Vehicles and People

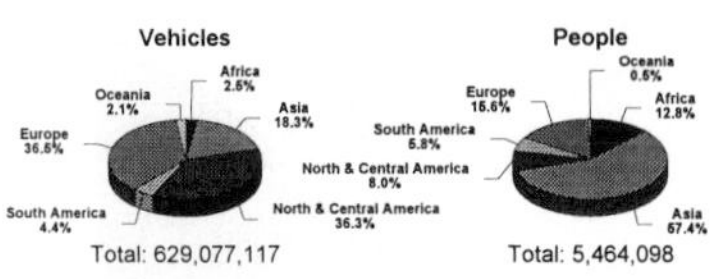

Global Trend In Motor Vehicles and People

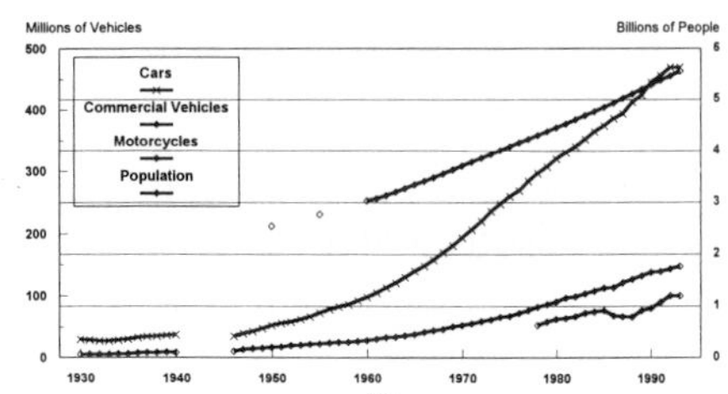

Global Trend In Motor Vehicles and People

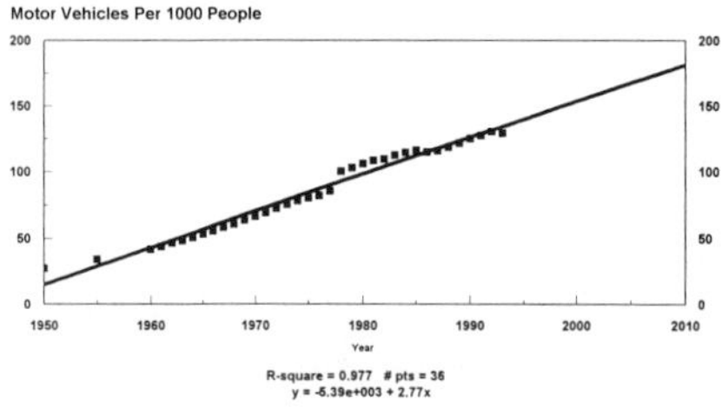

Vehicle Population Versus GNP

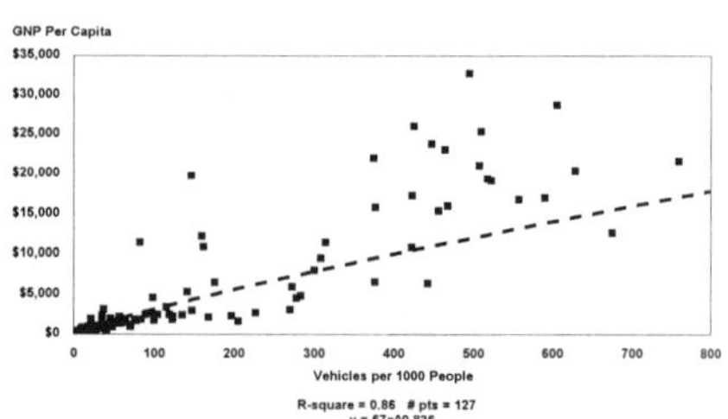

WHO Recommended Air Quality Guidelines For Europe

Compound	Guideline Value	Averaging Time
Ozone	120 Ug/m3	8 hours
Nitrogen Dioxide	200 Ug/m3	1 hour
Nitrogen Dioxide	40-50 Ug/m3	Annual
Sulfur Dioxide	500 Ug/m3	10 min
Carbon Monoxide	100mg/m3	15 min
Carbon Monoxide	60 mg/m3	30 min
Carbon Monoxide	30 mg/m3	1 hour
Carbon Monoxide	10 mg/m3	8 hours
Particulate	No Threshold	

Trends in Exhaust Emissions From US Cars Normalized to 1970

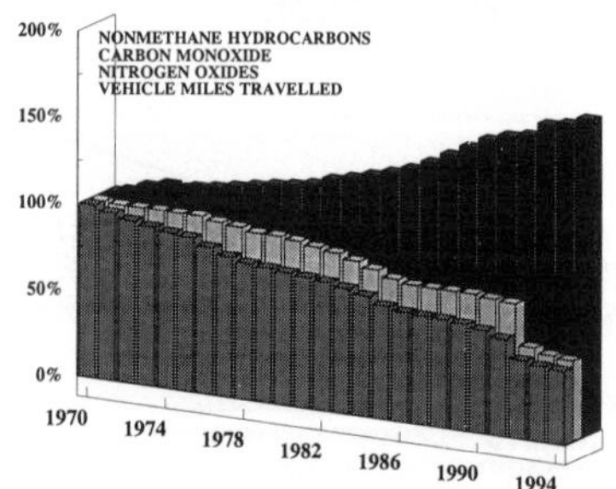

Emissions From Cars In Germany Normalized To 1980

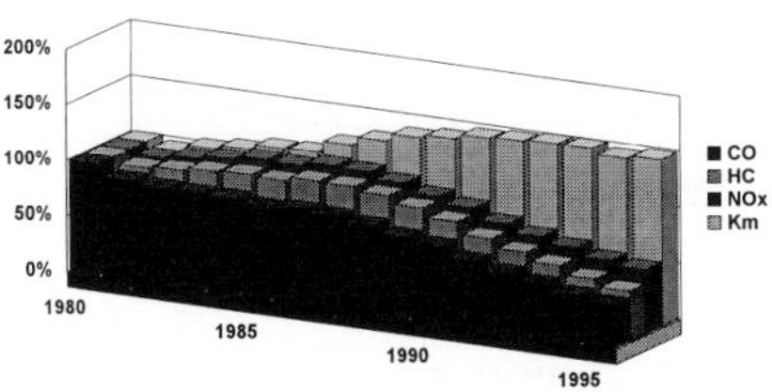

Source: Dr. Ulrich Hoepfner, IFEU, Heidelberg

Unleaded Gasoline Is Becoming Dominant

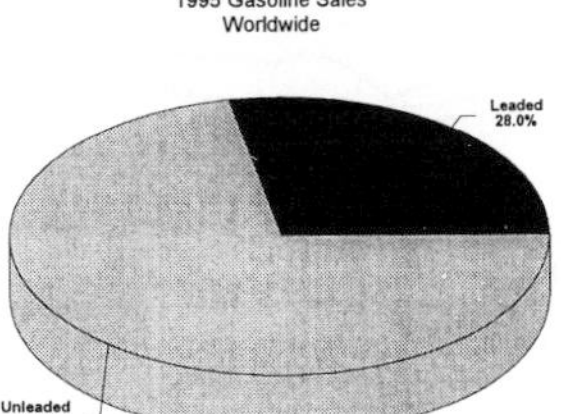

Light Duty Gasoline Vehicle Pollution Controls

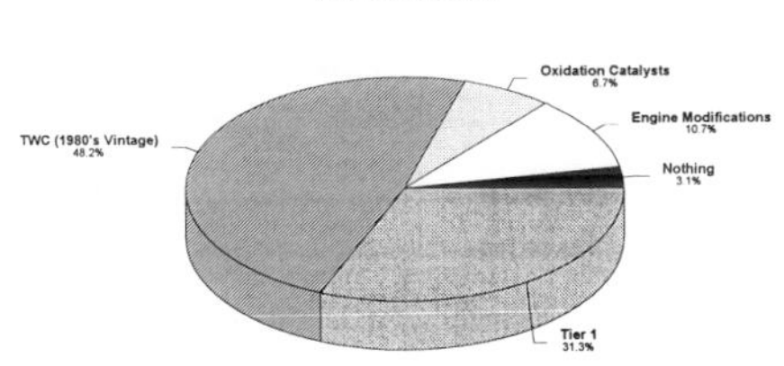

California Low Emissions Vehicle Program - (Grams Per Mile)

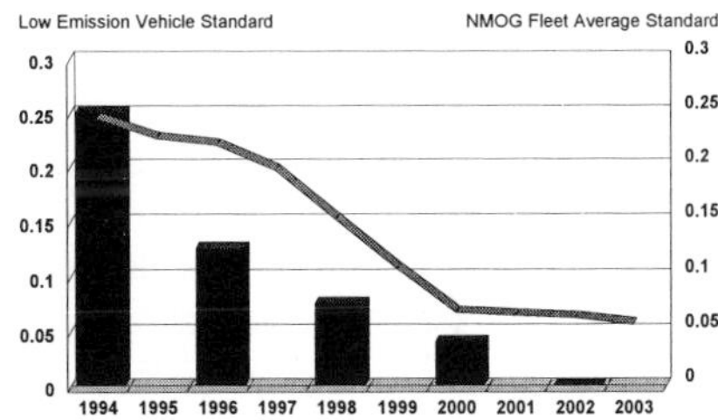

Global Atmospheric Concentrations of Greenhouse Gases - PPM

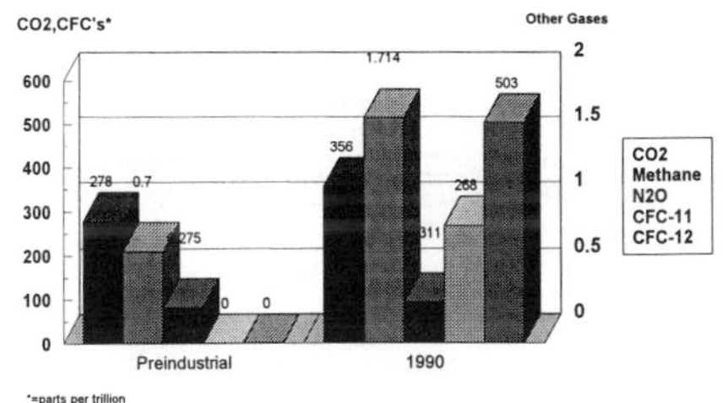

Recent Trends in the United States

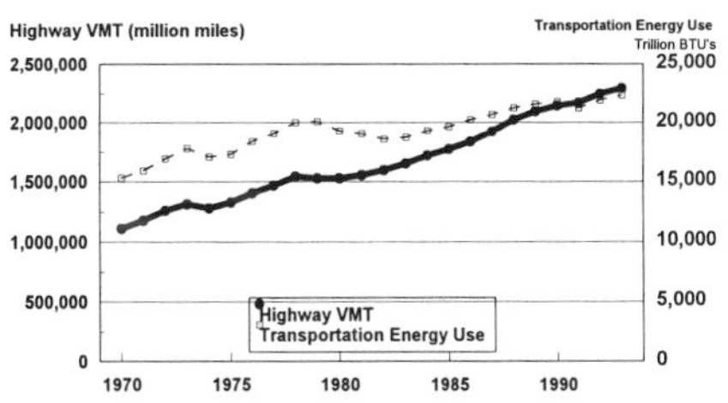

Fuel Economy of Gasoline Automobile Population for Selected Countries (MPG)

30
25
20
15
10
1970 1975 1980 1985 1990
Japan UK
France US
Sweden

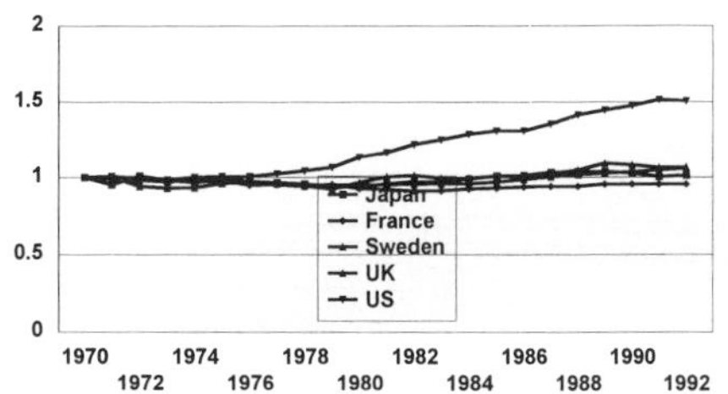

Recent Trends in the United States

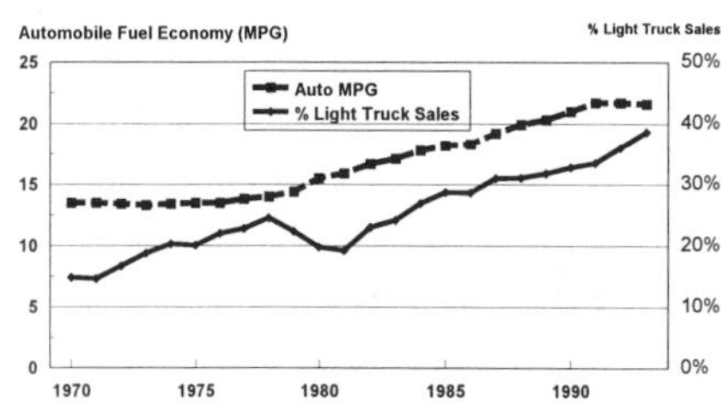

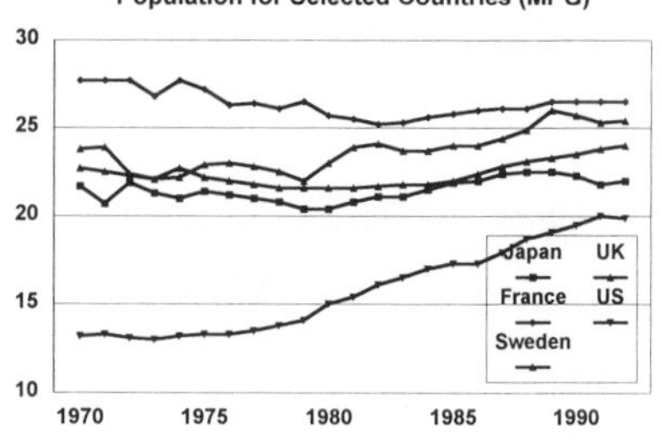

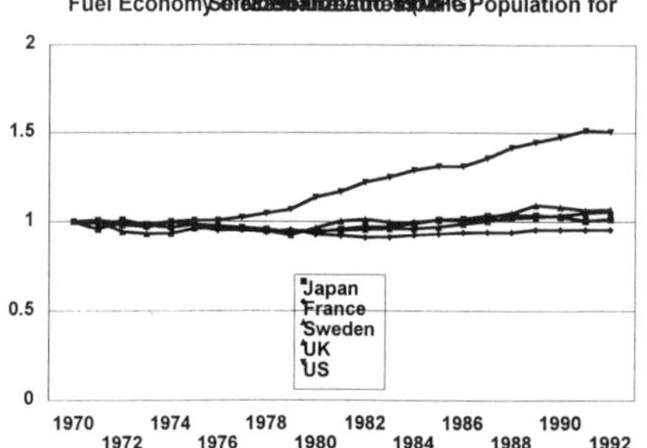

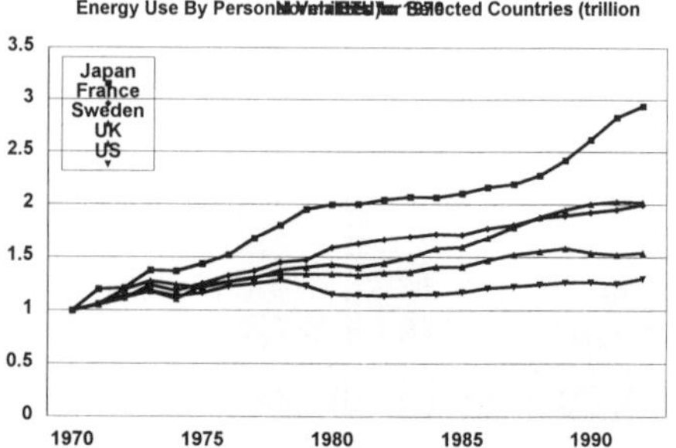

US Emissions of Greenhouse Gases From Transportation

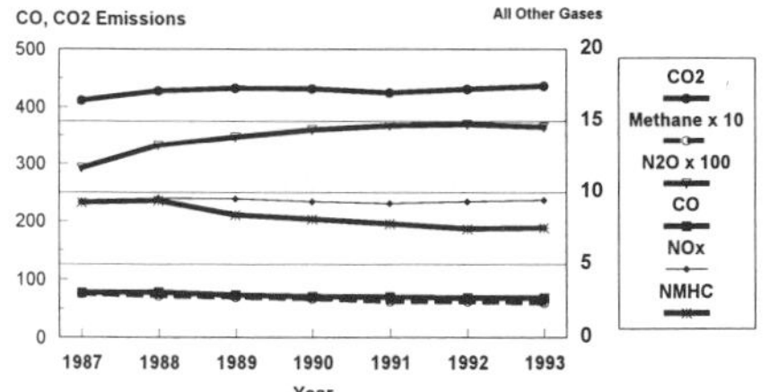

Transportation Contribution To Greenhouse Gases in the US - 1993

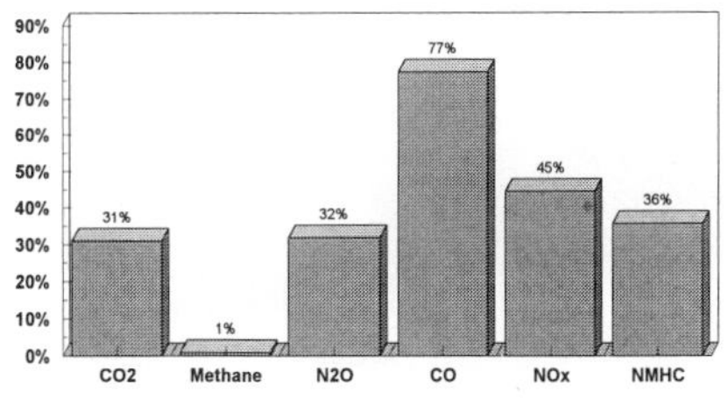

Average Source Contribution To Midtown Manhattan Site

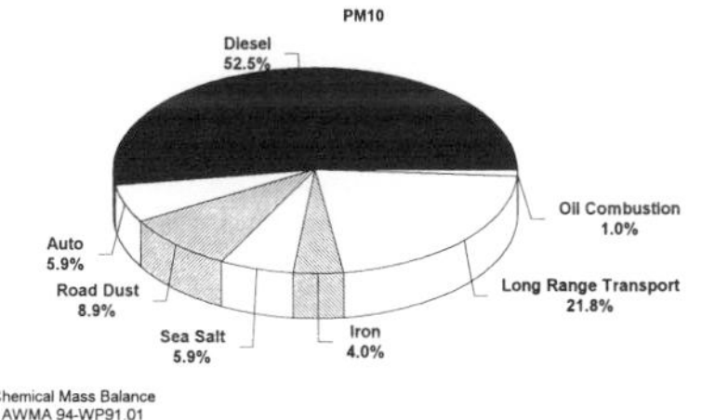

Europe Car Production By Fuel System Type

Trend In Diesel Vehicles by Category in Japan (000)

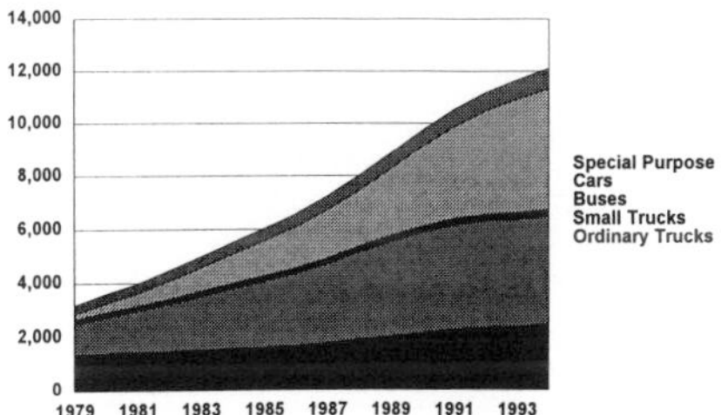

Effect of Diesel Penetration On Effective NOx Standard

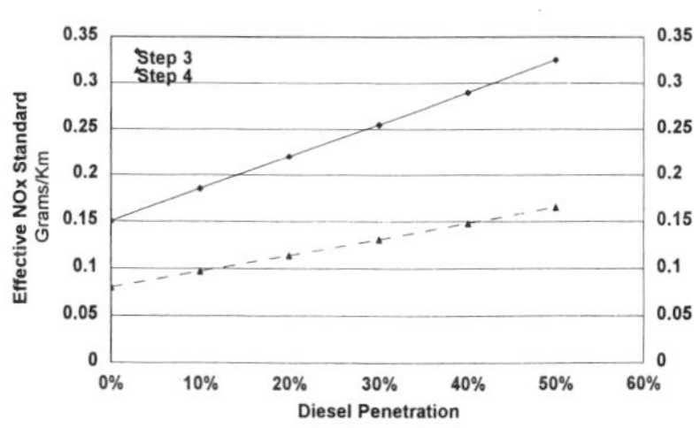

Total Fuel Cycle CO_2 Equivalent Emissions For Light Duty Vehicles

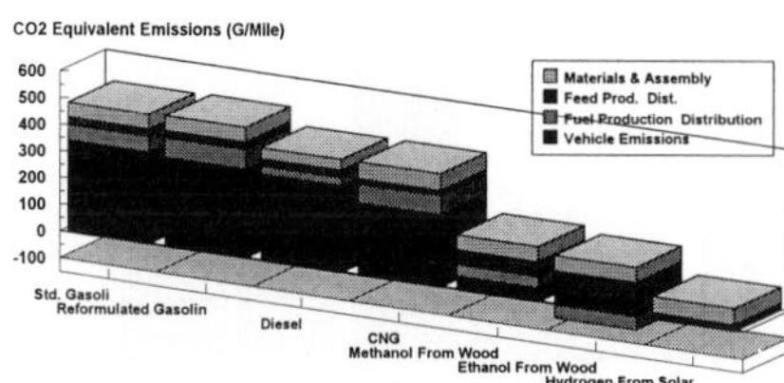

Conclusions

- Global Vehicle Population Continues To Grow Rapidly
- Greenhouse Gases From Transportation Sector Continue to Grow Rapidly
- Virtually No One Has An Adequate Program In Place To Address Greenhouse
- Focus on Diesels Without Tight NOX/PM Requirements Raises Health Risks Without Seriously Addressing Greenhouse Problem

European transport: emission trends and policy responses

A.N. Bleijenberg and J.M.W. Dings

Center for Energy Conservation and Environmental Technology,
Oude Delft 180, 2611 HH Delft, The Netherlands

ABSTRACT

This paper describes the trends in the main emissions to air by European road transport. Following this overview, policy options are discussed which can bring the emissions down to levels which meet environmental targets.

Of the so called "classical" pollutant emissions (NO_x, HC, CO, particulates and SO_2) HC, CO and SO_2 can both quite well be tackled by technical measures. The current trend towards lower emissions will continue, provided extra policy measures will be taken. NO_x is a more persistent problem. A strong technical effort improvement of diesel technology is required to bring this emission to acceptable levels.

Policy instruments are emission and fuel standards. Stringent fuel standards and check of conformity of vehicles in use become increasingly important. Special emphasis is required for cold start standards, two wheelers and off road machinery.

The trend in CO_2-emissions from road transport shows a different picture: an annual growth of 2 to 3%. This might slow down a little in the coming decades, because traffic growth could diminish somewhat. For CO_2 technical progress is clearly slower than traffic growth.

A strong government interference in the car sales market is suggested to achieve a reduction in CO_2-emissions. A 'downsizing' trend of the car can be forced and/or high additional costs are needed for the car. Both consequences are expected to reduce mobility growth somewhat.

For freight transport the answer is a drastic increase in efficiency of road haulage. Load factors are currently rather low and should go up. Next, the growth in transport intensity of economic production should be slowed down. Higher prices for road haulage seem the best policy instrument to achieve this.

1. CLASSICAL POLLUTANTS: NO_x, HC, CO, Pm, SO_2

1.1. Introduction

This chapter describes the development of the so called 'classical' emissions from mobile sources in Europe. These are the emissions that are covered by regulation: NO_x, HC, CO, Pm (particulate matter) and SO_2.

It describes the original emission situation, the current regulatory and technological developments and it gives an outlook to the future.

Volume developments will not be explicitly addressed in this chapter as they are treated in the next chapter on CO_2 emissions. It must be said that, as long as engine and fuel technology are not into play, measures to reduce CO_2 emissions mostly bring the same reduction of the 'classical' emissions. The measures discussed in this chapter must therefore be considered as *additional* to the measures presented in the 'CO_2' chapter.

1.2. Developments in new vehicle technology

Gasoline (petrol) cars

Historically the *gasoline car* has been the primary source of traffic pollution, due to the large share of total kilometers driven and the high emissions per kilometer driven.

Regarding the latter aspect, the reduction of emissions per kilometer from new SI (Spark Ignition, i.e. gasoline, LPG and CNG) vehicles is quite spectacular. The application of the three way catalyst on this type of engines has made over 90% emission reduction per km possible (except, of course, for CO_2). The first generations of catalyst equipped cars in the 80's faced quite some problems in the area of driveability and durability, increased fuel consumption and odd phenomena like a 'rotten egg' smell with certain catalyst types. The Dutch 'in use compliance' testing program shows that improvements in catalyst technology, fuel injection and engine management systems have largely eliminated these initial shortcomings.

The emissions of NO_x and HC from a modern well functioning gasoline car with a warm catalyst are practically negligible, a few *hundredths* of grams per kilometer (compared to a few grams per km in the early 80's). This equals to a few tenths of grams per liter of gasoline burnt. New emission standards as proposed by the European Commission for the year 2000 and 2005 are even 60 to 70% stricter.

However, the cold start emission behavior of gasoline cars is still poor, especially at low ambient temperatures, when an excess of fuel is injected into the cylinders to get enough evaporation. As a result even modern cars have excess HC and CO emissions in the first 2 kilometers after a cold start at lower ambient temperatures. The effect can be noticed very well on winter mornings in traffic peaks. Inclusion of a special cold test in the type approval procedure, which makes fuel pre-heating at low temperatures necessary, is therefore most desirable. This item is currently under discussion in Europe, but the outcome is still uncertain. The United States already have a low temperature test but the standards are fairly weak.

The costs of emission reduction of new gasoline cars amount to about 5 to 10 ECU's per kg of NO_x and HC emission reduced.

Diesel cars

The emissions from new diesel cars, especially NO_x and Pm_{10}, are generally considered more problematic than those from new gasoline cars. A modern passenger car diesel engine produces about 10 grams of NO_x per liter of diesel fuel burnt, about two thirds of the older generations' emissions. For the next five years only a 20% reduction in new cars is foreseen and this is not enough to compensate for the rapid growth in diesel car use.

Several NO_x after treatment technologies are currently under development but they are not expected to enter the market before about 2002. The other diesel car problem, the Pm_{10} emission, will be treated later on in this article. This emission gives especially rise to worries in countries with a large share of cars running on diesel, like France.

HC and CO emissions from diesel cars are not of great concern.

It must be said that the diesel car offers about 15% fuel consumption/CO_2 emission advantage compared to gasoline cars. However, careful analysis points out that this CO_2 advantage of diesel cars does not fully compensate for their NO_x and Pm_{10} disadvantage|1|.

The costs of emission reduction of new diesel cars amount to about 5 ECU's per kg of NO_x emission reduced.

Heavy trucks and buses

Without any new legislative measures the heavy truck will without competition be the main polluter on the road in the year 2010. The modern 'Euro 2' truck engine's NO_x emission is about 30 grams per liter of diesel fuel burnt, which is three times as much as the diesel passenger car and about 40% lower than the 80's generation of truck engines. Basic calculations indicate that traffic growth will more than compensate for this reduction, resulting in an unacceptable rise of total NO_x emissions from trucks and a 'market share' for trucks in NO_x emissions from road traffic in 2010 of about 70%. It is clear that either major technology steps or major volume reductions are necessary to move towards environmental targets for ozone formation, acidification and eutrophication.

The most promising concept currently under development to achieve a very substantial reduction of NO_x emission per unit of fuel burnt is Selective Catalytic Reduction or SCR. The technology is almost in the demonstration phase of development and seems to have about a 60 to 70% NO_x reduction potential over the current generation of trucks. It gives this performance without sacrifices in terms of fuel consumption or particulate emission.

Besides the uncertainty about the 'on the road' performance, this technology faces two main obstacles for a rapid and massive penetration on the market. The first obstacle is probably the need of an *extremely* low sulphur diesel fuel and the second obstacle is the need of a reducing agent like urea. This implies a substantial effort from the refining industry, the introduction of an extra product in the fuel distribution system and anti-tampering provisions. It must be ensured that a) the trucker will not 'forget' to fill the urea tank when driving in the EU, and b) is not prohibited from driving in non-EU states where no urea is sold.

Strong joint action of engine designers, catalyst manufacturers, oil industry and policy makers is needed to overcome these obstacles.

The costs of emission reduction of new trucks and buses amount to about 2 ECU's per kg of NO_x emission reduced.

Mopeds and motorcycles

Two wheelers (mopeds and motorcycles) deserve far more attention than they are currently given.

Mopeds, especially the two stroke versions, are a serious threat for urban air quality. Measurements show that HC emissions from mopeds can rise up to 10 grams per km and they can even double when the engine has been tampered with. This is about a hundred times more than a modern gasoline car with a hot catalyst. Combining this with the knowledge that mopeds mostly drive in urban areas it is clear that they are a key factor in future urban air quality improvement. This is especially the case in southern parts of Europe where mopeds are used extensively in cities.

The motorcycle problem is a bit different. They cover more kilometers in total but the emissions per kilometer are lower and they drive less in urban areas. BMW shows that low motorcycle emissions are very well possible by offering bikes equipped with injection engines and closed loop three way catalyst systems. Other manufacturers however still use carburetor systems and do not use any catalytic conversion.

Off road equipment

This extensive category of machinery includes ships, trains, tractors, shovels, forklift trucks, cranes etcetera. Altogether they have a surprisingly high emission 'market share' in the EU. They emit about 30% of NO_x emissions and 40% of Pm_{10} from mobile sources. For example, the NO_x emission from Dutch inland vessels is about one third of the NO_x emission from Dutch trucks.

Technically the off road diesel engines' emissions can be reduced in the same way as the truck emissions. However, these sources are not subject to any form of emission regulation yet. As volume growth rates are not likely to be smaller than those of road traffic it is expected that the share of total EU emissions will grow quite rapidly.

The first EU legislation on mobile machinery is expected to come into force from 1998. However, it does not cover ships, diesel trains and agricultural and forestry tractors. This leads to the odd conclusion that from the NO_x emission point of view it could in 2005 be better to transport goods by a new truck than by a new inland vessel !

This indicates that a lot of policy work will have to be done in the coming years.

The costs of emission reduction of new mobile machinery are relatively low. Even the first cheap emission measures have not been taken yet. However, cost effectiveness is highly dependent on the area of engine application. For example emission reduction of new engines of inland vessels is very cheap, about 0.4 ECU's per kg of NO_x emission reduced.

1.3. The issue of particulate emission

The issue of the emission of particulate matter (Pm) requires some extra attention. Pm emission is caused by poor fuel combustion, poor fuel quality, especially a high sulphur content, and by lube oil burn. Pm consists of soot (unburnt C atoms) and sulphate (from sulphur), often with very large polycyclic aromatic hydrocarbons (PAH) sticking to it.

In the earlier generation of trucks the problem is very much visible: thick black clouds of smoke coming from the exhaust pipes of diesel cars and trucks, especially in acceleration. The particulate emission factor from earlier generation diesel engines is about 3 grams per liter of diesel fuel burnt.

At first sight an enormous progress has been made in reducing particulate emissions. Current truck engines emit only about half a gram of particulates per liter of diesel fuel, which is 80 to 90% less than before. This progress is mainly caused by a spectacular improvement of combustion circumstances: a much higher injection pressure and electronic injection timing and quantity control, resulting in a much better air/fuel mixing. Also the lower diesel fuel sulphur content made a substantial contribution.

However, closer analysis points out that mainly the particulate *size* has been reduced, not so much the *number* of particulates. Evidence is growing that especially these small particulates can cause great health damage as they enter the lungs very deeply. Some people even say that gasoline engines' particulate emissions (which are fairly negligible in terms of mass) are therefore at least as harmful as the diesel engines' particulate emissions.

If the relationships as described above appears to be true the classical way of measuring and regulating emissions (by mass) would be not effective at all. The particulate problem becomes a very hard nut to crack. It is yet very unclear how emission policy is going to react to this phenomenon. Size related emission limits could be a consequence.

1.4. Control of 'in use' emissions

With increasing complexity of emission control equipment to be installed on new vehicles it becomes increasingly important to check whether the 'on the road' or 'in use' performance of the technology is still satisfactory. In the case of advanced emission technology (as applied in gasoline cars) the emissions from a malfunctioning vehicle can be ten or twenty times higher than the emissions from a correctly tuned and maintained vehicle.

There is a wide range of possible measures to ensure that every vehicle on the road remains well below the emission limits for that vehicle type. There are two main categories of measures.

In the first category every vehicle is inspected individually once in a while. These so called 'Inspection and Maintenance' (I&M) programs are generally quite costly. Mostly the costs are passed to the consumer. This makes these programs politically sensitive and therefore practically every country has got its own regime. It is obvious that these programs are potentially the key to ensure low emissions in every day reality. A major technology currently being implemented on new vehicles in the US and planned to be implemented in the EU in the year 2000 is the so called OBD (On Board Diagnostics) system. This system gives information on the functioning of the emission control devices. With this information checks and repairs in the I&M program become much easier. Probably the main effect of the OBD system is a more careful and robust design of emission control technology, as every manufacturer will try to avoid any fault indication.

In the second category a statistical sampling method is used to check a limited number of vehicles of every vehicle type. Generally this is called an In Use Compliance (IUC) procedure. If any systematic production faults appear to occur within a certain vehicle type appropriate measures can be taken. In the most extreme case a recall campaign could be the consequence with major costs for the manufacturer. This makes a IUC strategy a very effective tool to improve the quality of emission control devices, at the same time being a form of consumer protection.

Both types of, programs to control 'in use' emissions play a crucial role in actually *ensuring* low emission performance of vehicles 'on the road'. Limitations to the stringency of such programs are determined more by economics and politics than by technology.

1.5. The role of fuel quality

The role of fuel quality in attaining low emission performance is often not very well understood. Fuel quality influences a vehicle's emission performance in multiple ways.
The first way is the one-on-one link between fuel quality and vehicle emissions, *independent* of vehicle technology. This is the case for carbon/sulphur content of the fuel and CO_2/SO_2 emissions respectively.
The sulphur content of the EU main transport fuels has been reduced substantially to 0.03%m for gasoline and 0.04%m for diesel respectively. New desulphurization steps will be taken (see below). With these values the SO_2 emission from road traffic is not a major issue any more.
The carbon content cannot be influenced substantially. Therefore CO_2 is a very persistent problem which will be treated in the next chapter.
The second way is a direct influence of fuel quality on vehicle emissions, *dependent* on vehicle technology. Typical parameters here are benzene content, aromatics content, volatility (for gasoline) and cetane number, density and aromatics (for diesel fuel). The emission effects of changes in these parameters are different for every vehicle type but reductions achievable are as high as about 20%. These effects may be counterbalanced by increases in energy use (CO_2 emissions) by refineries. Studies have led to the introduction of cleaner burning RFG (reformulated gasoline) in certain zones in the United

States and the EU wide introduction of a set of environmental specifications for gasoline and diesel fuel by the year 2000.

A third way is the fuel quality being a prerequisite for proper functioning of emission abatement technology. This effect is most important when catalysts come into play. For example lead in gasoline used to be a main issue in relation to poisoning of the three way catalyst. At the moment sulphur in gasoline as well as in diesel fuel is widely discussed for its effect on the three way catalyst and on future diesel DeNO$_x$ catalyst systems. Especially the latter category of catalysts seems to be very sensitive to fuel sulphur, as described in an earlier paragraph.

1.6. Conclusions per emission

NO_x

From the description in the previous paragraphs it appears that in the future NO_x will be about the most persistent of the 'classical' emissions. Though gasoline vehicles' NO_x emissions are decreasing with several per cents per annum total NO_x emissions from traffic only slowly decrease. The diesel engine is responsible for this unsatisfactory development.

First, there still is a technical problem. The non-catalytic NO_x reduction techniques for diesel engines do not offer enough potential and there is quite some uncertainty about the future success of catalytic systems. Therefore it is not expected that NO_x emissions from diesel cars and trucks will decrease substantially until 2005.

Secondly, there is a political problem. A major part of today's heavy duty diesel engine sales does not face any emission legislation at all. Introduction of legislation for all types of diesel engines will at least take 5 more years. It will take about 10 more years before any substantial reduction of total NO_x emissions from these yet unregulated engines will be seen.

Summarizing, total NO_x emissions from mobile sources will decrease by only about 30% till the year 2005 and a 50% reduction will not be seen before 2010. Under the conditions that technology develops well and that continuous stringent regulation is applied to all new diesel engines NO_x emission targets (80% reduction) could be attained in the long term, about 2020 or 2025.

Particulate matter

This emission gives rise to worries for other reasons. In terms of tons of total emissions per annum a satisfactory improvement can be seen due to technical progress of the diesel engine. However, scientific evidence is growing that human health effects do not seem to correspond with the total load in tons of emission but merely with a combination of total load and particle size. Is therefore very uncertain whether the *effects* of particulate emissions are decreasing as well.

HC and CO

Currently the HC and CO emissions from historically the most important source, the gasoline car, are decreasing by about four per cent per annum as a result of the very low emissions of new gasoline vehicles entering the market. As a result, today's total HC and CO emissions from mobile sources are already about 50% lower than 15 years ago.

A lot of people are convinced that CO air quality targets will be met with the existing regulation only. HC emissions are more dangerous because these can have long term health effects and cause the quite persistent ozone problem.

Future strategies to reduce HC emissions from mobile sources are:

- controlling 'cold start' emissions from gasoline vehicles, especially at low ambient temperatures.
- keeping the existing fleet in good emission shape by a stringent inspection and maintenance policy and by 'in use compliance' monitoring.
- reducing the (urban) HC emissions from two wheelers.

If these issues will be tackled effectively the necessary 80% HC and CO emission reduction from 1990 levels could be met somewhere around 2010 or 2015.

SO_2

This issue has quite effectively been tackled by the desulphurization of diesel fuel. Emissions will be further reduced as a result from new desulphurization steps in 2000 and 2005. Emissions from combustion of high sulphur heavy fuel oil in marine diesel engines give most rise to worries.

2. CO_2-EMISSIONS

2.1. Introduction

The UN Framework Convention on Climate Change (FCCC - signed by most European countries - states as its main aim to stabilize the atmospheric concentration of green house gases on a level which avoids dangerous anthropogenic distortions of the climate system. The UN Intergovernmental Panel on Climate Change (IPCC) is a little more specific and concludes that the rise in global temperature should be limited to 2° C, with a maximum increase per decade of 0.1° C |2|. With these limits it is estimated that the industrialized countries need to reduce their emission of green house gases with 1 to 2% each year |3|. For this estimate it is assumed that developing countries have some scope to increase their emissions, while industrialized countries have to cut back. This paper takes the desired reduction in green house gases with 1 to 2% per year as starting point.
Current trends, however, reveal a growth in CO_2 emission in the industrialized world. For Europe the trend can be roughly estimated at an annual growth of 1% |4|. Road transport has a higher growth rate than average, which can be estimated at around 2-3% per year (see next sections).

2.2. Trends in passenger transport

The CO_2 emissions from passenger traffic increase by 2 to 3% each year. The main underlying trends in the distinguished factors will be discussed hereafter |4|.

Population size

Although the population in Europe has grown in the last decades - and centuries - this is not a major factor in explaining the fast growing environmental pressure from passenger transport. According to current trends for the coming years the European population grows with roughly 0.3% per annum, for both West and Central Europe.

Travel distance per person (per day)

Statistics from the ECMT show that the volume of passenger kilometers in Europe has grown on average with 3.1% per year in the period 1970-1994. Part of this growth, however, is caused by population growth, estimated at 0.7% per year in this period. This results in a trend in growth of travel distance per person of around 2.4%.

Modal split

The share of car traffic in total passenger traffic has increased from around 76% in 1970 to 82% in 1994 (excluding walking, cycling and aviation). This implies that the share of car traffic rose each year on average with 0.3 to 0.4%. Assuming that emissions per passenger kilometer of public transport are roughly half of those of cars, this modal shift corresponds with an annual increase in emissions of around 0.1 to 0.2%.

Occupancy

Because car traffic is dominant with respect to pollution from passenger transport, only the load factor of cars will be discussed.

It seems that the occupancy of cars tends to decline slowly. This can be illustrated by data for The Netherlands. In 1980 on average 1.73 persons were seated in a car, which went down to 1.65 in 1994. This correspond with an average reduction of the occupancy by 0.3% per year in this period. The average fuel consumption per passenger kilometer went up almost the same rate.

Efficiency

Again, only the efficiency of cars will be reviewed. Efficiency improvements of the cars have been achieved and resulted in a reduction of specific energy consumption of 0.5 to 1.0% per annum. This outcome is, however, a combination of two opposite trends. For it is well known that cars have become faster, heavier and more comfortable over time. This development can be summarized as upgrading. Several studies estimate that this upgrading trend results in 0.5% higher specific fuel consumption per year. This implies that technical progress is around 1 to 1,5% per year, but that roughly half of the technical achievements are offset by upgrading of the car.

Overview

Table 1 summarizes the main influences resulting in an annual growth of CO_2 emissions from passenger traffic of 2 to 3% per year. The dominant factors are the growth in travel distances and the efficiency improvement of the car. Next, the upgrading of cars is important as well. The other factors - population size, modal split and load factor - only explain the growth in CO_2 emissions for a small part.

Table 1

Estimated main trends behind the growing CO_2 emissions from passenger traffic.

Factor	Impact on CO_2 emissions
Population size	+0.3%/year
Travel distance per person	+2.0 - 2.5%/year
Mobility	*+2.3 - 2.8%/year*
Modal shift	+0.1 - 0.2%/year
Occupancy car	+0.2 - 0.3%/year
Upgrading car	+0.5%/year
Technological improvement car	-1.0 - 1.5%/year
CO2 emissions passenger traffic	*+2.1 - 2.8%/year*

Underlying forces

Two important underlying forces behind the growth in CO_2-emissions can be identified |4|. The first is increased travel speed. The introduction of the car and later of the aircraft allowed for higher travel speeds. This did, however, not result in time savings, but instead in increased travel distances, as economic research shows |5|. There appears to be a roughly Constant Travel Time Budget implying that a higher speed generates longer average travel distances. This is in the long run the main force behind the growth in mobility.

The second underlying force is income growth. Since 1965 real income per person has increased by around 2,5% a year in Western Europe, resulting in roughly doubling the total real income. The income growth generated both the large penetration of the (faster) car and allowed for a continuous 'upgrading' of the car. Maximum speed, acceleration, weight and size of the average car have increased over the last decades and are expected to continue to do so.

2.3. Trends in freight transport

The most familiar trend in the realm of freight transport is probably that the number of ton kilometers of all modes of transport is growing at the same rate as GDP. This means that the ton kilometers of road transport are growing faster than GDP, since the relative share of road transport is increasing. However, there are variety of factors behind this trend.

Growing transport distances

The underlying causes of the growth in freight transport can be illustrated with a description of trends in the UK food and drink sector |6|. The logistical changes taking place between 1983 and 1991 are shown in figure 1.

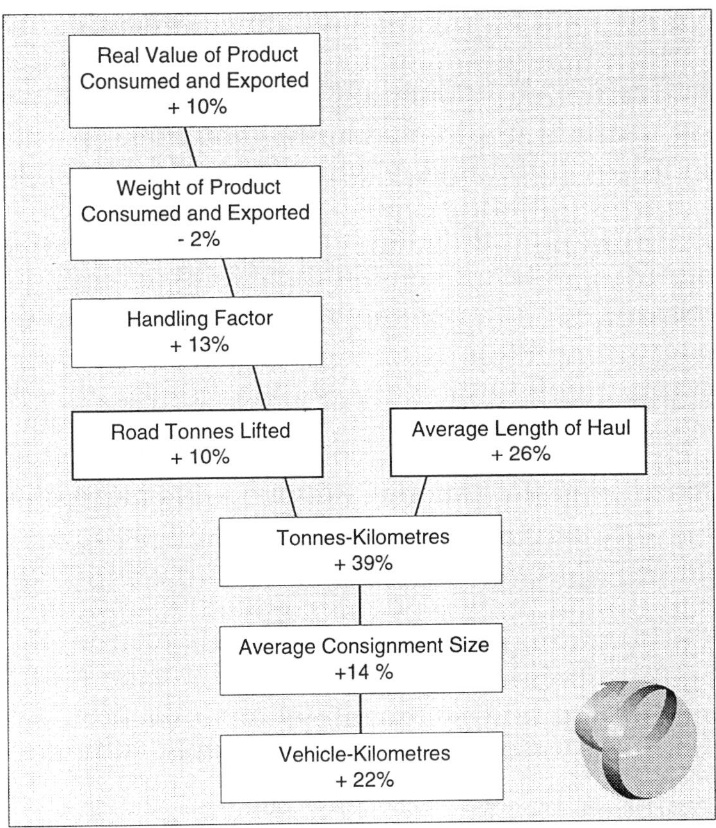

Figure 1. Logistical changes in the UK food and drink sector, 1983-1991 (Source: McKinnon and Woodburn, 1995)

It can be seen from the figure that the ultimate growth of 22% in the number of vehicle-kilometers is the result of trends in opposing directions. The main cause of the growth is a 26% increase in the average length of haul: an average increase of 3% per annum. A secondary cause of the growth in vehicle-kilometers is the increase in the number of hauls involved in the production chain as a whole, reflected in the 13% rise in the handling factor. This is due to the fact that the production process has become more complex and takes place at various different locations. Taking these two causes together, growth in freight transport is the result of more and longer hauls per unit end product.

This growth is tempered by two other trends, however. In the first place, the average load per haul has increased by 14%, so that fewer vehicle-kilometers are required for a given number of ton kilometers. In addition, the average value of a kilogram of product (food and drink) increased by 12%. The product value is tied to GDP (growth).

These trends in the food and drink sector provide a neat summary of the various factors behind the growth of freight transport. This sector is not representative of the economy as a whole, however, and similar analyses of other sectors are lacking. It is possible to give a rough indication, though. Over the past few decades, the number of ton kilometers has been growing at about the same rate as GDP. Growth in the number of tons of end product has been far lower, however: at an estimate, only one-quarter of GDP growth. This is not surprising, because the service sector is responsible for a growing share of economic activity. The share of industrial production is steadily declining and the economy is thus in a process of 'dematerialization'. However, this trend is not reflected in the flow of goods through the economy, because of the growth in length of supply and distribution lines and consequently transport distances. The growth in the number of links and associated hauls in the production chain and in the average length of haul are together responsible for an estimated three-quarters of the growth in ton kilometers observed. It is thus above all the spatial reorganization of the production process that has caused the increase in ton kilometers and far less the growth of GDP *per se*.

Several underlying economic and logistical trends contribute to the growth of transport distances|7|: The main factors are:

- less stocks (Just-in-Time)
- upscaling in production and distribution
- larger geographical market areas (both for resources and products).

Rising share of road transport

Another trend that can be discerned is the rising share of road transport. In the ECMT countries the share of road transport grew from 45% in 1970 to almost 60% in 1989. Assuming that road transport (long distance) uses 2 to 3 times more energy per ton kilometer than rail freight and inland waterways, this modal shift implies an annual 0.5 to 0.8% increase in CO_2-emissions. A further illustration is provided by the case of the

UK, where the share of road transport increased from 42% to 68% over the period 1953-1989 (measured in ton kilometers). Under the assumption of unchanged policy, all future projections indicate a continuation of this trend. An underlying reason is that little further growth is anticipated for bulk transport - much of which is transported by rail and waterway - while 'Other goods' will continue to grow substantially. Moreover, the aforementioned logistical trends also point towards further growth of the share of road transport.

Load factor

The average load factor of trucks has not changed substantially over the last decades and thus did not influence growth in CO_2 emissions.

Truck size

The average truck size tends to increase. Although statistical data are lacking, it is estimated that this development towards larger trucks reduces CO_2 emissions by around 0.5% per year.

Energy efficiency

Truck fuel consumption can be reduced by improving engine and drivetrain efficiency, reducing aerodynamic drag, lowering empty weight, and lowering rolling resistance. Statistical material gives not the right information to exactly calculate this technical improvement rate, but it can quite well be estimated from existing market knowledge. The fuel consumption per 100 km driven by a fully loaded 40 ton truck evolved from about 50 liters in 1970 to about 40 liters in 1980 (about 2% p.a.) to about 35 liters in the second halve of the 80's (1.5% p.a.). Since then progress has slowed down further to the current 1% per annum.

Overview

Table 2 summarizes the main trends behind the growth in CO_2-emissions from freight transport of 2 to 3% a year.

Table 2

Estimated main trends behind the growing CO_2-emissions from freight transport.

Factor	Impact on CO_2-emissions
GDP	+ 0.8% / year
Transport distance	+ 2.2% / year
Ton kilometers	*+ 3.0% / year*
Modal split	+ 0.6% / year
Load factor	0% / year
Truck size	- 0.5% / year
Efficiency truck	- 1.0% / year
CO_2-emissions freight transport	*+ 2.1% / year*

Table 2 shows that the growth in annual transport distance is the dominant factor. The main cause seems to be the decrease in the price of road freight transport. For The Netherlands the real price decrease is estimated at 1.6% per year for the period 1984-1992. In the same period other production factors, as labor and storage, have become more expensive. These changes in relative prices promote a shift towards transport and away from other production factors. The transport intensity of the economy increases.

2.4. Policy recommendations

Passenger traffic

A reduction in CO_2-emission from passenger traffic can only be achieved with the use of fierce policy instruments and thus after clear political choices. The reason for this is that a reduction of CO_2-emissions conflicts with two powerful trends: faster transport and growing income. The main effective policy instruments to counterbalance these forces are:

- strong government interference in the car sales market, to force the introduction of the very fuel efficient car against existing consumer preferences. Two main options are available. The first is a - possibly modified - version of the standards for Corporate Average Fuel Economy (CAFE), which have been successfully applied in the USA. The second policy option is a drastic increase in the fuel price, by e.g. doubling or tripling the current level of fuel excise duty in Europe. This will mainly promote the introduction of cars with a better fuel economy, but will also reduce mobility growth somewhat;
- reduced average speed of car traffic will bring specific energy consumption down by around 10%, counteracts the 'upgrading' trend in the car market and slows mobility growth. This combination of effects makes speed management a powerful policy instrument. Setting and enforcing proper speed limits - probably with the use of electronic speed limiters in the vehicle - is the main element of such a policy.

Other policy instruments can support these two main approaches, but can not replace them |4|.

It is clear that both policy lines are currently not political acceptable. Main reason for this is the 'price' attached to a reduction of CO_2 emissions: reduced growth in mobility and travel speed a substantial downgrading of the car and/or higher prices for mobility e.g., caused by the use of advanced environmental technology.

Freight transport

Reduction of CO_2 emissions from freight transport can mainly be achieved by some further improvements of the trucks, by changes in logistics (higher load factor) and by a higher transport efficiency of the economy (less transport per unit of GDP) |7|. Only for technical improvements direct regulation might be considered as policy instrument, although the effectiveness seems limited for the promotion of energy efficiency.

The other two options to reduce CO_2 emissions - logistics and transport efficiency - can mainly be promoted by the use of financial incentives. An increase in fuel price is attractive because it stimulates all types of CO_2 reduction, including driving behavior. It is estimated that the average price of diesel in the EU should be roughly doubled to come in line witch the total marginal costs of a truck kilometer in rural areas |8|. An alternative policy option is to introduce an electronic kilometer charge for trucks, based on registration of the number of kilometers driven in each country. This assures that the country where the truck has driven, and thus caused the costs, receives the revenues, which is not fully the case with the diesel tax.

Additional to these general cost related charges for trucks, it is needed to have extra charges in urban areas. Because congestion is higher and more people are exposed to air pollution and noise nuisance in urban areas, the marginal costs in urban areas are roughly twice as high as in rural areas |8|. Systems of urban road pricing can incorporate these additional urban costs.

REFERENCES

1. J.M.W. Dings, M.D. Davidson and G. de Wit, Optimale brandstofmix voor het wegverkeer (Optimal fuel mix in road traffic), CE, Delft (to be published in May 1997)
2. J.T. Houghton, G,J, Jenkins and J.J. Ephraums, Climate Change: the IPCC Scientific assessment, Cambridge University Press (1996)
3. Dutch Government, Second Memorandum on Climate Change, the Hague (1996)
4. A.N. Bleijenberg and J. van Swigchem, Efficiency and sufficiency, CE, Delft (1997)
5 P.B. Goodwin, Empirical evidence on induced traffic, in: Transportation 23: 35-54 (1996)
6. A.C. McKinnon and A. Woodburn, Logistical restructuring and road freight traffic growth: An empirical assessment, Herriot-Watt Business School, Edinburgh (1995)
7. A.N. Bleijenberg, Freight transport in Europe: In search of a sustainable course, CE, Delft (1996)
8. ECMT Task Force on the Social Costs of Transport, Draft report, ECMT, Paris (1997)

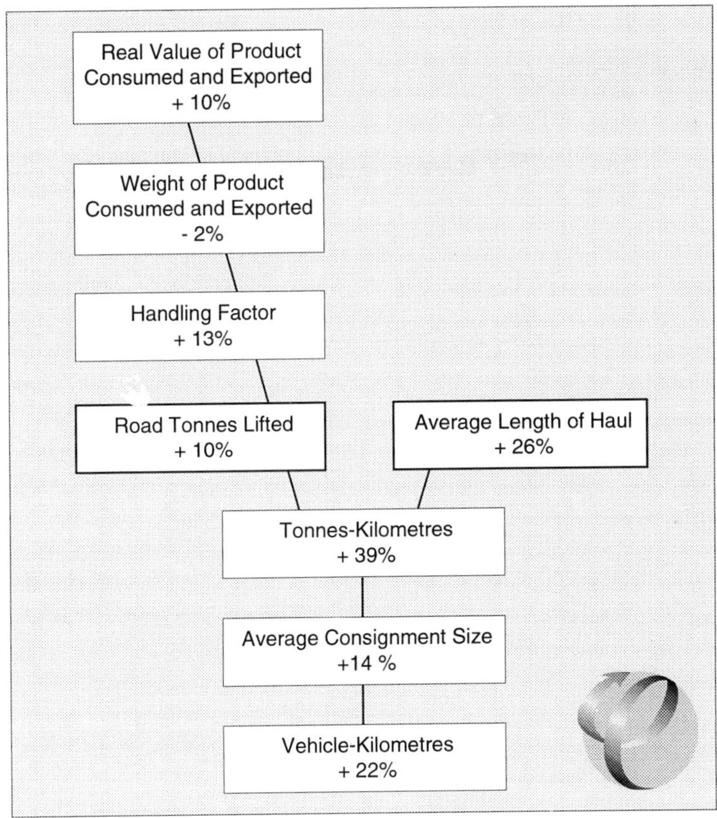
Real Value of Product
Consumed and Exported
+ 10%
Weight of Product
Consumed and Exported
- 2%
Handling Factor
+ 13%
Road Tonnes Lifted
+ 10%
Average Length of Haul
+ 26%
Tonnes-Kilometres
+ 39%
Average Consignment Size
+14 %
Vehicle-Kilometres
+ 22%

Aviation and Air Pollution

J.H.A.M Peeters

NL Ministry of Housing, Spatial Planning and the Environment, Directorate for Noise and Traffic, P.O. Box 30945, 2500 GX The Hague, The Netherlands

1. INTRODUCTION

This 5th US-Dutch international symposium deals with the priority issues and policy trends regarding air pollution in the 21st century. This paper will go into the issue of aviation and air pollution and is based on the Government Policy of The Netherlands on Air Pollution and Aviation [VROM, 1995].

The paper consists of the following parts:

- the current environmental impact of aircraft emissions;
- future emission trends;
- possible ways to reduce aircraft emission trends, and finally;
- a policy to reduce aircraft emission trends.

2. GENERAL

Modern aviation fuels are obtained from the refining of crude oil, and consist mainly of hydrocarbons. When the combustion of the fuel is complete, the combustion of aviation fuels gives rise to emissions of carbon dioxide (CO_2), water (vapor) (H_2O) and sulphur dioxide (SO_2). Although the combustion efficiency of jet engines is generally very high, combustion is not complete, and a number of other combustion products are also generated, particularly carbon monoxide (CO), volatile organic compounds (VOC) and 'particulates' (this term refers collectively to solid and liquid substances of diverse composition). In addition to incomplete combustion products, the high temperatures in the combustion chamber lead to the formation of oxides of nitrogen (NO_x). Aircraft engines also emit nitrous oxide (N_2O) and methane (CH_4). The emissions of these two substances are extremely small, and they are therefore not considered further.

Aircraft emissions contribute to climate change (depletion of the ozone layer and the greenhouse effect), acidification and disturbance (local air pollution and odors). The main contribution to air pollution is made by civil aviation, but military aviation also plays a role. Light aviation has an impact mainly at the local level.

The scientific understanding of the impact of aircraft emissions on the environment is still rather inadequate. Further research will be required in the coming years to remedy these deficiencies, such as the scientific research program that we currently undertake: the AIRFORCE project. The gaps in our understanding relate particularly to climate change, and the possible role played by NO_x emissions from aircraft in the upper atmosphere and chemical processes resulting in contrail formation and their effects on cloud properties and radiative forcing. The NO_x-issue has received a great deal of attention from the international scientific community in recent years. The contrail-issue has only recently become an area of interest for scientists. In the Special Report of the IPCC on Aviation and the Climate, that is scheduled for release in the end of 1998, further consideration will be given to these questions.

One reason behind the focus on the effects of NO_x emissions on the atmosphere is what is called the 'fuel-NO_x trade-off'. Historically, the fuel efficiency of jet engines has risen steadily. A higher efficiency will as a rule reduce the unit emissions of CO_2, H_2O, CO, VOC and SO_2. In increasing efficiency, however, temperature and pressure in the combustion chamber rise, whereby NO_x emissions will as a rule increase. This effect can by compensated through improvements in the combustion process in the combustion chamber. Examples are lean-burn and staged combustion techniques or revolutionary new combustion chamber concepts. If NO_x emissions would not be of environmental concern, aircraft and engine manufacturers would not be needing to invest in low NO_x engines.

3. THE CURRENT ENVIRONMENTAL IMPACT OF AIRCRAFT EMISSIONS

3.1. Composition of the atmosphere

The atmosphere, the ring of gases which girds our planet, can be divided into a number of layers, characterized by their temperature profile. The lowest layer of the atmosphere is the troposphere. In the troposphere, the temperature falls with increasing altitude. The troposphere is turbulent and the substances present in it undergo vertical mixing within a week or so. Above the troposphere is to be found the stratosphere. In this layer, temperature first remains constant and then rises with increasing altitude. This makes the stratosphere much more stable than the troposphere, and little vertical transport occurs. The boundary between the stratosphere and the troposphere is called the tropopause. The exact position of the tropopause depends on latitude and season, and is also influenced by weather systems; it fluctuates on a day-to-day basis. Near the poles the tropopause occurs at an average altitude of about 7 km and near the equator at an average altitude of about 18 km. It is exactly in this very complex region of the atmosphere where aircraft fly: both in the troposphere and in the stratosphere. The different characteristics of these two layers mean that the substances emitted by aircraft have different effects in them. Therefore, it is no simple matter to answer the question 'what are the atmospheric effects of aviation?'.

3.2. Depletion of the ozone layer

The aircraft pollutant which plays the most important role in depleting the ozone layer is NO_x. Model calculations indicate that this contribution is expected to be small in quantitative terms. Scientific understanding of the indirect effects of a number of aircraft pollutants is still incomplete, and the possibility that these effects may prove important cannot be excluded. On the basis of present knowledge the international scientific community estimates that the contribution of aircraft emissions to ozone depletion in the lower stratosphere is at present less than 1%. However, a new generation of supersonic airliners may have a major impact on ozone depletion.

3.3. Enhanced greenhouse effect

Aircraft emissions contribute to the greenhouse effect. The climate effects of aircraft CO_2 emissions are no different from those of other CO_2 emissions, and are relatively clear. The role of aircraft emissions of NO_x has become better understood in recent years. Changes in ozone concentrations due to aircraft NO_x emissions disturb the earth's thermal balance. Quantitatively speaking the effects depend on location and season, and are therefore difficult to compare with the global effects of persistent greenhouse gases such as CO_2. Nevertheless, it is required to gain a better understanding of the relative importance of the NO_x and CO_2 emissions from aircraft in the enhanced greenhouse effect. The international scientific community gathered together in the International Panel on Climate Change (IPCC) at present estimates that the indirect effect on the enhanced greenhouse effect of aircraft NO_x emissions, as a result of ozone formation, is of the same or a smaller order of magnitude as the direct effect of aircraft CO_2 emissions.

There is still a great deal of uncertainty about the effects of water vapor, SO_2 and soot particles. These pollutants emitted by aircraft could play an important part, because of their influence on the formation of clouds and aerosols, in contributing to the greenhouse effect. The radiative effect of aerosols and their ability to modify cloud properties are strongly influenced by the concentrations of the aerosol in the atmosphere, which exhibit very major local variations in magnitude and composition. Overall, changes in cloud cover and optical properties probably result in a net warming effect and the radiative effect in a net cooling. At present our knowledge does not allow us to quantify these climatic effects properly. It is assumed however that the indirect effects of H_2O, SO_2 and particulate emissions from aircraft are not greater in quantitative terms than the effects of aircraft emissions of CO_2 and NO_x.

3.4. Acidification

The contribution of aircraft emissions to acidification is in principle properly to quantify.

Of most importance are the NO_x emissions. SO_2 emissions are less important in this regard. On a global level, the aviation sector contributes about 0.7% to acidifying emissions of NO_x and SO_2, expressed in terms of acid equivalents.

3.5. Local air pollution in the vicinity of airports

The fourth environmental problem related to aviation emissions is the contribution to local air pollution in the residential areas around airports. In 1993 a study was carried out in The Netherlands for the purposes of the Project Schiphol Mainport and the Environment and the related integrated environmental impact assessment (IEIA). This study examined the nature and extent of air pollution and odor nuisance within a radius of 10 km from Amsterdam Airport Schiphol (AAS) in the period up to 2015.

The study used a dispersion model to calculate the concentrations of certain pollutants in 15 residential areas in the vicinity of Amsterdam Airport Schiphol. It concluded that the relative contribution of the 'Schiphol' emissions to these concentrations will not exceed several percent. It was also clearly demonstrated that in no case are any of the legal standards for NO_2, CO, SO_2, black smoke, benzene and benzo(a)pyrene exceeded. The concentrations of these substances in the various locations will decrease if, as assumed in the IEIA, the rise in emissions by air traffic can be offset by a fall in emissions by road traffic.

It is important to note that these results might not be applicable to other airports. For some airports the contribution of aviation to local air pollution is low, for other airports this is high and may cause severe problems. In addition, I must state that further research into this area is ongoing. In general, VOC, CO, SO_2, NO_x, particulates and odors are the emission products that are of importance with respect to local air pollution.

4. EMISSION TRENDS

4.1. Global

At present, CO_2 and NO_x can be regarded as being the most important aircraft pollutants. In both cases, aircraft emissions accounted for between 2 and 3% of total world emissions from the combustion of fossil fuels in 1990, as is shown in table 1.

Table 1

Aircraft emissions and their share of the total emissions due to the combustion of fossil fuels (coal, petroleum and gas) in 1990 [RIVM, 1995]

	CO_2 (Mton)	NO_x (kton)	VOC (kton)	CO (kton)	SO_2 (kton)
Aircraft	498	1,786	406	679	156
All sources (world total)	22,000	82,000	27,000	303,000	130,000
Percentage attributable to aircraft	2.3	2.2	1.5	0.2	0.1

There has been a continuous improvement in the efficiency of jet engines over the years. The specific fuel consumption, i.e. the amount of fuel required to generate one kiloNewton of thrust, has therefore fallen steadily. The emissions of CO and VOC per unit of fuel have generally fallen in line with the improvements in specific fuel consumption and with emission standards set by the International Civil Aviation Organization (ICAO). On the other hand the emissions of NO_x per kg. of fuel have increased. This is a result of the steadily rising temperatures in the engine combustion chamber which accompany the increasing fuel efficiency.

The efficiency trend was not enough, however, to offset the growth in emissions as a result of increased traffic volumes. Over the last two decades, air travel was the fastest growing mode of transport, and this trend is expected to continue.

Civil aviation is a growth market. It is expected to grow faster in future than the economy as a whole. This means that in the years to come the economic importance of air traffic will increase relative to other sectors. There will be a corresponding rise in the pollution caused by this sector, both in absolute and relative terms. That much is clear from calculations carried out for a white paper of the government of The Netherlands on Air Pollution and Aviation. These model calculations indicate that with current emission trends (including current international regulatory action) and without further policy measures, global aviation emissions in 2015 will be approximately three times those in 1990. Table 2 provides detailed information. Other forecast support these growth figures. According to a forecast by Environmental Defense Fund, the worldwide CO_2 figure for aviation could grow as large as 10 percent by 2050, depending on many factors associated with economic growth.

Table 2

Developments in world aviation emissions of CO_2 and NO_x for the period 1990 - 2015 for three economic scenarios [RIVM, 1995]

	CO_2		NO_x	
	Mton	Index (1990 = 1)	Mton	Index (1990 = 1)
Emissions 1990	498	1.0	1,786	1.0
Global Shift 2015	1,760	3.5	5,204	2.9
European Renaissance 2015	1,409	2.8	4,166	2.3
Balanced Growth 2015	1,678	3.4	4,964	2.8

4.2. European and national

In the near future similar high growth figures are expected in Europe. In the European Union the CO_2 growth figures associated with aviation will create an unbalance situation, since, under the influence of policy measures driven by the Framework Convention on Climate Change (FCCC), CO_2 emissions in other sectors will stabilize and/or decrease. I assume that a similar situation will arise in the US, depending on the goals, policies and measures that can be agreed in the light of a protocol under the FCCC for the greenhouse gas emissions after 2000.

At the national level, emissions attributable to flights related to The Netherlands will increase by a factor of two, as is shown in table 4. In 1995, the CO_2 emission of national and international civil aviation in The Netherlands was bigger than the CO_2 emission by heavy duty trucks.

Allowing for the effect that environmental policy will have on national emission sources, aircraft emissions under unchanged policy will become more significant. In 2010, it is estimated that emissions from flights related to The Netherlands will then account for 6% of national CO_2 emissions and 16% of national NO_x emissions. For other European countries and for the European Union as a whole, similar situations apply. It is clear that these numbers can not be neglected.

Table 3

Emissions of flights related to The Netherlands in 1990 and 2010 [VROM, 1995].

	1990	2010[1]	%-increase
CO_2 (Mton)	6,7	13,4	+100
NO_x (kton)	21,8	37,3	+71

4.3. Summary

In summary, aircraft emissions contribute to climate change (depletion of the ozone layer and the greenhouse effect), acidification and disturbance (local air pollution and odors). Although the aviation industry currently can not be regarded as a "climate ciller", its contribution is serious enough to develop a policy that is aimed at reducing the emission trends. The aviation industry is a growth market. It is expected to grow faster in future than the economy as a whole. There will be a corresponding rise in the pollution caused by this sector, both in absolute and relative terms. This should be of great concern.

Table 4 presents a systematic overview of the abovementioned findings.

Table 4

The importance of controlling emissions of the various aircraft pollutants for each of the relevant environmental problems [VROM, 1995]

Environmental problems	Control important	Control unimportant	Importance uncertain
Ozone depletion	NO_x	CO_2, VOC, CO	SO_2, H_2O, particulates
Greenhouse effect	CO_2, NO_x	VOC, CO	SO_2, H_2O, particulates
Acidification	NO_x	CO_2, SO_2[1], H_2O, VOC, CO, particulates	-
Local air quality[2]	VOC, CO, SO_2, NO_x, particulates, odors	CO_2, H_2O	-

1 SO_2 is an important acidifying agent. The contribution made by aircraft is small compared with that from other sources, however.

2 The impact of the various pollutants depends on local circumstances

5. POSSIBLE AVENUES TO REDUCE AIRCRAFT EMISSION TRENDS

Based on the results of various studies, a number of avenues can be pursued by policy-makers to mitigate the aircraft emission trends. Three categories of measures can be considered: technical measures, operational measures and mobility measures. Technical measures relate to the design of the aircraft engine, to the aircraft itself or to the aviation fuels used. Operational measures refer to measures taken while aircraft are cruising, or during the LTO cycle in and near the airport. Mobility measures can apply to substitution of traffic for shorter journey distances from air travel to more

environmentally-friendly alternatives, such as the high-speed train or the intercity train.
The technical avenue is aimed at development and introduction of new technology in the international civil aviation market. This new technology relates to changes in the overall design of aircraft engines, improvement of the combustion process of the combustion chamber, design of revolutionary new combustion chamber concepts, changes in existing aircraft design and, finally, designing new aircraft types that are optimized for lower cruising speeds that will emit less per passenger-kilometer for equal journey distances.
Operational measures can be changes in cruising altitudes, reducing cruising speeds, changing flight routes, improvement of air traffic control systems, modifying the distribution of airspace between civil and military aviation and, finally, measures related to the landing and take-off cycle at and around airports.
Mobility is a characteristic feature of our society. Civil aviation is one component of the transport network which meets our demand for mobility. The satisfaction of this demand is increasingly coming into conflict with environmental objectives, however. Where this transport network provides opportunities to substitute more environmentally-friendly alternatives there is an environmental gain to be had. Table 5 compares the emissions per passenger-kilometer for a journey of 500 km point-to-point for various transport modes. From an emissions point-of-view the high-speed train and the intercity train are, according to this table, the best alternatives.

Table 5

Emissions per passenger-kilometer for a journey of 500 km for various transport modes [VROM, 1995].

Mode	Occupancy rate	CO_2 (g/pkm)	CO (mg/pkm)	NO_x (mg/pkm)	VOC (mg/pkm)	SO_2 (mg/pkm)
Bus[1]	70%	22	75	479	56	23
High-speed train[1]	65%	48	2	87	12	60
Intercity[1]	44%	51	2	94	13	64
Private car (gasoline)	2 people	86[1]	250-1.600[2]	270-145[2]	45-220[2]	7[1]
Boeing 737-300[3]	71%	146	240	440	10	10

1 Mean for period 1988-1990

2 Data for 1990 car with three-way catalytic convertor; first value represents emissions under motorway driving conditions; second value represents emissions in city driving conditions [CBS]

3 Modelcalculation

6. A POLICY TO REDUCE AIRCRAFT EMISSION TRENDS

Let me now present to you an overview of our policy to reduce aircraft emission trends. This policy, which is described in the white paper 'Government Policy of The Netherlands on Air Pollution and Aviation', is predicated on an international approach and consists *inter alia* of the following elements:

- voluntary agreements with the Dutch aviation sector regarding the way that it can contribute to realizing the emissions policy;
- an international agreement between the partners to the Climate Treaty and the Chicago Convention that should provide as a minimum for a reduction in the growth of aircraft emissions on the basis of concrete international objectives for

air traffic and a methodology for calculating and apportioning international aircraft emissions;

- efforts within the International Civil Aviation Organization (ICAO) in order to achieve a substantial tightening of the existing NO_x standards for aero-engines, and the development of CO_2 standards for aircraft or the development of another system which will serve to reduce the growth in aircraft CO_2 emissions;
- introduction of a worldwide levy on kerosine;
- the discontinuance of exemptions on value-added tax for international flights;
- international Research & Development programs for clean and fuel-efficient aircraft and aero-engines;
- improvement of air traffic control systems.

From this policy package, I would like to emphasize two important elements. The first is *emissions regulation* that should preferably be a responsibility of ICAO's Committee on Aviation and Environmental Protection (CAEP). However, recent developments suggest that there consist different opinions between the United States and the European Union with respect to the need to tighten NO_x emissions regulatory rules for new aircraft engines. Although European members of the CAEP recommended the ICAO Council to increase the stringency of the current international NO_x requirement with 16 percent, the US opposed to this recommendation, arguing that there is no clear demonstrated environmental need and that the costs associated with the increased stringency are high. As a consequence, the ICAO Council decided not to increase the stringency and to refer this issue back to CAEP. This might drive Europe into a regional approach, thereby weakening the position of ICAO.

The second point I want to address are price measures, especially a *levy on aviation fuel*. Most countries follow the recommendation by the Council of the ICAO, that fuel used for international aviation should be tax-exempt. Aviation fuel taxation is precluded in most countries by provisions in the bilateral Air Transport Agreements which are the main legal framework for the operation of international civil aviation. High fuel prices drove airlines to achieve high energy intensity reductions during the seventies and the eighties through technical and operational changes. The primary aim of a levy on aviation fuel would be to reduce the consumption of aviation fuel, and hence reduce the CO_2 emission trend by aircraft. It is our strong belief that this policy measure must become an important element of an international approach towards a more sustainable civil aviation industry. The Netherlands therefore underlines the chairpersons' conclusion of the Informal Meeting of Environment Ministers in Dresden from 21 to 23 march 1997, proposing in paragraph 21 "that the Special Session of the General Assembly of the United Nations in june 1997 considers an initiative to introduce air fuel taxation at the international level".

REFERENCES

1. RIVM, 1995. Olivier, J.G.J., Scenarios for global emissions from air traffic. RIVM report 773 002 003, March 1995. National Institute for Public Health and Environmental Protection, Bilthoven.
2. VROM, 1995. Government Policy of The Netherlands on Air Pollution and Aviation, Ministry of Housing, Spatial Planning and the Environment (VROM), The Hague, The Netherlands.

Transportation & Market Incentives Group Strategic Plan for the 21st Century

L. Audette

Manager, Transportation Group, Office of Mobile Sources

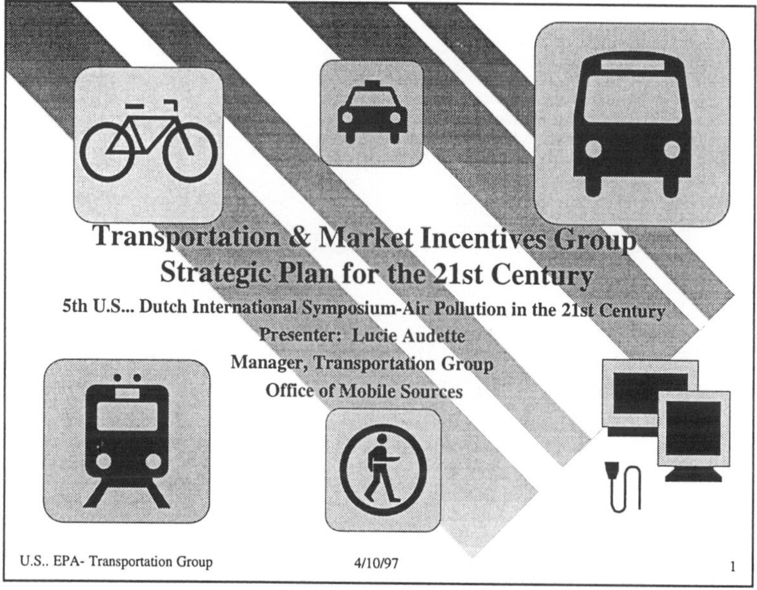

1. New Division and group created Sept. 1995 -

2. 9 month effort to develop a strategic direction and focus limited resources in most effective manner.

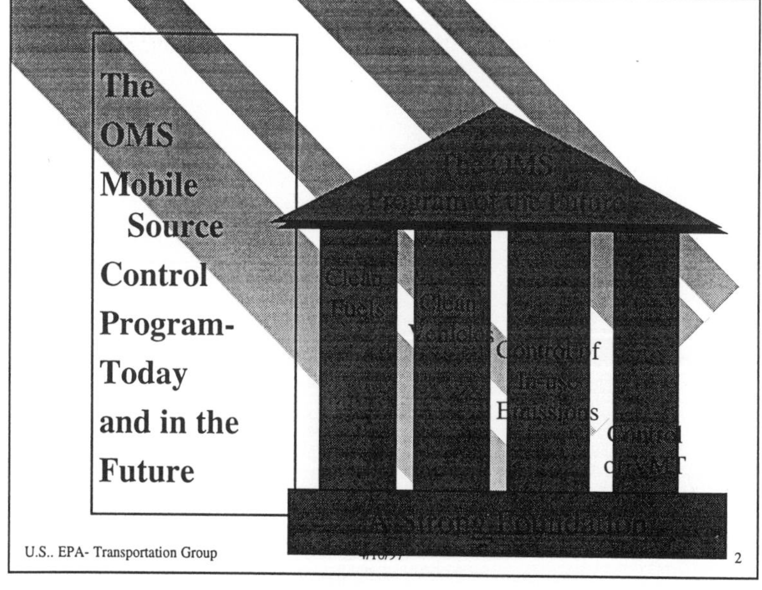

Transportation Background

- Significance of transportation is widespread
- Direct link between VMT increases and increased vehicle emissions.
- Consequences of continued VMT growth

U.S.. EPA- Transportation Group 4/10/97 3

1. One of the largest sector of the US economy when considering Number of resources consumed and people employed.

2. **Transportation contributes to many Environmental Problems:**

•Wetlands loss

•Urban Smog

•Poor Visibility

•Solid Waste

•Climate Change

•Ozone Depletion

•Air Toxic Emissions

•HazMat Spills

•Urban Water Contamination

•Ecosystem Destructions.

Transportation Challenge...
Changing OMS Role

- Past Role - Carrying out Clean Air Act Mandates, focus on command- and control regulations
- Cars are cleaner.... but transportation related air pollution continues to be a serious problem throughout U.S.

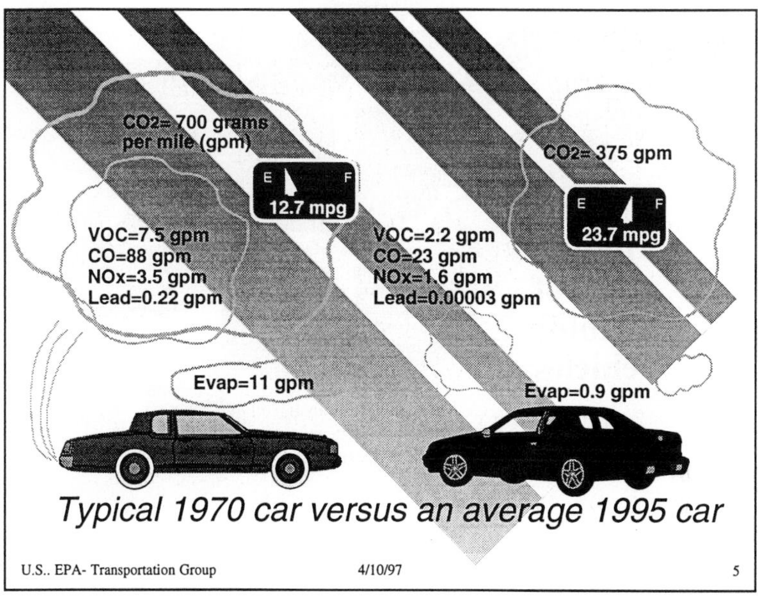

1. **National LEV Program**- FRM 1997- Tight exhaust emissions standards

>075 g/mile NMHC @ 50,000 Phase-in 1997 model year NE Ozone Transport Region, (benefits- 780 tons/day NMHC in 2015)

2. **Tier 2 Study-** Statute suggest: 0.125 g/mi NMHC and 0.22 g/mi NOx @ 100,000 Beginning 2004 model years. Study not complete.

3. **Urban Bus Retrofit-** Public Transit Buses, 1993 and Earlier

Urban areas over 750,000

Two kit certification options- 25% reduction in PM

Begin implementation this year

4. **Light Duty Vehicles and Light Duty Trucks** - Tier 1 Standards NMHC 0.31 g/mi and N)x 0.6 g/mi @ 100,000 miles for LDVs, Longer useful life, Phase-in model year 1994-1996 Benefits: 160,000 tones VOC nationwide in 2005)

5. On-Board Vapor Recover- refueling emissions prevented with on-vehicle controls, Phase-in model year 1998-2000. benefits: 300,000 - 400,000 tons VOC (95% reduction over uncontrolled levels)

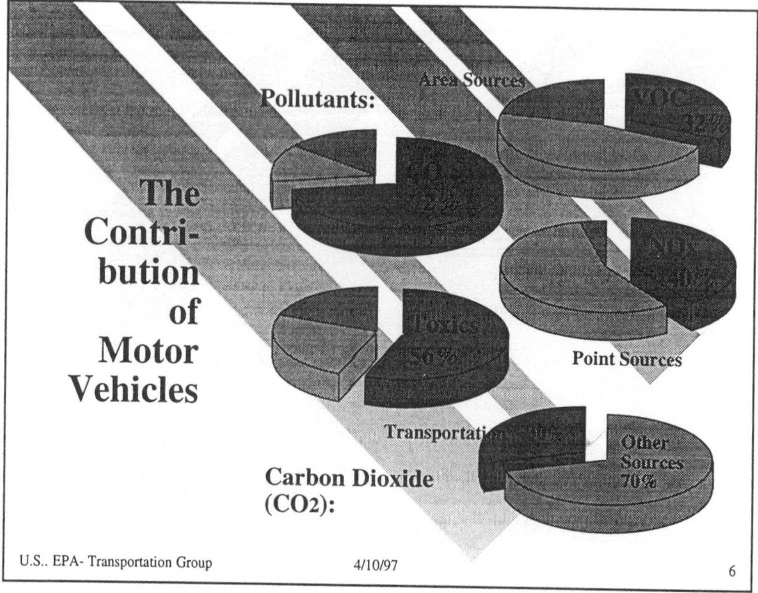
The Contri- bution of Motor Vehicles
Pollutants:
Area Sources
VOC
32%
NO
40%
Toxics
56%
Point Sources
Transportati
Other Sources 70%
Carbon Dioxide (CO2):
U.S.. EPA- Transportation Group
4/10/97
6

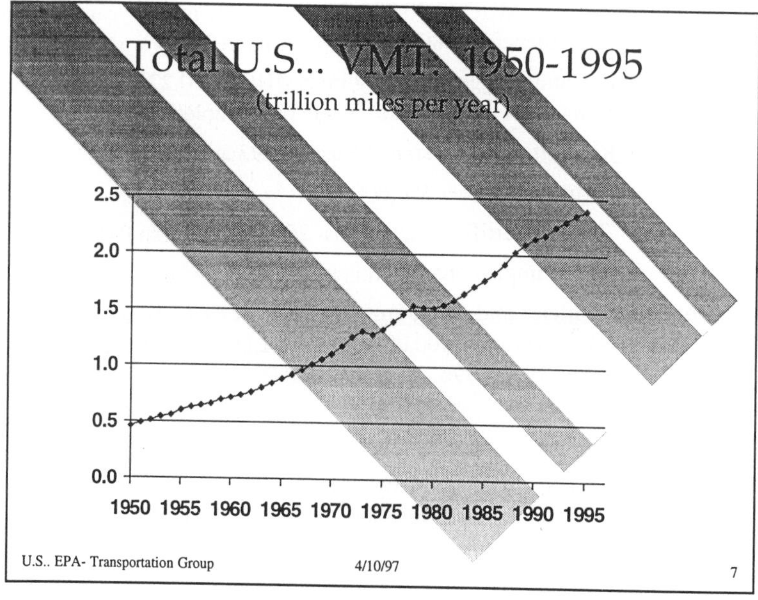
Total U.S... VMT: 1950-1995
(trillion miles per year)
2.5
2.0
1.5
1.0
0.5
0.0
1950 1955 1960 1965 1970 1975 1980 1985 1990 1995
U.S.. EPA- Transportation Group
4/10/97
7

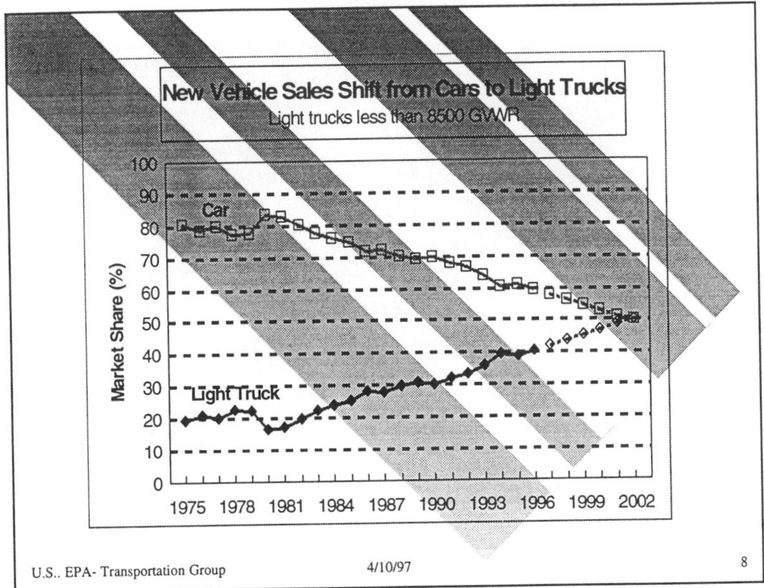
New Vehicle Sales Shift from Cars to Light Trucks
Light trucks less than 8500 GVWR
Market Share (%)
100
90
80
70
60
50
40
30
20
10
0
Car
Light Truck
1975 1978 1981 1984 1987 1990 1993 1996 1999 2002
U.S.. EPA- Transportation Group
4/10/97
8

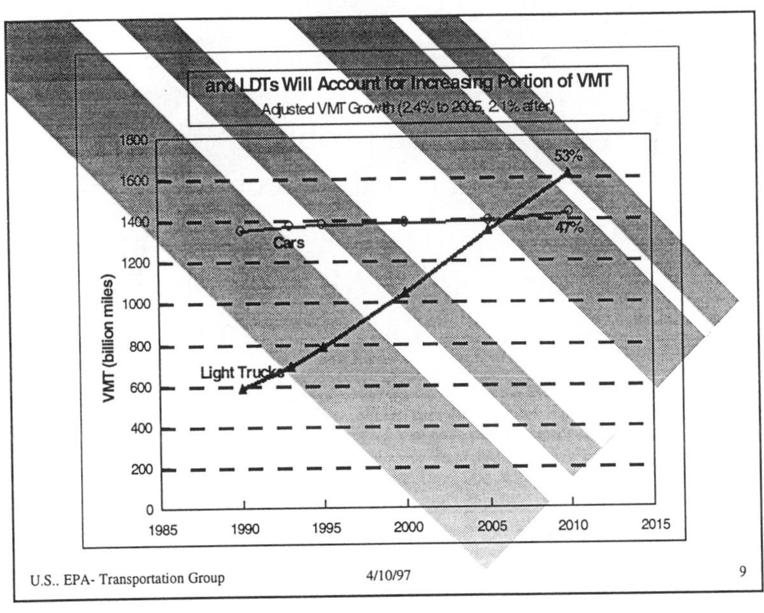
and LDTs Will Account for Increasing Portion of VMT
Adjusted VMT Growth (2.4% to 2005, 2.1% after)
VMT (billion miles)
1800
1600
1400
1200
1000
800
600
400
200
0
53%
47%
Cars
Light Trucks
1985 1990 1995 2000 2005 2010 2015
U.S.. EPA- Transportation Group
4/10/97
9

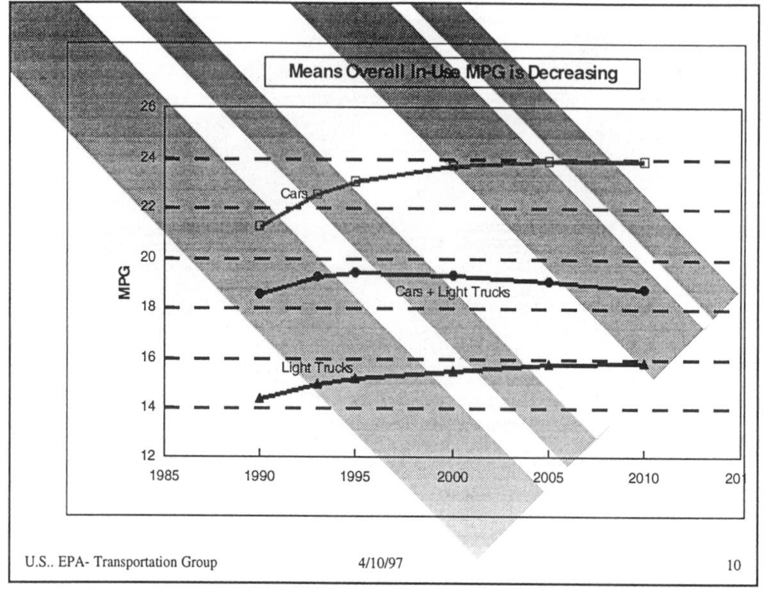

Transportation Challenge... Changing OMS Role

- Transportation problems complex and not subject to direct EPA control
- Changing Role- More emphasis on assisting States and Regions in developing a variety of measures/allowing for flexibility

U.S.. EPA- Transportation Group 4/10/97 11

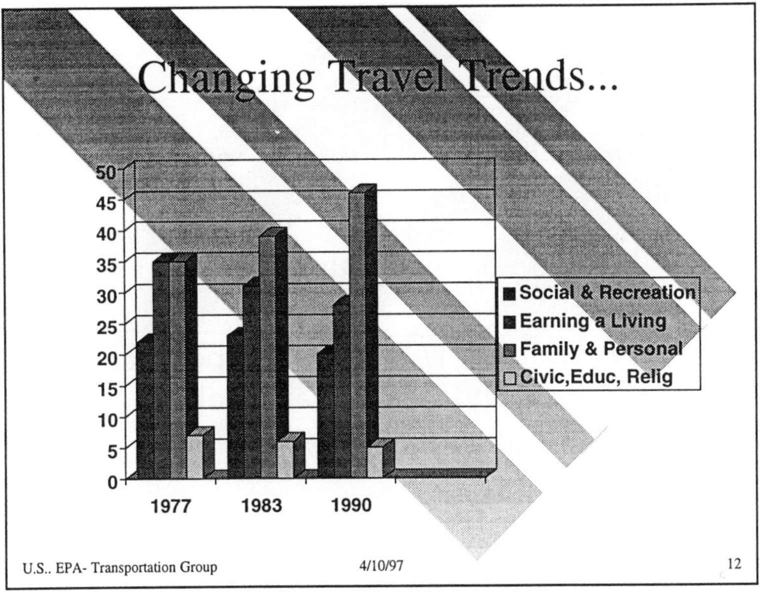

* From 1977 to 1990 work-related trips as a share of all trips taken decreased significantly, from 35% in 1977 to 28% in 1990, whereas the share of family and personal business trips grew from 35% to 46%.

Objective -- Transportation & Market Incentives Group

Reduce in-use emissions from transportation sector utilizing activity-based strategies.

⇨Reduce growth rate of VMT

⇨Reduce Trips

U.S.. EPA- Transportation Group 4/10/97 13

Focus -- Criteria Pollutants/Global Warming

Reducing the growth rate of VMT and trips supports...

- ⇨ EPA efforts to reduce criteria pollutant emissions and improve air quality
- ⇨ EPA global warming efforts
- ⇨ It's a double hit.

U.S.. EPA- Transportation Group 4/10/97 14

An Integrated Strategy

How will we reduce the historic growth rate of VMT?

By pursuing three track integrated approach...

U.S.. EPA- Transportation Group 4/10/97 15

An Integrated Strategy to Reduce VMT Growth Rate
➔Track 1: Properly merge air quality, transportation, and land use planning.
➔Track 2: Put in place and foster the use of viable transportation alternatives.
➔Track 3: Develop and implement transportation programs that incorporate market incentive measures.
U.S.. EPA- Transportation Group
4/10/97
16

1. Track 1: Inter-relation between urban design factors/ development patterns and transportation facilities and travel patterns.

2. Track 2: **The key here is providing flexibility and choice.**

Put in place viable transportation alternatives. This means developing a supply of alternatives that are realistic for a given community. this needs to address both low capital alternatives such as regional rideshare, subsidized transit, telecommuting, and high capital alternatives such as light rail, alternative fueled vehicles.

Foster the use of viable transportation alternatives. This requires providing accurate and timely information and conducting public education efforts both with the driving public and with school age children... the future.

3. Track 3: All our pilots and efforts should continually seek to utilize the creativity and power of the free market.

4. There is overlap between each of the tracks. We've separated for clarity but in reality there is merger.

Track 1 - Coordinate Air Quality / Transportation / Land Use Planning

- **Conformity reinvented**
  - » Air Quality/Transportation planning process merged.
- **Development of analytic tools**
  - » TMIP/TRANSIMs - a collaborative effort with DOT/DOE/EPA is fully coordinated. Other complementary modeling efforts are coordinated.
- **ISTEA Reauthorized**
  - » Incorporates environmental concerns
- **Urban Livability Initiative**

U.S.. EPA- Transportation Group 4/10/97 17

CONFORMITY

1. Final Rule will be published in April of 1997

2. CAP will issue initial report in early summer and Finalize in Fall of 1997/

3. Pilot projects- expressions of interest from 6 areas but they are waiting to see final conformity rule before submitting specific proposals.

ANALYTIC TOOLS

1. Transportation Model Improvement Program- short and long term effort to improve the modeling capability of MPOs (Metropolitan Planning Organization) and more fully integrate air quality and transportation modeling efforts.

2.. 5 track effort includes improving landuse modeling techniques and developing a long term transportation simulation model (TRANSIMS). This is a long term 10- 15 year effort.

ISTEA :

1. ISTEA of 1991 authorized $151 in transportation funding through FY97. Expires Sept. 30, 1996.

2. EPA has workgroup to develop key environmental positions and is working cooperatively within the administration to develop a proposal.

3. WHY we care:

(a) ISTEA through Conformity and Fiscal constraint provisions require that environmental impacts and resources must be considered in developing transportation programs, projects and plans (TPs and TIPs). Encourages local/public participation shifting decisions toward local/regional authorities.

(b) Provides funds that can result in environmental benefits:

- Flexible STP Funds

-$3 billion in planning funds

-$3 billion in enhancements funds

-$6 billion in CMAQ funds (Congestion Mitigation and Air Quality)

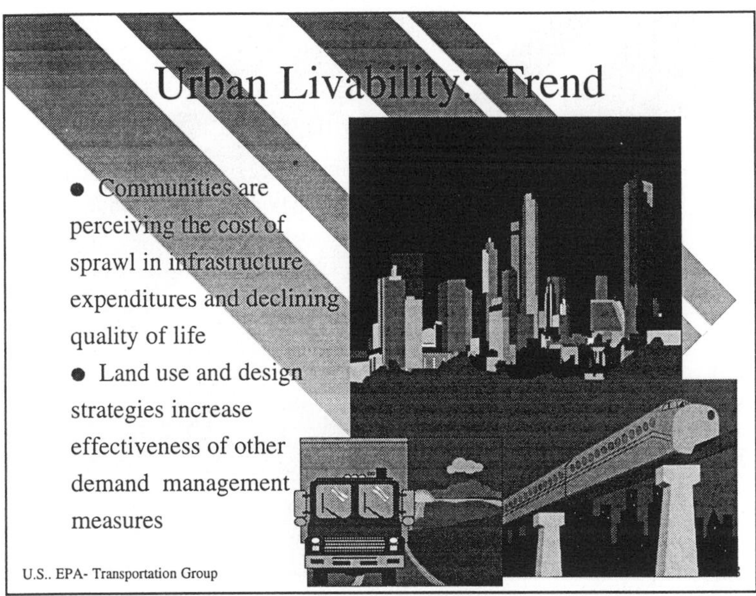

Urban Livability: Trend
● Communities are perceiving the cost of sprawl in infrastructure expenditures and declining quality of life
● Land use and design strategies increase effectiveness of other demand management measures
U.S.. EPA- Transportation Group

<u>Sustainable Development (cont.)</u>

There are a number of efforts currently underway within EPA to address sustainable development.

1. <u>OPPE-Urban and Economic Division</u>

Smart Growth Network is being created to (1) create coalitions to encourage infill development and provide information to local governments and developments on alternatives to sprawl.

2. <u>OPPE- Energy and Transportation Sectors Division</u>

Transportation Partners- teams with nongovernmental organizations (our Principal Partners) to develop innovative transportation solutions . Principal Patterns include Association for Commuter Transportation, Bicycle Federation of America, Local Government Commissions, Public Technology Inc., Renew America,

3. <u>REGIONS-</u> There is increasing sustainable development activity occurring throughout the U.S.. EPA regional staff are involved in a number of innovative pilot efforts... Region1- Visual Preference Survey, Regional 3- Sustainable Development Indicators work, Regional 9... involvement in several local general plan developments.

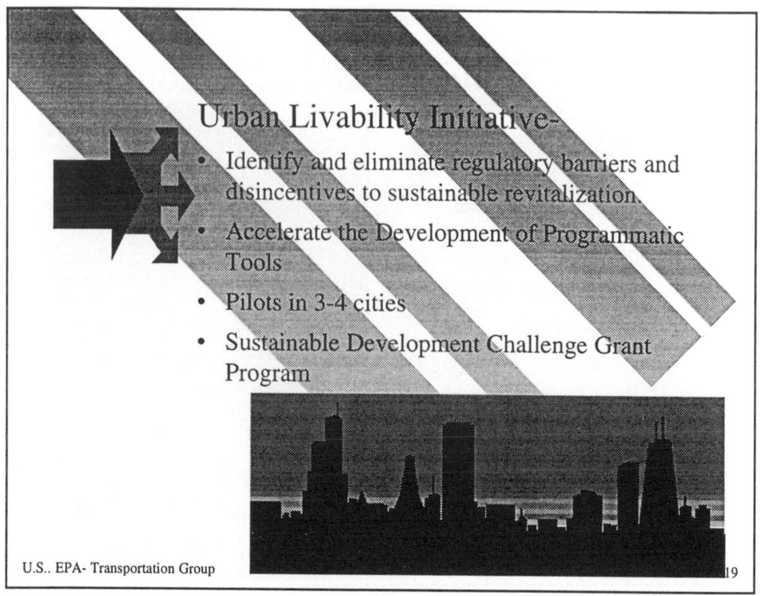

Urban Livability Initiative-
• Identify and eliminate regulatory barriers and disincentives to sustainable revitalization.
• Accelerate the Development of Programmatic Tools
• Pilots in 3-4 cities
• Sustainable Development Challenge Grant Program
U.S.. EPA- Transportation Group
19

4.**<u>FY97 Sustainable Development Challenge Grant Program</u>** -- OAR / OW

OMS sustainable development team will provide review and convey information to Regional transportation contacts.

<u>Purpose:</u> A total of $5 million will be awarded to 25 to 30 projects that link environmental management with sustainable development and economic revitalization. The FY96 program served as a pilot and 10 projects were chosen for grants. This year, the projects chosen will have more of a transportation / sustainable development / metropolitan focus, and EPA Regions will make the first selection cut.

<u>Status:</u> Federal Register announcement of the FY97 program is expected within a month.

5.**<u>The Urban Effort -- Joint EPA Office Initiative</u>**

The Urban Effort will create a small core staff of personnel detailed from offices within EPA.

<u>Purpose:</u> Develop and demonstrate strategies to pursue environmental and economic development goals of urban areas. EPA will fund pilots that have promise of replicable results in other areas.

Track 2 - Transportation Alternatives Are In Place and Used

- **Travel Smart Initiative**
- **Intelligent Transportation System (ITS) Support**
- **Technical assistance for transportation measures**
  - ✔ quantification of voluntary programs

U.S.. EPA- Transportation Group 4/10/97 20

Travel Smart Efforts

- DOT/EPA collaboration
- Youth targeted VMT reductio[n] Academy for Education. Development (AED)
- Smart Travel Resource Center
- Factsheet Development
- Product Distribution for States, MPOs and other stakeholders

U.S.. EPA- Transportation Group 4/10/97 21

DOT/ EPA Collaboration

What is it?

- Public Information Initiative
- Pilot site testing
- National and Local Campaign/ Coalition

Objectives

- Increase public understanding and awareness
- Encourage positive travel decisions
- Encourage positive life-work decisions

U.S.. EPA- Transportation Group 4/10/97 22

•IAG with DOT; DOT contractor-Global Exchange

•New name: DOT/ EPA Transportation/ Air Quality Public Information Initiative

•Target Audience: Driving Public

•Program begins with pilot site testing

•National/ Local Campaign

•National/ Local Coalition

1 - Diverse entities brought together: different perspectives; ideas

2 - Long-term sustainability == > funding, leadership shifts to partners

3 - Credible sources: assistance with message delivery

•Objectives: both short and long term

DOT/ EPA Collaboration: Pilots

Where:
- Dover, Delaware
- Albany, New York
- Milwaukee, Wisconsin
- San Francisco, California

What:
- Develop messages
- Test messages
- Establish baseline of:
  - awareness
  - attitudes
  - travel behavior
- Coalition building

Smart Travel Initiative Team // ##

Process for Pilot Selection

•**Determine criteria**: ozone non-attainment, range of transportation infrastructure, variety of demographic size, geographically diverse, etc.

•Input from Regions (approx. 25 recommendations)

•DOT/ EPA agreement on short list of 11 metro areas

•Request input from MPOs:

1- Ability, willingness to work with Feds

2- Resources (staff, financial, in-kind services) -- we are only giving $25 - $50 K

3- Potential Partners

4- Ability to survey/ evaluate; feds limited by ICR rules

Purpose of Pilots

•Develop program that is nationally applicable

•Every pilot will be different

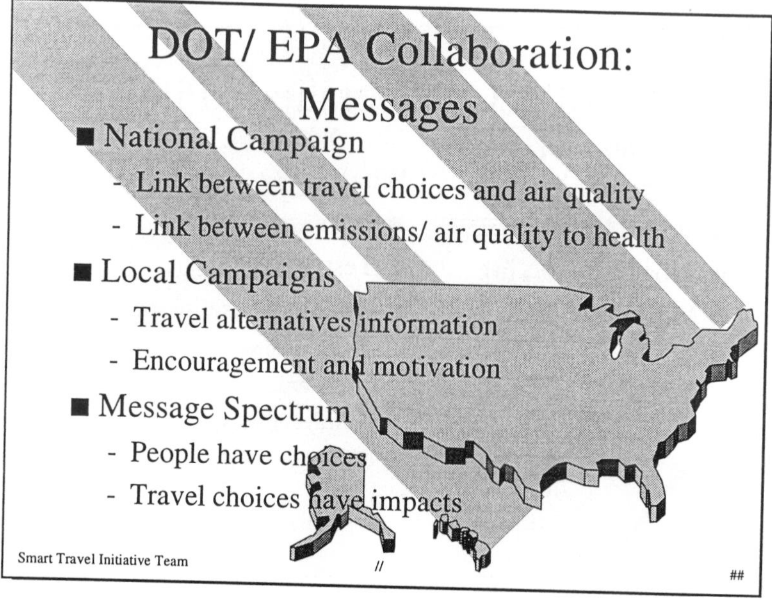

DOT/ EPA Collaboration: Messages
National Campaign
- Link between travel choices and air quality
- Link between emissions/ air quality to health
Local Campaigns
- Travel alternatives information
- Encouragement and motivation
Message Spectrum
- People have choices
- Travel choices have impacts
Smart Travel Initiative Team
//
##

•Consistent national message: Provides context and framework for individual responsibility in regards to transportation and air quality.

•Some examples of transportation alternatives

- Walking and biking
- Ridesharing
- Telecommuting (working at home)
- Trip Chaining

•Messages will not be designed to give drivers a guilt trip

•Messages will emphasize the "real" benefits of engaging in transportation alternatives

1- Reading and relaxation time while others drive

2- Cheaper to take mass transit than drive

3- More productive now that I telecommute

<u>Message Spectrum</u>

1- If "cooperative" pilot agreements consist of testing federal messages - EPA can have a role in message control

2 - If "cooperative" pilot agreements allows sites to develop their own messages - No EPA role in control of messages whatsoever

DOT/ EPA Collaboration: Timeline

- Pilot site announcement April 1997
- Pilot Kick-Offs begin September 1997
- Begin national coalition Winter 97-98
- Pilot evaluations on-going Fall 1998
- National Initiative roll-out 1999

			1	2	3	4
5	6	7	8	9	10	11
12	13	14	15	16	17	18
19	20	21	22	23	24	25
26	27	28	29	30	31	

U.S.. EPA- Transportation Group 4/10/97 25

Currently working with DOT to bring timeline forward

Pilot Site Announcement

•Internal EPA

•Letters to State & Local air officials/ directors and Stakeholders

•Press advisory, Trade Press

Pilot Kick-Offs: Invite Mary to be involved

National Coalition

•Begin recruitment after we can apply coalition building lessons learned from pilots

•Some examples of potential national partners: Air officials (STAPPA), grassroots transportation/ environmental (STPP), TP folks, environmental organizations (EDF), FACA folks - provide information and opportunity for participation, stationary source community, auto-related community (manufacturers, AAA, oil industry)

•Invite some to participate; others may be informative/ supportive role -- analogy of drunk driving campaigns

•Ask Margo/ Mary for suggestions and input

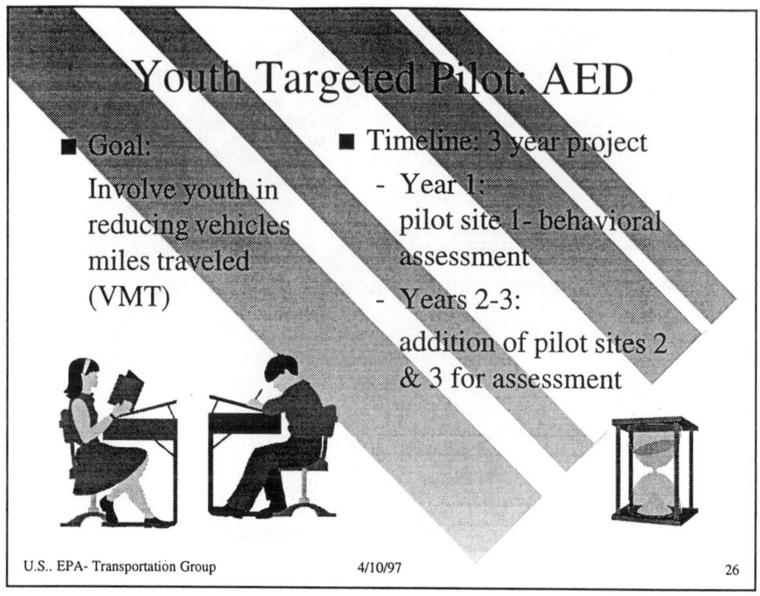

Youth Targeted Pilot: AED
Goal:
Involve youth in reducing vehicles miles traveled (VMT)
Timeline: 3 year project
- Year 1:
pilot site 1- behavioral assessment
- Years 2-3:
addition of pilot sites 2 & 3 for assessment
U.S.. EPA- Transportation Group
4/10/97
26

•Academy for Educational Development is: a non-profit service organization with expertise in youth development, social marketing and behavior change science.

• Partnership with Scholastic Magazine, one of the nation's premiere educational publishing houses.

•in-school and out-of school activities for youth:

Younger children: grades 5-7

Transition youth: grades 8-9

Young drivers: grades 10-12

Behavioral Asessment:

1) What's in it for me?, 2) Do my friends care?, 3) Can <u>**I**</u> do this?

Youth : can potentially have impact on: 1) their travel behave. now and in future and 2) their parents travel behave. now and in future.

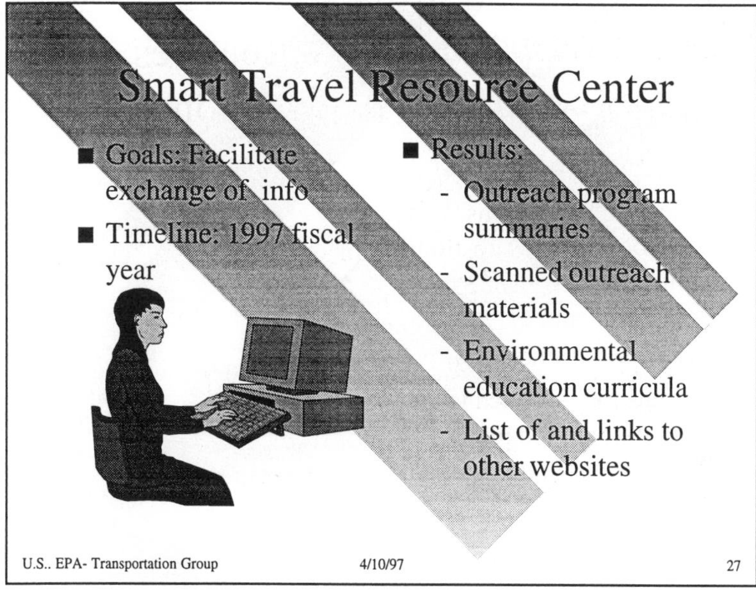

The STRC is a clearinghouse of tran-air quality education campaign information. (tool or mechanism for communication efforts). We expect that public decision makers, industry, consultants, public interest groups, as well as everyday individuals who support alternatives to driving alone will be among the many users of the STRC. The RC is basically an electronic depositiory of information assessable via the Internet and WWW.

The overriding goal of this effort is to facilitate the sharing of information on a variety of tran-air quality programs thru out the country and to improve the communications btw state and local officials regarding these efforts. This is especially valuable now because of the NAAQS revisions. We anticipate that many states and metropolitan areas will be interested in implementing these types of education and outreach programs in order to meet National Air Quality Goals.

The STRC will produce a variety of results. Program summaries will include pertinent information such as targeted pollutant, intended audience, sponsors, etc. The types of programs included will vary, for example, we will feature programs that have been operational for many years with significant budgets, and numerous sponsors. We will also feature programs that are more grass roots or community based. (read other results) The RC will link users to other websites, specifically those related to the environment, health and transportation. For example, there will be a link to a variety of episodic information which other members of TMIG are developing.

Transportation Alternatives Technology Applications

- Telecommunications technologies are substituting for vehicle travel, easing toll collection, and making transit more user-friendly
- 1-4 million Americans telecommute today; may rise to 29 million workers by 2010, reducing VMT by 4-5%

U.S.. EPA- Transportation Group 4/10/97 28

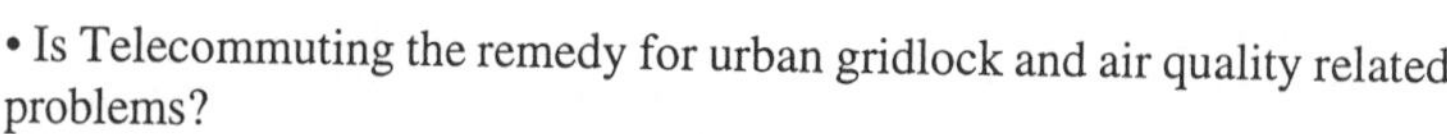

• Is Telecommuting the remedy for urban gridlock and air quality related problems?

•**Results to date in U.S.-** Falling short of expectations despite efforts by transit agencies, telecommunications firms to promote it. **Estimates** of telecommuters range from 4.1 million to 40 million, when work-at-home moonlighters are included.

•**Examples-** Summer Olympics in Atlanta- experiment in telecommuting- more than 200,000 workers telecommuted. BUT **those continuing after close have been too few to accurately count.**

•**Examples-** 1993 Congress appropriated $6 million--goal to get 60,000 federal workers to telecommute by October 1998(workforce 3 million) Additional $5 million appropriated in 1996.

•RESULTS TO DATE: 9000 government telecommuters

Has telecommuting peaked?

•1/3 of large companies in U.S. today let some employees work at least sometimes from home. Recent survey showed that only an additional 8% of those who currently don't allow considering- (May survey by Mercer Management Consulting)

•Annual growth in telecommuters- 6% today expected to decline to 1% growth in 2003 and beyond (According to Find/SVP, research firm which tracks telecommting trends) **Demise of 1990 ECO program.**

Intelligent Transportation Systems- Actions

ITS Goals

- reduce travel time
- squeeze more capacity out of existing roads
- Safer roads and faster emergency response
- reduce congestions and improve traffic flow (+50% VMT in 10 years)
- More efficient and productive commercial transport sectors

What is it?

- 9 Elements
  - traffic signal control systems
  - freeway management systems
  - transit management systems
  - multimodal traveler information systems
- 4 model Deployment Initiatives
- 75 largest Metro Areas in 10 years.

U.S.. EPA- Transportation Group 4/10/97 29

4 Model Deployment Initiative cities announced October 1996 by DOT

Seattle- $13.7M

Phoenix $7.5 M

San Antonio- $7.1 M

NY/NJ/Conn.- $10.43 Mm

Private sector fund "partner" >50%......Operational December 1997

ITS Program ASSESMENT- Crashes and Fatalities

Throughput and Time Savings

Customer Satisfaction

Cost Savings

energy and Emissions

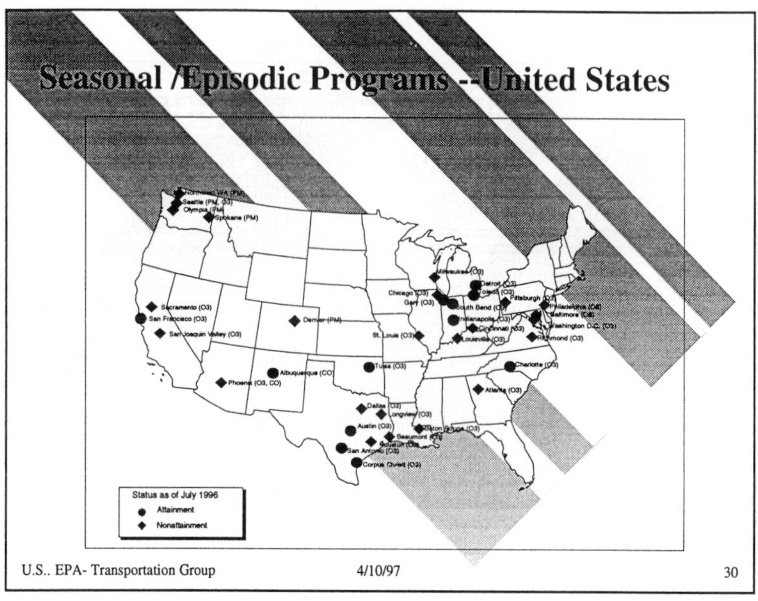

•Rapid increase in programs

•Goals- Public Education,

Attain or Maintain AQ Standards

Health Benefits

Congestion Managment

Reduce Emisisons

•Program Components-

Develop public and employer outreach tools

Coordinate with local community and businesses

Develop employer emissions reduction plans/measures

Develop accurate forecasting techniquest

Establish/operate media notification network

Collect data on impact of program

•Types of Control Measures- Focus on delay of emission producing activities

Mobile Sources- VMT reductions/shirting modes

Area Sources- Delay Activities (mow lawn, refuel)

Stationary Sources- Limit Ancillary Activities

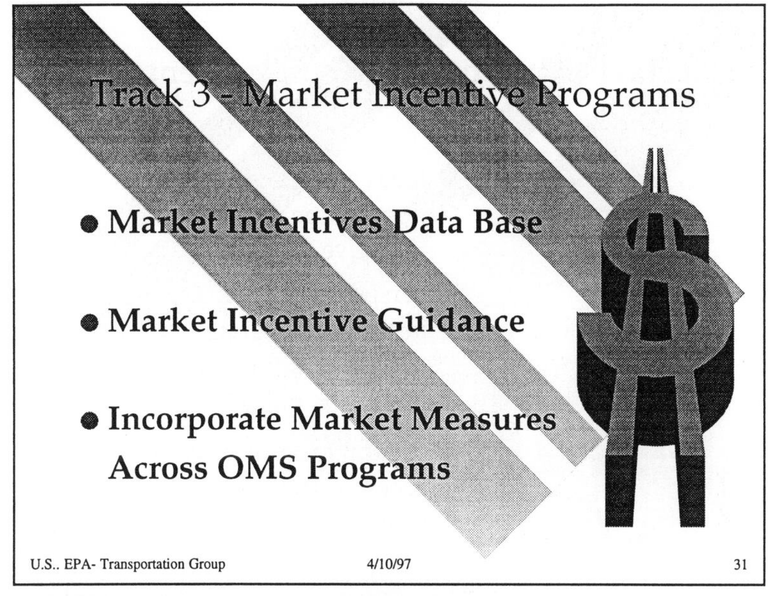

Track 3 - Market Incentive Programs
Market Incentives Data Base
Market Incentive Guidance
Incorporate Market Measures Across OMS Programs
U.S.. EPA- Transportation Group
4/10/97
31

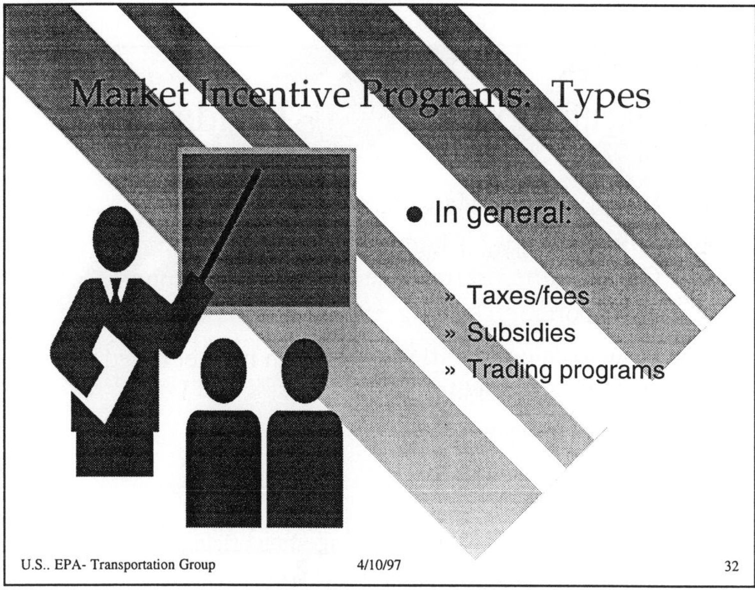

Market Incentive Programs: Types
In general:
» Taxes/fees
» Subsidies
» Trading programs
U.S.. EPA- Transportation Group
4/10/97
32

Transportation-sector specific:

Fee based on VMT and/or emissions

Road tolls, congestion pricing

Paying for parking/insurance

Transit fare subsidy

Scrappage

Market Incentives Resource Center

- Ideas and "how to" information on market incentive programs
- perspective of areas instituting programs
  - inclusive- Stationary and Mobile Source Programs
- Central location easy to access/ will be available on web

U.S.. EPA- Transportation Group 4/10/97 33

Transportation SIP Guidance

- EPA/FHWA
- Publication of "A Guide for Decision Makers: Opportunities to Improve Air Quality Through Transportation Pricing
- How to calculate SIP credits from market based strategies, including:
  - » At the pump charges
  - » Parking Pricing
  - » Emissions fees
  - » Roadway pricing (inc. congestion pricing)
  - » Subsidies

U.S.. EPA- Transportation Group 4/10/97 34

Implementing the Transportation Objectives

Collaborating with...

Intraagency
- OAQPS
- OPPE
- Regional Offices

Interagency
- FHWA/FTA
- DOE
- CARB

Regional/state/local governments
- STAPPA/ALAPCO
- NARC

Nongovernmental organization
- ACT
- STPP
- EDF

U.S.. EPA- Transportation Group 4/10/97 35

U.S.. EPA- Transportation Group 4/10/97 36

Transport, economy and air pollution in the 21st century

H.J. van Zuylen

Transport Research Center (AVV), P.O.Box 1031, 3000 BA Rotterdam, The Netherlands

1. INTRODUCTION

1.1. Looking into the future

The 20th century might be characterized as the century of diminishing distances and growing transport speed. The growth of distances travelled, travel speed and traffic volumes can be considered as a combined effect of technology, economy, social structures and national and international efforts to build more infrastructure with higher standards. The predictions of the future of the 20th century, made at the end of the last century or in the first decades of this century do not cover to a reasonable extent the real developments which have taken place. We should not expect that our expectations of what will happen in the next century will be more accurate or reliable. We can just do some wishful thinking, quantify our hopes and fears, extrapolate present growth processes, but we cannot expect that our views on the next century will be accurate, complete and relevant. Reviewing forecasts made in the nineteen-forties until the sixties for the second half of our century shows, that many forecasts were reliable, but that the emphasis in the many forecasted phenomena was seldom accurate: so even if we are able to forecast what will happen in the next decades, we will not be able to estimate whether the predicted trends will be important for the future of the world or that they will be accompanied by other, new trends which will overshadow the predicted ones. It is easier to look backwards and to (re)write history as a comprehensible, causal chain of events than to look ahead and do the same for the future. If we look forward, we can only see what we already know: extrapolation of the past. Really new developments can be identified, but their relevance and impact can not yet be assessed.

1.2. Creating the future

Air pollution due to transport is a technology-induced problem. It is reasonable to expect that technological innovation will be the first appropriate means to deal with the problem. However, technology is not an issue isolated from society. Technology is embedded in the society. Technology cannot be changed without taking into account what the reactions of the society will be. The time is yet not completely over when innovations came out of the laboratories, were brought to the market and became successful - or failed, without a careful consideration of its possible impacts. Most innovations are feasible and successful due to the fact that they satisfy certain apparent or hidden needs of users, match the existing infrastructure, have no conflicts with existing laws or division of competencies between relevant actors, match the existing preferences and habits. The innovation process is complex and forecasts are usually inaccurate and unreliable. The innovation process, as it is seen by many scientists, has a chaotic, unpredictable character: small factors apparently seem to determine the direction of the development process. It is already difficult to explain the process '*ex post*'. An '*ex ante*' prediction of success or failure is in most cases unrealistic.

1.3. Policies for the future

The role of governments in the innovation process is changing. The assumption that society can be constructed, planned, manipulated, changed to demand appears to be false. The dynamics of change processes is too complex to be fully understood, many factors which determine the dynamics cannot be controlled. Uncertainties can be coped with by using other approaches: scenario-development to get more grip on the - uncontrollable - influence factors, a more open steering concept with feed back and frequent readjustment of the steering and an approach where the government takes the role of facilitator, arranging the conditions for the innovation processes instead of steering the process. The new roles and approaches of the government require different skills and different structures in decision making.

Recently, in The Netherlands a project has been set up to develop option for a technology policy for transport [Slomovic et al. 1997]. The project INIT (Innovation Inland Transport Technology) identified about 15 innovation concepts which, according to expert views, will contribute to the policy objectives regarding accessibility and environment and which are feasible on a relative short term. For each innovation concept a preliminary survey has been done with focussing on *picture* (what are the characteristics of the concepts), *problems* (which problems will be solved by the concept) and *promises* (what will the concept contribute to the policy goals). Furthermore, the stakeholders have been listed and analyzed, the national and international initiatives which have a relation with the concept are enumerated. The strategy and necessary policy actions have been defined depending on the maturity of the technologies, the stakeholders and the existing R&D programs. The recommendations of INIT are now in a decision making phase and should lead to a coherent policy for the Ministry of Transport with respect to innovations. Most of this paper is based on the work done in INIT [INIT 1997].

1.4. Content of the paper

In section 2 the outlook on air pollution from transportation in The Netherlands is given. In section 3 the character of the transport system in relation with social and economic activities will be shortly discussed. In the approach of transport as a system, several ways can be found to solve the air pollution problem. Many of them involve technological innovations. In section 4 the character of the innovation process will be described and the possible roles of a government will be discussed. Section 5 describes how the innovations which are seen as promising and feasible, can be implemented in co-makership between private and public stakeholders. Chapter 6 gives the conclusions.

2. TRENDS AND SCENARIOS FOR AIR POLLUTION FROM TRANSPORT

In 1990, the Dutch Minister of Transport, Public Works and Water Management and the Minister of Housing, Physical Planning, and Environment published a policy statement on transport called *The Second Transport Structure Plan* (hereafter SVV-II, the Dutch acronym for *Tweede Structuurschema Verkeer en Vervoer*), which was approved by the Dutch government [Ministerie van Verkeer en Waterstaat 1992, Van de Hoorn 1993]. The SVV-II describes the current and foreseen problems associated with transport in The Netherlands and lists the goals and targets set by the government, as well as a wide range of measures and policy instruments proposed for reaching the goals. Although technology developments play an important role in the presented measures and instruments, the SVV-II paid no special attention to technology as a phenomenon in itself.

The SVV-II distinguishes three main problem areas:

- Economic centers are in danger of becoming inaccessible, and the national position of The Netherlands as a transport and distribution center is in jeopardy.
- Pollution continues to worsen: transport damages the environment through air pollution, noise nuisance and the fragmentation of the countryside.
- Road safety is deterioratinging after years of improvement.

The targets for 2010 have been made specific and quantitative in the Structure Plan. Recently, a similar exercise has been done for the OECD project on Environmental Sustainable Transport (EST) [van Wee et al. 1996]. Six quantitative criteria were given, where for The Netherlands these criteria have been extended with four additional ones. The criteria for SVV-II and EST which apply to air quality are given in table 1.

Table 1

Air quality criteria for SVV-II and EST

Component	SVV-II: Reduction between 1986 and 2010 (%)	EST: Reduction between 1990 and 2030 (%)
CO_2	10	80
NO_x	75 (cars) - 80 (commercial vehicles)	90
VOC (volatile organic compounds)	75	90
PM_{10}		90

Since the SVV-II was written, however, it became increasingly clear that the ambitious targets for many policy goals cannot be reached without an extra policy impulse. An assessment has been made of the intermediate situation and an estimation based on two scenarios. The most optimistic scenario with respect to economic growth was called the European Renaissance Scenario (ER) [Centraal Planbureau 1992], which assumes a dynamic, technological development, economic integration of Europe and a strategic role for the government as a co-ordinating mechanism. The second scenario, the Global Shift Scenario (GS) assumes a dynamic technological development, a free market, a stagnation in the European integration and an economic growth in Europe which is less than in Asia and North America. In the next two figures the estimated development of car kilometers and CO_2 emissions are given as estimated for two economic scenarios.

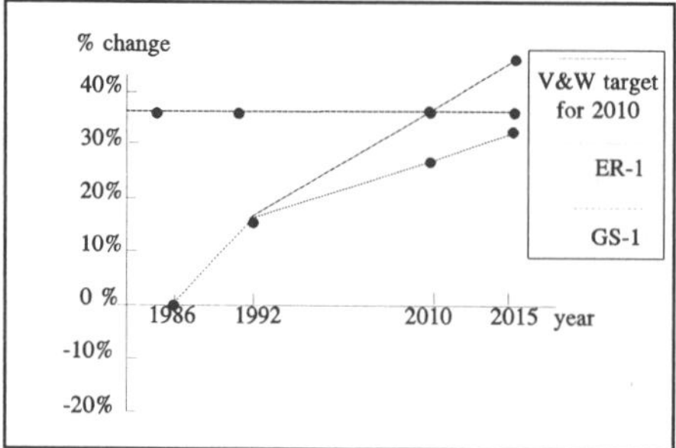

Figure 1 Observed and estimated development of car kilometers between 1986 and 2015

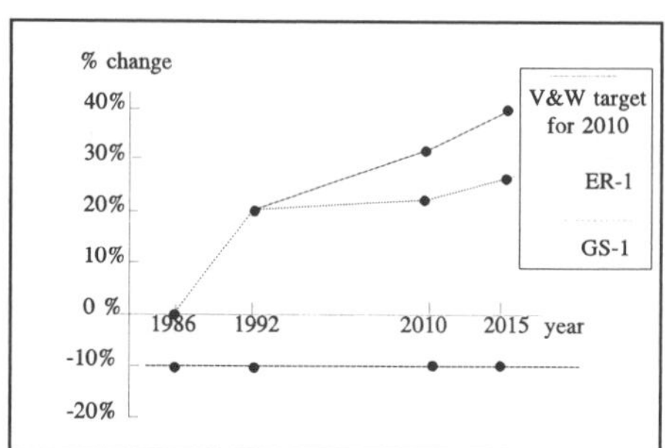

Figure 2 The expected development of emissions of CO_2

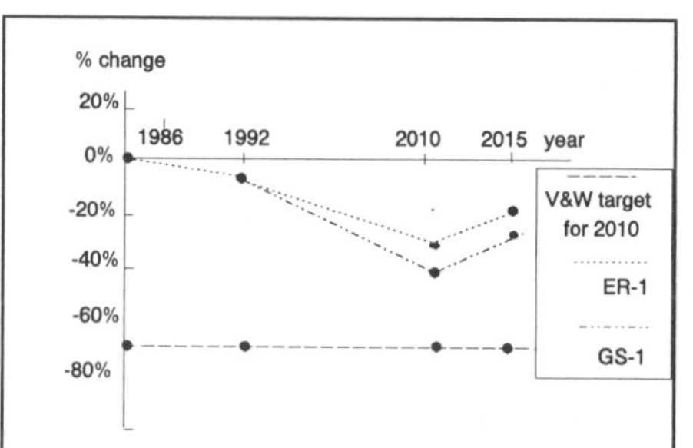

Figure 3 Percentage change in NO_x emissions from road traffic from 1986

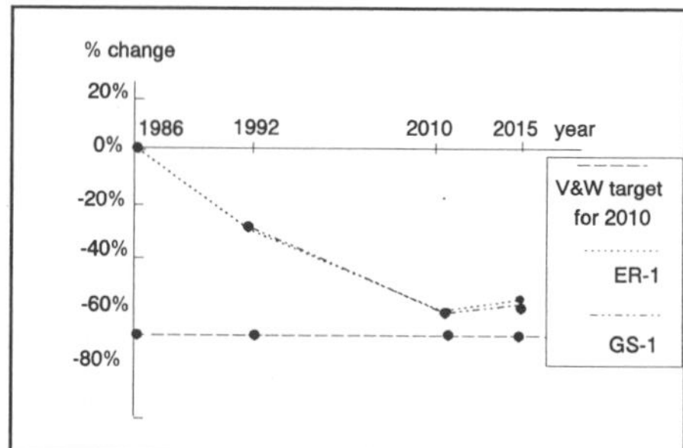

Figure 4 Percentage change in VOC emissions from road traffic from 1986

Preliminary studies suggested that part of this extra policy impulse could come from technology policy for traffic and transport. In the EST study [Van Wee et al. 1996] a detailed investigation has been made in which the impact technological innovations are compared with strictly restrained transport demand. This study shows, that the *Business As Usual* scenario does not lead to a situation which satisfies the requirements. A second scenario, in which technological innovations are supposed to be introduced, such as fuel cells, hydrogen and electrical and hybrid traction, shows sufficient reduction of emission to reach the targets for 2010 and 2030. A third scenario is the capacity constraint scenario. In this scenario it is assumed that the transport demand has been significantly reduced. Large changes in the society are needed, including a more efficient organization of transport, reduction of distances between production and consumption location, elimination of commuting and car travel, except for special services. The economic growth will be changed significantly from the present trend, because production has to be done more localized. A fourth scenario has been developed in the EST study, in which a mix has been made of technological innovations and the reduction of transport demand. Here less extreme assumptions had to be made, while the objectives still could be achieved.

The EST scenario study contains many assumptions which all represent uncertainties. However, the high technology scenario and also the fourth scenario show that extrapolation of the present economic grow does not necessarily conflict with high standards for air quality. The problem is, however, the realization of the necessary innovations, i.e. the changes in technologies and structures and culture in society and institutes the role politics has to play. INIT [Slomovic et al. 1997, INIT 1997] has developed policy options for the development and implementation of technological innovations which will help to achieve the necessary improvements in the transport system. In order to clarify the possibilities to improve the environmental impact of transport, a short discussion is given of the transport system. In this discussion it will become clear that technological innovations have more possibilities than just improving the direct source of pollution: the propulsion system of vehicles. Technology can also help to modify modal choice, facilitate intermodal transport and reduce transport demand.

3. THE TRANSPORT AND TRAFFIC SYSTEM

3.1. A systems view

Transport is a derived demand, where social and economical activities lead to the demand for mobility. In figure 1 a conceptual scheme is given of the transport system. Most environmental problems are located in the transport means part of the system, where vehicles give emissions, noise and risks. The directions for solutions can be on any level: e.g. demand reduction, alternatives for existing transport modes and improvement of the existing modes. The link between economy and transport is mainly that a growth of economic activities induces transport demand. Qualitative changes in the economy have also an important impact. For instance, the stimulation of recycling creates new transport patterns. Also changing logistic concepts and geographical shifts in economic activities have an impact on transport demand. Another influence of the economy on transport is, that a growing economy generates higher incomes which result in a higher travel demand.

An important factor which influences transport demand is regional planning. The geography of economic and social activities determines for a great deal the demand of transport. Optimization of the location of industries with respect to each other and of residential areas with respect to working places, cultural centers and recreation areas determine the distance for transport and often also the quantitative transport demand. Most important for transport demand are cultural patterns: habits, preferences and values determine the needs for transport, especially transport of people.

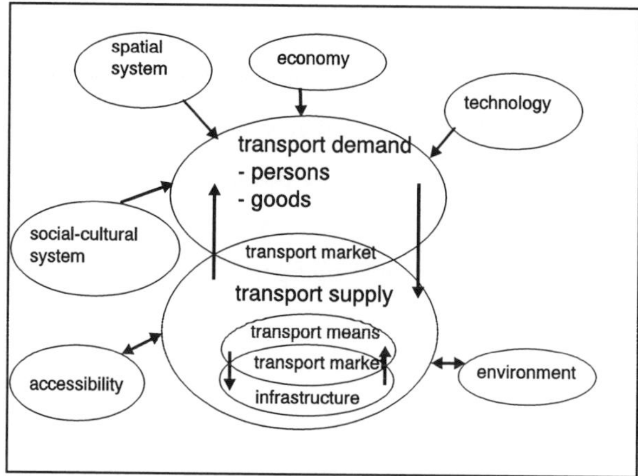

Figure 5 Conceptual model of the transport system [INIT 1997]

The environmental problems such as air pollution are caused at the transport means level of the transport system: vehicles and the way vehicles are used. The most obvious and logical solution is to develop clean engines and to manage traffic in such a way, that environmental problems are not caused at sensitive locations such as residential areas. However, it is often more effective to look also at the higher levels of the system. In The Netherlands the vision on the future of the transport system, in which both the environmental goals as well as the goals with respect to accessibility are realized, can be summarized as follows [INIT 1997]:

1. The demand for transport is restricted by making transport more expensive, provide alternatives for transport of people by stimulating tele-activities and improved land use patterns.
2. The character of the transport system is modified gradually into a service structure: transport is a integrated service from door to door, where the choice of route and vehicle is of less concern the ownership of vehicles is separated from its use. The choice of routes and vehicles is subjected to limitations, especial in urban areas, and the costs structure is such that the choices lead to a social acceptable situation.
3. The transport service is optimized such that a better utilization of vehicles and infrastructure can be realized. This implies that intermodal transport becomes normal: the mode which is the most cost-effective is chosen and interchanges between modes are no longer costly and give no unnecessary delays.
4. The transport chain has become transparent: for the supplier and user of transport services all the necessary information is accurate and complete to make the necessary decisions and to control the transport process.
5. The different links and nodes in the transport chain have been improved such that the infrastructure is utilized as good as possible and vehicles are efficient, safe and clean.

3.2. The concepts

In INIT the approach has been to investigate the potential of technology on all levels of the transport system, i.e. for innovation of vehicle traction as well as for demand reduction. INIT has identified 14 innovations which have the possibilities to contribute to the vision of the future. The innovations have a technological character and are assessed by experts as being feasible. The innovations are given as *concepts*: descriptions of a design or elaboration of an available set of measures and (technological) resources for a specific area of traffic and transport in the future. Concepts are more then just applied technologies. They bring technological innovations and changes in society and economic structures in relation with functions of the transport system.

The technology concepts that are formulated in INIT are technologically evident: they are no uncertain, creative inventions but applications of existing technologies and technologies which can be developed on a short term. They have been formed on the basis of discussions with experts about the technological possibilities and have grown into coherent future scenarios. They pose new challenges for society and the government. They hold promise, but sometimes bear risks as well; promise in terms of more efficient, safer and cleaner transport, risks in the context of, for example, changing economic relationships - some existing organizations will disappear or get a completely different role, and the rate of technological development in relation to social change - how much technological change can the society absorb in a few decades.

There are three categories of concepts:

1. tele-activities:

The use of information technology facilitates several activities where physical presence is no longer necessary: tele-working, tele-learning, tele-shopping etc. There are many possibilities to reduce physical transport. However, until now the applications are too limited to show a significant impact on transport and traffic. Further stimulation of experiments and removal of barriers may develop the full potential of this concept. The risk exists, that tele-activities will induce new transport demand, just as has happened with the telephone.

2. integrated goods transport chain:

Several concepts can help to facilitate (inter-modal) transport chains. The concepts apply to the infrastructure, vehicles, organization of transport, intermodal terminals and organizational issues:

- intermodal market linking system, an information system which links supply and demand for transport, supports the optimization of transport chains and provides the necessary information to follow goods along the transport chain,
- transhipment terminals, the optimization of intermodal nodes, such that costs and transshipment delays are reduced,
- integrated intermodal packaging infrastructure: standardized packaging and transport

units which are suited for different modes - also new modes like underground transport -and satisfy the requirements of producers, retailers, carriers and forwarders; the concept includes all the manufacturing and logistics to optimize the use of the standard units,

- underground (urban) haulage system, a new transport mode which eliminates the need for surface transport in sensitive areas, such as town centers, an improves the quality of life in those areas,
- automatic vehicle guidance, which will automate a part of the transport chain, reduce the costs, improve the utilization of vehicles and infrastructure,
- integration of inland shipping and coastal shipping: by technological innovations in ship building, waterborne transport in certain important transport niches can be made more competitive with respect to road transport,
- high-speed waterborne transport, using high frequency connections, fast transshipments and high-speed ships,
- improved designs for lorries/vans, which make them more suited for use in urban environments, cleaner and safer.

3. integrated passenger transport chain:

The separation between individual, private transport and collective transport has to become subordinate to the concept of the door-to-door transport service. Just as nobody is interested in the question whether the connection for a telephone call is made by a copper wire, glass fibre or wireless, the choice of vehicle, mode and route should in the future be determined by considerations like efficiency and social acceptability, without reducing the speed, efficiency and comfort of the trip. The transport system should be reliable, efficient and safe. Ownership of vehicles will no longer determine the preference for the use of a transport mode. Concepts which will support this vision of the future of passenger transport are:

- a fully integrated (public) passenger transport system
- a dynamic traffic management system
- a dynamic information system
- automatic vehicle guidance

- modular vehicles
- propulsion and fuel

The first concept primarily ensures the integration of different means of transport and connections between modes of transport. The goal is to achieve organizational improvement, supported by technological innovations.

A dynamic traffic management system ensures optimum utilization of existing infrastructure. A dynamic information system makes it possible to make transport choices on the basis of all the information available at any given time and supports the integrated passenger transport system.

4. THE VIEW ON INNOVATION

4.1. The innovation process

Technological innovations in the transport system are necessary. However, the direction of the innovation process is uncertain and difficult to steer. In the beginning of the innovation process everything is uncertain, there are plenty possibilities to steer the innovation process but there is little guidance for decisions in which direction to steer the process. As the innovation process goes on, the impact of the innovations and its possibilities become more clear. However, the possibilities to change the direction become less. Many forces have an influence on the innovation process. These forces work often in an independent, incoherent way. Cooperation between these forces is in most cases by accident.

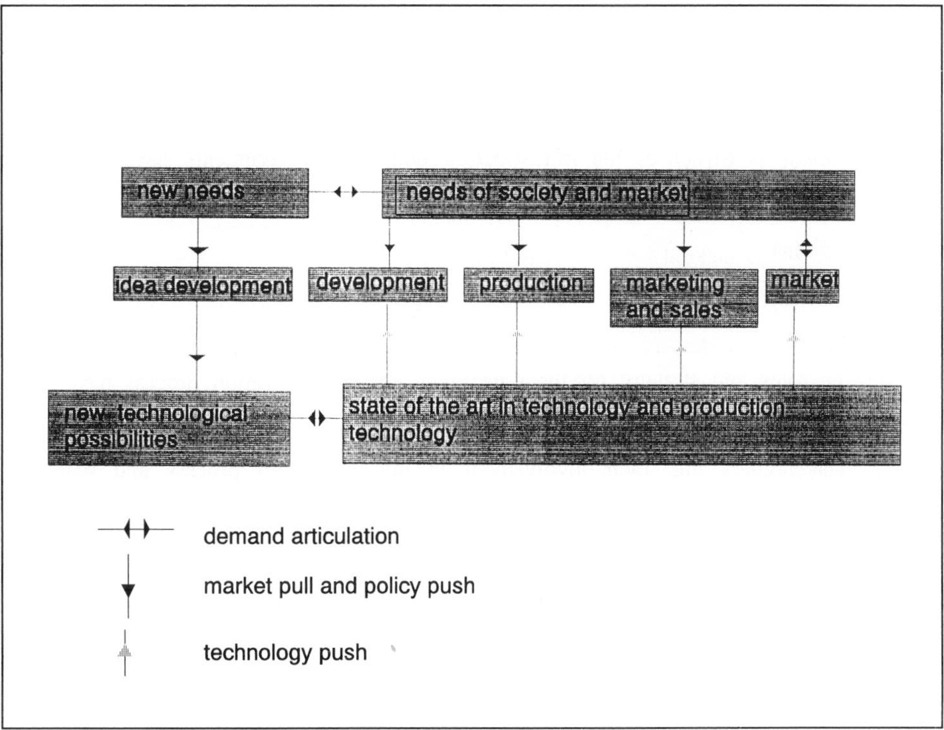

Figure 6 A model of the innovation process

The ways in which the government can have influence on the innovation process are:

- stimulating research and development by subsidies and research programs,
- facilitating the market uptake of innovation e.g. by tax measures, subsidizing the replacement of outdated equipment,
- strategic niche management: small niche markets are created where innovations can grow, build up the necessary infrastructure and networks without facing the competition of existing products,
- demand articulation, clarifying what the needs for innovation are and giving guidance to the industry indicating which directions will have the support of the government
- being an advanced buyer of new products,
- building networks of stakeholders to stimulate co-operation and removing barriers
- removing legal barriers,
- national and international co-ordination of innovation activities

- enforcing technologies by legal prescriptions,
- covenants between government and trade and industry.

In some cases the government can have different roles in the innovation process. Since the government is owner of the infrastructure and most governments have their own research and development organizations, the government is sometimes even found in the role of developer.

4.2. Strategies

In INIT the roles and strategy for the government in the innovation process have been determined for each concept, contingent with the maturity of the technology, the relation between stakeholders and their involvement in the innovation process and with the existing innovation programs. Some innovations are already in a mature state, many are already supported by initiatives of the government. Such initiatives should not be changed, unless more co-ordination is needed or the present initiatives are too limited in scope. Examples of such innovations are the *Dynamic Traffic Management System* and the *Travel Information System.*

Other innovations need technologies which are in a mature state and the innovation process and impact of the innovation is already reasonably clear. Such innovations require a kind of *Constructive Technology Assessment* [Rip et al. 1995]: stakeholders should come together, form a network and should identify their own goals with respect to the innovation. The government has to try and find a way to 'polarize' the different goals: the interests of the different stakeholders is taken into account and the goals which facilitate or fortify the common policy goals are supported. The process of the acceptance of a common shared goal, the growth of common understanding and the adjustment of different activities is first of all necessary. Examples of concepts that have to be dealt with in this way are:

- an intermodal market linking system
- terminal technology
- integrated intermodal packaging infrastructure
- an underground (urban) haulage system
- integration of inland and coastal shipping

- a fully integrated (public) passenger transport system.

The content of these concepts might change significantly during the development process. Little has been fixed yet and the future is still uncertain, which gives many possibilities for adjustment.

Finally there is the class of long-term innovations, which need several decades to be fully developed. Examples are *Fully automated vehicle guidance* and *Modular vehicles.* The approach for these concepts is, that an *Awareness Technology Assessment* [Smits and Leyten 1988] is executed, an analysis of possibilities, advantages and disadvantages, risks and opportunities, a map is made of stakeholders and a discussion is started whether such innovations are needed and what should be done to make a maximum profit out of the concept.

An analysis of the transport system is a precondition for an effective strategy. Experimental learning is a good approach supplementary to this analysis. Demonstration projects, pilots and niches are excellent ways to investigate the full potential of concepts, to find limitations and barriers.

4.3. Co-makership

Innovations in the transport system cannot be imposed, enforced or just left over to the market. The forces of the market have to be used to promote the innovations, but unfortunately many of these forces work in opposite directions. Market forces often lead to *social dilemmas*: short term, individual interests create a locked-in situation where optimization of other stakeholders become restricted or even impossible and the whole community ends in a worse condition than in the situation that some top-down, global optimization had been done and individual freedom of choice had been limited. Solving the social dilemma is a task for all stakeholders, but a government often has to take the initiative to change the structure of the market. The government is in most cases not able to steer the innovation process, but it is possible to change the characteristics of the dynamics of the process in such a way, that the innovation moves towards an attractive situation and will not be locked-in in a social undesirable state [van Zuylen 1995].

Innovations should be tailored to the specific application area. New technologies should contribute to the solution of existing problems and should be applied such, that they fit in existing patterns of behavior, competencies and equilibrium of power distribution. For example, in logistics several innovations will only be possible if they are introduced in collaboration with the stakeholders. The term 'acceptance is too weak to describe this aspect of innovation. The term 'co-makership' is more appropriate.

The practical approach proposed by INIT is, that for each concept a public-private organization is set up to develop implementation plans. These plans should be executed in collaboration between government, research and trade and industry, in order to get a guarantee of quality of the plan and a maximum support of the stakeholders. The co-ordination between concepts and the goals for the different concepts will be the care of a steering group according to a multi-echelon structure [Jantsch 1972].

The initial activities of these organizations should be to identify the stakeholders, to analyze the processes in which the concepts should play a role, look for existing initiatives which can support the development and introduction of the concepts. During this process it is possible and likely that the original concepts undergo an evolution and will be tailored to existing needs and problems of the stakeholders. This phase will be analytic as well as creative: analyzing subsystems of the transport system and creating new directions of solutions.

The next phase should focus on actions: *research* to reduce uncertainties, *development* of technologies, *pilots* and *demonstration projects*. In the following implementation phase the barriers have to be identified and removed and starting problems have to be solved. Suitable strategies are tax measures, legislation and strategic niche management. There has to be a coordination between actions of public and private stakeholders. Just as with goals, also action should be 'polarized': actions which work nearly in the same direction have to be fine-tuned and coordinated and conflicting actions have to be prevented. The basis for this 'polarization' is a good common understanding of all stakeholders [Eden en Vangen 1995].

This approach proposed by INIT for the innovation process is relatively new for the domain of technology policy in transport. It requires new skills and - at least for The Netherlands - new ways of co-operation between government, local authorities and private stakeholders. The dynamics of the innovation process in the transport system is not well understood yet. Research and experiments will be necessary to find effective ways for implementing innovations.

6. CONCLUSIONS

Transport has a negative impact on the environment. Without appropriate measures the pollution caused by transport will be unacceptable. The present situation has to be improved and the growth of transport demand have to be dealt with in such a way, that technological improvements of vehicles are not overshadowed by increased demand.

Transport is a derived demand where economy, the spatial system and social activities are determining factors. Choice behavior in often sub-optimal from the point of view of policy goals. Social dilemmas make market forces work in the direction of locked-in situations, where possibilities for optimization become very limited.

Several technological innovations have been identified, which can help to reduce the environmental impact of transport. Some innovations work on the level of the propulsion system, others have influence on the choice behavior, reduce transport demand, create new choices and make environmental friendly transport modes more attractive. Scenario studies show, that conflicts between economic growth and environmental quality can be solved.

Technological innovations have to be implemented very carefully. Transport is an organic system. Implementation of 'strange' technologies is difficult and often impossible due to repulsion reactions. Careful fine-tuning of the primary processes, identification of opportunities for innovation and 'polarization' of the goals and actions of the different stakeholders can lead to successful adaption of new technologies.

The approach which is proposed is based on co-makership in the innovation policy, where government, provincial and local authorities coordinate their strategies, private stakeholders are involved and share the responsibility for the implementation plans. The uncertainties in the dynamics of innovations and the emergence of new problems and technologies will be dealt with by monitoring the implementation process, executing technology surveys and periodic replanning.

INIT has recognized the fact that technological innovations have international dimensions. First of all is the automotive industry a worldwide complex that cannot easily be influenced by one single country. Furthermore, the competence of international organizations such as the European Union might be more suited to develop and implement an effective technology policy. There are many international activities in research and development which fit in the innovation concepts described in this paper. Duplication of these efforts is a waste of resources. Therefore, for each concept the international dimensions have been investigated and the implementation programs have to be set up in collaboration with existing international initiatives.

REFERENCES

1. Centraal Plan Bureau (1992) *Nederland in drievoud; een scenariostudie van de Nederlandse Economie 1990-2015*, SDU Uitgeverij, Den Haag, 1992.
2. Eden, C., Vangen S. (1995) An analysis of shared meaning in collaboration, *Second workshop on Multi-Organizational Partnership*, Glasgow
3. INIT, 1997 *Technologiebeleid in Verkeer en Vervoer: Samen werken aan innovatie*, Concept report Technology Policy in Traffic and Transport, Ministry of Transport, Public Works and Watermanagement, The Hague.
4. Jantsch, E. (1972) *Technological Planning and Social Futures*, Cassell, London
5. Ministerie van Verkeer en Waterstaat (1992) Dutch multi year programma on infrastructure and transport 1993-1997 The Hague
6. Rip A., Misa, Th. J., Schot, J. (1995) *Managing Technology in Society*, Pinter, London and New York
7. Slomovic, A., Smit S., van Grootveld, G., van Schaick, G.H.J., van Zuylen H.J. (1997) Developing a Technology Policy for Innovation in Traffic and Transport in The Netherlands. Int. Journal of Technology Management. To be published.
8. Smits, R. and Leyten, J. (1988) *Technology Assessment, Waakhond of Speurneus* Kerckebosch. Zeist
9. Van de Hoorn, A., (1993) The Dutch Transport Structure Plan 1986-2010, *Compendium of papers from ITE 63rd annual meeting*, ITE, Washington
10. Van Wee, G.J., Geurts, K.T., van den Brink, R.M.M., van der Waard, J. (1996) *Transport scenarios for The Netherlands for 2030.* Report 773002009. RIVM, Bilthoven
11. van Zuylen, H.J. (1995) The game of the rules. *The 23rd European Transport Forum* Seminar F, pp. 95-104, PTRC, London

'Hedging' Strategies for CO_2 Abatement

J.R. Ybema and A.J.M. Bos

Netherlands Energy Research Foundation ECN, unit Policy studies
P.O.Box 1, 1755 ZG Petten, The Netherlands

ABSTRACT

The future energy development of a country will differ substantially depending on the level of CO_2 emission reduction that is aimed at. To properly take the long term risk for drastic CO_2 emission reduction targets into account in the analysis of near term energy investment decisions, it is required to apply decision analysis methods that are capable to consider the specific characteristics of climate change (large uncertainties, long term horizon). Such decision analysis methods do exist. They can explicitly include evolving uncertainties, multi-stage decisions, cumulative effects and risk averse attitudes. The methods appear useful to select hedging strategies for CO_2 reduction. Hedging strategies for CO_2 reduction are sets of near term decisions which are most robust for various long term outcomes of climate change negotiations. The result of a hedging analysis gives a balance between the 'present' risk for costly premature emission reduction (when CO_2 reduction appears not needed) and possible 'future' risk for neglected CO_2 reduction in the past (when deep CO_2 reduction appears to be required). A stochastic version of a dynamic techno-economic energy model for The Netherlands was made. This model was used to quantify a CO_2 hedging strategy. Two outcomes of the climate negotiations were forecasted and probabilities were estimated for these outcomes. The results of the examples clearly showed that the calculated near term strategy differs from the results of conventional methods that do not have the capability to include uncertainty. The results of CO_2 hedging analyses indicate that it is better to take concrete action than to wait until uncertainty about CO_2 reduction targets is resolved.

1. INTRODUCTION

Energy consumption is the main source of greenhouse gas emissions. The choice between one energy technology or another determines to a large extent how much of a specific fuel is used, and thus how large the emissions of CO_2 will be during the active lifetime of the technology. Many energy technologies and the energy infrastructure have long technical lifetimes and long construction times. Therefore, energy is an area where long term planning is of crucial importance.

Since several decades scenario analysis is being used as an important decision support tool in this long term planning process. Various advanced modelling tools have been developed to support energy scenario analysis. In many countries, energy scenario analysis has also been applied to study the possibilities and consequences of reducing CO_2 emissions from the energy system. Such analyses have primarily been made on the national level, as energy policy mainly takes place at this level. Almost without exceptions these scenario studies followed deterministic approaches. This implies that uncertainty in reduction targets for CO_2 was not explicitly considered. Instead, a range of emission reduction targets was analyzed. In such an approach one analyses which measures and investments are required to achieve one or more 'certain' emission reduction targets. As such, scenario analysis remains oriented towards a 'learn-then-act' characterization of the decision problem: the uncertainty about the long term CO_2 reduction target is assumed to be resolved prior to the date at which action is taken [1].

However, the outcome of the international negotiations that take place over the next 10 to 20 years, is uncertain. Therefore, the national emission reduction allowances are also uncertain. They depend on the level of participation of developing countries in the convention, the total level of emission reduction and the use of flexibility increasing instruments. Regarding the current negotiations between countries under the Framework

Convention of Climate Change (FCCC), it is hard to predict what the outcome will be. Uncertainty about emission reduction targets is likely to remain for some time. In the meantime the most worthwhile thing to do is to find out what to do in the near term under this long term uncertainty; one has to 'act-then-learn'.

Being faced with the climate change problem, the best a country can do now is to strive for a flexible energy system in the near term at limited additional cost. Such a near term energy system configuration should be a good starting point to realize all possible long term CO_2 emission reduction targets. Such a strategy is called a CO_2 hedging strategy[10]. This paper presents elements of hedging strategies for CO_2 abatements. This is done in two ways. First, in Chapter 2, some practical ideas for hedging in concrete energy investment decisions are listed and explained. Further, the main body of this paper (Chapter 3, 4 and 5) presents an analysis of how a hedging strategy for an entire country could look like. This latter part includes a model based analysis of a CO_2 hedging strategy for The Netherlands. In Chapter 6 conclusions about CO_2 hedging analysis are drawn and the main limitations are listed.

2. IRREVERSIBILITY AND FLEXIBILITY IN THE ENERGY SYSTEM

At the moment when energy technologies, energy infrastructure and buildings are constructed, there is an opportunity to choose a less or more energy efficient type. After the construction has taken place, the energy consumption is more or less fixed for the lifetime of the equipment. One can of course modify or retrofit the original equipment but there are usually relatively high cost involved and there remains a limited potential for efficiency improvements. Thus, the initial construction of energy technologies, infrastructure and buildings create irreversibilities. The irreversibilities are of paramount

[10] 'Hedging' means securing oneself against possible losses or keeping one's options open. The term hedging originates from financial analysis and operations research. In financial analysis it implies the diversification of the risks of adverse financial shocks. Hedging is seeking the optimal path in an uncertain world. Implicitly, hedging approaches involve the protection against possible negative consequences by preserving future flexibility in courses of action.

importance for CO_2 reduction strategies. They determine for a large part the small size of low cost potentials for future emission reduction.

It is possible to reduce irreversibilities in the energy system to a certain extent by allowing more flexibility. This can be done by already anticipating at the moment of construction of the equipment that this equipment will later possibly be adapted. For many retrofit options it is indeed possible to comply with conditions that allow adaptation at relatively low cost, years after the original design. In this way the cost for abatement can be reduced and the potential for emission abatement can be enlarged. Many concrete flexibility increasing measures can be listed. Here, some examples are given to illustrate this concept.

> Orientate the roofs of houses to the south to have a higher electricity yield once photovoltaic systems or solar boilers might be installed on the roof.

> Change the design of products to increase the possibility for eventual recycling.

> Locate power plants close to areas with a high energy demand (industry, buildings). Eventually, the plant can be transformed to a cogeneration mode and the market to provide the heat to, is then nearby.

> Built a kitchen in a house with a slightly broader space for refrigerators than the currently common space to allow future replacement by a better insulated, and more space requiring, refrigerator.

The flexibility increasing measures are concrete 'hedging options'; they are means to keep options open. It is worthwhile to investigate the energy system in detail to identify the most prospective ones.

3. NATIONAL HEDGING STRATEGIES: A MODELLING APPROACH

This chapter informs how a CO_2 hedging strategy was constructed for the Dutch energy system. The model applied in this analysis, is a newly developed version of the MARKAL model. MARKAL (acronym for MARKet ALlocation) is a technology oriented model that has already extensively been applied to study the role of technologies in the future energy system, see e.g. [2]. MARKAL is a cost-minimizing model that becomes most often applied to analyze complete national energy systems. The stochastic model minimizes the expected net present value (NPV) of the energy system over the total time period considered. It is able to determine such a mix of energy technologies that the end-use demand for energy services is met at least cost, while the environmental and reliability conditions are taken into account. The model can calculate cost-effective strategies to abate CO_2 emissions when a dynamic CO_2 reduction path is imposed. The supply and demand side of the energy system are considered simultaneously when cost-effective CO_2 reduction strategies are calculated.

The version of MARKAL that has been applied for this study explicitly contains different uncertain emission targets for The Netherlands. It is hard if not impossible to give an objective assessment of CO_2 emission reduction targets and the probabilities for these reduction targets. Therefore this hedging analysis has based the probabilities on subjective assumptions. The model can be applied to include the time cumulated emission budgets for CO_2. For climate change this is important as CO_2 accumulates in the atmosphere. Within the climate negotiation budget approaches receive more and more attention. It is relevant to learn how a country can best use its emission budget over time under uncertainty.

The model is also capable to analyze multi-stage decisions. Hence, the uncertainty in national reduction targets will reduce over time, and thus more pointed reduction targets for CO_2 in the long term will appear. Multi-stages are also important as alternative energy investments will have different levels of flexibility to reach eventual future CO_2 reduction targets, and this flexibility needs to be valued.

The database that is applied represents the Dutch energy system in quite some detail for the time period 2000 to 2040 and in nine steps of five years. The energy demand projections for 45 kinds of energy end-use are roughly in line with recent energy demand projections [3]. It is noted that nuclear energy is not allowed as an option. CO_2 removal from coal power plants has been considered. Further, it has been assumed that the technologies considered improve over time. All energy technology data have been taken from a recent technology assessment study [4]. Energy demands and energy prices are exogenous to the model. Their projections have been taken from an earlier scenario study

[4]. The discount rate applied amounts to 5% per year.

It has been assumed that until the year 2020 it is uncertain by how much the Dutch CO_2 emissions have to be reduced. In 2020 it is assumed that the countries participating in the FCCC agree on long term national CO_2 reduction budgets. Then it appears that The Netherlands CO_2 emission budget for the period 2000-2040 is unlimited or the CO_2 emissions budget is equivalent to an annual reduction with 15% compared to the 1990 level for the time period 1997-2042. The associated probability assumptions are given in

Table 1.

Table 1: Distinguished states of nature and assumed corresponding CO_2 reduction budgets and probabilities attached.

State of nature	CO_2 emission budget for period 1997-2042 [% of 1990 level]	Assumed probability [%]
Unconstrained emissions	unconstrained	50%
Emission reduction	10%	50%

4. RESULTS OF CO_2 HEDGING ANALYSIS FOR THE NETHERLANDS

In order to be able to situate the effect of hedging, the results are compared with the results of three deterministic MARKAL calculations which correspond to the optimal energy system configuration for each of the three individual CO_2 emission reduction targets.

4.1. CO_2 emission levels

Until the period with 2015 as the central year, decisions to invest and/or to use energy technologies and primary fuels are taken without certainty about the time-cumulative CO_2 reduction target. The model will choose one optimal set of decisions which allows to achieve each of the long term emission reduction targets that have been distinguished.

This set of optimal decisions is the CO_2 hedging strategy. This strategy has the lowest expected cost and it takes into account that it possibly has to comply with both emission targets. After 2015 it becomes clear which time-cumulated CO_2 reduction target has to be met. The strategy for the period 2017-2042 will depend on the state of nature.

The total effect of the results of the model can be monitored by considering the total emissions of CO_2 over time, see Figure 1. The CO_2 emission linked with the calculated hedging path are presented by the solid line between 1995 and 2015 in Figure 2. After 2015 the CO_2 emission paths diverge from the common hedging path for the different realizations of emission reduction budgets (see the two solid lines after the year 2015. With the realization of the unlimited budget, the emissions of CO_2 increase rapidly after 2015 up to the same level as the deterministic unconstrained scenario. In case of realization of the restricted emission budget, CO_2 emissions will strongly decrease after 2015 (see lower solid line). Then the CO_2 emission level goes beyond the level of the deterministic scenario with restricted emissions to compensate the neglected reduction between 2000 and 2020.

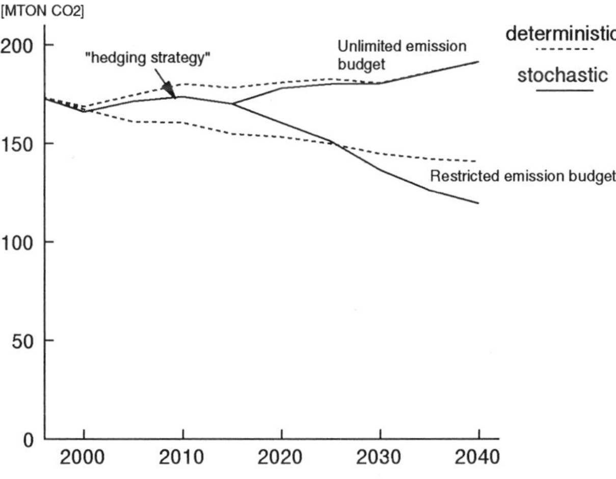

Figure 1: CO_2 emissions as calculated with stochastic and deterministic calculations.

The hedging strategy implies to adopt between 2000 and 2015 an emissions level that lies somewhere between the two deterministic cases. The emission level in the hedging strategy is closer to the case with unlimited CO_2 emissions than to the deterministic case with restricted emissions. In this example it appears preferable to achieve some emission reduction before 2015 to insure oneself against possible excessive cost after 2015 which would be linked to a strategy of 'waiting too long'.

4.2. Capacity expansion for electricity generation

Many energy technologies have technical lifetimes in the order of 30 years or more. Analysis of the energy investment decisions of long-lived equipment in a CO_2 hedging strategy is therefore very relevant. A few examples of the technology results of the scenario calculations are given in figures 2, 3 and 4. They presents the electricity production for three groups of technologies (coal fired electricity generation, gas-fired electricity generation and electricity generation with renewables) between 2000 and 2040. The hedging strategy has different effects for each group, as is illustrated below.

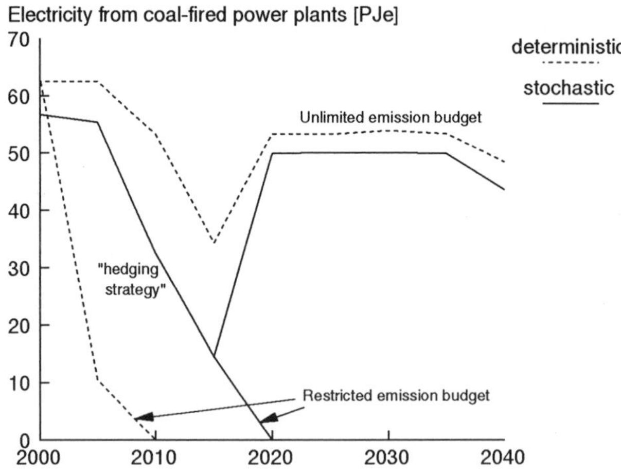

Figure 2: Electricity generation with coal fired power plants in the deterministic cases and in the CO_2 hedging strategy.

Figure 2 shows electricity production with coal fired power plants. The difference in results between the two deterministic scenarios is striking. With an unlimited emission budget, the production of electricity which is based on coal shows a small drop in 2015 (see upper dashed line) due to the normal retirement of some existing power plants, but after 2020 the contribution of coal to electricity generation is almost back at the same level as in the year 2000. With restricted CO_2 emission budgets, however, coal fired power plants are used with very low annual running hours in 2005 and by 2010 the coal plants will even be early depreciated. The hedging strategy is modest in comparison with the two deterministic cases. In the CO_2 hedging strategy, shutting down existing coal-fired power plants is not justified due to the existing uncertainty about the stringent reduction target. Instead, the model defers such drastic measures until uncertainty disappears. The existing coal fired power plants are kept in operation until 2015 although the plants are no longer running in base load mode but in intermediate load. New coal fired power plants are not built before 2020. When the emission budgets become certain in 2020, either new coal fired plants are constructed (with unlimited emission budgets) or the remaining coal fired plants are taken out of operation (in the case that restricted CO_2 emission budgets become certain).

For electricity generation with gas-fired STAG power plants the situation is the opposite (see Figure 3). In the deterministic scenario electricity generation with gas-STAGs is significantly higher with restricted CO_2 emission budgets than with an unlimited emission budget. Again the hedging strategy with the level of electricity production points at a more cautious strategy with the level of electricity production from gas-STAGs between the levels of the two deterministic scenario scenarios. The electricity production from gas-STAGs never achieves the same level as in the deterministic case with restricted CO_2 emissions, also not after 2020. This is due to the fact that other technologies than gas-STAGs, which have lower CO_2 emissions per kWh (such as renewables), are required after 2020 to keep the CO_2 emission within the budget.

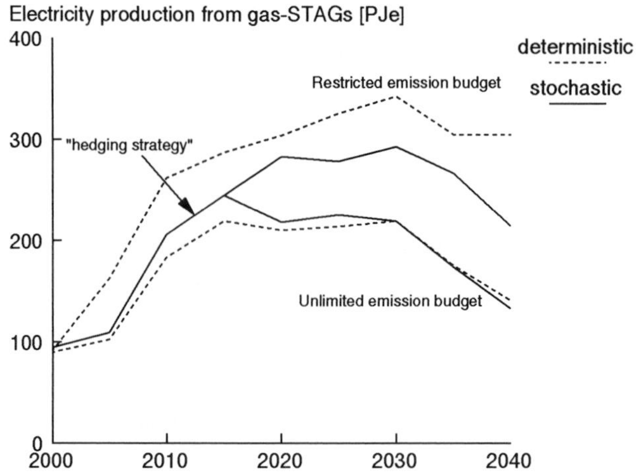

Figure 3: Electricity generation with gas fired power plants in the deterministic cases and in the CO_2 hedging strategy.

For electricity generation from renewables (wind turbines and solar PV systems), the results of the hedging strategy are equal to the results of the deterministic scenario with unlimited emission budgets (see Figure 4). After uncertainty unfolds in 2020, the contribution either remains low or the role of renewables increases rapidly. The level of electricity production from renewables is ultimately also much higher than in the deterministic scenario with restricted CO_2 budgets.

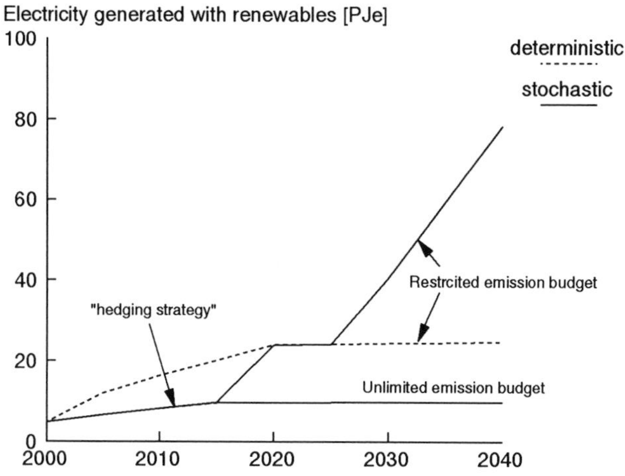

Figure 4: Electricity generation with renewable technologies (wind turbines and solar PV) in the deterministic cases and in the CO_2 hedging strategy.

4.3. Cost for CO_2 reduction

Uncertainty about CO_2 reduction targets is a cause of cost. If there was no uncertainty about the future CO_2 reduction target, it would be possible to make a plan of action for the energy system how the CO_2 emission target can be achieved as cost-effective as possible. The development paths of the energy system as calculated by each of the deterministic model runs follow such strategies. But as uncertainty does exist, the best one can do is to follow a flexible strategy with minimal expected cost, in other words to minimize regret caused by the uncertainties. The expected cost includes a weighing of possible strokes of luck and disappointments. Ex post (after the uncertainty disappears) the strategy is not likely to be optimal, however, ex ante the strategy reduces possible regret. This is always better than ignoring certain possible events. If that would be the case, it can be that a country is caught by surprise and that it faces very high cost linked with adjustment of the energy strategy within a short time span. Such a case can be referred to as an interrupted deterministic scenario. Hedging serves to avoid the high cost that arise from interrupted deterministic scenarios.

The MARKAL model calculates the annual cost of the energy system based on the cost of the technologies and the energy carriers with the application of a discount rate of 5% per year. The cost for CO_2 reduction have been calculated by comparing the annual cost of the energy system with the cost of the energy system in the unconstrained deterministic case.

The annual cost of the hedging strategy between 2000 and 2015 amount to a few hundred million guilders. After 2015 the annual cost for CO_2 reduction will depend on the 'state of climate' that will occur, and are shown to diverge very strongly. For the unconstrained case the cost get less. The annual cost rapidly increase if the stringent emission target has to be achieved. Then, the total cost are higher than in the deterministic stringent reduction case.

5. SENSITIVITY ANALYSIS

The results of the calculations are sensitive to the assumptions that had to be made. Several assumptions that affect the size of the CO_2 hedging have been analyzed:

> Selection of the range of possible outcomes of uncertainties.

> Probability assumptions. The probabilities assigned to the 'states of climate' determine the relative weights of the discerned states of nature in assessing the CO_2 hedging strategy.

> Technological progress and/or availability of technologies. The assumptions about the availability and maximum potentials of energy technologies affect the cost and boundaries for emission reduction. If a technology like CO_2 removal is available, flexibility increases to achieve far-reaching emission reduction targets in the long term. If additional policies would be assumed, like Joint Implementation (JI) or research and development (R&D) for energy technologies, the flexibility for CO_2 reduction would also increase and the optimal CO_2 hedging strategy may change.

> Annual emission constraint versus emission budgets for periods. In the discussion on targets and timetables for CO_2 reduction targets under the FCCC, the main focus is still on annual emission targets although emission budgets increasingly receive attention.

> Discount rate. The level of the discount rate determines the comparison of current and future financial flows.

> Moment in time that uncertainty about the emission budget disappears.

Table 1

Summary of results of sensitivity analysis.

Issue for which assumptions were made	Size of impact on the results
Annual CO_2 constraints vs. CO_2 budgets	Very high
Range of possible reduction targets	High
Availability of new technology	High
Moment that uncertainty disappears	Medium
Probabilities of constraints	Medium

Sensitivity analysis have been reported in [5]. Table 1 summarizes the sensitivity of the hedging results to the list of assumptions. The results appeared most sensitive to the assumptions to substitute emission budgets for annual emission constraints. If, after uncertainty has disappeared, one does not have to make up for neglected reductions before uncertainty about the target, the hedging strategy is much closer to the unconstrained emission case. The results are also highly sensitive to assumptions about the range of uncertain budgets under consideration and technology availability. Less sensitive are the results to assumptions on probabilities of targets and the moment in time that uncertainty unfolds.

6. CONCLUSIONS AND RECOMMENDATIONS

The following conclusions about the hedging approach can be drawn from the analysis presented in this paper:

> CO_2 hedging strategies provide a comprehensive way to analyze CO_2 abatement

strategies while properly accounting for uncertainties in future emission budgets.

> Analysis of CO_2 hedging strategies helps to define policy strategies for CO_2 reduction with minimum regret.

> The results of CO_2 hedging analyses suggest that it is better to take concrete actions soon than to defer them until uncertainty about CO_2 reduction targets is resolved.

> Four kinds of action cover the relevant elements of CO_2 hedging strategies:

- not investing in energy technologies with relative high CO_2 emissions;
- not investing in the short term in expensive CO_2 abatement technologies;
- increasing flexibility of the energy system, many ways are available to do this at low cost;
- research and development (R&D) for new low CO_2 energy technologies to facilitate long term emission reduction.

Analysis of CO_2 hedging strategies allows to make a trade-off between these kinds of action and to design an optimal portfolio of actions.

It is important to note that the hedging analysis represented in this paper can certainly not give the ultimate answer about CO_2 reduction strategies. The limitations need to be considered:

> For the application of the CO_2 hedging approach some critical assumptions need to be made. When a hedging method is to be explored, e.g. different CO_2 reduction targets need to be discerned and the probabilities of these targets need to be estimated and the process of unfolding of uncertainty over time has to be estimated. Currently, these assumptions can only be based on subjective judgements.

Until now, hedging methods have only to a limited extent been applied to address climate change considerations in energy investments. Applications of the hedging methods to more examples will increase insight in critical assumptions. One interesting direction will be to extend or replace the climate uncertainties with other uncertainties, e.g. uncertainties in energy prices or in energy technology development. Further, a more thorough analysis of a national CO_2 hedging strategies, guided by policy makers, can provide a more pointed answer to the question what actions should be taken now to prepare for uncertainty in long term CO_2 reduction. For such studies, it is recommended to also quantify the benefits of CO_2 hedging strategies.

REFERENCES

1. Manne, A.S. and R. Richels, Greenhouse insurance, the economic costs of carbon dioxide emission limits, MIT Press, Cambridge MA, (1992).
2. Kram, T., National energy options for reducing CO_2 emissions, Volume 1 and Volume 2, Netherlands Energy Research Foundation, report ECN-C--93-101 and ECN-C--94-024, Petten, Netherlands (1994).
3. Van Hilten, O. et al., The contribution of ECN to the Third White Paper on energy policy, Netherlands Energy Research Foundation, report ECN-C--96-014, Petten, 1996.
4. Ybema, J.R. et al., Prospects of energy technologies in The Netherlands, Netherlands Energy Research Foundation, report ECN-C--95-002, Petten, Netherlands (1995b).
5. Ybema, J.R. et al., Including climate change in energy investment decisions, Netherlands Energy Research Foundation, report ECN-C--95-073, Petten, Netherlands (1995a).

Towards a CO_2-free Energy System in the 21st Century

M. Bonney and W.J. Lenstra

1. INTRODUCTION

A world faced with the threat of climate change and characterized by an ever increasing demand for the services which energy can supply will need to develop and make use of CO_2-free forms of energy. Renewable energy sources such as wind, water and sun will ultimately have to play a major role. However, there are limits to the use of renewables. Some of the cost considerations and other barriers to renewable energy are discussed in section 2. Section 3 looks at the issue of fossil fuel availability, and concludes that the probability that mankind will exhaust the reserves at any time in the near future is remote. Section 4 analyzes the likelihood that the carbon constraint will become binding before the reserves are depleted. A CO_2-free energy system based on energy carriers which can be derived from either fossil fuels or renewables is presented in section 5, with special attention for the role that hydrogen might play in such a system. Section 6 describes the steps involved in recovering CO_2 from fuels and flue gases and storing it underground, and presents estimates of the costs associated with each step. Estimates of the costs of producing CO_2-free energy carriers in The Netherlands are presented in section 7. The paper closes with some thoughts about the impact that CO_2 recovery might have on fossil fuel price stability.

2. BRAKES ON THE INTRODUCTION AND UTILIZATION OF RENEWABLE ENERGY SOURCES

It is widely accepted that a sustainable energy system - one which can meet the needs of current and future generations without overtaxing the environment to the point where food supplies, ecosystems or economic development are at risk - will ultimately have to rely largely on renewable sources of energy such as wind, water and sun.

There is also general agreement that the transition to a renewables-based energy system must be made as quickly as possible in order to limit the build-up of carbon in the atmosphere. The greatest obstacles on the road to large scale implementation involve the relative costs of renewable-based energy carriers and the optimization of the necessary technology.

There are a great many measures that governments can take in order to remove these obstacles and most governments are taking them to a greater or lesser degree. In The Netherlands renewables currently account for about 1% of total energy use. The policy target is to increase this share to 3% by the year 2000 and to 10% by the year 2020. As shown in the following table, there are large differences in the costs of producing electricity from the different renewable options. All of them, however, are more expensive than electricity based on the fossil fuels used in The Netherlands, natural gas and coal.

Table 1: Comparative costs of electricity generation with fossil fuels and renewable energy sources, in NL*f*/kWh

fossil fuels	0.08
hydropower	0.15
gasification biomass	0.15
wind	0.17
heat pumps (non-residential buildings)	0.48
heat pumps (households)	0.68
photovoltaic (collective)	1.25
photovoltaic (individual)	1.50

Source: [9, pp. 13, 25]

The differences in the costs of producing electricity with various renewables reflect differences in the stage of development and market penetration. Wind and hydropower, for example, have already undergone a long development history and come the closest to being able to compete with traditional fossil fuel-based generation in The Netherlands. Photovoltaic solar energy, on the other hand, is still in a fairly early stage of development and has a long way to go before it can compete on a commercial basis with more traditional forms of energy. Technology development and experience gained through larger scale application will to some extent result in an autonomous decline in the costs of producing energy with these renewable sources.

The Netherlands uses two types of policy instruments to accelerate and strengthen the autonomous decline in the cost of renewable- based energy: those aimed at technological progress in order to improve the price/performance ratio drastically, and those aimed at fostering the actual application of those renewables for which the costs are only slightly higher than for fossil-based alternatives. Instruments in the former category are largely in the area of research, development and demonstration.

Fiscal instruments are used to foster the application of nearly competitive renewables. At the moment there are four fiscal programs in force in The Netherlands which have the effect of reducing the cost differential between traditional energy sources and renewables: a tax on small scale gas and electricity use with a refund for producers of renewable energy, free depreciation of investments and an investment tax credit in the corporate income tax, and a provision exempting earnings on investments in "green" facilities from the personal income tax. The following table indicates the effect that these fiscal instruments have on the costs of various renewable options.

Table 2: Effects of fiscal instruments in 1997, in NL*f*/kWh

	wind	heat pump (non-residential)	gasification biomass	landfill gas
cost	0.17	0.48	0.15	0.05
energy yield	0.08	0.29	0.08	0.08
tax advantage	0.04	0.10	0.04	0.03
remaining cost	0.05	0.09	0.03	-0.06

Source: [9, p. 25]

As Table 2 shows, even with tax advantages for renewables they are still costlier than their fossil-based competitors. In order to ensure that they are exploited despite their cost disadvantage, The Netherlands is instituting a system of "tradeable green labels". Agreements between the government and energy distribution companies set out each company's share in the total target for 2000 and each company commits itself to being responsible for providing that amount of renewable energy. The objective for an individual company is determined based on its share in the electricity and gas market in 1995. Since the potential for the different forms of renewable energy differs in different areas, the companies are introducing a system to make renewable energy tradeable without actually having to physically transport it.

They have developed a system of "green labels" which represent the renewable "value" of the energy, but not the energy itself. The companies can purchase green labels outside of their own service area, thereby making possible the production of renewable energy elsewhere, and crediting that energy, rightly, against their own commitment.

The energy companies have also found another way to cover the additional costs of producing electricity based on renewables. They have introduced a new product which they call "green current" on the market. The idea behind this approach is to offer a product to "green" customers for whom emission-free electricity has a higher value than regular electricity and who are therefore prepared to pay a premium for it. Customers are offered the possibility of signing a contract for a given amount of "green current" - that is electricity produced with renewables - within a given time period and they pay a higher price for it than they would for regular electricity. "Green current" was first offered on a trial basis about a year and a half ago and is proving much more popular than marketing studies carried out beforehand suggested it would be.

Besides the cost impediments to renewable energy there are also other barriers which are less sensitive to policy solutions. In The Netherlands, for example, the conflicting demands on land use form a particularly acute barrier. Windmill parks take up a lot of space and are noisy. The Netherlands is an extremely densely populated country, where residential, ecological, recreational, agricultural, industrial and infrastructural demands all compete for limited land resources. Other countries are confronted with other kinds of barriers to renewable energy, the most difficult to overcome being climatic conditions which limit the potential of wind and solar power, and geographic conditions which limit the potential of hydropower and biomass.

The combination of cost differentials and other kinds of barriers means that introduction of renewables on a large scale is likely to take a long time. The conclusion is that fossil fuels will have to continue to play a major role even as policy-makers strive to speed the transition to an energy supply system based on renewables.

3. FOSSIL FUEL AVAILABILITY

Having concluded that it will be necessary to continue using fossil fuels in order to meet the needs of current and future generations for energy services, the question arises of whether the remaining reserves are large enough to bridge the gap.

In attempting to answer this question, it is useful to distinguish between proven recoverable reserves and eventually recoverable resources. *Reserves* are estimated based on known geological occurrences, available technologies and current market prices, three elements which are strongly interdependent.

Low prices generally indicate that sufficient reserves are available, so fewer exploration activities are carried out. Moreover, when prices are lower there is no incentive to develop new and better extraction technologies to penetrate new reserves. *Resources* are estimated based on less certain geological occurrences and are not economically feasible to exploit under current market conditions. As market conditions change, technology evolves or advances are made in geology, however, resources can be "promoted" to reserves. Resources are generally expected to become economically exploitable in the near future.

The relationship between reserves and resources can be illustrated by the evolution in the so-called reserves-to-production (R/P) ratio. The R/P ratio provides an indication of how long current reserves will last given current production levels. The R/P ratios for conventional oil and natural gas have been stable or rising throughout most of the post-war period. The R/P ratio for oil hovered around 30 years during the mid-1970's; it is currently about 40 years. This demonstrates that it has been consistently possible to expand reserves faster than production has grown [12, pp. 35-37].

The following table summarizes various estimates which have been made of conventional fossil fuel resources in terms of tons of carbon as well as providing an indication of current annual consumption.

Table 3: Worldwide fossil fuel resources in GtC (10^9 tons carbon)

	WEC/93	Holdren	Skinner	WEC/IIASA	consumption per yr (1990) (WEC/IIASA)
coal	3600	4000	5300	3650	2.4
oil	170	380	300	250	2.6
gas	140	190	170	270	1.1
total	3910	4570	5770	4260	6.1

Source: [11, p. 77 and 12, p. 36]

As Table 3 shows, estimates of fossil fuel resources vary considerably. However, given current levels of consumption it does seem likely that oil and gas resources will be sufficient for at least the next century or so, while coal resources could last for several hundreds of years at the current rate of consumption. And these numbers do not take account of the unconventional oil and gas reserves, of speculative fuels such as the methane hydrates or of growth in the R/P ratio.

4. THE CARBON CONSTRAINT

The carbon present in the known fossil fuel resources is between about 5 and 8 times the present atmospheric carbon content of around 760 GtC. The question facing policy-makers is how much of the carbon in the resources should be allowed to enter the atmosphere, and how fast. Or framed differently: what is a "safe" level of atmospheric carbon? This is the issue which is being addressed by the scientists and policy-makers participating in the Intergovernmental Panel on Climate Change (IPCC) and in the negotiations under the Framework Convention on Climate Change (FCCC).

Policy-makers have not yet chosen a clear and unambiguous indicator of the environmental problem which would provide a starting point for determining a "safe" level of carbon in the atmosphere. Some scientists have suggested that limiting the increase in mean global temperature to 2°C relative to pre-industrial temperatures, and keeping the rate of increase below 0.1°C per decade would achieve the objectives of the Framework Convention on Climate Change. However, these indicators have not yet been embraced by policy-makers. If one assumes for the sake of argument that stabilization of carbon concentrations by the year 2100 at a level that is no greater than twice the pre-industrial level would provide adequate protection against the risks of climate change, then it is possible to put the fossil fuel resources into context.

The IPCC has determined that if concentrations are to be stabilized at about twice their pre-industrial level of about 280 ppmv, then cumulative emissions may not exceed something on the order of 880 to 1060 GtC between 1990 and 2100 [7, p. 16]. This is between 15 and 25% of the carbon present in the fossil fuel resources presented in Table 3, suggesting that the carbon constraint will likely become binding long before the resources are depleted.

In order to get maximum benefit from the remaining fossil fuel resources, there is a need for technologies which make it possible to use them in a carbon-constrained energy supply system. One such technology may be carbon dioxide capture and underground storage. This technology, described in greater detail in section 6, lends itself to use at large point sources of carbon dioxide, such as coal-fired power plants and installations for the production of hydrogen.

5. PRIMARY ENERGY SOURCES VERSUS ENERGY CARRIERS

Energy carriers such as heat, electricity, hydrogen and methanol can provide all the energy services which the world demands: heat, light, mechanical energy and motor fuels. They can be made from either fossil fuels or renewable energy sources, and can therefore furnish the necessary link between the current, fossil fuel-based energy system and the future, renewables-based energy system.

When fossil fuels are transformed into heat, electricity or hydrogen at large point sources, the CO_2 can be recovered and stored underground. Development of an infrastructure optimized to deliver these energy carriers and of products designed to use them (such as methanol-powered vehicles) should be a priority of government and private sector policies aimed at establishing a CO_2-free energy system for the 21st century and beyond.

Hydrogen has been "on the agenda" in the energy world for a long time. Its advantages as a fuel are well known. Since it can be made by electrolysis of water or by gasification of biomass, production can take place almost anywhere in the world. Since it does not emit any CO_2, particulates or sulphur when burned, it offers significant environmental advantages. NO_x emissions are also less of a concern than with fossil fuels since the constraints on lean-burn are much less stringent with hydrogen. Concerns about other environmental aspects (such as contrail formation from aircraft) and safety are lessening with additional research.

Since hydrogen does not occur naturally, it always has to be produced from another energy carrier: fossil or electricity. Production of hydrogen from fossil fuels is the least expensive method, but has not really been part of the discussion in any serious way in recent years because of the fossil CO_2 emissions which it generates. Hydrogen production based on nuclear energy was more or less scrapped from the agenda after the accident with the nuclear plant in Chernobyl and the accent was shifted to production through electrolysis of renewables such as hydropower, wind energy, or photovoltaic solar energy. The development of CO_2 removal and storage technologies adds a whole new dimension to the discussion about the potential role of hydrogen in the global energy supply system, opening up fossil fuels coupled with decarbonization as sources for hydrogen production [2,3].

Up to 15% hydrogen (by volume, 5% by energy share) can be added to the existing gas distribution pipeline without requiring modifications to the current system. New central heating units with ceramic boilers can take up to 50% hydrogen (by volume). In the longer term pure hydrogen could be distributed to final users, but this would require replacement of major elements of the distribution system such as compressors, older pipelines, meters and equipment for use [2].

Hydrogen could also be used in the transport sector. Car manufacturers in Germany are already researching hydrogen driven engines. Mercedes presented a new demonstration vehicle in 1996. Hydrogen is used to fuel busses in Canada. Hydrogen may offer even greater advantages as an aviation fuel. Not only the elimination of hydrocarbon emissions but also the lower weight per unit of energy makes hydrogen particularly attractive for aircraft. The Russian airframe manufacturer Tupolev and engine manufacturer TRUD demonstrated the feasibility of flying a liquid-hydrogen-powered aircraft in 1988. German, Russian and Canadian partners are currently working on a new prototype [8].

The U.S. Department of Energy has estimated comparative hydrogen production costs ranging from US$ 5/GJ for natural gas reforming, to US$ 10/GJ for coal gasification, US$ 15-25/GJ for water electrolysis and US$ 30/GJ for wind/electrolysis [1, Table 2]. In The Netherlands, where virtually all demand for heat is currently met with natural gas, hydrogen purchased on the spot market costs about NL*f* 9.00/GJ (US$ 5.00) and can therefore compete with natural gas purchased by energy distribution companies for delivery to small residential and commercial users, which costs about NL*f* 10.50/GJ (US$ 5.85). Gas purchased for delivery to large, industrial users costs the energy distribution companies something on the order of NL*f* 6.40/GJ (US$ 3.55), so hydrogen is still far from competitive in that part of the market [3].

6. CARBON DIOXIDE RECOVERY AND STORAGE

The first step in CO_2 removal involves its recovery from fuels or from flue gasses during energy conversion processes. It is beyond the scope of this paper to go into recovery technologies or processes in any detail. Extensive information about removal from coal-fired power plants is available in Hendriks [5] and about removal during hydrogen production from fossil fuels in Audus et.al [1]. Hendriks estimates the costs of capture from a conventional coal-fired power plant at between US$ 30 and US$ 70/avoided ton CO_2, which adds between US¢ 2.5 and US¢ 5.0 to the costs of producing a kWh of electricity [5, p. 221].

After the CO_2 has been recovered, it must then be made ready for transport to the storage site. This involves cleaning and dehydration in order to prevent corrosion of the transport system, as well as compression of the gas to a pressure of 60 to 80 bar. Investment costs for compression depend on the capacity of the compressor and the desired pressure. Estimates made in The Netherlands suggest that compression costs amount to something on the order of US$ 4.50/ton CO_2 [3, p.68].

Dry CO_2 can be transported through underground pipelines comparable to those used for natural gas. Pipeline costs are dependent on the length of the pipeline and the pipe diameter. Hendriks has estimated the cost of a 100 kilometer long pipeline which can handle 500 tons of CO_2 per hour at about US$ 2.00 per ton, including the costs of maintenance and extra compression during transportation [3, p.69].

Carbon dioxide may be stored permanently in deep geological formations such as gas and oil fields and aquifers, provided they have an impenetrable upper layer. Given the prevailing pressure and temperatures underground, only locations deeper than 800 meter are suitable for CO_2 storage. Gas fields offer a number of advantages relative to aquifers, one of which is that gas fields have already demonstrated that they have an impenetrable upper layer. Before CO_2 can be stored in aquifers the permeability of the top layer must be thoroughly studied. Another advantage of using gas fields is that much of the infrastructure put in place for exploitation of the field can also be used for CO_2 storage, such as the well head. An advantage of aquifers, on the other hand, is that they are more widely distributed throughout the world [3].

There is wide experience with underground storage of CO_2 in enhanced oil recovery projects in North America. In 1996 Norway's State oil and gas company, Statoil, started with injection of 1 million tons of CO_2 per year into an underground aquifer offshore in the Sleipner area of the North Sea, thereby reducing the CO_2 emissions from their gas operations and saving themselves US$ 55/ton CO_2 in Norwegian carbon taxes. Statoil is also contemplating longer term projects which would involve the export of hydrogen and electricity. Any CO_2 produced would be captured and stored in underground reservoirs [6].

There appears to be considerable potential worldwide for sites which are suitable for underground storage of CO_2. A study carried out under the auspices of the European

Commission's JOULE Program estimated the availability of offshore and onshore underground storage capacity in the European Union and Norway at more than 800 Gt CO_2, while the cumulative emissions of CO_2 from power plants in those countries over 25 years would amount to something on the order of 25 Gt CO_2 [4].

The costs of underground storage consist mainly of drilling and outfitting wells, with smaller additional costs for maintenance and management. Hendriks estimates the costs of storage in natural gas fields at between US$ 0.50 and US$ 3.00 per ton of CO_2, while his estimates of the cost of storage in aquifers vary between US$ 2.00 and US$ 8.00 per ton, excluding the costs involved in determining the suitability of a given aquifer [5, p. 209].

7. SITUATION IN THE NETHERLANDS

The Netherlands has nearly exhausted its No Regret measures for reducing CO_2 emissions. Possibilities for further reductions are quite expensive for a number of reasons.

Because natural gas already has a very large market share in The Netherlands (about 45% of total final energy consumption), there is very little potential for CO_2 reduction from fuel switching to other conventional fossil fuels. The Dutch economy is characterized by a very high proportion of energy intensive industries, which are already among the most efficient in the world. Any gains which could be made from further efficiency improvements would be quite small and very expensive, and due to the fact that these industries compete on a world market it would not be possible to pass on the costs of reduction measures in product prices. As explained in paragraph 2, the transition to a (largely) renewables-based energy system in The Netherlands will take time and the extent to which renewables can ultimately provide all of the energy services demanded in The Netherlands is limited by non-economic constraints which are not sensitive to policy solutions. CO_2 recovery and storage underground, coupled with electricity, heat and hydrogen production, and biomass-based methanol production will likely have to play a major part in the transition to a CO_2-free energy system in The Netherlands in the

21st century.

The capacity for underground storage of CO_2 in gas fields and aquifers in The Netherlands amounts to something on the order of 10 $GtCO_2$ [4, p.107], while CO_2 emissions are currently between 175 and 180 $MtCO_2$ annually. CO_2 storage could therefore provide a backstop for a very long time. There is already a dense system of pipelines in the country due to the high penetration of gas in heating and industrial markets and the ever-increasing application of district heating in most densely populated parts of the country. A great deal of the infrastructure needed to utilize CO_2-free energy carriers is thus already in place.

Table 4 shows some estimates of the cost of CO_2-free energy carriers in the Dutch situation.

Table 4: Comparative costs of traditional and CO_2-free energy carriers, estimates for the Dutch situation

	unit	production cost NL*f*/unit	internal rate of return
Electricity			
1. Reference: natural gas-combined cycle	kWh	0.08	10%
2. natural gas-combined cycle with CO_2 recovery/storage	kWh	0.13	10%
3. Reference: ICGCC	kWh	0.09	10%
4. ICGCC with CO_2 recovery/storage	kWh	0.15	10%
5. wind	kWh	0.17	10%
6. imported biomass	kWh	0.14	10%
Natural Gas			
1. Reference: natural gas	GJ	8	10%
2. hydrogen from gas with CO_2 storage	GJ	16	10%
3. hydrogen from imported biomass	GJ	18	10%
Motor fuels			
1. Reference: gasoline a. production cost excluding distribution and taxes b. retail price including distribution and taxes	GJ	a. 22 b. 60	market prices
2. Methanol from imported biomass excluding distribution and taxes (for use in gasoline engine also suited to methanol)	GJ	34	15%
Passenger cars			
1. Reference: car with gasoline engine a. costs per km excluding fuel taxes and fuel distribution costs b. costs per km including fuel taxes	km	a. 0,38 b. 0,44	15%
2. car with fuel cell based on hydrogen from natural gas with CO_2 recovery/storage	km	0.50	15%
3. car with fuel cell based on methanol from imported biomass	km	0.45	15%

The costs presented here have been calculated by the Ministry of Housing, Spatial Planning and Environment, except for the costs of integrated coal gasification combined cycle (ICGCC), which are based on Hendriks [5]. The numbers presented do not include the costs of modifications to central heating units, pipelines, gasoline engines etc. The energy prices used in the calculations are:

coal	NL*f* 4.00/GJ
natural gas	NL¢ 25.2/m^3
electricity	NL¢ 13.1/kWh
gasoline	NL¢ 72/liter (shadow prices)
	NL¢ 195/liter (retail prices)
biomass	NL*f* 100/ton

8. EFFECTS OF CO_2 STORAGE ON FOSSIL FUEL PRICES

In a carbon constrained world the use of fossil fuels will decrease more slowly if CO_2 recovery and storage is possible. Producing countries (not only OPEC, but also both The Netherlands and the United States) will be able to realize greater value added on their resources if fossil fuels can be made CO_2 free. This will contribute to stability in fossil fuel prices. Price erosion due to an international approach to the issue of climate change is underestimated by policy-makers and analysts. It seems likely that prices for fossil fuels will fall if demand declines. If producing countries react to falling prices by increasing production in order to generate revenue, then prices could fall even further and profit margins could erode entirely. Production of hydrogen from fossil fuels, coupled with decarbonization, could prevent this type of adverse impact from climate change policy.

REFERENCES

1. Audus, H., O. Kaarstad and M. Kowal, "Decarbonization of Fossil Fuels: Hydrogen as an Energy Carrier", undated paper.
2. Bergsma, C., "Waterstof levert nieuwe impuls voor Holland Gasland", in Gas Tijdschrift in de Energiemarkt, vol. 10, October 1996.
3. Center for Energy Conservation and Clean Technology, Waterstof voor kleinverbruikers en CO_2-opslag, Delft, 1996.
4. Elewaut, E. et.al., "Inventory of the Theoretical CO_2 Storage Capacity of the European Union and Norway", in The Underground Disposal of Carbon Dioxide, final report of JOULE II Project no. Ct92-0031, February 1996
5. Hendriks, C., Carbon Dioxide Removal from Coal-Fired Power Plants, University of Utrecht, Utrecht, 1994.
6. International Energy Agency Greenhouse Gas R&D Program,"Pioneering CO_2 Reduction", in Greenhouse Issues, Number 27, November 1996.
7. Intergovernmental Panel on Climate Change, Radiative Forcing of Climate Change, The 1994 Report of the Scientific Assessment Working Group, Summary for Policymakers, WMO/UNEP, 1994.
8. Klug, H., Daimler-Benz Aerospace Airbus GmbH, "Environmental Compatibility of Cryoplane, the Cryogenic Fuel Aircraft," paper presented at International Colloquium on Impact of Aircraft Emissions upon the Atmosphere, Paris, October 15-18, 1996.
9. Ministry of Economic Affairs, Duurzame Energie in Opmars 1997-2000, (Action Plan for Renewable Energy), The Hague, 1997.
10. Over, J. and J. Stork, "Ondergrondse opslag van CO_2: de mogelijkheden voor Nederland", in Energie- en Milieuspectrum, August 1996.
11. Scientific Council for Government Policy, Report to the Dutch Government Duurzame Risico's: een blijvend gegeven, The Hague, 1994.
12. World Energy Council and International Institute for Applied Systems Analysis, Global Energy Perspectives to 2050 and Beyond, London, 1995.

Better approaches to cleaner air from electricity generation

S. Napolitano [1]

B. McLean [2]

J. Bachmann [3]

L.R. Critchfield [1]

The Office of Air and Radiation, U.S. Environmental Protection Agency.

[1] Policy analyst for EPA's air office

[2] Director of the Acid Rain Program

[3] Senior policy advisor, responsible for EPA's development of revisions to the air quality standards under the Clean Air Act Amendments of 1990

ABSTRACT

The U.S. Environmental Protection Agency (EPA) developed the Clean Air Power Initiative (CAPI) as a more effective way to reduce air pollution from electric power generation. The Agency decided during the CAPI process to focus initially on nitrogen oxides (NOx) and sulfur dioxide (SOx) emissions. EPA developed a set of options that varied the level, timing, and type of approach to reducing emissions of NOx and SOx in order to meet the Clean Air Act Amendments of 1990 (CAAA) goals. The Agency analyzed the costs and emission reductions of these options using a new modeling approach. This paper describes the alternatives that EPA examined and summarizes its findings.

BETTER APPROACHES TO CLEANER AIR FROM ELECTRICITY GENERATION

During 1996, the U.S. Environmental Protection Agency (EPA) examined how the United States could use simple, market-based trading-and-banking approaches to control air pollution from electric power generation. The Agency wanted to see if these alternative approaches to pollution control could provide significant emissions reductions more cost-effectively than traditional regulatory actions -- often described as command-and-control options. This effort was EPA's Clean Air Power Initiative (CAPI).[1]

Initially, the Agency examined the significance of the air pollution problems that the United States faces and the electric power industry contribution to these problems. Then EPA considered the mandates and authorities the Agency has under the Clean Air Act Amendments of 1990 (CAAA) to control power industry emissions and considered how EPA and the States could work under the CAAA to design cost-effective ways of reducing the industry's most significant air emissions. After discussions with representatives of the electric power industry, state regulatory officials, fuel suppliers, environmental groups, and pollution equipment manufacturers, EPA developed a set of options that varied the level, timing, and type of approach to reducing emissions of NOx and SOx in order to meet the CAAA's goals.

The Agency analyzed the costs and emission reductions of these options and did a comparative analysis of a trading-and-banking approach versus the traditional regulatory approach. The analyses led to important conclusions about how EPA and the states may want to further control air emissions from electric power generation.

1. AIR POLLUTION AND ELECTRIC POWER GENERATION

Emissions from electric power plants contribute significantly to a number of important air pollution issues. These can be categorized as: 1) adverse effects on human health from ground level ozone, particulate matter, and persistent toxic air contaminants; 2) environmental impacts such as eutrophication of coastal surface waters, wide-spread regional haze that reduces visibility, acidification of surface waters from acid deposition, ecosystem and crop damage from ground level ozone, damage to various materials from ozone and acid deposition, and ecosystem damage from mercury and other persistent toxic pollutants; and 3) climate change due to the emissions of greenhouse gases. In 1994, power plants in the U.S. were responsible for 70% of all sulfur oxide (SOx) emissions, 33% of all nitrogen oxides (NOx) emissions, 23% of point source emissions of direct or "primary" particulate matter (PM), 23% of anthropogenic mercury

1EPA's paper on the Clean Air Power Initiative and supporting analyses that the Agency prepared from April 1996 to October 1996 can be found on the internet at the following web site address: http//www.epa.gov/capi.

emissions, and 36% of all anthropogenic carbon dioxide (CO_2) emissions[2].

For CAPI, EPA focused on the pollutants that are related to the first two categories of health and environmental effects noted above because these pollutants are associated with pressing regional health and environmental concerns in North America. In addition, EPA has clear statutory authority to regulate these pollutants. From a control/emissions perspective, the pollutants of greatest importance can be grouped into three categories: sulfur oxides, nitrogen oxides, and, potentially, mercury and other directly emitted toxic fine particles (see Table 1). Sulfur and nitrogen oxides emissions undergo complex atmospheric transformations that result in the formation of acidic fine particles and gases and ozone smog. The resultant mix, along with directly emitted mercury and fine particles, can be transported by weather systems over long distances and affect air and water quality and public health in areas far from where they were emitted. Because of this long range transport and the location of multiple power generation emission sources in the U.S., the resulting atmospheric and deposition problems affect broad multi-state regions.

2. REGULATORY CHALLENGE THAT CAPI ADDRESSES

The Clean Air Act Amendments of 1990 (CAAA) directed EPA to place tighter controls on power plant emissions of SOx and NOx to reduce acid rain, to reduce ground-level ozone, to consider the need to revise the air quality standards, and to work with states to meet revised standards. The Act directed EPA to revise its New Source Performance Standards (NSPS) and required EPA to address other issues such as air toxics and visibility, depending on the Agency's assessment of their severity. EPA will be acting on this agenda over the next 15 years.

The new controls on electric generation will occur during a very uncertain time for the power industry, compounding the challenges utilities face today in restructuring their operations to respond to the deregulation of their industry. If companies within the electric power industry could be told now what the Agency was going to do to further control their air emissions in the future and when new rules would go into effect, one major uncertainty in their future operations would be reduced. EPA's awareness of how clarity today about its future actions could alleviate some of the uncertainty that utilities face coupled with the Agency's objective to use its CAAA authorities cost-effectively to meet the Act's air quality goals led EPA to initiate CAPI.

During the CAPI process, EPA carried on a public dialog where the Agency worked with the electric power industry and other groups to develop cost-effective ways of reducing air pollution from electric power generation. A key element of that dialog was the completion of a set of analyses that provided estimates of emission reductions and costs that result from various approaches to controlling air emissions from the power industry.

U. S. Environmental Protection Agency, National Air Pollutant Emission Trends 1900-1994, Research Triangle Park, NC 1995.

TABLE 1
MAJOR REGIONAL AIR POLLUTANTS FROM POWER GENERATION

Sulfur Oxides
SOx and its transformation products (sulfates, sulfuric acid) are the primary contributors to acid deposition which makes lakes and streams acidic and damages man-made materials such as automobiles, buildings, and statues. Fine acid sulfates from SOx form a significant fraction of particulate matter and are the major contributors to regional haze in the Eastern United States. The national ambient air quality standards for particulate matter includes all particles smaller than 10 microns in diameter (PM_{10}), and are currently under review. A growing body of evidence suggests that early mortality and other serious health effects may occur at particulate matter levels below the current standard. EPA considers reducing particulate matter exposure to be one of its highest priorities in protecting public health.
Nitrogen Oxides
NOx contributes to ground level ozone, which is formed by the reaction of NOx and volatile organic compounds (VOCs) in the presence of sunlight. Ozone causes increased asthma attacks, reduced pulmonary function, coughing and chest discomfort, headache, and upper respiratory illness. In addition to human health impacts, ozone formed from NOx emissions adversely affects the growth of certain agricultural crops and forest species. In order to address widespread violations of the current health-based National Ambient Air Quality Standards for ozone, significant reductions in NOx emissions will be required from both mobile and stationary sources, including power plants. Revisions to the standards are under consideration. Nitrate particles contribute to PM_{10} nonattainment and visibility degradation, particularly in the Western United States. Deposition of nitrogen compounds into coastal estuaries (most notably the Chesapeake Bay) contributes to eutrophication and is also a significant component of acidic deposition. NOx transformation products also contribute to gaseous and particulate phase toxic air pollutants.
Mercury /Air Toxics
Mercury and certain fine particle toxic elements (e.g. arsenic) are of concern because they persist in the environment. Mercury, in particular, bioaccumulates (i.e., concentrates in animals and passes up the food chain through predatory animals). Current levels of mercury in the environment are of concern to health and ecosystems. Children can be affected as well as adults, and neonatal impacts from ingestion of methyl mercury by pregnant women have also been documented. Mercury believed to be primarily from air borne deposition is linked to reduced populations of the Florida panther and the common loon. Emissions of mercury and other air toxics from utilities are described in the Study of Hazardous Air Pollutant Emissions Generating from Electric Utility Steam Generating Units - Interim Final Report (EPA-453/R-96-013a - October 1996). A report to Congress on mercury is currently under review by EPA's Science Advisory Board.

3. CAPI'S FOCUS

Given the large share of total emissions that the electric generation sector produces, EPA believes that it is very important to consider further reductions of utility emissions of NOx and

SOx, if States and EPA are going to successfully address the ozone and fine particulate problems. The Agency also recognizes that there is sufficient information on utility emissions and knowledge of affordable pollution control techniques that are available to control NOx and SOx beyond what Title IV has required. Although fossil-fired generation units in the power sector produce a large share of the uncontrolled mercury emissions, the Agency is still evaluating the technologies that can be used to provide effective controls on these emissions.

In addition, effective strategies on controlling NOx and SOx could also help improve visibility, address eutrophication and acidification of water bodies, and contribute to the reduction in air toxics (specifically, mercury) and carbon (carbon dioxide) emissions. Therefore, EPA decided to focus initially on NOx and SOx controls in the Clean Air Power Initiative.

The Agency has also recognized through implementation of the Title IV SOx program that setting an emission cap to meet an environmental objective, allocating emission allowances to power generation units, and allowing the trading of these allowances by power plants has significant economic efficiency advantages over the traditional command and control type of regulation. The cost-effectiveness of the program is further increased by EPA allowing power generation units to "bank" early emission reductions for later use, or sale to other generators.

Because of the economic efficiency, the Agency focused on developing cap and trading/banking options for controlling SOx and NOx emissions. (See Table 2 for a list of the CAPI options.) Given the nature of the SOx problem and the existence of the Acid Rain Program, it made sense to design options that provided an emissions cap and allowed trading of emission allowances throughout the country. The summer cap would address ground-level ozone (formed primarily in the summer) while the winter cap would reduce eutrophication and acidification. For NOx, it made more sense to design options that provided summer and winter seasonal caps. EPA also hoped that a national trading scheme for NOx would keep the program simple and provide maximum liquidity for a trading market and maximum flexibility for the regulated community.

4. ANALYTIC APPROACH

To analyze the emissions reductions and costs of different options, EPA had ICF Resources Incorporated build a modelling system specifically for this purpose. The heart of the system was ICF's Integrated Planning Model (IPM), a dynamic linear programming model that selects the least-cost compliance approach for generation units providing electric power over a set time period subject to specified constraints -- in our case, pollution controls. IPM has been used by many large utilities to do capacity planning for the future. It also, has been used by the Edison Electric Institute, National Mining Association, the Department of Energy, the Tennessee Valley Authority, and EPA for similar national or regional analyses.

TABLE 2
CAPI EMISSION CONTROL OPTIONS

Each option is a stylized approach to air emissions control. One set of trading/banking options held the SOx cap constant and varied the NOx cap levels. Another group of options focused on alternative levels and timing of lowering the SOx cap. For comparative purposes, a traditional "command-and-control" approach was also evaluated.
NOx-Driven Trading and Banking Options
1. Trading/Banking for NOx at .15 and SOx with 50% Cap Reduction in 2010 - The existing CAAA Title IV SOx program continues, but in 2010 the current cap is lowered by 50 percent. In 2000, the Title IV NOx requirements form the basis for setting summer and winter season NOx caps. Fossil generating units are allowed to trade emission allowances and bank them for future use and/or trading (as occurs now in the Title IV SOx program). In 2005, the summer NOx cap is lowered. This new cap is based on a NOx "budget" equal to .15 per MMBtus of fossil fuel that EPA forecasts will be burned by generating units in 2000. 2. Trading/Banking for NOx at .20 and SOx with 50% Cap Reduction in 2010 - Same as Option 1, except that the summer NOx cap in 2005 is based on .20 pounds per MMBtus of fuel use. 3. Trading/Banking for NOx at .25 and SOx with 50% Cap Reduction in 2010 - Same as Option 1, except the summer NOx cap in 2005 is based on .25 pounds per MMBtus of fuel use.
SOx-Driven Trading and Banking Options
4. Trading/Banking for NOx at.15 and SOx with 60% Cap Reduction in 2010 - Same as Option 1 above, except the SOx cap in 2010 is lowered by 60 percent (as opposed to 50 percent). 5. Trading/Banking for NOx at .20 and SOx with 60% Cap Reduction in 2010 - Same as Option 2 above, except the SOx cap is lowered by 60 percent (as opposed to 50 percent). 6. Trading/Banking for NOx at .20 and SOx with 50% Cap Reduction in 2005 - Same as Option 1 above, except the SOx cap is lowered in 2005 (as opposed to 2010).
Traditional Regulatory Control
7. The current Title IV SOx program would continue in operation, but in 2010 the current cap is lowered by 50 percent and only trading is allowed (no banking). For NOx, in 2000 the proposed Title IV requirements go into effect. In 2005, all generating units using fossil fuels must meet standards for summer NOx emissions of .15 pounds per million Btus of energy consumed.

The modelling approach embodied in IPM allowed EPA to project future air emissions of selected pollutants and costs of electric production for a Base Case electric generation forecast that included consideration of existing environmental controls (a regulatory baseline). The Agency then ran the electric generation model after placing various types of pollution constraints on electric power generation to see the changes in selected air emissions and operating costs of electricity production after a control system was imposed.

The Agency focused on understanding the operations of fossil-fueled electric generation units in the future (i.e., where the greatest level of air emissions would result.) EPA's application of IPM considers the operations of nuclear and renewable energy sources, but does not attribute any air pollutants to them. EPA used the best available information on macro-energy and economic assumptions, electric generation technologies, air emissions under current and future federal and state regulatory requirements, and pollution control technologies. The Agency also considered how electric generation in the future is likely to occur in a more competitive environment. (See text box for more details.)

5. RESULTS

EPA's Base Case electric power generation and air emissions forecasts are provided below. These results are followed by the Agency's estimates of air emission reductions and costs that occur for different types of control options. Finally, we provide the results of sensitivity analyses of key assumptions.

5.1. Base Case Forecast

EPA's Base Case forecast for SOx and NOx emissions is provided in Table 3. The table also covers mercury and carbon emissions, which the Agency wanted to also track to see if significant side benefits result from the direct control of NOx and SOx. The Base Case shows EPA's forecast of air emissions from the electric power industry, if no new pollution controls beyond those in place occur. The table shows increases in NOx, mercury, and carbon emissions over the forecast period. EPA's implementation of the Title IV Acid Rain Trading Program of the CAAA lowers SOx emissions from 1990 to 2000 and holds them in check through 2010.

TABLE 3
BASE CASE FORECAST FOR THE US
OF SELECTED AIR EMISSIONS FROM ELECTRIC POWER GENERATION

POLLUTANT	Units	1990*	2000	2005	2010
Summer Nitrogen Oxides	Million short tons	3.3	2.60	2.76	2.79
Annual Nitrogen Oxides	Million short tons	7.5	5.97	6.39	6.46
Annual Sulfur Dioxide	Million short tons	15.9	10.22	10.44	10.00
Annual Mercury	Short Ton	51.3	60.9	66.1	67.4
Annual Carbon	Million Metric tons	477	551	606	639

* This estimate is based on EPA's National Air Pollutant Emission Trends 1990-1994, October 1995 (NOx and SOx) and Draft Mercury Report to Congress, August 1995 (Mercury) and EIA's Annual Energy Outlook 1994, January 1994 (Carbon).

CAPI Modeling Approach and Key Assumptions

ICF set up a modeling system that used four different existing models to implement the modeling approach that EPA required. They used the Integrated Planning Model (IPM) for estimating the type of electric generation capacity and regional distribution of power generation over the forecast period. This model examines power generation on a seasonal basis. IPM has a foresight feature that factors in how power generators would consider future electric demand, fuel prices, and other variables in making their decisions to build new capacity and to comply with environmental requirements.

ICF used a comparative cost model to screen out from consideration in IPM electric generation and pollution control technologies that would not be economic during the forecast period. This kept run time of the Model manageable. ICF used its Coal Electric Utility Model (CEUM) to develop delivered coal prices and the Gas Supply Assessment Model (GSAM) of the U.S. Department of Energy to estimate delivered natural gas prices. The supply curves of the two fuel price models were added to IPM so that it generated prices suitable for the fuel demand of each model run.

Coverage

IPM forecasts cover the 48 contiguous states and the District of Columbia. All existing utility power generation units are covered in the model as well as independent power producers and other cogeneration facilities that sell wholesale power, if they were included in the North American Electric Reliability Council (NERC) data base for reliability planning. The analysis considers future capacity additions by both utilities and independent power producers.

The air emissions forecasts are provided nationally and at the NERC region and subregion level. The air emissions forecasts cover sulfur dioxide, nitrogen oxides, mercury, and carbon (carbon dioxide) emissions.

Macro Energy and Economic Assumptions

EPA considered that the ongoing deregulation of utilities will make electric generation more competitive -- plant operators will be working harder to control or reduce costs and to cap, or lower customer prices. Therefore, the Agency used NERC's 1995 forecast, rather than the lower estimate of the Energy Information Administration (EIA), which did not fully reflect competition, to develop a generation forecast. The NERC forecast was modified to account for the electric demand reduction that should result from the President's Climate Change Action Plan. EPA assumed that utilities will build less new capacity in the future as they reduce planning reserve margins (extra capacity built to meet peak demands). In 2000-2010, they range from 15% to 20% in various NERC regions.

EPA also assumed that a 20-year trend of increasing operating availability for fossil units (less down time) will continue though 2005. Plant operators work to maintain heat rates at fossil units and those rates do not decline over time as they historically have. A small number of nuclear units that do not look like they can improve enough in the future to be competitive are retired. Power can be transferred between regions in the model at 75 percent of the First Contingency Total Transfer Capability, a sustainable transmission rate recommended by NERC staff.

Fuel Prices

EPA used recently revised coal and natural gas supply curves and transportation cost functions to develop delivered coal and natural gas prices during each model run. For the Base Case, the average delivered coal prices were $1.13, $0.98, and $.089 per million Btus for 2000, 2005, and 2010, respectively. The average delivered natural gas prices were $2.15, $2.37, and $2.42 per million Btus over the same years. Oil prices were $20, $22, and $24 per barrel over the same years.

Electric Generation Technology

EPA made assumptions on the efficiency and costs of installation and operation of all the electric generation technologies that it believed would be economically viable over the next 15 years. Most of the assumptions were based on the Energy Information Administration's estimates in the Annual Energy Outlook 1996 in conjunction with work done by the Tennessee Valley Authority. The Agency used EIA's approach to electric generation technology improvements that result from "learning-by-doing" to factor in performance and cost improvements over time. Repowering of coal and oil/gas steam to natural gas combined-cycle and integrated gasification (of coal) combined-cycle and life extension of fossil facilities were considered.

Regulatory Baseline

Electric generation units are constrained to meet applicable State NOx and SOx emission rate requirements, such as Reasonably Available Control Technology (RACT), but not Phases II or III of the Ozone Transport Commission's NOx Memorandum of Understanding. New capacity must comply with EPA's New Source Review and New Source Performance Standards. Compliance with the existing requirements of the Title IV SOx program and the NOx Phase I rule standards are also part of the baseline.

Pollution Control Technology

EPA used engineering estimates of performance and costs prepared by Bechtel Corporation for low NOx burner, selective catalytic reduction (SCR), selective non-catalytic reduction (SNCR), and gas reburn technologies that could be used at coal-fired and oil/gas-fired units for NOx control. No improvements in performance or costs over time were assumed. The Agency used information from the Electric Power Research Institute to update its performance and cost functions originally used in analysis supporting the Title IV program of the CAAA in 1991 for SOx controls (scrubbers). An assumption was made that control costs would decline 20 percent by 2000 in keeping with the recent trend. To reduce SOx emissions, coal-fired units can either employ scrubbers, or switch to lower sulfur content coals, or to natural gas.

Figure 1 shows the electric generation forecast on which the Base Case air emissions are based. Coal will continue as the dominant fuel for electric power generation. Natural gas use in combined-cycle units will have an increasing role in power generation to meet growing electricity demand. Nuclear power generation will decline. There will be little change in renewable sources of power.

5.2. Emission Reductions and Costs of Control Options

EPA analyzed three groups of options: trading and banking options with alternative summer NOx emission caps and a fixed SOx emissions cap, alternative SOx caps and alternative timing of the lowering of the SOx cap with the NOx summer cap held in most cases to the same level,, and a traditional regulatory approach. EPA conducted sensitivity analyses for key assumptions. The results from each set of analyses are presented below.

5.3. NOx-Driven Trading and Banking Options

Figure 2 shows the national summer NOx emission reductions from each option compared to the Base Case for the NOx-driven options. Each of the options provides substantial early reductions as plant operators bank emissions for later reductions. In 2005 and 2010, summer NOx reductions from the Base Case projections range from 34 percent to 56 percent and 54 percent to 66 percent, respectively. Figure 3 shows how the emissions reductions are widely dispersed geographically in 2005 (a similar pattern also exists for 2010). There are substantial reductions that occur in the mid-West, mid-Atlantic, and Southeast states. Reductions in these areas are critical to addressing the ozone problem in non-attainment areas.

The annual costs of each option are shown in Table 4. The jump in costs between the NOx .20 pound option and .15 pound option is due to the increase in the use of selective catalytic reduction technology to meet the lower emissions cap. However, the vast majority of units will not be using this technology to comply. This is illustrated in Figure 4.

TABLE 4
ANNUAL COSTS FOR NOx-DRIVEN OPTIONS
(Billion 1995 $)

Basis for Summer NOx Cap	2000	2005	2010
.15 pounds of NOx per MMBtus	$ 1.3	$ 2.5	$ 4.0
.20 pounds of NOx per MMBtus	$ 0.9	$ 1.8	$ 3.2
.25 pounds of NOx per MMBtus	$ 0.7	$ 1.3	$ 2.9

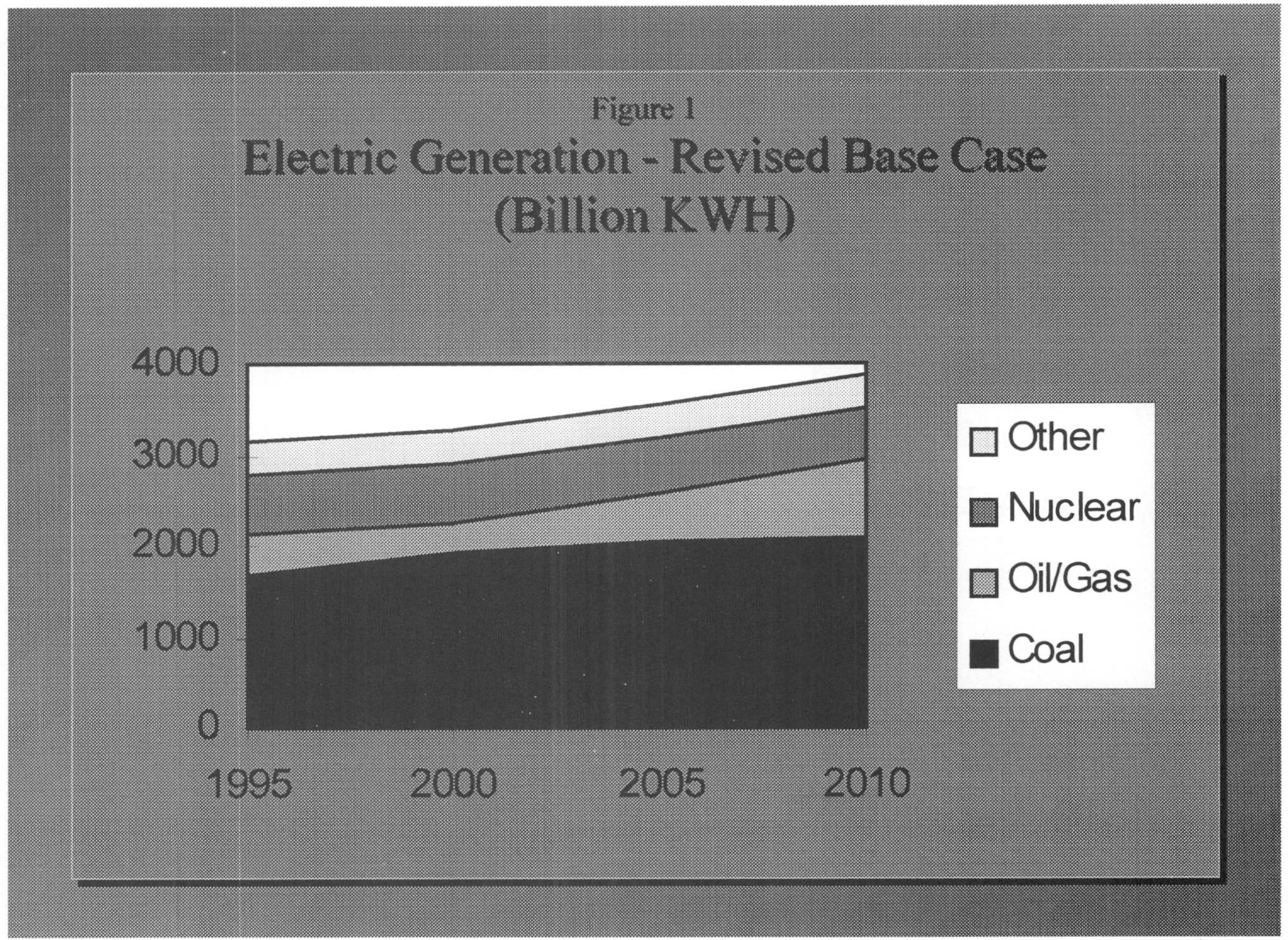
Figure 1
Electric Generation - Revised Base Case
(Billion KWH)
4000
3000
2000
1000
0
1995
2000
2005
2010
Other
Nuclear
Oil/Gas
Coal

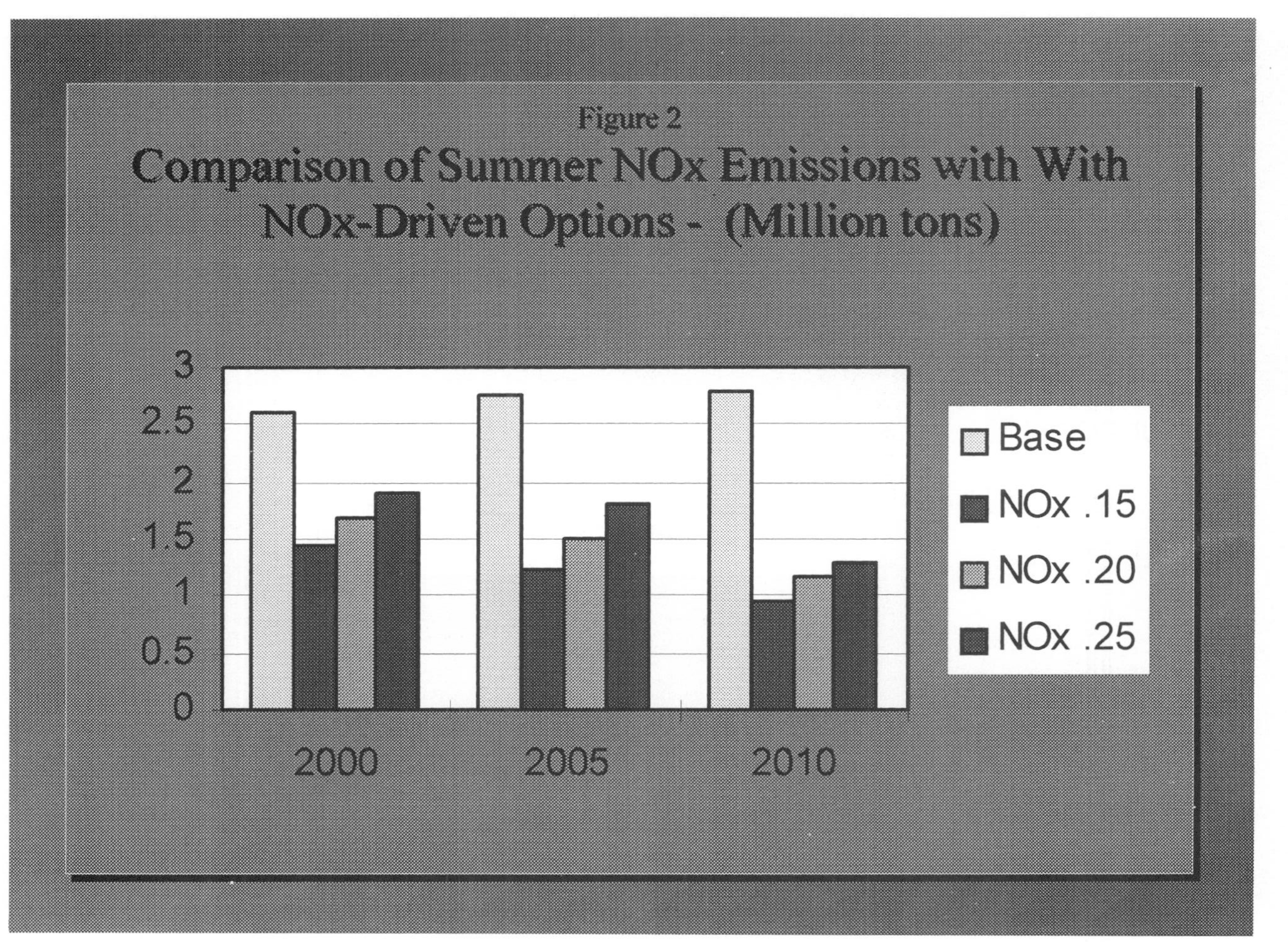
Figure 2
Comparison of Summer NOx Emissions with With NOx-Driven Options - (Million tons)
3
2.5
2
1.5
1
0.5
0
2000
2005
2010
Base
NOx .15
NOx .20
NOx .25

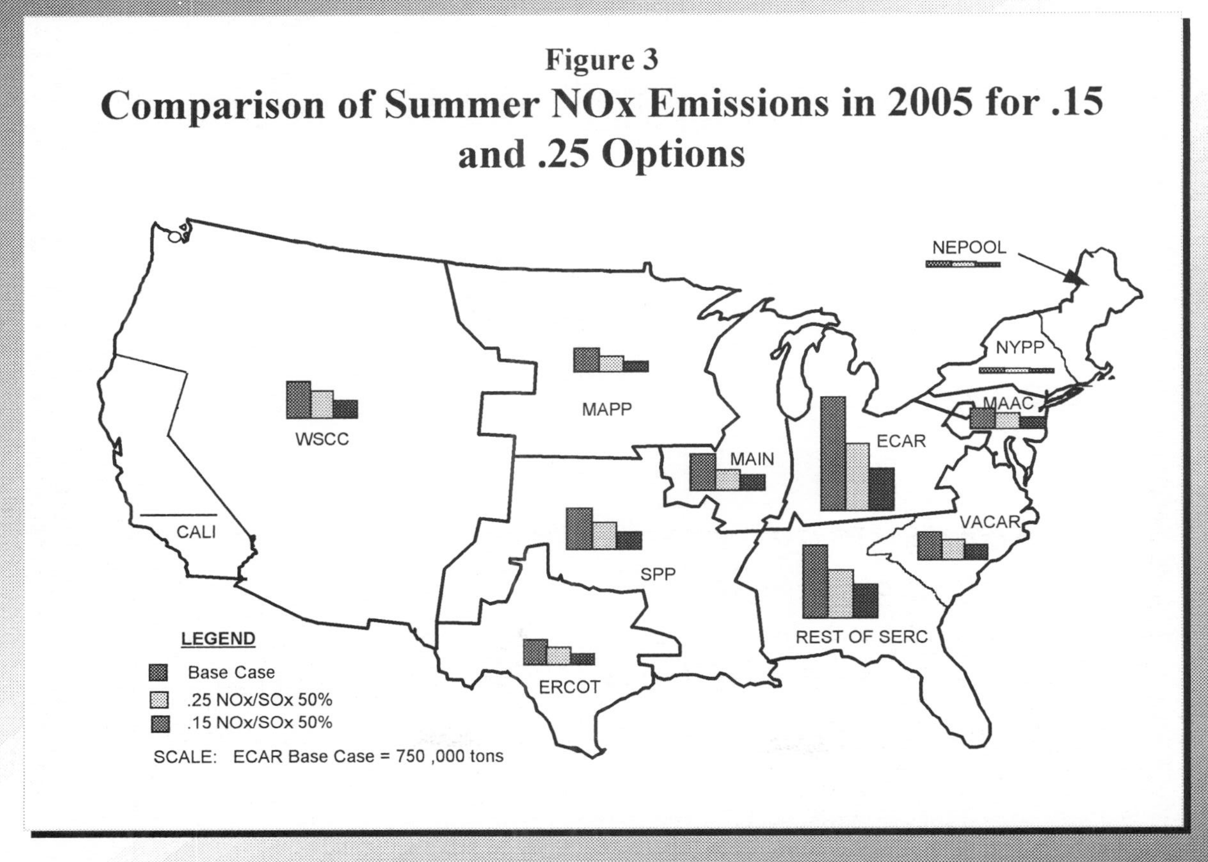
Figure 3
Comparison of Summer NOx Emissions in 2005 for .15 and .25 Options
NEPOOL
NYPP
MAAC
ECAR
VACAR
REST OF SERC
MAIN
MAPP
SPP
ERCOT
WSCC
CALI
LEGEND
Base Case
.25 NOx/SOx 50%
.15 NOx/SOx 50%
SCALE: ECAR Base Case = 750 ,000 tons

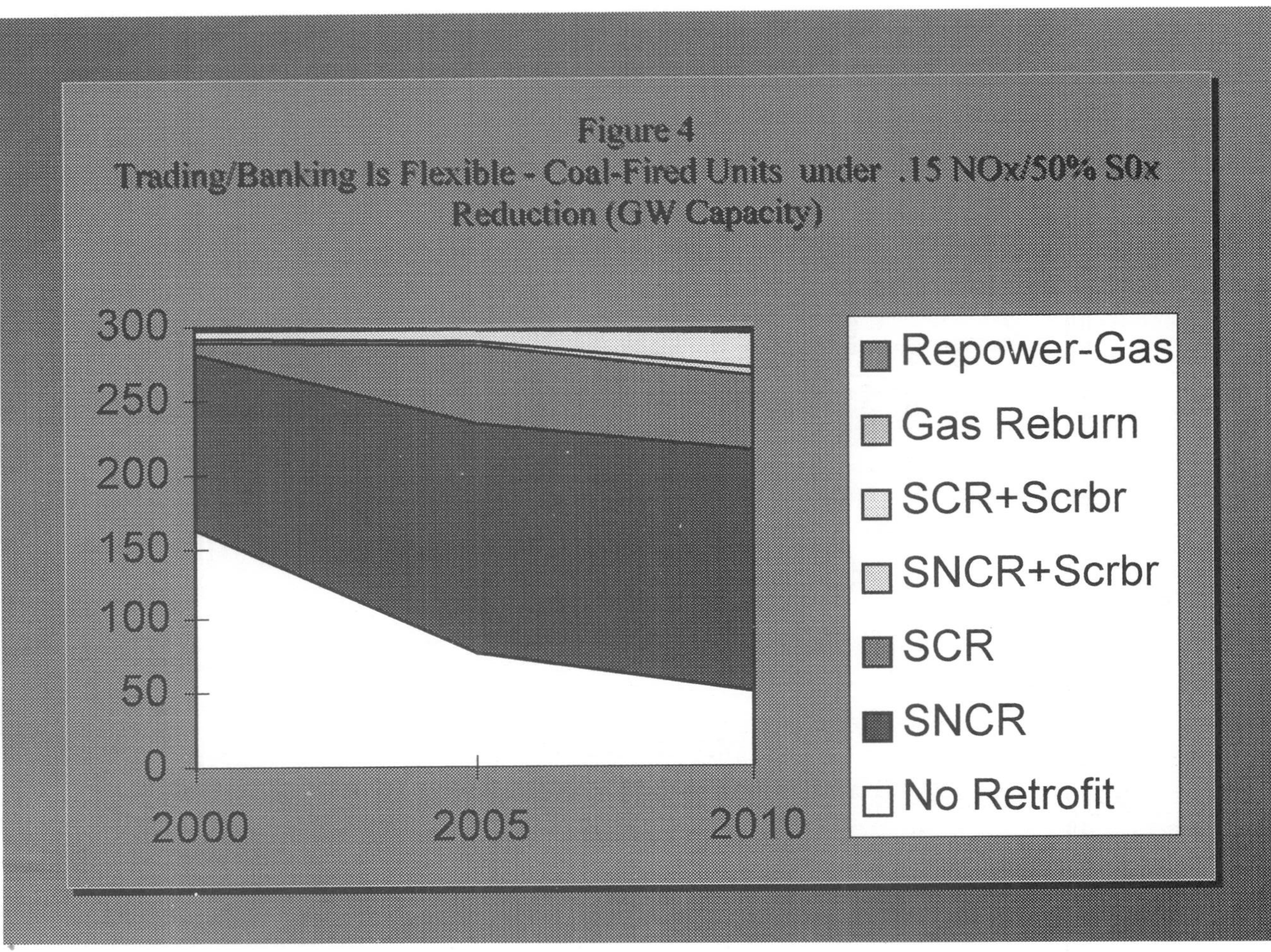
Figure 4
Trading/Banking Is Flexible - Coal-Fired Units under .15 NOx/50% S0x Reduction (GW Capacity)
300
250
200
150
100
50
0
2000
2005
2010
Repower-Gas
Gas Reburn
SCR+Scrbr
SNCR+Scrbr
SCR
SNCR
No Retrofit

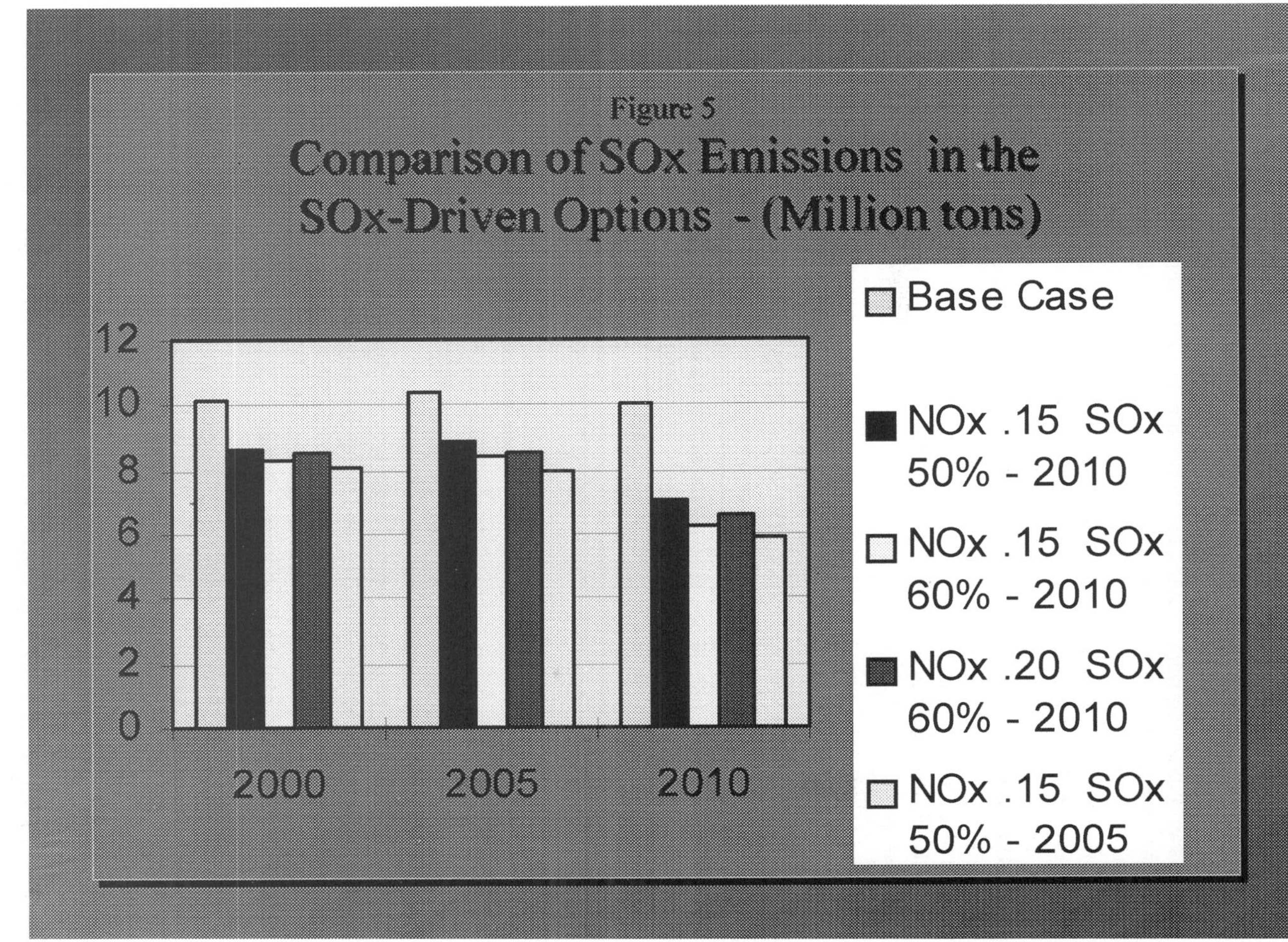
Figure 5
Comparison of SOx Emissions in the
SOx-Driven Options - (Million tons)
12
10
8
6
4
2
0
2000
2005
2010
Base Case
NOx .15 SOx 50% - 2010
NOx .15 SOx 60% - 2010
NOx .20 SOx 60% - 2010
NOx .15 SOx 50% - 2005

5.4. SOx-Driven Trading and Banking Options

Figure 5 shows the national SOx emission reductions from each SOx-driven option compared to the NOx .15 and SOx 50 percent cap reduction in 2010. Although in one option the SOx cap is moved forward to 2005, the emission reduction is not that dramatic in that year. This is because the freedom to bank and trade emissions reductions leads generation units to make and bank early reductions and withdraw them in a manner that smooths out SOx emission reductions over time. This is a more economical compliance strategy than a making a more abrupt reduction in 2005 when the capped emission level is reduced.

The annual costs for the SOx-driven options appear in Table 5. The table shows that its more costly to move the cap reduction forward (from 2010 to 2005 for a 50 percent reduction) than to reduce the cap level (from 50 percent to 60 percent in 2010). As we saw with NOx, the SOx emission reductions in these options are widespread throughout the country. The greatest reductions occur in the Eastern US where most of the emissions that affect urban areas exist.

TABLE 5
ANNUAL COSTS OF SOx-DRIVEN OPTIONS
(Billion 1995 $)

Summer NOx Cap/SOx Cap Reduction	2000	2005	2010
.15 NOx/60% Reduction in SOx Cap in 2010	$ 1.4	$ 2.6	$ 4.7
.20 NOx/60% Reduction in SOx Cap in 2010	$ 1.0	$ 2.0	$ 3.7
.15 NOx/50% Reduction in SOx Cap in 2005	$ 1.5	$ 2.9	$ 5.0

5.5. Traditional Regulatory Approach

The best way to appreciate the benefits of the trading-and-banking options is to compare them to the traditional command-and-control approach. EPA did this by constructing a traditional option that was based on the same type of objectives as one of the trading and banking options. The Agency compared the Trading/Banking for NOx at .15 and SOx with 50% Cap Reduction in 2010 (Option 1) to an option where unit-specific rates of .15 pounds of NOx per MMBtu are set in 2005 for the summer and the Title IV SOx emissions cap is reduced by 50 percent in 2010 and banking is no longer allowed.

The trading and banking option has 27 percent lower summer NOx emissions than the traditional regulatory approach in 2000. In 2005 and 2010, the traditional approach produces 38 percent and 21 percent lower summer NOx emissions, respectively. For SOx, the trading/banking approach has 22 percent and 19 percent lower emissions than the traditional option in 2000 and 2005, but is 56 percent higher than the traditional approach in 2010.

The annual costs of the two options are compared in Table 6. In 2000, the traditional approach is less costly as some plant operators decide not to add scrubber units and switch coals to make as many early emission reductions, since they can not use the emission allowances banked after 2010. However, in 2005 and 2010, the trading and banking option is much less expensive. On a net present value basis, from 2000 through 2010, the total costs of the traditional approach are more than 50 percent greater than the total costs of the trading-and-banking approach.

TABLE 6
ANNUAL COSTS OF TRADITIONAL VERSUS TRADING/BANKING APPROACH FOR A NOx SUMMER EMISSIONS CAP BASED ON .15 AND A 50 PERCENT REDUCTION IN THE TITLE IV SOx CAP IN 2010
(Billions 1995 $)

Approach	2000	2005	2010
Traditional Regulation	$ -.1	$ 3.9	$ 9.3
Trading/Banking	$ 1.3	$ 2.5	$ 4.0

5.6. Mercury and Carbon Emissions

EPA estimated mercury and carbon emissions for each of the trading/banking options. In 2000, reductions in mercury emissions ranged from 1.7 tons to 2.5 tons of the Base Case emissions of 60.9 tons. In 2010, reductions in mercury ranged from 6.7 tons to 10.2 tons of the Base Case emissions of 67.4 tons. Carbon emission reductions in 2000 ranged from 2 million metric tons to 5 million metric tons from the Base Case emissions of 551 million metric tons. In 2010, carbon emission reductions ranged from 29 million metric tons to 43 million metric tons from a Base Case of 639 million metric tons of carbon emissions.

5.7. Sensitivity Analyses

EPA conducted sensitivity analyses of several key assumptions in the Base Case to see the difference the changes made in the forecast of NOx, SOx, carbon, and mercury emissions. If natural gas prices were 15 percent higher, there would not be much difference in emissions except in 2010 where there would be 30,000 additional tons of NOx emissions. If transmission capability between NERC regions were 12 percent higher in 2000 and 50 percent higher in 2005, mercury and carbon emissions would not be much different than in the Base Case. NOx emissions were increased by 20,000 tons, 55,000 tons, and 35,000 tons in 2000, 2005, and 2010, respectively. If the Clinton Administration' Climate Change Action Plan proved to be 40 percent less effective, the most significant change would be the additional emissions of NOx of 122,000 tons and 63,000 tons in 2000 and 2005, respectively.

The Agency also developed a Higher Emissions Case to see how much higher future emissions would be in the future under other plausible assumptions that would lead to higher emissions. This case was premised on the potential for higher electric demand in the future, greater changes in operations to be more competitive, greater ability of generating units to send power to other regions, greater reductions of nuclear capacity, higher gas prices, and gas combined-cycle technology improving at a slower rate than assumed in the Base Case. Due to the Title IV program, the SOx levels are similar to the Base Case. For the other pollutants, there is an increase of annual NOx, carbon, and mercury emissions of close to 6 percent, 6 percent, and 3 percent in 2000 and close to 6 percent, 12 percent, and 13 percent in 2010, respectively.

EPA also considered how much more expensive it would be to implement one of the more aggressive trading/banking options if the Higher Emissions Case did occur. The results are provided in Table 7. The results suggest that controlling NOx and SOx at levels higher than our best estimate is not likely to be prohibitively expensive.

TABLE 7
COMPARATIVE ANNUAL COSTS OF THE TRADING/BANKING FOR NOx AT .15 AND SOx WITH 60% CAP REDUCTION IN 2010 FOR THE BASE AND HIGHER EMISSIONS CASES
(Billion 1995 $)

Emissions Case	2000	2005	2010
Base Case	$ 1.4	$ 2.6	$ 4.7
Higher Emissions Case	$ 2.1	$ 3.4	$ 5.4

6. CONCLUSIONS

From the results of the analysis, there are five major conclusions that we believe can be drawn:

- National trading and banking approaches for summer NOx and annual SOx control can provide large emission reductions. By 2010, the CAPI options produced a 54 to 67 percent reduction in summer NOx emissions and a 27 to 41 percent reduction in annual SOx emissions below the Base Case projection.

- The NOx and SOx emission reductions will occur throughout the U.S. Furthermore, EPA's analysis suggests that there will be large NOx and SOx reductions in the areas where it can do the most good--areas that make significant contributions to the air quality of large urban areas.

- Trading and banking options offer electric generators compliance flexibility in terms of timing and options. They can install pollution control technologies, switch fuels, change dispatch/utilization, and add cleaner generation units. They also can make early "bankable" reductions and/or purchase allowances.

- The annual costs of the trading and banking options range from $1.3 to $2.9 billion in 2005 and $2.9 to $5.0 billion in 2010. If higher emissions occur in the future, there will also be higher control costs. However, our analysis suggests that these costs are unlikely to increase to levels that should limit actions to address these problems.

- A traditional regulatory approach to NOx and SOx control is likely to cost considerably more than a trading and banking approach. EPA's comparative analysis in CAPI showed that from 2000 through 2010, the total costs of a traditional regulatory approach are more than 50 percent greater than the costs of a trading-and-banking approach.

REFERENCES

1. U.S. Environmental Protection Agency, "EPA's Clean Air Power Initiative", October 1996.
2. U.S. Environmental Protection Agency, "Supporting Analysis for EPA's Clean Air Power Initiative", October 1996.
3. U.S. Environmental Protection Agency, "Analyzing Electric Power Generation under the CAAA, July 1996.
4. Bechtel Power Corporation, "Cost Estimates for NOx Control Technologies Final Report", March 1996.
5. Bechtel Power Corporation, draft technical study on the use of gas reburn, June 1996.

ACKNOWLEDGMENTS

ICF Resources Incorporated used its Integrated Planning Model (IPM) to prepare the extensive cost and emission reduction analyses for the CAPI under the direction of Mr. John Blaney, vice-president of ICF Resources. The authors also wish to acknowledge Mary Nichols, EPA's Assistant Administrator for Air and Radiation, who initiated CAPI as part of her effort to develop better approaches to controlling pollution from electric power generation. We wish to thank the U.S. Energy Information Administration for supporting our efforts to develop a forecast future electric power generation to use in our analysis. EPA also received valuable input from the Edison Electric Institute, the Electric Power Research Institute, the National Mining Association, the Gas Research Institute, Utility Air Regulatory Group, the Center for Clean Air Policy, and other outside groups.

Trends in ammonia emissions from agriculture in Europe and in The Netherlands

ir. N.J.P. Hoogervorst

Laboratory of Waste Materials and Emissions (LAE), National Institute for Public Health and Environment (RIVM), P.O.Box 1, 3720 BA Bilthoven, The Netherlands

1. INTRODUCTION

The scientific and political attention for ammonia emissions originated in the 1980's when acidification and large scale forest die-back were a public concern. At first, acidification was attributed to SO_2 and NO_x, largely emitted by non-agricultural sources. Agriculture emits them in relatively small quantities, as a by-product of combustion of fossil fuels. By the end of the 1980's it became apparent that also ammonia (NH_3) contributed to acidification. For this agent the contribution from agriculture to the total emission is relatively high. In most European countries 90-100% of national NH_3-emissions originate from animal husbandry[11]. On a European scale, NH_3-emission densities are highest in The Netherlands reaching an average level of over 100 kg/ha while the average in the EC-12 and in Europe is roughly only 25 kg/ha (see table 1). The high emission levels stimulated the Dutch government to be the first to launch a national ammonia abatement program in 1990. The results of this program will be discussed in chapter 2 of this paper. Chapter 3 provides an overview of trends in ammonia emissions on a European scale.

---

[11] Exceptions are Sweden (60%), Luxembourg (76%), Poland (79%), Austria (81%), Slovenia (85%) and the UK (87%), according to Corinair data for 1990.

Table 1

Average NH_3 emission from agriculture in 1990 at six regional scale levels: the Northern Hemisphere, Europe, the EC-15[a)] and -12, The Netherlands, and the Dutch municipality of Noordwijk.

	average (kg/ha)	lowest	highest
Northern Hemisphere	n.a.	Canada: 5	China: 50
Europe	26	USSR: 7	Netherlands: 95
EC-15	25	Spain: 10	Netherlands: 95
EC-12	25	Spain: 10	Netherlands: 95
The Netherlands	109	Almere: 20	Deurne: 340
Noordwijk municipality	48	bulbs: 5	dairy: 200

[a)] the 15 memberstates from the European Community include EC-12 and Austria, former DDR, Finland and Sweden.

Source: computed from Corinair and FAO data; Netherlands from RIVM data, N.Hemisphere from EMEP-East.

2. AMMONIA EMISSIONS IN THE NETHERLANDS

2.1. Historic development of ammonia emissions

The emissions of NH_3 from Dutch agriculture have grown exponentially since the end of the nineteenth century and reached their peek level in 1987. These emissions originate largely from animal manure (see figure 1) and for a minor share (1-5%) from the application of chemical nitrogen fertilizers.

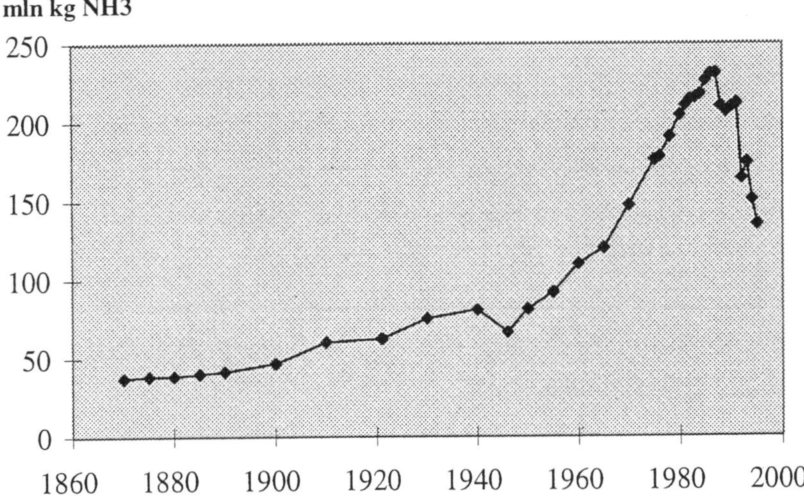

Figure 1. Ammonia emissions from livestock in The Netherlands, 1870-1995.

2.2. Environmental damage from ammonia deposition

The first notions about the contribution of ammonia to soil acidification in nature areas stem from 1982, when Van Breemen and his colleagues described forest soil acidification from atmospheric ammonium sulfate (Van Breemen et al., 1982). In 1984 a large coordinated research program was launched, reflecting the growing concern of policy makers with effects of air pollution, primarily on forest. The first two phases of this Dutch Priority Program on Acidification (Heij en Schneider, 1991) resulted in understanding of the causes and effects of deposition of acidifying substances on forest and heathland, which was substantial enough to implement an Acidification Policy (VROM, 1989) and an Ammonia Policy (LNV and VROM, 1990).

By then, ammonia (NH_3) was accepted to be a *potentially* acidifying substance. In the air it neutralizes acidity by transforming into ammonium (NH_4^+) but when this reaches the ground it is either taken up by plants or transformed into nitrate (NO_3^-), leaving 2 ions of H^+ for every ion of ammonium. Actual soil acidification occurs to the extend that nitrate leaches to the groundwater instead of being taken up by plants. This occurs when the potential acid and nitrogen supply is abundant, compared to the availability of other (neutralizing) minerals. Thus, acidification can be seen as a result of direct acid inputs *plus* eutrophication, an oversupply of nitrogen. This oversupply leads to rapid growth of stems and leaves but retards the development of the root system. This increases the susceptibility of trees to insects, fungi, drought, frost and storms. In addition, the diversity of (forest) ecosystems is reduced by eutrophication as grasses, stinging nettle and blackberries become dominant species in forest undergrowth and lichen disappear. Heathlands are converted into grassland and fens loose their oligotrophic conditions and associated flora and fauna. Groundwater is polluted by nitrate and aluminum, both toxic to humans (Heij and Schneider, 1991).

Dutch farmers are not as convinced as scientists and policy makers that ammonia emissions are harmful to the environment. It is very likely that the restrictions they experience from the ammonia policy and their inability to discern the visual effects from NH_3-emissions lead them to distrust the abundance of scientific evidence. They rather believe statements of Heidelberg Appeal Nederland, an organization that claims to represent Nobel laureates, publicly challenges the findings of 10 years of acidification research but avoids a scientific discussion on the issues they raise. This has increased public uncertainty about the harmfulness of ammonia emissions. Even the Minister of Environment seems to question the gravity of the ammonia problem. Despite big efforts, scientists and policy makers have not succeeded to eliminate these feelings which is a potential threat to a full implementation of the ammonia abatement policy.

2.3. The Dutch ammonia abatement policy

Setting targets

The Dutch Priority Program on Acidification provided a scientific justification for the Dutch government to develop a policy aiming to reduce ammonia emissions (LNV en VROM, 1990). The long term emission targets were derived from critical loads for various types of ecosystems and from targets for nitrate leaching in nature areas (max. 25 mg/l). It turned out that the protection of forests and heathlands required a maximum deposition of 1400 mol H^+/ha containing 1000 mol N/ha. This meant a reduction of national NH_3-emissions by 80-90%, assuming a reduction in the surrounding countries by 60% with respect to 1980 emission levels. For SO_2 and NO_x similar reductions were needed (RIVM, 1989). Protection of natural ecosystems and fens would require sharper deposition levels (down to 400 mol H^+/ha) and larger emission reductions. Calculations showed that a 70% emission reduction in The Netherlands was technically feasible but relatively costly. This was chosen as a target for the year 2000 since it was thought to be necessary to reach a minimum level of protection to national forests. The costs of a 30% emission reduction appeared to be very limited and would not create financial problems to the farming sector (Oudendag en Wijnands, 1989). This target was set for the year 1994, 5 years after the Policy Plan was put in action. Based on cost calculations, farmers didn't want to go beyond 50% reduction by the year 2000. Policy documents in the early 1990's still mentioned a target of 50-70% for 2000 but by 1995 the targets had been adjusted to 50% reduction in 2000 and 70% reduction in 2005.

Instruments

On the national level, policy instruments include mandatory coverage of manure storage and mandatory incorporation of slurries into the soil. Policy makers are preparing legislation which prescribes that new stables and major stable adaptations meet certain emission requirements (Green Lable stables). On the local level, farmers have to comply to production permits, based on calculated ammonia deposition on acid-sensitive areas. In addition to the nationally prescribed measures for emission reduction, farmers can invest in low ammonia housing systems to meet the requirements for a production permit or move the location of their stables away from the sensitive areas.

2.4. Effects of the Dutch ammonia policy

Without the ammonia policy, NH_3 emissions from livestock would have stabilized roughly at the 1986 level. This stabilization was due to other policies such as the national manure policy from 1986 (which put a hold to the systematic expansion of the pig and poultry industry) and the EU dairy policy from 1984 (which put a maximum on the national milk production and thus results in annual reductions of the number of dairy cows). Although the manure production per animal continued to grow, the net result was a stabilization of the national manure production (measured in nitrogen) since 1986.

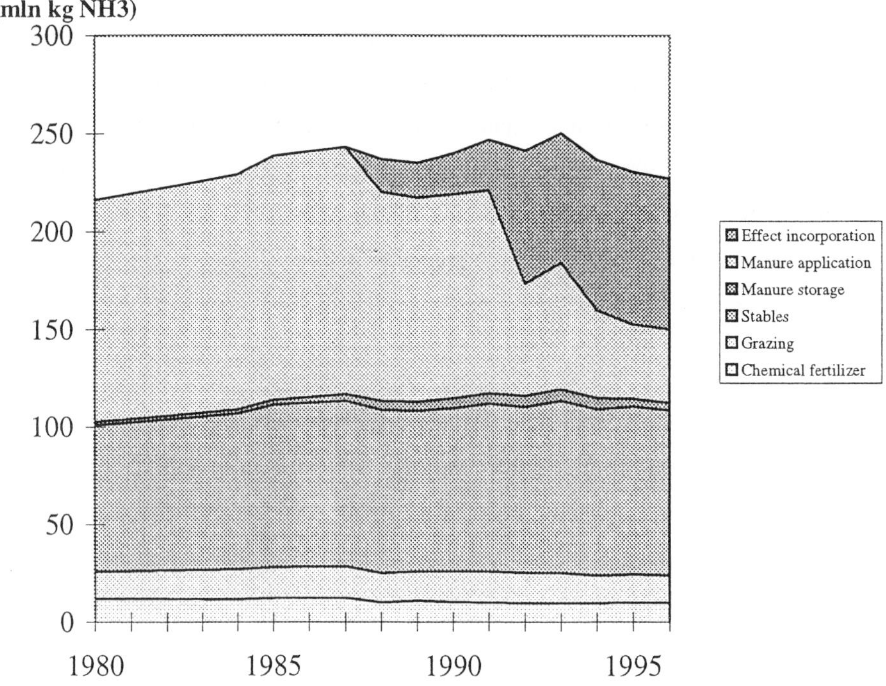

Figure 2. Effects of incorporating manure on NH_3-emissions between 1980 and 1996.

The largest contribution to NH_3 reduction came from incorporation of manure into the soil, which was made compulsory in a stepwise fashion between 1987 and 1995, see figure 2. This effect may be overestimated since we observe a growing gap between measured and calculated ammonia concentrations in ambient air since 1992. The emission from storage facilities grew slightly but would have grown substantially if they had not been covered. That is because the manure policy prescribed a restricted period for manure spreading in order to reduce nitrate leaching. This meant that more manure was stored for a longer period of time before spreading on the fields.

2.5. Projections of NH_3-emissions from Dutch agriculture

Dutch scientists are developing a tradition of scenario studies on pollution from Dutch agriculture. These studies are used in ex-ante policy evaluations and provide a basis for discussion on policy adaptations. The first scenario study was carried out in 1989, as part of the 'Concern for Tomorrow' study (RIVM, 1989). It predicted that the acidification policy from 1989 would reduce NH_3-emissions between 1980 and 2000 by 33%, mainly by adapting housing systems and by incorporating and injecting animal manure. With additional effort a reduction of 70% was thought to be achievable, basically by applying stronger versions of the same kind of measures but than on a larger scale. The study also mentioned the alternative of large scale processing of animal manure to a full substitute of chemical fertilizer, reflecting the believes of optimists about the potentials of technological innovation. Five years and 50 million guilders later (FOMA, 1995) it was concluded that that option might have been technically feasible but not compatible with the rules of a free market economy.

Table 2

Overview of RIVM scenario studies on NH_3-emissions from Dutch agriculture: application rates of measures and overall emission reduction in 2010 compared to the 1980 level.

Emission reducing measures	Concern for Tomorrow (1989)	2nd National Environmental Outlook (1991)	3rd National Environmental Outlook (1993)	4th National Environmental Outlook (1997)
	(%)			
Change N% in feed				
- cattle	+25	+10	+ 4	+5 / - 0
- pigs	- 1	- 10	- 20	- 10 / -20
- poultry	- 1	- 10	- 7	+ 22
Low NH_3 housing				
- cattle	0	0	0	25
- pigs	53	100	100	75
- poultry	72	100	100	70-90
Cover manure storage	0	100	100	100
Incorporate slurry on grassland	60	100	100	100
Incorporate slurry on arable crops	100	100	100	100
Manure processing	25-100	11-18	26/187	5
% Emission reduction 1980-2010	70	63	74	60

Subsequent scenario studies (see table 2) all came to the conclusion that all available technical measures need to be applied in order to reach emission levels that are compatible with targets for ecosystem protection. As more field data became available on the effectiveness of emission abatement techniques, the scenario studies became less optimistic about the actual emission reductions that each technique could achieve. This implies that larger cuts in the livestock production sector are needed to reach the emission targets. This means that farmers' resistance will grow and that the likelihood of reaching the targets diminishes as time goes by. The long term target, 80-90% emission reduction or a return to 1920 emission levels, seems to be out of reach forever.

2.6. Scientific developments in The Netherlands

Dutch policy makers have a number of concerns with respect to the ammonia problem, some of which can be dealt with by conducting more research. Policy makers are looking for new approaches to the ammonia problem, which is more attuned to region specific circumstances, such as the type of nature that is to be protected and the types of farming that are regionally dominant. A regional approach requires accurate region specific data on emissions, atmospheric transport, deposition and dose-effect-relations on vegetation. This means that research should pay more attention to uncertainty analysis, and verification of emission factors. For this a two year Nitrogen research program (STOP, Erisman et al., 1996) was financed. It addresses two main topics: (1) improvement of source-receptor relations on a local and regional scale, and (2) better understanding of causal ecological relations to improve critical load estimates for different ecosystems. Furthermore, RIVM currently runs a research project in which measured NH_3 concentrations in ambient air are compared to calculated concentrations based on estimated NH_3-emissions. Our uncertainty analysis seems to indicate that emission estimates are highly sensible to variations in emission factors of cattle stables. To establish better emission factors we need new and more accurate measuring techniques for stable emissions. A second area of research deals with the regional distribution of sources and receptors of ammonia. Since emission targets, which were derived from the present regional distribution patterns, seem unattainable in the next decades, it becomes interesting to examine alternative distribution patterns of the two which require less ambitious emission targets to protect ecosystems. These alternative patterns may become interesting policy options, especially when the public costs of relocating nature areas and/or farms are lower than the costs of increasing national emission reductions from 80 to 90%.

3. AMMONIA EMISSIONS IN EUROPE

Ammonia emission levels in Europe are generally much lower than in The Netherlands. But emission levels per hectare in the Po-valley (Italy), Brittany (France), Sleeswijk-Holstein (Germany) and Flanders (Belgium) are also far above the European average. The high concentration of animals in these regions is accountable for this situation.

3.1. Historic development of ammonia emissions in Europe

Historic data on NH_3-emissions in Europe are hard to come by and they show a large variation in accuracy, both over time and between countries. EMEP has compiled timeseries for all European countries from 1980. These data show that total European NH_3-emissions have remained stable around 5.6 billion kg NH_3 until 1989 and have then dropped 0.9 billion kg until 1994 (see table 3). Part of this reduction can probably be ascribed to improved estimation procedures but a substantial part is caused by reductions in the number of animals. In the EC the quotation of milk production led to a downward trend in cattle numbers. In former East-European countries animal numbers diminished as a result of the decay of the communist system. In both parts of Europe emission levels dropped by roughly equal amounts between 1989 and 1994.

Table 3

Development of NH_3-emissions in Europe between 1980 and 1994.

	1980	1985	1990	1994	'94-'80
	(mln kg NH_3)				
Netherlands	234	256	236	172	-62
EC-12	3217	3254	3031	2819	-398
EC-15	3683	3726	3475	3212	-471
Europe	5894	5936	5621	4936	-958

Source: EMEP/MSC-W, 1996.

3.2. Ammonia deposition and public awareness in Europe

The public awareness of the ammonia problem is probably highest in The Netherlands but even there it is not overwhelming. Most European countries have been reluctant to take action against ammonia emissions, simply because they are not aware of the problem. Research over the past five years, however, has led to new combinations of existing knowledge which gives a better view on the nature and size of the ammonia problem. This is related to the development of the concept of critical loads, which are indications for the maximum annual deposition of a certain agent that will not harm a chosen ecosystem. These critical N loads take account of physical and chemical conditions of soils, regionally specified targets for protected ecosystems and depositions from other acidifying (and atrophying) agents such as SO_2. The environmental damage from ammonia can be derived from critical loads for nitrogen, derived simultaneously from critical loads for acidification and eutrophication. Generally, the critical N load for eutrophication is lower than the one for acidification (Hettelingh et al., 1995). In the largest part of Europe (with the exception of Scandinavia, Italy and the east coast of the UK) ammonia depositions take more than 50% of total nitrogen deposition. In high emitting regions this share exceeds even 70%. This means that in large parts of Europe ammonia deposition contributes significantly to the excedence of critical nitrogen loads. Presently, total nitrogen depositions exceed critical N loads all over Europe, except in Ireland, the northern part of Scandinavia, southern Italy and the largest part of Greece. In the EC-15 more than 34% of the ecosystems (38 million hectares) were unprotected against eutrophication in 1990 (IIASA, 1996). The exceedence is highest in the central-western part of Europe (Germany, Denmark, Poland, Czech Republic, Austria, Switzerland, Luxembourg, Belgium, The Netherlands, and central England), with levels of more than 10 kg/ha (or 1 g/m^2) per year (Posch et al., 1997). More than half of this can be attributed to ammonia. This means that ammonia deposition is a substantial threat to endangered ecosystems all over Europe.

3.3. Ammonia policies in Europe

Emission abatement policies are always a response to a widely established public concern about possible harmful effects of emissions. Since ammonia emissions are generally low in Europe and people are generally unaware of its harmful effects, NH_3 abatement policies are rare in Europe. At present, only Flanders (Belgium), Denmark and The Netherlands have a national policy aiming to reduce NH_3 emissions. All countries prescribe special manure application methods which reduce volatilization of reduced nitrogen in manure. In The Netherlands this is supplemented by obligations to cover outside manure storage facilities and subsidies to built low emitting housing systems (see par. 2.3).

The effects of these measures are not clearly visible in reductions of NH_3 emissions because simultaneous changes in the number of livestock occurred. In Flanders livestock expansion outweighed the effect of low emission manure application. This increased average NH_3 emission from 70 to 80 kg/ha between 1985 and 1994. In Denmark the abatement measures were only partly counteracted by a 30% increase in pig and poultry numbers, resulting in a net reduction from 55 to 45 kg/ha in the same period (RIVM, 1996).

The lack of national ammonia policies in Europe does not imply that there is no political attention for the ammonia problem. The UN-ECE currently facilitates negotiations for a new Nitrogen Protocol, covering emissions from both oxidized (NO_x) and reduced nitrogen (NH_3) to the air. Information on critical nitrogen loads (see par. 3.2) and on technical potentials and costs of emission reductions for NO_x and NH_3 (see par. 3.5) provides the basis for these negotiations, which are expected to be completed in 1998.

3.4. Future trends in European ammonia emissions

Future developments of ammonia emissions are dependent on developments in livestock production and in application of emission control measures. Table 4 gives an overview of current reduction plans in the EC-15 countries and in non-EC Europe. These data have been derived from an inventory of officially declared national emission ceilings, collected on a routine basis by the Secretariat of the Convention on Long-range Transboundary Air Pollution (UN-ECE, 1995). In cases where no projection was supplied by a country for the target year 2010, emission ceilings were derived in accordance with the practice used for modeling work under the Convention. This means that target levels were derived from projections of livestock numbers and control measures.

Compared to the base year 1990, ammonia emissions in the EC-15 would be lower by about 15% and by 17% in the non-EC countries, see table 4. Reductions in Denmark (26%), Finland (27%), Germany (29%), and The Netherlands (66%) are far above the EC average. The reductions in Finland and (East) Germany, as in many other countries, are based on expected reductions in animal numbers only. Realization of these CRP's will increase the protection of ecosystems against eutrophication. The combined effect of the CRP's for NO_x and NH_3 are estimated to reduce the unprotected area of ecosystems in Europe from 18% in 1990 to 11% in 2010 (IIASA, 1996). Within Europe large regional differences will remain, however, with unprotected area of ecosystems ranging from more than 80% in Belgium, The Netherlands, West Germany, Belarus, Czech Republic, Lithuania, and Poland to less than 20% in, Ireland, UK, Norway, Sweden, Finland, Albania, Greece, Romania and Russia.

Table 4

Current reduction plans for ammonia emissions in EC-15 countries and in non-EC Europe in 2010, compared to emissions in 1990[a]).

Country	base year	target 2010	Change	Index	control
	(mln kg NH_3)			(index)	
Austria	91	93	2	102	
Belgium	95	106	11	112	yes[b])
Denmark	140	103	-37	74	yes
Finland	41	30	-11	73	
France	700	669	-31	96	
Germany(incl.DDR)	759	539	-220	71	
Greece	78	76	-2	97	
Ireland	126	126	0	100	yes
Italy	416	391	-25	94	
Luxembourg	7	6	-1	86	yes
Netherlands	236	81	-155	34	yes
Portugal	93	84	-9	90	
Spain	353	373	20	106	
Sweden	61	53	-8	87	yes
United Kingdom	320	270	-50	84	
EC-15	3516	3000	-516	85	
non-EC	4213	3484	-729	83	
EUROPE	7729	6484	-1245	84	

[a]) base year data, collected by UN-ECE, differ from EMEP data as presented in previous tables.

[b]) not mentioned in the IIASA study.

Source: IIASA, 1996

Table 4 also shows that very few countries are expected to take control measures to realize their reduction plans. According to a IIASA study all countries mentioned will apply more efficient application techniques for most types of manure, see table 5. This is a logical choice since it is the most cost-effective way to reduce NH_3 emissions. Stable adaptations are expected for laying hens in all countries and for other poultry and dairy cattle in Ireland, Luxembourg and The Netherlands. It is remarkable that no stable adaptations are foreseen for pigs. These stables are presently being built in The Netherlands and they prove to be a fairly cost-effective way to reduce NH_3 emissions. Biofiltration of stable air, on the other hand (and envisioned for The Netherlands), is still relatively expensive and might sooner be applied in cattle stables than for pigs. Covers on manure storages are applied to all manure types in The Netherlands but only to cattle and pig manure in Ireland and Sweden and to pig manure in Luxembourg. This might also be an interesting measure for countries which have limited time periods for manure spreading and thus need substantial manure storages. It is also seen as an appropriate (and thus not necessarily cost-effective) measure to offset potential acidifying side-effects of a eutrophication abatement measure. Given the overview of table 5 we see a large variation in control measures between the countries listed. This might suggest that there is no standard best approach to reducing ammonia emissions and that each country (andregion) needs to find its own appropriate measure mix.

Table 5

Ammonia emission control measures by country, assumed to realize the CRP's for 2010.

CONTROL MEASURES	Dairy cows	Other cattle	Pigs	Laying hens	Other poultry
Low Nitrogen Feed			Ireland		
			Luxembourg		
	Netherlands		Netherlands	Netherlands	Netherlands
Stable adaptation				Denmark	
	Ireland			Ireland	Ireland
	Luxembourg			Luxembourg	Luxembourg
	Netherlands		Netherl.[a])	Netherlands	Netherlands
				Sweden	
Biofiltration			Netherl.[b])		
Coverings for manure storage		Ireland	Ireland		
			Luxembourg		
	Netherl.[a])	Netherlands	Netherlands	Netherlands[a])	Netherlands[a])
	Sweden	Sweden	Sweden		
Low emission manure application	Belgium[a])	Belgium[a])	Belgium[a])	Belgium[a])	Belgium[a])
		Denmark	Denmark	Denmark	Denmark
	Ireland	Ireland	Ireland	Ireland	Ireland
	Luxembourg	Luxembourg	Luxembourg	Luxembourg	Luxembourg
	Netherlands	Netherlands	Netherlands	Netherlands	Netherlands
				Sweden	Sweden

[a]) my addition; not mentioned in the IIASA study.

[b]) considering technical circumstances I question this assumption.

[c]) the source also mentions application to sheep manure in Luxembourg, The Netherlands and Sweden but I question the applicability of this technique to this manure type.

Source: IIASA, 1996.

3.5. Scientific developments

Recently, new protocols have been developed to arrive at uniform NH_3-emission estimates throughout Europe (EEA, 1996) but the results still have to find their way to the various emission databases. It is clear that the availability of statistical data form the largest obstacle to more adequate emission estimates, not insufficient knowledge of underlying chemical and physical processes.

Accurate data bases on current ammonia emissions in European countries are needed as a starting point for multilateral negotiations on emission reductions. Corinair provides such a data base but harmonization is still needed to attain comparable data. Countries apply different estimation procedures to arrive at national emission estimates. The simplest procedure uses fixed emission factors (obtained from literature) per animal, distinguishing only between crude categories such as cattle, sheep, horses, pigs and poultry. The second stage procedure uses nation specific emission factors, and possibly increasing the number of animal categories. In the third stage emission factors are applied to a mathematical description of the nitrogen flow in farming systems. This allows calculation of the effects of applied emission abatement techniques. As application of these techniques increases over time, average emission factors will also differ between years. In a fourth stage, emission factors take account of the N composition of feed rations or excreta. Most European countries currently apply first or second stage estimating procedures for NH_3-emissions. Some countries, like Belgium (Flanders), The Netherlands, and the UK use third stage procedures. The fourth stage procedure is still only being developed at research stations in The Netherlands and it may take several years before we will see field application.

Table 6

Application of estimating procedures for NH_3 emissions in Europe since 1980.

Estimating procedure	before 1985	1985 -1990	1990 - 2000	after 2000
1. general emission factor, few animal categories	Netherlands		most European countries	
2. nation specific emission factors		Netherlands		
3. time specific emission factors			Belgium Netherlands United Kingdom	
4. feed ration specific emission factors			research stations	Netherlands ?

Various research centers in Europe study emission processes of ammonia, especially in the United Kingdom, Denmark, Germany and The Netherlands. There is more than average attention in The Netherlands and the UK for new techniques to measure NH_3-emissions from naturally ventilated stables. Scientists in the UK specialize in modeling evaporation of NH_3 after manure spreading in the field. German researchers seem more interested in measurements of NH_3 emissions from pig and poultry stables. Actually, scientists all over Europe continue to improve their techniques to better measure ammonia volatilization at each stage of the nitrogen cycle. This provides the basis for improved emission inventories in the (near) future.

REFERENCES

1. Breemen, N. van, P.A. Burrough, E.J. Velthorst, H.F. Dobben, T. de Wit, T.B. Ridder, H.F.R. Reinders (1982) 'Soil acidification from atmospheric ammonium sulfate in forest canopy throughfall', in: *Nature*, 299: 548-550.
2. EEA (1996) Joint EMEP/CORINAIR Atmospheric Emission Inventory

Guidebook, first edition. Vol. 1-2. Copenhagen, Denmark.

3. Erisman, Jan-Willem, Roland Bobbink and Ludger van der Eerden, eds. (1996) Nitrogen pollution in the local and regional scale; the present state of knowledge and research needs. Bilthoven: RIVM, report no. 722108010.

4. FOMA (1995) Jaarverslag 1994 [Annual Report]. Wageningen: Financierings-Overleg Mest- en Ammoniakonderzoek.

5. Heij, G.J. and T. Schneider (eds.) (1991) Dutch Priority Program on Acidification. Eindrapport tweede fase additioneel programma verzuringsonderzoek. Bilthoven: RIVM, report no. 200-09.

6. Hettelingh, J-P, M. Posch, P. de Smet and R.J. Downing (1995) *The use of critical loads for emission reduction agreements in Europe*, in: Water, Air and Soil Pollution **85:2381-2388**.

7. IIASA (1996) Cost-effective Control of Acidification and Ground-Level Ozone, Second Interim Report to the European Commission, DG-XI. Working Document, December 1996.

8. LNV en VROM (1990) Plan van Aanpak Beperking Ammoniak Emissies van de landbouw [Ammonia Policy]. Tweede Kamer, vergaderjaar 1990-1991, 18225, nr.42 en 43. Den Haag: SDU-uitgeverij.

9. MSC-W (1996) MSC-W Status Report 1996, part One; Estimated dispersion of acidi- fying agents and of near surface ozone. EMEP/MSC-W Report 1/96, Oslo: MSC-W.

10. Oudendag, D.A. en J.H.M Wijnands (1989) Beperking van de ammoniak-emissie uit dierlijke mest; een verkenning van mogelijkheden en kosten. [Reducing ammonia emissions from manure; an exploration of possibilities and costs] Den Haag: LEI, Onderzoeksverslag 56.

11. Posch, M, J-P. Hettelingh, P.A.M. de Smet, R.J. Downing (eds.) (1997) Calculation and mapping of critical thresholds in Europe; status report 1997. Bilthoven, The Netherlands: RIVM-CCE (in preparation).

12. RIVM (1989) Concern for Tomorrow, National Environmental Outlook 1985-2010. Bilthoven: RIVM, pp.456.

13. RIVM (1991) Second National Environmental Outlook 1990-2010. Alphen a/d Rijn: Samsom H.D. Tjeenk Willink, pp.550.

14. RIVM (1993) Nationale Milieuverkenning 3, 1993-2015. Alphen a/d Rijn: Samsom H.D. Tjeenk Willink, pp.167.

15. RIVM (1996) Milieubalans 96, Het Nederlandse milieu verklaard [Environmental

Balance 1996, explaining the Dutch environment] Alphen a/d Rijn: Samsom H.D. Tjeenk Willink, pp.142.

16. VROM (1989) Bestrijdingsplan Verzuring [Acidification Policy] Tweede Kamer, vergaderjaar 1988-89, 18225, nr.31. Den Haag: SDU-uitgeverij.

New Perspectives: Sustainable Technological Development in Agriculture.

Oskar de Kuijer, Hans Linsen and Jaco Quist [1]

[1] Interdepartmental research program "Sustainable Technological Development", P.O. Box 6063, 2600 JA Delft, The Netherlands

ABSTRACT

New long term perspectives on fulfilling our (basic) needs for food and short term actions to work on these perspectives are necessary. In agriculture in The Netherlands and elsewhere there are currently many economic and environmental problems. This situation is further aggravated by trends towards liberalization of world trade, a demanding environmental policy, technological progress and fewer price subsidies for agricultural production. The next four decades will bring a considerable growth in the world's population and an increasing welfare. This will lead to a sharp increase in the demand for food.

If sustainable development is to become a reality, the environmental efficiency (i.e. the use of energy, space and raw materials) of current agricultural production methods must be increased with a factor of 20 by 2040. This is a challenge for a technological breakthrough.

In the contribution 'New Perspectives: Sustainable Technological Development in Agriculture' two sustainable perspectives and action agendas, which are studied at the Dutch research program Sustainable Technological Development (STD), will be discussed. The first perspective is the development of protein foods which are attractive to both consumers and manufacturers. These products must be able to meet in the future consumer demands in the same way as meat does, but their production in environmental terms must be at least 20 times more environmentally efficient than current meat production in The Netherlands.. Research points out that in 2040 novel protein foods will occupy 40% of the meat market.

The second perspective, sustainable land use, is focusing on the rural area. The essentials of this perspective are the integration of "new" functions like the gaining of water and energy, the processing of organic waste and the management of nature and culture into the production of food and raw materials for industry. This should lead to new business opportunities both for farmers and other parties like energy and water companies and a production which in environmental terms is 10 times more efficient.

Market parties, research-institutes, government bodies and non governmental organizations are working together to realize the action agendas and try to achieve these perspectives.

1. INTRODUCTION

The next four decades will see a growth in the world's population from 6 billion to some 10 billion. Welfare will increase at the same time by an average of 2% per year in OECD countries and by 4-6% in Third World countries, the latter rise leading to a more evenly spread demand for food in the world as more people eat meat. And to produce one kilo of meat requires four kilos of plant food. The quotation of figure 1 shows the consequences.

	EB	=	EB/Pr	*	Pr/P	*	P
1990	1	=	1	*	1	*	1
2040	0,5	=	1/20	*	5	*	2

EB = Environmental Burden
EB/Pr = Environmental Burden per unit of prosperity
Pr/P = Prosperity per capita
P = Population

Figure 1.1. The relations that determine the environmental burden

Twice as many people, a five-fold increase in worldwide prosperity and a reduction in environmental pollution, all by the year 2040. This means that we are going to need to become twenty times as efficient in our treatment of energy, raw materials, space and other environmental factors. Thus we need to become twenty times as efficient in meeting our food supply needs. This amounts to a substantial technological challenge. (Weterings en Opschoor, 1992).

If in the twenty-first century we wish to be assured of an economically feasible and ecologically sustainable food supply, now is the time for us to be developing a new outlook, and daring to invest in the future. We need to be making long-term investments, in policy-making, in research and in development programs.

In the "Choosing the Opportunities" memorandum the government of The Netherlands puts a strong emphasis on technological development in order to harmonize economic and ecological objectives. In the memorandum and the subsequent debate reference is made to a three-way approach: continual improvement, redesign of products and processes, and reconsideration.

Continual improvement refers to the small forward steps like the catalysator or, in food terms, the reduction of the use of remedial substances. Redesign should draw our attention to plants which can themselves counter insects and fungi. Reconsideration is the rethinking of functions that products and services perform, for instance the obtaining of necessary protein by means other than the eating of meat. The interdepartmental research program Sustainable Technological Development has targeted the route of reconsideration by formulating the future needs of society: a leap forwards not only in the technological sense but especially in terms of social change.

2. PROGRAM SUSTAINABLE TECHNOLOGICAL DEVELOPMENT

In the last four years, research in the context of interdepartmental research program Sustainable Technological Development (or STD) has been carried out with the aim of ascertaining (Jansen en Vergragt, 1992; Vergragt en Jansen, 1992):

> widely supported possibilities there are for becoming twenty times as efficient in meeting social needs;

> to what extent it is possible to halt current trends and initiate a long-term sustainable technological development.

At this stage, the program is clearly research-oriented, rather than being concerned with policy-making. Learning by doing. The process of carrying out research and learning from its results has been organized into twenty projects, which in turn fall under the sub-programs Food, Transport, Housing, Water and Chemistry.

The ultimate goal of STD is to have started with sustainable technological development before the end of 1997 in order to give substance to the process of reconsideration. Based on a number of credible examples STD wants to demonstrate that it can be economically attractive to make provisions on the long term with much less pressure on the environment than is now the case. Other criteria for success are not only financial and other support but also the adoption of the results by companies, governmental organizations, research institutes and the government. A practical derivative of the program is a manual for the development of sustainable technology projects.

The program STD is intended to function as a starter motor: it is a question of looking at long-term goals, and with those in mind, taking the initial steps in the direction of sustainable technological development. This initial steps will include policy-forming, research activities and investment plans. In so doing a method, it is referred to as

Backcasting, is developed to ensure that the starter motor will work effectively.

The long-term goal, the social implications and the large number of parties involved in developing sustainable technology tend to mean in practice that it is the government which often has to take the initiative.

3. CURRENT SITUATION IN AGRICULTURE

The Food sub-program of the Sustainable Technology Development research program aims to contribute to the development of sustainable technology in food production. Sustainable in many respects.

Sustainable means for the consumer that he can count on an uninterrupted supply of healthy safe and tasty foods which fulfil his physiological and emotional needs. From the economic point of view sustainable means that food provision has a firm economic base and contributes to incomes. Ecologically, sustainable means that environmental pressure from food production is significantly reduced.

There are a number of problems inherent in the current food supply situation in The Netherlands. Not only environmental problems, but also socio-economic problems, such as a fall in the income of farmers and threats to the quality of rural subsistence. Table 2.1 shows the entire food production chain for various product groups. The vertical line shows the different product groups, and the horizontal line shows each link in the chain. The spots indicate the degree of environmental damage caused by a particular link. The more spots, the more serious the environmental problem.

Product groups and their respective contributions to environmental strain

Product group	primairy production	transport & storage	processing	transport distribution	prepa-ration	waste
meat	●●●●●●	●◐	●●◐	●◐	●●	●●●●●●
fish	●●●●	●◐	●●●	●◐	●●	●●
beverages			●●●●	●		●●
sugar	●●●●	●◐	●●●●	●		●●
potatoes	●●●●●	●●●	●●●●	●	●	●●
grains	●●	●●●	●●●		●	
vegetables	●●●●	●●●	●●●●	●◐	●●	●
fruit	●●●	●●	●●●●	●		●
oils en fats			●●●		●●	●
dairy	●●●●	●●	●●●●	●●	●●	●●●●

●● Current situation
● Autonomous trends

Source: Basic document product group tables

Interdepartemental Research Programme
Sustainable Technological Development

Table 2.1 Product groups and their respective contributions to environmental strain in The Netherlands

In the case of meat production, the problems related to matters such as the processing of manure and the amount of physical space required for feed production. Space, not only in The Netherlands, but also in foreign countries where cattle feed is grown. In the case of the production of greenhouse vegetables, the problems relate to energy use and related issues, such as the greenhouse effect. In primary production as a whole, problems include acidification and manure surpluses.

Taking into account forecasts of consumer demands in 2040 and the current environmental and economic problems for which a solution has to be found, the sub-program Food has developed the possible options Sustainable Land Use, High-Tech Agroproduction and Novel Protein Foods.

4. WHAT WILL BE EATEN IN 2040?

To develop technology for sustainable food and the associated networks we have to gain a picture of where we are heading. Not a blueprint but a charcoal sketch of the needs of society in forty years. This assumes that these needs can be fulfilled while environmental pressure is substantially reduced. However, to prevent the perspective from being pie in the sky, we first have to see the attraction of the first step before we invest.
In the area of food supply, a vital factor in developing a sketch of the needs of society is the anticipation of consumer behavior. Much is known about current consumer behavior in the Western world. Certain aspects of consumer behavior are extremely difficult to predict, since it is subject to capricious, spur-of-the-moment decisions. But it is possible to pick out a number of general trends.
Consumer research has shown that the consumer of the twenty first century will be increasingly inclined to demand high quality food which tastes good, and is safe, healthy and reliable. Any food which is considered to be either unhealthy or unreliable will be forced out of the western dietary package. Health, and therefore products promoting good health, will become increasingly important. In The Netherlands, the rising number of older people in the population will make this trend even more marked.
Demands regarding the way in which food is produced will also play an increasingly important role in purchasing decisions. These demands relate, for example, to environmental and Third World concerns. Consumers will wish to be supplied with information regarding where products come from and the production methods used. Information technology and extensive knowledge of the entire production chain will therefore become extremely important.

Alongside an increase in the consumption of ready-to-eat meals, it is also reasonable to assume that demand will increase for quality products such as locally produced, ecologically grown products.

5. PERSPECTIVES ON SUSTAINABLE TECHNOLOGICAL DEVELOPMENT IN AGRICULTURE

5.1. Sustainable Land Use

In the Sustainable Land Use project a perspective has been developed for the rural area, the core of which is the integration of "new" functions like the gaining of water and energy, the processing of waste and the management of nature and culture with the production of food and produce for industry.

In the Sustainable Land Use project several main functions per plot have been targeted. Multiple land use saves space and benefits both nature and the environment. Moreover, it gives the land user more possibilities to earn income. The integration begins at plot level but has consequences for other levels too. Corporate structures and control mechanisms will change as investors other than farmers and conservationists vie for the rural areas.

Multiple land use makes the rural areas attractive for new investors like drinking water and energy companies. If ground and surface water is no longer subject to contamination and if this can be maintained, it will become possible to obtain more and cleaner water, on top of which incidence of drought will decrease. Gaining energy can also have its place in multifunctional land use. Think of the biomass crop and solar and wind energy. A further function of the rural area is waste processing that offers the farmer savings (less fertilizer) and is a compensation for the reduction of residues. In terms of the value of nature the quality of the countryside will increase as pressure on the environment and the threat of drought are reduced. A more varied landscape is also more interesting from a recreational point of view.

The Sustainable Land Use vision is currently being applied to an area of just under twenty thousand hectares around the town of Winterswijk in the eastern part of The Netherlands. Ten practical projects are being worked in order to stimulate the acquisition of the knowledge and technology we currently lack, and also to implement new forms of cooperation. Fundamental research questions will be imbedded in research programs in the knowledge infrastructure.

It is expected that by combining existing functions, total environmental pollution could be reduced by a factor of ten, and the environmental problems related to primary production could be alleviated. At the same time is expected that there will be more rural incomes then autonomous developments predict. (IP-DL 1997, p.m.).

5.2. High Tech Agroproduction

The High Tech Agroproduction project fits in with the knowledge and experience of the Dutch horticulture sector. Various interested parties, including the farming community, the processing industry and providers of energy from the growing base of the project. The focus of the project is on the demand-focused production of fresh vegetables in enclosed production units, under controlled conditions and in close geographical proximity to the consumer. It will be possible in the future, for example, for sprouts to be produced in this type of 'factory', the advantages being that production will be more than twenty times more environmentally friendly than is currently the case within glasshouse horticulture and that it will be possible to respond to the precise demands of the consumer.

The aim of the High Tech Agroproduction project is to design an enclosed production system. It is important to develop systems in which the only raw materials are sunlight, water and residues. All other materials must be recyclable. This will require the acquisition of a broad international knowledge base, which can then be coordinated by the Dutch centers of expertise. To this end contact has been made with research organizations and businesses in the United States, Japan, Germany, Australia and the United Kingdom. (IP-HTA 1997, p.m.).

5.3. Novel Protein Foods

In this project, research has been carried out into the future development of new protein foods which are attractive to both consumer and producer, and which meet the same requirements as meat, but the production of which is twenty times as environmentally efficient as current meat production. It is not the intention to replace meat altogether, but to achieve a significant market share for environmentally friendly new protein foods (Novel Protein Foods or NPF's) by the year 2035. These products should be able to occupy their own position in the food market. In the next chapter the case Novel Protein Foods will be worked out more in detail.

6. THE NOVEL PROTEIN FOODS CASE

At the moment, meat is a very important source of protein in our diet, and it meets a wide range of requirements in such areas as taste, habit and status. Unfortunately production places a heavy strain on the environment in The Netherlands through a surplus of manure, harmful emissions from pesticides and large use of energy, space and raw materials. Apart from the growing problem of manure disposal and recent discussions surrounding diseases like BSE and swine fever, the Dutch meat sector is also under pressure from increasing foreign competition and a growing demand on the part of the consumer for new meat-replacement products. The latter in particular is proving a challenge to Dutch businesses. They are faced with a choice: either they can remain passive, and leave foreign competition to capture and develop these new markets, or they can decide to take up the challenge and exploit the opportunities presented by these developments. In real terms, there is no choice: the challenge must be taken up.

In order to achieve a significant market share for Novel Protein Foods instead of meat, a research and development plan was drawn up. This plan had to be attractive to both consumers and producers and must be accepted as feasible by scientists and technology developers, if the relevant market parties are to continue down the path set by this project (STD, 1996). In order to develop such a plan for NPF's in the twenty-first century research was carried out in five different lines environmental aspects, consumer, technology, business economics and macro-economics. Therefore, five research questions have been posed as an integral part of this project (see Figure 6) (De Haan et al, 1995).

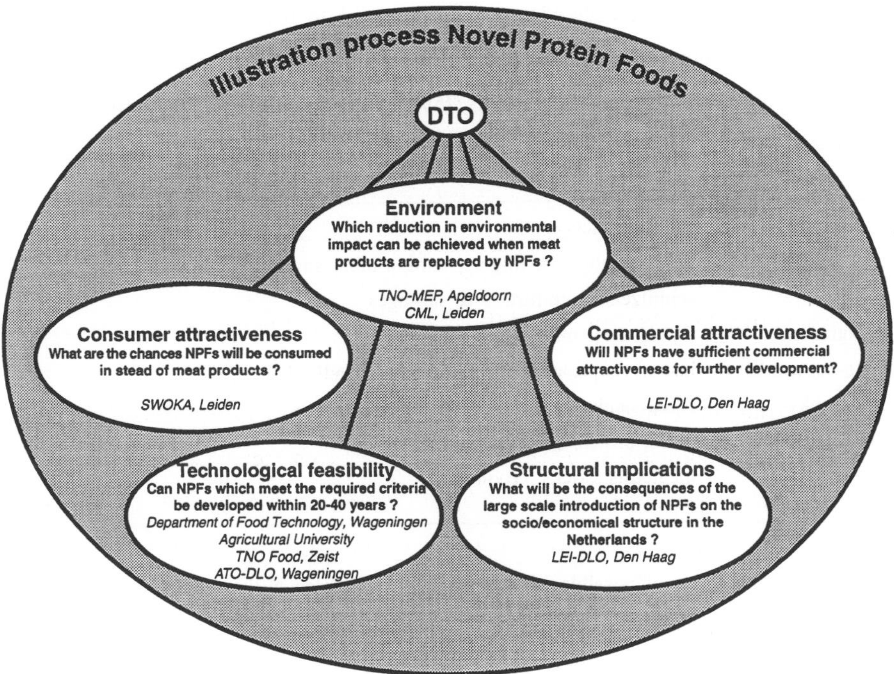

Figure 6. The research questions and structure of the Novel Protein Foods project

In the following paragraphs the main results of the Novel Protein Foods project and the next steps in NPF-development are discussed.

6.1. The consumer: the deciding factor is taste

It became evident in the course of consumer research that it is the taste of NPF's which will determine whether they are purchased and consumed or not (Hamstra en Verhoeven, 1995; Baggerman en Hamstra 1995; Fonk en Hamstra 1996). In addition a trend analysis was carried out including a prediction of consumption and purchasing behavior in 2035. An important assumption in this trend analyses is that in the future it will be possible to manufacture NPF products the taste, smell and structure of which are considered by the consumer to be as good as, or better than meat.

Consumer research also revealed that consumers will be reluctant to give up quality cuts of meat. NPF's are therefore most likely to function as a replacement for compound meat products and in the processed sector as opposed to the unprocessed sector, which consists of straight cuts of meat (see Table 6.1). This led to forecasts that the processed sector will grow from its present 45% to 75% in 2035, whilst in the same period the share of straight cuts of meat will fall from 55% to 25% (Baggerman en Hamstra 1995).

meal category	NPF products
meal component	burgers, nuggets, fingers, cordon-bleu, souffles, patties, schnitzels, frankfurters & sausages
meal ingredient	frying meat mix, strips, cubes
ready-to-eat-meals	hearty soups, vegetables, rice and noodle dishes
savory snacks	filled sandwiches, hot dogs, dried sausages
sandwich fillings	pates, pastes, spreads, cold cuts
appetizers	meatballs, sticks, croquets

Table 6.1. A number of possible product concepts in which NPF's could be included as an ingredient (Quist et al. 1996).

On the basis of these same consumer trends and assumptions, it has been predicted that it must be possible for NPF's to gain more than half of the processed segment, reducing consumption of products such as sausage meat and minced meat in particular. This represents a market share of 40%. Our assumptions regarding the behavior of the consumer of the future were of course of crucial importance in carrying out trend analysis and making predictions about diet and purchasing behavior in the future.

Several parties were involved in seminars lasting several days, extensive desk studies and interviews with consumers, the results of which helped us to draw up a picture of the consumer of the future. These parties included representatives of the business world (both research and marketing), research organizations (both institutes and universities), government and non governmental organizations (representing the interests of the environment and the consumer) (Fonk and Hamstra, 1996).

In the short-term, there are plenty of options for the inclusion of NPF's in main meals. In the longer term, changing eating patterns will lead to an increase in the importance of meat substitutes in snacks and ready-to-eat products (Fonk and Hamstra, 1996; STD

1996) In actual fact, NPF's will not be introduced onto the market as meat substitutes, but as new protein products. Products which meet the demands of the consumer of the future: healthy, easy to prepare and reliable!

6.2. New technologies make NPF's possible

In principle hundreds of different NPF's could be produced from around 20 protein-source/technology combinations. A stepwise selection process was followed to identify those NPF's most likely to succeed in the market. These NPF's scored well in areas such as technological attractiveness, consumer attractiveness, commercial attractiveness, the degree to which they would reduce environmental damage and macro-economic and institutional effects (de Haan et al., 1995).

The selection process finally led to the identification of 3 ingredients which can be produced in 7 different ways (referred to as the 7 options) from peas, lucernelupin, the fungus *Fusarium* and the cyanobacteria *Spirulina*.

The three ingredients which have been given fancy names, are Protex, Fibrex and Fungopie. Protex is a product resembling to minced meat in structure and made from either Spirulina, Pea or Lucerne. Fibrex is a fibrous ingredient and made by the continuous fermentation of *Fusarium*. Fungopie is a fermented ingredient and resembling temper in structure. It can be made from either Pea or Lupin by fermentation with the fungus *Rhizopus*. In Table 6.2 is shown the 7 high-potential options, each representing the combination of a particular protein source with a particular technology.

Ingredients		Protein source
Protex	1	Spirulina (cyano bacterium)
an ingredient resembling	2	Pea
can be made form bacteria,	3	Genetically modified pea
yeasts and plants	4	Lucreme
Fibrex		
a fibrous ingredient		
produced by continuous	5	Fusarium
fermentation of fungi		
Fungopy		
an ingredient produced by	6	Pea with the fungus Rhizopus
fermenting plants with fungi	7	Genetically modified lupin with the fungus Rhizopus

Table 6.2 The selection of 7 high-potential NPF-options, each representing the combination of a particular protein source with a particular technology

Although most of the technologies involved are already being used in current production techniques, it has not been possible yet to produce products which are sufficiently attractive to the consumer to cause a significant reduction in the amount of meat consumed. By relating consumer demands to product quality demands it was possible to identify the areas in which new - and in some cases, basic - knowledge must still be gained (Sijtsma et al 1996a). This areas are sensory sciences (Sijtsma et al, 1996b), nutritional value (Jansen et al., 1996a), an increase in scale to large-scale production (Jansen et al., 1996b) and a further reduction in environmental strain (Linnemann et al., 1996). Research programs have been planned to fill these gaps. They are to be carried out between now and the year 2010.

6.3. NPF's are far more environmentally friendly than meat

Life Cycle Analysis (LCA) was carried out in order to be able to calculate and compare the environmental strain imposed by the production of NPF's on the one hand and pork on the other (Van den Berg et al, 1996). This method takes into account and quantifies - as far as is possible - all the factors in the entire production process which are in any way damaging to the environment.

For every environmental theme in the result was standardized in relation to the world situation. All standardized effects were added together to produce an environmental index. It was then possible to determine the degree of environmental damage caused by each of the selected NPF options, and to compare them with the environmental damage caused by the production of pork, based on current technological conditions. In Table 6.3. the result is shown of the environmental register for the selected NPF options compared to pork.

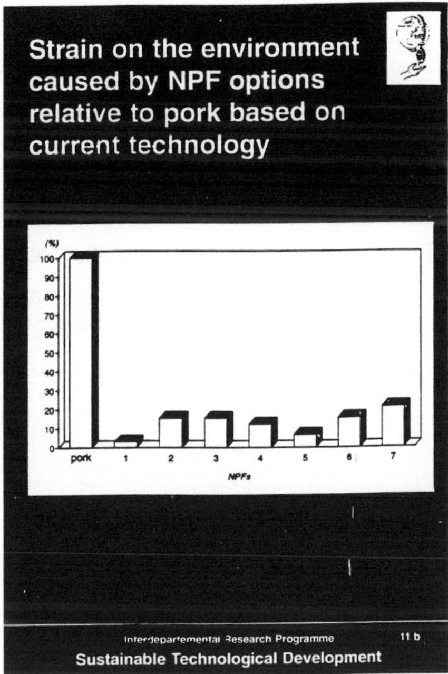

Table 6.3. The environmental register for the selected NPF options compared to pork as an ingredient (pork is 100%) based on the production processes prevailing in 1995 (Van den Berg et al, 1996).

In Table 6.3. the numbers 1, 2, 3 and 4 refers to options for the Protex ingredient, with respectively a cyanobacteria, a normal pea, a genetically manipulated pea and alfalfa as protein sources. The option 5 refers to a Fibrex ingredient, using a fungus as protein source and 6 and 7 are options for the Fungopie ingredient, with respectively, a pea and a genetically manipulated lupin as the protein sources, both in combination with a fungus.

According to the latest estimates, the NPF options Spirulina for Protex (option 1) and Fusarium for the ingredient Fibrex (option 5) are expected to achieve a reduction factor of more than 20, whilst the remaining options are expected to achieve a reduction factor of between 6 and 13.

6.4. NPF's are commercially attractive

Research in the area of business economics has revealed that NPF's are far less expensive to produce than meat (Reinhard et al., 1996). Depending on the precise NPF, production is expected to be between 20% and 50% less expensive (See Table 6.4. Estimated costs per ton NPF in 2035). These relatively low production costs, combined with the fact that the market share is expected to increase from 5% in 2005 to 40% in 2035 make NPF's an extremely attractive business proposition. Especially for those food companies which are already active in that part of the chain which leads from protein source to end product.

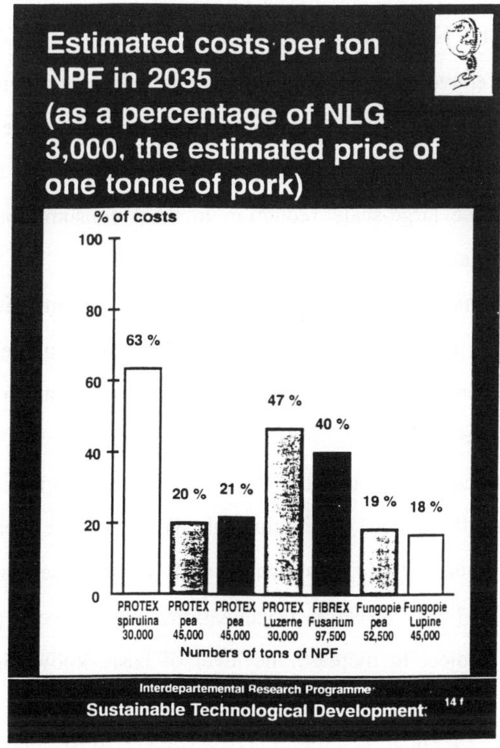

Table 6.4. Estimated costs per ton NPF in 2035 (as an percentage of DFL 3,000 the estimated price of one ton of pork), (Reinhard et al., 1996)

The development costs entailed in meeting the high quality standards and building up the expertise required in the areas of sensory sciences, nutritional value, upscaling and a reduction in environmental strain make it impossible for any one business to develop any given option single-handed. But where the development costs relate to more than one option and can be carried by a number of businesses throughout the chain working together, investment in NPF development can be a commercially attractive option.

Research has shown that if 40% of the meat will be replaced by NPF's that will not have a dramatic effect on future economic structure compared to possible autonomous developments. There will be nevertheless a negative effect on part of the meat sector. On the other hand, the environmental benefits will be considerable (Jahae et al., 1996, De Vlieger et al., 1996).

6.5. Next steps: First NPF's on the market

The NPF project has shown that the development and large-scale introduction of NPF's would be possible in the future, and would contribute to sustainable development. NPF's are attractive to both consumers and producers; the structural effects are relatively limited and if they cause large-scale reduction in meat consumption they will lead to a significant reduction in environmental damage.

A process to replace meat as an ingredient to a large extent with alternative protein products has now been initiated. It is now up to the market parties involved to ensure that the process is carried through. Fortunately that does appear to be happening as the following three examples will show.

The first example is that most elements of the research program have been taken on board by the recently founded Technological Top Institute for Food Sciences in Wageningen. Within this institute, companies such as Unilever and Gist-brocades are working together with those research organizations which were also involved in the Novel Protein Foods project to increase the level of basic knowledge concerning Novel Protein Foods.

The second example is that two companies are working together with research organizations on the development of NPF products. Their aim is to develop knowledge to a sufficient extent to allow the introduction of NPF products onto the market within the foreseeable future. The last example is the cooperation between research bodies such as TNO-Nutrion, the DLO-institute for Agrotechnological Research and Wageningen Agricultural University in a number of follow-up research projects. Next to these joint projects, the various institutes have also started work on individual projects. TNO-Nutrion, for example, has embarked on an extensive six-year research project into Future Protein Foods.

These and other follow-up activities which fit within the development plan being carried out are supporting the supposition that the program STD has succeeded in fulfilling its role as a starter motor. Research, development work and the introduction of new products can now proceed through existing channels.

7. CONCLUSION

This article leads to the following three conclusions. The first conclusion is that sustainable technological development is not a luxury but a necessity, if we wish to avoid problems in the future and if we wish to ensure that in the twenty-first century the world's environmental problems and other problems related to food supply are not as great as they are now. The second conclusion we would like to draw is that STD has development and tested a method by which concrete steps can be taken towards the initiation of sustainable technological development. Now Sustainable technological development has become possible. The third and final conclusion is that projects carried out within the STD framework have demonstrated that there are a number of high-potential visions. These and other visions point the way to an economically feasible and ecologically sustainable means of supplying food.

In order to ensure that the initiatives taken by STD are carried through the organizations involved, such as businesses, government, social organizations and centers of expertise must be prepared to invest in and become involved in long-term development. There must be also an organization which can initiate and facilitate new projects independent from existing structures, until such time as they are strong enough to be incorporated into existing research or investment programs - this comes down once more to the starter motor function. And there must be sufficient funding for the carrying out of new projects.

The learning time is now over, and the time for harvesting has arrived. The government would be wise to make use of the lessons learned within the context of the STD program. The STD program should in fact be continued, not in the form of an experimental learning program but as a long-term policy program. A program aimed at the continuous development and stimulation of new ideas and the stimulation of technical and social research. It is a tool for redesigning products and processes and for reconsideration of the way we fulfill our needs. The result of the NPF and other projects shows it will be an effective and efficient program to realize more synergy between environment and economy.

REFERENCES

1. Weterings and J.B. Opschoor, 'The eco-capacity as a challenge to technological development' Advisory Council for Research on Nature and Environment, 1992.
2. Vergragt and L. Jansen, 'Sustainable Technological Development: the making of a long-term oriented technology program', Project Appraisal, vol.8, no. 3 pp. 134-140, 1992.
3. Jansen and Ph.J. Vergragt, "Sustainable Development: a Challenge to Technology!", proposal for the STD-program, 1992.
4. STD-report Novel Protein Foods in 2035: lekker eten in een duurzame toekomst, STD-program, Delft, The Netherlands, 1996.
5. de Haan, H. Hermans, O. de Kuijer, I. Larsen, H. Linsen and J. Quist, 'Kansrijke NPF's als ingrediënten voor toekomstige eiwithoudende voedingsmiddelen', STD-publication VN10 (in Dutch), 1995.
6. Hamstra en P. Verhoeven, 'De betekenis van vlees en andere eiwitprodukten voor consumenten, STD-publication VN8 (in Dutch), 1995.
7. Baggerman and A. Hamstra, 'Motieven en perspectieven voor het eten van Novel Protein Foods in plaats van vlees, 1995. STD-publication VN9 (in Dutch), 1995.
8. Fonk en A. Hamstra, "Toekomstbeelden voor Consumenten van Novel Protein Foods", STD-publication VN12 (in Dutch), 1996.
9. Quist, O. de Kuijer, A. de Haan, H. Linsen, H. Hermans, Ivo Larsen, 'Restructuring meat consumption: Novel Protein Foods in 2035', paper at the Greening of Industry Network Conference, Heidelberg, November 24-27, 1996.
10. Sijtsma, M. Rabenberg, R. Janssens, A. Linnemann, 'Sensorische aspecten van Novel Protein Foods', STD-publication VN13 (in Dutch), 1996.
11. Janssens, L. Sijtsma en A. Linnemann, 'Voedingswaarde van Novel Protein Foods', STD-publication VN14 (in Dutch), 1996.
12. Janssens, A. Linnemann, M. Rabenberg, L. Sijtsma, 'Proceseisen bij grootschalige produktie van NPF's', STD-publication VN15 (in Dutch), 1996.
13. Linnemann, L. Sijtsma, R. Janssens, 'Milieutechnische potentie van grondstoffen voor NPF's', STD-publication VN16 (in Dutch), 1996.
14. Sijtsma, R. Janssens, A. Linnemann, 'R&D-programma's voor de ontwikkeling van NPF's' STD-publication VN17 (in Dutch), 1996.
15. Reinhard, R. Koster en G. Boers, 'Ontwikkeling van NPF's door bedrijven', STD-publication VN21 (in Dutch), 1996.

16. van den Berg, G. Huppes, B. van der Ven, B. Krutwagen, 'Novel Protein Foods: milieu-analyse van de voortbrengingsketen', STD-publication VN18 (in Dutch), 1996.

17. Jahae en K. Wijnen, 'Belang en ontwikkelingen Nederlandse vleessectoren, STD-publication VN5 (in Dutch), 1995.

18. de Vlieger, I. Jahae, T. van Gaasbeek, J. van der Hoek, M. van Leeuwen, K. Wijnen, 'Substitutiescenario's en modelontwikkeling voor NPF's', STD-publication VN19 (in Dutch), 1996.

Global policy on sustainable agriculture: a 2020 vision

P. Pinstrup-Andersen and
Rajul Pandya-Lorch [1]

Recent research by IFPRI and others show that the extent of future food insecurity, hunger, and unsustainable exploitation of the environment will depend on appropriate economic and social policies rather than on absolute limitations of the earth's carrying capacity. However, failure to design and implement appropriate policies and other action now to assure sustainable management of natural resource may lead to a situation where the earth's carrying capacity is reduced to a level where it does become the limiting factor to the well-being of future generations. The current large number of food-insecure and malnourished people combined with the risk to future generations call for action now. This paper focuses on the policies and related action required.

Enough food is being produced around the world today that nobody should have to go hungry. Yet, more than 800 million people go hungry, 185 million preschool children are seriously underweight for their age, and diseases of hunger and malnutrition are widespread (FAO 1996; UN ACC/SCN 1992). During the next quarter century, about 80 million people will be added every year to the world's population (UN 1996), the largest annual population increase in history. Assuring food security for the current population as well as for future generations without degrading the environment is a fundamental challenge confronting farmers and policymakers around the world. The widespread food insecurity, unhealthy living

[1] Director General and Special Assistant, respectively, of the International Food Policy Research Institute (IFPRI), 1200 Seventeenth Street, N.W., Washington, D.C., 20036-3006, U.S.A.

conditions, and abject and absolute poverty in many developing countries today are already threatening global stability. Poor, hungry people who are marginalized in economic processes and disenfranchised in political processes are desperate people. Failure to assure sustainable food security will foster the very conditions that will further destabilize and polarize the world in the years to come, with tremendous consequences for all people.

Prospects for reducing malnutrition among the world's children are grim. One-third of all children under the age of five years are malnourished (UN ACC/SCN 1992). Close to 100 million of these 185 million malnourished children are in South Asia, while about 30 million are in Sub-Saharan Africa (Table 1). Projections to the year 2020 (Rosegrant, Agcaoili-Sombilla, and Perez 1995) suggest that, under the most likely or baseline scenario, the number of malnourished children could decrease to 155 million or 25 percent of the preschool children population. This baseline scenario incorporates the best assessment of future growth in population, income, and productivity of staple crops and livestock. Large decreases in the number of malnourished children are expected in South and East Asia, but in Sub-Saharan Africa their number could increase by 50 percent to reach 43 million.

While the baseline may be the most likely outcome, action that is taken now and in the next few years will greatly influence the outcome in 2020. Two scenarios illustrate the opportunities associated with alternative actions. In a pessimistic scenario of slow growth and low investment — a situation where nonagricultural income growth has been reduced by 25 percent, international investment in national agricultural research systems has been eliminated, and direct core funding of the international agricultural research system has been phased out — the number of malnourished children could increase to 205 million in 2020

(Table 1). The increase will be particularly pronounced in Sub-Saharan Africa. In an optimistic scenario of rapid growth and high investment — a situation where nonagricultural income growth has increased by 25 percent, national and international agricultural systems have been strengthened, and investments in public goods such as health and education have increased by 20 percent — the number of malnourished children could decline to 109 million, about 50 million less than in the baseline scenario. However, even in this more optimistic scenario, the number of malnourished children in Sub-Saharan Africa is projected to increase relative to the 1990 level.

We do not have to accept the "most likely" scenario of large magnitudes of child malnutrition and food insecurity. The conditions that will assure that all people are fed without damaging the environment can be created if we take the necessary action. Continuing with business as usual is certain to lead to persisting hunger and poverty and to continued degradation of the environment, catalysts for an increasingly unstable world.

It is, in the best of cases, going to be a tremendous challenge to achieve the 2020 Vision. In the quarter century between 1995 and 2020, world population is expected to increase by about 35 percent to a total of 7.7 billion people (UN 1996). About 98 percent of the population increase is expected in developing countries, whose share of world population will exceed 80 percent by 2020. While the absolute increase will be largest in Asia (1.2 billion), the rate of growth will be most rapid in Sub-Saharan Africa where the population could double to 1.2 billion in 2020. Although the global population growth rate is slowing down and is projected to reach 1.0 percent by 2015–2020, compared to about 1.5 percent in

1990–1995, Africa's projected population growth rate of 2.2 percent will be more than twice that of other regions.

Most of the population increase in the next 25 years is expected in the cities. Rapid urbanization could more than double the urban population in developing countries to 3.6 billion by 2020, by which time urban dwellers could outnumber rural dwellers (UN 1995). While the rural population will continue to grow, the growth rate will be much greater in urban areas. It is of critical importance that investment in rural areas be accelerated. About 80 percent of the developing world's poor live in rural areas. There is still a window of opportunity to solve the poverty and nutrition problems in rural areas before they become urban problems, but that window is gradually closing.

Developing countries are projected to increase their cereal demand by about 80 percent between 1990 and 2020, while the world as a whole will increase its cereal demand by about 55 percent (Figure 1). Meat demand in developing countries will increase by a staggering 160 percent, and world meat demand will increase about 75 percent. The percent increase in demand for roots and tubers will be slightly lower than that for cereals. These large increases will put tremendous pressures on future agricultural production and marketing and, unless current policies are changed, on the environment.

The projected increase in the demand for cereals, meat, and roots and tubers varies significantly among developing-country regions (Figure 2). Sub-Saharan Africa is projected to increase its demand for these three commodity groups by at least 150 percent. Of note is the very rapid increase in meat demand in Asia.

So, how much of the demand is likely to be fulfilled through developing-country production? In the early 1990s, developing countries had net cereal imports—the difference between consumption and production—of around 90 million tons. Rosegrant, Agcaoili-Sombilla, and Perez (1995) project that these will increase to about 190 million tons by 2020. Because Sub-Saharan Africa is expected to continue its poor production performance relative to population growth, its net import requirements for cereals are projected to triple during this period.

The composition of these additional imports is shown in Figure 3. IFPRI research suggests that the net cereal import requirements of developing countries in 2020 will consist primarily of wheat and maize. There will also be a very large increase in net imports of meat in response to more rapid economic growth in developing countries, especially Asia.

Assuming that the projected production and import requirements are correct, per capita food availability will increase in all regions, but the increase will be very small in Sub-Saharan Africa (Figure 4). By 2020, average daily calorie consumption per person in Sub-Saharan Africa will still be only about 2,100 as compared to 3,000 calories in Asia and 3,500 calories in the developed countries. The largest improvement is likely to occur in Asia, and there is—unfortunately—strong evidence to suggest that some of this improvement will result in increasing obesity and related health problems in that part of the world.

Notwithstanding the rapid increases in maize and wheat prices during 1995 and the first half of 1996, we believe that the long-term trends for real food prices will continue to fall. In fact, maize and wheat prices decreased very significantly during the second half of 1996.

As Figure 5 shows, prices for wheat, rice, maize, beef, and roots and tubers are projected to fall significantly in real terms between now and 2020.

World grain stocks have decreased markedly during the last 10 years (Figure 6) although they have recuperated slightly from a low of about 13 percent of annual world consumption in mid-1996. Rapidly falling cereal prices during the 1980s and early 1990s have contributed to the falling stock levels. Changes in the European Common Agricultural Policy and the GATT agreement have also contributed to lower stocks, and world grain stocks will be considerably lower in the future than they have been in the past. This is likely to be reflected in the availability of food aid, which is currently about one-half of the level it was four years ago. Lower future grain stocks may imply larger price fluctuations in the future, because the buffer available in periods of bad weather and production shortfalls in general will be smaller.

FUTURE FOOD PRODUCTION

While the world is far from approaching the biophysical limits to food production, there are indications that growth in food production has begun to lag. For instance, food production increases did not keep pace with population growth during the 1980s and early 1990s in 49 developing countries with a population of one million or more (FAO 1995). The annual rate of growth of global grain production also dropped from 3 percent in the 1970s to 0.7 percent during 1985–95. In addition, yields of rice and wheat have been constant for the last few years in Asia, which is a major producer (Pinstrup-Andersen 1994).

World cereal production is projected to grow on average by 1.5 percent per year between 1990 and 2020, meat production by 1.9 percent, and production of roots and tubers by 1.4 percent (Rosegrant, Agcaoili-Sombilla, and Perez 1995). Production growth rates are expected to be substantially higher in developing countries than developed countries. Cereal production is projected to grow at an average annual rate of 1.9 percent in developing countries (compared to 1.0 percent in developed countries), meat production at 2.9 percent (compared to 0.9 percent in developed countries), and production of roots and tubers at 1.7 percent (compared to 0.8 percent in developed countries). Aquaculture production, which doubled between 1984 and 1992, is projected to increase at a slower rate between 1990 and 2020, and marine fish catches are likely to be no higher than current levels in 2020 (Williams 1996).

REQUIRED POLICIES AND RELATED ACTION

The 2020 Vision initiative[3] has identified six priority areas of action, including policy changes, in order for global food needs to be met without damage to the environment (IFPRI 1995). These are:

(1) Selective strengthening of the capacity of developing-country governments;

(2) Investing more in poor people;

(3) Accelerating agricultural productivity;

(4) Assuring sound management of natural resources;

---

[3] Information about the 2020 Vision initiative is available from IFPRI.

(5) Developing competitive markets; and

(6) Expanding and realigning international development assistance.

The action needed will require changes in behavior, priorities, and policies, and it will require developing and strengthening the needed relationships between individuals, households, farmers, local communities, nongovernmental organizations (NGOs), national governments, and the international community. Each country must design its action program; the six priority areas of action identified here should serve as a point of departure for designing country-specific strategies.

The first priority area of action is to selectively strengthen the capacity of developing-country governments to perform appropriate functions such as establishing and enforcing property rights, promoting private-sector competition in agricultural markets, and maintaining appropriate macroeconomic environments. Predictability, transparency, and continuity in policymaking and enforcement must be assured. Governments must also be assisted to get out of areas that are best handled by the private sector or civil society. In many countries, NGOs have come to play a much more important role in areas traditionally covered by government, such as poverty relief, health care, nutrition, and management of natural resources. For the 2020 Vision[4] to be realized, the efforts and contributions of NGOs and other elements of civil society must be fully recognized and supported, and a more effective distribution of labor between government and civil society, including NGOs, be achieved.

---

[4] The 2020 Vision is *a world where every person has access to sufficient food to sustain a healthy and productive life, where malnutrition is absent, and where food originates from efficient, effective, and low-cost food systems that are compatible with sustainable use of natural resources* (IFPRI 1995)

The second priority area of action is to invest more in poor people. One billion people lack access to health services. 1.3 billion people consume unsafe water. Almost 2 billion do not have access to adequate sanitation systems. One-third of primary school enrollees drop out by Grade 4. For a large share of the world's population to be malnourished, illiterate, sick, and without resources is not only unethical but wasteful. Governments, local communities, and NGOs must assure access to primary education, primary health care, and clean water and sanitation for all people. They must work together to improve access by the poor to productive resources and remunerative employment.

The rate at which population grows in developing countries is one of the key factors conditioning when and whether the 2020 Vision is realized. Strategies to reduce population growth rates include providing full access to reproductive health services to meet unmet needs for contraception; eliminating risk factors that promote high fertility, such as high rates of infant mortality or lack of security for women who are dependent on their children for support because they lack access to income, credit, or assets; and providing young women with education. Female education is among the most important investments for realizing the 2020 Vision.

The third area of action is to accelerate agricultural productivity. Agriculture is the lifeblood of the economy in most developing countries. In the lowest-income countries, it provides up to three-quarters of all employment and half of all incomes. Agriculture has long been neglected in many developing countries, resulting in stagnant economies and widespread hunger and poverty. Yet, there is considerable evidence that East Asia's rapid economic growth in recent years has been facilitated by a vibrant and healthy agricultural sector that supported the nonagricultural sector.

Agricultural growth and development must be vigorously pursued in low-income developing countries for at least four reasons: (1) to alleviate poverty through employment creation and income generation in rural areas; (2) to meet growing food needs driven by rapid population growth and urbanization; (3) to stimulate overall economic growth, given that agriculture is the most viable lead sector for growth and development in many low-income developing countries; and (4) to conserve natural resources. Poverty is the most serious threat to the environment in developing countries: lacking means to appropriately intensify agriculture, the poor are often forced to overuse or misuse the natural resource base to meet basic needs.[5]

Existing technology and knowledge will not permit production of all the food needed in 2020 and beyond. National and international agricultural research systems must be mobilized to develop improved agricultural technologies and techniques, and extension systems must be strengthened to disseminate the improved technologies and techniques. Developing countries must increase their national agricultural research expenditures in the near term to 1 percent of the value of agricultural output with a longer term target of 2 percent. Interaction between public-sector agricultural research systems, farmers, private enterprises, and NGOs must be strengthened to assure relevant of research and appropriate distribution of responsibilities. A clear policy on and agenda for biotechnology research that focuses on the problems of developing-country farmers must be developed.

---

[5] See Per Pinstrup-Andersen and Rajul Pandya-Lorch (1995), 2020 Brief No. 15, for further discussion of these issues.

The fourth priority area of action is to assure agricultural sustainability and sound management of natural resources. Governments, NGOs, and local communities must work together to establish and enforce systems of rights to use and manage natural resources, to improve the way water is allocated and used, to reverse land degradation where it has occurred, to reduce the use of chemical pesticides, and to implement integrated soil fertility programs in areas with low soil fertility. Local control over natural resources must be strengthened and local capacity for organization and management improved. Investments in less-favored geographical areas, that is, areas with agricultural potential, irregular rainfall patterns, fragile soils, and many poor people, must be expanded. About one-half of the world's poor people reside in less-favored areas. Yet, most investment, including agricultural research investment, is still focused on the more-favored areas. If reducing poverty and protecting the environment are serious goals, the balance between less-favored and more-favored areas must be redressed. Poverty and environmental degradation are closely linked, often in a self-perpetuating negative spiral in which poverty accelerates environmental degradation and degradation results in or exacerbates poverty.

Continuing to neglect the less-favored, vulnerable areas where many of the world's poor live will make degradation worse and perpetuate poverty. Whereas the long-term solution for some of these areas may be outmigration, most countries cannot accommodate the movement of large numbers of mostly poor and uneducated people in the short term. While failure to address the problems effectively in the less-favored areas themselves will accelerate degradation, outmigration transfers poverty and population pressures to urban areas and rural areas with better natural resources. There is growing evidence that agricultural intensification in fragile lands is possible and that degraded natural resources can be rehabilitated. Accelerated investments in agricultural research and technology, rural infrastructure, family planning, education, primary health care, and appropriate policies are urgently needed to eradicate extreme poverty and associated food insecurity and environmental degradation. Agricultural research and resulting technologies can simultaneously increase food production and protect the environment. There does not have to be a trade-off between meeting future food demands and maintaining the natural resource base.[6]

The fifth priority area of action is to develop competitive markets. As a result of inefficient markets and poor infrastructure, the cost of bringing food from the producer to the consumer is very high in many low-income countries, particularly in Africa. Governments should phase out inefficient state-run firms, invest in or facilitate private sector investment in developing and maintaining infrastructure, especially in rural areas, and provide technical assistance to help strengthen small-scale competitive rural enterprises.

The sixth priority area of action is to expand and realign international development assistance. Many years ago, industrialized countries agreed to allocate at least 0.7 percent of their GNP to foreign assistance. Most countries do not maintain this target. Besides increasing international development assistance to reach the 0.7 percent target, it must be realigned to low-income developing countries, primarily in Sub-Saharan Africa and South Asia where the potential for further deterioration of food security and degradation of natural

---

[6] See Per Pinstrup-Andersen and Rajul Pandya-Lorch (1995), 2020 Brief No. 29, as well as Peter Hazell (1996) for a fuller discussion of these issues.

resources are great. Developing countries in turn must seek measures to diversify sources of external funding and to stem capital flights.

CONCLUSION

If the global community does not get its act together soon, hunger and malnutrition and resulting illnesses will persist, natural resources will continue to be degraded, and conflicts over scarce resources such as water will become even more common. For most of humanity, the world will not be a pleasant place to live. Yet it does not have to be this way. With foresight and decisive action, we can create a better world for all people. We have the resources to do so; let us act while we still have choices.

REFERENCES

Alexandratos, N. (ed.). 1995. *World agriculture: Towards 2010.* Rome, Italy: Food and Agriculture Organization of the United Nations.

FAO (Food and Agriculture Organization of the United Nations). 1996. Food, agriculture, and food security: Developments since the world food conference and prospects. World Food Summit. Technical Background Document No. 1. Rome: FAO.

________. 1995. FAO agrostat-pc, production domain. Rome, computer disk.

Hazell, P. 1996. Sustainable development of less-favored lands: IFPRI's research agenda. Notes prepared for International Centers Week. Washington, D.C.: International Food Policy Research Institute, October.

IFPRI (International Food Policy Research Institute). 1995. *A 2020 vision for food, agriculture, and the environment: The vision, challenge, and recommended action.* Washington, D.C.: International Food Policy Research Institute.

Pinstrup-Andersen, P. 1994. *World food trends and future food security.* Food Policy Report. Washington, D.C.: International Food Policy Research Institute.

Pinstrup-Andersen, P., and R. Pandya-Lorch. 1995. *Poverty, food security, and the environment.* 2020 Brief No. 29. Washington, D.C.: International Food Policy Research Institute, August.

________. 1995. *Agricultural growth is the key to poverty alleviation in low-income developing countries.* 2020 Brief No. 15. Washington, D.C.: International Food Policy Research Institute, April.

Rosegrant, M. W., M. Agcaoili-Sombilla, and N. D. Perez. 1995. *Global food projections to 2020: Implications for investment.* Food, Agriculture, and the Environment Discussion Paper No. 5. Washington, D.C.: International Food Policy Research Institute.

UN ACC/SCN (United Nations Administrative Committee on Coordination—Sub-committee on Nutrition). 1992. *Second report on the world nutrition situation*, Volume 1. Suffolk, England: The Lavenham Press Ltd. for the United Nations ACC/SCN Secretariat.

UN (United Nations). 1996. *World population prospects: The 1996 revision.* New York: United Nations.

________. 1995. *World urbanization prospects: The 1994 revisions.* New York: United Nations.

USDA (United States Department of Agriculture). 1995. *Grain: World markets and trade.* Foreign Agricultural Service Circular Series FG 8-95, August. Washington, D.C.: United States Department of Agriculture.

Williams, M. J. 1996. *The transition in the contribution of living aquatic resources to food security.* 2020 Discussion Paper No. 13. Washington, D.C.: International Food Policy Research Institute.

Table 1—Number of malnourished children in developing regions, 1990 and 2020

	1990	2020		
		Baseline[a]	Slow Growth & Low Inv.[b]	Rapid Growth & High Inv.[c]
		(millions)		
West Asia and North Africa	6.76	6.30	11.05	1.87
Latin America	11.71	8.12	13.23	3.12
Sub-Saharan Africa	28.61	42.67	52.75	33.61
East Asia	41.45	24.70	35.58	14.26
South Asia	95.81	72.94	92.54	56.01
Developing countries	184.34	154.73	205.15	108.87

Source: Rosegrant, Agcaoili-Sombilla, and Perez (1995).

[a] The baseline scenario incorporates the best assessment of future trends in population, income growth, urbanization, rate of increase in food production due to technological change and productivity growth, commodity prices, and response of supply and demand to prices.

[b] The slow growth and low investment scenario simulates the combined effect of a 25-percent reduction in non-agricultural income growth rates and reduced investment in agricultural research and social services.

[c] The rapid growth and high investment scenario simulates the combined effect of a 25-percent increase in non-agricultural income growth rates and higher investment in agricultural research and social services.

Figure 1—Percent increase in total demand, 1990-2020

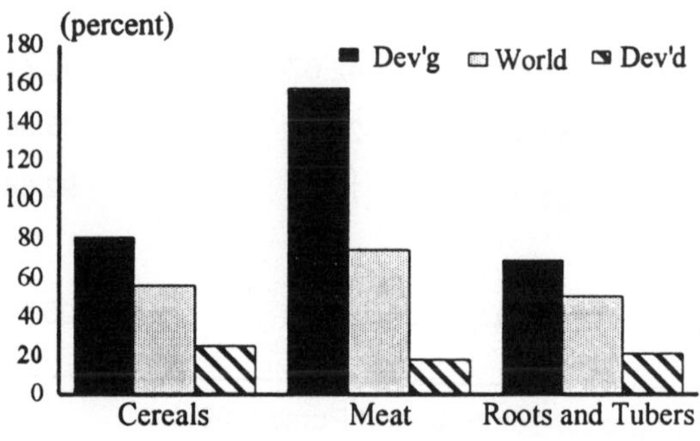

Source: Rosegrant, Agcaoili-Sombilla, and Perez (1995).

Figure 2—Percent increase in total demand in developing regions, 1990-2020

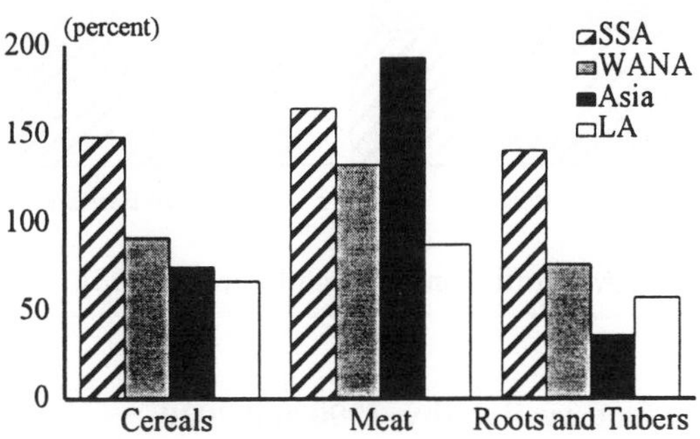

Source: Rosegrant, Agcaoili-Sombilla, and Perez (1995).

Figure 3—Composition of imports

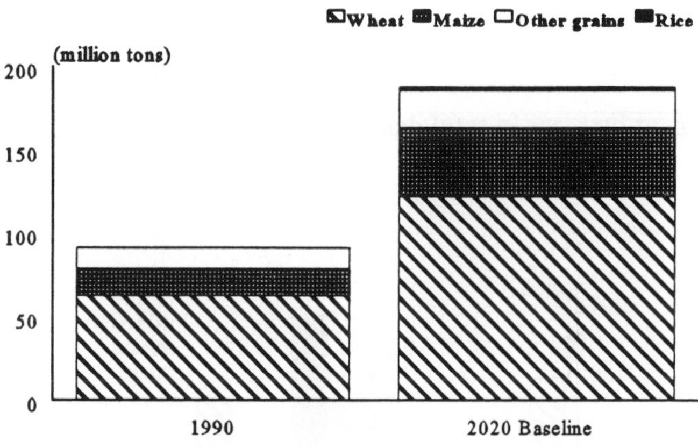

Source: Rosegrant, Agcaoili-Sombilla, and Perez (1995).

Figure 4—Per capita food availability, 1990 and 2020

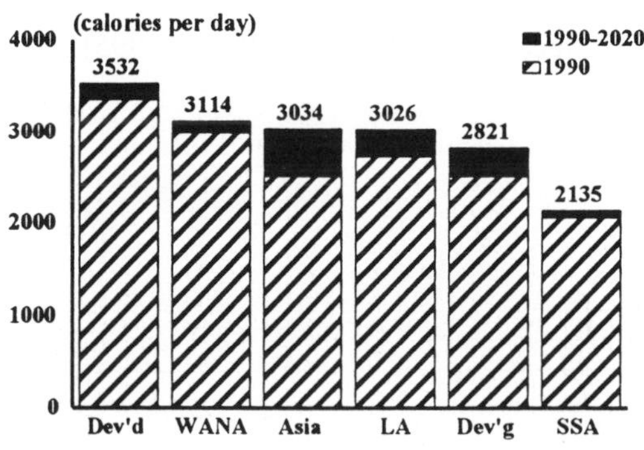

Source: Rosegrant, Agcaoili-Sombilla, and Perez (1995).

Figure 5—Projected world prices (in 1990 dollars)

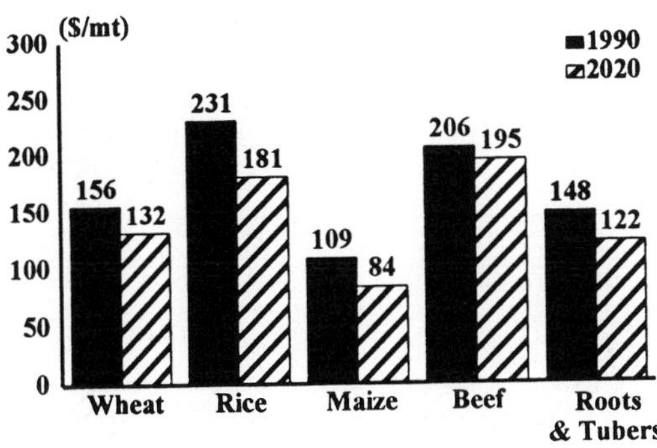

Source: Rosegrant, Agcaoili-Sombilla, and Perez (1995)
Note: Beef prices are \$/100 kilograms.

Figure 6—Global grain stocks: Level and percent of consumption, 1968/69–1996/97

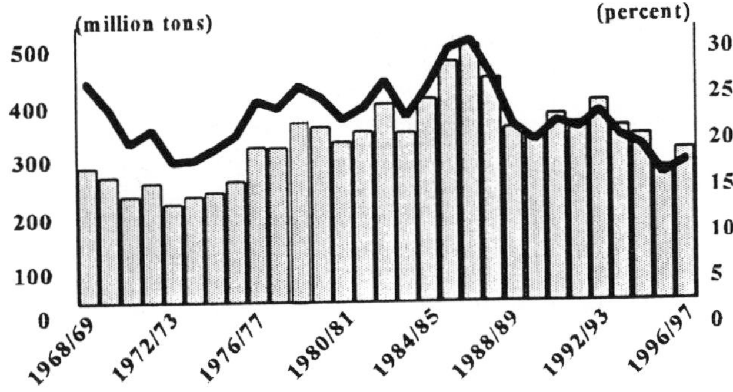

Source: USDA (1996).

Strategies for future agricultural control policies from the Dutch perspective

drs. R.J.T. van Lint and drs. J.F. Tanja

Ministry of Housing, Spatial Planning and the Environment,
Directorate Water Supply, Water, Agriculture

ABSTRACT

The Dutch agricultural sector is of significant importance to the Dutch economy, but is also one of the major contributors to the total environmental pollution of the country. At the moment the environmental policy for this sector is centered around two major issues, manure/ammonia and pesticides. The manure-surplus causes pollution of soil, air, ground- and surface water with phosphates, nitrates and ammonia. And although emissions of pesticides are being reduced, the quality of the surface-water is not improving.

Under the influence of an increasing market-orientation on both national and international level, a growing insight in the need of regional differentiation and a growing environmental awareness outside the Ministry of the Environment, the way environmental policies are developed is changing in The Netherlands. More and more, parties concerned are involved in an early stage of policy-development, resulting in a policy oriented on objectives that requires a high level of selfregulation instead of a policy based on strict rules and set measures.

1. INTRODUCTION

In the past decades environmental policy regarding the agricultural sector has been developed. This policy has been successful to a certain extent in certain field, but the objectives that have been set will be difficult to meet. That is the reason why an evaluation of these policies is necessary. In the perspective of some major developments on both national and international scale, new strategic lines are formulated.
This paper presents a short overview of the main environmental issues in the agricultural sector, a set of specific developments on both national and international level that have a specific influence on the Dutch environmental policy-development, new strategic lines that have been formulated and the results in some branches of the agricultural sector so far.

2. MAIN ENVIRONMENTAL PROBLEMS

The Dutch agricultural sector consists of 100.000 companies, covering 1.8 million hectares of land, more than half of it covered with grass. Production and incomes differ per branch, but with an export balance alone of over 10 billion a year the agricultural sector makes a significant contribution to the Dutch economy.
Unfortunately, they also make a significant contribution to the total environmental pollution of the country; for example, 90 % of the ammonia-emissions and 50% of the emissions of methane and di-nitrogen oxide can be attributed to agricultural sources.
At this moment the environmental policy for the agricultural sector is centered around two major issues: manure/ammonia and pesticides. A third important issue is energy. Other issues, like contamination with heavy metals, non-CO_2 greenhouse gases, NO_x, noise, odor and waste-management are either dealt with on a larger scale or have less political priority at the moment. This does not mean the agricultural sector is not doing anything about them.

Manure and ammonia

Considering the size of the country, The Netherlands have an enormous cattle-population (mainly cows, chickens and pigs) that far exceeds the human population. Besides producing milk, meat and eggs, these cattle also produce an abundance of manure, that cannot possibly be disposed of in an environmentally sound way. Over-fertilization has led to levels of phosphates and nitrates in the ground that are far too high, causing leaching to ground- and surface water.

Other sectors, like the industry and wastewater treatment plants, also cause pollution with nitrates and phosphates. In the past few years, however, environmental measures have strongly reduced the emissions from these sources.

The agricultural sector has also taken steps to reduce emissions, for example by banning the spreading of manure in autumn and winter and reducing the maximum amounts of manure used per acre. This resulted in reductions of 25% for phosphates and 16% for nitrates in 1995, compared to 1985. Compared to the industry and wastewater plants, however, they have been less successful, and a major effort has to be made to meet the objectives set in the 1995 'Policy Document on Manure and Ammonia'.

As said before, 90% of the total ammonia emission of The Netherlands can be attributed to agricultural sources. Although the negative effects of ammonia are known, it is difficult to make a direct connection between the ammonia-emissions of one particular farm and the ammonia-deposition on a nearby natural area. Without scientific proof, ammonia-reducing policies are not widely supported by the agricultural sector, making them very hard to enforce.

A reduction of 25% has been reached in the past few years, as a result of a set of specific regulations, e.g.:

- Covering of manure-storage tanks.
- Low-emission application of manure, preferably by injecting the manure into the ground. This can reduce emission of ammonia by as much as 95%.
- Low-emission housing, using technical improvements to reduce emission from stables. The techniques are rather expensive, but in due time such stables should replace the old ones.

It has already become clear that the objectives set out in the 'Policy Document on

Manure and Ammonia' (a reduction of 70% compared to the levels of 1980 in the year 2000-2005) will not be reached.

As said before, the objectives for nitrates, phosphates and ammonia will be very hard to meet. The result, on the other hand, will not be good enough to meet environmental standards. The current objectives for nitrate, for example, secure the quality of newly formed groundwater under 85% of the agricultural acreage, but European objectives for surface-water will not be reached.

Pesticides

In 1993, a management-agreement with the sector was reached on the multi-year crop protection plan. This long-term agreement sets different objectives for different years, allowing the sector enough time to take the necessary steps to reduce both the amount used and the level of emission. Evaluation of the results so far seems positive; the set reductions for the use of pesticides for 1995 have been met. Unfortunately, these figures are flattered, a closer look gives a far less positive result.

It is true that the use of pesticides has decreased by 45%, whereas the target for 1995 was only 30-35%. This decrease, however, is mainly caused by tightening up the rules on soil-detoxication. In other areas levels have reduced as well, but not enough to meet the targets. The emission of herbicides decreased 13%, and not the 31% government and sector agreed on. The use of insecticides, fungicides and other agents decreased 4% instead of 23%, and even those 4% are flattered, since mineral oils have been taken into account; the use of fungicides has actually not decreased at all since the introduction of the multi-year plan.

Although emissions have decreased considerably towards soil, air, ground- and surface water, the quality of the surface-water has not significantly improved.

Energy

In developing environmental policy, energy-reduction is increasingly taken into account as one of the environmental objectives that have to be met. As a result, energy-efficiency has grown considerably in the last few years in specific branches like the glasshouse horticulture-branch. The energy-use of the sector as a whole, however, is still rising, making CO_2-reduction objectives very hard to meet.

3. SETTING

The way in which the Dutch government develops and executes her environmental policy is influenced by a number of developments on both national and international scale. Although these developments are different in their line of approach, they are in a sense closely linked, and result in a "system" where target groups more and more take their own responsibilities, allowing the government to limit its role in the longer term more to the establishment of the framework within which the target groups can exercise their selfcontrol, and facilitating others in taking up their responsibilities.

Market-orientation

On both national and international level orientation on the world market is a rising phenomena. This has considerable consequences for the agricultural sector. In the past, farmers were strongly production-oriented, working for a good harvest that would fetch a good price on auction or with the cooperative. In the dairy branch, for example, an individual farmer, a cooperation, a dairy factory, they would all be just links in a chain of production who didn't really bother about the rest of the chain.

More and more they become market-oriented. This leads to a strong tendency of all parties concerned to concentrate on the total chain of production and to find a way to control this chain. To gain a position on the market and to keep that position, the image of the total branch becomes more important. Some branches turn out to be better equipped for this than others, like the dairy farmers and the glasshouse horticulture-branch. Although economically not always well off, they are better organized and more flexible, enabling them to react faster to the demands of the market.

Environmental legislation plays a complex role in this process. Environmental measures are expensive, thus putting pressure on the prices. It is hardly possible to pass the extra costs for environmental measures on in the prices they get for their products. On the contrary, prices are falling while costs keep rising. environmental measures tend to be the last drop, and therefore a cause for unrest and protest. The only way farmers are able to meet environmental demands made by both customers and government, is if they are given the scope to solve the problems in their own way, and fit them into their own pace of investment. The more market-oriented a branch becomes, the better it responds to a policy of selfregulation.

Regional differentiation

Up till now, environmental legislation has been made to fit the entire country. More and more it's becoming clear that this might not be the best way to do it, for the simple reason that circumstances differ too much from region to region. A national norm on nitrates that could be met easily on the claysoils in the north might turn out to be nearly impossible to meet on the dry sandsoils in the east of the country, and even if it is met it might not have enough result to reach environmental objectives.

Legislation has to be feasible for a farmer, without losing sight of the economic basis of his enterprise. If environmental measures will cost too much they cannot be implemented.

The farmer is not the only one asking for scope, so are the local and regional authorities like provinces, municipalities and water boards. They want to have the possibilities to adapt the general rules to fit their special situation, and are willing to take responsibility for those adaptations. More and more, the central government enables them to take this responsibility, indicating objectives and frameworks rather than strict rules and measures. For example, an Ammonia Reduction Plan gives a region a certain amount of freedom to indicate how the objectives set out in the Interim Act on Ammonia and Stock Farming will be met. This might result in stricter rules in some areas that are more vulnerable to environmental pollution. A certain limitation of freedom should be made by the national government, to reduce the risk of extreme differences between rules and regulations from region to region and to secure equality of rights.

Regional differentiation is also something that will have to be considered on an international level. At this moment, Europe tends to make general legislation and set general objectives that have to be met by all the member states. The geological and environmental differences between those states are of course enormous; standards that might work out fine in the dry interior of Spain will probably turn out disastrous in the wet parts of The Netherlands. This results in rules that will have to be far more strict in one country than another, which will have an effect on the competitive position of the countries.

"Environment-included" thinking

In the past decades, care for the environment and environmental legislation have found a place in society. It becomes more and more self-evident to consider environmental issues in all fields and actions. Houses are built with environmentally friendly materials, cars drive on unleaded gasoline, and waste is segregated for recycling in over 80% of the household kitchens. The object of the Ministry of the Environment is that environment is included in the way we think and the way we live. It should become included in the policies of the government, on national, provincial and local level. Other departments should take the responsibility to include environmental issues in their policies and legislation, without direct intervention by the Ministry of Environment. The Ministry of Agriculture, Nature Management and Fisheries, for example, has got the responsibility for the legislation on manure, including the environmental issues involved. What is more, the agricultural sector itself starts to include environmental issues and objectives in their plans and policies.

This changes the role of the Ministry of the Environment. Instead of initiating and developing environmental policies in different fields, the Ministry has to participate, facilitate, stage-manage if necessary, but to allow others to take the responsibility.

In itself this is of course a very positive development, but it does contain a risk. Environment is not the only issue taken into consideration, and interests will be weighted. If a choice has to be made between a certain level of economic growth and the protection of the environment, environmental measures might be considered too expensive. It will have to be the task of the Ministry of the Environment to take care of the formulation of clear objectives and to monitor and evaluate environmental results, to make sure a balance is kept and environmental issues will not be sacrificed for economic growth and financial security.

While the Ministry of the Environment takes on another role, the traditional negotiating partner of the environmental movement disappears, forcing them to change their approach. This results in an increasing amount of direct contact between environmental pressure groups and agricultural organizations or small groups of farmers, to foster environmental action. Sometimes the environmental pressure-groups more or less force the farmers around the table, threatening with lawsuits or consumer-oriented actions, sometimes the farmers themselves invite the environmental pressure groups to come and talk and try to solve the differences. In all cases, the objective is to find solutions that will benefit both farmer and environment. This might lead to rather small-scale projects, like the protection of birdsnests, it might also result in large scale research project (for example into environmentally friendly ways to grow flower bulbs) or collective plans on regional environmental protection.

Internationalization

The international aspect has been mentioned before. There is a growing recognition of the fact that we have a collective responsibility for sustainable development on a global level. This causes an increasing tendency to tackle environmental issues on an international scale. Sometimes this might mean cooperation with one or two countries on a particular issue, like the cooperation between Belgium, Germany and The Netherlands to tackle ammonia-problems. In other cases, the scale is larger, like the ECE protocols on NO_x and POP's, or the international negotiations on climate change.

Agricultural environmental issues don't have high priority yet on the international stage, but this is bound to change, considering: 1) the amount of transboundary pollution caused by the agricultural sector, 2) the need of harmonization of environmental policy to guarantee a minimum level of environmental quality to all citizens an 3) the need of an equal competitive playing field.

Being a member of the EU, the Dutch government is committed to follow European legislation and objectives and fit them into the national legislation. For example, the European Directive on Nitrates forces the Dutch to take extra measures that will reduce the leaching of nitrate from dry sandsoils, since the objectives set in this Directive cannot be met by the current Dutch legislation.

4. STRATEGIC LINES

The social and political processes set out in the previous paragraph have a strong effect on the environmental policy for the Dutch agricultural sector. This is a field where the call for regional differentiation grows louder, just as the call for more influence on the objectives and means to get there.

In the past few years, ideas about how to develop environmental policies have changed considerably. This resulted in a new way of working, that combines two main elements: 'external integration' and 'selfregulation'. At the basis of these two concepts are the growing environmental awareness and the call for more scope to adapt rules and regulations to specific situations, as described above. Integral approaches and the use of new market opportunities are essential.

'External integration' basically means integrating environmental issues into the policies of other 'parties', both on the side of the government and the agricultural sector, and making them take the responsibility to provide for the quality of the environment. In order to do that, they will need the opportunity to 'regulate themselves', finding ways to fit environmental issues into daily routines and regulations. The best way to do this is to involve all parties concerned in an early stage, so the policy can be develop as a kind of co-production.

The parties involved usually include, apart from the Ministry of the Environment, several other ministries like the Ministry of Agriculture, Nature Management and Fisheries and the Ministry of Transport, Public Works and Watermanagement. Local and provincial authorities, water boards and the like are the ones responsible for granting the permits, making local plans or maintaining a certain environmental quality standard. They want to be included so they have the possibility to adapt rules to the local situation, asking for extra measures when necessary and the option to be more lenient when possible.

The agricultural sector will have to deal with the objectives and regulations, and find a way to fit them into the way they run their business.

To ensure a balance between economical and environmental interests, environmental pressure-groups are more and more invited to take part in this process.

In such a system of policy-development by co-production, the role of the Ministry of the

Environment will change. Instead of initiating and developing environmental policies, the Ministry has to participate, facilitate and stage-manage the process if necessary. The Ministry has to take care of the formulation of clear agreements on measurable results, and of the monitoring and evaluation of the environmental results. In the perspective of the market-orientation, the awareness and use of the total chain of production from producer to consumer may be stimulated by instruments like an environmental certificate, and agreements between agricultural organizations and environmental pressure groups or consumer-organizations.

The development of a policy on an international scale will also be the responsibility of the central government.

5. TWO CASES

To give a good idea of how an open process might work and what the results can be, two short cases will be presented.

First, the voluntary agreement with the glasshouse horticulture-branch, where the process until now has been very successful.

Second, the policy on pesticides. This is an example with a different character, since the process so far has not been an entirely open one. It did however contain certain elements of co-production. Here an agreement with the sector has lead to certain results, but the objectives for 2000 will not be met. Additional measures are necessary and it's not clear yet if negotiations on these measures will be successful.

Covenant glasshouse horticulture-branch

For the past two years municipal, provincial and national government, water boards, agricultural and environmental organizations have been working on a covenant for the glasshouse horticulture-branch.

The first initiatives for this covenant came from the branch itself. They found themselves confronted with an increasing amount of environmental legislation and environmental targets they could not fit into their individual conduct of business. Instead of general

legislation, they wanted a tailor-made system that would suit individual companies.
The authorities on the other hand had to deal with a sector that causes a high level of environmental pollution and a set of rules that was difficult to enforce. Together they've tried to improve the system of environmental legislation for this particular sector in such a way that both environment and sector will benefit.

At the basis of the system stands a set of 'Integral Environmental Targets' (IET), to be met in 2010. Covering issues ranging from pesticides and water management to light-pollution and the use of energy, these are the targets that have to be met by the sector as a whole. How they are going to do that is more or less up to them; the government does not dictate any specific environmental measures that have to be taken, unless there's really no other option. This IET will then be translated into targets for individual farmers, based upon their individual circumstances and the crop they grow. Farmers who choose to take part in this system make a Company Environmental Plan (CEP). In this CEP the farmer sets out what measures he is going to take to tackle the different environmental problems in his own company, and the results these measures will have on the amount of environmental pollution. The CEP forms the basis for the company's environmental permit. Results have to be monitored, and the farmer has to report these results to the authorities. Some environmental problems will be easier to improve on in his particular company than others. He will have to produce results on all environmental issues, but can make up for small results at one issue with better results on another one. If it turns out he does not reach his personal targets, the CEP will have to be adjusted. This system enables farmers to concentrate on their own strong points, and to fit environmental measures into their business-planning. It also links up with quality-control systems already used by the sector, reducing the amount of paperwork for both farmer and government and enabling the farmers to promote themselves as environmental-friendly.

There are still some juridical discussions going on about how to fit this system into the existing legislation. Only a few years ago a system had been set up with general rules for the entire sector. These rules might not be ideal, but neither is a system of individual permits for every company. This brings high costs for both local authorities and farmers, not to mention the amount of paperwork.

The idea right now is to apply the general rules, unless a farmer indicates he wants a more tailor-made approach; in that case he can make a CEP, and apply for a permit. The general rules will function as a safe-guard, to make sure the targets will be met even if the system would not work out.

Pesticides

From the 1950's onward, the use of pesticides has greatly increased all over the world. Thanks to chemical pest control, the Dutch crop cultures had become more intensive and consequently more susceptible to pests and diseases. As a result, Dutch agriculture became highly dependent on pesticides. In the second half of the 1980's, an average of 10 kg of pesticides per hectare was used. Some substances were starting to loose there efficacy, and resistance problems were increasing.

The necessity of an active restrictive policy became even more evident when in the second half of the 1980's traces of pesticides were found in drinking-water. All of a sudden, pesticides became an issue of public concern and was raised on the political agenda.

This resulted in the joint development of a new pesticides policy by the ministries of Agriculture, Nature Management and Fisheries; Housing, Spatial Planning and the Environment; Welfare, Health and Cultural Affairs; Social Affairs and Employment, while the ministry of Transport, Public Works and Water Management was involved when necessary. They aimed to develop a policy that would take both socio-economic and environmental factors into account. This resulted in the Multi-Year Crop Protection Plan.

The main objectives of this plan were the reduction of the dependence on pesticides, a reduction of the use of pesticides with 50% and of pesticide- emissions to air of 50%, to soil and groundwater of 75% and to surface-water of 90% by the year 2000, with interim targets for the year 1995. On top of that authorization of the most harmful products would have to be withdrawn.

The Multi-Year Crop Protection Plan was sanctioned by Parliament in 1991, and by 1993 the government and agribusiness concluded a management agreement on the execution of this plan. With this agreement, the agricultural sector subscribed to the objectives and targets set out in the plan, and to an active contribution to various implementation activities.

In 1996, the results were evaluated for the first time, and - as mentioned before - at first glance the policy seemed successful. A closer evaluation, however, showed that the objectives set for 1995 were not met at all, and that it would not be likely that current measures were enough to reach the objectives set for 2000. It is clear that extra measures will have to be taken to improve on these results, either by the agricultural sector or - if they will not or cannot do that - by the government. At this moment, the sector has promised to work out new initiatives. If they appear not to be satisfying, the central government will take it's own responsibility. Regulations and / or financial instruments may be considered.

An effective national solution is complicated by a number of factors, particularly in the area of reducing the range of authorized substances. One of the main objectives of the Crops Protection Plan is to reduce the dependence on pesticides. In order to do that it is essential that the farmer has a wide range of substances to choose from, so he can solve emergencies with as little use of chemicals as possible. The criteria used in the authorization process, however, are rather strict and will lead to a range of authorized substances that is too small to meet this objective.

Plant diseases and pests are transboundary, spreading both actively and passively. Passive spread has strongly increased as a result of international trade in agricultural products. Strict national legislation is not only ineffective in this field, it also influences the competitive position of the Dutch farmers. Considering the disparities among the EU member states in climate, production conditions and environmental requirements, international legislation does not seem to be forthcoming in the near future. To accelerate this process an active role in the EU is necessary.

FUTURE OUTLOOK

Despite all the rules and regulations set for the agricultural sector in The Netherlands, solving the environmental problems caused by this sector turned out to be rather difficult in the past years. The strategic lines set out in the previous paragraphs might be a solution, if these lines are continued and improved in the future.

Integration of environmental values in the policies of other Ministries and local authorities will have to be stimulated even more. Selfregulation, as worked out for the glasshouse horticulture-branch, should be extended to other branches where it might have a good chance to solve the problems, like the dairy-farming branch. The continuing market orientation and growing focus on the total chain of production could be used, and environmental agreements between the agricultural sector and environmental organizations should be fostered. Instruments like an environmental certification system deserve more attention. The same holds true for the us of financial instruments like levies.

Environmental issues will have to be tackled on a larger scale, both national and international, in connection with policies on spatial planning, renovation of the countryside and reorganization of the agricultural sector.

REFERENCES

Along with a certain amount of non-published information, the following documents have been used in composing this paper:

1. RIVM, "Milieubalans 1996. Het Nederlands milieu verklaard", Alphen ad Rijn, 1996
2. Ministry of Agriculture, Nature Management and Fisheries, "Policy document on manure and ammonia", The Hague, 1995
3. Ministry of Agriculture, Nature Management and Fisheries, "Environmental policies in agriculture in The Netherlands. Nutrients and pesticides", The Hague, 1993
4. Ministry of Agriculture, Nature Management and Fisheries, "Essentials: Multi-Year Crop Protection Plan", The Hague, 1990
5. Ministry of Housing, Spatial Planning and the Environment, "The Netherlands' national environmental policy plan 2", The Hague, 1994

New environmental guidelines for IFC projects

A.D. Fitzgerald

Consultant, Environment Division, International Finance Corporation, Washington, DC

Introduction

The International Finance Corporation (IFC)—the private sector investment arm of the World Bank—and the World Bank[1] use environmental guidelines when financing projects in developing countries. Guidelines were written in 1982-'83 and published in 1984, and were republished in 1988 but with no substantial changes. Other guidelines were written by IFC in the early '90s because by that point in time IFC felt that the 1984/88 guidelines were out-of-date and did not meet the needs of the private sector projects in which it was investing—essentially the 1984/88 guidelines were based on late '70s and early '80s production and environmental control technologies, and reflected the command-and-control regulatory approach. A commitment was made in 1993 to start the process of writing new environmental guidelines that would eventually replace both the 1988 guidelines which are still used by the Bank, and the series of guidelines that IFC had written for its projects, and which continue to be used by IFC. When completed, the new guidelines will apply to all operations of the World Bank Group.

Methodology Used in Developing the Guidelines

The guidelines specify the maximum emission values for new projects that are financed by the World Bank Group. In developing these maximum emission levels it was agreed that the following factors would be considered: the protection of human health; acceptable mass loadings to the environment; the use of commercially available/proven technologies (production and control); regulatory trends; cost effectiveness of pollution prevention and control systems; and the use of Good Industrial Practices. Brief comments on each of these follow:

- **The Protection of Human Health:** The guidelines are designed to protect human health to the extent that current technology allows. In some cases emissions levels selected for the guidelines were derived by back-calculating from target ambient

[1] The World Bank consists of the International Bank for Reconstruction and Development (IBRD) and the International Development Agency (IDA), however IBRD is commonly referred to as the World Bank or the Bank. IBRD, IDA, IFC and MIGA (the Multilateral Investment Guarantee Agency) together constitute the World Bank Group.

levels recommended by internationally recognized organizations such as the World Health Organization (WHO). Extensive literature on Good Industrial Practice was reviewed to determine maximum emission levels which are considered acceptable to protect human health. In addition, it was recognized that site-specific conditions may warrant a lowering of the maximum emission numbers in order to protect human health—the environmental assessment for the project would address this situation. In cases where short term exposure is of concern, the guidelines recommend actions that can be taken. For example, short term effects from sulfur dioxide might be controlled by the burning of cleaner fuel at a power plant.

- **Acceptable Levels of Mass Loadings to the Environment:** The World Health Organization's publication "Management and Control of the Environment", the US EPA's "Compilation of Air Pollutants Emissions Factors (AP-42)", the European Union's "Atmospheric Emission Inventory Guidebook", and other sources were used to develop an emissions inventory of the major pollutants of concern. The guidelines stress good environmental management programs which, at a minimum, will maintain loading rates in the air and water sheds below the removal rates for these major pollutants.

- **Commercially Proven Technologies:** The guidelines recommend production and control technologies which are commercially available, which are resource and energy efficient, and which minimize the type and quantity of pollutants generated. In situations where pollutant release levels are greater than acceptable health based levels then best demonstrated technology may be required.

- **Current Regulatory Trends:** In developing the guidelines the regulatory trends of OECD member countries and developing countries were reviewed to determine levels of pollutant releases in air emissions, liquid effluents, and solid/hazardous wastes which are considered acceptable by these regulators.

- **Cost Effectiveness:** Pollution prevention and control systems were reviewed to determine levels of control which promote productivity and optimize resource utilization to achieve sustainable development. Pollution prevention is stressed in the guidelines as a means of cost-effective environmental management.

- **Good Industrial Practices:** For each of the industrial sectors covered in the guidelines, documents from several organizations were reviewed to determine those Good Industrial Practices which prevent pollution and which promote energy efficiency.

Who Will Use the Guidelines?

At the outset, the guidelines were intended to inform the staff of the World Bank Group, and consultants to the Bank, what level of emissions and discharges are acceptable for new projects. This is still the primary audience. However, interest in the guidelines from outside the World Bank Group has surfaced and has grown dramatically during the time of their preparation. It is now the case that the following stakeholders have indicated interest in referring to the guidelines:

- Other multilateral financial institutions (e.g. Asian Development Bank, African Development Bank)

- Commercial banks and on-lending institutions

- Industry associations

- Equipment manufacturers and suppliers

Representatives of the NGO community have commented that the guidelines are expected to draw a global audience. Moreover, developing countries that may be in the process of establishing, or improving on, their own regulatory requirements are expected to consider the requirements of the Bank's guidelines. With the global interest in the guidelines the Bank is perceived to be assuming leadership in establishing emission requirements for projects in developing countries.

The Pollution Prevention and Abatement Handbook

The industry specific guidelines are included in a forthcoming Bank publication—the Pollution Prevention and Abatement Handbook. This Handbook has three parts:

- **Part I—Pollution Management: Key Policy Issues:** This part is aimed at government decision makers and other policy makers in developing countries. It will guide decision makers in adopting appropriate policies by providing lessons learned from previous development experience. It outlines the incentives needed to promote pollution prevention and control with greater emphasis on pollution and adoption of Good Industrial Practices which in turn lead to sound environmental management.

- **Part II—Implementing Policies in Practice:** Part II identifies the basics of good environmental management, including the use of indicators for pollution management, implementing the environmental assessment process, and use of sound environmental standards. A methodology for setting the right priorities is also presented. Other subjects covered include air, water, and solid/hazardous waste management, environmental audits and monitoring, global and transboundary environmental issues, and financial incentives/disincentives needed to promote good environmental management.

- **Part III—Project Requirements:** Part III focuses on the requirements for industrial projects financed by the World Bank Group. This part of the Handbook includes discussions of the characteristics of major pollutants and their environmental and health effects, as well as technologies for their control. Included are papers on sulfur oxides, nitrogen oxides, particulate matter, arsenic, cadmium, lead, mercury and ground level ozone. And finally, Part III contains the Industry Sector Guidelines.

A table of contents for the Handbook is attached.

A Typical Industry Sector Guideline

An attempt was made to include in each industry sector guideline contained in Part III of the Handbook sections on: industry description, waste characteristics, pollution prevention and control measures, pollution reduction targets, emissions requirements, monitoring and reporting requirements, and key issues. Not every guideline includes every one of these sections. A description of the contents of each section follows:

- **Industry Description and Practices:** The major manufacturing processes and the energy consumption associated with each of the processes are described in this section. If there are processes which increase productivity and energy efficiency together with minimizing environmental impacts, then these are recommended. Raw materials used in the process and manufactured products are discussed to assist in evaluating the processes for their potential environmental impacts.

- **Waste Characteristics:** Here the major pollutants associated with the manufacturing process, their range of concentrations, and their loading levels in air emissions, liquid effluents, and solid/hazardous wastes are described. Information on the quantities of wastewater and solid/hazardous waste generated together with the quantity of pollutant per unit of output are presented for the major processes.

- **Pollution Prevention and Control:** Measures which reduce or eliminate pollutant loads are described in this section—these are measures that, if implemented, minimize pollutants at source.

- **Target Pollution Loads:** The levels achievable through adopting pollution prevention and control measures and Good Industrial Practices are specified. management would be able to achieve these before the addition of any add-on pollution control devices or treatment systems. In some cases one can argue, for example, that a fabric filter for capturing particulate matter is part of the process or is an add-on pollution control device. In cement production a fabric filter would be argued as part of the process in order to capture valuable product that would otherwise be lost, rather than a pollution control device.

- **Emissions Guidelines:** Although this section has "guidelines" in its title the Bank requires projects to discharge not more than the maximum values given in this section of each guideline unless an environmental assessment fully justifies a variance from the values and the variance(s) is/are approved by Bank management. Values are normally expressed as concentrations to facilitate monitoring and in many of the guidelines these concentrations are accompanied with values expressed as units of pollutant per unit of production. These maximum values are based on the level or pollutant reduction that is achievable through inclusion of cleaner production processes and appropriate pollution prevention and control technologies.

- **Monitoring and Reporting:** The frequency of sampling for each parameter is specified—sampling and testing methodologies are not included here since a separate document elsewhere in the Handbook addresses this. Reporting requirements are also specified in this section of the guideline.

- **Key Issues:** This section captures the key points and recommendations that are embedded in the various sections of the guideline—adopting these will help management meet the requirements of the guideline.

One of the new guidelines (Iron and Steel Manufacturing) has been attached to this paper to illustarte the content of the above described sections.

Status of the Handbook and the Guidelines

IFC and Bank staff have agreed that the Handbook will be a "living" document. As such, other guidelines will be added from time-to-time. At present about 35 industry sector guidelines have reached final draft stage, although it is expected that additional comments will be received after its release to the public and these comments will be reviewed for there potential inclusion. An additional 15 guidelines are in various stages of drafting and are expected to reach final draft status by mid-June, 1997. The Handbook will be available for wide distribution by the first of July. There are also plans to eventually make the complete Handbook available through the Internet.

Use of the Guidelines by IFC

As stated earlier, IFC is the private sector investment arm of the World bank Group. When considering a loan to a private sector project or investment in a project in a developing country, IFC uses four criteria for making an investment: (1) the project must be in the private sector and it must be profitable; (2) the project must contribute to the national economy; (3) the project must be technically feasible; and (4) the project must be environmentally sound.

IFC's Environment Division staff review the project for environmental issues, including the results of the environmental assessment, and approve the project when they are satisfied that environmental issues are being appropriately addressed. The industry sector guidelines in the Handbook are critical to IFC's involvement in a project—the legal agreement between IFC and the project sponsor will specify that the project must meet the maximum emissions numbers in the guideline, and exceptions or variations to these numbers would only be made in rare and compelling circumstances and on approval by IFC's management.

IFC further requires that project sponsors provide to IFC an annual performance report summarizing emissions information. As well, IFC monitors the environmental performance of its investments through visits to the project site—large heavy industry projects might warrant an annual visit while smaller less polluting projects might warrant a supervision/monitoring visit from an IFC environmental specialist on a less frequent basis, perhaps once every two years. The guidelines and the maximum emission values are criteria that the environmental specialist would refer to during his/her visit.

In Closing

Industry is moving forward with inclusion of cleaner production processes that use less resources, that are energy efficient, and that emit fewer pollutants. As well, new management practices such as those found in ISO 14000 are quickly falling into place, including in projects in developing countries. Older World Bank Group environmental guidelines were based on late 1970s and early 1980s production and control technologies and the command-and-control approach. The new guidelines are justified and timely and are based on a sound approach. This approach has been described and includes: recommending cleaner production processes and pollution prevention methods which minimize pollutants at source and which are energy efficient, coupled with, as necessary, additional treatment and control systems to meet specified maximum emission levels that are attainable and which take into account the impact on human health and the environment.

IFC believes that the pollution prevention approach taken in developing the guidelines is correct for its investments. A recent case study by IFC confirmed this—a heavily polluting cement plant was modernized with IFC investment to substantially reduce the discharge of air emissions and IFC followed up with a study of the benefits of the modernization. The results are published in: "Cost Benefit Analysis of Private Sector Environmental Investments—A Case Study of the Kunda Cement Plant". Here is a summary of the findings:

- The investment to reduce air pollution resulted in significant net economic benefits—the economic rate of return has been calculated at 24.7%, and sensitivity analysis confirmed that even if benefits are significantly reduced the economic rate of return is more than 16%.

- The significant environmental benefits that were identified include:
  — reduced global effects of SOx and NOx emissions
  — better health and health-related costs
  — reduced soiling and material damage
  — increased tourism income
  — greater real estate values
  — increased forestry and agricultural yields

The advantages and benefits of pollution prevention are clear in this example. It is interesting to note that the approach taken in the study and the findings that came out of the study have many applications in the private sector. They include: (a) improved project investment analysis; (b) clearer demonstration of the development impact of environmental investments; (c) help in developing investment plans, by estimating the returns to specific environmental investments; (d) improved corporate image and public relations; and (e) higher environmental awareness.

IFC identified as well that an environmental cost-benefit analysis can also help public policy makers with respect to: (a) funding investments; (b) providing incentives; (c) justifying environmental regulations; and (d) establishing penalties for non-compliance of regulations.

IFC's environmental staff have been a full partner with the Bank's environmental staff in developing the guidelines. The guidelines respond to IFC's project needs and will be used by IFC as they are released—expected to be in mid 1997.

Appendix 1
CONTENTS OF POLLUTION PREVENTION AND ABATEMENT HANDBOOK

Pollution Prevention and Abatement Handbook - Part III

Iron and Steel Manufacturing

Industry Description and Practices

Steel is manufactured by reducing iron ore using an integrated steel manufacturing process or a direct reduction process. In the conventional integrated steel manufacturing process, the pig iron from the blast furnace is converted to steel in the basic oxygen furnace (BOF). Steel can also be made in an electric arc furnace (EAF) from scrap steel and in some cases, from direct reduced iron. BOF is typically used in high tonnage production of carbon steels, while EAF is used to produce carbon steels and low tonnage specialty steels. An emerging technology, direct steel manufacturing, produces steel directly from iron ore. **This document only addresses integrated *iron and steel manufacturing*; a separate document on mini mills addresses the electric arc steel process and steel finishing processes. Steel manufacturing and finishing processes which are discussed in the document on Mini Steel Mills are also employed at integrated steel plants.** The manufacturing of coke is also dealt within a separate document.

When making steel using a BOF, coke making and iron making precede steel making; these steps are not necessary with an EAF. Pig iron is manufactured from sintered, pelletized, or lump iron ores using coke and limestone in a blast furnace. It is then fed to a BOF in molten form along with scrap metal, fluxes, alloys and high-purity oxygen to manufacture steel. In some integrated steel mills, sintering (heating without melting) is used to agglomerate fines so as to recycle iron-rich material, such as mill scale.

Waste Characteristics

Sintering operations can emit significant dust levels of about 20 kg/ton of steel. Pelletizing operations can emit dust levels of about 15 kg/t of steel. Air emissions from pig iron manufacturing in a blast furnace include: particulate matter (PM - ranging from less than 10 to 40 kg/t of steel manufactured), sulfur oxides (SO_x - which are mostly from sintering or pelletizing operations) (1.5 kg/ton of steel), nitrogen oxides (NO_x - which are mainly from sintering and heating) (0.5 kg/ton of steel), hydrocarbons, carbon monoxide, in some cases, dioxins (mostly from sintering operations), and hydrogen fluoride.

Air emissions from steel manufacturing using BOF may include PM (ranging from less than 1t5 to 30 kg/ton of steel), chromium (0.8 mg/Nm^3), cadmium (0.08 mg/Nm^3), lead (0.02 mg/Nm^3), and nickel (0.3 mg/Nm^3). In the desulfurization step between the blast furnace and the BOF, the particulate matter emissions are about 10 kg/ton of steel manufactured.

In the conventional process without recirculation, waste waters (including those from cooling operations) are generated at an average rate of 80 m^3 per ton of steel manufactured. Major pollutants present in untreated waste waters generated from pig iron manufacture include total organic carbon (typically 100 to 200 mg/l), total suspended solids (7,000 mg/l), dissolved solids, cyanide (15 mg/l), fluoride (1,000 mg/l), COD (500 mg/l), and zinc (35 mg/l).

Major pollutants in waste waters generated from steel manufacturing using the BOF include total suspended solids (up to 4,000 mg/l), lead (8 mg/l), chromium (5 mg/l), cadmium (0.4 mg/l), zinc (14 mg/l), fluoride

(20 mg/l), and oil and grease. The process generates effluents with high temperatures.

Process solid waste from the conventional process, including furnace slag and collected dust, is generated at an average ranging from 300 to 500 kg per ton of steel manufactured, of which 30 kg may be considered hazardous based on the concentration of heavy metals present.

Pollution Prevention and Control

Where technically and economically feasible, direct reduction of iron ore for the manufacture of iron and steel is preferred because it does not require coke manufacturing and because it has fewer environmental impacts. Wherever feasible, pelletizing should be preferred over sintering for the agglomeration of iron ore. The following pollution prevention measures should be considered:

Pig Iron Manufacturing

- Improve the blast furnace efficiency by using coal and other fuels (such as oil or gas) for heating instead of coke thereby minimizing coke guidelines.
- Recover thermal energy of the blast furnace off-gas before using it as fuel.
- Increase fuel efficiency and reduce emissions by improving blast furnace charge distribution.
- Improve productivity through screening of the charge and better tap-hole practices.
- Take action to reduce dust emissions at furnaces, such as covering iron runners when tapping the blast furnace, using nitrogen blankets during tapping.
- Use pneumatic transport, enclosed conveyor belts, or self-closing conveyor belts, wind barriers, and other dust suppression measures to reduce the formation of fugitive dust.
- Use low NO_x burners to reduce NO_x emissions from burning fuel in ancillary operations.
- Recycle iron-rich materials such as iron ore fines, pollution control dust, and scale *sintering* in a sinter plant.
- Recover energy from sinter coolers.
- Use dry sulfur oxide removal systems (such as carbon absorption for sinter plants or lime spraying in flue gases).

Steel Manufacturing

- Use dry dust collection and removal systems to avoid the generation of wastewater.
- Use BOF gas as fuel.
- Use doghouse enclosures for BOF.

Other

Use blast furnace slag in construction materials.

Target Pollution Loads

The recommended pollution prevention and control measures can achieve the following target levels:

Liquid Effluents

Over ninety percent of the wastewater generated can be reused. Discharged waste waters should in all cases be less than 5 m^3 per ton of steel manufactured and preferably less than 1 m^3 per ton.

Solid Wastes

Blast furnace slag should be normally generated at a rate less than 320 kg/ton iron. Slag generation rates from BOF should be between 70 and 170 kg/t of steel manufactured but this will depend on the impurity content of feed materials. Approximately, 65 percent of BOF slag from steel manufacturing can be recycled in various industries such as building materials. Zinc recovery may be feasible for collected dust.

Treatment Technologies

Air Emissions

Air emission control technologies for the removal of particulate matter include scrubbers (or semidry system), baghouses, and ESPs; the latter two technologies can achieve 99.9 percent removal efficiencies for the PM and associated toxic metals.

Sulfur oxides are removed with scrubbers with a 90 percent or better removal efficiency. However, the use of low sulfur fuels and low sulfur ores may be more cost-effective.

The acceptable levels of nitrogen oxides can be achieved by using low NO_x burners and other combustion modifications.

For iron and steel manufacturing, the following emissions levels should be achieved:

Load Targets per Unit of Production

Parameter	*Maximum value*
Particulate matter (PM_{10})	100 g/t of product (blast furnace, basic oxygen furnace), and 300 g/t from sintering process
Sulfur oxides (SO_x)	1,200 g/t (500 mg/Nm^3) (for sintering)
Nitrogen oxides (NO_x)	500 g/t (200 mg/Nm^3)
Fluoride	1.5 g/t (5 mg/Nm^3)

Wastewater Treatment

Wastewater treatment systems typically include sedimentation to remove suspended solids, physical/chemical treatment such as pH adjustment to precipitate heavy metals, and filtration.

The following target levels can be achieved for steel making processes:

Target Load per Unit of Production

Parameter	*Blast Furnace*	*BOF*
Wastewater	0.1 m^3/t steel	0.5 m^3/t steel
Zinc	0.6 g/t	3 g/t
Lead	0.15 g/t	0.75g/t
Cadmium	0.08 g/t	N/A

N/A = Not applicable.

Solid Waste Treatment

Solid wastes containing heavy metals may have to be stabilized using chemical agents before disposal.

Emission Guidelines

Emission levels for the design and operation of each project must be established through the Environmental Assessment (EA) process, based on country legislation and the *Pollution Prevention and Abatement Handbook* as applied to local conditions.[1] The emission levels selected must be justified in the EA and acceptable to the World Bank Group.

The following guidelines present emission levels normally acceptable to the World Bank Group in making decisions regarding provision of World Bank Group assistance; any deviations from these levels must be described in the World Bank Group project documentation.

The guidelines are expressed as concentrations to facilitate monitoring. Dilution of air emissions or effluents to achieve these guidelines is unacceptable.

All of the maximum levels should be achieved for at least 95% of the time that the plant or unit is operating, to be calculated as a proportion of annual operating hours.

Air Emissions

For integrated iron and steel manufacturing plants, the following emission levels should be achieved:

Air Emissions from Integrated Iron and Steel Manufacturing

Parameter	*Maximum value*
Particulate matter (PM)	50 mg/Nm^3
Sulfur oxides (SO_x)	500 mg/Nm^3 (sintering)
Nitrogen oxides (NO_x)	750 mg/ Nm^3
Fluorides	5 mg/Nm^3

Liquid Effluents

The following effluent levels should be achieved:

[1] For reference, see the Tables in the Statement of Principles (first section of Part III of the Handbook).

Effluents from the Integrated Iron and Steel Manufacturing

Parameter	*Maximum value*
pH	6 - 9
Total suspended solids	50 mg/l
Oil and grease	10 mg/l
COD	250 mg/l
Phenol	0.5 mg/l
Cadmium	0.1 mg/l
Chromium (total)	0.5 mg/l
Lead	0.2 mg/l
Mercury	0.01 mg/l
Zinc	2 mg/l
Cyanides (free)	0.1 mg/l
Cyanides (total)	1 mg/l
Temperature increase	less than or equal to 3°C[1]

[1] The effluent should result in a temperature increase of no more than 3 degrees Celsius at the edge of the zone where initial mixing and dilution take place. Where the zone is not defined, use 100 meters from the point of discharge.

Note: Effluent requirements are for direct discharge to surface waters. Discharge to an **offsite wastewater treatment plant** should meet applicable pretreatment requirements.

Sludges

Sludges should be disposed of in a secure landfill after the stabilization of heavy metals to ensure that heavy metal concentration in the leachates do not exceed the levels presented for liquid effluents.

Ambient Noise

Noise abatement measures should achieve the following levels, measured at noise receptors located outside the project property boundary, with an increase in existing ambient level of L_{dn} 10 dB(A) or less where background levels are less than L_{dn} 55 dB(A), and with an increase in existing level of L_{dn} 3 dB(A) or less where background levels are above L_{dn} 55 dB(A).

Ambient Noise

Receptor	Maximum dB(A)
Residential; institutional; educational	L_{dn} 55
Industrial; commercial	L_{eq} (24) 70

The emission requirements given here can be consistently achieved by well-designed, well-operated and well-maintained pollution control systems.

Monitoring and Reporting

Air emissions should be monitored continuously after the air pollution control device for particulate matter on (or alternatively opacity level of less than 10 percent) and annually for sulfur oxides, nitrogen oxides (with regular monitoring of sulfur in the ores), and fluoride. Wastewater discharges should be monitored daily for the listed parameters, except for metals, which should be monitored at least on a quarterly basis. Frequent sampling may be required during start-up and upset conditions.

Monitoring data should be analyzed and reviewed at regular intervals and compared with the operating standards so that any necessary corrective actions can be taken. Records of monitoring results should be kept in an acceptable format. These should be reported to the responsible authorities and relevant parties, as required.

Key Issues

The following box summarizes the key production and control practices that will lead to compliance with emissions guidelines:

- Prefer direct steel manufacturing process where technically and economically feasible.
- Use pelletized feed instead of sintered feed where appropriate.
- Substitute a portion of the coke used in the blast furnace by injecting pulverized coal or by using natural gas, or oil.
- Achieve high energy efficiency by using blast furnace and basic oxygen furnace off-gas as fuels.
- Implement measures (such as encapsulation) to reduce the formation of dust (including iron oxide dust) and where possible, recycle collected dust to a sintering plant.
- Recirculate waste waters. Use dry air pollution control systems where feasible. Otherwise treat waste waters.
- Use slag in construction materials to the extent feasible.

Further Information

The following are suggested as sources of additional information (these sources are provided for guidance and are not intended to be comprehensive):

British Steel Consultants. 1993. Research Study, International Steel Industry. Prepared for the International Finance Corporation.

The Government of The Netherlands. 1991. "Progress Report on the Study of the Primary Iron and Steel Industry." Third Meeting of the Working Group on Industrial Sectors. Stockholm: January 22-24.

Paris Commission. 1991. *Secondary Iron and Steel Production An Overview of Technologies and Emission Standards Used in the* PARCOM Countries.

World Bank. Environment Department. 1996. "Pollution Prevention and Abatement: Iron and Steel Manufacturing". Technical Background Document.

This technical guideline is part of the World Bank Group's *Pollution Prevention and Abatement Handbook* and should be read in conjunction with the Introduction as well as Parts I and II which spell out important policy and procedural considerations. The information is intended for use by World Bank staff and consultants in carrying out the policies set out in the Operational Policy on Environmental Assessment (OP4.01) and related documents.

Renewable technologies and their role in mitigating greenhouse gas warming

F.T. Princiotta

Director, Air Pollution Prevention and Control Division
National Risk Management Research Laboratory
U.S. Environmental Protection Agency (MD-60)
Research Triangle Park, NC 27711

Human activity has led to an increased atmospheric concentration of carbon dioxide (CO_2), methane (CH_4), and other gases which resist the outward flow of infrared radiation more effectively than they impede incoming solar radiation. This imbalance yields the potential for global warming as the atmospheric concentrations of these gases increase. For example, before the industrial revolution, the concentration of CO_2 in the atmosphere was about 280 ppm, and it is now about 360 ppm. Similarly, CH_4 atmospheric concentrations have increased substantially, and they are now more than twice what they were before the industrial revolution, currently about 1.8 ppm. Recent data also suggest that airborne particulates have increased significantly in the post-industrial period and have contributed to a counteracting cooling impact.

In this paper we will discuss the role that renewable and other mitigation approaches could play in ameliorating such projected warming. In order to put this issue in context, the following issues will be discussed:

-What is the range of projected warming?
-What is the relative importance of the various greenhouse gases?
-What are the major and projected sources of CO_2?
-What emission controls achieve what level of greenhouse gas warming mitigation?
-What are candidate mitigation technologies - on both the end use side and the production side?
-Focusing on one particular renewable technology, the Hynol process, what are some of the economic, institutional, and other barriers that hinder commercialization?

A model (Glowarm 3.0) that the author has developed to help evaluate these questions is a spreadsheet (Lotus 1-2-3) model which calculates global concentrations and their associated global warming contributions for all the major greenhouse gases. The model calculates atmospheric concentrations of greenhouse gases based on projected emissions in 10-year increments. For CO_2, look-up tables are used to relate the fraction of CO_2 remaining in the atmosphere as a function of time after emission for two alternative CO_2 life cycles. For the other gases, an inputed lifetime value is used. Average global equilibrium temperatures are calculated by adding contributions of each gas, using lifetimes and radiative forcing functions described in Intergovernmental Panel on Climate Change (IPCC), 1990, along with an assumed input atmospheric sensitivity. Realized (or actual) temperature is estimated using an empirical correlation algorithm we developed based on general-circulation model (GCM) results presented in IPCC, 1992. This approach uses a correlation which relates the rate of equilibrium warming over the period between the target year and 1980 to the ratio of actual to equilibrium warming. The greater the rate of equilibrium warming, the smaller is the ratio of the actual to equilibrium ratio. Note that it is much easier to calculate average global warming than it is to estimate warming on a geographical or seasonal basis. Such geographical or seasonal projections require more complex models which are subject to a much greater degree of uncertainty.

Figure 1 shows fields for the model. Note that equilibrium and transient (realized or actual) warming can be calculated for any year (to 2100) for a variety of emission and control scenarios, two CO_2 life cycles, an assumed atmospheric sensitivity to a doubling of CO_2 concentration, CH_4 lifetime, and both sulfate cooling and CFC phaseout assumptions. Under the same assumptions, the model output temperatures fall generally within 10% of values calculated by other more complex models (IPCC, 1996b; NAS, 1991; Krause, 1989).

UNCERTAINTIES IMPACTING DEGREE OF WARMING EXPECTED

There are many uncertainties associated with the expected magnitude of global warming. The following are major uncertainties which will be considered and quantified:

1. Atmospheric Sensitivity. This critical variable is generally defined as the equilibrium temperature rise associated with a doubling of CO_2 concentration. GCMs are utilized by climate modelers to forecast the impact of CO_2 warming. Unfortunately, the range of their results is wide and not converging (Dornbusch and Poterba, 1991). The IPCC (IPCC, 1996a) has concluded this range to be between 1.5 and 4.5°C.

2. CO_2 Life Cycles. The Earth's carbon cycle, which involves atmospheric, terrestrial, and oceanic mechanisms, is complex and not completely understood. Yet, in order to estimate CO_2 atmospheric concentrations and subsequent warming, it is necessary to assume a relationship between CO_2 remaining in the atmosphere and time after emission. For this analysis, two CO_2 life cycles were utilized, one based on IPCC (1992) and the other described by Walker and Kasting (1992). The Walker model yields longer atmospheric lifetimes leading to higher CO_2 concentrations.

3. Projected Growth of CO_2 Emissions Over Time for a "Business as Usual" Case
Attempting to predict the future is a risky business, at best. Yet, to scope the magnitude of the warming issue, it is necessary to estimate emissions of greenhouse gases as far in the future as one wishes to project warming. As we discussed and quantified in a previous paper (Princiotta, 1994), the following are key factors which will determine a given country's emissions of CO_2, the most important greenhouse gas:

    - current emission rate
    - population growth
    - growth of economy per capita
    - growth rate: energy use per economic output
    - growth rate: carbon emissions per energy use unit

Since future global CO_2 emissions will be the sum of an individual country's emissions, all subject to varying factors listed above, it is clear that even for "business-as-usual" (or base case) there is a large band of uncertainty.

4. Methane Lifetime. A variety of investigators have provided a range of estimates for the atmospheric lifetime of CH_4. The longer the lifetime, the greater is CH_4's contribution to global warming.

5. Projected Growth of Methane Emissions. There is an incomplete understanding of the current contributions of the major anthropogenic sources of CH_4. They include: landfills, rice production, coal mines, natural gas production and distribution systems, and the production of cattle. There is even more uncertainty regarding the likely growth of such emissions over time as population grows, industrialization accelerates in developing countries, and agricultural practices change.

6. Use of High Global Warming Potential Compounds (e.g., HFC-134a) to Replace Chlorofluorocarbons (CFCs).
As the international community phases out of CFC production, due to concerns associated with stratospheric ozone depletion, hydrofluorocarbon (HFC)-134a and other compounds with significant greenhouse warming potential are being utilized as replacements. The importance of the extent to which compounds such as these are utilized will be evaluated.

7. Actual Temperature Response Versus Calculated Equilibrium Warming. GCMs often calculate projected equilibrium warming rather than transient or actual warming. Equilibrium warming can be defined as the temperature the Earth would approach if it were held at a given mix of greenhouse gas concentrations over a long period of time. Transient (also called realized or actual) temperatures are those that would actually be experienced at a given point in time, taking into account the thermal inertia of the Earth, especially its oceans. There is only an incomplete understanding of this thermal inertia effect and its quantitative impact on actual warming.

8. Aerosol (Sulfate) Cooling. A recent development (IPCC, 1992) has been the availability of evidence that emissions of sulfur dioxide (SO_2), other gases, and aerosols have contributed to a significant cooling impact, counteracting greenhouse gas warming. There is significant uncertainty over the magnitude of the direct impact of such fine particles and even more uncertainty over their secondary impact on clouds (generally thought to be significant and in the cooling direction).

In order to attempt to understand the impact of these variables, we have estimated warming for five scenarios spanning what we believe are reasonable ranges of values for these variables. For certain factors, such as atmospheric sensitivity, there is a reasonable consensus regarding the possible range of values. For other factors, there is no such consensus. It should be recognized that the credibility of this uncertainty analysis is only as good as the variable ranges assumed. Table 1 shows the assumed range of values from the "lowest" scenario, which assumes that all of these variables are at values which will yield the lowest degree of warming, to the "highest" case, which assumes those values which will yield the highest projected warming. These can be characterized as representing best versus worst case scenarios, respectively. In the middle is the base case which is generally consistent with the IPCC (1992, 1996b) and represents current conventional wisdom regarding the most likely scenario.

Figure 2 graphically summarizes the results of model calculations for the five scenarios examined. Also included in this figure is the actual warming estimated in 1980 relative to the pre-industrial era (NAS, 1991). As indicated, the range of projected global warming varies from significant to potentially catastrophic. We believe a more likely range of uncertainty is represented between the low and high scenarios. The predicted warming at 2100 for these cases is 2.1 and 5.7C°, respectively. The magnitude of these values and the difference between them support the contention that we are dealing with an issue not only of unprecedented potential impact, but also of monumental uncertainty. It is noteworthy that, even for the "low" scenario, temperature increases of 2.1C° over pre-industrial values (1.6C° over 1980 levels) are projected by 2100. According to Vostock ice core measurements (Dornbusch and Poterba, 1991), the last time the Earth experienced such an average temperature was 125,000 years ago.
As a basis for comparison, recently the IPCC (IPCC, 1996b) has projected warming at 2100 to range from 1.8°C to 3.0°C depending on the projected emission scenario, with the base case warming at 2.5°C. This warming includes the 0.5°C warming experienced from the pre-industrial era to the current time. On the same basis, the Glowarm model calculates a base warming of 2.6°C.

It is important to note that uncertainty influences not only the predicted degree of future warming, but also the effectiveness of a given mitigation strategy. Figure 3 illustrates this point. Realized warming versus time is plotted for the "low," "high," and base scenarios. In addition, two stringent mitigation cases are included. Both assume that, by the year 2000, worldwide mitigation is imposed to decrease emissions of all greenhouse gases by 1% annually. However, the first mitigation case assumes all of the "high" variables summarized in Table 1 . The second, imposes a mitigation program assuming base (or "most likely") variables. The results are dramatic. They show that, even with a stringent emission reduction program, if the "high" case values are assumed, warming will be **greater** for all years before 2100 than for the uncontrolled base case! Note that, if a mitigation program (1% per year reduction for this "high case") were initiated further in the future, 2010 for example, the results would be even more dramatic. In this case, the controlled temperature at 2100 is now about 2.4°C versus the 2.1°C for the uncontrolled base case.

WHICH GASES ARE IMPORTANT?

Let us now examine the important greenhouse gases and their potential warming contributions. Figure 4 shows the projected contribution by greenhouse gas over the period 1980-2050 for the base scenario. CO_2 and CH_4 are clearly the most important contributors to warming, with CFCs and their substitutes, nitrous oxide (N_2O), and tropospheric ozone (O_3) [1] playing small but significant roles. Noteworthy, is the projected cooling impact of aerosol sulfates.

However, again, uncertainty is significant, this time in determining the relative contributions of the greenhouse gases. Such uncertainties are considered in Table 2. For each greenhouse gas, this table summarizes: atmospheric lifetime, the ratio of current to pre-industrial atmospheric concentrations, projected contributions to realized warming, and the projected impact of mitigating emissions. Also included is a judgment regarding the relative confidence of the predicted warming impacts, along with major uncertainties and the major human sources. Uncertainty is important for all gases, but especially for aerosols and tropospheric ozone.

When one considers the importance of a given greenhouse gas, it is informative to evaluate warming prevented for a given mitigation scenario. Figure 5 shows results of model calculations for the period 1980-2050 comparing equilibrium base scenario warming to warming prevented assuming a stringent mitigation program. In this case, a 1% annual reduction in emissions is assumed for each gas (or its precursor), exclusive of sulfates, starting in the year 2000. The main result here is that a higher fraction of their base warming can be mitigated for the short-lived gases such as CH_4 and O_3. For example, whereas less than half of CO_2's base warming is mitigated in this case, about three-quarters of CH_4's base warming is mitigated. When viewed from a mitigation (or warming prevented) viewpoint, CH_4 is about half as important as CO_2; whereas, from an emission viewpoint, it is less than a third as important.

Figure 6 shows additional model results to help shed light on this point. In this case, the effect of annual mitigation rate (starting in 2000) on equilibrium warming mitigated by gas is illustrated. An interesting observation that can be made is that a stringent 2% per year mitigation program for CH_4 could have almost as much benefit by the year 2050 as capping (0% growth) CO_2 emissions. Of course, such conclusions are subject to the uncertainties previously discussed.

WHICH COUNTRIES ARE MAJOR CONTRIBUTORS TO EMISSION OF GREENHOUSE GASES? WHAT ARE LIKELY TRENDS?

It is useful to look at recent histories of CO_2 emissions for key countries. Figure 7 derived from NAS, 1991, illustrates growth in CO_2 emissions from 12 key countries between 1960 and 1988. As indicated, the U.S., USSR (now Russia, Ukraine, and other independent countries), and China are by far the major sources of CO_2. However, when one considers the recent (1980-1988) growth rate, China and India are especially significant since this portends

---

[1]This value assumes volatile organic compound (VOC), nitrogen oxide (NO_2), and carbon monoxide (CO) precursors contribute to O_3 formation. However, the small component of O_3 warming associated with CH_4 emissions is included in the CH_4 value.

future contributions to CO_2 emissions. Table 3 summarizes 1988 CO_2 data (NAS, 1991) for key countries listed in order of overall emissions, per capita emissions, and per gross national product (GNP) emissions. Although, the U.S. leads the world in overall and per capita emissions, China easily has the largest per emissions GNP.

In order to provide insight into the various sectors contributing to 1990 CO_2 emissions for key countries, Figure 8 was generated based on Oak Ridge National Laboratory (ORNL) calculated data (Bowden, et al., 1993). This figure illustrates that each country has a distinctive mix of activities yielding CO_2 emissions. In the case of the U.S., coal combustion (for electricity and steam), petroleum for transportation, and natural gas combustion (primarily for power generation and space heating) are the three most critical contributors. The pattern is similar in the former Soviet Union with the major difference in the automobile sector; much less CO_2 is generated by a much smaller fleet of vehicles. In China, coal combustion is the dominant source of CO_2 emissions, helping to explain why China's CO_2 unit of GNP is so high; coal is by far the most CO_2-rich fuel source per unit of useful output energy. Germany, the fourth most important source of CO_2, is also dominated by coal use: in their case, brown coal (lignite) is indigenous to their country. Japan, with few indigenous fossil fuel resources, is heavily dependent on imported coal and residual oil for power generation. It is interesting to note that India, the second most populous country in the world and likely a major future contributor, has a pattern similar to China, with steam coal the dominant source.

We have already discussed the uncertainties associated with future emissions of CO_2. Such emissions will depend on country-specific factors: population growth, rate of industrialization, energy use per economic output, and carbon use per energy utilized. Table 4 (Princiotta, 1994) shows a projection of growth of these factors for the developed (Organization for Economic Cooperation and Development--OECD) and relatively undeveloped Asian countries for the period 1990-2025. This projection is derived from information presented in IPCC, 1992. For the OECD countries, the key driver yielding increased CO_2 emissions is expected to be economic growth, whereas population growth is projected to be quite modest. For the Asian countries, the key driver is likely to be economic growth, with population growth also significant. For both regions, in the absence of a CO_2 mitigation program, energy efficiency gains and a decrease in carbon-intensive energy use are projected to be modest over this time period.

It is useful to examine the likely results of these drivers on projected emissions of CO_2 from selected countries. Figures 9 and 10 show such a projection assuming economic, population, and energy use trends summarized in IPCC, 1992. Projections for the years 2030 and 2100 are combined with actual CO_2 emission data (NAS, 1991) from the 1960-1988 time period. These graphics show that the Asian countries, especially China and India, driven by high projected economic growth and large populations, will be dominant CO_2 emitters by the middle of the next century.

MITIGATION: HOW MUCH AND WHEN TO START?

Figure 11 illustrates the projected results of two hypothetical mitigation scenarios compared to the base case which assumes current expectations for greenhouse gas emissions. If emissions were held constant at year 2000 levels, the rate of projected warming could be slowed substantially; although significant warming would continue for the foreseeable future. However, if emissions for all greenhouse gases were reduced 1% annually, post-1980 warming could be stabilized below about 1° C by the year 2100. Therefore, in order to mitigate warming over the long term, it will be necessary to reduce greenhouse gas emissions substantially over time. This will be a difficult goal, considering projected rates of economic and population growth which are key drivers for greenhouse gas emissions. Figure 12 illustrates this point by showing on one graphic, projected economic activity, population, base case CO_2 emissions, and mitigated CO_2 emissions unitized at 1990.

Figure 13 illustrates the impact of the year control starts on realized warming projected in 2050 for two mitigation scenarios (1% of annual control and an emission cap). As indicated, early emission control allows for a larger degree of climate stabilization. These results suggest that there can be major stabilization benefits for early initiation of mitigation.

MITIGATION: WHICH SOURCES/WHICH TECHNOLOGIES?

It is useful to examine recent CO_2 energy use patterns in order to ascertain which sectors and fuels are significant CO_2 emitters and candidates for mitigation. Table 5 (adapted from IPCC, 1996a) illustrates that all major energy categories are important emitters of CO_2, so that all major energy sectors will require major improvements in end use efficiency and, in the longer term, migration away from fossil fuels if stabilization efforts are to be successful.

Since it is clear that fossil fuel use is the key driver for greenhouse gas warming, a relevant question is: how much fossil fuel is available and how long will it last?

Table 6 adapted from IPPC, 1996a, and augmented by the author, summarizes the prevailing view on this subject. Basically, oil appears to be the least abundant fossil fuel with reserves plus most likely discovered conventional oil estimated to be 8500 EJ. Such an amount would be depleted by about 2035, if oil use rate were to increase by 1.5% per year. If unconventional reserves which, include heavy oil, oil shale, and oil tar deposits, are included, the availability of oil could be extended to about 2080, again assuming 1.5% increase in use per year until depletion. For conventional gas, the reserves plus expected discovered resources are estimated to be 9200 EJ . If gas use rate would increase 1.5% annually, these resources would be depleted by 2065. If unconventional gas sources are considered, gas resources wouldn't be depleted before about 2135 under the assumptions described above. As indicated, known reserves of coal are much larger, with depletion estimated at 2195 under the 1.5% annual growth assumption. Taken together all fossil fuel resources appear to be sufficient to last until about 2150. From a greenhouse warming viewpoint, gas is the most desirable fossil fuel since it has a high hydrogen to carbon ratio and generates substantially lower quantities of CO_2 than do oil and coal.

A reasonable scenario, then, is that sometime during the first half of the next century conventional sources of oil and later gas will become scarce and more expensive. Unconventional sources will likely become available but at substantially higher prices than for more easily extracted conventional sources. At the same time, depending on economic conditions, political policies, and technology availability, coal and/or alternative sources of energy, (e.g., biomass, solar) will fill the gap left by the depletion of relatively inexpensive oil and gas resources.

Since fossil fuels are the key driver for greenhouse gas warming, and their resources are limited, especially for oil, a key question is what renewable resources are potentially available to displace oil, gas and coal. Table 7, adapted from IPCC, 1996a, and augmented by the author, shows the potential renewable resource available in the 2020-2025 time frame and for the longer term. Also included is information estimating the fraction of total projected global energy use that these resources could supply. As can be seen, in the nearer term horizon (2020-2025) only biomass, hydro, and solar (in that order) appear to be available in sufficient quantities to displace a major component of fossil fuel use. In the longer term only solar and biomass appear to offer the potential for wide scale displacement.

Any successful mitigation program dealing with the critical energy sector must aggressively deal with the two fundamental components of the energy cycle: end use efficiency and production. Tables 8 and 9, adapted from NAS, 1991, list and briefly summarize candidate mitigation options for the end use and production sectors, respectively. Since electricity production and use, residential, commercial and industrial combustion, and transportation energy use are all major current and projected generators of CO_2, all these sectors must make fundamental end use and production improvements if a stabilization program is to be successful.

The author is convinced that meaningful mitigation can be achieved only with an aggressive program aimed at using less energy in all sectors in the near term, supplemented by new technologies capable of displacing fossil fuels in the longer term. This contention is supported by one of the most detailed assessments of its kind (EPA, 1990), in which a multitude of options were evaluated for their quantitative potential in mitigating greenhouse warming in the 2050 and 2100 time frame. Table 10 is adapted from that study and compares the mitigation potential of those options which can reduce emissions of CO_2. As can be seen, both end use and production strategies can be effective in mitigating greenhouse gas warming in these time frames. Of particular potential importance are end use efficiency in transportation and stationary source combustion systems, and in the production side via extensive displacement of fossil fuels by biomass. Also, potentially significant would be a forest sequestration strategy to reverse the current trend of deforestation with wide-scale reforestation.

When one considers the potential problem posed by long term fossil fuel use from a greenhouse warming viewpoint, and the likely depletion of cheap fossil fuels, especially oil and gas, within a few generations, one might expect a massive worldwide effort to develop renewable alternatives and energy conservation technologies. This is not the current situation. Table 11 (IPCC, 1996a) summarizes energy research in the IEA countries (industrialized) from 1983 to 1994. As can be seen, R&D expenditures have been generally decreasing during this period, especially when calculated as a fraction of Gross Domestic Product (GDP). Also, by far the largest component of such research has been focused on nuclear fission, a commercial technology with many economic and political problems not likely solved with research. It is interesting to note that the U.S. military research budget alone in the post cold war era is about 3.5 times greater than the combined energy research for all the IEA countries!

Focus on Hynol Process Utilizing Biomass and Methane to Yield Transportation Fuels

In order toconsider some of the real world difficulties of developing and commercializing a potentially significant CO_2 mitigation technology, we will discuss the Hynol process. This process, which was innovated at DoE's Brookhaven National Laboratory, has been under development via EPA sponsorship with contributions from the California Energy Commission and DoD's Strategic Environment Research and Development Program (SERDP).

Bench scale work over a 4 year period has been performed at Brookhaven and at EPA's Research Triangle Park, NC, facility to provide fundamental design information. This process could be used to provide fuel to dedicated light and heavy duty vehicles designed for methanol fuel as well as fuel-cell powered vehicles. Figure 14 is a schematic of the process.

Analysis of technological options for converting biomass to liquid fuels showed that methanol, produced by the Hynol process, could displace more gasoline at lower cost--and with greater effect on the net CO_2 emissions--than other process options (Borgwardt, in press). Methanol from the Hynol process cost is estimated at \$0.48/gallon for a 7870 tonne/day plant with 15.45% Capital Recovery factor, \$61/tonne biomass cost, and natural gas at \$2.50x10^6 Btu. It is currently estimated to be competitive with current equivalent gasoline prices in conventional vehicles.

A patent for the Hynol process was issued on September 6, 1994. A 50 lb/hour pilot test facility has been constructed for testing the critical gasifier and will commence operation soon. University of California, Riverside, has the lead research role via a cooperative agreement with EPA. Figure 15 illustrates the projected cost of Hynol methanol as a function of natural gas price.

The following are the author's observations and opinions regarding the difficulty of developing and ultimately commercializing such a process under the current economic and political situation.

> Despite the potential of a process such as Hynol to displace oil and/or to reduce greenhouse gas emissions, there is no commercial incentive to develop biofuels as long as their cost exceeds, or even is equivalent to, that of fossil fuels. Situations where biomass

can compete economically with fossil fuels are very few and have insignificant potential for affecting global greenhouse gas emissions.

- If greenhouse gas emissions is the only factor justifying biofuel utilization, development of biofuel technology is improbable without support by the government for the R&D that is necessary to demonstrate the technology and for providing incentives for renewable energy use.

- In the U.S., despite a robust economy, funding for renewable energy R&D is constrained to a modest level because of concerns about budget deficits and the absence of any imminent energy or environmental emergency. In the case of Hynol, it has been difficult to convince federal and private research sponsors to provide the resources for comprehensive testing of an integrated pilot of the process. The current pilot program is limited to gasifier evaluation.

- As long as petroleum is one of the lowest-cost sources of energy, and there is no global commitment to greenhouse gas reduction, only market forces will determine the fate of any effort to develop a biofuel alternative. The current basic cost of petroleum production is so low that it could undercut any attempt to start a major biofuel industry.

- If either petroleum displacement or greenhouse gas reduction is to be appreciably affected by biofuel, a very large plant must be considered, like 9000 tonnes/day of biomass feed. This is simply a matter of the number of plants that would be required to displace a significant portion of the current consumption of fossil fuel. The logistics of producing and delivering 9000 tonnes of biomass per day is formidable, given the land area, transport system, storage, etc., that are required. Capital investment in the plant alone would be over $1billion; raising such an amount would be quite difficult unless risks were very low and potential profits high.

- Convincing landowners of the merits of investing and establishing dedicated energy plantations on a large scale, even before a conversion plant is built, will be difficult. Building a conversion plant before the energy crops are in production, will also be a risk. Government guarantees would likely be necessary.

- Even at 9000 tonnes/day, leveraging of the yield of liquid fuel from biomass will be necessary for practical consideration, given the amount of fuel needed, the number of plants required, and the production cost. Hynol methanol provides a means of such leveraging by use of natural gas as cofeedstock. Further leveraging will be achievable when high efficiency fuel cell vehicles become commercialized, probably about the same time as a viable biofuel industry could be established.

- Energy companies have billions of dollars invested in infrastructure for the petroleum fuel cycle; therefore, there is a tremendous amount of inertia to make fundamental changes in this area. Energy companies have a considerable vested interest in the status quo, considering this investment.

- As a new fuel to be potentially used in unprecedented quantities and in locations all around the country, the following issues will have to be evaluated and resolved before such widescale use is practical: (1) potential toxicity; (2) potential for groundwater contamination; and (3) corrosiveness to vehicle components.

SUMMARY AND CONCLUSIONS

- A spreadsheet model has been utilized to calculate both equilibrium and realized greenhouse warming as a function of key variables including: greenhouse gas emission growth rates, CO_2 life cycles, CH_4 lifetime, current aerosol cooling, and CFC phaseout assumptions.

- Model calculations for the three most credible cases, assuming a varying range of assumptions, yield projected warming at 2100 from a substantial 2.1C° to a potentially catastrophic 5.7C°. The most likely case yields 2.6C° projected warming from pre-industrial values; such warming is consistent with the most recent IPCC (IPCC, 1996b). Such uncertainty also impacts the estimates of the effectiveness of a mitigation program. Model results suggest that, even assuming a stringent mitigation program, if key uncertainties all align toward maximum greenhouse warming, warming will be greater than it would be for a business-as-usual case assuming the mid-range of the key variables contributing to uncertainty.

- Aerosol/sulfate cooling is an important phenomenon, with recent data suggesting cooling comparable to the warming associated with CH_4, the second most important greenhouse gas. Again, uncertainty in current and projected cooling is substantial.

- CO_2 is the largest potential contributor of the greenhouse gases, with CH_4 the second most important contributor. Warming associated with tropospheric ozone could be important, but the underlying science allowing a quantitative judgment is weak.

- Mitigating CH_4 emissions can achieve substantial benefits, in the near term, in light of its relatively short atmospheric lifetime. In fact, a 2% per year CH_4 mitigation program can be almost as effective as placing a cap on CO_2 emissions, assuming mitigation started in 2000 and the target year is 2050.

- The United States, the former Soviet Union, China, Germany, and Japan are the largest emitters of CO_2 (in rank order). Each has a distinctive profile with regard to contributions per fuel-use sector. Developing countries in Asia, such as China and India, are expected to have exponential growth in greenhouse gas emissions, driven primarily by projected economic growth and dependence on coal as a major fossil fuel.

- Model analysis shows that the time mitigation is initiated has an important impact on the degree of mitigation achievable. For example, a program to cap (hold constant) greenhouse gas emissions can be equally effective as a more stringent mitigation program initiated 10 years later.

- Mitigation of greenhouse gas emissions will be a major challenge, since it may be necessary to dramatically decrease emissions over time. This would run counter to very strong trends toward progressively increasing emissions, driven by projected economic and population growth and widescale use of coal. Such mitigation may require major enhancements in end use efficiency in the short term and a major transition to renewables in the longer term.

- Fossil fuels are a finite source of energy. Oil and gas are projected to become scarcer and much more expensive during the middle portion of the next century. Among the renewable energy resources, only biomass and solar appear to have the potential for large scale fossil fuel displacement. Despite this, research on renewable technologies is at a constrained level, and, in the author's view, unlikely to provide technology capable of displacing large quantities of fossil fuel at competitive costs anytime in the foreseeable future.

- The Hynol process is a potentially attractive technology generating methanol (or hydrogen) for the transportation sector. However, as for other renewable technologies, a host of political, economic and policy factors inhibit commercialization.

References

Borgwardt, R.H. "Biomass and Natural Gas as Co-Feedstocks for Production of Fuel for Fuel-Cell Vehicles", U.S. EPA, National Risk Management Research Laboratoy, Research Triangle Park, NC, in press.

Bowden, T.A., et al. "CDIAC Catalog of Numeric Data Packages and Computer Model Packages." Carbon Dioxide Information Analysis Center. Oak Ridge National Laboratory, Oak Ridge, TN. Document No. ORNL/CDIAC-62. May 1993.

Dornbusch, R., and Poterba, J.M. "Global Warming Economic Policy Responses." The MIT Press, Cambridge, MA, 1991.

Environmental Protection Agency. "Policy Options for Stabilizing Global Climate," Report to Congress, 21P-2003.1, U.S. EPA, Office of Policy, Planning and Evaluation, Washington, DC, December 1990.

Intergovernmental Panel on Climate Change (IPCC), "Scientific Assessment of Climate Change," Cambridge, UK, June 1990.

Intergovernmental Panel on Climate Change (IPCC), "Climate Change 1992 - The Supplementary Report to the IPCC Scientific Assessment," Cambridge, UK, 1992.

Intergovernmental Panel on Climate Change (IPCC), "Technologies, Policies and Measures for Mitigating Climate Change", Cambridge, UK, 1996a.

Intergovernmental Panel on Climate Change (IPCC), "Climate Change 1995, The Science of Climate Change", Cambridge, UK, 1996b.

Krause, F. "Energy Policy in the Greenhouse." International Project for Sustainable Energy Paths (IPSEP), El Cerrito, CA, 1989.

National Academy of Sciences (NAS). "Policy Implications of Greenhouse Warming." National Academy Press, Washington, DC, 1991.

Princiotta, F.T. "Greenhouse Warming: The Mitigation Challenge." In Proceedings: The 1992 Greenhouse Gas Emissions and Mitigation Research Symposium, EPA-600/R-94-008 (NTIS PB94-132180), U.S. EPA, Air and Energy Engineering Research Laboratory, Research Triangle Park, N.C., January 1994.

Walker, J.C.G., and Kasting, J.F. "Effects of Fuel and Forest Conservation on Future Level of Atmospheric Carbon Dioxide." Paleoclimatol., Palaeocol, (Global and Planetary Change Section). 97 (1992) pp. 151-189. Elsevier Science Publishers.

Table 1: Five Scenarios Impacting Degree of Global Warming

Variable Impacts on Predicted Warming	*Range of Impacts——>Greater Warming*				
	Lowest	Low	*Most Likely*	High	Highest
Atmospheric Sensitivity	1.5	2	*2.5*	3.5	4.5
CO2 Life Cycle	IPCC	IPCC	*IPCC*	Kasten	Kasten
CO2 Growth Rate:1990-2030	1.4%	1.6%	*1.85%*	2.00%	2.2%
Co2 Growth Rate:2030-2100	0.5%	0.65%	*0.78%*	1.85%	2.2%
Methane Lifetime	7	8	*11*	12	13
CH4 growth Rate:1990-2030 / 2030-2100	0.67%/0.32%	0.77%/.52%	*1.17%/.82%*	1.27%/.92%	1.37%/1.02%
Penetration of HFC-134a	15%	25%	*35%*	45%	55%
Actual/Equil. Temp.Ratio @ 0.35 degree/yr	0.3	0.4	*0.505*	0.6	0.7
Current Sulfate Cooling	-2.5	-2	*-1.65*	-1	-0.1
Sulfate Cooling Emission Exponent	1	0.9	*0.8*	0.7	0.6
OUTPUT Calculations,Degree C.					
Equilibrium Temperature @ 2050	0.5	1.2	*2.3*	5.1	7.8
Realized Temperature @ 2050	0.5	0.9	*1.2*	2.6	4.4
Equilibrium Temperature @ 2100	1.1	2.4	*4.3*	10.3	15.9
Realized Temperature @ 2100	0.9	1.7	*2.2*	5.2	9.1

Table 2: Greenhouse Gases -- What is Known and What is Not

CHARACTERISTIC	CARBON DIOXIDE	METHANE	AEROSOLS	HFC-134a	TROPO. OZONE	N2O
1. Atmospheric Lifetime (yrs)	50-100	10-12.5	<<1	16	<<1	150
2. Current Concentration/	1.26	2.15	Uncertain	New CFC Substitute	>1, But Poor Data	1.08
Pre-Industrial Concentration						
3. Projected Realized Warming/						
By Gas at the Year 2100	+1.8	+0.5	-0.5	+0.2	+0.1	+0.1
Most Likely Case: Total Warming = 2.2		Incl. Indirect Effects			(Excludes CH4 source)	
4. Impact of 1% /Yr Mitigation:	60%	31%	-	-	4%	4%
Control starts at 2000, the impact at 2050:						
Calculated as % of total mitigation						
5. Confidence in Warming Calculations	Fair/Good	Fair	Poor	Good	Poor	Fair
for Items 3. and 4. Above						
6. Major Uncertainties	Carbon Cycle Influence on	1. Quantification of Natural and	1. Current Extent of Cooling	Extent to Which Will	1. Atmospheric Chemistry	Atmospheric Concentration
	CO2 Atmospheric Lifetime	Human Sources and Sinks	2. Relationship of Emissions	Substitute for CFCs	Models Insufficient	Rising Faster Than Known
		2. Explanation Needed for Decelerat-	to Atm. Aerosols		2. Data on Tropo. Ozone	Sources/Sinks Predict
		ing Growth in Atm. Concentrations	3. Impact on Cloud Formation		Trends Poor	
					3. Emission Data for NOx,	
					Hydrocarbons and CO	
					Precursors Poor	
7. Major Human Sources	Fuel Combustion	Coal Mining	Fossil Fuel Combustion	Refrigeration Cycles	Mobile Sources: VOCs, NOx,	Biomass Burning
	- Electric Power	Natural Gas and Oil Production and	Biomass Combustion		and CO	Adipic Acid and HNO3 Prod.
	- Mobile Sources	Transportation			Stationary Combustion:	Mobile Sources
	- Industrial Deforestation	Landfills			NOx and CO	Farming
		Rice Paddies			Biomass Burning: CO and	Stationary Source
		Ruminants			VOCs	Combustion
		Biomass Burning & Decomposition				

Table 3: 1988 CO_2 Data for Key Countries

CO2 Emissions-1988		CO2 per capita		CO2 per GNP	
(Million of Tons)		(tons per person)		(Mt CO2 per $1000 GNP)	
United States	4804	United States	19	China	6.0
USSR	3982	Canada	17	South Africa	3.6
China	2236	Czechoslovakia	15	Romania	2.8
Germany	997	Australia	15	Poland	2.7
Japan	989	USSR	14	India	2.5
India	601	Germany	13	Czechoslovakia	1.9
United Kingdom	559	Poland	12	Mexico	1.7
Poland	459	United Kingdom	10	USSR	1.5
Canada	438	Romania	10	Korea	1.2
Italy	360	South Africa	8	Canada	1.0
France	320	Japan	8	United States	1.0
Mexico	307	Italy	6	Australia	1.0
South Africa	284	France	6	United Kingdom	0.8
Australia	241	Korea	5	Germany	0.7
Czechoslovakia	234	Spain	5	Brazil	0.6
Romania	221	Mexico	4	Spain	0.6
Korea	205	China	2	Italy	0.4
Brazil	202	Brazil	2	Japan	0.3
Spain	188	India	1	France	0.3

Table 4: Assumed Annual Growth Factors Influencing CO_2 Emissions (1990 - 2025)
(Derived from IPCC, 1992)

FACTOR	OECD	ASIA
Growth of Economy Per Capita	2.2%	3.5%
Population Growth Rate	0.3%	1.5%
Growth Rate: Energy Use Per Economic Output	-1.1%	-0.8%
Growth Rate: Carbon Emissions Per Energy Use Unit	-0.7%	-0.3%
Annual CO_2 Growth Rate (Sum of above factors)	+0.7%	+3.9%

Table 5: **1990 Global Energy Use and CO2 Emissions from Energy Sources**

Carbon expressed in Gt C; Energy as EJ

	Energy Used	**CO2 Emitted**
Electric Generation	96	1.3
Direct Use of Fuels by Sector		
Resid./Comm./Inst.	47	0.9
Industry	68	1.4
Transportation	51	0.9
TOTAL	**262**	**4.5**
Demand Side		
Resid./Comm./Inst.	86	1.4
Industry	123	2.1
Transportation	53	1.0
TOTAL	**262**	**4.5**
By Source		
Solids	77	1.9
Liquids	90	1.7
Gases	61	0.9
Other	34	0.0
TOTAL	**262**	**4.5**

Table 6: Global Fossil Energy Reserves and Resources, In EJ

	Consumption		Reserves Identified	Conventional Resources Remaining to Be Discovered at Probability			Unconventional Resources		Resource Base [a]	Year Resource Is Depleted (at 1.5% annual growth rate)
	1860-1990	1990		95%	50%	5%	Currently Recoverable	Recoverable W/Techno-Logical Progress		
Oil										
Conventional	3343	128	6000	1800	2500	5500			8500	2035
Unconventional	-	-	7100					9000	16100	2080
Gas										
Conventional	1703	71	4800	2700	4400	10900			9200	2065
Unconventional	-	-	6900				2200	17800	26900	2135
Coal	5203	91	25200				13900	86400	125500	2195
Total	10249	290	50000	>4500	>6900	>16400	>16100	>113200	>186200	2150

Notes: All totals have been rounded: - = negligible amounts: blanks = data not available

[a] Resource base is the sum of reserves and resources. Conventional resources remaining to be discovered at probability of 50% are included for oil and gas

Table 7: Global Renewal Energy Potentials by 2020-2025, and Maximum Technical Potentials in EJ Thermal Equivalent [a]

	Consumption 1860-1990	Consumption 1990	Potential by 2020-2025[b]	Fraction Global Energy By 2050 [c]	Long-Term Technical Potentials[d]	Fraction Global Energy by 2100 [e]
Hydro	560	21	35-55	5-8%	>130	>9%
Geothermal	-	>1	4	0.5%	> 20	>1%
Wind	-	-	7-10	1-1.5%	>130	>9%
Ocean	-	-	2	0.2%	>20	>10%
Solar	-	-	16-22	2-3%	>2600	100%
Biomass	1150	55	72-137	9-19%	>1300	>87%
Total	1750	76	130-230	18-32%	>4200	100%

Notes: All totals have been rounded; - =negligible amounts; blanks = data not available

[a] All estimates have been converted into thermal equivalent with an average factor of 38.5%.

[b] It represents renewable potentials by 2020-2025, in scenarios with assumed policies for enhanced exploitation of renewable potentials.

[c] Based on potential by 2020-2025 and assuming 709 EJ utilized in 2050.

[d] Long-term potentials are based on the IPCC Working Group II. This evaluation is intended to correspond to the concept of fossil energy resources, conventional and unconventional.

[e] Based on long-term potentials by 2100, assumed 1492 EJ in 2100.

Table 8: Brief Descriptions of End Use Mitigation Options For the United States (NAS, 1991)

END USE: RESIDENTIAL AND COMMERCIAL ENERGY MANAGEMENT	
Electricity Efficiency Measures	
Residential Lighting	Reduce lighting energy consumption by 50% in all U.S. residences through replacement of incandescent lighting with compact fluorescents.
Water Heating	Improve efficiency by 40 to 70% through efficient tanks, increased insulation, low-flow devices, and alternative water heating systems.
Commercial Lighting	Reduce lighting energy consumption by 30 to 60% by replacing 100% of commercial light fixtures with compact fluorescent lighting, reflectors, occupancy sensors, and day lighting.
Commercial Cooling	Use improved heat pumps, chillers, window treatments, and other measures to reduce commercial cooling energy use by 30 to 70%.
Commercial Refrigeration	Improve efficiency 20 to 40% through improved compressors, air barriers and food case enclosures, and other measures.
Residential Appliance	Improve efficiency of refrigeration and dishwashers by 10 to 30% through implementation of new appliance standards for refrigeration, and use of no-heat drying cycles in dishwashers.
Residential Space Heating	Reduce energy consumption by 40 to 60% through improved and increased insulation, window glazing, and weather stripping along with increased use of heat pumps and solar heating.
Commercial and Industrial Space Heating	Reduce energy consumption by 20 to 30% using measures similar to these for the residential sector.
Commercial Ventilation	Improve efficiency 30 to 50% through improved distribution systems, energy-efficient motors, and various other measures.
Oil and Gas Efficiency	Reduce residential and commercial building fossil fuel energy use by 50% through improved efficiency measures similar to the ones listed under electricity efficiency.

(continued)

Table 8 (continued)

Fuel Switching	Improve overall efficiency by 60 to 70% through switching 10% of building electricity use from resistance heat to natural gas heating.
END USE: INDUSTRIAL ENERGY MANAGEMENT	
Cogeneration	Replace existing industrial energy systems with an additional 25,000 MW of co-generation plants to produce heat and power simultaneously.
Electricity Efficiency	Improve electricity efficiency up to 30% through use of more efficient motors, electrical drive systems, lighting, and industrial process modification.
Fuel Efficiency	Reduce fuel consumption up to 30% by improving energy management, waste heat recovery, boiler modifications, and other industrial process enhancements.
New Process Technology	Increase recycling and reduce energy consumption primarily in the primary metals, pulp and paper, chemicals, and petroleum refining industries through new, less energy intensive process innovations.
END USE: TRANSPORTATION ENERGY MANAGEMENT	
Vehicle Efficiency	
Light Vehicles	Use technology to improve on-road fuel economy to 25 mpg with no changes in the existing fleet.
	Improve on-road fuel economy to 36 mpg with measures that require changes in the existing fleet such as downsizing.
Heavy Trucks	Use measures similar to those for light vehicles to improve heavy truck efficiency up to 31 mpg.
Aircraft	Implement improved fanjet and other technologies to improve fuel efficiency by 20% to 130 to 140 seat-miles per gallon.
Transportation Demand	Reduce solo commuting by eliminating 25 % of the employer-provided parking spaces and management placing a tax on the remaining spaces to reduce solo commuting by an additional 25 %.

Table 9: Brief Descriptions of Production-side Mitigation Options for the United States (NAS, 1991)
(Note 1 Quad = 1.055×10^{18} J)

Option	Description
ALTERNATIVE FUELS FOR TRANSPORTATION	
Methanol from Biomass	Replace all existing gasoline engine vehicles with those that use methanol produced from biomass
Hydrogen from Nonfossil Fuels	Replace gasoline with hydrogen created from electricity generated from nonfossil fuel sources such as nuclear and solar energy directly in transportation vehicles.
ELECTRICITY AND FUEL SUPPLY	
Heat Rate Improvements	Improve heat rates (efficiency) of existing plants by up to 4% through improved plant operation and maintenance.
Advanced Coal	Improve overall thermal efficiency of coal plants by 10% through use of integrated gasification combined cycle, pressurized fluidized-bed, and advanced pulverized coal combustion systems.
Natural Gas	Replace all existing fossil-fuel-fired plants with gas turbine combined cycle systems to both improve thermal efficiency of current natural gas combustion systems, and replace fossil fuels such as coal and oil that generate more CO_2 than natural gas.
Nuclear	Replace all existing fossil-fuel-fired plants with nuclear power plants such as advanced light-water reactors.
Hydroelectric	Replace fossil-fuel-fired plants with remaining hydroelectric generation capability of 2 quads.
Geothermal	Replace fossil-fuel-fired plants with remaining geothermal generation potential of 3.5 quads.
Biomass	Replace fossil-fuel-fired plants with biomass generation potential of 2.4 quads.
Solar Photovoltaics	Replace fossil-fuel-fired plants with solar photovoltaic generation potential of 2.5 quads.
Solar Thermal	Replace fossil-fuel-fired plants with solar thermal generation potential of 2.6 quads.
Wind	Replace fossil-fuel plants with wind generation potential of 5.3 quads.
CO_2 Collection and Disposal	Collect and dispose of all CO_2 generated by fossil-fuel-fired plants into the deep ocean or depleted gas and oil fields.

Table 10: Selected CO_2 Emission Global Mitigation Policy Strategies: Decrease in Projected Warming (Equilibrium) Relative to Base Case (Adapted from EPA, 1990)

Strategy	Assumptions	Potential Emission Reductions 2050	2100	Comments
End Use Strategies Improved Transportation Efficiency	See Footnote a	6%	9%	Recent trends in US moving in opposite direction, many low mpg vans, light trucks replacing autos
Residential, Commercial Industrial Efficiency Gains	See Footnote b	9%	15%	Such reductions would require major marketing campaign, carbon taxes and other economic incentives
Production Strategies More Nuclear Power Use (Electricity production)	See Footnote c	2%	4%	Such increased use would need to be accepted by public. Marketing incentives as well as assumed cost reduction needed
Solar Technologies (Electricity Production)	See Footnote d	2%	4%	Breakthrough in technology would be necessary for such penetration
Natural Gas Incentives (Electricity Production)	See Footnote e	<1%	<1%	Natural gas generates about half the CO_2 per output relative to coal. Availability of natural gas limits option.
Commercialized Biomass (Transportation & Stationary Source)	See Footnote f	8%	12%	Largest potential impact of renewable technologies; feasibility dependent on large areas dedicated to energy crops and available production technology and end use of infrastructure
Sequestration Strategies Reforestation	See Footnote g	7%	5%	Would require a massive turn around toward net forest gain relative to current rapid deforestation

a.The average efficiency of cars and light trucks in the U.S. reaches 30 mpg (7.8 liters/100 km) by 2000, new cars achieve 40 mpg (5.9 liters/100 km). Global fleet-average automobile efficiency reaches 43 mpg by 2025.

b.The rates of energy efficiency improvements in the residential, commercial, and industrial sectors are increased about 0.3-0.8 percentage points annually from 1985 to 2025 compared to the base case and about 0.2-0.3 percentage points annually from 2025 to 2100.

c. Assumes that technological improvements in the design of nuclear powerplants reduce costs by about 0.6 cents/kWh by 2050. In the base case nuclear costs in 1985 were assumed to be 6 to 10 cents/kWh (1988 $).

d. Assumes that low-cost solar technology is available by 2025 at costs as low as 6.0 cents/kWh. In the base case these costs approached 8.5 cents/kWh but these levels were not achieved until after 2050.

e. Assumes that economic incentives to use gas for electricity generation increase gas share by 5% in 2000 and 10% in 2025.

f. Assumes the cost of producing and converting biomass to modern fuels reaches $4.25/GJ (1988 $) for gas and $6.00/GJ (1988 $) for liquids. The maximum amount of liquid or gaseous fuel available from biomass (i.e., after conversion losses) is 205 EJ.

g. The terrestrial biosphere becomes a net sink for carbon by 2000 through a rapid reduction in deforestation and a linear increase in the area of reforested land and biomass plantation. Net CO_2 uptake by 2025 is 0.7 Pg C. In the base case, the rate of deforestation continues to increase very gradually, reaching 15 Mha/yr in 2097, and no reforestation occurs.

Table 11: Total Reported IEA Government R&DBudgets (Columns 1-7; US$ Billion (10^9) at 1994 Prices and Exchange Rates) and GDP (Column 8; U.S.$ Trillion (10^{12}) at 1993 Prices)

Year	(1) Fossil Energy	(2) Nuclear Fission	(3) Nuclear Fusion	(4) Energy Conservation	(5) Renewable Energy	(6) Other	(7) Total	(8) GDP	(9) % of GDP
1983	1.70	6.38	1.43	0.79	1.05	1.08	12.40	10.68	0.12
1984	1.60	6.12	1.44	0.70	1.02	0.99	11.88	11.20	0.11
1985	1.51	6.26	1.42	0.70	0.85	1.04	11.77	11.58	0.10
1986	1.51	5.72	1.31	0.59	0.66	0.94	10.74	11.90	0.09
1987	1.37	4.36	1.23	0.65	0.62	1.04	9.27	12.29	0.08
1988	1.46	3.64	1.13	0.53	0.62	1.19	8.58	12.82	0.07
1989	1.30	4.42	1.07	0.45	0.57	1.33	9.13	13.23	0.07
1990	1.75	4.48	1.09	0.55	0.61	1.15	9.62	13.52	0.07
1991	1.52	4.45	0.99	0.59	0.64	1.39	9.57	13.58	0.07
1992	1.07	3.90	0.96	0.56	0.70	1.28	8.48	13.82	0.06
1993	1.07	3.81	1.05	0.65	0.71	1.38	8.66		
1994	0.98	3.74	1.05	0.94	0.70	1.30	8.72		

INPUT INFORMATION		CO2 Gwth	CO2 Decay Option			Sulfate	Input		Tropo.O3
1ST YR:	1980	1990-2030	IPCC-1992}		1	1980 Sulfate Impact			PPM-O3/KG:
; END YR:	2100	1.85%	Kasting Model}		2	-1.65	watts/sq.m		CH4
;IMPACT YR:	2100	2030-2100	CO2 Option =		1	Average for North Hemi.			3.50E-15
CONTROL CASE NO:	1	0.780%	OUTPUT SUMMARY			0.8	{SO4 effect.exponen		NOx
1=BASE, 2=Control, 7=Cap		CH4 Gwth	Equil.Warming			For 1850 to 1980			3.80E-14
START CONTROL	2000	1980-2030	4.24	Deg.Celcius		EQUIL.WARMING			CO
;ANN.EMIS.CONTROL:	1.0%	1.170%				0.54	Deg.Celcius		6.40E-15
CFC PHASEOUT?	1	2030-2100	Transient Warming=						NMHC
(1=YES,2=NO,3=see B9)		0.820%	2.12	Deg.Celcius		Transient Warming=			3.60E-14
Effect. of CFC Warming	0.5	SO2 Gwth				0.48	Deg.Celcius		Actual/Derwent=
% CFCs to H FC-134a	0.35	1990-2030	CH4 ppm	N2Oppm	CO2ppm		1980		0.3
METHANE LIFETIME	*11*	1.20%	4.55	0.388	812		CH4 ppm	CO2ppm	Transient Respons
ATM.SENSITIV.2X,degre	2.50	2030-2100	Rate of Eq.Warm.		0.035	pre-ind.	0.8	280	1
		0.35%	(Degree C per Year)			1980	1.65	338	4=slowest,5=slow
			Allowable C02:IPCC 1994						1=base,2=fast
			450	ppm	eq.				3=fastest

Fig. 1: Glowarm Model Input and Output Screen

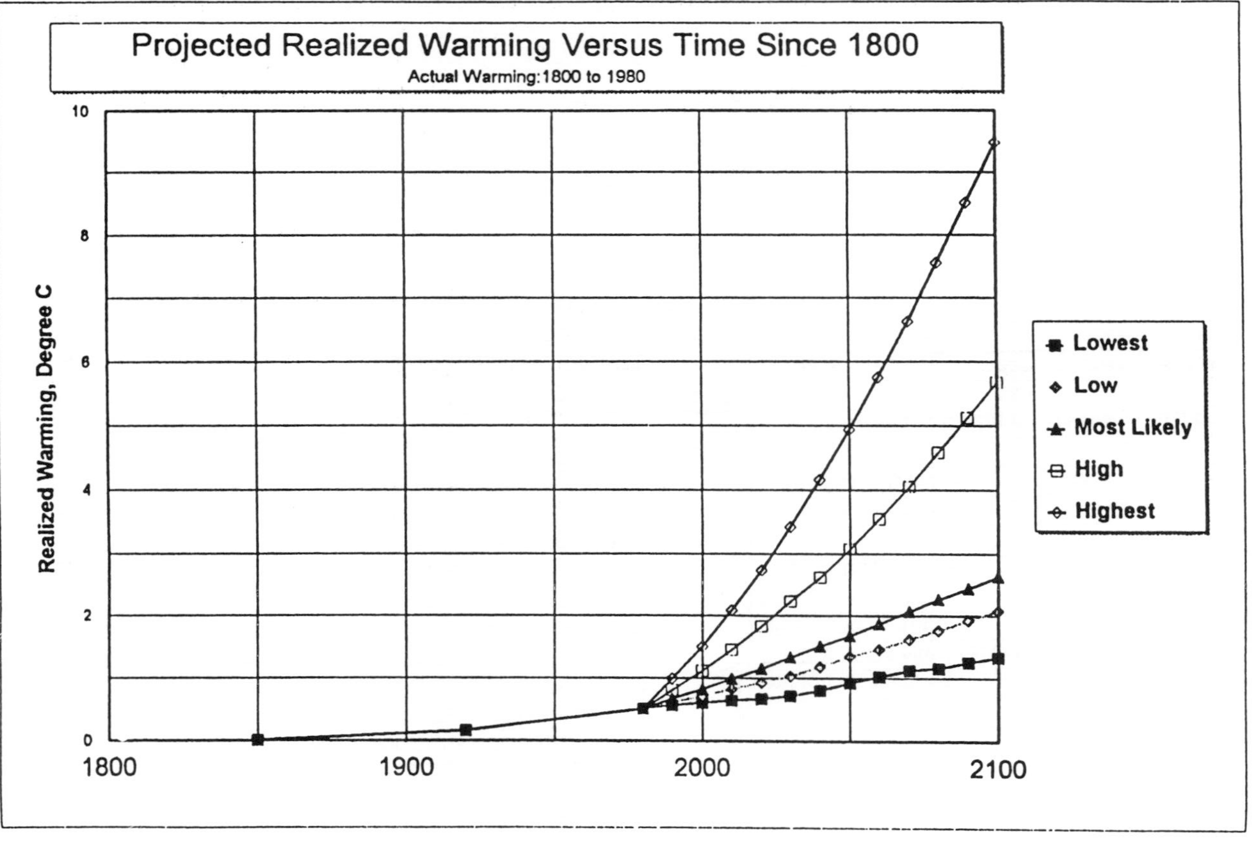

Fig. 2: Projected Realized Warming Vs. Time Since 1800 (actual warming 1800 to 1980)

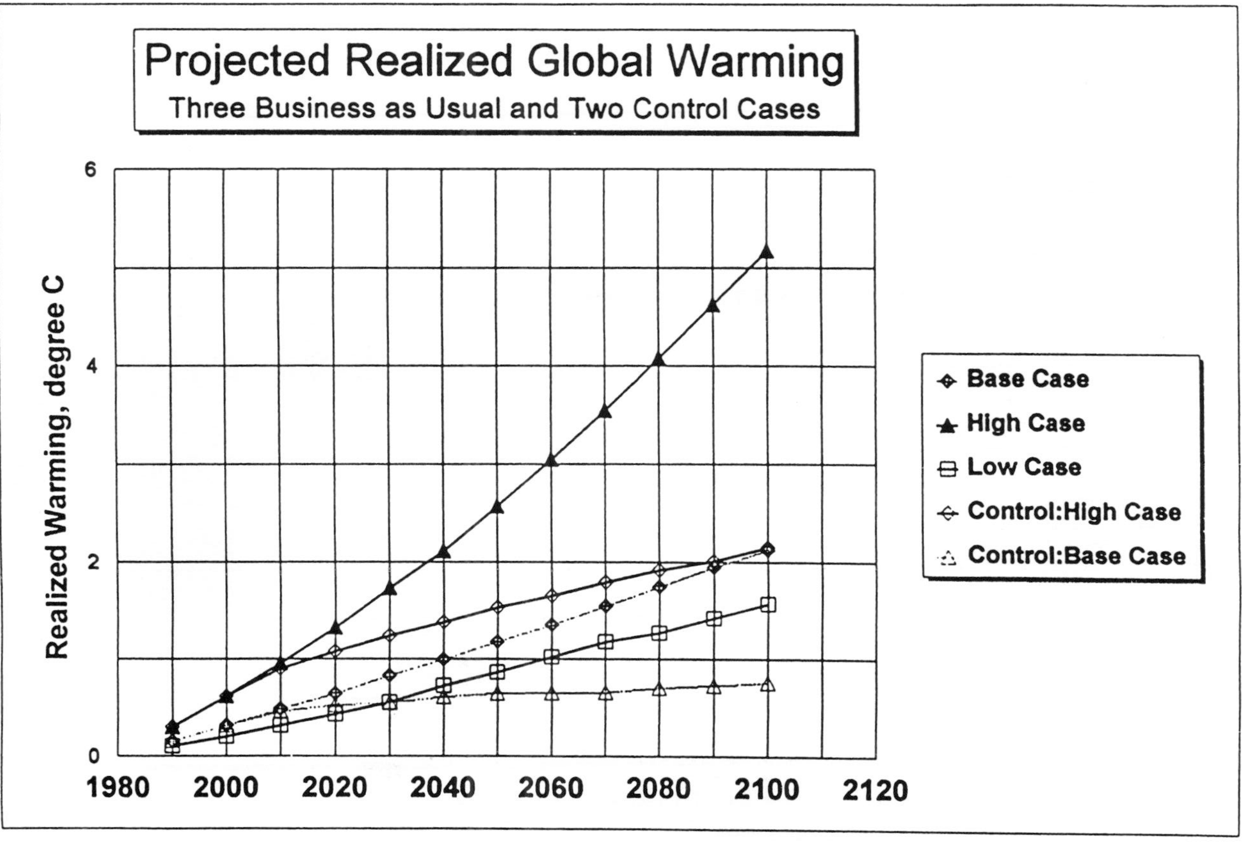

Fig. 3 : Projected Realized Global Warming for Three Business as Usual and Two Control Cases

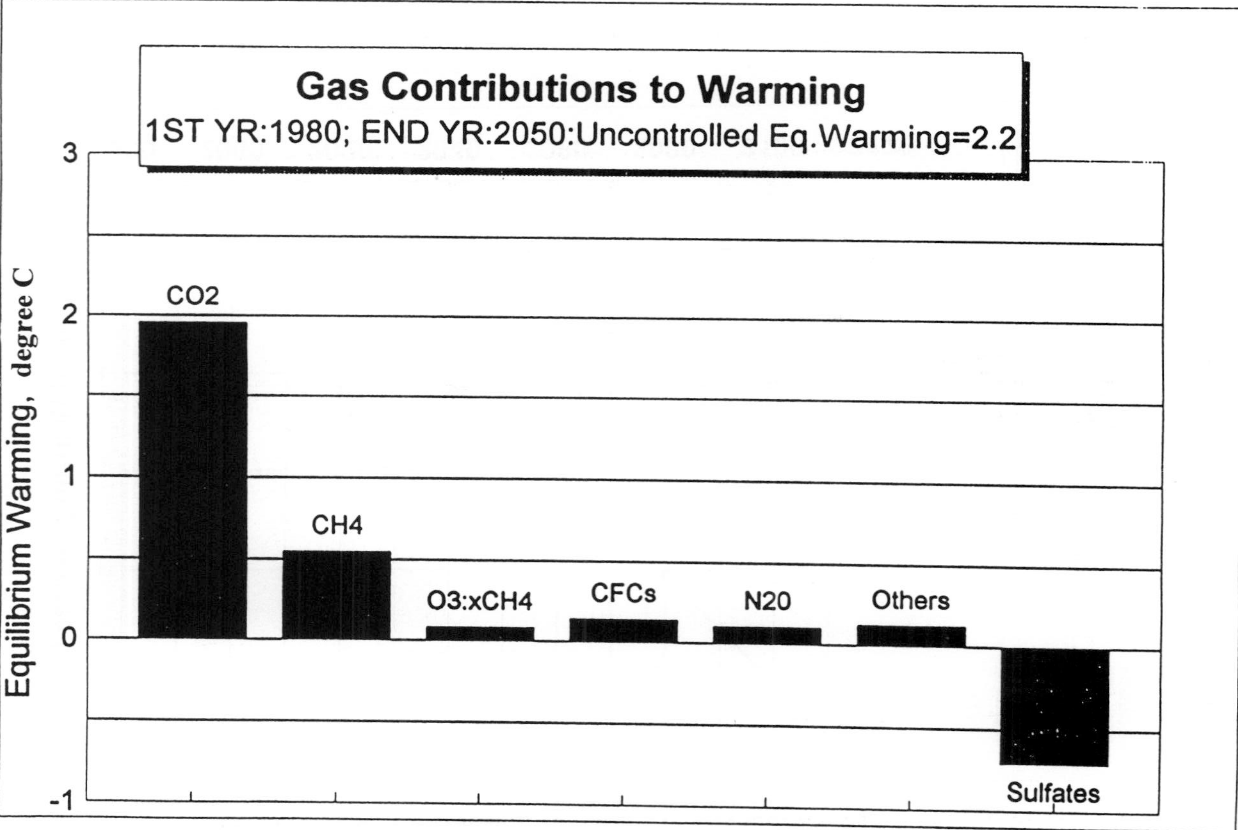

Fig. 4: Gas Contributions to Warming

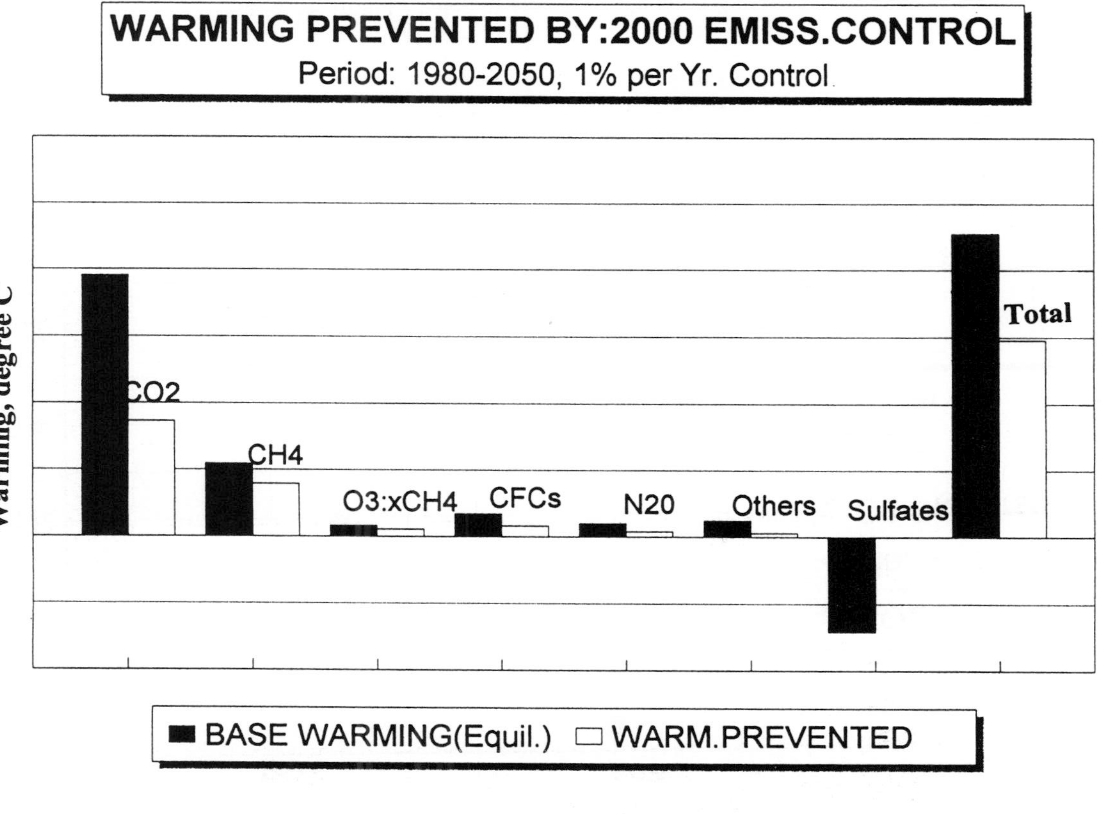

Fig. 5: Warming Prevented by 2000 Emission Control (period 1980 - 2050, 1% per yr. control)

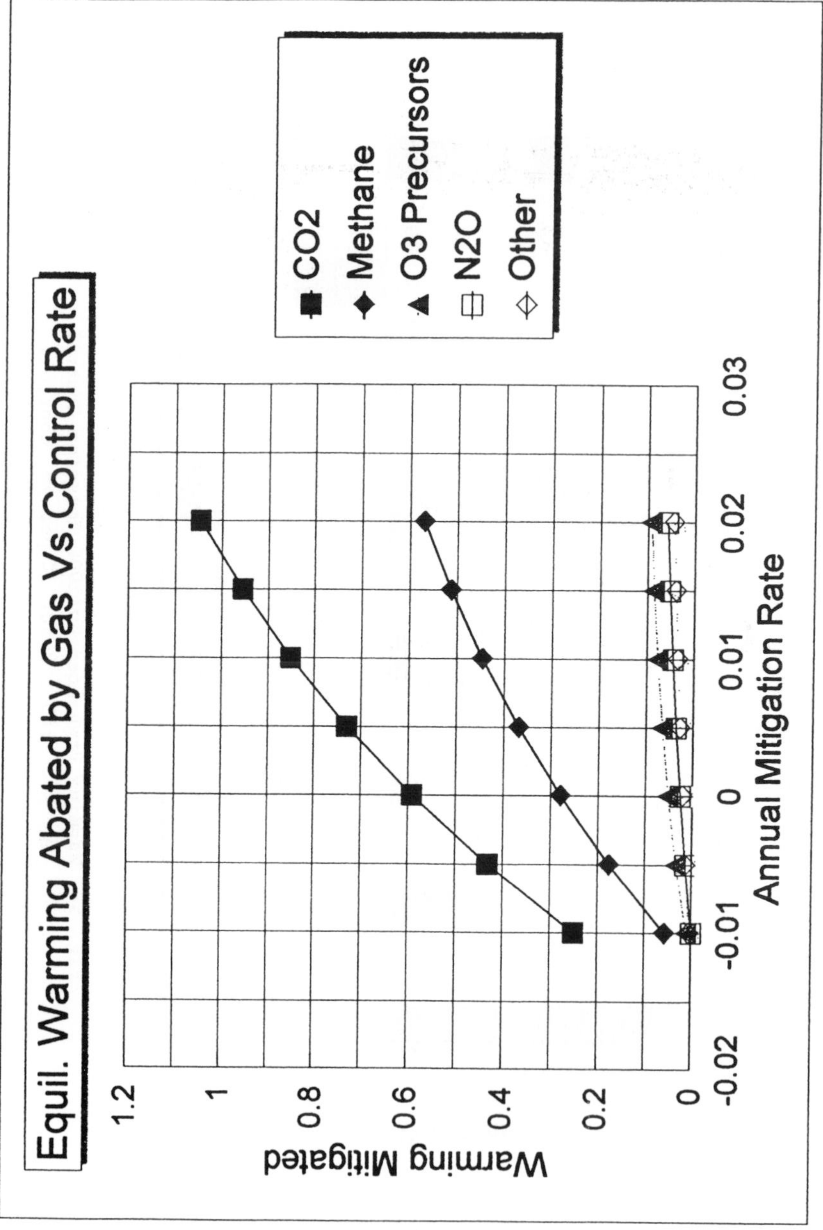

Fig. 6: Warming Abated by Gas Vs. Mitigation Rate

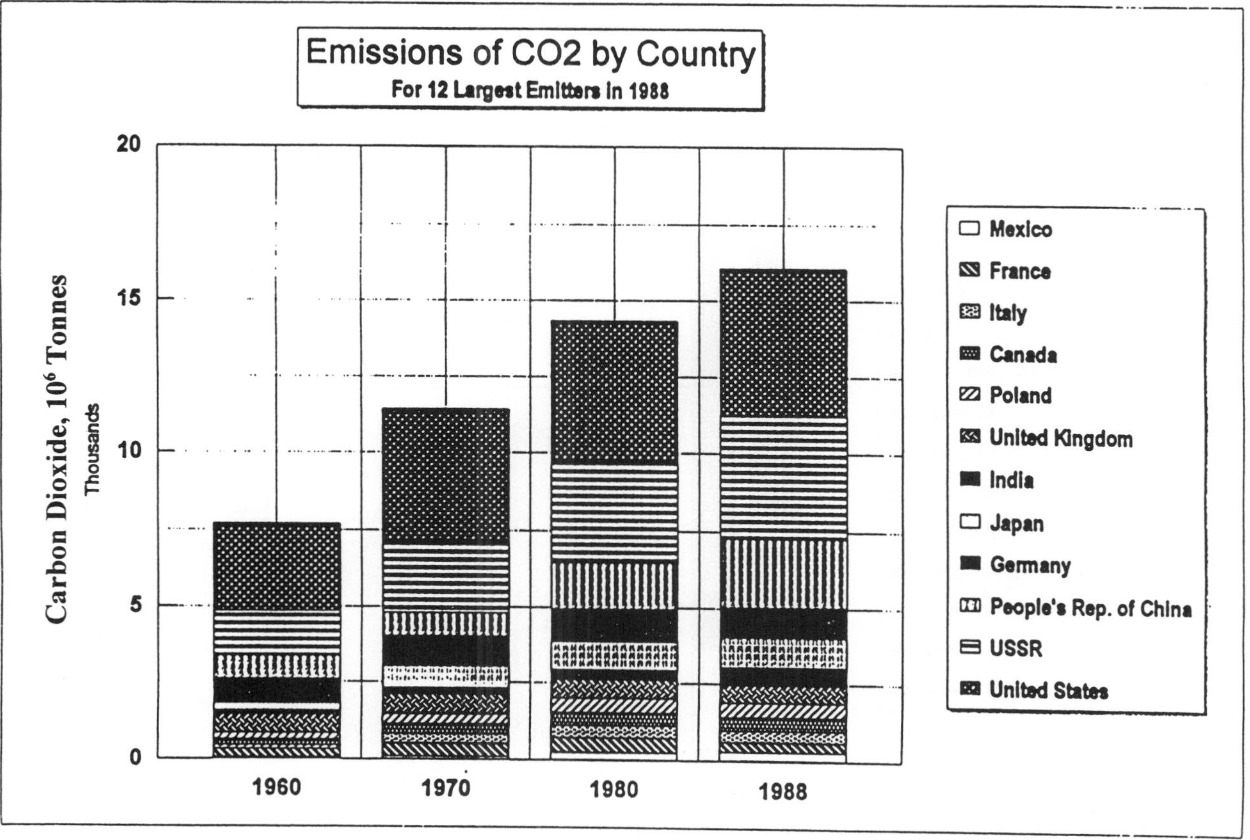

Fig. 7: Historical Emissions of CO_2 by Country (12 largest emitters in 1988)

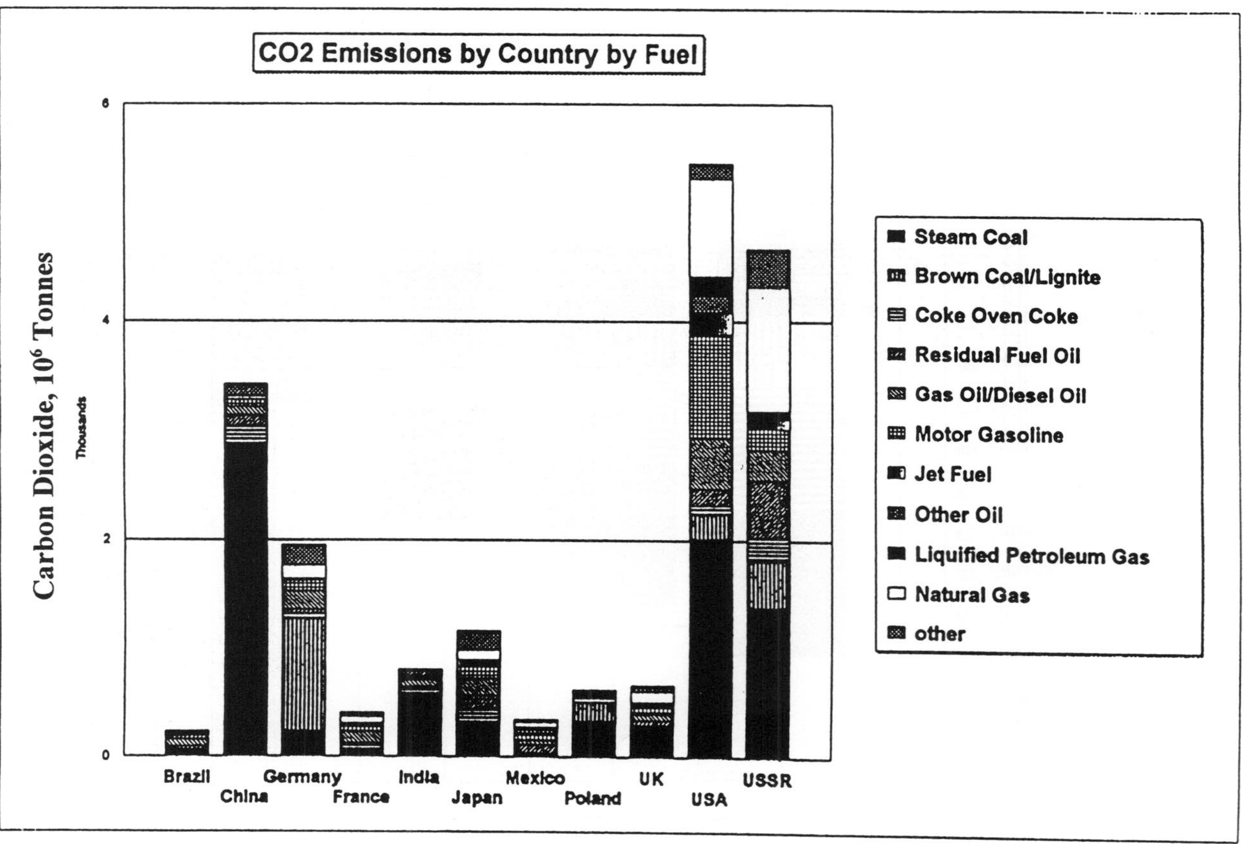

Fig. 8: Recent CO² Emissions by Country by Fuel

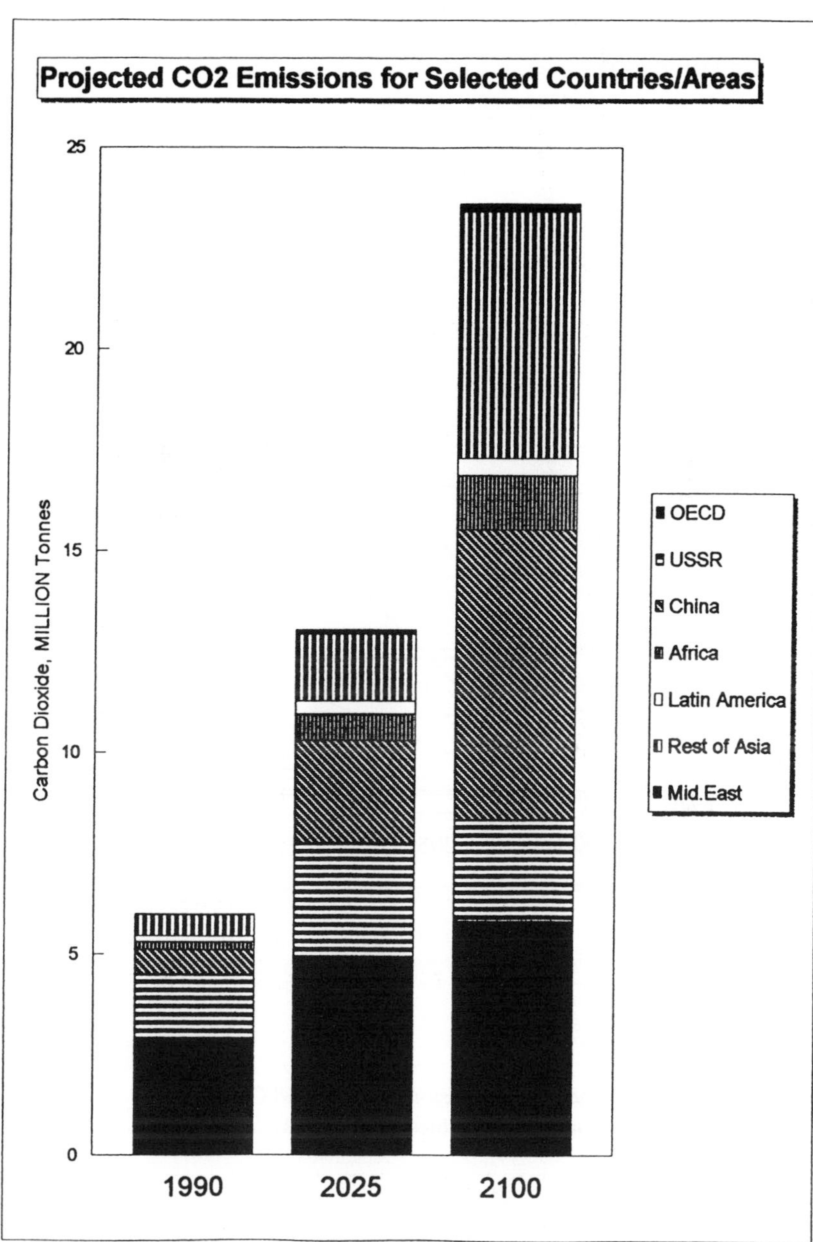

Fig. 9: Projected CO_2 Emissions for Selected Countries
(Cumulative bar chart; actual values 1960-1988)

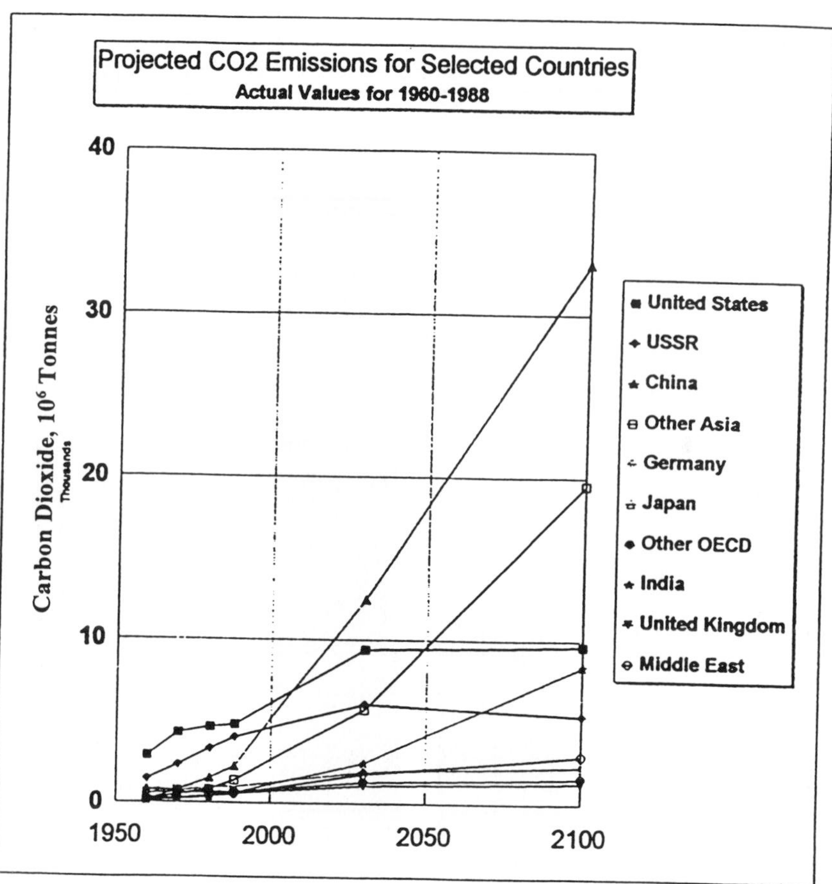

Fig. 10: Projected CO_2 Emissions for Selected Countries
(line chart; actual values 1960 - 1988)

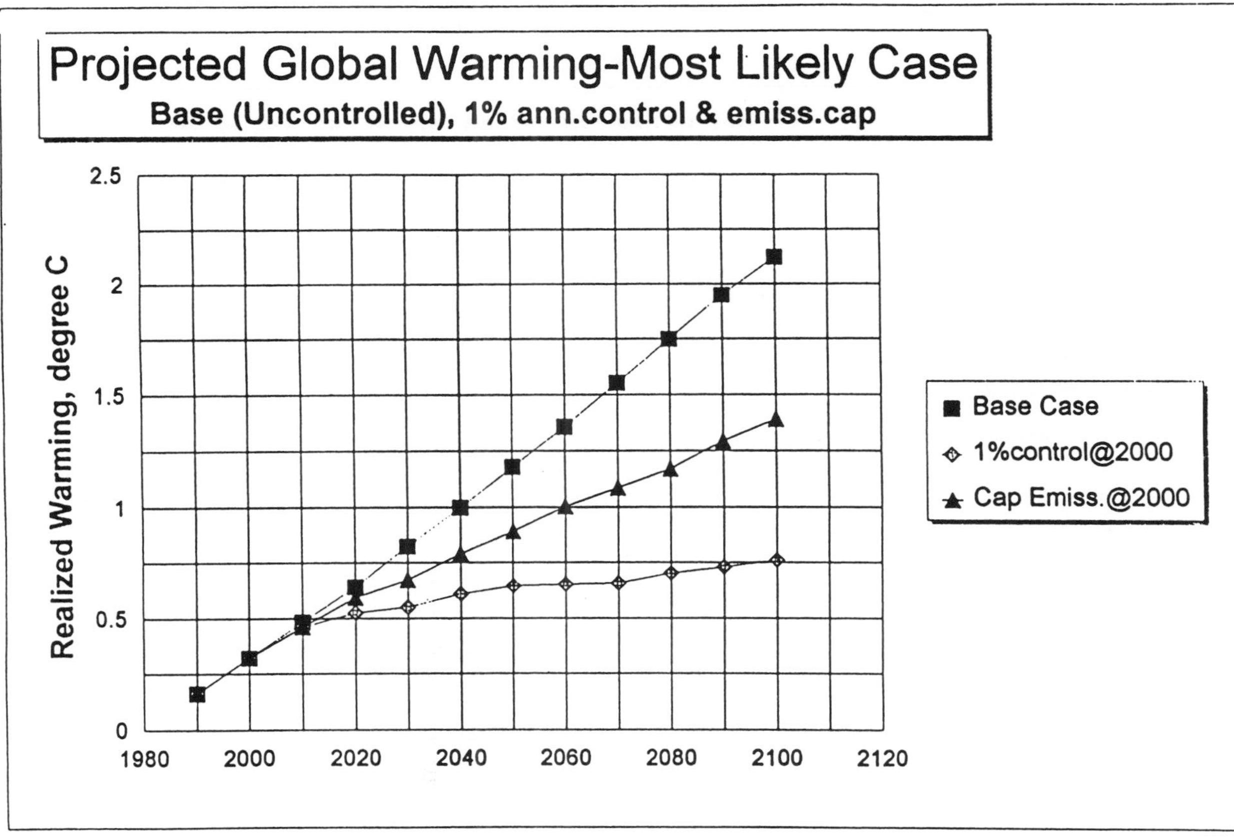

Fig. 11: Projected Global Warming for the Base Case and Two Mitigation Scenarios

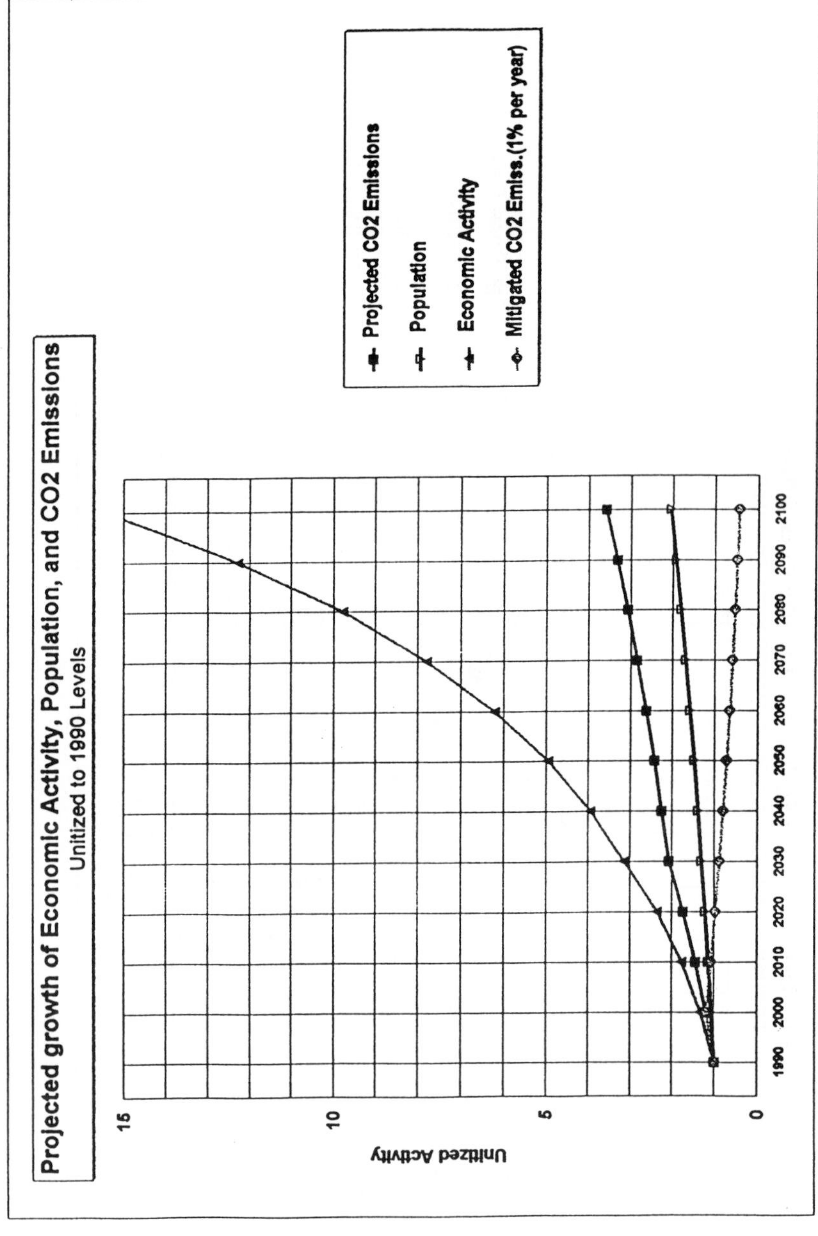

Fig. 12: Projected Growth of Economic Activity, Population, and CO² Emissions (unitized at 1990 levels)

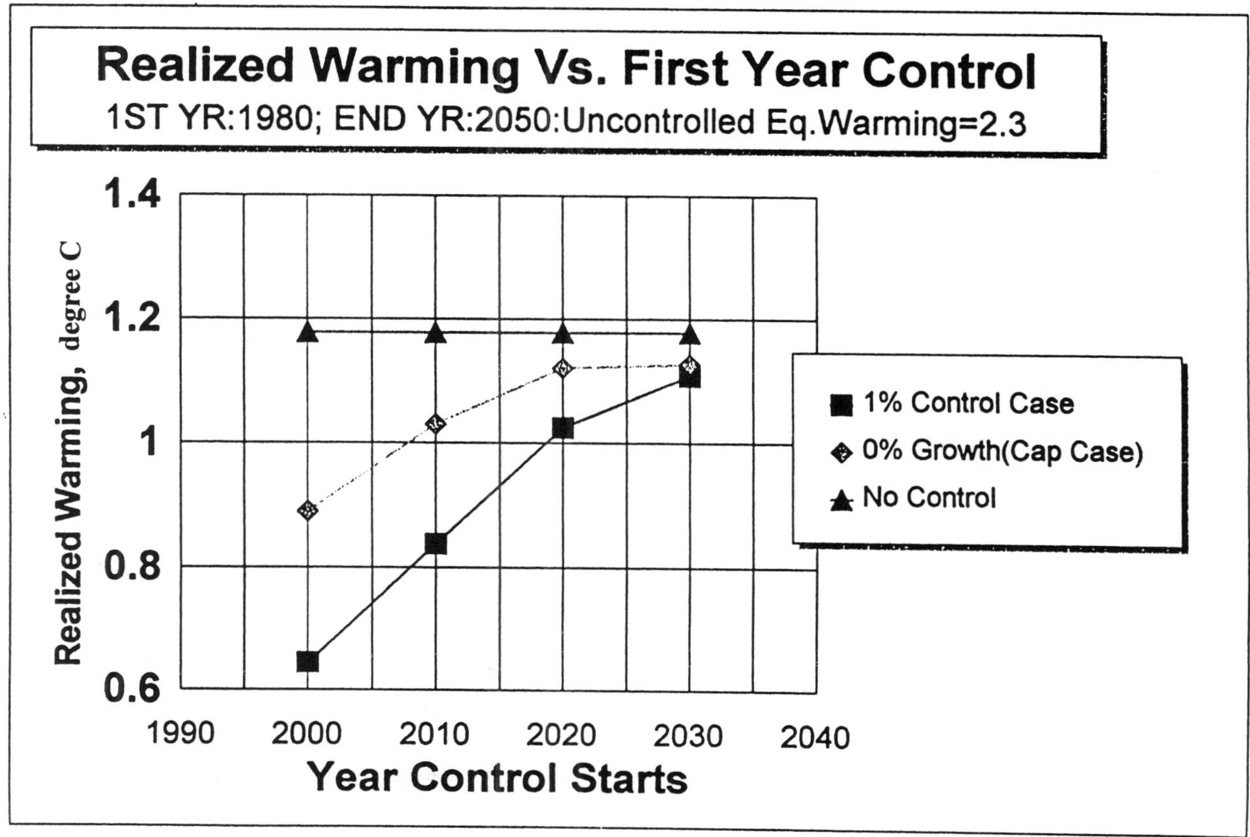

Fig. 13: Projected Realized Warming Vs. First Year Control

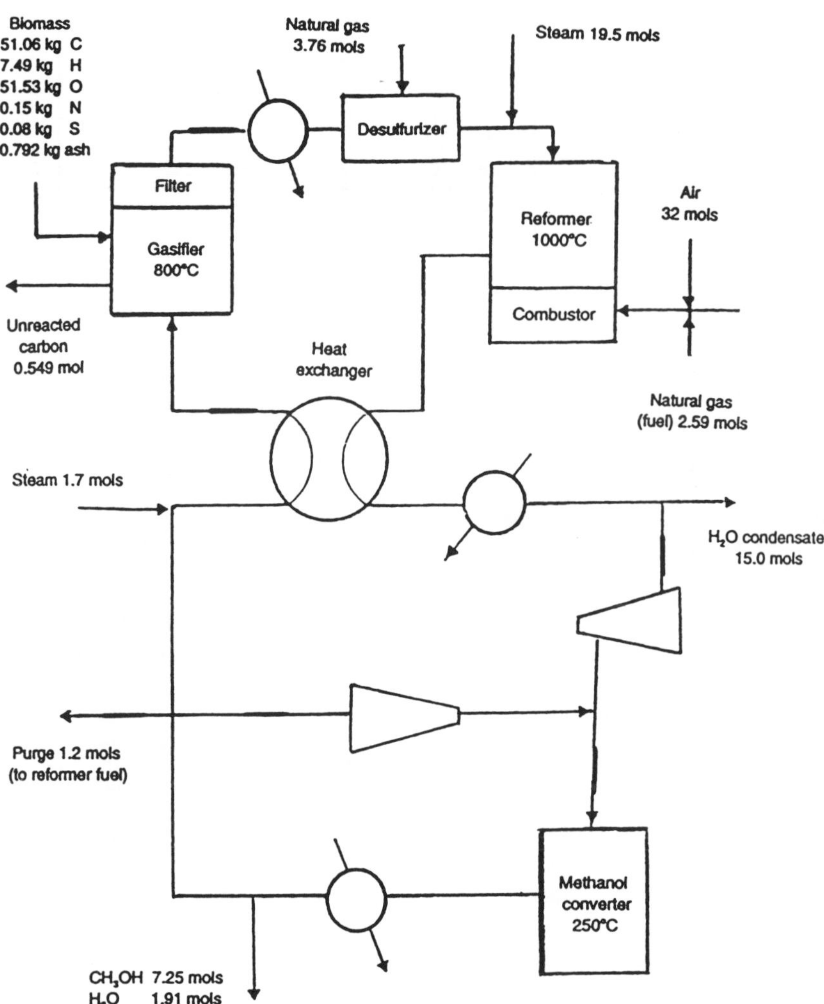

Fig. 14: The Hynol Process

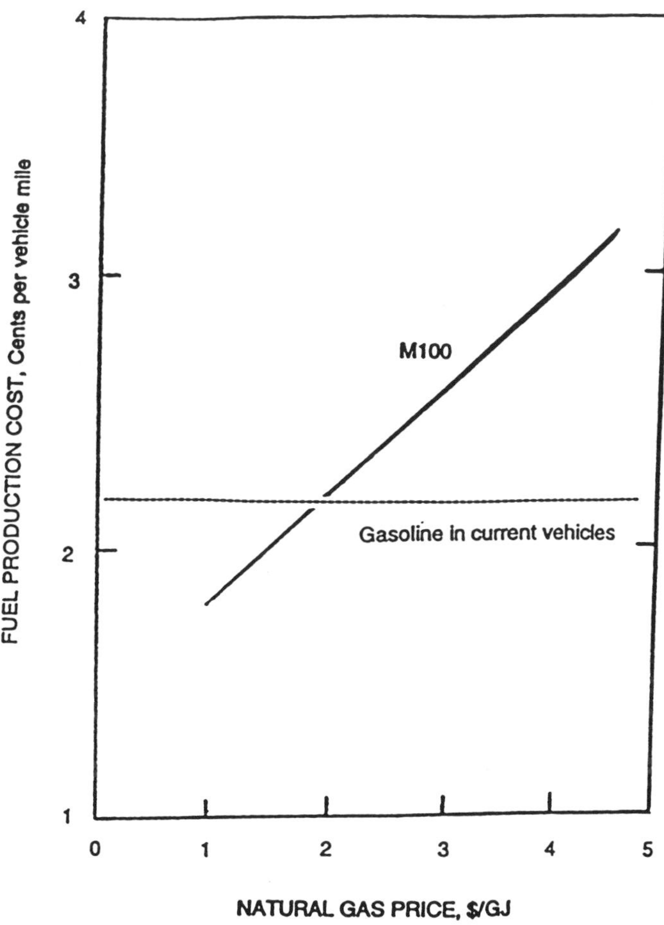

ASSUMPTIONS

1. Production cost of gasoline is assumed constant at $0.60/gal
2. Fuel economy of conventional gasoline vehicles is assumed to be 27 miles/gal
3. Optimized methanol vehicles using M100 are 27% more fuel efficient than gasoline vehicles
4. Hynol plant size is 7900 tonnes/day
5. Biomass is delivered at $61/tonne
6. Optimized Hynol process produces methanol at $0.42/gal

Fig. 15: Hynol Methanol Vs. Gasoline Used in Vehicles With Internal Combustion Engines

Canada's environmental technology verification program

Ray Klicius, P. Eng. and Andrew Spoerri

Environmental Technology Advancement Directorate, Environmental Protection Service, Environment Canada

1. THE ETV PROGRAM

The Environmental Technology Verification (ETV) Program is an initiative designed to accelerate the growth and marketability of Canada's environment industry. The Program builds on Canada's reputation by emphasizing our capabilities and credibility in the environmental market.

The ETV Program provides validation and independent verification of environmental technology performance claims. This initiative has been developed to promote the commercialization of new environmental technologies into domestic and international marketplaces and thus provide industry with the tools to address environmental challenges efficiently, effectively and economically.

Environmental technology vendors (suppliers) apply to the Program for verification of the claims they make concerning the performance of their environmental technologies. If the claim is verified, the company is issued three documents: a Verification Certificate; a Verification FactSheet; Final Verification Report, and; is entitled to use the ETV logo (on the specified documentation) to market their technology in Canada and abroad. As each claim has very specific parameters, the logo may only be used in conjunction with the three verification documents.

This is a voluntary Program to provide buyers with an assurance that vendor's claims of performance for an environmental technology are valid, credible and supported by suitable demonstration test information. Besides suppliers of environmental technologies, suppliers of equipment-based environmental services (where performance can be verified) are eligible to apply for verification.

The Government of Canada federal department, Environment Canada, was the lead department in the conceptualization and development of this initiative in cooperation with Industry Canada and with direction from the ETV government/industry Steering Committee. The Program originated from one of the key initiatives of the *Canadian Strategy for the Environment Industry.*

2. BACKGROUND

The *Canadian Strategy for the Environment Industry*, announced by Environment Canada and Industry Canada in 1994, called for the examination of the certification of products, processes and services (Initiative 5). In response, Environment Canada and Industry Canada, in partnership 6 with the Canadian Environment Industry Association (CEIA) and other stakeholders, embarked on an examination of options in October 1994 for a national certification program. Based on national consultations of various options in October - November 1995, the initiative was refocused towards the concept of an Environmental Technology Verification Program, modelled after a successful program in the American state of California.

During the Program's development since its conceptual stage in 1994, environment industry representatives and the CEIA have been continually consulted and generally very supportive. A *Draft Business Plan* was prepared and distributed for review in April 1996 for a series of national consultations held in May 1996. There was strong endorsement for the Program and positive feedback on the *Draft Business Plan.*

A strong message was heard from the environment industry participating in these consultations on the need for greater industry representation on the ETV Program Steering Committee. Consequently in July 1996, the Steering Committee was restructured to include a majority industry representation. The Steering Committee was consulted on all major issues during the development of the ETV Program and played an integral part in all key Program milestones.

Canada faces the challenge of enhancing the international credibility and competitiveness of our environment industry and pursuing reciprocal agreements with international programs. A key message heard from stakeholders during the consultation process was the need for reciprocity with similar American (and international) initiatives. For reciprocity to be effective, Canada's ETV Program must have comparable elements and compatibility with other programs.

Environment Canada and the U.S. Environmental Protection Agency signed a Cooperative Agreement seeking harmonization of respective Programs. Furthermore, California Environmental Protection Agency, who is currently operating a similar environmental technology verification program, signed a Memorandum of Understanding with Environment Canada to strive for reciprocity. The agreement, designed to facilitate the exchange of information and reciprocity on respective verification programs, will set a precedent for similar agreements and international cooperation and open doors for Canadian environmental companies doing business abroad.

3. DEFINITIONS

3.1. Environmental Technology

For the purposes of the ETV Program, environmental technologies are ***products and processes*** that offer an environmental benefit or address an environmental problem. This definition includes products and processes whose primary purpose is environmental protection or remediation. It also includes products or processes that contribute to environmentally sound production, including alternative production processes and materials. The focus is on environmental technologies and equipment-based services for industrial and institutional applications. (Note: the ETV Program does not deal with "green" consumer products which are addressed by Canada's Environmental Choice Program™.)

Environmental technologies address a wide range of environmental protection and conservation needs, including:

- Pollution prevention
- Pollution detection and monitoring
- Environmentally-related human health protection
- Pollution control and treatment
- Energy efficiency/management
- Emergency response
- Non-hazardous and hazardous waste management
- Site remediation and restoration
- Land and natural resource management.

3.2. Equipment-based Service

For the purposes of the ETV Program, equipment-based environmental services are services that can make claims based solely on measurable performance of the equipment or technology used. Such services can be verified in the same manner as technologies. Excluded from consideration are "people-based" environmental services; essentially any service for which a strict performance-based verification would not be possible. Also excluded is the certification of individual environmental practitioners.

3.3. ETV FactSheets

Two ETV FactSheets will be made available through the Program. One is the ETV *Program* FactSheet, providing a general overview of the ETV Program. The second is a *Verification* FactSheet, providing vendor-specific information when a vendor successfully completes the ETV process.

The ETV Program FactSheet is a one page (double-sided) information sheet containing recent developments and updates on the Program and published in English and French. The 'Government of Canada' insignia and the Canada wordmark is in the bottom left and bottom right corners respectively. The Program FactSheet is published and updated approximately every 3 months by ETV Canada Inc. with input from Environment Canada and contains a brief description of the ETV Program, its benefits and advantages to industry, and application information. It may also outline verifications completed to date and in which technology sectors.

Each successful vendor is also issued a *Verification* FactSheet. This contains the same template markings as the Program FactSheet (ETV Logo and government wordmarks). It states that the company has successfully completed the ETV Program and describes the verification claim in detail, including specific parameters, operating conditions and applications. A brief statement about the nature of the Program is also included.

3.4. Verification Certificate

The Verification Certificate is awarded to vendors by ETV Canada Inc. upon successful validation of their performance claim. The Certificate is the vendor's authenticated proof of having successfully completed the ETV Program. This is the vendor's primary credential proving that the verified performance claim is recognized by the Government of Canada and will serve as a integral marketing document. It contains the same template markings as the FactSheet as well as: the successful vendor's full corporate/identifier, address and contact numbers; performance claim; statement that the vendor has successfully completed the ETV Program; signature by ETV Canada Inc.; Certificate No.; effective date and expiry date, and; a brief statement about the nature of the ETV Program.

3.5. Verification Report

The Verification Report is issued to vendors by ETV Canada Inc. upon successful validation of their performance claim. The Report contains: a detailed description of the technology; a detailed description of the performance claim including specific parameters, operating conditions and applications; the results of data assessment and claim validation, and; a disclaimer.

The ETV Logo is not intended to serve as stand-alone proof of verification, but rather must be used in conjunction with a specific performance claim. Vendors, having successfully obtained a verification through the Program, may only use the ETV Logo on the ETV FactSheet, Verification Certificate and Final Verification Report issued to them with respect to a validated claim.

3.6. Verification Entity

Verification Entities (VE) are third party, impartial, specialized and accredited laboratory and testing facilities, technical review services, and various technical specialists sub-contracted by ETV Canada Inc. to supply technical data generation, assessment and validation expertise. As the basis for the credibility of the vendor's verified claim is the data assessment and verification, the VE sub-contracted by ETV Canada Inc. (and approved by the vendor) must have the technology-specific and specialized expertise, capability and facilities to either conduct the data assessment for the verification **or** conduct necessary demonstrations and testing and generate the required technical data to support the claim. However, a VE may not both generate the required data and then assess/validate that same data for any one performance claim, unless it is acceptable to both ETV Canada Inc. and the vendor.

4. PROGRAM STRUCTURE & STRATEGY

The ETV Program, conceptualized and originating as a federal government initiative, is delivered and administered by a private sector partner - ETV Canada Inc. ETV Canada Inc. operates the ETV Program on behalf of the Government of Canada and is permitted for a specified period of time under a license agreement to use the ETV Logo and issue Verification Certificates, FactSheets and Final Reports.

Environment Canada is responsible for Program policy and general direction through provisions in the license agreement. The Canadian Environment Industry Association and private sector representatives, also under provisions of the license agreement, would provide input to Environment Canada on Program oversight and direction through participation in semi-annual performance reviews of the license agreement.

When a vendor comes to ETV Canada Inc. and wants their technology "verified", this involves the confirmation of a quantifiable performance claim supported by reliable data. The verification of particular technology performance claims are based on stringent technology-specific protocols. These protocols serve as the template and manual for the complete verification procedure to validate the specified performance claims of a particular technology. The Verification Entity assesses the integrity of supplied data and the validity of associated performance claim(s) based on the data provided and following the detailed, rigid, technology-specific protocol. For a claim to be verified, ETV Canada Inc. must be satisfied that all of the following criteria have been fulfilled:

- The technology is based on sound scientific and engineering principles.
- The claim is fully supported by peer-review quality data, which are supplied by the applicant or generated upon applicant's request through an independent test program conducted by a qualified testing agency.
- The conditions of performance for the claim are clearly defined.

The process of having a claim verified through the ETV Program consists of four stages: **Pre-screening; Application Review; Verification,** and; **Award**.

- In the **Pre-screening stage**, for a technology to be eligible, it must be an environmental technology or an equipment-based environmental service, meet minimum Canadian environmental standards and/or national guidelines for the specific technology or claim, and be currently commercially available or commercially ready for full-scale application. If the technology meets these criteria, the applicant submits a Pre-screening Application to ETV Canada Inc. which is reviewed to assess eligibility and feasibility and to resolve any conflict of interest which may exist between the applicant and ETV Canada Inc.
- During the **Application Review stage**, if the technology is eligible for application, the applicant submits a Formal Application which requests additional information about the technology, the claim to be verified, and the information (or data) that is currently available to support the claim. ETV Canada Inc. reviews the Formal Application for completeness and determines if it can be accepted into the ETV Program. If the application isn't acceptable, the applicant may choose to modify and resubmit it. If the application is accepted, ETV Canada Inc. receives detailed data and information from the applicant to substantiate the claim. It reviews the information and proposes a verification process for the claim, including identification of a Verification Entity (i.e. approved technical expert(s)) and a cost estimate for the program. ETV Canada Inc. discusses the scope and cost of the proposed program with the applicant, and reaches

agreement on the Verification Entity (including resolution of any conflict of interest between the applicant and the Verification Entity) and associated costs.

- During the **Verification stage**, the Verification Entity reviews the supporting data to determine if the claim is adequately substantiated or if additional testing is required to substantiate the claim. If additional testing is required, an independent testing of the technology is conducted by another Verification Entity or approved testing facility, with costs paid by the applicant. The Verification Entity prepares a report on the results of the verification program, and submits it to ETV Canada Inc. and the applicant for review and approval. If the claim cannot be substantiated, the applicant may choose to modify the claim such that it is substantiated with the existing data.
- In the **Award stage**, if the applicant's claim is substantiated, ETV Canada Inc. prepares a FactSheet to accompany the Verification Report and awards a Verification Certificate to the applicant. The applicant is then entitled to use the Certificate, FactSheet and Final Report in domestic and international marketing activities.

The cost of verification through the ETV Program is predicted to be approximately $10,000 - 15,000. These figures are estimates and will vary depending on the complexity of the performance claim. Verification costs are separate from demonstration and/or testing costs. As every applicant is expected to submit supporting test data, the costs to generate this data is typically required for technology commercialization regardless of whether the company chooses to verify a technology performance claim. In addition, if additional test data is required to substantiate the performance claim, these costs are incurred by the technology vendor and are above and beyond the verification cost.

ETV Canada Inc. and the Government of Canada encourage buyers and regulatory authorities to accept the verification certificate as support for suppliers' claims concerning product performance. This increases the marketability of the verified products, expedites provincial permit approvals, and reduces the need for repeated demonstration testing. The benefits include enhanced international marketability, accelerated commercialization of innovative technologies, and potentially greater inter-provincial acceptance of new environmental products and processes. The result should be a healthier environment and increased domestic and international trade.

Pilot verifications are important to ensure adequate technology-specific protocols are created and established for new technologies. As vendors apply to the Program with new and innovative technologies, protocols must be developed and implemented for the verification of these technologies. These pilots must be conducted and protocols developed by a qualified facility with appropriate expertise. Pilot activities and protocol development are scheduled to continue for at least the first year of the Program's operation.

Efforts to achieve reciprocity with the U.S. programs and other similar international initiatives are ongoing. Efforts are also underway to investigate the establishment of a new work item under ISO and pursue recognition from the United Nation's Economic Commission for Europe and NAFTA's Commission for Environmental Cooperation. Endorsement and buy-in from all Canadian provinces and territories is important to the success of the ETV Program. The goal is to obtain formal recognition by the provinces/territories by engaging them into the ETV Program through joint development of standardized environmental technology evaluation procedures that provide Canada-wide acceptance of technology evaluations. Negotiations are ongoing with provincial representatives to solicit their participation on an Inter-provincial Working Group on the Standardization of Environmental Technology Evaluations and to sign bilateral agreements. The goal of the Working Group is to facilitate cooperation among all provinces/territories for the testing, demonstration, evaluation and verification of environmental technology performance claims in a manner that allows for mutual acceptance among all parties.

5. CURRENT STATUS

The license agreement between Environment Canada and ETV Canada Inc. will be signed in March 1997. ETV Canada Inc. will be open for business and accepting applications by April 1, 1997. Also in preparation for program operation, the first phases of the pilot verifications and development of the general verification protocol will be completed by April. The ETV Program will be officially launched by the Minister of

Environment Sergio Marchi in June 1996. Efforts to solicit the endorsement and buy-in by the provincial governments and confirm the participation of the provinces/territories on the proposed Inter-provincial Working Group on the Standardization of Environmental Technology Evaluations is an on-going activity.

The development of a new ISO work item, as well as pursuing recognition from international organizations such as the United Nation's Economic Commission and NAFTA's Commission for Environmental Cooperation are also ongoing.

6. ROLES / RESPONSIBILITIES

As the ETV Program will be a government / private-sector partnering initiative, there are certain Program development and operational activities which Environment Canada is best suited to undertake and some which ETV Canada Inc. should undertake. The respective roles between the Environment Canada and the ETV Canada Inc. are outlined below.

6.1. ETV Canada Inc.

ETV Canada Inc. is operating under license from Environment Canada to deliver and administer the ETV Program. Specifically, the licensee is accountable for:

- assuming day-to-day responsibility for ETV Program delivery;
- overall management and leadership functions;
- financial management on a cost recovery basis;
- marketing and promotion;
- coordination of technical expertise to complete verifications;
- issuing of awarded verification certificates;
- monitoring use of verification certificates;
- addressing conflict of interest and confidentiality issues;
- general administration of the Program.

ETV Canada Inc. is responsible for the final development, implementation and ultimate success of the Program.

The contact for ETV Canada Inc. is:

Mr. John McMullen, President, 2197 Riverside Dr., Suite 300, Ottawa, ON K1H 7X3, (613) 247-1900 ext.228, Fax: (613) 247-2228

6.1. Government of Canada

Environment Canada's role is to ensure the credibility and quality oversight of the ETV Program, particularly for the ETV Logo, is maintained. The mechanism for this oversight will be through the semi-annual review of the license agreement.

Initially, Environment Canada will take the lead with respect to international reciprocity activities with input from ETV Canada Inc.

Efforts to obtain endorsement and recognition from the provinces and the Canadian Council of Minister's of the Environment (CCME's) Environmental Protection and Planning Committee are the responsibility of Environment Canada with input from ETV Canada Inc. This includes activities to establish, solicit membership and provide secretariat services for the Inter-provincial Working Group for Standardization of Environmental Technology Evaluations.

The facility currently conducting the pilot/protocol development project under contract to Environment Canada is Water Technology International (WTI) Corp. This pilot/protocol development will continue under the direction and supervision of Environment Canada with input from ETV Canada Inc. WTI's activities include: further general protocol development; technology-specific protocol development; pilot verifications of technologies not investigated in earlier activities; round-robin verifications with U.S. programs; assistance in the design and implementation of a program to train proposed Verification Entities; technical input for ISO new work item development and Inter-provincial Working Group activities, and; refinement of the ETV database's interface with the marketplace.

European IPPC BAT Reference Documents

P. Wicks

Directorate-General for Environment, Nuclear Safety and Civil Protection
Unit XI.E.1 'Industrial Installations and Emissions'
European Commission, Rue de la Loi 200, B-1049 Brussels

1. INTRODUCTION: THE IPPC DIRECTIVE

In 1996, the EU Council of Ministers adopted Council Directive 96/61/EC on Integrated Pollution Prevention and Control, commonly known as the 'IPPC Directive'. Its purpose is to achieve integrated prevention and control of pollution arising from the activities listed in its Annex.

The Directive sets general principles governing the basic obligations of the operators of industrial installations. First and foremost among these is the obligation to take 'all the appropriate preventive measures against pollution', which is defined as

> 'the direct or indirect introduction as a result of human activity, of substances, vibrations, heat or noise into the air, water or land which may be harmful to human health or the quality of the environment, result in damage to material property, or impair or interfere with amenities and other legitimate uses of the environment.'

The other obligations of the operator involved waste management (prevention, recovery and disposal), efficient use of energy, accident prevention and the return of the site of operation to a satisfactory state upon definitive cessation of activities. Measures to prevent pollution must in particular involve 'application of the best available techniques' (BAT).

Fulfillment of these obligations is ensured by means of an integrated permitting procedure, in which permit applications must include information on the installation and its activities, the substances and energy used or generated, emission sources, conditions of the site, the nature and quantities of the foreseeable emissions as well as the likely environmental impact, proposed abatement techniques, measures taken for the prevention and recovery of waste, and the measures planned to monitor emissions. Likewise, the permit issued by the competent authority must contain conditions, and in particular emission limit values based on BAT. Member States must ensure that the competent authority follows or is informed of developments in BAT.

Although the IPPC Directive does not itself set uniform Community-wide emission limit values for any substances, it leaves in force emission limit values provided for by existing Directives and provides for new emission limit values to be set in the future where a need for such action is identified.

2. BAT INFORMATION EXCHANGE: DESCRIPTION OF THE ACTIVITIES

2.1. Objectives, timeframe and legal status

In addition to the permitting procedure, the Directive requires the European Commission to organize 'an exchange of information between Member States and the industries concerned on best available techniques, associated monitoring, and developments in them', and to published the results of the exchanges of information.

The primary objective of such an exercise is to support the competent authorities in their implementation of the Directive, and in particular in their obligation to follow developments in BAT. In addition, the participation of industry and the general availability of the published results should in itself stimulate the uptake by industry of cleaner production techniques.

The information exchange will be a continuous operation, with the Commission required by the Directive to publish the results of the exchange of information every three years. The multi-annual work program drawn up by the Commission foresees coverage of all sectors included in Annex I of the Directive by the year 2001.

Although the information exchange process is a legal obligation, the published results will not themselves be legally binding, although it is expected that they will have a considerable influence on permitting practice. However, the Directive does specify that the future Community-wide emission limit values mentioned above must be set on the basis of the information exchange.

2.2. Scope

The information exchange will include all industrial activities covered by the IPPC Directive. The list of such activities, given in Annex 1 of Directive, comprises a total of 33 industrial sectors in seven groups: energy industries, production and processing of metals, mineral industry, chemical industry, waste management, and "other activities".

2.3. Organization

The approach taken is sector-by-sector, and mostly follows the structure of Annex 1, with some minor regrouping. The result is that documents containing the results of the information exchange - called BAT Reference Documents or BREF's - will be published for each of 30 sectors. In addition, a number of cross-sectoral issues have been identified as worthy of particular attention and will result in 'horizontal BREF's'. These are vacuum and cooling systems, monitoring techniques, and emissions from storage.

The Commission unit responsible is XI.E.1 'Industrial installations and emissions', within the Directorate-General for Environment, Nuclear Safety and Civil Protection (DG XI). Nevertheless, much of the technical work, including initial drafting of the BREF's, will be carried out by the European IPPC Bureau, which has recently been established at the Institute for Prospective Technological Studies (IPTS) in Seville, Spain. IPTS is an institute of the European Commission's Joint Research Centre.

The parties to the information exchange - EU Member States and industry, and also environmental NGO's - participate on two levels. At a 'technical' level, Technical Working Groups (TWG's) consisting of expert representatives will be set up for each sector covered. At a 'political' level, an Information Exchange Forum consisting of official representatives meets two or three times a year. Its role is to oversee the information exchange process and to provide official comments on the draft BREF's. Final responsibility for publication of the BREF's rests with the Commission.

2.4. Definition of BAT

The term 'best available techniques' is defined in some detail in the Directive, and it is worth citing the full definition here:

> " 'best available techniques' shall mean the most effective and advanced stage in the development of activities and their methods of operation which indicate the practical suitability of particular techniques for providing in principle the basis for emission limit values designed to prevent and, where that is not practicable, generally to reduce emissions and the impact on the environment as a whole:
>
> - 'techniques' shall include both the technology used and the way in which the installation is designed, built, maintained, operated and decommissioned,
> - 'available' techniques shall mean those developed on a scale which allows implementation in the relevant industrial sector, under economically and technically viable conditions, taking into consideration the costs and advantages, whether or not the techniques are used or produced inside the Member State in question, as long as they are reasonably accessible to the operator,
> - 'best' shall mean most effective in achieving a high general level of protection of the environment as a whole."

Of particular note is relatively wide definition of 'techniques' to include such aspects as operation and maintenance, the integrated nature of the concept ('protection of the environment as a whole'), and the need the take into consideration costs and advantages. In addition to the above definition, the Directive contains a list of 12 items to be considered when determining BAT, one of which is the published results of the information exchange and also information published by international organizations. The list also includes the consumption and nature of raw materials (including water) and the length of time needed to introduce the best available technique, as well as other items that follow more directly from the BAT definition or from the basic obligations of the operator.

The definition can perhaps be seen as the result of two influences: firstly, the desire to take a genuinely integrated approach and therefore to include all relevant factors, environmental, economic and practical; secondly, the need for political compromise between Member States having more or less advanced industrial practices. Although the resulting definition is coherent, rational and widely accepted, it may well be argued that the different interpretations will be possible when it comes to concrete applications. Such differences will presumably reveal themselves during the course of the information exchange.

3. ISSUES TO BE ADDRESSED

3.1. Contents of the BREF's

The contents of the BREF's will be determined to a large extent by the definition of BAT described above. Thus, BREF's should not only contain information on technologies to be used but also cover design, construction, maintenance, operation and decommissioning. In particular, practical experience shows that the quality of operation and maintenance of an installation is frequently more important than, for example, the process route. Also, the need to take costs and advantages into consideration when determining BAT implies that the economic performance, as well as the environmental performance, of a given technique should be addressed during the information exchange, and that information about this should be included in the published results. The extent to which the focus is more on end-of-pipe technologies or on integrated solutions will presumably depend on the nature of the sector to be treated.

The financial and human resources available and the vastness of the scope of the exercise will not permit the drafting of handbooks containing an in-depth study of each sector; nor will it be possible for local authorities to apply directly the results without first taking into account local economic and environmental conditions. The BREF's will not stipulate a list of best available techniques to be prescribed as the only permissible techniques throughout Europe. Instead, the idea is to draw up a list of reference techniques to define the environmental performance that is achievable in the sector. These reference techniques are agreed on the basis of an evaluation of various 'candidate BAT's', identified at an early stage of the process.

An outline for the BREF's has been discussed in the Information Exchange Forum. This outline involves the following elements: general information about the sector, a description of the currently applied processes and techniques, present consumption and emission levels, a selection of 'candidate BAT's' together with an evaluation of their environmental and economic performance, a selection of BAT's, and finally a description of emerging techniques.

3.2. A vertical or a horizontal approach?

It has already been mentioned that the approach taken will be a sectoral one, supplemented by three BREF's on horizontal themes. It seems fairly clear that neither a purely vertical nor a purely horizontal approach is ideal. The advantages of a sectoral approach are that different sectors have distinctive characteristics requiring a different approach, and that existing information and expertise is often organized sectorally. On the other hand, certain issues such as the three identified - vacuum and cooling systems, monitoring techniques and emissions from storage - are essentially of a cross-sectoral nature and are therefore best treated as such.

It is important to note that the information exchange process will be a dynamic and evolving one, and in particular that more issues requiring a horizontal approach may come to light as more experience is gained. The vertical/horizontal question is one of many issues which will require a continuous learning process.

3.3. The integrated approach and environmental trade-offs

Another such issue is the question of how to address environmental trade-offs. One of the major reasons for an integrated approach is to address the problem of abatement techniques merely shifting pollution from one environmental medium to another. A genuinely integrated BAT will be one achieves 'a high general level of protection of the environment as a whole': in other words, it needs to be optimized for the overall environmental impact rather than for a specific environmental effect (e.g. air pollution).

This cannot be done with out some method for deciding on environmental trade-offs. Broadly speaking, three approaches can be identified. The first would be to refrain from addressing environmental trade-offs in the BREF's, except to mention that they exist. The idea would be to give information on the performance of candidate BATs concerning all the relevant environmental impacts, while leaving permitting authorities free to make trade-off decisions on the basis of local conditions and sensitivities. Although permitting authorities are in any case free to take local conditions into account when determining BAT, this approach would seem to be too limited.

A second approach would be to try to develop fully quantitative weightings for the different environmental criteria. This would be a much more ambitious plan and has considerable appeal as a truly rational approach. However there seems to be a general consensus amongst parties to the information exchange that it is unrealistic to adopt such an approach since the weightings would contain a large amount of subjectivity and are too dependent on local conditions.

The third option, and the one most likely to be taken, is therefore to adopt a qualitative approach, refraining from establishing quantitative weightings but nevertheless giving some guidance at least on which of the environmental effects are more important. In any case such a judgement will be implicit in the selection of reference techniques.

3.4. Economic considerations

As mentioned above, economic considerations are included in the legal definition of BAT, and will therefore need to be taken into account explicitly in the selection of reference techniques. An evaluation of the economic performance of candidate BAT's is included in the BREF outline. It should cover both the investment costs and the operating costs of a particular abatement technique, and also take account of economic savings due to increased efficiency. A distinction is to be made between new and existing installations.

3.5. Interaction between organizations

The European Commission is not alone in conducting activities concerning best available techniques or related concepts. Much information is available, whether from industry,

Member States and other nations, or international organizations. UN-ECE and Parcom are examples of fora involved with similar activities. It will be important to ensure that the information exchange process is well-coordinated with such activities and benefits from all relevant information.

There are no obvious obstacles in principle to the sharing of information - on the contrary, countries and regions with more developed environmental legislation and practices generally have an interest in encouraging similar practices elsewhere. The greatest obstacles are therefore likely to be operational: lack of awareness of related activities, or lack of time and resources to co-ordinate adequately. There is therefore much to be gained from a comparative discussion on the different activities that exist, and on possible modalities of co-operation.

4. CONCLUSIONS

The BAT information exchange being organized by the Commission is a legal obligation of Council Directive 96/61/EC on integrated pollution prevention and control (IPPC). Its main objective is to provide guidance for the permitting authorities within the EU when determining BAT as required by the Directive. It may also prepare the ground for future Directives setting EU-wide emission limit values.

The information exchange will cover a wide range of industrial sectors - all those covered by the Directive - as well as some horizontal issues, and will be an ongoing process lasting several years. Within the Commission, DG XI is responsible for the exercise but will be assisted by a European IPPC Bureau, Technical Working Groups and an Information Exchange Forum.

The published results - called BREF's - will contain list of reference techniques that can be considered BAT, but which are not to be seen as exclusive lists. The definition of BAT to be used is given in the Directive.

An outline for the contents of BREF's has been established, and much thinking has already been done on questions such as the amount of detail to be included, the usefulness of making separate BREF's for horizontal issues, how to deal with environmental trade-offs and how to include economic considerations. The issues are complex, however, and it will require flexibility and a continuous learning process. Interaction between different organizations carry out similar activities is important, and the sharing of information between them will need to be well-coordinated.

Environmental Technology Verification Program Verification Strategy

F. Princiotta

Office of Research and Development
U.S. Environmental Protection Agency
Washington, DC 20460

Background

Throughout its history, the U.S. Environmental Protection Agency (EPA) has evaluated technologies to determine their effectiveness in monitoring, preventing, controlling, and cleaning up pollution. Since the early 1990s, however, numerous government and private groups have identified the lack of an organized and ongoing program to produce independent, credible performance data as a major impediment to the development and use of innovative environmental technology. Such data are needed by technology buyers and permitters both at home and abroad to make informed technology decisions. Because of this broad input, the President's environmental technology strategy, *Bridge to a Sustainable Future,* and the Vice President's *National Performance Review,* contain initiatives for an EPA program to accelerate the development of environmental technology through objective verification and reporting of technology performance. In 1994, EPA's Office of Research and Development formed a workgroup to plan the implementation of the Environmental Technology Verification Program (ETV). The workgroup produced a Verification White Paper that guided the initial stages of the program. This document, *Verification Strategy,* updates the earlier paper based upon the evolution of the program over the last two years. It outllnes the operating principles and implementation activities that are shaping the program, as well as the challenges that are emerging and the decisions that must be addressed in the future. The program will continue to be modified through input from all parties having a stake in environmental technology, through further operational experience, and through formal evaluation of the program.

The goal of ETV, which remains unchanged, is to verify the environmental performance characteristics of commercial-ready technology through the evaluation of objective and quality assured data, so that potential purchasers and permitters are provided with an independent and credible assessment of what they are buying and permitting.

Important Definitions

A clear definition of the words "evaluate" and "verify", along with the word "certify", is important to establish at the outset of the program. The technology development community, the regulated community, and those charged with executing environmental standards at all levels of government require a precise understanding of what EPA means and does not mean by the activities to be undertaken through ETV.
EPA intends to sponsor the evaluation of environmental technologies through adequate testing and verify that they perform at the levels reported. By evaluate and verify we mean:

Evaluate / Evaluation

To carefully examine and judge the efficacy of a technology; to submit technologies for testing under conditions of observation and analysis; *syn.*, measure, estimate, classify, test.

Verify / Verification

To establish or prove the truth of the performance of a technology under specific, predetermined criteria or protocols and adequate data quality assurance procedures; *syn.*, confirm, corroborate, substantiate, validate.

EPA does not intend to certify that a technology will always, or under circumstances other than those used in testing, operate at the levels verified. By certify we mean:

Certify / Certification

To guarantee a technology as meeting a standard or performance criteria into the future; *syn.*, ensure, warrant, guarantee.

EPA understands that the word certify can have a variety of meanings, but the Agency believes that the above definition is the one most commonly understood. Misuse of the term, could cause confusion among the public.

Operating Principles

Several important operating principles have defined the basic ETV program structure and remain fundamental to its operation. These are briefly outlined below.

1. Performance Evaluation Goal

Under ETV, environmental technologies are evaluated to ascertain and report their performance characteristics. EPA and its partners will not seek to determine regulatory compliance; will not rank technologies or compare their performance; will not label or list technologies as acceptable or unacceptable; and will not seek to determine "best available technology" in any form. In general, the Agency will avoid all potential pathways to picking "winners and losers". The goal of the program is to make objective performance information available to all of the actors in the environmental marketplace for *their* consideration and decision making.

2. Commercial-Ready Technologies

The ETV program is a service of EPA to the domestic and international marketplace in order to encourage rapid acceptance and implementation of improved environmental technology. ETV, therefore, focuses its resources on technologies that are either in, or ready for, full-scale commercialization. The program does not evaluate technologies at the pilot or bench scale and does not conduct or support research. Participation in ETV is completely voluntary.

3. Third-Party Verification Organizations

ETV leverages the capacity, expertise, and existing facilities of others through third-party partnerships in order to achieve universal coverage for all technology types as rapidly as possible. Third-party verification organizations are chosen from the both the public and private sector, including states, universities, associations, business consortia, private testing firms, and federal laboratories. EPA designs and conducts auditing and oversight procedures of these organizations, as appropriate, to assure the credibility of the process and data. In order to determine if EPA participation is important to the commercialization process, ETV is testing the option of one totally unstructured and independent, private sector pilot in which EPA's role will be solely fiduciary. In addition, the Agency will continue to publish the results of commercial-ready technology evaluations that it conducts in the normal course of its business.

4. Pilot Phase

The program will begin with a three to five year pilot phase to test a wide range of partner and procedural alternatives, as well as the true market demand for and response to such a program. Throughout the pilot period, EPA and its partners will operate in a flexible and creative manner in order to identify new and efficient methods to verify environmental technologies, while maintaining the highest credibility standards. The operational objective will be to actively look for ways to optimize procedures without compromising quality. The ultimate objective of the pilot phase is to design and implement a permanent verification capacity and program within EPA by 2000, should the evaluation of the effectiveness of the program warrant it.

5. Pilot Technology Areas

ETV has begun with pilots in narrow technology areas in each of the major environmental media and will expand as appropriate, based on market forces, availability of resources, and the willingness of the marketplace to pay for third-party verification. For example, the drinking water technology pilot has started with a focus on microbial and particulate contaminants, and disinfection byproducts in small systems (less than 3300 users), an obvious and very large domestic and international market with pressing environmental problems. In fiscal year 1997 (FY97), the program will be expanded to the wider area of nitrates and synthetic organic chemicals and pesticides in all drinking water systems. Success in particular technology areas will allow the program to have a "pump-priming" effect to bring new technologies to the marketplace. Selection criteria for ETV pilot programs and other verification focus areas are discussed in a subsequent section of this paper.

6. Stakeholder Groups

ETV is guided and shaped by using the expertise of appropriate stakeholder groups in all aspects of the program. These groups consist of representatives of all verification customer groups: buyers and users of technology, developers and vendors, and, most importantly, technology "enablers", i.e., the consulting engineering community that recommends technology alternatives to purchasers, and the state permitters and regulators who allow it to be used. Stakeholder groups must be unique to each technology area in order to capture the important individual aspects of the different environmental media and to get buy-in from affected groups. For example, state drinking water permitters are necessary to participate in development of testing protocols for *cryptosporidium;* air pollution regulators are needed to evaluate innovative compliance monitoring devices; metal production parts manufacturers need to help design testing procedures for new coating compounds. In general, the role of stakeholders will be to assist in the development of procedures and protocols, prioritize types of technologies to be verified, review all important documents emerging from the pilot, assist in defining and conducting outreach activities appropriate to the particular area, and, finally, to serve as information conduits to the particular constituencies that they represent. As of June 1996, over 80 individuals are serving in the three stakeholder groups formed to date.

7. Private Sector Funding

Over the three to five year pilot phase of the program, the costs of verifying technologies in many pilots will move from a primarily government funded effort to a primarily private sector funded effort. At least two pilots will be vendor supported from the beginning. The original goal, as articulated in the 1994 strategy, called for complete private sector sponsorship within three years. A recent review (1995) of the program by a distinguished panel of outside experts convened by the EPA Science Advisory Board (SAB) concluded that such a goal was probably not achievable in so short a time-frame (they suggested five to eight years) and that some level of government support (10 to 20% of ongoing costs) would remain necessary to keep the activity viable. Conclusions on this issue will have to be reached as data emerge on the economic value-added of the program and the level of cost that the private sector is willing to bear in the various technology sectors.

8. Pilot Evaluation and Program Decisions

The Agency will collect data on operational parameters, e.g., number of participants; cost and time required to perform tests and report results, and on outcomes, e.g., use of data by the states and public; sales reported by vendors, in order to evaluate all aspects of the program. EPA will use this information to make long-term recommendations to the Congress on the future and shape of the program in December 1998. Among the choices at that time will be the formulation of a permanent, broad scale program; the narrowing of efforts to certain areas in which ETV appears to be effective; or the discontinuance of verification efforts. The latter conclusion could be reached either because state regulators/permit writers and the technology innovation industry are not assisted by ETV or because the cost of verification proves to be prohibitive.

9. Outreach and Information Diffusion

As was pointed out by the SAB in its 1995 review of ETV, verification alone will not move better, cheaper, faster technologies to success in the marketplace. Substantive and substantial interface with the permitters of environmental technology (primarily at the state level) will be necessary to have any chance of rapidly implementing innovative approaches. To date, the outreach activities of the program have been limited to assuring substantial state representation on the Stakeholder Groups that are designing the protocols and procedures for each pilot; developing informational fact sheets about the program; and placing a Web page on the Internet. In 1997, the Agency intends to develop an overarching outreach strategy with the help of a "corporate board" of major organizations in the technology area, e.g., National Governors Association, Western Governors Association, Environmental Council of the States, National Pollution Prevention Roundtable, appropriate corporations, and others. State permitter training, a national conference and other efforts will be included.

10. Market Gap Definition

Lastly, EPA will track applications and expressions of interest on the part of technology developers who come to all parts of the Agency that do not fit into the present suite of verification activities. This universe will be characterized during the initial stages of the pilot period and a strategy to address gaps will be developed.

ETV Pilot Process

Although a wide degree of flexibility will characterize the pilot projects (see above), each ETV pilot generally will go through two periods of development: an organizational phase and an operational phase (see Figure 1).

Figure 1. ETV Pilot Process

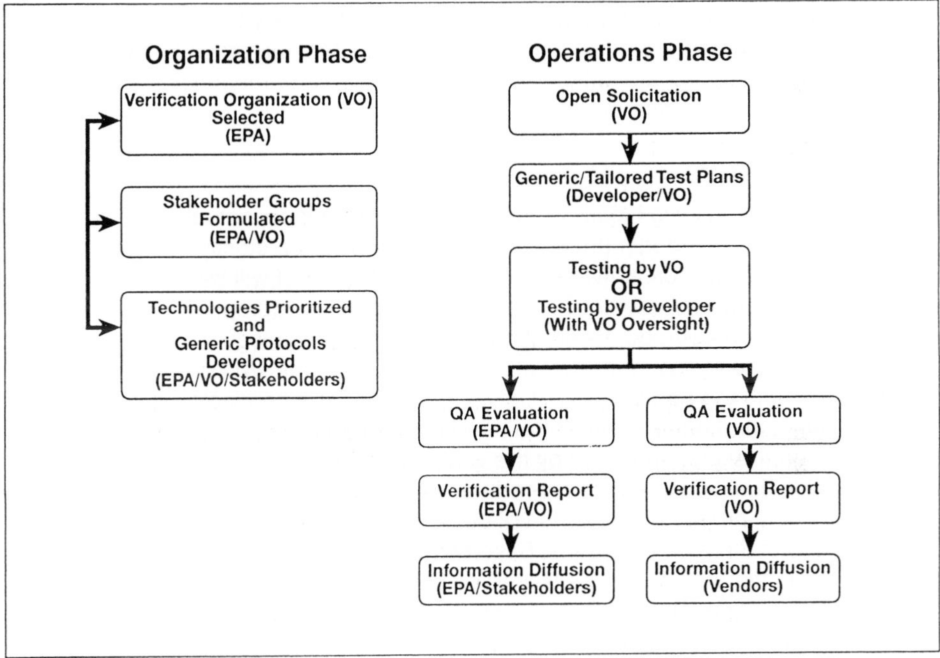

Organizational Phase

During the organizational phase, EPA will select one or more partner organizations to oversee and conduct verification activities. This important step will usually occur through an open solicitation process, although some exceptions will be appropriate. All partner proposals will be peer reviewed. EPA and its partner(s) will then select approximately 25 appropriate participants for the Stakeholder Group (see #6) that will guide the progress of the program. EPA may formulate the Stakeholders Group prior to selecting the partner organization if the procurement process is protracted. Stakeholder Groups will then begin the important process of establishing priorities and defining procedures and protocols appropriate to that particular type of environmental technology and customer group.

Operational Phase

Once the basic building blocks of the program are in place, actual verification activities begin (these steps will not always be strictly sequential). Verification activities in each technology area will be announced in the Commerce Business Daily and other appropriate publications to encourage maximal participation by technology developers, and to assure a level playing field. Test plans will be developed with the participation of developers and tests conducted by independent third parties (either the verification organization or other testing organizations approved by the verification organization). Appropriate quality assurance procedures will be incorporated into all aspects of the project and reports will be peer reviewed. Verification statements of three to five pages, based on the performance data contained in the reports, will be issued by EPA and appear on the ETV Internet Web page. Other outreach activities, as defined by the Stakeholder Group, such as state permitter training, will be conducted.

Selection Criteria for ETV Pilot Projects

The selection of verification pilot programs and other ETV verification activities to be carried out under the program is critical to its ultimate success. The ETV pilot programs are designed to meet the needs of the many stakeholders in environmental technology, while allowing EPA to experiment with a variety of procedures and partnership arrangements to determine the optimal form of program implementation. The following criteria are being used to choose ETV verification pilots and other verification activities. The first three criteria are applied in cascading order, i.e., each pilot must pass the previous criterion in order to go on to the next.

1. Address Important Environmental Needs

All programs conducted by the EPA have environmental improvement and protection as their ultimate objective. It is particularly important that this program, which is designed to assist the environmental technology industry, select pilot verification activities that have clear and positive environmental benefits. Such benefits may improve the environment by achieving higher levels of pollution reduction or by accelerating the rate of technology implementation through lower cost or simplified operation.

2. Present Substantial Business Opportunities for the Private Sector

Pilot areas selected will have clear market niches, both domestically and internationally, that present the potential for a substantial increase in technology sales and use. Although voluntary environmental improvement technologies, e.g., indoor air filtration systems, and existing regulatory program technologies, e.g., hazardous waste monitoring devices, are of interest to the program, particular attention will be paid to technology areas that stress the Agency's commitment to pollution prevention or address upcoming regulatory requirements and deadlines that present obvious and major opportunities for innovation and increased conomic activity.

3. Involve Multiple Developers and Vendors

To clearly demonstrate the potential of a successful verification program, the ETV pilots must benefit the largest possible number of technology developers and users. Pilot programs will focus on areas in which a number of technologies and companies are active.

The last two criteria are programmatic and allow the Agency to test operating program parameters in the pilot phase.

4. Address the Full Range of Environmental Media

Assuring that all major environmental media and program areas are included in the pilots will allow ETV to begin interaction with interested parties, e.g., technology developers, technology buyers, regulation writers and state permitters, across a broad range of environmental areas. This will assist the program to identify particularly fruitful technology focus areas for the future, spread the word about the EPA verification function widely, identify regulatory barriers that inhibit the use of verified technologies, and assist developers across a wide spectrum of technology.

5. Test a Variety of Verification Organization Types

As described above, EPA is interested in evaluating all possible verification organization alternatives, including federal government laboratories, state verification programs, universities, industry associations, independent testing organizations, and developer conducted testing (with verification organization or EPA oversight). Pilot programs selected are expected to cover all of these alternatives.

Table 1 contains the application of these criteria to the pilots that were selected for implementation in FY96 using FY95 funds. Table 2 contains those that have recently been approved (8/16/96) for implementation in FY97 using FY96 funds.

Table 1. Selection Criteria and Customer Drivers — ETV FY95 Pilots

Five Selection Criteria and Customers	Small Package Drinking Water Systems	P2 (Pollution Prevention)/ Waste Treatment Systems	Site Characterization and Monitoring Technologies	Indoor Air Products	Independent Entity (See page 12)
1. Address Important Environmental Needs	*Cryptosporidium* Other DW problems	Industrial waste reduction	Lower cost of site characterization and monitoring	Toxic compounds & biocontaminants in the indoor environment	Unknown
2. Present Substantial Business Opportunities to the Private Sector	Thousands of communities worldwide	Hundreds of companies	Superfund sites, Brownfields	Thousands of offices, homes	Assumed
3. Involve Multiple Developers and Vendors	Numerous	40 applicants to first solicitation	Growing universe	Hundreds of products	Assumed
4. Address Full Range of Environmental Media	Drinking water	P2/waste treatment	Remediation	Air	Unknown
5. Test Variety of Verification Organization Types and Processes	Private sector	State	DOE, Federal Laboratories	Private sector, university	Private sector association
Targeted Customers	State regulators; small communities	CSI (Common Sense Initiative) industries	Federal & state regulators	Consumers, consulting engineers, industry	Unknown

Table 2. Selection Criteria and Customer Drivers — ETV FY96 Pilots

Five Selection Criteria and Customers	Advanced Monitoring Systems	Air Pollution Control Technologies	Wet Weather Flow Technologies	P2 (Pollution Prevention) Coatings and Coating Equipment
1. Address Important Environmental Needs	Basic to regulatory reinvention	Needed to implement Clean Air Act (CAA)	Leading cause of water quality impairment, Clean Water Act (CWA)	Pollution prevention for Volatile Organic Compounds (VOCs), CAA, others
2. Present Substantial Business Opportunities to the Private Sector	Regulatory reinvention growing area of environmental focus	Thousands of businesses in regulated universe	Over 1 M businesses, thousands of municipalities	Large number of small businesses use coatings
3. Involve Multiple Developers and Vendors	Dozens of new technologies emerging	Hundreds of vendors	Dozens of vendors & developers, 5 companies identified	22 vendors identified to date
4. Address Full Range of Environmental Media	Air, water, soil monitoring	Air pollution control	Water, soil, air	Air pollution prevention
5. Test Variety of Verification Organization Types and Processes	Private sector	Private sector	Private sector associations, universities	DoD Laboratory
Targeted Customers	Federal & state regulators; industry	Federal & state permitters; industry	Small & large municipalities & businesses	Small business (metal & plastic)

Program Implementation To Date

FY95 - $7 million[1]

In September 1995, EPA initiated four pilot programs that were selected through its competitive Environmental Technology Initiative process. These were: Small Package Drinking Water Systems, with NSF International, a private sector testing and standards organization as partner; Pollution Prevention and Waste Treatment Systems, with the State of California EPA as partner; Site Characterization and Monitoring Technologies, with Sandia National Laboratories as partner; Indoor Air Products, with Research Triangle Institute, a private sector organization, as partner. In addition, the Agency solicited proposals for a fifth pilot to test the option of a private sector, non-technology specific, independent entity. Through a peer reviewed process, the Civil Engineering Research Foundation has been selected and will shortly initiate a pilot effort. ETV activities during the first year of the program focused on program planning, the selection of stakeholder group participants, procedures and protocol development, priority setting, and solicitation of technologies for testing. The Site Characterization and Monitoring Technologies pilot, the first to become operational, completed tests of five technologies. (See Table 3 for budget summary.)

FY96 - $10 million

In FY96, Congress directed that all ETI funds be focused on verification. Activities for this year include expansion of the scope of the first year verification pilots under ETV and initiation of four new pilots. The technology categories selected for the new pilots are: Advanced Monitoring Systems to encourage regulatory reinvention; Air Pollution

---

1 FY95 funds were distributed at the end of that fiscal year and were expended in FY96. Similarly, FY96 funds were distributed at the end of FY96 and are being expended in FY97.

Table 3. ETV FY95 and FY96 Pilot Program Funding

Pilot Area	Partner	EPA Contact Telephone #)	Amount ($million)		
			FY95	FY96	Total
Drinking Water Systems	NSF International Ann Arbor, MI	Jeff Adams (513-569-7835)	1.4	1.1	2.5
Site Characterization & Monitoring Technologies	Sandia National Lab. Alburquerque, NM Oak Ridge National Lab. Oak Ridge, TN	Eric Koglin (702-798-2432)	1.1*	0.6	1.7
Pollution Prevention (P2)/Waste Treatment Systems	State of California Sacramento, CA	Greg Carroll (513-569-7948)	1.8	0.8	2.6
Indoor Air Products	Research Triangle Institute RTP, NC	Les Sparks (919-541-2458)	1.0	0.2	1.2
Independent Entity	CERF Washington, DC	Norma Lewis (513-569-7665)	1.8	0.0	1.8
P2/Innovative Coatings and Coating Equipment	CTC Johnstown, PA	Mike Kosusko (919-541-2734)	0.0	0.6	0.6
Advanced Monitoring Systems	Open Solicitation	Robert Fuerst (919-541-2220)	0.0	1.6	1.6
Air Pollution Control Technologies	Open Solicitation	Ted Brna (919-541-2683)	0.0	1.4	1.4
Wet Weather Flows Technologies	Open Solicitation	Mary Stinson (908-321-6683)	0.0	0.8	0.8
US TIES (International Verification)	Open Solicitation	Steve James (513-569-7877)	0.0	1.0	1.0
Total			7.1	8.1	15.2

*FY94 and FY95 funding

Control Technologies aimed at facilitating the Clean Air Act Amendments of 1990; Innovative Coatings for Pollution Prevention; and Wet Weather Flow Technologies. Remaining resources will focus on verification activities under the Design for the Environment pollution prevention program conducted by the Office of Pollution Prevention and Toxic Substances, and USTIES (U.S. Technology for International Environmental Solutions) verification activities abroad with EPA's Office of International Affairs as an active partner (See Table 3). Important support activities will include the ongoing ETV program evaluation; a study of the fate of technologies for which no verification pilots exist ("market gap" study); and a substantially expanded information diffusion/outreach/ technology transfer program. These activities are summarized on the next page.

Other FY96 ETV Funded Activities - $1.9 million

ETV Outreach and Support ($0.5M)

Evaluation Process ($0.1M)

Technology Data Collection ($0.1M)

Design for the Environment (DFE) ($1.0M)

Small Business Innovation Research (SBIR) set aside ($0.2M)

Total ETV FY95 and FY96 Funding—$17.1 million

FY97 - $10 million

Investment of FY97 resources will follow the pattern of FY96 for verification activities, with additional emphasis on communication, outreach and evaluation. The existing pilots will be entering the full scale operational phase of the program and will be expanded to include new technology areas as the market dictates. Candidate technology categories for one or more new pilots will be based upon the results of the "market gap" studies conducted during 1997. These new areas would be created to support segments of the environmental technology industry that are not adequately addressed by the existing ETV program. Outreach and information diffusion on the results of the earlier evaluations will be expanded, and program results to date will be evaluated for preparation of decision options on the future of the program.

ETV Program Vision and Projected Development Scenario

Figure 2 lays out the projected ten year development scenario for the ETV Program in which assumptions are made concerning: (1) the amount of funding received by the program; (2) the number of viable technology areas (defined as pilots in the early stages) needed; (3) the

Figure 2. ETV Program Scenario

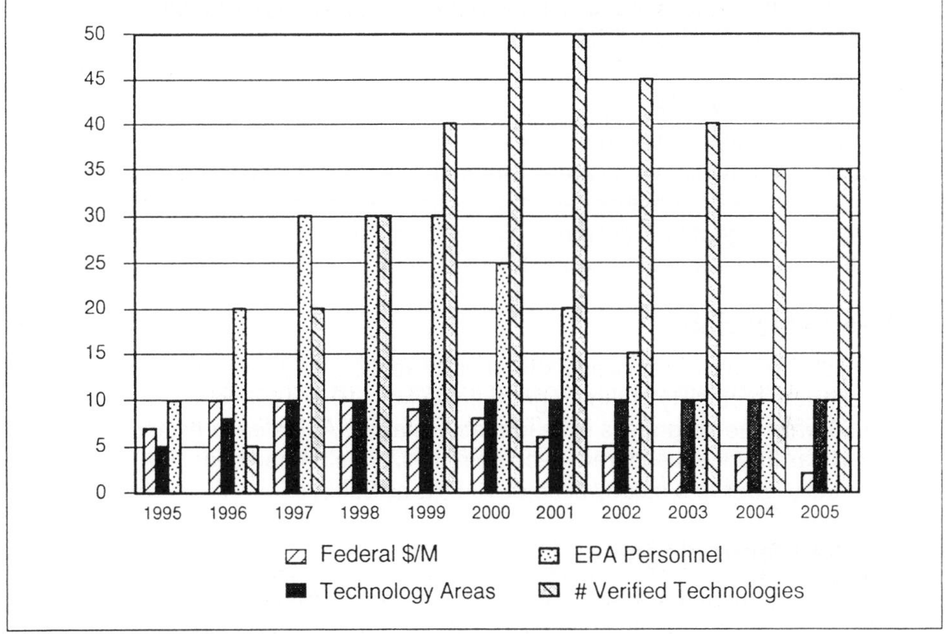

number of EPA staff assigned to quality assurance and technical oversight; and (4) the number of new technologies that come to the program for verification. All of these assumptions are projections based upon the experience of technology verification programs in other fields and the level of resources needed to accomplish the work projected. Changes to any of the assumptions would, of course, change all of the projections.

1. Federal Funding

The projected budget moves from $7M in 1995 to $10M for three years of development and building market acceptance. It then begins a gradual decline as the private sector pays more and more of the costs. By 2005 the steady-state cost of the program for all environmental areas is projected to be $2M.

2. Technology Areas

Starting in 1995, pilot programs gradually expand to cover all appropriate environmental areas (areas defined primarily by customer groups). By 1997, ten areas are defined and implemented and remain in place throughout the program, although their technology focus can shift over time, based on market forces.

3. EPA Personnel[2]

EPA staff necessary to oversee third-party verifiers and maintain the quality and credibility of the program move from 10 in 1995 to as many as 30 while the program is in its formative stages. As procedures and quality assurance measurers become routinized, staff demands gradually lower to a steady state of 10 FTE by 2003.

4. Verified Technologies

Due to a demand backlog, the number of technologies verified rapidly rises from five in 1996 to 50 in 2000, stays at this level for a few years, and then declines to a steady state of about 30 to 35 technologies.

If executed as projected in this scenario, the Environmental Technology Verification Program could reasonably be expected to verify the performance of approximately 350 innovative technologies in a decade.

---

[2] EPA personnel are assigned to ETV from appropriate media areas (e.g., drinking water, advanced air monitors) within ORD's national laboratories and centers based upon programmatic requirements. Six EPA staff are assigned to ETV directly for management coordination, information dissemination, and oversight.

Dutch experience with integrated approaches/covenants with industry

Ir. W.C.J. Quik - Technical Director VNCI

The Dutch chemical industry covers a broad spectrum of activities carried out by national as well as multinational companies. At approximate 150 sites around 80.000 staff are employed. In 1996 total production amounted to $ 30 billion of which 75% was exported.

			Perc. of industry
Employment		78.600	10
Production	b.Dfl	49,4	15
Export	b.Dfl	35	20
Investment	b.Dfl	3,1	25
R & D	b.Dfl	1,9	30

Roughly 60% of the turnover is based on the manufacture of petrochemicals. Comparing it to the size of the US chemical industry the Dutch chemical industry represents roughly 10% of the US industry's turnover.

The member companies of the Association of the Dutch Chemical Industry (VNCI) cover over 98% of the total turnover of the chemical industry.

Out of the range of instruments for environmental policy until the end of the 1980's legislation was virtually the only instrument used to realize environmental objectives in The Netherlands.

Instruments for environmental policy

- Legislation
- Fiscal instruments
- Non-fiscal instrument
  - > economic
  - > managerial

At the end of the 80's the Dutch Government developed the National Environmental Policy Plan aiming to achieve sustainable development in one generation. Before completion of the National Environmental Policy Plan industry was asked what position it would take. There were two options for industry i.e.:

- being unable to accept the policy aims as long as there was uncertainty about its organizational, technical and financial consequences (i.e. opposition):
- accept the aims on condition that industry were involved in implementing them and only if it was felt that they were feasible in practice (i.e. voluntary cooperation).

The last option was chosen and a new instrument for implementation of environmental policy developed.

Government, authorities and industry embarked on a route for environmental management. Basically we are moving away from a command and control approach by authorities and a defensive, reactive attitude by companies towards a pro-active industry and positive, cooperative and trusting position of government and authorities.

By this cooperative approach internalization of environmental issues is being fostered and thereby integrated as a normal business issue in company management.

In Holland we have over 100 negotiated agreements in operation for a wide variety of environmental issues and most of them produce the required results.

The main elements in the negotiated agreements can be summarized as follows:

- quantified objectives
- staged targets
- obligations of parties
- monitoring
- reporting
- verification

Participation by industry is on a voluntary basis but once the agreement is signed companies are expected to honor their obligations like a business contract. The agreement also specifies that non-participating companies -free riders- will have their operating license reviewed and adjusted by the authorities.

The major agreement for the chemical industry is the one related to the execution of the National Environmental Policy Plan with emission reductions as main element.

The agreement contains a wide variety of environmental objectives upto 2010. It basically calls for emission reductions for the entire sector, between 50 and 70% for the period 1985-2000 and reductions between 70 and 90% for the period 1985-2010.

DIFFUSION TO AIR		
REDUCTION TARGETS* (PERC.)		
	2000	2010
SO_2	77	90
NO_x	60	90
VOC	58	80
1985 = 100		

Participating individual companies will have to draw up a 4 year site environmental plan stating the measures they will implement or consider to implement on the basis of best available technologies not entailing excessive costs (BATNEEC).

Commitments

Industry association

- promote participation
- facilitate execution
- environmental diplomacy

Individual company

- 4-year site environmental plan
- annual progress report

The plan is negotiated with the permit issuing authorities and upon approval used as the basis for permit application.

Commitments

Authorities

- review environmental plan
- approve plan
- approved plan basis permit
- consistent policies
- international harmonization

Over 90% of the Dutch chemical industries have signed the agreement, the remaining companies participate as if they have signed.

STATUS COVENANT SEPTEMBER 1996			
	Locations	Site env.plan	Progress report 1995
VNCI-members	108	101	103
Others	17	11	11
Total	125	112	114

The chemical industry envisages a number of potential advantages in undertaking the agreement i.e.:

- certainty about long term environmental targets;
- attuning environmental/normal investments;
- international harmonization environmental policies;
- integrated approach air/water/waste;
- level playing field for industry;
- consistent government policies;
- consultation platform industry/authorities.

Results obtained over 1995 demonstrate that most of the objectives will be achieved including a 20% energy efficiency improvement over the period 1990-2000.

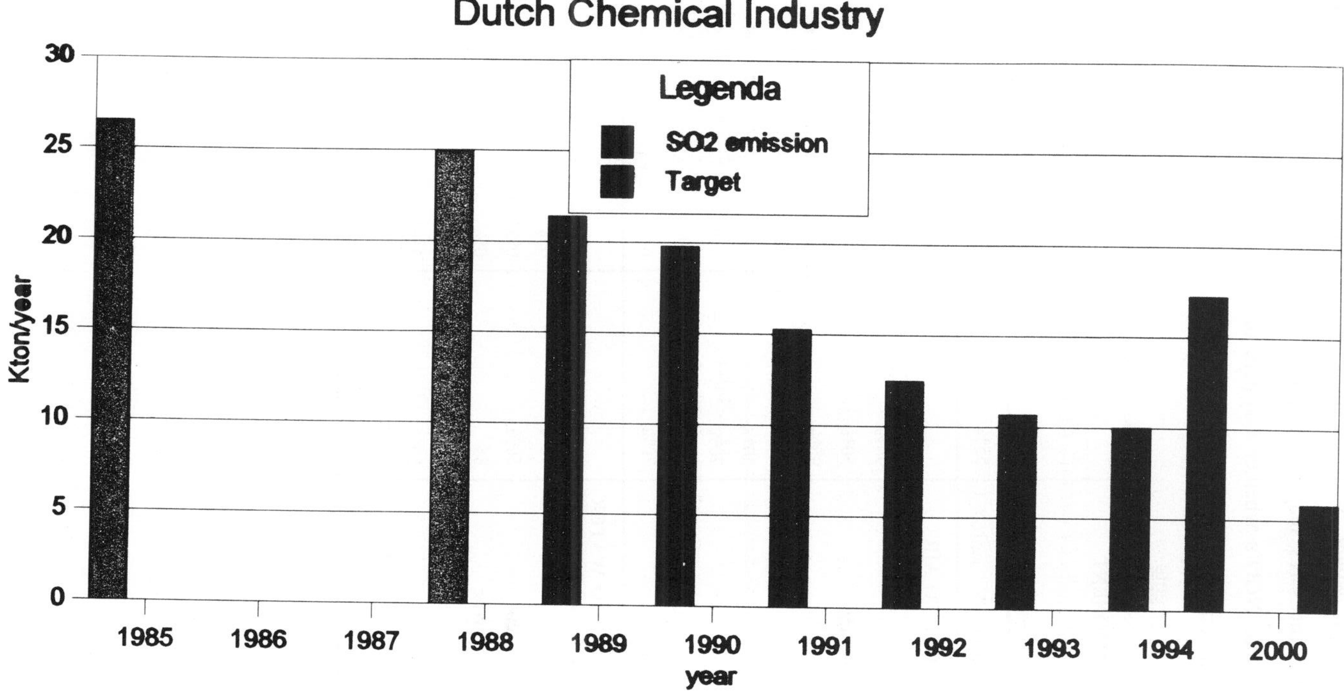
SO2 emission
Dutch Chemical Industry
Legenda
SO2 emission
Target
Kton/year
30
25
20
15
10
5
0
1985
1986
1987
1988
1989
1990
1991
1992
1993
1994
2000
year

For one target which is likely not to be met (NO_x) installation of BAT rather than BATNEEC measures may be required. We are now studying with the government and environmental NGO's whether a system of tradeable emission rights for our sector could offer an attractive cost-effective way to resolve the shortfall.

Non-acieved targets

COMPONENT	TARGET REDUCTION 2000	CHECK SECTOR	NUMBER COMPANIES
NO_x - combustion	55%	46%	81
DISPERSION TO AIR			
Vinylchloride	90%	77%	7
Carbonmonoxide	50%	30%-	39
Fluorides	95%	61%	12
Cadmium	70%	43%	7
Chromium	50%	31%	8
Mercury	70%	49%	5
Copper	50%	39%	8
Dioxins/Furans	70%	63%	5
DISPERSION TO WATER			
Phthalate esters	50%	0%	2
Thichlorobenzene	50%	0%	1
Hexachlorobenzene	85	0%	1
Hexaclorobutadiene	599%	0%	1
Dioxins/Furans	50%	38%	2
WASTE			
Gypsum			p.m.
Halogenated HC			p.m.
Polymers			p.m.

Non-achieved targets

• SECTOR	> NO_X > VINYLCHLORIDE > CARBONMONOXIDE
• INDIVIDUAL • COMPANIES	BALANCE SITE ENVIRONMENTAL PLAN 1997

Companies participating in the execution of the negotiated agreement and having an adequate environmental management system (ISO 14001 or EMAS-level) can opt for a simplified permit application and an adjusted inspection regime.

Industry experience so far is in general positive, for many small/medium sized companies it was the first time that an integral environmental plan -a strategic document- was produced. Execution of the agreement perfectly fits within the chemical industry Responsible Care program. Furthermore these efforts have contributed to an improved environmental image of the chemical industry.

For the next site environmental plan (start mid 1997) the emphasis is likely to shift from end of pipe to in process measures, more attention to the economic situation of the company with respect to the possibility for environmental investments and cost effectiveness of emission reductions. Furthermore companies are being asked to pay attention to the utilization of feedstocks and energy as well as to their products in the light of sustainable development.

In our view the use of negotiated agreements in the chemical industry has demonstrated to offer an attractive alternative to conventional command and control. No complex legislation is required and agreed environmental improvements can be realized in a shorter time frame in a more cost effective manner.

CONCLUSIONS

> Learning process

> Positive experience

> Business contract

> Management tool

> Improved efficiency

The development and implementation of STRETCH

(Selection of sTRategic EnvironmenTal CHallenges)

J. Cramer, University of Tilburg and TNO-Center for Technology and Policy Studies, P.O. Box 541, 7300 AM Apeldoorn, The Netherlands

1. INTRODUCTION

The need to achieve sustainable development represents an enormous challenge to society. It means that within just a few decades we must learn to deal much more efficiently with energy and raw materials [1]. According to some estimates, within the next 50 years the burden on the environment will have to be reduced to an average of one-tenth of the current levels (this means an increase in eco-efficiency by a factor of ten) in the highly industrialized, Western countries [2], [3]. As a first step into this direction Von Weizsäcker, Lovins and Lovins promote an increase in eco-efficiency by a factor of four (this means one quarter of current levels) [4].

Steps have already been taken within industry to increase the average eco-efficiency of products. Most of these efforts focus on *step-by-step*, cost-effective environmental improvements of existing working methods, products and services within a time scale of 1 -3 years. Various techniques and methodologies have been developed to analyze and assess the environmental merits of such product improvements. Incremental improvements provide significant progress in the early stages by capitalising on 'low-hanging fruit' (the easy improvements). After that first period, incremental changes become less profitable in terms of both economic and ecological efficiency. Then more far-reaching environmental improvements begin to deliver a higher reduction in environmental impact at relatively lower costs [5].

If one wishes to reach the target of a tenfold increase in the average eco-efficiency mentioned above, *more far-reaching* improvements are therefore necessary. Contrary to incremental improvements, relatively little experience has been gained within industry with the implementation of such product improvements. Within Philips Sound & Vision more and more attention is being paid to these more far-reaching improvements. This paper describes the way in which Philips Sound & Vision has set up this approach and how involvement has been created across the organization.

2. ENVIRONMENTAL POLICY WITHIN PHILIPS SOUND & VISION

Philips Sound & Vision is part of the Philips Sound & Vision/Business Electronics division. This division is one of the eight divisions of Philips Electronics. Philips Sound & Vision consists of three business groups: BG TV, BG Audio and BG IR3 (=VCR). Every BG has its own environmental coordinator and most industrial facilities have also appointed an environmental coordinator. The Environmental Competence Center (ECC) was established in the early 1990's to coordinate environmental activities within the whole Sound & Vision/Business Electronics division.

Every Philips' division, including Philips Sound & Vision/Business Electronics, has built up experience in the environmental field since the 1970's. In the 1970's and 1980's, the emphasis in the environmental policy of Philips Sound & Vision was on incremental improvements, especially in its production *processes.* One of the major driving forces behind this was legislation and regulation, and the associated rules concerning licensing.

Since the early 1990's, the focus has widened to encompass improvements in the consumer electronics *products* themselves. An initial driving force for this was the corporate environmental policy formulated by Mr Timmer. Another reason was the growing public pressure to find socially responsible ways of disposing of used consumer electronics goods. Additional factors were the (professional) customers' requirements with respect to the use of certain chemical substances and the short-term cost-effectiveness of some environmental improvements (e.g. through material saving;

application of recycled material).

In recent years, Philips Sound & Vision has initiated numerous activities to improve its products from an environmental perspective. A manual on environment-oriented product development ('eco-design') has been produced for designers. The manual includes mandatory environmental requirements for design, and voluntary guidelines to stimulate creativity for eco-design. For instance, a major project is being carried out to reduce the number of environmentally harmful substances in consumer electronics products. One example of this is the decision to stop using flame retardants in the plastic housing of televisions (which, in contrast to other brands, has been the case with Philips televisions since 1987). In addition, the manual contains guidelines concerning the best ways of designing consumer electronics products so that they can be reprocessed in environmentally sound ways at end-of-life. Training programs and workshops are organized to transfer environmental expertise to those responsible for product development.

All these activities have learned the organization that environmental improvements can lead to a win-win situation, in which business opportunities can also be created.

Based on this learning process, attention within Philips Sound & Vision is now turning to more far-reaching and complex solutions, aimed at radical redesign based on existing concepts and at product alternatives. In that context, it has developed the concept of the 'green television', which incorporates all the accumulated environmental know-how of the moment. This concept will be used as a measure prove for future generations of the product.

After gaining some experience with the design of these more far-reaching environmental product improvements, the need was recognized to structure the way in which decisions about strategic environmental product planning were prepared. No guidelines or rules of thumb existed for determining how to select promising environmental opportunities.

The question arose of how the company could systematically elaborate its strategic environmental opportunities and decide which ones to take on board. Until recently, this had not been a prominent issue within Philips Sound & Vision.

Thus, originally the business strategy to be followed by Philips Sound & Vision was relatively simple: a defensive strategy in order to meet existing environmental regulation and covenants, or a cost-reduction strategy aimed at improving the environmental performance in a way that realized short-term costsavings. However, the strategy became more complex as Philips Sound & Vision began to introduce more far-reaching environmental improvements. The growing interest in this latter type of product improvements went hand in hand with the adoption of a third, more offensive strategy. This latter strategy aimed at a better competitive market position through increasing its market share and improving its public image. Identifying promising environmental opportunities and selecting those options turned out to be much more complicated in this case. It required clear strategic choices with regard to the environmental issues which it wants to boast in the market. Not only Philips Sound & Vision but also most other companies had little experience with such an offensive strategy.

3. THE 'STRETCH' METHODOLOGY

To generate and select green opportunities in the context of an offensive strategy a methodology called 'STRETCH ('**S**election of s**TR**ategic **E**nvironmen**T**al **CH**allenges) has been designed and tested within Philips Sound & Vision. The basic questions that needed to be answered were: What opportunities or threats does the environmental issue present for a company such as Philips, particularly for Sound & Vision? What technological options are available for dealing as adequately as possible with environmental problems? And finally, the most crucial question: *which environmental opportunities should be selected to enhance the business and improve the environmental performance of its products?*

In order to address these questions, data is needed on the key drivers that will determine the future business strategy in general. For instance, in the case of Philips Sound & Vision the collection of data consisted of information about economic factors (i.e. future market perspectives of the consumer electronics sector in general and of the company itself) and the technological innovations to be expected. Moreover, some information was needed about cultural trends and the possible set of environmental issues at stake in the future. On the basis of this information, a limited number of plausible scenarios can be formulated related to possible future product market strategies. These scenarios are used to help priorities, select and finally implement the most promising environmental challenges to be adopted by the company.

In total, the 'STRETCH' methodology consists of the following five activities:

Step 1: the identification of the crucial driving forces that will influence the business strategy in general
Step 2: the design of a limited number of plausible scenarios that the company can adopt on the basis of step 1, leading to a list of potential product market strategies
Step 3: the specification of potential environmental opportunities and threats for each scenario on the basis of a checklist of environmental design options
Step 4: the selection of environmental challenges per product leading to a substantial improvement of its environmental performance
Step 5: the implementation of the environmental challenges ultimately selected

To illustrate the step-by-step plan sketched above, a description of how the methodology has been applied within Philips Sound & Vision is given below.

Step 1: Identification of crucial driving factors

Data collection about the market perspectives of consumer electronics products and their technological perspectives is no easy job. Due to rapid multimedia development, the consumer electronics products' business environment is changing quickly. The sector is moving from analog to digital signalling and to an integration of modes (such as text, sound, and visuals) that used to be completely separate. Moreover, a variety of once separated businesses are now starting to converge and compete against each other. This holds in particular for the following sectors: communications, entertainment and information and business/consumer electronics. Finally, the multimedia development gives to end consumers the opportunity to be less passive than before and have more control, choice of selection and/or interactivity.

At the moment it is very hard to predict the future trends in multimedia. Roughly speaking, four aspects can be distinguished in the realization of multimedia [6]:

1. hardware: the development and production of multimedia technology or multimedia hardware (including standardized software tools)
2. software: the development and the provision of multimedia services and applications
3. distribution: the transportation or distribution of multimedia services and applications to the users
4. application/use: the real use of multi media applications and services. This can be professional use within or between organizations, and consumer use.

At the moment the producers of multimedia hardware (especially the computer hardware and consumer electronics companies) are active in new growth markets due to stagnating turnovers and price erosion in their traditional markets. At the same time, the development and distribution sectors are growing. The application/use of multimedia seems to be the greatest bottleneck in the take-off of multimedia on a large scale. All kinds of questions are still unanswered. For instance, what the future home will look like in relation to multimedia developments and how multimedia will penetrate professional circles.
On the basis of a literature study and interviews the following three plausible scenarios can be formulated for thinking about future developments within the consumer electronics sector:

Scenario 1: The consumer electronics companies continue to focus on making the hardware

Scenario 2: The consumer electronics companies shift their emphasis towards the development of software/multimedia

Scenario 3: The consumer electronics companies concentrate on providing services

Philips Sound & Vision has recently chosen a particular mix of the three scenarios as a starting point for its strategic planning for the coming 5-7 years. On the basis of this particular scenario, concrete product market strategies have been formulated.
Information on cultural trends has been provided by Philips Corporate Design (PCD), a department that works closely with trend labs to spot future cultural trends. Experience has shown that people can no longer be so easily classed as having a particular lifestyle. The consumer combines various lifestyles. However, according to these trend lab studies,

positive drivers for environmental awareness are in particular: 1. time; 2. quality instead of quantity; 3. health; 4. growing consciousness of waste and how a product is made; and 5. homeliness. The trend seems to be that the consumer will become more critical, ethical, spiritual, emancipated, demanding and creative. In order to gain a more detailed indication of the influence of cultural trends, focused marketing research at the level of individual products is necessary.

Step 2: Design of plausible scenarios

Within Philips Sound & Vision one particular scenario had already been chosen and elaborated in detail to distinguish between activities which are already mainstream for Philips, new to Philips and new to the outside world. For each of these three categories a list of products had already been defined which represents the particular category of activities. This list of products formed a good starting point to analyze the environmental opportunities and threats for the coming 5 - 7 years with the support of various Philips' experts. The most interesting in this respect were products in the categories: new to Philips and new to the outside world.

Step 3: Specification of potential environmental opportunities

After collecting and integrating available data in steps 1 and 2, the Environmental Competence Center (ECC) of Philips Sound & Vision/Business Electronics identified a number of promising environmental opportunities (step 3). The particular environmental issues which will be headline news in the coming 5 years, or even beyond that, cannot be predicted with great precision. The Environmental Competence Center (ECC) of Philips Sound & Vision has therefore developed *a general checklist of environmental product design options* that serves as a guideline for prioritization (see table 1).

This checklist has been compiled on the basis of various sources [7], [8], [9]. The list of environmental design options has served as a tool to assess the environmental challenges at stake when a company implements the product market strategies formulated in step 2. To priorities these potential challenges, the ECC organized two sequential brainstorming sessions. One of the first companies to try structuring such brainstorming sessions is D-OW [10]. The way in which this company designed the brainstorming process has been an inspiring example in developing our own methodology within Philips Sound & Vision and is now being used within Philips too. The brainstorming sessions were held with representatives of various key persons within Philips, namely representatives of strategy development within Sound & Vision, representatives of Philips Corporate Design and environmental experts from Sound & Vision. This group of people then formulated a number of criteria to guide the process of prioritization.

Table 1: CHECKLIST OF ENVIRONMENTAL DESIGN OPTIONS

Minimisation of production impact

* Minimisation of waste, emissions and energy use
* Respect for biodiversity

Minimisation of product impact

* Reduction of toxic substances
* Minimisation of materials consumption (e.g. through miniaturization, weight reduction; systems integration)
* Minimisation of use of non-renewable resources
* Minimisation of fossil energy consumption (e.g. through energy efficiency and durable energy use)

Efficient distribution and logistics

* Produce where you consume
* Direct distribution to consumer

Intensity of use

* Lease vs sell
* Collective use

Durability of products

* Reuse
* Technical upgrading
* Longer lifetime
* Reparability
* Refurbishing
* Aging with quality

Recyclability of materials

* Reduction of materials diversity
* Materials cascading
* Design for disassembly
* Selected, safe disposal

These criteria were:

1. environmental improvements should provide a business opportunity or competitive advantage
2. projects should have clear environmental relevance
3. environmental improvements should preferably be quantifiable
4. environmental problems directly related to health and safety issues require more attention
5. implementation should not be hampered because of difficulties in cooperation with third parties or because of lack of expertise within the company.

With the help of the criteria mentioned above, the brainstorming group made an initial selection of nine promising projects.

Step 4: Selection of environmental challenges per product

In step 4 the environmental coordinators of each of the three main business groups (BG's) within Sound & Vision selected those items out of the list of nine division priorities that were considered relevant for their BG. Each BG selected five environmental priorities. Together, the BG's covered all nine priorities. Within the framework of the elaboration of each priority, brainstorming sessions and interviews were held with relevant persons from the particular BG, including product managers, marketing people and technical experts.

The elaboration of the various environmental challenges was tailor-made to each BG and each priority item.

To illustrate this point the following three examples will be given:

- the reduction in the energy intensity of Consumer Electronics products
- the reduction of the material intensity of Consumer Electronics products
- the development of potential strategies to enhance the durability of products

With respect to the item 'reduction in the energy intensity', an intensive brainstorming session was held in the BG's TV, Audio and VCR in order to generate and select more far-reaching environmental improvements in the energy consumption during use and standby. As improvements could be made in various parts of the product (e.g. in the components or in the printed circuit board), experts from various backgrounds were present at these workshops. The options that these experts proposed are being elaborated in a technical, economic and marketing sense.

Secondly, the 'reduction of the material intensity of Consumer Electronics products' was also elaborated in a specific tailor-made way. In order to generate options for the reduction of material intensity, close cooperation was established between Philips and one of its main suppliers of materials. Various brainstorming sessions were held to identify promising alternative materials that are lighter, but at the same time have the appropriate functionality for fulfilling the demands on the product. The results of these brainstorming sessions are currently being elaborated in R&D projects.

The project related to 'the development of potential strategies to enhance the durability of products' was elaborated in a slightly different way. First, a summary of the potential options for optimizing the life of products was made on the basis of a literature survey. Next, the capability of Philips Sound & Vision in meeting these options as a way to achieve further optimization of the life of its products was assessed. At this stage it was found important to gauge the view of the outside world on this matter.

To this end, the Environmental Competence Center organized a brainstorming session with external stakeholders in The Netherlands which was attended by 15 representatives from environmental, consumer and women's groups, from the Dutch Ministry of Housing, Physical Planning and the Environment and the Dutch Ministry of Economic Affairs, from relevant research institutes and from Philips.

The participants at this session were asked which five (not more) activities they thought Philips Sound & Vision should give the highest priority in the context of the theme of 'optimizing product life'.

The reactions of the participants suggested a clear prioritization [11]. Particular attention was given to the following topics:

- making more robust constructions
- designing modular constructions
- selling the use of products/leasing

These results were presented in brainstorming sessions with the BG's Audio and VCR. Establishing which additional methods stand a good chance of success in the future of Philips Sound & Vision is currently part of further internal consultation and investigation.

Initial results show that products usually break down due to thermal problems (too high temperature) or defective components or joints. Only after more information has been gathered on the various advantages and disadvantages of improving the durability of the products will Philips take concrete action.

A PhD student from Delft University of Technology was appointed to the ECC/S&V in 1996 to elaborate concrete designs related to the durability issue.

The three examples clearly show that it usually takes a number of brainstorming sessions and specific R&D initiatives before a final assessment is made of the most promising environmental opportunities to be implemented. Through these sessions and specific projects, learning experiences are built up that are used to reduce the present uncertainties about environmental opportunities and market perspectives. When the company has learned more about these more far-reaching environmental improvements, it becomes easier to integrate these endeavors into the regular product development process.

Step 5: Implementation of the environmental challenges ultimately selected

The promising environmental challenges are presently being elaborated in particular R&D or product development projects. Before final decisions can be made, data should still be collected on market perspectives and consumer trends. If necessary, results of specific consumer tests assessing the interest in the new product should also be available. Because the further elaboration of promising environmental challenges takes time, the final implementation of the results has not been effectuated yet.

4. STRUCTURAL EMBEDDING OF STRETCH WITHIN PHILIPS SOUND & VISION

The selection of promising environmental challenges, as described above, is one of the two main pillars of strategic environmental product planning. The other pillar concerns the structural embedding of this endeavor within the organization. In practice, this is an even harder job than identifying and selecting strategic options. It requires a strategic way of thinking about environmental issues within the organization, especially at senior management level.

The implementation of this strategic approach can only be successful if the environmental aspects are incorporated into the process of product planning as a structural component. Companies usually structure this whole process, from generating to ultimately realizing new products, in a more or less similar fashion. Within Philips Sound & Vision the product creation process is divided into two main phases: first, the strategy & planning (including "know-how" planning) phase and next the product realization process (from concept start to commercial release).

In the first phase, a product/marketing strategy is formulated, and the architecture and standard design planning is derived from this strategy. In the second phase, various quality controls and validation procedures are carried out by implementing numerous go/no-go decisions. Each step in the product realization process must conform to a set of standards and release criteria before the next step can be made. In this second phase, major changes in the product design cannot be implemented. Such decisions need to have been made in the first phase.

In the context of the structural embedding of STRETCH the *first action* within Philips Sound & Vision was the integration of environmental goals into an early phase of the product creation process. First of all, the written procedures already in place were evaluated with regard to environmental aspects. Where necessary, these procedures were reformulated in order to incorporate the environmental items to be taken into account. After having incorporated environmental aspects into the written procedures, the next step is to *deploy the environmental responsibilities.* This process, currently taking place, is the most difficult part of the integration process. It requires people at various levels within the organization to take environmental aspects into account. This often involves substantial cultural changes to the way people think and act. Mental changes of this kind take time.

The *second action* to be undertaken in the context of the structural embedding of STRETCH concerns the attuning of the selected environmental challenges to the general marketing strategy of the company. In order to bring about these challenges, a *fundamental change is required in the company's marketing strategy on the environment.* The company needs to adopt a more offensive strategic attitude. This is quite different from following a defensive strategy that is designed to comply with all the relevant environmental legislation and regulations. A direct cost-reduction strategy which focuses mainly on measures which provide short-term solutions is strongly supportive but not the heart of the matter. A company that wants to introduce more far-reaching environmental measures derived from STRETCH must make strategic choices as to how the company wants to strengthen its market position by means of a better green profile than its competitors.

Not only Philips Sound & Vision but also most other companies hardly gained practical experience with such an offensive strategy yet. Exceptions in this respect have been market entrants with market-shaping strategies. They profile themselves from the start as environmentally responsible companies. Good examples of the latter (outside the area of consumer electronics) are companies such as Ben and Jerry's ice cream and the Body Shop. Most companies that have already built up a particular image and tradition have more difficulty in adopting an offensive environmental strategy. Such companies cannot just change their deeply rooted image, culture and knowledge overnight for the sake of the environment.

5. CONCLUSIONS

Until now no structured methodology existed for attuning environmental considerations to the business strategy of companies. The Philips Sound & Vision Environmental Competence Center has developed *a methodology* for this purpose. This methodology is called 'STRETCH' (**S**election of s**TR**rategic **E**nvironmen**T**al **CH**allenges). The objective of STRETCH is to incorporate environmental considerations into the business strategy and select strategic environmental challenges in an early phase of business development. The application of STRETCH provides the possibility of meeting *three main objectives:*

Firstly, focusing on long-term environmental product design strategies can elicit innovations that may enhance the competitive position of the company. Through the integration of eco-efficiency goals into product innovation in general, a company does not aim to beat the competitors purely on environmental grounds, but on its innovative product strategy in general. In this way economy and ecology can go hand in hand.
By taking environmental aspects into account at an early stage of product development, more far-reaching improvements can be made in future consumer electronics products compared with the current range of products. The first strategic environmental efforts, like those taken by Philips Sound & Vision, are still more the exception than the rule [12]. This approach, however, could provide a way forward to substantial improvements in eco-efficiency.

Secondly, the environmental opportunities and threats to be expected in the future can be anticipated in an earlier phase. Through this early warning system an attempt can be made to diminish the negative consequences in an early stage and a response is not required when it is actually too late. In this way actions are more pro-active rather than defensive. The company can even be one step ahead of all kinds of government demands and public pressure by redirecting product development in the context of sustainability in a more fundamental way. By pro-actively integrating environmental aspects into the earlier phases of the product creation process, external criticism can be avoided and the lead taken in environmental priority setting.

Thirdly, as a result of more far-reaching environmental improvements even higher eco-efficiencies are expected to be reached than through incremental improvements. At this stage, the exact data on eco-efficiency gains to be realized within the nine strategic projects currently being carried out within Philips Sound & Vision cannot be provided;

these will be collected during the execution of the projects.

On the basis of the STRETCH methodology, Philips Sound & Vision has prioritized nine projects for further investigation. Through the performance of these projects, learning experiences are built up that can reduce present uncertainties about the environmental opportunities and market perspectives. Once the company has learned more about the more far-reaching environmental improvements, it becomes easier to integrate these endeavors into the regular product creation process.

From initial experiences with the application of STRETCH within Philips Sound & Vision it could be learned that *environmental objectives can be attuned very well to the business strategy.* Moreover, it became clear that the implementation of environmental challenges is not only the task of product development departments but of the whole business.

REFERENCES

1. Schmidheiny, S., Changing Course; A Global Business Perspective on Development and the Environment, MIT Press, Cambridge, 1992.
2. Weterings, R.A.P.M. and J.B. Opschoor, The Environmental Utilization Space as a Challenge for Technological Development, RMNO No. 74, Rijswijk, 1992.
3. Schmidt-Bleek, F. et al., Carnoules Declaration of the Factor Ten Club, Wuppertal, Oct. 1994.
4. Weizsäcker von, E.U., A. Lovins and L. Lovins, Faktor Vier, Doppelter Wohlstand-Halbierter Naturverbrauch, Droemer Knaur, München, 1996.
5. Arthur D. Little, Sustainable Industrial Development; Sharing Responsibilities in a Competitive World, Conference Paper, The Hague, 1996.
6. Hertog, P. den, Multimedia als Innovatie: Fake or Future? (Multimedia as Innovation: Fake or Future?), STB-TNO, nr. 94/002, Apeldoorn, 1994.
7. Business Council for Sustainable Development, Getting Eco-efficient, Report of the First Antwerp Eco-Efficiency Workshop November 1993, Geneva, 1993.
8. Boer, M. de, Milieu, Ruimte en Wonen (Environment, Space and Housing), Ministry of Housing, Physical Planning and Environment, The Hague, 1995.

9. Fussler, C. with P. James, Driving Eco-innovation; A Breakthrough Discipline for Innovation and Sustainability, Pitman Publishing, London, 1996.
10. Ibid. note 9.
11. Cramer, J., Pros and Cons of Optimizing the Life of Consumer Electronics Products, in: Proceedings of the First International Working Seminar on Reuse, organized by the Eindhoven University of Technology, Eindhoven, November (1996) 73-84.
12. IVA (Royal Swedish Academy of Engineering Sciences), Environmental Management; From Regulatory Demands to Strategic Business Opportunities, Stockholm, Sweden, 1995.

Sustainability in the chemical industry by 2050

(The outlook for a sustainable chemical industry in 2050)

J.J.M. Mulderink

It was Albert Einstein who, long before the awareness of depletion of ores and fossil fuels and pollution of the environment gathered momentum, had already stated that: "The world we created today as a result of our thinking thus far faces problems that cannot be solved the way we thought when we created them".

Everybody agreed with him at that time although the message was hardly understood. But in the early seventies the consciousness grew that conservation of energy and protection of the environment deserved particular attention, thanks to the Report of Rome and Dennis Meadows' doomsday scenario in "Limits to Growth".

Of course, there were other problems Einstein warned about, but we should concentrate on the two issues that were emphasized by Meadows.

Especially environment established its place in the new thinking. Only in the early seventies was energy an issue when the Organization of Oil Exporting Countries (OPEC) was able to raise the price of a barrel of oil to a level above US$ 40. Later the price stabilized at a much lower level and price worries subsided.

It was not until 1992, that the Dutch government launched an initiative for a "Sustainable Technology Development" and asked for *the* guidelines. The basis of their questions to an interdepartmental program group was: What will the world look like in about fifty years time?" and "Are there any avenues that should be taken into now in order to arrive without major shocks in the society of 2050?".

It was realized by the program group that these simple questions had many aspects for areas as diverse as: water, housing, transport and food.

One area was chemistry, recognizing the fact that chemistry is the major driving force in all life processes and a major contributor to many products that enhance the quality of life. Think of fibers, plastics, coatings, food-ingredients, pharmaceuticals, paper, detergents and many others.

The mission for "DTO-Chemie", the group responsible for working out the chemistry contributions part, was two-fold:

- Sketch the contribution chemistry could make to a sustainable society in the middle of next century by looking back from 2050. This "looking-back" with backcasting Techniques was crucial in the approach to answering the questions.
- Formulate the avenues of developments and changes that have to be taken at this very moment.

The goal was evident: Increase the quality of life with the help of chemistry.
The study initially centered on the availability of energy carriers. That is: finding substitutes for oil, petrol, kerosene, coal, natural gas and promoting the use of electricity based on solar radiation from a chemical point of view.
Another study dealt with the possibilities of raw material supplies for the production of chemicals with particular notice to the large polymers: polyethylene, polypropylene and PVC. At present, these polymers are responsible for half of the global output of industrial organic products.
Also, in dematerialization (reduction of material use) chemistry should play an important role. For instance, in lighter materials in transport vehicles in order to save energy; and in more efficient reaction technology in order to reduce emissions to water and air. Or in the construction of laminates in solar cells in order to trap solar energy for conversion into electricity.

Every proposed technology should be:

- Environmentally neutral as well as
- CO_2 neutral

The process was started by asking groups consisting of representatives from industry, science, politics, government and banks;

- What will be the shape, the motivation and the need of the world in the year 2050?
- Will industry and science be capable to fulfil the needs of that world?
- Which conditions should be satisfied from now on to implement all the necessary activities?

The backcasting sessions supplied many expectations and, of course, many uncertainties. However, some conclusions for the year 2050 were supported by nearly everyone involved.

- ▸ There will be roughly 10 billion people living on our planet.
- ▸ The energy consumption will, despite major energy conservation, at least have doubled compared with that of 2000. The share of electricity in the energy supply will be much larger than in the preceding century.
- ▸ The economy will have grown exponentially.
- ▸ A much larger part of the global population will be living in cities.
- ▸ The mobility of the people will have strongly increased.
- ▸ The strong individualism of people will lead to many more personal vehicles.

The parties that must satisfy these conditions are: government, industry, universities and development institutes world-wide.

The members of the backcasting groups stated that the parties should work in close cooperation but also realized that their possibilities especially for the long term, would be limited.

The driving force will be the realization that fossil reserves are finite and that prices will go up. Moreover, one wants to become less dependable on distant suppliers.

Before we reach that point large quantities of oil, natural gas and coal are going to be used. This will initially result in a further increase of the CO_2 content in the atmosphere till the years 2020-2030. According to the calculations of the IPCC (International Panel on Climate Control) this will have a modest impact on the rise of the sea level and the global temperature.

After 2030 a gradual stabilization of the CO_2 content by an increased use of electricity from wind and photovoltaic cells and by using liquid fuels from biomass conversion will taken place. In the second part of the next century the CO_2 concentration will decrease. This will be the result of the scrubbing effect of the oceans as mentioned before.

An activity that will precede the CO_2 reduction is removal of detrimental quantities of certain persistent organic pollutants in the atmosphere by better controlled chemical processes. Not only moral considerations but also financial reasons will lead to reducing the spilling of these chemicals.

In the more developed countries legal constraints will have an accelerated modifying effect.

Last but not least the coating industry with their emissions of solvents.

There is already enough knowledge developed to switch to either waterbased coatings or to high-solid paints with only a fraction of the solvents used in more antiquated coating materials. Another technology is the take-up of monomers, used as solvent, in the coating recipe.

It is only a matter of time to bring about the needed cleaning effect in the atmosphere. Too drastic actions now could be counter productive.

In a changing society in which sustainability is going to play an increasing role the effects of an increase in land use for agriculture to produce food and energy, should not be overlooked.

Plants have their own emissions and too large plantations of a particular crop can be detrimental to human and animal health.

Another issue that deserves attention is the water vapor content in the atmosphere. It is too easily accepted that all water vapor coming from combustion is going to condense soon afterwards.

In the equilibrium situation that existed before the industrial revolution this was hardly a problem. Evaporation of sea and landwater and the closed loop of biomass burning were the major processes in water vapourforming. Water vapor however, that originates from combustion of large quantities of natural gas is not part of the equilibrium system.

It can be imagined that especially water vapor originating from natural gas burning could increase the average relative humidity of the global atmosphere by a few percent.

If this should be the case, and there are hardly any measurements to check this assumption, the greenhouse effect would be influenced as much by this fossil water vapor increment as by the increment of CO_2 from the same natural gas. The greenhouse effect of water vapor is namely around 30 times stronger than that of CO_2.

I informed you about the program that we proposed to the Dutch government. It is not at first sight a quantum leap in completely new technologies. The inventions that will be leading to a large leap forward should be following as a matter of course when this program is going to be tackled with vigor. What is new is the connection of the different avenues.

Chemical products are dependent on energy carriers and agriculture plays at long range the key part. Cooperation between agriculture and the chemical industry in this field cannot be avoided. New studies of the methodology of chemical processes play an essential role in this scenario.

I want to finish by expressing my optimism that through intensified programs for sustainability, especially aimed at energy supply and methodological studies of chemical processes, air pollution will be greatly reduced too.

Technology Vision 2020: The U.S. Chemical Industry

Steven C. Weiner

Pacific Northwest National Laboratory[1], 901 D Street, SW, Suite 900
Washington, DC, 20024-2115 USA

The U.S. chemical industry's future and its key technology needs and pathways are the subject of a joint effort of the Chemical Manufacturers Association, American Chemical Society, American Institute of Chemical Engineers, Council for Chemical Research, the Synthetic Organic Chemical Manufacturers Association and the chemical industry to:

- Provide technology vision and establish technical priorities in areas critical to improving the chemical industry's competitiveness.
- Develop recommendations to strengthen cooperation among industry, government and academe.
- Provide direction for continuous improvement through step-change technology.

Through the Technology and Manufacturing Competitiveness Task Group, more than 200 people participated in this look at the future for the chemical industry which resulted in the publication of *Technology Vision 2020: The U.S. Chemical Industry* in December 1996.

---

1. The Pacific Northwest National Laboratory is operated for the U.S. Department of Energy by Battelle Memorial Institute under Contract DE-AC06-76RLO 1830.

1. DIMENSIONS

The chemical industry provides the building blocks of a modern society, and its products are essential to a broad range of manufacturing industries and service sectors. These products meet our most fundamental needs for food, shelter, and health, and they are vital to the high technology world of computing, telecommunications and biotechnology. Chemicals are a keystone of U.S. manufacturing, essential to a large range of industries, such as pharmaceuticals, automobiles, textiles, furniture, paint, paper, electronics, agriculture, construction, appliances, and services.

More than 70,000 different products are registered in three broad and diverse categories: (1) basic chemicals; (2) chemical products used in further manufacture (intermediates); and (3) finished chemical products used for ultimate consumption or used as materials in other industries. More than 9,000 corporations develop, manufacture, and market products and processes in the following categories:

- industrial inorganic chemicals
- plastics, materials, and synthetics
- drugs
- soap, cleaners, and toilet goods
- paints, varnishes and allied products
- industrial organic chemicals
- agricultural chemicals
- miscellaneous chemical products.

The U.S. chemical industry is the world's largest producer of chemicals (over $375 billion worth shipped in 1996), representing over 10% of all U.S. manufacturing on a value-added basis, contributing the largest trade surplus of any nondefense-related sector to the U.S. economy ($16 billion in 1996), and employing over one million Americans enjoying a relatively high standard of living. The energy dimensions of the industry are substantial: over 5.8 quads are used for both feedstocks and fuel/power, even recognizing that energy consumption per unit of output has dropped by over 30% since 1970.

2. THE CHANGING BUSINESS ENVIRONMENT

The driving forces that are changing the nature of the business environment and impacting other industries, as well, stimulated this effort to think about the future of the chemical industry.

- **Increasing Globalization**. The accelerating interdependence of nations and companies challenges the chemical industry to "develop competitive, advanced manufacturing technologies, improve logistics and management of supply chains, and create new products that support the needs of customers globally.
- **Sustainability**. Societal demands for higher environmental performance present significant challenges and opportunities, and accelerate the development of technologies that use resources - both raw materials and energy - more efficiently.
  The industry's commitment to Responsible Care® positions it to be a welcome neighbor and meet the challenges of sustainability: meeting "the economic and environmental needs of the present while enhancing the ability of future generations to meet their own needs.
- **Financial Performance**. Increasingly, the $17 billion spent by the industry on R&D is being shaped by shorter term performance. This emphasis on near-term profitability improves stockholders' satisfaction, but is a significant barrier to funding long-term R&D.
- **Customer Expectations**. Customers are increasingly asking for products and services that are "faster, better, cheaper, and cleaner". Thus, opportunities are provided to forge new and stronger relationships. Challenges abound to remove inefficiencies in supplier/customer relationships and operations through reduced product cycle times, improved product quality and continuously improving service.
- **Changing Work Force Requirements**. More highly skilled workers will be required for tomorrow's work force given the rapid pace of technological change and the complexity of technology. Computers and automation will make plants easier to rum, but will require a more technically advanced understanding of those processes.

3. THE VISION

In order to encompass the chemical enterprise, working groups dealing with new chemical sciences and engineering technology, manufacturing and operations, computer utilization and information management, and supply chain management examined key technological challenges and critical success factors for the industry as it enters the next century. Smaller groups dealt with issues such as partnering and sustainable development, and those themes are embodied in the report. The elements of the vision highlight where the industry wants to be in 2020 and have been characterized as a 'call to action, innovation and change':

- The U.S. chemical industry leads the world in technology development, manufacturing, and profitability. *Profitability pays for technology development and the next generation of manufacturing excellence.*
- The U.S. chemical industry is responsible for breakthroughs in R&D that enhance the quality of life worldwide by improving energy use, transportation, food, health, housing, and environmental stewardship. *The role that chemicals play in our quality of life is not broadly appreciated and it is hoped that Technology Vision 2020 will help with that message.*
- The U.S. chemical industry leads the world in creating innovative process and product technologies that allow it to meet the evolving needs of its customers. *Emphasis is placed on both process and product technologies. "Responding to needs of customers in real time increasingly defines financial success for both the customer and the chemical producer."*
- The U.S. chemical industry sets the world standard for excellence of manufacturing operations that are protective of worker health, safety and the environment.
- The U.S. chemical industry is welcomed by communities worldwide because the industry is a responsible neighbor who protects environmental quality, improves economic well-being, and promotes a higher quality of life. *These two elements are directly related to the industry's Responsible Care® program in which companies are "committed to support a continuing effort to improve the industry's responsible management of chemicals" according to an accepted set of guiding principles.*

- The U.S. chemical industry sets the standard in the manufacturing sector for efficient use of energy and raw materials. *For example, in 1993, the industrial sector accounted for 86% of all operating cogeneration capacity in the U.S. The chemicals sector alone accounted for 31% of that capacity.*
- The U.S. chemical industry works in seamless partnerships with academe and government, creating "virtual" laboratories for originating and developing innovative technologies. *The emphasis on partnerships and collaboration recognizes that companies alone will be unable to meet the challenges of the 21st century and that the capabilities and competencies of the academic community and federal laboratories represent a vital resource.*
- The U.S. chemical industry promotes sustainable development by investing in technology that protects the environment and stimulates industrial growth while balancing economic needs with financial constraints. *Sustainable development represents two balances - firstly, environmental protection and industrial growth; secondly, economic need and financial constraints - that ultimately will have global impacts.*

4. STEPS TO GETTING THERE

A series of steps are outlined for the roadmap toward achieving *Technology Vision 2020*:

- **Generate and use new knowledge** by supporting R&D focused on new chemical science and engineering technologies to develop more cost-efficient and higher performing products and processes.
- **Capitalize on information technology** by working with academe, federal and national laboratories, and software companies to ensure compatibility and to integrate computational tools used by the chemical industry; and developing partnerships for sharing information on automation techniques and advanced modeling.
- **Encourage the elimination of barriers** to collaborative precompetitive research by understanding legislation and regulations that allow companies to work together

during the initial stages of development.

- **Work to improve the legislative and regulatory climate** by reforming programs to emphasize performance rather than a specific method of regulatory compliance, and increasing consideration of cost, benefits, and relative risk.
- **Improve logistics efficiencies** by developing new methods for managing the supply chain and by sponsoring an effort to shape information technology and standards to meet the industry's manufacturing and distribution needs.
- **Increase agility in manufacturing** by planning manufacturing facilities capable of responding quickly to changes in the marketplace using state-of-the-art measurement tools and other technologies for design, development, scale-up, and optimization of production.
- **Harmonize standards**, where appropriate, by working with governments within the United States and internationally, and with independent standards groups on nomenclature, documentation, product labeling, testing, and packaging requirements.
- **Create momentum for partnering** by encouraging companies, government, and academe to leverage each sector's unique technical, management, and R&D capabilities to increase the competitive position of the chemical industry.
- **Encourage educational improvements** by strengthening educational systems and encouraging the academic community to foster interdisciplinary, collaborative research and provide baccalaureate and vocational training through curricula that meet the changing demands of the industry.

5. MEETING THE CHALLENGE

One means of summarizing is to reflect on the conclusions of the Task Group regarding the five broad goals that the chemical industry must accomplish over the next 25 years in order to meet the vision.

- Improve operations, with a focus on better management of the supply chain;
- Improve efficiency in the use of raw materials, the reuse of recycled materials, and the generation and use of energy;

- Continue to play a leadership role in balancing environmental and economic considerations;
- Aggressively commit to longer term investment in R&D; and
- Balance investments in technology by leveraging the capabilities of government, academe, and the chemical industry as a whole through targeted collaborative efforts in R&D.

These goals are clearly interrelated. For example, an unanswered plea to both the public and private sectors to aggressively commit to longer term investment in R&D could severely impact the ability to improve efficiency in the generation and use of energy for future generations. The answer may lie in balancing investments through targeted collaborative efforts in R&D that create a "multiplier effect" on the pool of resources: people, facilities and capital. More broadly, collaboration, whether it be among chemists, engineers, physicists and biologists; or suppliers and customers; or public and private sectors; or other diverse groups, may be the most important message of *Technology Vision 2020.*

6. ACKNOWLEDGEMENTS

The author wishes to thank the U.S. Department of Energy, Office of Industrial Technologies, and its Chemical Industry Team for the support of this work and participation in the 5th U.S.-Dutch International Symposium on Air Pollution in the 21st Century.

REFERENCE

1. Technology Vision 2020: The U.S. Chemical Industry. American Chemical Society, American Institute of Chemical Engineers, Chemical Manufacturers Association, Council for Chemical Research, Synthetic Organic Chemical Manufacturers Association. December 1996. Washington, DC.

Industry's role in air quality improvement: environmental management opportunities for the 21st century

D.A. Rondinelli [a] and M.A. Berry [b]

[a] Glaxo Distinguished International Professor of Management and director of the Center for Global Business Research, the Frank Hawkins Kenan Institute of Private Enterprise, University of North Carolina, Campus Box 3440, Chapel Hill NC 27599-3440, USA.

[b] Adjunct associate professor of management, the Kenan-Flagler Business School, and senior research associate, the Kenan Institute of Private Enterprise, University of North Carolina, Campus Box 3440, Chapel Hill, NC 27599-3440, USA.

1. INTRODUCTION

More than 200 years ago Adam Smith, the founder of modern economics, wrote in the *Wealth of Nations* that "...consumption is the sole end and purpose of all production; and the interest of the producer ought to be attention to...that of the consumer." In recent years the rapid growth of the world economy has given Adam Smith's maxim new meaning. The age-old hunger for consumption grows unabated and increasing numbers of consumers around the world are attaching new value to the environmental quality of goods and services and expressing concern over the environmental impacts of industrialization. That concern is growing because more than 70% of the world's urban population live in areas where the air is seriously polluted and as many as 750,000 people - the majority in developing countries - die each year of ailments caused by air pollution.[1]

Over the past 25 years corporations throughout the world have made dramatic changes in the way they do business as more people come to understand how the ecological system works and how polluted air and water endanger human health. The key to increasing industry's participation in the drive for higher standards of air quality is the growing realization that effective environmental management, technological development, and technology dissemination are cost-effective and profitable business strategies. Global competition is making firms around the world more customer-conscious and, to the extent that consumers demand products that minimize environmental degradation and enhance the quality of their lives, businesses in every industry must respond in order to survive.[2]

This paper examines how changes in business practices, driven by a better understanding of how natural environments function, are converging to provide new opportunities for environmental management that goes beyond regulatory compliance to reduce air pollution. Although sound and well-enforced environmental regulations are an essential foundation for improving air quality, command-and-control systems alone are unlikely to achieve the lower levels of pollution that will be necessary to achieve sustainable development in the 21st century. In cooperation with government, businesses in every industry can play crucial roles in achieving higher standards of air quality while at the same time maintaining acceptable levels of economic growth. We explore three ways in which corporations can contribute to environmentally sustainable development: (1) by adopting proactive environmental management systems that focus on air pollution prevention; (2) by developing new technologies for air pollution control and reduction; and (3) by transferring air pollution control and prevention technologies through international trade and investment.

2. TOWARD A HOLISTIC VIEW OF BUSINESS AND ENVIRONMENT

Recent developments in the global economy are pushing firms in every industry to develop new strategies of competition and new processes for managing their environmental impacts. These trends include: (1) a growing awareness of the relationships between economic and environmental sustainability; (2) a better understanding of the business opportunities—both potential cost reductions and higher profits—in adopting quality environmental management practices; (3) a growing realization in government and the private sector that regulatory controls, while necessary, are not sufficient to achieve pollution prevention; and (4) growing international pressures on corporations to adopt voluntary standards for environmental management that go well beyond regulatory compliance as a precondition for participation in global trade and investment.[3]

2.1. Changing Business Perceptions of the Environment

We should not for a moment think that environmental performance is, or ever will be, the dominant concern of business. In market economies the primary purpose of business is to create wealth for shareholders who risk capital in anticipation that it will appreciate sufficiently to give them a fair return on investment. Customers, in addition, demand goods or services of value, the sales of which allow companies to earn profits, pay their employees' wages, and reward their shareholders. Profit-making will always be the primary objective of corporations, but an increasing number of companies understand that they also share the common benefits of the natural environment. Beyond creating monetary wealth, businesses have a second "bottom line" that reflects the ethical system in which they operate. Firms must develop core corporate values and adopt quality-based management systems in order to satisfy consumers. This second bottom line accounts for things of social value such as environmental health and safety.

Wealth-creation remains a fixed business objective, but the manner in which wealth is created changes continuously. In the 21st century nearly every aspect of the business environment affecting corporate competitiveness will change dramatically. Three trends currently offer hope that businesses will continue to adopt positive cost-effective environmental performance policies that will significantly reduce global air pollution. They include, first, broad changes in business practices, especially those related to quality-based management, that have built-in environmental protection attributes; second, changing attitudes in the business community toward the environment that are making environmental management an integral part of business strategy; and third, new philosophies and technical approaches to environmental management that are making pollution prevention programs more effective and efficient.

Business leaders have become accustomed to thinking about a firm's environment; the social, political, and economic conditions that directly impinge upon its operations; and ways of adapting their internal organizations to changing external conditions in order to survive and grow. Only recently, however, have corporations defined their environments to include the *biosphere*, the life supporting system from which industries draw their raw materials and deposit their wastes. In the rapidly shrinking world of commerce, corporations must come to grips with the complex web of relationships between the natural environment in which we live and the economic system in which we work. Companies that perceive of environmental systems and economic systems integrally see a *biospheric market place* where environmental, economic, and human life systems are tightly bound. In such a system both producers and consumers begin to understand the broader context of production and distribution. Consumers begin to demand goods and services produced through processes that do not pollute the air they breathe or degrade the natural resources from which they derive their livelihoods.

To make the biospheric marketplace function smoothly, government has traditionally provided infrastructure and imposed regulations that sustain the natural system, just as it manages diseconomies and preserves a minimum level of order in the economic system. Similarly, management in the private sector seeks to guarantee and shape the future of individual businesses. Increasingly the problem that both governments and businesses face is to balance the financial health (cash flow and profitability) of wealth-creating firms with the efficient and responsible use of natural and environmental resources.

Public policies in most countries assume that corporations make environmental decisions and management decisions differently and that the two are basically irreconcilable at the organizational level. They further assume that corporations cannot compare the costs and benefits, for example, of improving air quality with those of production changes or product design options. Some have concluded that the mere act of making environmental decisions automatically forces a conflict of values. And in the past corporations have too often accepted these assumptions, opposing environmental controls with the argument that it is impossible to conserve ourselves into prosperity. However, decades of conflict over the environment have produced a new economic model. Enlightened corporations have begun to recognize that environmental resources, including the quality of air, must be treated as a unique form of productive capital with properties that make them different but no less important to the economy than manufactured capital. In the words of Frank Popoff, former CEO of Dow Chemical Company, "there can be no economic development without environmental responsibility." Natural resources such as air, water, and forests are capital resources and their wise use benefits industries, employees, and communities.

Industry's role in conserving and, indeed, enhancing environmental resources is now widely recognized. The McKinsey Corporation's survey of more than 400 top executives of companies around the world found that 92% agreed with Sony Corporation president Akio Morita's assertion that the environmental challenge will be one of the central issues of the 21st century.[4] A growing number of U.S.-based multinational corporations, 3M, DuPont, Monsanto, Xerox, Johnson & Johnson, Procter & Gamble, SC Johnson, and others are beginning to view the natural environment and their business environment holistically.

A recent survey of 256 large and small manufacturing firms in the United States found that nearly 78% of the respondents ranked pollution prevention as "very important" or "important" to corporate performance.[5] About 84% of the companies were pursuing reduced emissions strategies and 16% were seeking zero emissions levels. Clearly, regulations were a significant factor in their corporate environmental strategies, but respondents also listed corporate citizenship, improving technologies, service to key customers, and improving productivity as critical reasons for adopting proactive environmental management strategies.

Enlightened companies begin with a policy and a plan that reflect sound environmental goals and secure top management commitment and long-term funding. More than 79% of the executives responding to the McKinsey survey reported that their firms had written company environmental policy statements. Good policies identify environmental protection as a priority and are reinforced with specific goals, target dates, and issue-specific policies and procedures. The policies are backed up by a long-term strategy. They mandate a strong program to monitor performance and take corrective action when necessary. Many firms have adopted environmental programs that focus employee and public attention on their objectives—"Pollution Prevention Pays" at 3M, "Waste Reduction Always Pays" at Dow, "Priority One" at Monsanto, and "Save Money and Reduce Toxics" (SMART) at Chevron. But beyond slogans and symbols, successful companies declare clear goals and measurable targets. General Electric adopted a program to decrease toxic emissions by 90% between 1988 and 1993. Xerox reduced hazardous waste generation by 50% between 1990 and 1995. Nortel, the Canadian-based telecommunications MNC, for example, established clear and specific targets for the years 1993 to 2000: a 50% reduction in pollutant releases, a 50% reduction in solid wastes, a 30% reduction in paper purchases, and 10% improvement in energy efficiency.[6] Kodak uses its customer satisfaction objective as its environmental goal: "to create customer confidence with on-time delivery of defect-free, reliable products and services exactly as ordered with no wasted material or labor."[7]

These corporate policy statements are not merely meant for public image-building; they reflect a new understanding of the economics of environmental resources. The unwise use of environmental resources can deprive industries of the materials they need for production and eventually drive them into bankruptcy. To profit and grow companies must be able to regenerate natural resources and maintain an ecosystem capable of sustaining both production and consumption. Proactive environmental management and pollution prevention are being adopted by major corporations because they understand that it is a profitable strategy. AT&T saved $3 million a year by redesigning its circuit board cleaning process to eliminate ozone-depleting chemicals.[8] Canadian chemical companies that adopted "Responsible Care" practices have saved millions of dollars in insurance costs as insurers began to consider them highly protected risks.[9] In 1996, Dow Chemical announced that would spend $1 billion in training, R&D and new facilities to achieve an ambitious set of environmental objectives, including a 75% reduction in emissions of 29 priority compounds and a 50% reduction in emissions of other chemicals, by the year 2005. Dow fully expects a 30% to 40% return on its investment.[10]

Proactive corporate environmental policies reflect a growing realization that the objectives of sustainable environmental development and industrial progress are much the same—they both include maintaining growth and improving quality; satisfying the basic needs of life such as jobs, food, energy, water, and sanitation; conserving and enhancing the natural resource base from which many industries derive inputs; reorienting technology; and managing risk. As world population grows, resource use must be redefined and redirected, a mandate with both fundamental environmental implications and enormous business opportunities.

2.2. The Limitations of Regulatory Controls

A second force driving industry's search for solutions to air pollution problems is the growing realization in both government and the private sector that regulation alone is unlikely to achieve the air quality objectives that are essential for sustainable development in the 21st century. Until recently most people believed that environmental management was mainly the responsibility of government and environmental regulations proliferated as public demands on government grew to remedy the adverse environmental effects of industrialization. In the United States and many other industrialized countries, however, environmental legislation was adopted piecemeal, creating a complex regulatory process. In 1970 there were about 2,000 federal, state, and local environmental rules and regulations in the United States; today there are more than 100,000. Environmental regulations are listed in over 789 parts of the Code of Federal Regulations. A command-and-control system for environmental management became the foundation for scores of environmental, health and safety programs and thousands of federal, state and local standards, regulations, and guidelines within which businesses must operate. This complex system imposes tremendous compliance burdens on corporations and especially on small- and medium-sized firms.

In the United States, the Federal government has steadily increased its enforcement of environmental regulations making business executives and owners liable for environmental pollution. The U.S. Environmental Protection Agency takes hundreds of enforcement actions against businesses every year leading to prison sentences and heavy fines. But regulatory enforcement is expensive for both government and business. The total costs of complying with environmental laws over the past 25 years have easily exceeded $1 trillion. About $120 billion is spent annually for pollution abatement and control. Current estimates of compliance costs under the new Clean Air Act

Amendments alone are on the order of $50 billion a year. Many companies will spend hundreds of millions of dollars on environmental projects over the next few years simply to stay abreast of current environmental regulations.

Clearly, sound and well-enforced regulations have brought tremendous progress in reducing air and water pollution and toxic hazards in North America, Western Europe, and other regions of the world. As Table 1 indicates, air pollution has been reduced significantly in the United States over the past 25 years. These reductions in air pollution are all the more remarkable because they occurred during a period in which population increased by 28%, vehicle miles traveled rose by 116%, gross domestic product grew by 99%, and most Americans attained a higher quality of life.

Table 1
Common air pollutants U.S. air pollution emissions 1970-1995

Pollutant	Million Short Tons Annually		
	1970	1995	Change
Carbon monoxide	127	90	-28%
Nitrogen dioxide	24	25	+6%
VOC's	31	24	-25%
PM-10	15	3	-79%
Sulfur dioxide	30	21	-41%
Lead	225	10	-98%

Source: U.S. Environmental Protection Agency, National Air Quality and Emissions Trends Report, Washington, DC: U.S. EPA, 1995.

But the steady proliferation of regulations and the continuing expansion of an already complex regulatory system has made enforcement more expensive and marginally less effective. The continuous tightening of environmental laws in piecemeal fashion in the United States has placed more sophisticated and costly burdens on state and local governments. Rarely, however, have national or state lawmakers fully understood or reassessed the cumulative effects of their environmental regulations.

Unfortunately this reliance on constantly changing regulations may simply produce a temporary false sense of security, since controls require add-on technologies that must be replaced or modified every time regulations change or new ones are added. Under a command-and-control system businesses are constantly struggling to comply. And both scientists and corporate environmental managers are discovering that the complex command-and-control approach to environmental protection, which often addresses only one environmental medium at a time, can merely be an expensive way of moving pollutants around. Water treatment often results in the collection of hazardous wastes or toxic sludge that must be deposited on land or incinerated and returned to air in some form. Air controls often depend on liquids and gases which must be treated and disposed of through water or land.

The growing complexity of environmental regulation and the all too common technical inefficiencies and administrative weaknesses of command-and-control systems have spawned increasing interest in pollution prevention. Governments everywhere have limited financial resources to monitor pollution and enforce regulations even in the best of economic times, but the continued reliance on regulatory controls keeps bidding up the costs. In every country governments must struggle to oversee and enforce environmental laws. Consequently, governments are under growing pressures to motivate the private sector to become proactive through market mechanisms and pollution prevention programs.

2.3. Changing Business Practices

A third force that is creating positive pressures for industry to reduce air pollution are changes in business practices (see Table 2). The traditional business traits of *individuality, independence, hierarchy, local markets,* and *limited communication* are giving way to corporate strategies that emphasize *community, interdependence,* that they cannot survive unless they adopt the environmental values of the community.[11] Although every business has a fiduciary responsibility to shareholders, it must also answer to a broader audience of stakeholders, including owners, employees, customers, regulators, suppliers, competitors, community interest groups, and the media.

Table 2
Changing profile of general business practice

Traditional Business Traits	Emerging Business Traits
Individual------------------------------------>	Part of a Community
Independence------------------------------->	Interdependence
Hierarchy------------------------------------->	Networks
Local or Regional Markets---------------->	Global Markets
Crisis and Cost----------------------------->	Proactive-Investing in the Future
Maximized Profits-------------------------->	Sustained Business
Quantity-Price------------------------------>	Quality-Value
Limited Communication------------------->	Immediate Global Communication
Supply and Inventory---------------------->	"Just in Time" Delivery and Service

As the global economy becomes more integrated, corporations inevitably become part of a network or a "value chain" that constitutes a worldwide system of exchange. In the network, information, raw materials, goods, services, and even pollution and wastes are moved around in the process of generating revenue and creating wealth. For a business to be part of a network, it must abide by broader environmental protection standards—examples in the United States include the chemical industry's adoption of the "Responsible Care" Program, the carpet industry's testing and labeling program, and the electronics industry's move away from chemicals that contribute of stratospheric ozone depletion and global warming.

Because of the enormous improvements in communication and transportation systems, business networks are expanding rapidly. This network expansion connects virtually every business to the "global village" and its market. The rapidly expanding global consensus that environmental protection is crucial for survival is reflected in the environmental requirements of international trade agreements that have steadily grown over the past 50 years. Those corporations that seek access to the global market must be prepared to meet international environmental performance expectations.

A basic principle of the market asserts that when people have access to information they make better choices. The tremendous advances in communication technology make it possible to observe and communicate with every region of the world almost instantly. The information age makes it possible to acquire information on virtually any subject, including environmental pollution, in a matter of minutes. More information means more choices and consumer demands. A business must keep up with changing demands for better environment performance or risk the loss of market share. One of the benefits of rapid communication and transportation, for example, is just-in-time delivery of goods and services, a practice that is good for both business and the environment. JIT systems reduce the environmental costs associated with over-production, inventory management, spoilage, and disposal.

Business' traditional objective has been to *maximize profits*. As global competition becomes more intensive, corporations must develop quality-based management competencies that are now essential for long-term business success. The strengths emphasized in a quality-based management system are conducive to positive and proactive environmental performance and overall business success.

Leadership. Quality management begins at the top of successful organizations in the form of strong leadership. Leaders set the tone, commitment and high standards for environmental responsibility and performance. When there is a strong and consistent vision of environmental performance from the top, quality-based businesses almost always demonstrate exemplary environmental performance.

Intense Customer Focus. In quality-based management there is always an intense focus on customer needs and values. If customers value the quality of the environment, so will the quality- based business.

Proactive Environmental Behavior. Almost always in a well-managed organization there is a clear environmental policy and a process within the company of complying with environmental regulations and using pollution prevention strategies.

Valuing Human Resources. In quality-based organizations there is a tendency of management to use fully the creative abilities of employees, suppliers, and customers to arrive at strategies and solutions to all problems including environmental management issues. The various environmental values held by stakeholders motivate improved environmental performances of the organizations that allow them to participate in decision-making and problem solving.

Management By Fact. In quality-based organizations decisions are based on good science and factual information. Quality-based organizations know where they stand in relation to the business world and the natural environment. Products and services are fashioned through careful analysis and by building on knowledge and experience within the company.

Creativity and Flexibility. In quality-managed organizations there is much thinking "outside of box". Managers realize that the business faces different threats and opportunities at different times. The quality-based organization excels at maximizing the different inherent capabilities of the organization to be successful.

Quality Over Cost. Although cost-efficiency is always a concern, well-managed organizations emphasize quality-seeking ways of managing environmental pressures, internalizing environmental values, and meeting customer demands simultaneously.

Constancy of Purpose. Businesses that focus on quality constantly improve the business process including their environmental performance. Environmental performance is often used as a service to customers, as a means of protecting existing the existing customer base, and as a way of diversifying and expanding market share

Win-Win. In quality-based organizations all stakeholders are winners. The win-win management philosophy produces a culture based on civility, where environmental values are respected. The win-win strategy often leads to economies and opportunities both for stakeholders and for the corporation.

Quality Tools, Technology, and Systems. The success of a company's environmental management program depends on using and developing economically viable operating systems. Logistics, for example, is critical to business efficiency but is also critical to reducing pollution. Less wasteful new technology and quality systems make businesses more accessible to existing customers, more attractive to potential new customers, and more efficient in pollution prevention.

Quality management principles require firms to shift their goals from maximizing profits by manipulating quantity and price to sustaining profit growth by adopting an intense customer focus, continuous improvement, and error and waste reduction.

2.4. International Standards of Environmental Performance

The fourth factor driving greater industry participation in finding solutions to air pollution problems in the 21st century are international pressures to adopt voluntary standards that emphasize the integration of environmental management and corporate strategy. For example, the American Society for Testing and Materials (ASTM) is making headway in standardizing environmental auditing, assessment, and criteria for investment and insurance. British standard BS7750 was an industrial response to the adoption of the 1990 Environmental Policy Act in Great Britain that has been widely adopted internationally. The European Community has issued a Standard Eco-Management and Audit Scheme (EMAS), which member nations are expected to implement. Through these international standards, firms are finding that it is better to make a product right the first time and save the cost of re-work later; that it is cheaper to prevent a spill in the first place than to clean it up; and that it is more cost-effective to prevent air pollution rather than to control emissions.

The ISO 14000 series is likely to become the dominate international standard for environmental management systems. These standards seek to integrate environmental and corporate management systems. Integrated management systems help firms identify the causes of environmental problems and eliminate them. The 14000 series includes standards for environmental management systems, environmental auditing, environmental labeling, performance evaluation, life cycle inventory and assessment, and environmental aspects in products.[12] As companies become certified under the new standards they are finding many business benefits, including improved environmental performance, reduced liability, lower costs, better access to capital, fewer accidents, more employee involvement, improved public image and enhanced customer trust.

3. INDUSTRIAL RESPONSES TO IMPROVING AIR QUALITY AND ENVIRONMENTAL SUSTAINABILITY

All four of these forces - changing business perceptions of the environment, the limits of regulatory control, changing business practices, and international standards of environmental performance - are stimulating corporations in a wide range of industries to find more effective ways to reduce or eliminate air pollution. Three responses are likely to intensify in the 21st century: (1) the adoption of proactive environmental management systems focused on pollution prevention; (2) innovations in pollution reduction and control technology, and (3) dissemination of pollution prevention and control technology through international trade and investment.

3.1. Proactive Environmental Management

Truly effective environmental protection requires the prevention of air pollution rather than the control of emissions. Pollution prevention uses materials, processes, or practices that reduce, minimize, or eliminate pollutants or wastes at the source. Pollution prevention technologies in manufacturing include materials substitution, process modification, materials reuse within existing processes, materials recycling to a secondary process, and materials reuse within a different process.[13]

Increasing legal liabilities and rising costs of emission control have become driving forces for corporations to find more effective ways of preventing pollution. Cutting-edge corporations in the United States and around the world are using five major approaches to proactive environmental management that in combination form a comprehensive pollution prevention program: (1) full cost accounting; (2) waste minimization; (3) demand-side management; (4) design for environment; and (5) product stewardship.

The use of full cost accounting (FCA) is beginning to reshape the concept of environmental accounting. Corporations like Dow Chemical, DuPont, and Ciba Geigy use FCA to identify, quantify, and allocate the direct and indirect environmental costs of ongoing operations.[14] FCA identifies and quantifies environmental performance costs for a product, process, or project including direct costs, the hidden costs such as monitoring and reporting, contingent liability costs, and intangible costs such as public relations and good will.

In the 1980's many corporations began focusing on, anticipating, and preventing waste problems before they occurred.[15] By the end of the 1980's waste minimization programs had been adopted by a diverse group of U.S.-based MNC's such as Allied Signal, General Dynamics, Dow Chemical, Chevron, Boeing, AT&T, Amoco, General Electric,

IBM, Polaroid and Xerox.[16] Most successful businesses were voluntarily performing internal environmental compliance audits to identify and correct their environmental liabilities, demonstrate good faith effort, and reduce government pressures. More important, the voluntary audits forced businesses to evaluate operating systems, identify the actual cost of controls, and develop environmental performance strategies to eliminate liabilities altogether. Waste minimization is a powerful business strategy because it encourages the efficient use of raw materials and reduces the costs of waste. But most companies minimize wastes because it provides competitive advantages and satisfies customers' needs. In the process, companies often learn how to control pollution better than the regulators and at lower cost.

Demand-side management is an approach to pollution prevention that originated in the utility industry, but has spread to other industries as well. It focuses on understanding customers' needs and preferences and on their use of products. It seeks to minimize or eliminate wasted product, to sell customers exactly what they demand, and to make the customer more efficient in the use of the product. Demand-side management forces an industry to look at itself in a new light which often leads to the discovery of new business opportunities. In the case of the utilities, demand-side thinking emphasized that companies are not primarily in the business of selling electricity or gas, or even light or heat, but are really in the business of selling environmental conditions such as comfort, brightness, and conveyance.

Design for environment (DFE) is also becoming an integral part of pollution prevention in proactive environmental management. Corporations such as AT&T, Xerox, Hewlett Packard and Baxter International are finding that it is far more efficient to design products for disassembly, modular upgradeability, and recyclability at the outset than to deal with disposal problems at the end of a product's life.[17] Procter & Gamble's objective is to "design manufacturing waste out" of business areas that account for at least 50% of its production volume by the beginning of 1998. DFE reduces reprocessing costs and returns products to market more quickly and economically.

Product stewardship is yet another concept taking hold in industrial countries seeking to curtail air pollution and solid and liquid wastes. Companies such as Dow Chemical, Procter & Gamble, and Scott Paper are responding by using product life cycle analysis

(LCA) to determine ways of reducing or eliminating wastes at all stages -- from raw materials acquisition, production, distribution, and customer use to waste reclamation, recycling, reuse, and disposal.[18] Japanese universities and research institutes are applying LCA to a wide range of products from aluminum cans, automobiles, and office buildings to vending machines, washing machines and steel alloys.[19] Firms serious about product stewardship seek alternative products and applications that are less polluting and alternative materials, energy sources, or processing methods that eliminate waste; compare the cost of managing for conformance versus for assurance; and adapt to customers' needs, preferences and uses of products.[20]

3.2. Technological Innovation

Inherent in proactive environmental management has been the search for new technologies, processes, systems, equipment, and know-how that reduce or control pollution more effectively. Air pollution in urban areas, for example, reflects more than poorly maintained automobiles; it points to infrastructure and market place breakdown. Entrepreneurial firms see business opportunities in the form of less polluting durable goods especially cars and trucks, mass transit, cleaner fuels, and efficient materials handling. Diesel engine makers such as Varity Perkins in the United Kingdom, for example, are developing new sealing technologies and using computer simulations to market a zero-pollutant engine.[21]

Technological innovation is the key to the environmental progress required for sustainable development. Effective regulatory compliance depends on businesses having access to effective and affordable processes and equipment to meet environmental standards. Pollution prevention is based on process or technological changes that reduce and eliminate waste. By assessing its product and processing technologies, SC Johnson & Son, for example, was able to cut manufacturing waste nearly in half, reduce the use of virgin packaging materials by more than 25%, and reduce volatile organic compound (VOC) use by 16% between 1990 and 1995.[22] Its product stewardship projects produced environmental benefits in its plants around the world.

Corporations in a wide range of industries are developing and adapting commercially viable new technologies for pollution control and prevention that are not only used internally but developed into new products and services. Enlighten firms view pollution reduction and control as business opportunities. New technologies and processes are being developed for input substitution, product reformulation, product unit redesign, product unit modernization, improved operation and maintenance, and internal recycling and reuse.[23]

Many firms are developing and adapting process innovations to reduce noxious and toxic emissions. The survey of 256 large and small firms in the United States mentioned earlier found that more than 60% of the firms used improved process technology or new process technology, and about 58% reported using new product technology to prevent pollution.[24] The Olin corporation, a specialty chemicals, metals and aerospace products corporation, for example, substantially reduced air emissions of carbon tetrachloride by applying technologies that reclaim the material for reuse in several of its production processes. It also achieved 80% reductions of 1,1,1-trichloromethane by altering its overall production processes to wash parts using water-based cleaners instead of chlorinated solvents.[25] Dow Chemical is replacing CFC's and other volatile organic compounds in the manufacture of high volume commercial foam products. Canadian industries have found ways of eliminating halogenated degreasing solvents to reduce the volume of hazardous wastes requiring disposal and eliminating VOC emissions.[26]

Much of this technological innovation has been driven by regulatory requirements, but in many industries corporations are seeing strong business opportunities in the demand for environmentally sustainable development. The adoption of waste minimization strategies has stimulated computer design processes and new software, expert systems, and comprehensive data bases that integrate waste minimization requirements into process design and plant restructuring. Firms are developing new retrosynthesis- and computational chemistry-based techniques that will make pollution prevention far easier in the chemical industry.[27] The technological innovation accompanying the search for environmentally sustainable development is not constrained to North American and Western European firms. Russian scientists have created an air pollution control called Pulsatech that can destroy polluting molecules from industrial smokestacks using specially generated high-frequency, high-voltage pulses within a reaction chamber.[28]

3.3. Technology Diffusion and Transfer

One of the most effective ways of controlling air pollution and progressing toward environmental sustainability is for companies with effective and efficient pollution prevention and control technologies to commercially diffuse them in international markets. In many of the former socialist and developing countries of the world, significant reductions in air pollution will occur only when indigenous industries have access to highly-effective technologies at a reasonable cost. The Brazilian state petroleum company (Petrobras), for example, by investing $1 billion in sulfur reduction technologies at five refineries, is now able to distribute low-sulfur (maximum 0.3 wt %) diesel fuel that can improve air quality and vehicle performance in highly polluted Brazilian cities.[29] The new low-sulfur diesel will cut sulfur emissions by 900 tons a month in the State of Sao Paulo alone. Diffusing and transferring these technologies from industries that have developed them in North America and Western Europe to developing nations and emerging markets is a win-win solution; it earns revenues for the exporting firms and assists importing firms to reduce noxious or toxic air emissions.

The worldwide market for environmental technology and services is large and growing rapidly. In 1994 the global market for environmental goods and services was estimated at $408 billion and it was projected to grow to more than $572 billion by the year 2001.[30] The air pollution control segment of the market was more than $26 billion in 1994 and if it retains its share of the overall market can be expected to reach nearly $38 billion by the year 2001. In 1997, the world market for particulate systems and parts was forecast to exceed $6.7 billion, for gas treatment systems and parts to exceed $8.3 billion, and for other air pollution equipment and services to exceed $4.7 billion.[31]

Additional opportunities are inherent in other process and prevention technologies, instrumentation, monitoring, and energy technologies not reflected directly in air pollution control market estimates. Although in North America and Western Europe the annual growth in markets for environmental protection technologies and services have stabilized at from 4% to 5% on average, demand in Asia is estimated to be growing at 17% to 20% a year, in Latin America from 12% to 15% a year, and in Africa at about 10% annually. As the economies of Central and Eastern European countries begin to recover and expand, and as their governments seek membership in international trade organizations, demand for air pollution and other environmental technologies is likely to grow steadily.

4. CONCLUSIONS

As we enter the 21st century, many old business attitudes toward the environment are changing rapidly (see Table 3). The traditional business attitude that industries could "*conquer nature*" and use the environment without limit is rapidly giving way to a strategy that "*conserves nature*." Not too long ago pollution was viewed as a symbol of a strong industrial base and as an unimportant diseconomy. Today pollution is used to measure error and waste. Pollution is increasingly seen as a symbol of poverty, ignorance, mismanagement and industrial inefficiency. Command-and-control regulations framed environmental management as a cost to industry that reduced profit margins. Today many businesses advertise environmental protection as a corporate *value* and operate *beyond compliance* often by *voluntary standards*.

Table 3
Changing business attitudes toward the environment

Traditional Attitude	Emerging Attitude
Command and Control Regulation------->	Voluntary Standards
Regulatory Compliance-------------------->	Beyond Compliance
Conquer Nature------------------------------>	Conserve Nature
Pollution: An unimportant diseconomy-->	Pollution: A measure of error and waste
Business uses the environment------------> without limit	Business depends on a healthy environment to sustain business
Environment is an "Issue"----------------->	Environment is a 'Value'
Environmental protection as a business cost--------------------------------->	Environmental protection as a business opportunity
Pollution as a symbol of a-----------------> strong industrial base	Pollution as a symbol of poverty, ignorance, industrial inefficiency, social instability, and mismanagement

Many businesses are embracing environmental management as a rapidly growing business opportunity. By integrating business objectives and environmental values, corporations can enhance their environmentally responsible reputation, create more valuable goods and services, and attain a stronger market position.

There is no longer much disagreement about whether or not the environment should be protected; the dialogue now centers on how to achieve environmentally sustainable development. As we embark on a new century, corporations are building on the creative potential of the market to find new solutions to environmental problems (see Table 4). While there will always be the need for an appropriate level of environmental regulation, proactive corporate environmental management and pollution prevention can make the greatest contributions to sustainable development. Progress will depend on cooperation among governments, corporations, environmental groups, and consumers to convince a larger number of businesses that it is far less costly to prevent pollution than to clean it up.

The spread of pollution prevention and proactive environmental management practices will also depend, however, on the adoption of new and more flexible government regulations. In order to advance the cause of pollution prevention, governments and the public must clearly recognize that there is no such condition as "zero risk." Risk-free management philosophies must be replaced with "acceptable risk" concepts; the principle of "control regardless of cost" must be replaced with "control in relation to benefit." As more corporations use pollution as a measure of waste and business inefficiency, regulations mandating the use of control technology must give way to those allowing companies to seek less polluting process modifications or replacements. Source by source control must be replaced by more efficient and less expensive industrial performance standards.

Table 4
Changing environmental management strategies

Past or Present Strategy	Present or Emerging Strategy
Environmental quality management------>	Pollution prevention
Local ambient environmental------------> concerns and effects	Environmental Effects:Indoor,local, regional, and global
Control regardless of cost----------------->	Control in proportion to benefit
Control technology-------------------------->	Process replacement or modification
Air, water, land assessed and-------------> managed as independent media	Air, water, land assessed and managed as an integrated system
Local and national control ----------------> strategies for pollution and effects	Global pollution control strategies and international agreements
Standards and emission limitation--------> permits	Tradable emission allowances
Source by source control------------------>	Industry performance standards
Zero risk------------------------------------->	Acceptable risk

Government has an important role to play in helping businesses find a economic advantage in operating beyond regulatory compliance. It can help firms discover that when manufacturing processes generate less waste, profit margins often increase and pollution control costs decline. Regulatory permitting should be changed so that it works more efficiently, encourages innovation, and creates more opportunities for stakeholder participation. Regulations must be reconstructed to give industry the incentives and flexibility to develop innovative technologies that meet or exceed environmental standards while cutting costs.

The U.S. Environmental Protection Agency's "Common Sense Initiative" is an example of the types of programs government can use to support proactive corporate environmental management.[32] Under the initiative, the EPA reviews existing environmental regulations to identify opportunities to get better environmental results at less cost and seeks improved rules through increased government, business, and stakeholder coordination. The program promotes pollution prevention as a standard business practice and the central ethic of environmental management. It establishes systems that make it easier to create, find, use, and disseminate pollution and environmental management information. It creates incentives to assist businesses that want to obey or exceed legal requirements and, at the same time, it applies harsh penalties to competitors who seek an advantage in breaking the law.

Governments must also adopt market-driven policies for reducing pollution. These include tradable permits and emission trading through which companies can buy and sell, for example, sulfur emission allowances on the open market, a mechanism that is currently being used in the United States to reduce acid rain. More flexibility is needed to allow companies to replace one form of pollution control with another when it achieves overall environmental goals.

In brief, the intensification of all three emerging trends—the adoption of proactive environmental management and pollution prevention practices, the pursuit of technological and process innovation in industry, and the transfer and diffusion of pollution control and prevention technologies through international trade and investment—can give business a central role in improving air quality and attaining environmentally sustainable economic development during the 21st century. But to accomplish this goal, governments and businesses must work together in new partnerships that seek to attain sustainable environmental objectives in ways that benefit both the public and the private sector and that allow businesses to satisfy the needs of consumers.

REFERENCES

1. Jane Vise Hall, "Air Quality in Developing Countries," Contemporary Economic Policy, Vol. 13, No. 2 (1995): 77-85.
2. Peter Bartelmus, Environment, Growth and Development -- The Concepts and Strategies of Sustainability, New York: Routledge, 1994.
3. Michael A. Berry and Dennis A. Rondinelli, "Proactive Corporate Environmental Management: A New Industrial Revolution?" Working Paper, Chapel Hill, NC: Kenan Institute of Private Enterprise, University of North Carolina at Chapel Hill, 1997.
4. McKinsey & Company, "The Corporate Response to the Environmental Challenge," Summary Report, Amsterdam, The Netherlands: McKinsey &

Company, 1991.

5. Richard Florida, "Lean and Mean: The Move to Environmentally Conscious Manufacturing," California Management Review,Vol. 39, No. 1 (1996): 80-105.
6. Margaret G. Kerr, "Looking at Environmental Management through the Lens of Quality," press release, Mississauga, Ontario, Canada: Nortel, 1995.
7. Eastman Kodak Corporation, Health, Safety and Environment 1995 Report, Rochester, NY: Kodak, 1996.
8. Amal Kumar Naj, "Some Companies cut Pollution by Altering Production Methods," The Wall Street Journal, (December 24, 1990): 1, 28.
9. Dave Lenckus, "Taking Initiative to Prevent Pollution Can Reduce Premiums," Business Insurance, (November 11, 1996): 30.
10. William H. Miller, "Making Pollution Prevention Pay," Industry Week (May 20, 1996): 136L.
11. Michael A. Berry, Dennis A. Rondinelli and Gyula Vastag, "Nitrokemia's Environmental Communications Strategy: Building Public Trust through Effective Communications," Corporate Environmental Strategy, Vol. 4, No. 1 (1996): 73-79.
12. See Dennis A. Rondinelli and Gyula Vastag, "International Environmental Management Standards and Corporate Policies: An Integrative Framework," California Management Review, Vol. 39, No. 1 (1996): 106-122.
13. Thomas W. Zosel, "Pollution Prevention in the Chemical Industry," in David S. Edgerly (ed.) Opportunities for Innovation: Pollution Prevention, (Washington, DC: U.S. Department of Commerce, National Institute of Standards and Technology, 1994):13-25.
14. Elizabeth Kirschner, "Full Cost Accounting for the Environment," Chemical Week, Vol. 154, No. 9 (1994): 25-26.
15. Kathi Futornick, "Government and Industry Programs for Reducing Pollution," in Thomas E. Higgins (ed.) Pollution Prevention Handbook, (Boca Raton, FL: Lewis Publishers, 1995): 15-42.
16. Robert P. Bringer and David M. Benforado, "Pollution Prevention and Total Quality Environmental Management: Impact on the Bottom Line and Competitive Position," in Rao V. Kolluru, Environmental Strategies Handbook, (New York: McGraw Hill, 1994): 165-197.
17. Robert B. Shelton and Jonathan Shopley, "Improved Products Through Design for Environment Tools," Prism, First Quarter (1996): 41-49.
18. Patricia S. Dillon and Michael S. Baram, "Forces Shaping the Development and

Use of Product Stewardship in the Private Sector," in Fischer and Schot, Environmental Strategies for Industry, pp. 329-341.

19. David J. Hunkeler and Ellen A. Huang, "LCA in Japan: A Survey of Current Practices and Legislative Trends and Comparison to the United States," Environmental Quality Management, Vol. 6, No. 1 (1996): 81-91.
20. Carl C. Henn and James A. Fava, "Life Cycle Analysis and Resource Management," in Rao V. Kolluru (ed.) Environmental Strategies Handbook, (New York: McGraw Hill, 1994): 541-641.
21. Unsigned article, "Diesel Makers Take Next Step toward Sealed-for-Life Zero Emissions Engine," Machine Design, Vol. 69, No. 1 (1997): 32-35.
22. SC Johnson & Son, Inc., 1990-1995 Sustainable Progress Report, Racine, Wisconsin: SC Johnson and Son, 1996.
23. Christopher J. Keyworth, "Integrating Pollution Prevention and Environmental Management Programs," Environmental Manager, Vol. 1 (August 1995): 18-24.
24. Richard Florida, "Lean and Green: The Move to Environmentally Conscious Manufacturing," California Management Review, Vol. 39, No. 1 (Fall 1996): 80-105.
25. U.S. Environmental Protection Agency, "Design for the Environment (DfE) Current Projects," Washington, DC: U.S. EPA, 1995.
26. Brian Kishbaugh and Orlando Martini, "Companies Get Proactive About Pollution Prevention," PEM: Plant Engineering & Maintenance, Vol. 18, No. 5 (1995): 16.
27. G. Sam Samdani, "Cleaner by Design," Chemical Engineering, Vol. 102, No. 7 (1995): 32-37.
28. Unsigned article, "Russian Device Zaps Smokestack Pollution," Machine Design, Vol. 68, No. 10 (1996): 28-29.
29. Unsigned article, "New Low-Sulfur Diesel Debuts in Brazil," Oil & Gas Journal, (December 16, 1996): 22-23.
30. Environmental Business International, "The Global Environmental Market and United States Environmental Industry Competitiveness," Washington: U.S. Environmental Protection Agency, 1996.
31. Robert W. McIlvaine, "The Air Pollution and Municipal Waste Forecast in 1997," EM (January 1997): 34-38.
32. U.S. Environmental Protection agency, "The Common Sense Initiative: A New Generation of Environmental Protection," Policy Paper, Washington, DC: U.S. EPA, 1994.

SESSION F

ENVIRONMENT & ECONOMY

Sustainable Economic Development

L.B. Lave

Graduate School of Industrial Administration

Pittsburgh, PA 15213, Carnegie Mellon University

Three issues are central to an examination of sustainable economic development and public policies to foster this development:

1. Who speaks for the environment and future generations?
2. How rosy is the future?
3. Are environmental resources unique in production?

The first issue is at the heart of the sustainability debate. What will future generations desire? What do we mean by "environmental quality?"

The people currently alive can articulate what they regard as good or bad futures, but we have no idea what future generations will desire. For example, suppose that your great, great grandfather, at great sacrifice, preserved the best spot for a privy for your generation. The point is that technologies change and what was a wonderful gift at one time becomes useless at another. This doesn't mean that the current generation has no obligation to preserve opportunities for future generations, but it does mean that we must examine our sacrifices in view of likely changes in technology, population size, and tastes.

Environmental quality is a complicated vector, not a simple scalar. Acidified lakes in northern New York are clear and inviting, like swimming pools. In contrast, "living" lakes have muddy bottoms and are filled with plants and animals of varied size, including harmful bacteria and leaches. How hard should we work to preserve harmful bacteria and leeches? How much effort should we put into preserving wetlands that are breeding grounds for mosquitoes and disease?

We all desire environmental quality, but few of us agree on what constitutes quality. One extreme would be to minimize the environmental footprint of human actions. The other extreme would be to shape the environment to offer more services, e.g., drain swamps to prevent disease, clear brush to make hiking trails, and stock trout in lakes to provide better fishing. Still more extreme are those who regard ultimate environmental quality as a place completely undisturbed by humans - no human access allowed. As extreme in the other direction are those who regard the wilderness as attractive principally because it gives space for motorized craft, such as snow mobiles, four wheel drive trucks, and speed boats. The point is that there is no agreement about what constitutes environmental quality.

The attractiveness of our future, issue 2, is an unstated assumption in most discussions. Pessimists regard the future as spiraling ever downward, with environmental quality and the well-being of future generations continually declining [U. S. Department of Commerce, 1980, Ehrlich & Ehrlich, 1981]. If so, major action is needed now to prevent this disaster. Optimists see an improving future that gives humans more opportunities and choices [Simon & Kahn, 1985, Simon, 1995]. Will those living in 100 years have a small fraction of the income we currently enjoy or will they be richer than we are?

If the future generations can be expected to be worse off than we are, we need to consider making large sacrifices in order to increase their welfare. We need to preserve more natural resources and save more to increase the capital stock. In this view, the current generation is unwilling to recognize the perilous state of future generations and unwilling to make the sacrifices necessary to preserve the lifestyles of future generations.

If future generations can be expected to be richer than we are, additional sacrifices today would make future generations still better off, but there is little reason to make these sacrifices. Why should I lower my consumption so that people alive in 100 years will have three times my income, rather than twice my income?

Differences in how rosy the future will be are fundamental to discussions about the future. Unfortunately, these expectations about the future are rarely stated. Nonetheless, they are perhaps the primary reason for the different proscriptions concerning the future.

The substitutability of resources, issue 3, is important: Every barrel of oil and every ton of rich iron ore used by the current generation means less resources for future generations. If environmental resources are unique, the only sustainable economy is one that uses no nonrenewable resources. Thus, the path to sustainability requires that we phase out our consumption of nonrenewable natural resources as quickly as possible.

If environmental resources are not unique in production, the classical economic view is that society should ensure that the value of resources passed on to future generations should be as great as our generation received. I will elaborate below, but in this view, the economy is sustainable as long as the value of resources passed on to future generations is at least as great as that given to the generation.

Managing A Modern Economy for Sustainability

A modern economy, such as that of the United States or the Netherlands, is large and complicated, with many interactions. The economy is too large and complicated for a central authority to be able to control it in detail, as the Soviet Central planners attempted to do. In a market economy, no one is responsible for making sure that there is bread and milk in the supermarket for consumers or indeed for insuring that consumers don't go hungry for weeks because of production or delivery problems. No one is responsible for ensuring that adequate supplies of coal, petroleum, and other fossil fuels are available for production and consumption.

Modern economic theory could be thought of as a way of decomposing the complicated structure of an economy so that workers, firms, and consumers who are each trying to do the best for themselves wind up doing what is best for the economy. Prices are the signaling devices by which consumers indicate that they want more of some good or service; they are a way of signaling producers of products and raw materials to increase production. If there is insufficient petroleum at the current price, price rises signaling consumers to purchase less and producers to increase the quantity they are offering. If petroleum resources are being depleted so that petroleum will be much more valuable in the future, the current price will rise to reflect the future high prices.

Assuming the economy is competitive and there are no externalities, economic theory shows that a competitive market produces an efficient outcome. When markets exist for all future goods and services, the framework can be extended to show that a market will account for resource production and use in the future, bringing about efficient outcomes over time.

Three major challenges have arisen to the desirability of this competitive outcome. The first is that markets must be competitive. Both the USA and the European Union have antitrust policies to preserve and enhance competition.

The second challenge comes from externalities, both positive and negative. Environmental discharges are an example of a negative externality. Without ways of charging firms for the environmental discharges, the market will not produce an efficient outcome. Both the European Union and the USA have erected an elaborate array of environmental laws and regulations to internalize this externality.

The third challenge arises from the fact that markets do not exist for goods and services in future markets. I cannot purchase one million barrels of crude oil for delivery in Pittsburgh on June 1, 2050. Of greater importance, I cannot purchase a contract that, if petroleum costs more than $100 per barrel in 2050, I will have a technology available to me that gives me the services I desire without using petroleum.

I leave the first of these difficulties to the U.S. Department of Justice Antitrust Division, the U.S. Federal Trade Commission and their European Union counterparts. Challenges two and three are my focus here.

The Role of Life Cycle Analysis in Environmental Regulation

The landmark legislation for environmental protection is the 1970 Clean Air Act which defined a central role for the U.S. federal government. The newly created Environmental Protection Agency was given the mission of getting emissions standards for a range of pollutants. Subsequent legislation expanded EPA's role to setting discharge standards for a wide range of pollutants in water and solid waste.

Both for a regulatory agency and for the firms being regulated, selecting an approach to meeting the standards control technology is difficult. Decisions, especially those concerning the environment, have wide ramifications and unexpected consequences. For example, installing flue gas desulfurization equipment on a coal burning power plant results in more energy per kilowatt hour. In a world concerned with resource depletion and greenhouse gas emissions, the lower efficiency is an unfortunate consequence.

In selecting a technology, design, or material, we need to know the life cycle implications of the alternatives - the implications for resource and energy use and for environmental discharges from the extraction (or growing) of the material through its production, use, and recycle/disposal.

Producers strive to lower their production costs in choosing among materials, designs, and capital equipment. Market prices reflect only private costs and demand. Social costs, such as environmental discharges, enter principally through regulation. For example, the price of steel reflects the demand for steel and the cost of making it, including the ore, coal and coke, labor, and capital. The costs will also reflect environmental regulations. However, if environmental regulations are too lax or too stringent, are not enforced uniformly, or do not cover important areas, prices will not reflect social costs.

Current market prices reflect the scarcity of raw materials over the next decade or so. High interest rates mean that material scarcity a century or more in the future will have almost no effect on current prices. Since greenhouse gas emissions are an externality, they are not reflected in current prices.

The 1970 Clean Air Act deals with externalities by a centralized, command and control approach. EPA is assigned the task of setting detailed emissions standards for producers in each industry based on available control technology and environmental problems. Defining "best available control technology," "new source performance standards" and the other emissions regulations for each industry have proven to be a difficult, controversial task.

The 1990 Clean Air Act tried a different approach to regulating sulfur dioxide emissions. Congress decided to experiment with a market based approach. The decided on the total amount of emissions that would be allowed, allocated these emissions among the electricity generating plants, and then allowed each plant to sell some of its "allowances" if it reduced emissions more than the stipulated amount or to buy allowances if reducing emissions was extremely costly. This market based approach resulted in saving about 2/3 of the estimate abatement cost and achieve the reductions more quickly than mandated by law.

The success of sulfur dioxide allowances increased support for a much greater use of market mechanisms in achieving environmental goals. However, at present, prices contain only a distant aspect of the social costs of depleting stratospheric ozone, contributing to air pollution and water pollution or otherwise harming human health and the environment. Command and control regulation does affect the prices of produces and materials through the costs of control. However, the current system is sufficiently imperfect that EPA regulations are not close to optimal levels. As a result, the resulting control costs can either overstate (as for Superfund) or understate (as for materials that contribute to indoor air pollution) the social costs of improving environmental quality.

Since market prices don't reflect social costs, the market cannot perform its decomposition function properly. This means that a separate analysis is required of the life cycle implications of choosing among alternative materials, designs, processes, and recycling/disposal options.

EPA and the Society of Environmental Toxicologists and Chemists (SETAC) developed the current method for life cycle analysis. It is a straight forward analysis of production processes beginning with the production of the product of interest. It then goes backward to the most important processes that provide inputs to the product production. Perhaps the most important step in the analysis is to draw a boundary around the problems to decide what will be considered. This approach has been criticized as inadequate and arbitrary [Portney, 1993].

The Carnegie Mellon Green Design Initiative has developed a different approach [Lave et al., 1995, Hendrickson et al., 1997]. We take the input-output table for the United States and append to it environmental and raw material vectors to describe the energy and other inputs to each process and the environmental discharges that result. This approach allows us to avoid setting arbitrary boundaries on the problem. This is a general equilibrium economic model and so all of the inputs and all of the environmental discharges, both direct and through supplies, are calculated.

One application of this model is to the requirement of the California Air Resources Board that a specified proportion of new cars sold, beginning in 1998, had to be zero emissions cars. CARB is concerned with reducing emissions of the gases that form ozone over much of California. A zero emissions car would lower air pollution emissions because it has no tailpipe.

The cars designed to meet the CARB requirement are battery powered. In particular, the cars currently marketed contain 500 kg or more of lead-acid batteries. The cars have no tailpipe, but require electricity to recharge the batteries. While electricity generation produces air pollution, a modern electricity generation plant with its stringent environmental controls will produce less pollution than the comparable number of gasoline fueled cars that it replaces.

However, each car has 500 kg or more of lead acid batteries. A large amount of lead must be mined, smelted, made into batteries, and then must be recycled into new batteries. We calculate that the most advanced electric car, General Motors' EV1, would result in much more lead being discharged into the environment, per vehicle mile, than a comparable internal combustion engine car burning gasoline with tetraethyl lead [Lave et al., 1995]. In a second analysis, we calculated that half a million EV1s in Southern California in 2007 would lower peak ozone from 200 ppb to 199 ppb, a reduction too small to be measured reliably [Lave et al., 1996]. At the same time, half a million EV1s would increase total lead use in the USA by 20%, presumably increasing total lead discharges by 20%. Although we did not make the direct calculations about health implications, it seems evident that a 1 ppb reduction in peak ozone in one region is much less of a health benefit than a 20% increase in lead discharges throughout the USA.

This calculation illustrates the importance of a life cycle framework in considering decisions concerning alternative designs, materials, processes, and recycle/disposal decisions.

Sustainable Development: The Role of Technology

The Bruntland Commission defined sustainable development to mean satisfying current needs without diminishing the ability of future generations to satisfy their needs. This statement is wonderfully crafted to cover a wide range of meanings, depending on the goals of the listener. This is perhaps the principal reason why there has been such wide spread agreement on satisfying the goal: Without further definition, it is a principle largely without content.

Some environmentalists take a static view of technology and economics in interpreting the phrase. If one assumes that one cannot depend on technology improving and that people in the future will have the same tastes as those alive today, the USA must drastically cut consumption by ending all use of nonrenewable resources. The implications of this view are that all countries, both rich and poor must restructure their economies and curtail consumption. Furthermore, current population is too large for sustainability.

Some industrialists take a dynamic view of technology and resources.
With rapidly advancing technology, resources could be stretched to provide much more consumer goods with less damage to the environment. In this world, we can be less concerned about using up raw materials since technology will find ways of using less of the scarce resources and will find substitute resources. Thus, curtailing US economic activity is needlessly painful. A better way to achieve sustainability is to develop new technology to accomplish the goal.

Important examples support each of these views. On the one hand, soil loss in the USA from cultivating unsuitable areas combines with the damage from strip mining and the superfund sites resulting from thoughtless manufacturing to produce an unattractive future. On the other hand, Current cars have twice the fuel economy of 1974 models and computing is millions of times more efficient than it was 40 years ago.

One's individual interpretation of "sustainability" leads to a wide range of policies. Technological-economic optimists can insist that the government keep its hands off the economy, since a free market will produce technology and resources for sustainability. Optimists with a more interventionist bent might insist on increased expenditures for R&D to increase the efficiency of resource utilization. Interventionist pessimists might insist on government programs to promote zero population growth and a reduction of consumption among the rich. Less interventionist pessimists might advocate educating consumers to want fewer products and services and to share wealth with poor countries.

We have no way of knowing whether future technology will require only a tiny proportion of today's resources. Whenever the price of some raw material has increased sharply, alternative supplies have been supplied within a short period of time. For example, a vast amount of oil was discovered after the 1973 OPEC oil embargo and subsequent price increase. An optimist might have no hesitation in continuing to increase the use of petroleum. A pessimist might want to see the substitute resources and new technology first.

The point is that viewing sustainability in static terms is too narrow and limited. Great progress has been made in getting more of the goods and services that people desire out of raw materials whose price has increased. Assuming that no further technological progress will occur is an extreme assumption. However, assuming that technology will become available when desired is an equally extreme assumption. Investors in the USA and other nations demand a high return on their investments. In practice, this means that they and the companies they own have extremely short time horizons: If we won't run out of petroleum in the next few years, we don't worry about the problem today. This viewpoint neglects the time required to develop viable new technologies or put them into place.

Perhaps the one point of agreement is that technology is the key to the future. Society could suffer a large loss if the technology is not available when needed.

A Neo-Classical Economic Growth Model

Elena Shewliakowa, Eduardo Vergara and I [Shewliakowa et al., 1997] have been exploring sustainable development in a neoclassical growth model. This highly stylized model is of a single nation that has a stock of cheap resources, a level of technology for utilizing available renewable sources, and a level of productivity in accomplishing R&D. This is a simple world where a single commodity is produced with labor and energy. People consume the commodity; however, they can decide how much of the labor force to devote to R&D. R&D workers don't contribute to current production and so each worker switched from production to R&D lower the total consumption of all workers. Thus, switching workers from production to R&D sacrifices current consumption for better technology that will increase production (and consumption) in the future.

Energy is needed in both the production and R&D activities. The inexpensive oil stock increases production - while it lasts. When the oil is gone, the economy will have only solar energy. The current technology for using solar energy is not efficient. However, R&D can make the technology better, eventually increasing the efficiency of solar energy to the point where productivity is as high as with oil.

In our model, people must decide how much of their consumption to devote to R&D. If they devote too little, the technology for using solar energy will still be primitive when the oil runs out and their consumption level would decline precipitously. However, that event is in the future, long after the current generation has died. Thus, the current generation could decide that this is not their problem and do no R&D. At the other extreme, they might decide to cut their consumption to the subsistence level and switch almost all workers to R&D.

The model produces interesting results. There are two sustainable futures. One path has low consumption based on inefficient technology for using solar energy. The other path has high consumption based on efficient use of solar technology.

A nation with large stocks of cheap oil and high R&D productivity (USA) will be able to develop excellent technology for using solar energy with hardly any sacrifice in current consumption. A nation with no oil reserves and little ability to do R&D (Namibia) has low income initially because it has little oil. Since it cannot afford to allocation many workers to R&D, and since workers are not productive at R&D, the sustainable path is at this low consumption level.

More interesting are nations with a moderate level of oil and moderate R&D productivity (Great Britain). Such a nation could decide to have high consumption today, leaving future generations with the choice of curtailing their consumption stringently, or again deciding to let the future generations fend for themselves. If each generation decides to allocate few workers to R&D, their consumption will be high until the oil runs out. When the oil is gone, the sustainable path will be at a low consumption level because it is based on primitive solar technology.

Nations willing to sacrifice a great deal so that future generations will be better off can reach the high sustainability path with little oil. Such a nation could lower current consumption, devoting all possible effort to R&D (post war Japan).

For market economies, our willingness to sacrifice for the future is indicated by the amount of income that is saved and invested for the future. The greater is savings, the lower will be the interest (discount) rate. Thus, societies, that care little about the future, will have high discount rates and low savings rates. The USA savings rate is very low, both in comparison with our historical savings rates and with other nations. Americans don't appear to be willing to sacrifice for the future. However, Technology is advancing rapidly in the USA, leaving me optimistic about the future.

Modeling the Role of Uncertainty

Policy making on energy conservation and greenhouse gas emissions tends to be paralyzed by uncertainty. What are the stocks of petroleum and natural gas that can be produced at low costs? What is the quantitative effect of increasing concentrations of greenhouse gases and how bad will the resulting climate change be for people and the environment?

For a decade, Europeans have advocated the "precautionary principle" in dealing with this uncertainty: When the world is at risk, don't take chances. Americans have been much more optimistic about the future

Making decisions under uncertainty has received major attention in the fields of decision theory and game theory. We might think about greenhouse warming as a game against nature [Lave & Dowlatabadi, 1993]. Humans select a policy that determines the concentration of greenhouse gases. Then nature decides one the amount of climate change and ensuing disruption. In this framework, Europeans are assuming that nature will be malevolent, choosing a bad outcome. If so, humans should employ a "minimax" strategy of minimizing the maximum loss - choosing the strategy that gives the smallest loss, even when nature is malevolent.

The opposite assumption is that nature is benevolent, showering gifts on humans. If nature were conceived to be accommodating, we would choose a strategy that assumes that nature will move to make the outcome as good as possible fore humans.

Finally, we could decide that nature is indifferent to humans, employing a random strategy: If so, nature chooses outcomes using a probability distribution. In this model, we would pick the strategy that minimized expected cost/maximized expected benefit.

The precautionary principle is equivalent to a minimax strategy. Americans are not so pessimistic as Europeans and tend to favor a strategy that minimizes expected cost. The expected cost criterion leads to quite a different strategy than the minimax criterion. Thus, the difference between Europeans and Americans is not so much in their understanding of the greenhouse effect and its implications. Rather the difference is due more to their method of making decisions under uncertainty.

Summary

I have discussed several aspects of sustainable development, beginning with the three fundamental issues: 1. Who speaks for the environment and future generations?
2. How rosy is the future?
3. Are environmental resources unique in production?
Answering these questions is inherently difficult.

The second topic was an examination of economic theory as giving a decomposition of the economy such that a free market can produce outcomes as efficient as the best central planner can. However, the decomposition runs afoul of the assumption that externalities cannot be present. Internalizing externalities, such as environmental pollution, has been done by command and control regulation. Market mechanisms, such as effluent fees and cap and trade schemes provide more efficient mechanisms for controlling the externalities. The experiment with trading allowances for sulfur dioxide emissions showed that these mechanisms could achieve the goal at must less cost and more quickly.

In order to have efficient regulation or to set effluent fees, decision makers need to know the full life cycle implications of a decision. For example, I commented that California's requirement for zero emissions cars likely harms environmental quality by substituting a large amount of lead discharges to the environment for a tiny reduction in ozone concentrations. I described a better tool for calculating the life cycle of a product, process, or material.

Sustainable consumption paths depend on the level of renewable resources and the technology for using them. I described a neo-classical growth model the examined the interrelationships among the endowment of fossil fuels, the ability to accomplish R&D, and the willingness to sacrifice current consumption in order to improve the well-being of future generations.

Finally, I discussed the role of uncertainty in making decisions about greenhouse policy. The differences between American and European policies on greenhouse policy are more closely related to their view about what to do under uncertainty than their perceptions of the seriousness of the problem.

Although long term environmental issues are difficult to model and put into policy analysis, there are some helpful approaches and tools that have been developed.

REFERENCES

P. Ehrlich, and A. Ehrlich, Extinction, New York: Random House, (1981).

C. T. Hendrickson,, A. Horvath, S. Joshi, and L. B. Lave, "Introduction to the Use of Economic Input-Output Models for Environmental Life Cycle Assessment," Technical Report, Carnegie Mellon University, July (1997).

L. B. Lave,., E. Cobas-Flores, C. T. Hendrickson, and F. C. McMichael, "Using Input-Output Analysis to Estimate Economy-Wide Discharges," Environmental Science & Technology, 29: 420-6, (1995).

L. B Lave,. and H. Dowlatabadi. "Climate Change: The Effects of Personal Beliefs and Scientific Uncertainty," Environmental Science & Technology, 27: 1962-71, (1993).

L. B., Lave, Christ T. Hendrickson, & Francis C. McMichael, "Environmental Implications of Electric Cars," Science, 268:993-5, (1995).

L. B. Lave, ,A. G. Russell, C. T. Hendrickson, and Francis C. McMichael, "Battery-Oiwered Vehicles: Ozone Reduction Versus Lead Discharges," Environmental Science and Technology: 30:402-7, (1996) .

P. R., Portney, "A Critique of Life-Cycle Analysis," Issues in Science & Technology, 9: 69-75, (1993).

E. Shewliakowa, E. Vergara, and L. B. Lave, "How Much Resources for Future Generations? A Model of Sustainable Growth with Costly Innovation, Exhaustible & Inexhaustible Resources, and Environmental Amenities," working paper, Carnegie Mellon University, April 10, (1997).

J. L. Simon, (ed), The State of Humanity, Cambridge, MA: Blackwell Publishers, Inc, (1995).

J. Simon, and H. Kahn (eds), The Resourceful Earth, New York: Basil Blackwell, (1985).

U.S. Department of Commerce, Global 2000: Report ot the President, Washington, DC: U.S. Government Printing Office, (1980).

Dual goal: economic growth along with environmental improvement

H. Verbruggen

Institute for Environmental Studies, Vrije Universiteit Amsterdam
De Boelelaan 1115, 1081 HV Amsterdam, The Netherlands

The concept of sustainable development conceals an inherent tension. It pursues a dual goal: economic growth along with environmental improvement. Is that feasible? On the one hand, some studies portray a black future if population and economic growth continue on the present footing and pace (cf. Wetzel and Wetzel, 1995). Continuous economic growth characterized by a growing volume of throughput through the economy, will contribute ever less to welfare. Many maintain that there is something like a carrying capacity of the Earth that cannot be surpassed on pain of welfare losses due to over-exploitation of the Earth's environmental capital. Beyond that point, environmental degradation is irreversible. And as a result of the phenomenon of entropy, technological solutions are ultimately also not feasible. On the other hand, there are studies indicating at a brighter future: environmental problems will more or less be automatically resolved in the process of economic growth. Studies into the so-called Environmental Kuznets Curve claim that it is true that environmental pressure increases with income growth, but decreases after a particular level of income has been reached. However, there are important pitfalls that undermine this hypothesis, both from a methodological and empirical point of view (Heintz and Verbruggen, 1997). Moreover, recent studies in this field find again a relinking between income growth and pollution (De Bruyn and Opschoor, 1994).

A new perspective in the environment-economic growth debate is opened up by the

Factor 4 discussion. Technological options and cleaner forms to satisfy human needs are available so that welfare can double and environmental pressure can be halved. Von Weiszäcker et al. identified more than 100 of these so-called win-win options (Von Weiszäcker, et al., 1995).

In my opinion, the environment-economic growth debate cannot be settled on the basis of scientific knowledge. And the introduction of the concept of sustainable development in the 1980's has also not been very helpful. Sustainable development indicates a favorable direction by introducing the condition that the needs of future generations have to be taken into account in present-day economic decision making; neither more, no less. It does not provide a blueprint, nor does it specify the means by which a sustainable economy is to be achieved. This is not to say, however, that the concept of sustainable development gives no policy indications at all, as many have alleged. Sustainable development provides a framework within which the use of, or investments in different forms of capital can be compared. In the economic process, different forms of capital are being combined to produce goods and services to meet human needs. In this connection, the World Bank distinguishes four forms of capital, namely environmental capital, physical capital, human capital and so-called social capital (Serageldin and Steer, 1994). The latter refers to the social cohesion and the problem-solving abilities of government and society. Sustainable development is all about the question how much of these different stocks of capital are invested to produce welfare, and in which combinations. To put it in economic terms, to what extent can these capital stocks be substituted for one another. This is the central question in operationalizing sustainable development. In doing so, the needs of future generations have to be taken into account. This means that the trade-offs have to fulfil only one condition: a per capita welfare which does not decrease over time. This condition implies a supplementary one, namely that the total stock of capital goods with which our welfare is produced must, at the very least, be maintained. And if we take account of the increasing needs of a growing world population, then there has to be an enlargement of this total stock, or, at any rate, in total welfare that the capital stock is able to produce. This will require as yet unimagined forms of technological development and also drastic changes in current production methods and consumption patterns. In my opinion, sustainable development can only be brought closer by means of an unprecedented economic and technological dynamism. Seen in this light, sustainable development, or to put it differently, the reconciliation of economic growth and environmental improvement, actually becomes a sort of 'meta-growth'.

By indicating that a sort of meta-growth is needed, we still have no answers as to the substitution possibilities of environmental capital for other forms of capital. Another important question concerns the spatial level at which sustainability has to be achieved. Do we have, for instance to close substance cycles at each and every spatial level to become sustainable, or is it possible to leave a particular cycle open at one spatial level to be closed at another spatial level. And how can these spatial sustainability questions be coordinated. A final question is, of course, how -by which policies and measures- a sustainable development can be pursued.

I shall address these three questions.

With respect to the substitutability of the different forms of capital, it has to be realized that the environmental capital stock must clearly be distinguished from the other three forms. Environmental capital makes, via the production process, a twofold contribution to welfare: it furnishes primary inputs and space, and it also provides a sink for the waste and pollutants our production processes generate. Remarkably, environmental capital also possesses an autonomous capacity for self-regeneration, though this capacity is rather vulnerable. Equally remarkably, we are far from having exhausted every possible application of the world's environmental assets; consider, for example, the richness of its genetic diversity. Finally, environmental capital contributes directly to our welfare, that is, without transformation in a production process, in scientific, cultural, recreational and aesthetic terms. These aspects, taken together, mean that there is a limit to the replaceability of environmental capital. However, these characteristics of environmental capital, or put differently, the nature and extent of ecological-economic relations, or again in other words: what is the carrying capacity of the various components of environmental capital, all this is surrounded by uncertainty. This means that the substitutability of environmental assets by other forms of capital is open to dispute. And as this is the case, it has to be admitted that several forms of sustainable development can be envisaged, according to the assessment of these uncertainties. These assessments may well include assumption about future technological possibilities.

In addition, various subjective preferences also play a role; for example, the conservation of an eco-system may be considered to be a higher priority than the strengthening of social capital by creating jobs for the long-term unemployed. The choice between the numerous substitution options - and that is what I would like to stress - is made on the grounds of individual and societal risk analyses and preferences. The operationalization of sustainable development represent a series of value judgements and normative choices which eventually lead to political decisions.

There is, then, no shortage of alternatives, notwithstanding the requirement I have made to achieve a sustainable development, namely, that per capita welfare provided by the sum of different capital stocks may not decease over time, and my warning to be prudent with the use of environmental capital. And this is why so many definitions of sustainable development have been formulated; definitions that vary so much that they can be divided into different categories. The best-known division is between strong and weak sustainability. The adherents of the strong sustainability concept are very reluctant to substitute environmental capital for other capital stocks, whereas the followers of weak sustainability combine a high degree of flexibility with technological optimism. So, as long as these different interpretations exist, there will be a tension between economic growth, that is fuelled by weak sustainability, and the environment which is better protected in the strong sustainability concept.

In discussing sustainability, no spatial distinction has been made yet. But that is rather abstract, and only relevant for genuine global environmental problems. At lower spatial levels, the degrees of freedom further increase, and hence, the tension between economic growth and the environment becomes more manifest. This has to do with the fact that at lower spatial levels, the external effects of environmental degradation can be shifted to other and/or higher spatial levels. It also opens new opportunities: unsustainable practices in one country can be compensated by sustainable practices in another country. It is even thinkable that economic benefits can be reaped from these, what I would like to call, sustainability interactions between different spatial levels. Take, for instance, the case of recycling or re-use of materials. It can be sustainable to have our waste or scrap be recycled in India. We leave our materials cycle open to be closed in another country. In other words, if we take into account cross-country differences in environmental endowments and natural conditions, it is possible to realize environmental gains through trade and specialization, deliberately leaving open specific cycles at specific spatial levels, but closing them at higher spatial levels.

Not only the degrees of freedom increases if different spatial levels at which environmental capital can be used, are taken into account. It also poses an entirely new question: namely, how do we have to measure, value or compare sustainability at different levels? Can importing countries be held responsible for the depletion and externalities in exporting countries. Why should we grow wheat in The Netherlands when it can be grown substantially less environmentally harmful in Spain? Is it sustainable to reallocate polluting industries outside the country, without changes in the pattern of consumption?

Just to give you a concrete example. At the Institute for Environmental Studies, Vrije Universiteit Amsterdam, a research project has been carried out on different scenarios for sustainable economic structures for The Netherlands for the year 2030 (Verbruggen, 1996). The study was commissioned by the Ministry of Housing, Spatial Planning and the Environment. The objective of the research was to construct four economic structures for the Dutch economy that could be labelled sustainable depending on a chosen perspective. These perspectives ranged from a world-wide pursuit of strong sustainability with limited substitution possibilities between various forms of capital and ambitious environmental objectives, to a world-wide market-led weak sustainability.

The economic structures in the different scenario's were shifted such that value added was optimized subject to different environmental objectives and other conditions. The Ministry also asked us to construct a so-called ecological trade balance. This balance is defined as the difference between the ecological impact related to exports, and the ecological impact related to foreign production for domestic consumption, that is import. The results are shown in the table.

The ecologic trade balance in 1991 and the four scenarios for 2030, The Netherlands

	1991	Strong (Together) *Sustainability*	Strong (Alone) *Sustainability*	Negotiated *Sustainability*	Weak *Sustainability*
Climate Change (10^6 carbon equivalents)	21.572	3.833	-10.910	-8.653	3.234
Acidification (10^6 acid equivalents)	3.183	808	-305	92	1.348
Eutrophication (10^6 manure equivalents)	34	-4	-25	33	9
Waste (10^9 kg)	2.469	-830	-1.150	-624	-409

Source: H. Verbruggen (project leader, 1996. Duurzame Economische Ontwikkelings-scenario's voor Nederland in 2030 (Sustainable Economic Structures for The Netherlands in 2030), Publikatiereeks Milieustrategie 1996/1, VROM, The Hague.

In 1991, the Dutch economy has a surplus in all four identified environmental themes, indicating that, on balance, environmental impacts are accepted on behalf of foreign consumers. This ecological trade balance "detoriates" if economic structures in The Netherlands are made more sustainable: the surplus reduces and even becomes negative in a number of cases. But what does that tell us? In case of international environmental problems like climate change and acidification, it can be argued that economic activities are re-allocated from a high energy-efficient country to countries with low energy-efficiency. This is bad for the global environment. For national environmental problems like eutrophication and waste, the inference is ambiguous. Environmental quality in The Netherlands will improve, whereas the environmental impact of these reallocated activities in other countries and the appreciation thereof depends on local environmental circumstances and preferences.

Finally, I would like to address the issue on policies and measured to achieve sustainability, or to reconcile economic growth and environmental care. As it presently stands, and that is relevant for all developed economies, environmental policy is largely being implemented by regulating the sources of pollution, through rules and regulations, permits and so on and so forth. In this approach, the polluters have to bear the cost of complying to the regulations, but for the rest, the use of the environment is still free. That is to say, that the environment is not managed as a stock. Hence, the use of the environment has no price. The tension between economic growth and the environment is exactly due to this fact: regulated polluters who get the environment for free. They feel themselves restricted; they may be restricted in their growth prospects, they have to bear extra cost, and they might lose their competitive edge.

This conflict can only be overcome, if the environment becomes a genuine economic good, that means scarce and adequately priced. Irrespective the interpretation of sustainable development at whatever spatial level, environmental capital has become a very scarce resource indeed. A scarce resource is an economic resource, and should be treated accordingly. This implies that property and use rights should be defined and allocated for all the components of environmental capital. I am all too aware of the political resistance to, and the practical difficulties and problems involved in an actual transition to a system which allocates environmental property and use rights, and creates markets to trade these rights. A great deal of creativity will be needed in the development of such markets for environmental goods. The already existing examples of tradeable emission permits and tradeable fish quota should be extended on a much wider scale (for other emission, CO_2 in particular and other natural resources). For The Netherlands, a system of tradeable manure rights, tradeable parking lots and road pricing are interesting possibilities.

To conclude, the use of the environment should become a hundred percent economic good, that can be traded on the market. Only then will environmental capital be put to an efficient use and only then can economic growth and the environment be reconciled.

REFERENCES

1. Heintz, R.J. and H. Verbruggen (1997). Meer groei en toch een schoner milieu? De groene Kuznets-curve (Do economic growth and a clean environment go hand in hand? The environmental Kuznets curve), Milieu, Tijdschrift voor Milieukunde, 12, 1997/1, pp. 2-9.
2. Wetzel, K.R. and J.F. Wetzel (1995). Sizing the Earth: Recognition of Economic Carrying Capacity, Ecological Economics, 12, pp. 13-21.
3. Bruyn, S.M. de, and J.B. Opschoor (1994). Is the economy ecolizing? Discussion paper TI 94-65, Tinbergen Institute, Amsterdam.
4. Weiszäcker, E.U. von, A.B. Lovins and L.H. Lovins (1995). Faktor vier, doppelter Wohlstand-halbierter Naturverbrauch, Droemer Knaur, München.
5. Seralgeldin, I. and A. Steer (eds.) (1994). Making Development Sustainable: From Concepts to Action, Environmentally Sustainable Development Occasional Paper Series Nr. 2, World Bank, Washington D.C.
6. Verbruggen, H. (project leader) (1996). Duurzame Economische Ontwikkelingsscenario's voor Nederland in 2030 (Sustainable Economic Structures for The Netherlands in 2030), Publikatiereeks Milieustrategie 1996/1, VROM, The Hague.

Environmental policy and fiscal instruments in The Netherlands

Pieter Hamelink

Ministry of Housing, Spatial Planning and the Environment in The Netherlands

1. INTRODUCTION

Environmental policy in The Netherlands started in the 1960's. Water and air pollution had become urgent problems, which needed to be tackled. Later also other environmental problems emerged. Problems which were related to the economic development and the growing population of the country.

In the past thirty years new policies were developed and a number of environmental laws were installed to deal with the various environmental problems. Next to the development of these classic command-and-control instruments, a lot of attention was - and still is to-day - given to the role of other instruments like economic instruments, social instruments and communication instruments. More in particular, the use of a combination of these instruments, tailormade for specific environmental problems and situation, has played a major role in the implementation of Dutch environmental policy.

Among the various economic instruments, fiscal instruments tend to become more and more important. And not only the straightforward 'environmental' taxes, but especially the development of 'greener' general taxes, either as a burden or as an incentive. One has to keep in mind however that, being a very open economy and member of the European Community, there are limitations as to what can be done in the field of greener taxes.

2. FINANCING ENVIRONMENTAL POLICY

The question of how to finance the costs of the environmental policy has been one of the main topics from the very start of the environmental policy making. The basic principle of the financing system of the costs of the environmental policy in The Netherlands is the so called Polluter Pays Principle, which was adopted in the early seventies by the European Union and by the Organization for Economic Cooperation and Development. According to this principle everyone who causes damage to the environment should pay for the costs related to that damage. In doing so, environmental costs would be internalized in the prices of products and services and - if other countries should follow the same policy - the (international) competition would therefore not be distorted as a result of necessary environmental measures.

This should be, of course, the ideal situation. But, although the concept of the principle is in essence very clear, it is in practice not always possible to implement the principle in full. The relation between pollution or effects of pollution and certain activities of polluters is not always clear to establish. Besides that, not all environmental damage can easily be expressed in financial terms. The amount of the costs of environmental damage and who should pay for these costs can therefore not always be established.

A major element of the financing system in the Dutch environmental policy is closely related to the concept of direct regulation. Where a clear relation exists between environmental damage and its causes, the responsible polluters themselves have to take measures to avoid or diminish their pollution. Those measures can be imposed on the basis of direct regulation, such as licensing and permits, or can be agreed on in so called voluntary agreements or covenants. In those cases the polluters concerned must also bear the costs of these measures. One could say that the Polluter Pays Principle is here applied in a most direct sense. The environmental costs on the basis of direct regulation amount to almost 10 billion guilders ($ 5.4 billion) in 1997, which is about 48 percent of the total environmental costs in The Netherlands.

So almost half of the total costs of environmental policy in The Netherlands is directly paid for by the polluters involved.

This means that the other half is financed by other, more indirect ways. In several cases specific charges are used, while a part of the environmental costs also is financed out of the general budget of the government. Different forms of taxation play therefore also a role in the financing system of the environmental policy in The Netherlands.

3. DIFFERENT TYPES OF ENVIRONMENTAL TAXES AND LEVIES

Discussions about environmental taxation often distinguish between different types of levies. The Netherlands' Scientific Council for Government Policy, for example, distinguishes them according to their primary purpose and the way the revenue is used. Following this interpretation one can distinguish between earmarked charges, regulating charges and environmental taxes.

3.1. Earmarked charges

The primary purpose of earmarked charges is to generate revenues to pay for certain environmental activities or investments. The rate of the charge is determined by the funding needed and the volume of the tax base. The environmental effect is primarily a consequence of the use of the revenue. Dependent on the rate, however, the charge itself may also have a regulating effect.

Earmarked charges can be used where it is not possible to make a direct link between the causes of pollution and the measures to deal with them on an individual level or where this may not be an efficient approach. Instead of individual measures of polluters themselves, collective measures may be a more appropriate solutions for the environmental problems. The costs of such collective measures can be covered by charges to be paid by the polluters concerned.

Earmarked charges in The Netherlands are mainly used to finance environmental policies of regional and local governments, like provinces, water boards and municipalities. Charging systems for waste collection and disposal and for sewerage systems of municipalities and for the waste water treatment by provinces and water boards are examples in this respect. Total revenue of the charges of the regional and local governments is currently almost 6 billion guilders ($ 3.2 billion).

Central government uses only in a few cases earmarked charges to finance some specific policy fields. This concerns charges related to aircraft noise, manure surplus and water pollution regarding the state waters. The total revenue of these charges is a relatively insignificant amount of 160 million guilders ($ 85 million).
Although it is not possible to fill in the "Polluters Pay Principle" in full, in the case of earmarked charges one might speak of "Polluters Pay" to give a general description. The total of the revenues of the earmarked charges in the country covers about 30 percent of the total environmental costs.

3.2. Regulating charges

The purpose of regulating charges is to give price incentives in order to change economic behavior, irrespective of any funding which might be required for abatement measures. The rate of the regulating charge is in principle determined by the environmental objective related to the base of the charge. To avoid any second thoughts on behalf of the taxpayers, revenues of the regulating charges could be returned for example by lowering other taxes. Regulating charges are not supposed to have a structural role in the financing system of the environmental policy.
A recent example is the regulating **energy tax,** which was introduced in The Netherlands in 1996. On the introduction was decided when it became clear that an european wide CO_2/energy tax - a tax measure the Dutch government always strongly had supported - could not expected to be implemented by that time. The European discussion about such a tax was - at least for the time being - at a dead end. At the moment, however, the European Commission is working on new proposals for energy taxation in the European Union. The discussion on the european level can therefore be expected to go on very soon.
The Dutch regulating energy tax is focusing on small scale energy consumption: households, small commercial establishments such as restaurants and shops, office buildings, schools and so forth. These are target groups which are difficult or impossible to reach with policy instruments such as long term agreements or environmental permits. The regulating energy tax provides an important addition to the set of policy instruments currently being used to encourage energy conservation and reduction of CO_2-emissions in

The Netherlands.

Some industries are exempted from the tax in order to avoid the economic risks which would result from unilateral imposition of an energy tax on large industrial energy users facing competition from countries where such a tax is not in force. However, this observation does not mean that the industry has no obligations regarding energy saving. Other instruments than taxation are more suitable for that purpose. Industrial energy conservation is being realized in the context of long term agreements that industrial sectors have signed with the government. In these agreements sectors commit themselves to take measures to improve their energy efficiency by an average of 20 percent over the period 1990-2000. Since these policy instruments are proving to be effective in inducing large energy consumers to save energy, exposing them to the economic risks of a unilateral tax was felt to be unwarranted.

The Dutch energy tax features some interesting provisions to promote environmentally friendly options. First, there is a tax-exemption for heat supplied via district heating. This improves the competitive position of district heating relative to individual central heating, thus contributing to the expansion of the district heating in The Netherlands. The second provision is one which exempts natural gas used for electricity generation. This might contribute to a greater use of cogeneration of heat and electricity. And third, there is a special provision for energy generated from renewables (water, wind, solar and biomas), stimulating their use.

The tax is expected to raise about 2.1 billion guilders, including VAT to be paid over the tax, ($ 1.1 billion) structurally in 1998 when the rates reach their ultimate level, as foreseen in the tax law. These revenues are recycled through relief in other taxes paid by households and businesses. Since one of the ideas behind the tax was to shift taxes from labor to the environment, labor taxes were the first choice as a recycling instrument.

3.3. Environmental taxes

Environmental taxes combine funding and regulation by introducing environmental considerations into the tax system. They can thus be a contribution to the "greening of the tax system", a subject which has become an important issue in The Netherlands as well as in other countries of the European Union and at the Organization for Economic Cooperation and Development. Environmental value is added as a tax basis to the traditional system of taxing income, capital and consumption. And since the prime purpose is funding, this also determines the rate. The regulating effect of the price increase is a secondary effect, albeit it can be a desirable one. To prevent the overall tax base from being eroded, environmental taxes can best be levied on environmental resources for which demand is fairly constant (or in economic terms: when there is a low price elasticity of demand).

The revenues of environmental taxes accrue to the general budget of the government. As mentioned previously, a part of the total environmental costs are paid out of the general budget, because a lot of expenses of the government for environmental policy could only be linked very indirectly to certain kinds of pollution. In general it would not be possible to say which specific polluters are to be held responsible for the need of the expenses involved and in relation to that who should pay for those expenses. The Polluter Pays Principle is here difficult to apply, or may be only in a very indirect way by means of very general environmental charges. In that case there is not a specific link any more between pollution and its effects and the taxpayer.

Important categories of expenses by the central government, where a clear link with specific causes of pollution do not exist or where the polluters can not be traced, are for instance expenses for soil sanitation (especially in cases of "historic" pollution), research expenses, expenses for the promotion of environmental technology, energy conservation and sustainable energy, payments to other governments (provinces, municipalities), bilateral and multilateral international cooperation and of course the costs of the government organization itself. The environmental expenses out of the general budget will be an ample 4 billion guilders ($ 2.2 billion) in 1997. This is about 22 percent of the total environmental costs.

4. DEVELOPMENT OF ENVIRONMENTAL TAXES

In the 1970's and 1980's the Dutch environmental policy was characterized by a number of sectoral environmental laws. These laws together created possibilities for at least 15 different levy systems, of which only some were actually in force at that time. Each levy was supposed to finance the costs of the corresponding policy field, in accordance with the Polluter Pays Principle.

This system of different levies for different programs resulted in fragmentation and lack of transparency for tax payers. It was feared that continuing along the road of "different levies for different programs" would lead to tax collection systems in which the administrative costs were disproportionately high compared to the rather small revenue raised. There arose a need for a more integral system for financing environmental policy expenditures. In 1988, these levies were therefore replaced by one fuel charge, earmarked to finance the costs of the governments' environmental policy. Fuel was chosen as the tax base because it was felt that this would provide a general link with the Polluter Pays Principle. A great deal of pollution is directly related to fuel use. There is also a more indirect linkage, namely that fuel use can be seen as a rough indicator of activities which result in pollution. Nevertheless, the relationship between the tax and (the financing of the cost of) the environmental problem decreased, for the sake of simplicity of the tax system.

In the period 1988-1992 environmental policy intensified and the revenue of the fuel charge had to rise accordingly. This higher revenue amplified ongoing discussions about the relation of the fuel charge with the Polluter Pays Principle. Why should, in the end, users of energy have to pay for soil purification or chemical waste policy? In 1992 it was therefore decided that the revenues raised would no longer be earmarked for environmental expenditures but would accrue to the general budget. Consequently, the fuel charge became a fuel tax. Environmental expenditures of the central government were from that moment on to be paid from the general budget. Needless to say that this change of names did not alter the effect of the tax on energy prices at all.

The government proposal in 1991 to raise the fuel tax substantially for budgetary reasons led to fierce resistance by the largest energy-intensive companies. In reaction, Parliament asked the government to look for other taxes with an environmental base. Taxes on the extraction of groundwater, the dumping of waste and on the use of uranium for the production of electricity were consequently introduced in 1995. The revenue of these taxes goes, like the revenue of the fuel tax, to the general budget of the government. The government argued that by introducing these taxes other taxes (like those on income from labor) did not have to rise. Introducing environmental taxes lead in this reasoning to an implicit tax shift to the environment, by increasing the share of environmental taxes in total tax revenue.

The total revenue of the environmental taxes - on fuels, waste, groundwater, uranium - and the regulatory tax on the small use of energy is expected to rise to an amount of about 4 billion guilders ($ 2.1 billion) in 1998, which will than be about 2.5 percent of the total tax receipts of the Dutch government. This percentage may seem not very high, but with the introduction of these environmental taxes, we have made some important steps in the process of "greening the tax system". The Dutch government has set up a special Green Tax Commission to study further possibilities and to make proposals in this respect.

5. FISCAL INCENTIVES AND SUBSIDIES

Subsidies financed from the budget are still used in The Netherlands, especially in the field of research and development and the stimulation of new techniques. However, their role is increasingly taken over by fiscal incentives. These incentives consist of amendments to existing tax laws, like the personal income and corporate tax. The introduction of such instruments can also be seen as a contribution to the "greening of the tax system".

5.1. Accelerated depreciation

In 1991 the scheme for accelerated depreciation for investments in environmentally friendly equipment was introduced on the initiative of Parliament. This scheme is not supposed to trigger certain investments as such, but it should stimulate that already planned investments will go into a more environmental friendly direction. This accelerated depreciation give companies the freedom to choose how to write off the costs of investing in a piece of equipment.

Accelerated depreciation keeps taxable income down so that in the year concerned companies pay less income tax or corporate tax. Of course in later years there will be less to write off, but it is precisely the deferral of tax payments which is of benefit to companies' cash and interest position. Depending on the write-off period and the interest rate this advantage can be calculated at 5 to 10 percent of the investment sum.

Accelerated depreciation is only possible for equipment that is specified on a so called environmental list, which is updated every year by the Ministry of Environment. The list includes equipment in the field of water pollution, soil pollution, waste, noise and energy saving.

To qualify for the list, equipment should meet some specific criteria. For example, for reasons of controllability and feasibility, the equipment or asset must be quickly and unambiguously definable both in technical terms and in terms of cost structure. Moreover, the equipment or asset should not yet be in common use in The Netherlands. The penetration rate of the market should be below 30 percent of the potential outlets for a certain application. This feature makes the accelerated depreciation scheme an important instrument in promoting the dissemination and market introduction of new developed technology.

There is a close link between government sponsored R&D and the accelerated depreciation scheme. About 60 percent of the technologies placed on the list were sponsored by the Dutch government in an earlier phase of development, most through demonstration projects for new technologies. Thus the accelerated depreciation scheme complements the R&D programma by facilitating market penetration and encouraging diffusion.

5.2. Green investment

Possibilities of investing in green investment funds tax free have been available in The Netherlands since 1995. This means that private investors are not taxed on their interest and dividend payments, provided that these payments derive from investment in certain green investment funds. These green investment funds, in turn, have to invest in certain green projects. The aim of this tax concession is to encourage investment in major environmental projects, involving forests and nature areas, sustainable energy supplies and environmental technology.

The initiative for tax free green investment was taken by Parliament in The Netherlands, which considered it desirable to get Dutch citizens more involved in investing in green projects. The reasoning was that by offering fiscal incentives, more savings would be made available for these green projects. Such projects are often difficult to finance, since they do not always provide the lavish returns the market expects. By ensuring that investors' returns on such projects are untaxed, this allows them to compete with the returns of regular investments funds on the market, which are normally taxed. So green projects offer not just a sound environmental yield, they also provide a sound financial return as well.

Returns on investment are only free from income tax if they come from green investment funds. Green investment funds are obliged to invest at least 70 percent of their total assets in green projects. Green projects are labelled as such by the Dutch government. Green projects should be new projects which are important for environment and nature, generate some return but cannot be funded in a normal commercial way. Categories of green projects are:

- Projects in the field of nature, forestry, landscape and organic farming.
- Green mortgages. These mortgages can then be granted at a lower interest rate for houses which score high environmentally because they are economical with energy, have been built from environmentally friendly building materials or using an ecologically benign building processes.
- Projects in the field of renewable energy. These can be wind, solar, or geothermal energy, energy from hydro power, from timber or energy rich crops, the use of heatpumps, the storage of heat or cold in aquifers, and the heat distribution networks for district heating and the heating of greenhouses for market gardens.

At the moment, all large Dutch banks offer green investment funds to their clients. The largest share of the projects these funds invest in consists of projects in the field of organic farming and energy.

5.3. Energy investment tax allowance

On 1 January 1997 a special energy investment allowance was introduced. This allowance is a provision in the personal and corporate income tax. It is comparable to the investment allowance for investments that already existed in The Netherlands. The energy investment allowance makes it possible for firms to subtract a percentage of the investment sum from the profit of the firm. In this way, the firm gets an advantage because less profit tax has

to be paid to the Treasury. Depending on the modalities that are ultimately chosen, the advantage will amount to about 10 to 15 percent of the investment sum.
Investments that qualify for the energy allowance are specified on a list. This list contains in the first place all energy items that are also on the list for accelerated depreciation among which renewable sources like wind, water and solar energy. For these innovative investments there is thus a double incentive. In addition, the list contains investments in specified equipment for energy saving, but which are not innovative and therefore not on the list for accelerated depreciation. Finally, the energy investment allowance facilitates energy saving investment in production processes satisfying a certain energy saving requirement.

6. FUTURE DEVELOPMENTS

The use of economic instruments in The Netherlands is not so much the result of a deliberate choice of the Dutch government at a specific point in time, but it is the result of a continuing political process taking place over the years. Political preferences of governments change, especially after elections. This is one of the reasons that the weighing of environmental effectiveness, revenue requirements, feasibility, competitiveness of firms and purchasing power of the citizens can lead to a diffuse outcome. Each economic instrument has its own history and background. Without understanding of the instruments genesis, it may often be difficult to see the rationale for the modalities chosen.
Nevertheless, the above mentioned fiscal instruments all contribute to the Dutch effort to further 'green' the tax system. Existing fiscal instruments will be amplified by differentiated tax rates and target groups. New instruments will very much be geared to transportation, the use of automobiles in particular, and to consumption. The Dutch VAT (Value Added Tax) could be a very useful tool to stimulate or discourage the consumption of particular goods and services. One must recognize however, that the Dutch membership of the European Community brings its limitations as well as new chances.
In its efforts to create uniformity between tax systems and to avoid unlimited

taxcompetition between members, members of the EC have only limited playground to use their national tax system to pursue national goals. This in particularly the case for differentiating the VAT on a national level.

On the other hand the EC-members recognize that fighting unemployment is the biggest challenge facing the EC today. One of the options under consideration is to lower the taxation of labor. An alternative tax base, to prevent erosion of public expenditure, could then be consumption, or factors of production other than labor. At this point, an increased use of environmental and energy taxes and differentiating the VAT could support national budgets and at the same time help the EC in its task to promote sustainable economic growth which respects the environment.

Health damage of air pollution: an estimate of a dose-response relationship for The Netherlands [1]

Thijs Zuidema [a] and Andries Nentjes [b]

[a] Thijs Zuidema,
University of Groningen,
P.O.Box 800, 9700 AV GRONINGEN
The Netherlands

[b] Andries Nentjes,
University of Groningen,
P.O.Box 716, 9700 AS GRONINGEN
The Netherlands

ABSTRACT

This paper estimates the dose-response relationship between air pollution and the number of work loss days for The Netherlands. The study is based on illness data (work loss days) for the Dutch labor population and average year concentrations of air pollution in 29 districts. The dose-response relationship has been estimated by means of two different techniques: the ordinary least squares method (OLS) and the one-way fixed-effects method (OWFEM), which we consider to be more adequate. In general health effects are much smaller when OWFEM is applied than if OLS is used.

With OWFEM a significant relationship is found between sulphate aerosol (SO_4), ammonia (NH_3) and the number of work loss days (WLD's). Particulates (TSP), O_3 and SO_2 have no significant effect on the number of WLD's. These results differ from those

[1] We wish to thank Shelby Gerking (University of Wyoming, U.S.A.) for his advice on the application of the one-way fixed-effects method. We are very indebted to Che Wah Lee, who did the computerwork. (In the period he worked at the University of Groningen.)

obtained in studies in the United States, which indicate that particulates (TSP) and other small particles, ozone (O_3) and to a lesser extent SO_4 and SO_2 significantly influence the number of WLD's.

1. INTRODUCTION

To calculate the monetary health benefits of reducing air pollution one needs firstly an estimation of the relationship between the pollutant and its physical impacts; secondly a translation of physical effects in money terms. This paper concentrates on the first part of the problem. It discusses the physical impacts of a number of air pollutants on potential output, in particular labor input. We do this by estimating the dose-response relationship (DRR) between air pollution and the number of work loss days owing to illness of respiratory organs for The Netherlands. Studies of this type have been done in the US since the early seventies. They have identified a significant relationship between specific air pollutants and morbidity (and also mortality) in a number of cases.

Until now no DRR has been estimated on the basis of Dutch data. Only a few rough estimates have been made, in which relationships found in the US were applied to calculate the health impacts of SO_2 concentrations in The Netherlands. This means that it was assumed that relationships which were estimated for the United States also held true for The Netherlands. Of course, this is very dubious. This study is original in the sense that we have estimated a DRR purely on the basis of Dutch data. A second interesting element is that we compare the results obtained with the ordinary least squares method with a more advanced method: the so-called one-way fixed-effects method.

The paper is structured as follows. In section 2 a review of the research literature is given. Subsequently, section 3.1 explains which DRR is used for The Netherlands. The data used in this study are described in section 3.2. Then section 3.3 describes the estimation method, namely the one-way fixed-effects method. Subsequently, the empirical results are presented in section 3.4. Finally some conclusions are drawn in section 4.

2. REVIEW OF THE RESEARCH LITERATURE

Ideally the DRR should be based on a health model that specifies the factors that determine the health situation of a person. Clearly, health status is affected by many factors; see, among others, Zuidema [1992, paper 27.06].
A DRR can be presented by the following formula:

$$WLD = f(P_1, \ldots, P_N, X_1, \ldots, X_M) \qquad [1]$$

WLD = annual work loss days;
P_i = air pollutant i;
Xj = other variables j.

In the literature on the health impacts of air pollution, a DRR is usually a function of a large number of variables, such as air pollution: nitrogen oxides, sulphur dioxide, sulphate aerosol, black smoke, particulates, ozone and ammonia. In addition to these variables, the number of WLD's is influenced by other variables as well. The literature on epidemiology mentions, among others: education and occupation, income, job situation (employed or not), race, sex, age, and habits such as drinking and smoking.
For empirical research we have to confine ourselves to an equation with a small number of variables, because only then the relationship can be estimated. In the literature on research in the US the emphasis lies on the estimation of the relationship between air pollution and the number of WLD's. Air pollution, however, can also result in an increase in mortality. Research in this field has been conducted, for example by, Lave and Seskin [1971; 1977]; Mendelsohn and Orcutt [1979], Chappie and Lave [1982], and Lipfert [1984]. In this article and survey of the literature we shall not deal with mortality but concentrate on morbidity (=WLD's).
The research for the estimation of DRR's has mainly been done for the US. Table I classifies the seven major studies of a number of air pollutants for the United States. It shows that the DRR's have been estimated by applying different techniques. Apparently, no standard technique exists for the estimation of a DRR. The studies were published in

the eighties and early nineties, but refer to different regions in the U.S. in the seventies. The types of pollutants that have been distinguished and the other explanatory variables are shown in table II. The variable measuring WLD's is based on the response to the survey question asking how many days in the past 2 weeks did illness prevent one from working. Other data on background variables has been obtained from the same set of panel data.

Table I. Estimation techniques, place and period in studies about morbidity

Studies	Estimation techniques	Place[1]	Period
Cropper [1981]	- Tobit model	US	1970, 1974 en 1976
Krupnick, Harrington and Ostro [1990]	- logit estimation procedure	Los Angeles	1978 en 1979
Ostro [1983a]	- OLS[1]; - Tobit model; - logit linear combination	US	1976
Ostro [1983b]	- OLS; - logit linear combination	US	1976
Ostro [1987]	- Fixed-effects method by using a Poisson distribution	US	1976-1981
Ostro [1990a]	- logit estimation	US	1979-1981
Portney and Mullahy [1986]	- maximum likelihood method by using a Poisson distribution	US	1979

[1] OLS = ordinary least squares method

Table II. Dose-response relationship for morbidity[1]

Studies	Pollutants								Health Situation	Socio-economic variables		Demographic variables				Habits	Other variables
	NO2	SO_2	SO_4	O_3	TSP[2)]	IP[2)]	FP[2)]	COH[2)]	existence of chronic condition	income	educa tion	age	marital status	race	sex	smoking	population density
Cropper [1981]		*							-	-	-		-	-			
Krupnick, Harrington and Ostro [1990]	-	-		*				*	*		*	*		-	*	*	
Ostro [1983a]			-		*				*	-		*	-	-	*	-	-
Ostro [1983b]			-		*			*	-		*	-	-		-		-
Ostro [1987]							*		*	*	*	*	*	*	*		
Ostro [1990a][4)]			*		*[3)]	*[3)]	*[3)]										
Portney and Mullahy [1986]			-	*					-	*	-	-		*	-	-	

1) * = significant;
- = in significant;

2) TSP = total suspended particulates;
IP = inhalable particles;
FP = fine particles;
COH = coefficient of haze (= a surrogate for fine particles);

3) Only significant if lagged by one 2-week period

4) Several socio-economic and demographic variables are included in the DRR. It is, however, not known which variables are significant.

Table II shows that the variables included in DRR's can differ between the studies. Which variables are actually taken into account is to a large extent determined by pragmatic reasons, such as the availability of data.

The pollutant NO_2 is only examined in the study by Krupnick et al. [1990], but it is not statistically significant. SO_4 is significant in only one of the four cases. Statistically significant in all cases are ozone, total suspended particulates and the coefficient of haze. The conclusion can be drawn that the most important variables are ozone and the four types of particulates.

It appears from table II that the variables chronic condition, education, age, and sex are usually statistically significant. Smoking is only significant in one of the three studies.

Next, the figures about air pollution are presented in table III.

Table III. Air pollution concentration[4]

Studies	Pollutants							
	NO_2	SO_2	SO_4	O_3	TSP	IP	FP	COH
Cropper [1981]		- [3)]						
Krupnick, Harrington and Ostro [1990]	O $<NO_2 < 31$ pphm[1)]	O $< SO_2 < 6$ pphm		$2 < O_3 < 43$ pphm				$4 <$ COH < 26 μ gm -3
Ostro [1983a]		8 μ gm^{-3}			78 μ gm^{-3}			
Ostro [1983b]			-		-			
Ostro [1987]							22 μ gm^{-3}	
Ostro [1990a]			8 μ gm^{-3}		69 μ gm^{-3}	44 μ gm^{-3}	24 μ gm^{-3}	
Portney and Mullahy [1986]			11 μ gm^{-3}	0,042 ppm [2)]				

1) pphm = parts per hundred million;
2) ppm = parts per million;
3) - = the study does not present the figures.
4) Krupnick et al. present minimum and maximum concentrations. The other studies present average values.

It is important that studies present data about the interval of air pollution concentration for which the DRR is estimated. This is done only by Krupnick et al. Most other studies mention only the average pollution and Ostro [1983b] does not even mention this. The authors hardly ever compare their results with the results of other studies. As Ostro has done a substantial amount of the research, it would have been interesting if he had examined why SO_4 is statistically significant in Ostro [1990a], but not in Ostro [1983a] and [1983b]. The influence of the estimation technique on the empirical results is not discussed in the various studies. This effect can be very large, however, as will be shown in section 3.

Dutch research

Although the relationship between WLD's and air pollution has not been estimated for The Netherlands, research has been done on health effects of air pollution. Hoek et al. [1990] and [1993] researched the effects of air pollution on children living in The Netherlands. They measure the effect of short term exposure to relatively high concentrations of the air pollutants ozone, sulphur dioxide and total suspended particulates on the pulmonary function of a group of children. The research was conducted in the eighties in a small number of communities in The Netherlands. Hoek et al. found a significant negative association between O_3, SO_2, TSP and the pulmonary function of the children. Our research differs from this approach because we have estimated long run effects of air pollution on work loss days of adults by using aggregated data.

Of a quite different category is the research of Jansen [1974], the OECD [1981] and Ostro et al. [1990b]. They estimated the monetary health damage due to air pollution for The Netherlands. Since an empirical DRR for The Netherlands was not available they applied a DRR which was estimated by Lave and Seskin [1971; 1977] for the US. In another publication, Jansen [1980] used Crocker's so-called Wyoming study. If one takes the American DRR for granted the damage to health in The Netherlands is considerable. According to Jansen [1974] the annual damage is about one billion guilders (1967 prices). However, the estimation of the actual damage is very uncertain, for the OECD [1981] annual estimates range from 100 million to 2.5 billion guilders.

3. A DOSE-RESPONSE RELATIONSHIP FOR THE NETHERLANDS

3.1. The model for a dose-response relationship for The Netherlands

As a starting point for estimating an empirical DRR we have formulated the following health model.

$$WLD = f(P_1, \; .... \; ,P_6,X_1,..... \; ,X_4) \qquad [2]$$

WLD = annual work loss days;

P_1 = sulphur dioxide (SO_2);

P_2 = sulphate aerosol (SO_4);

P_3 = black smoke;

P_4 = particulates;

P_5 = ammonia (NH_3);

P_6 = ozone (O_3);

X_1 = unemployment percentage in a region;

X_2 = percentage of labor force in a region receiving a pension under the Dutch Disablement Insurance Act;

X_3 = population density as an indicator for the urbanization rate of a region;

X_4 = average annual gross income per capita in a region.

The choice of explanatory variables is partly inspired by theoretical considerations and partly influenced by availability of data. In this study aggregated cross-sectional data have been used. We would have prefered to use microdata because these data contain person-specific information. Unfortunately no suitable dataset is available in The Netherlands. When compared with the models used in section 2 (table II), it is striking that in equation 2 the following background variables are lacking: chronic condition, age, race, sex, smoking and drinking. The reason for this is that there is no information available on a regional level for The Netherlands. Next, we discuss the variables in equation 2.

Pollutants

The selection of air pollutants has been influenced by availability of information. In further defence of this choice we recall that US studies suggest that the following pollutants may have a significant positive effect on WLD's: SO_2, SO_4, particulates, black smoke and ozone. To this we have added ammonia, which has not been researched in US studies. It, however, is a major air pollutant in The Netherlands.

Work loss days

The question can be asked how WLD's should be defined. Since we are interested in the health impacts of air pollution only WLD's caused by illnesses of the respiratory system, such as chronic bronchitis, asthma and chronic non-specific lung disease (CNSLD) has been used in this research. In the empirical study only the WLD's of employed persons are used. WLD's of the unemployed and disabled are not taken into account.

Unemployment

Four socio-economic variables are used. The first one is unemployment. We expect that with increasing unemployment the WLD's of the employed will decrease. This can be argued in the following two ways:

- When the fear of loosing one's job increases, people will not report themselves sick so readily because they are anxious not to be fired;
- It is possible that unhealthy people are more likely to lose their jobs than healthier people.

Disablement

The percentage of disabled workers might also affect the WLD's of the employed people. Two influences can be distinguished:

- The labor force may become healthier when more people are considered as disabled. This may reduce the number of WLD's of the remaining labor force.
- The disablement percentage can be seen as a health indicator for the labor force: the higher it is the lower the health situation of the labor force may be. In this view a high disablement percentage correlates with a large number of WLD's.

The above-mentioned two influences work against each other. Therefore it is a priori not clear what sign this variable will have in equation 2.

Population density

The labor force in urban areas is different from that in rural areas. The socio-cultural climate is also different. Generally speaking, urban areas have greater problems than rural areas. Therefore we expect a positive relationship between the population density and the number of WLD's.

Income

It is well known that people who are well educated and belong to the higher occupations have fewer WLD's than people who belong to the lower classes. Income can be considered as an indicator of the schooling and occupational level. Therefore we expect a low average regional income per capita to have a negative influence on the number of WLD's.

The empirical results will show whether these hypotheses are falsified or not.

3.2. Data

The following data are used for the estimation of the DRR: work loss days, air pollution, and socio-economic figures.

Work loss days

The Netherlands consist of 29 administrative health districts. For each district we have information on WLD's of employed persons due to illnesses of the respiratory system for the years 1987, 1988 and 1989. This means that aggregated data (cross-sectional and time series) had to be used and we lack the person-specific information (smoking habits, age, sex, etc.) that are contained in microdata that have been used in the US studies. On the other hand our sample contains the complete employed labor population and the whole territory of The Netherlands. The US studies lack this scope. For estimation purposes WLD's per district have been expressed as a percentage of the district's employed labor population. Its variation between district's is large with the maximum more than three times the minimum.

Air pollution

The data refer to outdoor air pollution. Although the influence of occupational exposure on the health situation of people is possible, this influence is not included in the DRR. The reason for this is that there is no data set available.

Data about outdoor air pollution come from the annual reports of the National Institute of Public Health and Environmental Protection (RIVM) for 1987, 1988 and 1989. Air pollution is monitored at a number of monitoring sites. The air pollution in a "district" is estimated by taking an average of the air pollution at the monitoring sites in that "district". We use annual averages since WLD's have also been measured on a per year basis.

The pollutants SO_2, SO_4, BS, Part., and O_3 are measured in μgm-3 and NH_3 in mol ha-1. Figures on concentration are presented in table IV. They show that differences across the regions are considerable. The maximum concentration is about two to three times higher than the minimum value. In general the northern part of The Netherlands is the least polluted and the southwestern part the most.

Table IV. Minimum and maximum concentration of air pollution in The Netherlands

(measured in μg m-3, except NH_3 (mol ha^{-1})						
	year					
air pollution	1987		1988		1989	
	min.	max.	min.	max.	min.	max.
SO_2	10	36	7	24	7	26
SO_4	7	9	4	6	7	10
BS	11	30	7	24	10	28
Part.	42	57	33	55	36	61
NH_3	570	1190	320	1060	490	1000
O_3	27	56	26	57	28	58

Part. = particulates;
BS = black smoke;
NH_3 = ammonia;
min. = region with lowest annual pollution;
max. = region with highest annual pollution.

A comparison with air pollution concentration of the US studies, as shown in table II, reveals that the concentration of SO_4 in The Netherlands in the late eighties is between 4 and 10 μg m^3 and in the areas in the US between 8 and 11 μg m^3 during the seventies. The concentration of particulates in The Netherlands is lower than the concentration in the study by Ostro [1983a] and in the range of Ostro [1990a].

Socio-economic data

The socio-economic data on unemployment percentages, disablement percentages, population density and average gross nominal income per capita come from the Dutch Central Statistical Office. These figures are available for 40 areas, the so-called "Coropregions", which differ from the health districts. Therefore all socio-economic data about Coropregions had to be assigned to the 29 health districts. In doing this the number of inhabitants per Corop region has been used as weights.

The unemployment and disablement percentages are calculated as percentages of the labor force per district. The population density is calculated by dividing the population of a district by the number of square kilometers. The variance of the population density, however, is very large. Amsterdam has more than 4000 inhabitants per km^2 and Assen has less than 170. The variance in the unemployment and disablement percentages across the regions is considerable. The variance of the incomes across the regions is small. The maximum income is about 10 percent higher than the minimum income.

3.3. Estimation technique: ordinary least squares method versus one-way fixed-effects method

We have estimated the DRR both by application of the least squares method (OLS) and by the so-called one-way fixed-effects method (OWFEM). It is not necessary to explain OLS, but this may be necessary for OWFEM, because this method is not so widely known.

The DRR for The Netherlands is estimated by means of data that refer to 29 regions in The Netherlands. When the relation is estimated by application of the ordinary least squares method it is actually assumed that the same DRR holds for each region. This is, however, very unlikely. Some regions may have a specific population composition which influences the number of WLD's. Urban areas could therefore have health models that differ from those for rural areas. If these differences are not taken into account the effect of air pollution on the number of WLD's cannot be estimated in an unbiased way. A method which can take into account the fact that the DRR's differ across the regions is the *one-way fixed-effects method* (OWFEM), see, among others, Judge et al. [1988,chapter 11.4] and Greene [1990, chapter 16.4]. The essential difference between the least squares method and OWFEM is that in the latter method the intercepts are different across the various regions while in the former method they are all equal. On the other hand, it is assumed that the independent variables and their impacts on health are the same for the various regions. Because specific regional differences in the WLD's are reflected in the intercepts it is not necessary to use different DRR's for the various regions.[2] The method is called fixed-effects because the differences across the regions

---

[2] Besides the one-way fixed-effects method there is also a two-way fixed-effects method. In the two-way the intercepts differ both across the regions and across time intervals. Because the data are related to a short period, namely three years, it is not likely that the DRR's will

can be considered as shifts of the functions.

The computer program LIMDEP 6.0 has been used for the estimation of the equations, see Greene, [1991]. Assuming we have i=1,2,..., N regional observations and t = 1,2,..., T time-series observations, the (i,t) observation can be written as:

$$Yit = \beta 1i + \sum_{k=1}^{K} \beta k.\ Xkit + eit \qquad [3]$$

K = number of explanatory variables;

The fact that each region has a different DRR can be adequately captured by specifying a different intercept coefficient for each region. All the information is used in estimating equation 3. Output consists of X-effects (that is β2,...,βk) and the N values for intercepts β1. Only when the intercepts differ significantly is there a basis for applying the fixed-effects method. The testing is as follows:

Hypothesis 0: β11 = β12 = ... = β1N

The null hypothesis can be tested by calculating the following value for F.

$$F = \frac{(e'e - \hat{e}'\hat{e}) / (N-1)}{\hat{e}'\hat{e} / (NT-N-K')} \qquad [4]$$

e’e = the residual sum of squares from the restricted model. This is the model in which all the intercepts are equal.

ê’ê = the residual sum of squares from the unrestricted model. This is the model that is used here, which means that the intercepts are not equal.

K’= K -1

Under the null hypothesis the statistic in [4] has the F distribution with [(N-1),(NT-N-K)] degrees of freedom. The critical value of F can be found in a table.

It is also of interest to examine which part of the variance is explained by the explanatory variables. This is done by means of an analysis of variance. Total variation

---

differ over the various years. In addition to the fixed-effects method there is also a random-effects method. In the One-way Random-effects method it is assumed that the intercepts are randomly distributed across the regions. We can use OWFEM because we do not assume this.

is equal to the sum of within group variation and between group variation. The explanatory variables only give an explanation of the variance within the regions. The differences between the regions are caught by the differences between the constants. Before presenting the empirical research, an analysis of variance is given for the dependent variable, that is the number of WLD's for each of the pollutants. Total variation is equal to the sum of within group variation and between group variation. Systematic differences between the regions that cannot be related to differences in explanatory variables are caught by the differences in the intercepts. Although the number of WLD's per region is the same each time, there are important differences because the available information is different. For SO_4, for instance, data on air pollution exist for only 9 out of 29 regions, so use can be made of the information concerning the number of WLD's in these 9 regions only. The result is that the analysis of variance shows differences.

Table V. Analysis of variance

N		variance within regions	variance between regions	total variance
29	SO_2	460.23	9436.71	9896.94
9	SO_4	97.82	2440.51	2538.33
14	BS	261.51	5180.98	5442.49
5	Part.	48.38	1738.95	1787.32
11	NH_3	136.73	3588.56	3725.29
20	O_3	349.21	6428.11	6777.32

Table V shows that total variance of the number of WLD's consists, for more than 95 %, of variance between the regions. This means that the explanatory variables explain only a small part of the total variance in the number of WLD's. This indicates that it is especially useful to apply a technique like the fixed-effects method in which this is taken into account.

3.4. Empirical results

The DRR's for The Netherlands are estimated, with each equation containing only one pollutant variable. The reason for this is that multicollinearity exists between the various pollutants. A consequence of this estimation procedure, however, is that the effect of a specific air pollutant on the number of WLD's may be overestimated.

Firstly, empirical results which were obtained by application of OLS and subsequently by means of OWFEM are given. We first estimate a simple model with pollution as the only explanatory variable and then a model with more explanatory variables.

Ordinary least squares method

The following equation is estimated for the period 1987-1989:

$$WLD = a.Pi + c \qquad [5]$$

WLD = number of work loss days per 100 working persons a year;

P_i = pollutant i.

Table VI. Relationship between air pollution and the number of work loss days (WLD = a.P_i + c) OLS

N		constant	Pi	R^2
29	SO_2	34.363 (12.279)***	0.338 (1.988)**	0.044
9	SO_4	26.142 (3.001)***	0.035 (1.416)	0.074
14	BS	26.442 (4.626)***	0.946 (2.700)**	0.154
5	Part.	20.527 (0.940)	0.337 (0.779)	0.045
11	NH_3	29.918 (4.601)***	0.011 (1.284)	0.051
20	O_3	65.774 (11.074)***	-0.659 (-4.637)***	0.270

t-statistic is in parentheses;
* = coefficient is significant at a 10 per cent level;
** = coefficient is significant at a 5 per cent level;
*** = coefficient is significant at a 1 per cent level;
N = number of regions.

Table VI gives the results of the ordinary least square method with only one pollutant as explanatory variable. The following conclusions can be drawn:

1. The coefficient that shows the influence of air pollution on WLD (a) is positive, except for ozone which has a negative impact;
2. The coefficient a differs significantly from zero in three cases;
3. The constant c is significant at a 1 per cent level in five of six cases.

One can try to improve these results by adding socio-economic variables to explain differences in WLD's. Table VII presents results with population (Pd) and unemployment percentage (u) added. The disablement percentage has been left out since adding it did not improve empirical results. For the same reason average income per capita was left out. The reason for it is the small variation of this variable across the regions.

Table VII. Relationship between the number of work loss days and air pollution, population density and unemployment. (WLD = a.Pi + b. Pd + d.u + c) OLS

N		constant	Pi	u	Pd	R2
29	SO_2	30.849 (7.905)***	0.337 (1.869)*	0.372 1.299	-0.000,12 (-0.163)	0.063
9	SO_4	23.799 (2.662)**	0.031 (1.191)	-0.048 (-0.081)	0.005,72 (1.631)	0.174
14	BS	19.108 (2.569)**	0.955 (2.676)**	0.668 (1.392)	0.000,60 (0.363)	0.214
5	Part.	11.365 (1.239)	0.193 (1.144)	-0.188 (-0.477)	0.015 (7.564)***	0.880
11	NH_3	19.900 (2.505)**	0.019 (2.147)**	-0.218 (-0.477)	0.009 (2.426)**	0.216
20	O_3	71.336 (8.837)***	-0.74 (-4.669)***	-0.019 (-0.063)	-0.002 (-1.219)	0.290

The following conclusions can be drawn from table VII:

1. As in table VI, the effect of air pollution (a) is in most cases positive, except for ozone.
2. Adding u and Pd in the DRR does not substantially change the effects of air pollution as a comparison of the columns under Pi in table VII and VI shows. The difference is that next to SO_2 and BS, ammonia has a positive significant effect on the number of WLD's.

3. The influence of the population density (Pd) is also usually positive (as expected).
4. The influence of unemployment is usually negative (as expected), but it is never significant.
5. In general R^2 is hardly improved by adding u and Pd. The exception is the equation for particulates. This may be caused by the effect of Pd.

One-way fixed-effects method

The simple equation, similar to 5, is now estimated by means of fixed-effects. The constants - as many as there are regions and for each pollutant - are not shown in this publication, but they differ considerable, with a minimum of 24 for SO_2 for region 7 and a maximum of 62 for SO_2 in region 19. These base levels are influenced by variables which are not included in the model. The test of whether the regional constants differ significantly is done on the basis of F-values. If the hypothesis is rejected, application of fixed-effects is not useful. If the F-value exceeds the critical value, the hypothesis is accepted. The highest critical value holds for particulates, namely 6.42, at a significance level of 1 per cent. Since all F-values from table VIII are much higher than 6.42 it is even more than 99 % certain that the constants (the intercepts for various regions) differ significantly.

Table VIII. Relationship between work loss days and air pollution
(WLD = a. P_i + c) OWFEM

N		Pi	R^2	F-value
29	SO_2	0.054 (0.803)	0.954	40.27
9	SO_4	0.007 (1.101)	0.964	52.57
14	BS	0.221 (1.445)	0.955	37.31
5	Part.	0.005 (0.031)	0.973	77.18
11	NH_3	0.001 (0.315)	0.963	52.48
20	O_3	-0.067 (-0.633)	0.949	27.31

Table VIII shows:

1. Just as in the ordinary least squares method, the effect of air pollution on the number of WLD's is positive for all pollutants except ozone (column 3);
2. The value of the coefficient, however, is much lower than the same coefficient in table VI. This can be explained by the fact that in the one-way fixed-effects method, the other influences on the number of WLD's are represented in the constants. Owing to this, the estimation of the effect of air pollution is improved.
3. Finally it appears that R^2 is now much higher. In table VI the correlation coefficient R^2 was usually less than 0.1.

The explanation for the high R^2 is that the regional constants explain a large part of the differences between regions. The variance that is left can be associated largely with differences in air pollution between the regions. It should, however, be admitted that the impacts of air pollution on the number of WLD's does not differ significantly from zero in table VIII for each pollutant.

Next, equation 6 is estimated by application of the one-way fixed-effects method.

$$WLD = a.Pi + b.Pd + d.u + c \qquad [6]$$

Table IX: Relationship between number of work loss days and air pollution, population density and unemployment
(WLD = a. Pi + b. Pd + d. u + c) OWFEM

N		Pi	Pd	u	R^2	F-value
29	SO_2	0.081 (1.207)	0.017 (0.676)	-0.185 (-1.975)*	0.958	42.18
9	SO_4	0.021 (3.456)***	0.322 (2.850)***	-0.149 (-1.166)	0.982	86.60
14	BS	0.203 (1.206)	0.024 (0.815)	-0.125 (-0.731)	0.959	35.06
5	Part.	-0.186 (-0.903)	-0.202 (-0.977)	-0.551 (-1.705)	0.982	9.63
11	NH_3	0.007 (1.880)*	0.267 (1.848)*	-0.187 (-1.272)	0.975	58.53
20	O_3	-0.189 (-1.639)	0.015 (0.577)	-0.251 (-1.986)*	0.955	29.03

The results are presented in table IX and the following conclusions can be drawn.

1. The influence of air pollution is usually positive, as in table VIII. The exception is again ozone, and this time also particulates: both have a negative coefficient (but neither of them is significant).

   The main difference with table VIII is that two pollutants are now significant: SO_4 and NH_3. The t-values have improved for all the pollutants compared with the simple model presented in table VIII.
2. The effect of population density on work loss days is usually positive.

3. Just as in table VII the effect of unemployment is negative and confirms our expectation that higher unemployment is associated with lower WLD's.
4. R^2, which was already high in the simple specification of OWFEM (table VIII) has improved slightly.

It is interesting to compare the results obtained by OLS with those by OWFEM by comparing the tables VII and IX. The following conclusions can then be drawn.

1. The effect of air pollution on health is in the fixed-effects method much lower than in the least squares method. The reason for this is that the intercepts explain a large part of the WLD's. The estimated effect of air pollution is then lower.
2. With OLS the pollutants SO_2, BS, NH_3 and O_3 are statistically significant, but with OWFEM SO_4 and NH_3 are statistically significant.

4. A COMPARISON WITH OTHER STUDIES AND CONCLUSIONS

In our quest for a DRR between air pollution and diseases of the respiratory system it has been shown how much the results are influenced by the specification of the model and the statistical method that is applied. As to the statistical method: if one were to simply look at t-values, the least squares method is better than OWFEM, but as we explained, it tends to overestimate the health impacts of air pollution. OWFEM is the most advanced method and its results are more reliable. It has appeared from the analysis of variance and from the calculated F-values that each region has its own health model. This fact is taken as a starting point in OWFEM.

By not taking into account these structural regional differences, the estimation of the health effects of air pollution is too high when the least squares method is used. Therefore we consider OWFEM with multiple explanatory variables as presented in table IX to give the most reliable results.

The empirical results of this study show important differences with results of research in the US. As table II shows, the significant pollutants in the US are: TSP and other small particles, O_3 and to a lesser extent SO_4 and SO_2. In our OLS version for The Netherlands

SO_2, BS, O_3 and NH_3 are significant, but in OWFEM only pollutants SO_4 and NH_3 significantly influence the number of WLD's.

The pollutant NH_3 does not occur in the US studies because it is of minor importance. For The Netherlands, however, this is different. Ammonia concentrations in specific regions of The Netherlands are very high so it need not come as a surprise that they have a negative impact on health. Ammonia concentrations are caused by the emissions from manure from the intensive cattle-raising sector, which is concentrated in rural areas in the middle, east and south of the country.

The injurious influence of particulates has been proved various times in the United States. The reason that this cannot be found for The Netherlands with OWFEM may be that the number of observations for this pollutant is too low (N = 5; see table IX).

Ozone has a negative influence on the number of WLD's which is highly significant in OLS. As has already been said the results with OLS are not reliable to our opinion. The data show that the level of ozone is highest in the northern part of The Netherlands. This means that the level of ozone is high in areas that are rather clean with respect to other types of air pollution and might explain the negative relationship between ozone and WLD's. With OWFEM the DRR is also negative but not significantly different from zero.

The research in the field of health effects of air pollution has been developing in the past two decades. The results, however, are still rather uncertain and we have shown how sensitive they are to the model specification and the estimation technique. As the results up to now of both US research and of our own study are rather uncertain it is questionable whether the monetary valuation of health damage based on such dose-response relationships makes sense.

REFERENCES

1. Chappie, M. and L. Lave, 1982, The Health Effects of Air Pollution: A Reanalysis, Journal of Urban Economics, 12, Bristol, pp. 346-376.
2. Crocker, T.D., W.D. Schulze, S. Ben-David and A.V. Kneese, 1979, Methods Development for Assessing Air Pollution Control Benefits, Vol.I, Experiments in the Economics of Epidemiology, EPA-600/5-79-001a.
3. Cropper, M.L., 1981, Measuring the Benefits from Reduced Morbidity, American Economic Review, 71, pp. 235-240.
4. Dickie, M. and S. Gerking, 1987, Benefits of Reduced Morbidity from Air Pollution Control: A Survey (paper for congress in Wageningen, The Netherlands).
5. Greene, W.H., 1990, Econometric Analysis, MacMillan, New York.
6. Greene, W.H., 1991, LIMDEP User's Manual and Reference Guide, Version 6.0, New York.
7. Hoek, G. et al., 1990, Effects of Air Pollution Episodes on Pulmonary Function and Respiratory Symptoms, Toxicology and Industrial Health, **6**, pp. 189-197.
8. Hoek, G. et al., 1993, Acute Effects of Ambient Ozone on Pulmonary Function of Children in The Netherlands, American Review of Respiratory Disease, **147**, pp. 111-117.
9. Jansen, H.M.A., G.J. van der Meer, J.B. Opschoor and J.H.A. Stapel, 1974, An Estimate of Damage Caused by Air Pollution in The Netherlands in 1970, Institute for Environmental Problems, Free University of Amsterdam (IvM-VU no. 8a), Amsterdam.
10. Jansen, H.M.A. 1980, Luchtverontreiniging en gezondheid - enige dosis effectrelaties voor gebruik in een economische waardering, Amsterdam (IvM-VU no. 116 and 116 A).
11. Judge, G.J., R. Carter Hill, W.E. Griffiths, H. Lutkepohl and T.C. Lee, 1988, Introduction to the Theory and Practice of Econometrics, second edition, John Wiley & Sons, New York.
12. Krupnick, A.J., W. Harrington and B. Ostro, 1990, Ambient Ozone and Acute Health Effects: Evidence from Daily Data, Journal of Environmental Economics and Management, 18, pp. 1-18.
13. Lave, L.B. and E.P. Seskin, 1971, Health and Air Pollution, Swedish Journal of Economics, 73, pp. 76-95.
14. Lave, L.B. and E.P. Seskin, 1977, Air Pollution and Human Health, Baltimore.

15. Lipfert, F., 1984, Air Pollution and Mortality: Specification Searches using SMSA-based data, Journal of Environmental Economics and Management, 11, pp. 208-243.
16. Mendelsohn, R. and G. Orcutt, 1979, An Empirical Analysis of Air Pollution Dose Response Curves, Journal of Environmental Economics and Management, 6, pp. 85-106.
17. OECD, 1981, The Costs and Benefits of Sulphur Oxide Control, Paris.
18. Ostro, B., 1983a, The Effects of Air Pollution on Work Loss and Morbidity, Journal of Environmental Economics and Management, 10, pp. 371-382.
19. Ostro, B., 1983b, Urban Air Pollution and Morbidity: A Retrospective Approach, Urban Studies, 20, pp. 343-351.
20. Ostro, B., 1987, Air Pollution and Morbidity Revisited: A specification Test, Journal of Environmental Economics and Management, 14, pp. 87-98.
21. Ostro, B., 1990a, Associations Between Morbidity and alternative Measures of Particulate Matter, Risk Analysis, **10**, pp. 421-427.
22. Ostro, B., D. Robert and L.G. Chestnut, 1990b, Transferring Air Pollution Health Effects Across European Borders, paper presented at Congress of European Association of Environmental and Resource Economists, Venice (Italy).
23. Portney, P.R. and J. Mullahy, 1986, Urban Air Quality and Acute Respiratory Illness, Journal of Urban Economics, 23, pp. 21-38.
24. Rijksinstituut voor Volksgezondheid en Milieuhygiëne, Luchtkwaliteit, Jaarverslag 1987, 1988 en 1989, Bilthoven.
25. Zuidema, Th., 1992, A Determination of the Benefits of Environmental Goods. An Application to the Abatement of Air Pollution, Proceedings of the 9th World Clean Air Congress in Montreal in 1992, Vol. 5, Pittsburgh.

SESSION G

INTEGRATED URBAN PLANNING

Influence of Land-use on Traffic and Environmental Impact of Traffic

B. van Wee [a] and B. van Bleek [b]

[a] Dutch National Institute of Public Health and the Environment
PO Box 1, 3720 BA Bilthoven, The Netherlands

[b] Dutch Ministry of Housing, Spatial Planning and the Environment
National Spatial Planning Agency
P.O.Box 30940, 2500 GX The Hague, The Netherlands

ABSTRACT

From the literature we know that the influence of land-use planning alternatives on passenger traffic and transport is potentially large. However, strong governmental policy will, especially at the 'higher than urban level', be required to arrive at a land-use situation where car use will be relatively low. One of the key issues in Dutch land-use policy at the urban level is the question about compact or diffuse urbanization patterns. A more compact pattern results in a lower level of car use and in some forms of environmental pressure on car use than a more diffuse pattern does. However, to have a significant impact, compact building has to be the guideline for urbanization over a long period of time. Apart from the direct effects, compact building will also offer better opportunities for future measures, such as car-free zones in (central) urban areas, improved public transport, stimulation to cycle and reduction of parking problems. In doing so local environment and liveability in cities are expected to improve.

1. INTRODUCTION

This paper will overview the possible impact land-use can have on traffic as well as the environmental impact of traffic itself. Specifically, Section 2 will survey the categories of land-use measures to reduce environmental pressure while Section 3 will review the literature on the possible influence of land-use on traffic. Section 4 will present scenario studies carried out by the Dutch Spatial Planning Agency of the Ministry of Housing, Spatial Planning and the Environment, with Section 5 giving possible effects of two scenarios ('compact' and 'diffuse') on traffic volumes and the environment. Finally, Section 6 will present the main conclusions.

2. LAND-USE MEASURES TO REDUCE ENVIRONMENTAL PRESSURE

Land-use measures can reduce environmental pressure in several ways, as outlined in Table 1.

Table 1
Categories of measures to reduce environmental pressure

Categories of measures	Example
1. Measures that reduce the volume of human activities producing environmental pressure, and so emission levels, or the structure of these activities.	Volume: the number of car kilometers can be reduced by a comprehensive planning of residential areas, working areas and infrastructure. Structure: land-use measures can influence the modal split (e.g. share of car and of public transport).
2. Land-use measures that influence the locations of human activities in such a way as to cause less environmental pressure.	If industrial activities are located far away from residential areas they will contribute less to local air pollution than if they are located within residential areas.
3. Land-use measures to influence the locations of the receptor.	If houses are built far away from existing motorways, concentrations of pollutants will be lower than if houses are built close to motorways.
4. Land-use measures related to the area between the source and the receptor.	If offices are built between a main road and a residential area, noise levels in this residential area will be lower than without these offices. Noise barriers; city parks and reducing concentrations of pollutants are also measures.
5. Another distribution of households within the same housing stock.	Students in city centers; families in suburban areas.
6. Spatial/technological measures to decrease pressure of human activities.	Building below the land surface, e.g. roads; vertical mixing of land-use; bedrooms in houses away from the roadside.
7. Compensating measures.	A city park close to areas with high environmental pressure.

The fifth point in the table deserves some explanation. It is related to the fact that not all people are affected by environmental pressure the same way. For example, students might experience less noise nuisance from cafes and bars in the city center than families with children, which means that changing the distribution of households within the housing stock could result in less nuisance.

To date we have not seen a comprehensive inventory of possible land-use measures for lowering environmental pressure, considering all the categories listed in Table 1. Although it is not the aim of this paper to fully describe the possibilities of these categories, we can conclude that because of the broad scope of the measures cited, both researchers and policy-makers can make more use of them in future (studies on) land-use policies.

Land-use can be studied on different spatial levels. Based on Van Wee (1993) we distinguish between the macro/meso level and the micro level. The macro/meso level is related to items like the position of the Randstad[3] within The Netherlands (macro) and the location of new urban areas (meso). The micro level is related to the location of individual households, companies and other institutions (forewith to be known as 'actors'). Besides this, we distinguish between a change in the function of land (e.g. the change from agricultural use of land to urban use) and a change of actors within a given land-use pattern (e.g. another distribution of households within the given housing stock or another distribution of companies within the given number of office buildings). See Table 2.

Table 2
Combining scale (macro or meso) with changes in land-use

	Change in function of land	Changes within a given land-use pattern
Macro/meso level	1	2
Micro level	3	4

[3] The highly urbanized western part of The Netherlands, including the provinces of Noord-Holland, Zuid-Holland and Utrecht.

Traditionally, Dutch land-use policy focuses mainly on cell 1 in Table 2. The so-called Location Policy that tries to get 'the right business at the right location' is an example of cell 3 (see Van Wee and van der Hoorn, 1996). Most literature on the influence of land-use on traffic and transport focuses on the macro or meso level, with hardly any on the micro level.

3. INFLUENCE OF LAND-USE ON TRAFFIC: AN OVERVIEW OF THE LITERATURE

3.1. Introduction

We used to distinguish between the macro/meso and the micro level. However, because so little literature related to land-use measures on the micro level is found[4], it is better to distinguish between land-use alternatives at the urban level and higher (macro/meso level) and land-use alternatives within the urban environment (meso level). This categorization has some overlap. Studies that consider urban forms and types of urbanization as part of land-use on a higher level are described in the first category, whereas studies only considering urban forms within a given urban environment is part of the second category. We have based Section 3 on mainly two references reviewing the literature[5].

3.2. An overview of scenario studies on the macro/meso level

Van Wee and Van der Hoorn (1997) review four land-use - transport scenario studies cited below that were carried out for (parts of) The Netherlands between 1986 and 1995:

> Strategic Study Randstad (De Jong *et al.*, 1986)

> Working group EROMOBIL (EROMOBIL, 1990)

> Land-use scenarios for The Netherlands (Clerx and Verroen, 1992)

---

[4] The work of Van Wee and Van der Hoorn (1996) as previously mentioned holds one of the expectations.

[5] Other recent references showing the influence of land-use on transport are Bolt (1982), Cervero (1996a, b), Cervero and Radisch (1996), Curtis (1996) and Newman (1996).

> Model calculations of Randstad Scenarios (Verroen *et al.*, 1995)

The studies were carried out using more or less 'traditional' transport models. In these models the study area contains many zones, each of which has (at least) inhabitants / household data and labor data. These models also include road and public transport infrastructure by nodes and links between these nodes by which networks are formed. The behavior of people/households is modelled by means of a mathematical formula. The relationships between residential and employment locations depends mainly on distances and/or travel times per mode, car ownership levels and travel costs. The models calculate the so-called long-term equilibrium.

Van Wee and Van der Hoorn (1997) state that it is difficult to compare the results of the scenario studies because of the differences in base year and future (scenario) years, differences in the assumed 'default' scenario and differences in the character of the land-use scenarios. Nevertheless, they were able to draw some general conclusions:

> Potential influence of land-use planning alternatives on traffic and on passenger transport is large. The differences in car use between the scenarios can be more than 20 per cent when related to homes and employment locations with different locations in the scenarios.

> Differences in mobility between these alternatives are caused mainly by the extent to which there is a balance between working and living in regions.

> The alternatives that result at the lowest level of car use differ greatly from the situation brought about by the Dutch policy of the past 20 years. A strong governmental policy is therefore required to arrive at a land-use situation where there is a relatively low level of car use. Such a policy will result in strong opposition from some of the relevant 'actors'.

> Further urbanization of the so-called 'Green Heart' of The Netherlands can be beneficial from a transport point of view.

> Urbanization on the Ring of the Randstad and along the axes results in a relatively low level of car use, especially in the case of mixed land-use.

> Most scenario studies use only 'traditional' indicators to compare scenarios, such as car kilometers and public transport passenger kilometers. The studies do not use indicators for the total 'costs' (including external costs) and the 'benefits' of the

transport system.

To demonstrate the possible impact of land-use on transport, we will use the study of Verroen and Hilbers (1995). Although only six per cent of houses have different locations in the scenarios of 2015, the difference in car use is two per cent. This difference is not the largest possible: scenarios that are seen as 'undesirable' (e.g. scenarios with a very diffuse pattern for new urbanization or scenarios with the urbanization of areas that lack proper public transport) were not considered. So, relatively speaking, the possible impact of land-use on car use is very large.

3.3. An overview of the literature on the meso level: urban form, traffic and the environment

We will use a recent article of Anderson *et al.* (1996) for this section, with all the references taken from this work. Anderson *et al.* have reviewed a great deal of literature on urban form, energy and the environment. Some of the conclusions of the studies cited are listed below:

> Compact urban land-use patterns result in less car use than more disperse patterns.

> The least-desirable form of urban development from an environmental perspective is low-density sub-urbanization (see Roseland, 1992).

> Mixed land-use patterns result in less car use than non-mixed land-use patterns.

> Land-use patterns significantly influence possibilities for public transport and public transport use.

> It is desirable to preserve open spaces with relatively dense vegetation within urban areas. In addition to their recreational and aesthetic value, these spaces play a significant role in improving local air and water quality (see Spirn, 1984).

> There are large differences between the possibilities expected to reduce transport by land-use policies; some authors are very optimistic, others very pessimistic. The differences are mainly related to what the observers consider to be realistic changes and policies.

> Many of the studies are or can be criticized on grounds of methodology.

We conclude that the studies reviewed by Anderson *et al.* show that urban land-use patterns can significantly influence mobility in general, and car use too, especially if not only 'traditional' urban land-use patterns are concerned.

4. RECENT DUTCH SCENARIO STUDIES

In this section we will discuss two recent studies, RUIMPAD and The Netherlands 2030, in which the National Spatial Planning Agency of the Ministry of Housing, Spatial Planning and the Environment is involved.

RUIMPAD

RUIMPAD (VROM/V&W, 1997) is a long-term survey for the period up to 2050. It is aimed at support of future national land-use and transport policy of the Dutch government. Measures under category 1 in Table 1 and cell 1 in Table 2 dominate the survey. The most important measures having positive environmental and/or spatial effects are:

- urbanization in compact medium-sized towns near public transport centers.

Compact building and mixed land-use result in short travel distances for daily to weekly services. Cycling and walking can be stimulated in combination with a high level of infrastructure for slow transport modes. This strategy can also be adopted in existing urban areas and cities.

- spatial integration and connection of different modalities.

Spatial and organizational integration of infrastructures for different modes allow people to make easy use of different transport modes, e.g. building transferee, combining infrastructure for cars, public transport (national rail system; local/regional public transport) and slow modes (cycling, walking) at the edges of cities. All this can significantly reduce local car use and so improve local environmental quality.

- land-use measures to improve transport to and from the main public transport mode.

Public transport will have a higher market share if transport to and from the railway stations (and other high-level public transport nodes) does not take much time (short egress and access times). Compact building can reduce egress and access times. Improving infrastructure for slow-traffic modes and local public transport (resulting in reduced travel times and better quality) can improve the competitiveness of public transport. Besides, mixing land-use near public transport nodes gives better opportunities to combine longer distance trips by train with other destinations, such as shops.

The Netherlands 2030

At the time of writing this paper 'The Netherlands 2030', a survey to show the possible impact of long-term spatial planning dilemmas, was still in progress. Dilemmas on air pollution are:

1. Compact of scattered settlement patterns?
2. How to deal with demand for mobility without a loss in environmental quality?
3. How can we find the right balance between international economic developments and the environment?

To answer these (and other) questions, four scenarios have been created, with some of their characteristics shown in Table 3. An evaluation of the scenarios is to take place.

Table 3
Some characteristics of the scenarios for 'The Netherlands 2030' categorized according to urbanization, mobility, sustainable economy and setting

	Urbanization	Mobility	Sustainable economy	Setting
'Land of cities'	compact	public transport	knowledge and culture	large-scale
'Landscape park urbanization'	interwoven	individual transport	footloose activities	'cultural' natural
'Urbanization along streams'	water, road and rail axes as leading principles	public and multi-model	mainports and distribution	wetlands
'Scattered urbanization'	freedom of settlement	individual transport	freedom of settlement	consumer natural

Results for 'transport and the environment' are not available yet.

5. COMPACT OR DIFFUSE LAND-USE PATTERNS: WHAT'S THE DIFFERENCE?

5.1. Introduction

Section 4 concludes that one of the main current policy issues is the one on compact or diffuse building. On the basis of the literature and policy discussions as mentioned above, we calculated possible effects of two land-use scenarios in comparison with a reference scenario[6]. The scenarios are:

> a reference scenario, representing current land-use policies for the period up to 2020
> a compact scenario: 10% of new dwellings to be built between 2010 and 2020 are additionally located within existing urban areas. New residential areas outside existing urban areas have relatively high densities (in the Randstad provinces: 30 dwellings per acre; outside the Randstad provinces: 25 dwellings per acre)
> a diffuse scenario: 14% of new dwellings to be built between 2010 and 2020 and located in the Randstad in the reference scenario will be built in the provinces of Noord-Brabant and Gelderland (southwest and west of the Randstad provinces, respectively), with a density of 25 dwellings per acre.

The study area contains the Randstad provinces, Noord-Brabant and Gelderland. According to current demographic projections we assume that between 500,000 and 550,000 dwellings have to be built in the period 2010 to 2020 (about 7.5% of all dwellings in The Netherlands in 2020). The scenarios differ with respect to both the location of dwellings and the so-called 'critical design dimensions', as presented by Verroen (1994) and Verroen and Hilbers (1996) and outlined below:

- Single-core as opposed to multi-core orientation: Further urbanization through the development of separate metropolitan areas or assuming the development of networks of cities with the idea of spreading the daily patterns of activities out over several metropolitan areas.
- Clustered as opposed to dispersed: Development of large-scale, urban extensions with emphasis on the main infrastructure vs. a development of small-scale urban extensions with emphasis on the secondary.
- Mixing as opposed to separating: Mixing or keeping all kinds of spatial functions separate, such as living, working and facilities in various sub-areas within the urban areas (thus as the 'small municipality' or as 'metropolitan district/central

[6] The authors would like to thank Mariëlle Damman, Hans Eerens, Karst Geurs, Marianne Kuijpers, Hans Nijland and Brigit Staatsen for their contribution.

municipality').

These design dimensions are combined with the compact and diffuse scenarios. The compact scenario has a single-core orientation, clustered urban development and mixed land-use. The diffuse scenario has a multi-core orientation, dispersed urbanization and activities which are not mixed.

Mobility effects of the scenarios are based on the TNO-INRO reports that describe the study 'Model calculations of Randstad Scenarios' (Verroen *et al.*, 1995), using interpolation and extrapolation techniques. Calculations of pollutant concentrations are based on the method as described in Van Wee *et al.* (1996). Calculations of noise nuisance levels are based on dose-response functions as given in the literature. Other variables in the table below are based on expert judgement.

Table 4 gives indicative results of the comparison between the compact and diffuse scenarios using the reference scenario.

Table 4
Indicative results of the scenarios
(index reference scenario = 100)

			Compact	Diffuse
Urban land-use	national		< 100	> 100
	Randstad		< 100	< 100
	Brabant/ Gelderland		< 100	>> 100
Car use	national		99	101
	Randstad	within built-up area	101	98
	Brabant/ Gelderland	within built-up area	99	106
Noise nuisance	national		100	100
	Randstad	within built-up area	100	100
	Brabant/ Gelderland	within built-up area	100	>100
Local air pollution (concentrations of pollutants)	national		101	99
Safety (number of accidents)	national		> 100	< 100
	Randstad	within built-up area	> 100	< 100
	Brabant/ Gelderland	within built-up area	< 100	> 100

Source: RIVM, partly based on Verroen *et al.* (1995)

The compact scenario results in a lower level of car use and related emissions on a national scale, and in a lower level of overall urban land-use. Urban land-use is higher in the diffuse scenario, but the claim for urbanization moves towards Noord-Brabant and Gelderland, where pressure on land-use is lower than in the Randstad. The scenarios hardly differ with respect to noise nuisance. In the diffuse scenario noise nuisance is marginally higher in Noord-Brabant and Gelderland. Compact building leads to more road accidents than diffuse building if the traditional form of urbanization is assumed. However, additional measures can improve road safety in such areas (Hilbers, 1996).

The overall impression might be that from an environmental point of view it hardly makes any difference whether a more compact or a more diffuse form of urbanization is implemented. However, we think this conclusion is incorrect for two reasons: Firstly, the variation between the locations and densities of all dwellings in 2020 will be only a few percentage points, so that even with large relative implications, the overall differences cannot be high. Secondly, compact building offers better opportunities for future measures, such as car-free zones in (central) urban areas, improved public transport, stimulating cycling, reducing parking problems etc. If such measures are implemented, compact building will probably be rated better for all indicators in Table 2. In other words, compact building does not necessarily reduce environmental pressure, but gives better opportunities for future measures to improve the local environment and liveability in cities.

6. CONCLUSIONS

1. There are several categories of possible land-use measures: measures to reduce volume of activities, measures related to the polluter, the receptor or the intermediate zone and measures to redistribute population within the existing housing stock. We have not yet seen a comprehensive inventarization of possible land-use measures that can contribute to lower environmental pressure taking all categories into consideration.
2. To date, national Dutch land-use policy has mainly focused on changes in land-use at the meso or macro level (the urban level and higher). Hardly any attention has been paid to the micro level (individual actors, such as companies and households) and to possibilities for redistribution among individual actors within the existing urban structure.

3. Potential influence of land-use planning alternatives on traffic and transport of persons is large. However, strong governmental policy is required to arrive at a land-use situation that will result in a relatively low level of car use. Such a policy will result in strong opposition of some actors.
4. Further urbanization of the so-called 'Green Heart' of The Netherlands and urbanization of the Randstad Ring, and along the axes, results in a relatively low level of car use, especially for mixed land-use.
5. It is desirable to preserve open spaces in urban areas with relatively dense vegetation. In addition to their recreational and aesthetic value, these spaces play an important role in improving local air and water quality.
6. A more compact form of urbanization and mixed land-use will result in a lower level of car use and some kinds of environmental pressure from car use than with a more diffuse pattern of urbanization. However, to have a significant impact, compact building has to be the long-term guideline for urbanization.
7. Compact building offers better opportunities for future measures, such as car-free zones in (central) urban areas, improved public transport, stimulating cycling, reducing parking problems etc. and in so doing will improve local environment and liveability in cities.

REFERENCES

1. Anderson, W.P., P.S. Kanaroglou, E.J. Miller (1996), Urban Form, Energy and the Environment: A Review of Issues, Evidence and Policy, Urban Studies, Vol. 33, No. 1, 7-35
2. Bolt, D. (1982), Urban form and energy for transportation. Delft: Research center for physical planning, Organization for Applied Scientific Research (TNO).
3. Cervero, R. (1996a) Transit-Based Housing in the San Francisco Bay Area: Market Profiles and Rent Premiums, Transport. Quar., Vol. 50, No. 3, 33-49
4. Cervero, R. (1996b), Mixed land-uses and commuting: evidence from the American housing survey, Transportation Res.-A, Vol. 30, No. 5, pp. 361-377
5. Cervero, R., C. Radisch (1996), Travel choices in pedestrian versus automobile

oriented neighborhoods, Transport Policy, Vol. 3, No. 3, 127-141

6. Clerx, W.C.G., E.J. Verroen (1992), Land-use scenarios for The Netherlands: transport patterns and environmental impact. Delft: INRO-TNO (in Dutch)
7. Curtis, C. (1996), Can strategic planning contribute to a reduction in car-based travel? Transport Policy, Vol. 3, No. 1/2, 56-65.
8. De Jong, M.A., L.H. Immers, J.W. Houtman, C.W.W. van Lohuizen (1986), Strategic Study Randstad. Delft: INRO/TNO (in Dutch).
9. EROMOBIL (1989), Interaction between land-use and mobility in the Randstad (Report of the EROMOBIL working group - in Dutch).
10. Hilbers, H. (1996), Urbanization and road safety, Delft: TNO-INRO (in Dutch).
11. Newman, P. (1996), Reducing automobile dependence, Environ. Urban., Vol. 8, No. 1, 67-92.
12. Roseland, M. (1992), Toward Sustainable Communities: A Resource Book for Municipal and Local Governments. Ottawa: National Round Table on the Environment and the Economy.
13. Spirn, A.W. (1984), The granite Garden. Urban Nature and Human Design. New York: Basic Books.
14. Van Wee, G.P. (1993), Location Policy and Land-use: the Effects on Traffic and Transport. Literature Study. Bilthoven: National Institute of Public Health and the Environment. Report nr. 251701010 (in Dutch).
15. Van Wee, G.P., K. Van Velze, H.C. Eerens (1996), Traffic and local air pollution: a forecasting method applied to The Netherlands, Paper prepared for the symposium 'Verkeer en Milieu: een duurzame tegenstelling?', Amsterdam, 29-10-1996.
16. Van Wee, B., T. van der Hoorn (1996), Employment location as an instrument of transport policy in The Netherlands. Fundamentals, instruments and effectiveness. Transport Policy, Vol. 3, No. 3, 81-89.
17. Van Wee, G.P., T. van der Hoorn (1997), The Influence of Land-use on Traffic and Transport: A Comparison of Scenario Studies. Tijdschrift Vervoerswetenschap, 1/97, 43-61.
18. Verroen, E.J. (1994), Accessibly proximity. An investigation of mobility friendly forms of urbanization of the ring of towns in the Center of The Netherlands after 2005. Delft: INRO-TNO (in Dutch).
19. Verroen, E.J., H.D. Hilbers, C.A. Smits (1995), Model calculations of Randstad Scenarios. Delft: TNO-INRO (in Dutch).

20. Verroen, E.J., H.D. Hilbers (1996) Promising mobility friendly urbanization strategies: time for a paradigm shift? Paper presented at the PTRC summer annual meeting, London, 1996.
21. VROM/V&W (1997) (Ministry of Housing, Spatial Planning and the Environment (VROM), Ministry of Transport and Public Works and Water Management (V&W), Choosing for space to move. Exploring the future of mobility, urbanization and transport networks. The Hague: Ministry of Housing, Spatial Planning and the Environment, Spatial Planning Agency.

Environmental Planning and the Compact City
A Dutch Perspective

Gert de Roo

Department of Planning and Demography, Faculty of Spatial Sciences
University of Groningen, P.O. Box 800, 9700 AV Groningen, The Netherlands

1. INTRODUCTION

In The Netherlands the compact city policy was welcomed as a spatial concept during the mid eighties. The compact city concept meant a strengthening of the city as a place to live and to work in. It should also be the answer to the two most important problems the economic hart and denser populated part of The Netherlands is facing; a fast urbanization of open space and a continues increase of mobility. Therefore it was hoped that the compact city will not only contribute to the spatial quality of the urban area and the country side. It should also have a positive effect on the environmental quality, partly because of an expected reduction of traveling distances. Although the compact city concept was embraced as the answer to major issues urban planners had to deal with in the eighties, urbanization and mobility are today more than ever issues to be dealt with! Now, in the nineties, Dutch planners have to locate more than a million houses which have to be build in the near future. This enormous task confronts Dutch planners with dilemmas of the compact city concept, which is basically a conflict between spatial and environmental policy making.

2. THE COMPACT CITY CONCEPT

During the mid eighties 'concentration' became a leading factor in Dutch urban planning. Concentration was preceded by periods of deconcentration and concentrated deconcentration (Faludi and Van der Valk 1994). Deconcentration during the sixties proved to be too encouraging to (sub)urbanization of the country side. In the seventies concentrated deconcentration had to be the answer, but two major oil crises proved different. The concept of the seventies was too expensive to maintain, and was converted into a policy of spatial concentration (Van der Cammen and De Klerk 1993). In The Netherlands and beyond, the concentration policy became known as the compact city policy.

The National Physical Planning Council, the planning department of the Dutch Ministry of housing, physical planning and environment, defined the compact city policy as "(more than before) aiming at concentration of functions (living, working, provisions) in the city" (NPPC 1985). The compact city concept was expected to influence not only spatial developments. A positive effect on the environmental quality was expected as well.

For historical reasons Dutch cities are already reasonably compact. Till mid nineteenth century Dutch cities were enforced by law to have defence walls, which contributed highly to the compactness of inner cities. Also the long tradition of city planning in The Netherlands explains the relative compactness of Dutch cities. In the eighties scarcity of space made planners (again) aware of the continues need for compactness in urban planning, which meant in fact a revival of the compact city.

Although scarcity of space is an important reason to adapt a compact city policy, the concept deals with the structuring of functions within the city as well. The compact city concept can be characterized as follows (Bartelds en de Roo 1995):

- intensive use of existing urban areas;
- concentration instead of dispersion of functions;
- mixing instead of separating functions;
- building in high densities.

Dutch urban planning policy is "aiming for a good living and production climate, the utilization of the available capacity of the urban area for housing, working, recreation and care taking and the mix use of these functions" (VROM 1983; 10). Compactness is the key word in this policy, and is focussed on:

"- support of urban capacity;
- mobility reduction;
- support of bike and public transport;
- limiting the urbanization of the country side" (VROM 1993; 6).

Closeness and accessibility became important criteria for new urban developments. In 1990 the Fourth physical planning report Extra part one (VROM 1990) was published, giving a perspective of new urban developments till the year 2015.

Although compactness was seen as the only acceptable answer to stop further urbanization of the country side the number of houses that has to be build till the beginning of the next century is phenomenal and it will be impossible to have this all located in existing urban areas.

Part one of the Fourth report talks about 835.000 houses to be build between 1995 and 2015 (VROM 1990; 25). After the Fourth report part one was published the estimated number of houses to be build only increased. The latest estimates are above one million houses! (VROM 1993) Knowing that the population size in The Netherlands is about 15 million inhabitants and is growing only marginally one might conclude that the Dutch are building mainly for smaller households. In Amsterdam, the capital of The Netherlands, most of the households are single person households.

Closeness is probably the most important compact city criteria (Bartelds and De Roo 1995). Based on closeness to the city center priority is given to where new developments have to take place. In order of priority housing developments have to take place within existing cities, next to existing cities, and further away in connection to existing urban areas (Needham et al. 1994; 29). About one third of the housing program has to be worked out within cities. Roughly another third of the program will lead to a further expansion of the city, because housing developments will lead to the construction of new neighborhoods as well. Also the last one third will contribute to a further urbanization of the country side.

The four main urban areas accommodating the cities Amsterdam, Rotterdam, The Hague and Utrecht are all located in the spatial and economic heart of The Netherlands called the Randstad. The Randstad is located around the so-called Green Heart, which is preserved open space and used for agricultural, ecological and recreational purposes (Faludi and Van der Valk 1994). The four urban areas being part of the Randstad have the biggest concentration of population and are the economic centers of The Netherlands. Commuting issues and accessibility difficulties in these areas are hard to deal with and are daily topics in the news. And, of course, the main part of the housing program has to be worked out here.

3. DILEMMAS OF THE COMPACT CITY

The more compact a city will be the closer functions are located near each other. From a spatial point of view this might be seen as a contribution to the variety and multifunctionality of a city. Though, from an environmental point of view a compact city might lead to considerable conflicts between environmentally intrusive activities and environmentally sensitive functions. The more intrusive a function or activity is the further away it should be located from sensitive functions. This means that intrusive (urban) activities simply need more space than the site where these activities take place. Some distance should be taken into account to keep these functions away from environmental sensitive areas, such as residential areas. Nowadays there is political pressure to build as much as possible within the existing boundaries of a city. Understandably urban planners will find it difficult to include environmental constraints in their spatial plans, facing the task to build 'compact'.

Dutch compact city policy is confronted with spatial and environmental policy dilemmas which can be a constraint to spatial and economic developments in urban areas, and which will effect the environmental quality of these areas in a negative way (Bartelds and De Roo 1995, VROM 1993b). Issues that are specially related to these dilemmas are inner city activities, such as industry and recreation, the restructuring of former industrial sites, soil pollution, inner city traffic and transport, the use of inner city open space, and so on. Most of these issues are about environmental spillovers which are in conflict with environmental rules to maintain a reasonable quality in residential areas (De Roo 1993).

Far the most of the dilemmas of the compact city are more or less related to environmentally intrusive activities and their (environmentally sensitive) surroundings or are related to (constraining) intentions to planning sensitive functions on locations facing an amount of environmental load.

Amsterdam, the capital of The Netherlands, is asked by the Dutch government to build 20.000 houses in the coming years (VROM 1990; 45). Amsterdam is surrounded by a green belt, water, industrial sites and the Amsterdam Schiphol Airport, and is therefore limited to expand. To prevent urban sprawl a new neighborhood (15.000 houses) will be build by reclaiming land from the IJ lake opposite to the city center (Amsterdam 1994). Another attractive site for city development are the old and neglected harbors called the Houthavens on walking distance from downtown Amsterdam (see figure 1). The restructuring of the Houthavens from low quality industrial and transport activities into multifunctional and housing activities will improve the spatial quality of this part of Amsterdam to a great extent (Amsterdam 1994b). It will also be a positive response to the compact city policy, and it will give Amsterdam a new waterfront which will effect the image of Amsterdam as a whole. Unfortunately opposite to the Houthavens the industrial site Westpoort is responsible for a number of environmental spillovers. According to the current environmental standards both odor and noise in the Houthavens are too high to allow residential use of the area, and are stagnating the execution of the spatial plans (Amsterdam 1995). At the moment Amsterdam is suggesting to redefine environmental quality (and therefore environmental restrictions) by using more indicators in addition to environmental load, and to introduce 'compensation' as a possibility to create at least an acceptable quality to live with (Amsterdam 1994b).

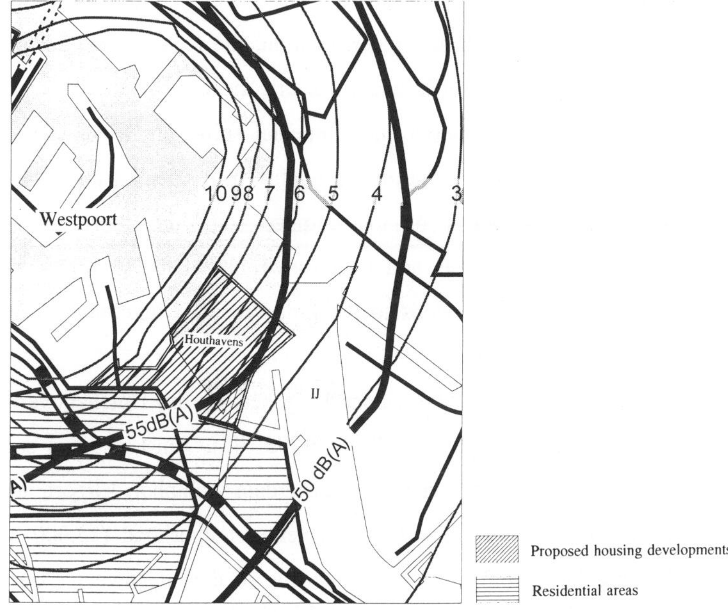

Figure 1. The Amsterdam Houthavens Housing project is located within the environmental zones for noise (< 50 dB(A) = acceptable, > 55 dB(A) = unacceptable) and odor (< 1 g. 99.5% = acceptable, > 10 g. 98% = unacceptable).

More or less the same situation can be found in Rotterdam, having one of the biggest harbors in the world within its city boundaries. Major inner city developments are on going to locate about 16.000 houses inside Rotterdam its boundaries (VROM 1990; 49). Most of the developments are concentrated in locations like Kop van Zuid and Katendrecht, old harbor areas near and opposite to the city center. In part of these harbors industrial and transport activities are still taking place. These activities are interfering with the plans of the municipality, who now wants to relocate most of these activities to the Waal-/Eemhaven, a site further away and more suitable for harbor related activities. Unfortunately the environmental zone for noise around the Waal-/Eemhaven does not allow a further rise of the noise level produced in the Waal-/Eemhaven (see figure 2). The municipality of Rotterdam now argues that an increase of noise by 5 dB(A) should be allowed because the net results would make it worthwhile

(Rotterdam 1994). Rotterdam was hopeful that the increase of noise, which has a negative effect on residential areas located near the Waal-/Eemhaven, will lead to a reduction of environmental spillovers in the redevelopment areas, allowing thousands of houses to be build. Although a number of uncertainties are still unanswered the national government accepted this reasoning, and changed the Noise abatement act for this and similar cases (Second Chamber 1991).

Figure 2. Rotterdam wants to expand activities in the Waal-/Eem harbor, which will influence the noise output of the harbor area and effect residential areas (55h = actual load, 55t = expected noise level).

Not only Amsterdam and Rotterdam, the largest cities in The Netherlands, are having difficulties to find suitable locations for their urban development plans. In Arnhem, a city of 200.000 inhabitants, environmental zones are such that construction of housing should be prohibited everywhere within the city (Boei 1993). That is according to environmental standards for toxic substances and odor. This situation is seen as highly unacceptable, because Arnhem too has to take its part in the immense housing program of The Netherlands. The Drechtsteden, a cluster of cities south east of Rotterdam and one of the main economic centers of The Netherlands, has a number of potential locations which might be turned into attractive residential areas. A well though-out integrated environmental scan of the area showed however major 'black areas'. The research was based on a number of environmental aspects, such as noise, odor, carcinogenic and toxic substances, external safety and soil pollution (Voerknecht 1994). Groningen, the capital of northern Netherlands, also wants to take its share in the Dutch housing program. And as other municipalities Groningen is favoring the compact city policy. A number of abandoned sites within its city boundaries are appreciated as locations where housing development can take place if not the soil was heavily polluted. A lack of financial means is one of the main reasons preventing spatial developments in the near future. At the moment Groningen has started building houses in the country side, on the edge of the city (Groningen 1996).

This five examples are given as examples of Dutch dilemmas of the compact city policy. From these examples the following conclusion can be made: In almost all of the cases where conflicts arise between spatial wishes and environmental restrictions a densening and/or a change of functions or a change of (size of) activities takes place or is at hand.

In The Netherlands about all spatial change of functions takes place through government planning. When planning spatial developments environmental issues have to be taken into consideration. The measuring and mapping of environmental loads (mostly from industry, traffic and leisure activities) might lead to the conclusion that The Netherlands is far more polluted than expected. Unfortunately most of the pollution can be found in and around cities. And cities are at the moment seen as places where major developments should take place to prevent urbanization of the country side and to fulfill the need for housing in the near and far future.

A number of developments which lead to a densening and/or a change of functions or an expansion of activities give rise to stagnation or can't take place at all, because the actual or expected environmental quality is seen as unacceptable. Often these developments are desirable from a spatial point of view.

Dilemmas of the compact city do occur because seemingly two policy intentions can't be fulfilled easily at the same time. One of the intentions is to keep cities as compact as possible. The other intention is to maintain at least an acceptable environmental quality, a quality that is based on the actual or expected environmental load.

4. THE TRADITIONAL WAY OF DUTCH ENVIRONMENTAL POLICY MAKING

In the beginning of the seventies Dutch policy makers became convinced that corrective and ad hoc policies were no longer enough to deal with environmental issues. In the Urgency report from 1972 (VM 1972) a strategy was developed based on compartment wise sanitation of the environment. This more or less meant a sectoral policy approach for soil, water, air, noise et cetera. For all these environmental sectors aims, goals and objectives were created, including quantitative standards based on dose/effect relations.
During the eighties the compartment wise approach to deal with environmental issues was put aside. This approach was found not effective enough. Instead of a reduction of environmental pollution a diversion of pollution was seen between the compartments. A more integrated approach had to be the answer to this problem. This integrated approach got shape on the basis of environmental themes, such as acidification, climate change, drought, eutrophication, disturbance and so on. Based on these themes or environmental topics policy was made to deal with sources of pollution and its negative effects. This meant policy making concentrating on the polluters or target groups (source), such as industry, agriculture, consumers etc. and policy making to protect environmentally sensitive areas (effect) (VROM 1989, VROM 1993b). This integrated approach of environmental policy making proved to be more effective than a compartment wise approach.
What didn't change during the eighties was the use of environmental zoning. On the contrary, after a more or less successful introduction of the Noise abatement act in 1979, the national government was convinced that an expansion of the environmental standard program could be helpful to separate environmentally intrusive activities and environmentally sensitive land uses.
Quantitative environmental standards are relatively easy translated in spatial zones,

pointing out areas where the environmental load is seen as unacceptable (within the zone) and where environmentally sensitive areas, functions or activities can be accepted (outside the zone).

The comfort of using standards and zoning almost logically leads to a new and innovative initiative at the beginning of the nineties. An integrated program for environmental zoning was introduced by the national government, taking into account different environmental loads, such as noise, odor, external safety and air pollution, at the same time (VROM 1990b, Van der Gun and De Roo 1994).

The popularity of the environmental standard was not solely based on the easy, almost mathematical way it can be translated into spatial zones. There is definitely another reason at least as important. Standards and their translation into zones fitted very well in the environmental policy hierarchy that existed already from the moment environmental policy was introduced in The Netherlands. Since the beginning environmental policy was initiated and worked out on national level, while the implementation was left to local and region authorities. Environmental standards therefore were introduced on national level, leaving the translation into spatial zones, the abatement of local intrusive sources, the issuing of permits and the implementation of the spatial environmental zones in the local land use plans to the provincial and municipal government (Borst et al. 1995).

In this centrally based system environmental standards became a strict framework where other policy sectors such as spatial or physical planning should remain within.

This all goes well as long as there is not too much difficulty on local level implementing the standards, which will lead to a sustainable separation between intrusive and sensitive functions. From the beginning of the nineties on it became clear that - unfortunately - this was no longer the case (Miller and De Roo 1996). The compact city policy was interfering.

5. NEW STRATEGIES IN DUTCH ENVIRONMENTAL PLANNING

Although the number of critics against the centrally oriented environmental standard system is growing, the analysis of a number of cases proved that these standards can be implemented most of the time without a decline of spatial quality (Borst et al 1995). There are also a number of cases where environmental standards can be implemented with a bit of struggle and an acceptable portion of financial input. Knowing these conclusions one might conclude that the critics are maybe less interested in the environmental quality of an area, and that their critique should be explained by the wish to gain more political control on local level; more political control on local level means a less tight environmental framework and more freedom for local decision making. There can be doubts about the way critics are against the system of environmental standards, on the other hand there is no doubt that there are a handful projects in The Netherlands where standards are not the right answer to deal with the unacceptable environmental situation.

In these projects, like Arnhem and the Drechtsteden, large parts of residential areas have to be demolished if standards and zones are implemented. Seen from a spatial, social and economic point of view this situation is unacceptable (Borst et al 1995).

Looking at the analysis above, where three types of conflicts are identified based on the usability of standards, *complexity of the local situation* might be the key to new and workable strategies solving spatial and environmental conflicts in compact areas. The more complex a situation becomes the more difficult it will be to implement standards without effecting other than environmental qualities of an area, such as spatial, social and economic qualities.

Environmental conflicts can be seen as complex when there is a multiple mixture of loads, from different sources, having an above local dispersion, which will have consequences for existing functions and spatial developments if no substantial actions are taken.

When environmental conflicts in compact cities are seen as extremely complex, and when costs/benefit analysis makes clear that actions to be taken from an environmental point of view based on centrally imposed standards are far beyond reason, and when other than environmental qualities will be effected negatively, some excessive load should be accepted. In these cases the excessive load in an environmentally sensitive area should be compensated effecting the livability of that area in a positive way. This is the latest change in Dutch environmental policy system, which means in a way the acceptance of the complex reality of the compact city by the national government (VROM 1995).

Environmental standards will remain an important instrument for creating a sustainable separation between intrusive and sensitive functions, but no longer it will dominate spacial developments entirely. By this change of national policy the idea is accepted that local authorities rather then the national government have a better idea of spatial and environmental ins and outs of locally complex and often unique situations. This also means that local authorities should decide, in consultation with higher authorities, which environmental framework is acceptable taking into account the developments that are needed in the compact city.

Lately, this change in responsibility has become more or less official. A few more, but less official changes dealing with environmental policies can be seen in the Dutch political landscape as well. One of the changes is a widening of the definition of environmental quality. Environmental quality as used above can be seen as the direct translation of the negative environmental load in an area. Policy based on this kind of quality measurement is highly restrictive.

The idea is that additional measures taken to improve the livability of an area or which will contribute to sustainability should be taken into account as well. This change in defining environmental quality should lead to a more progressive form of environmental policy making.

Also the introduction of an objectives approach next to the standard approach can be seen as an interesting innovation (Miller and De Roo 1996). So far environmental issues with a spatial effect were seen as a source/effect relationship where standards are used to translate the environmental issue into space consuming zones. Every source/effect relationship was seen independent from other conflicts and policy programs. When policy is developed on the basis of an objectives approach each environmental issue is no longer standing on its own, but will be taken into a wider perspective (Amsterdam 1994b). In this case standards are no longer strict frameworks for spatial initiatives, but are replaced by broadly defined objectives determining the environmental outcome. This could mean that the environmental quality is not a target to be reached in the short run, but could very well be a target that can be reached through spatial planning, taking more time but creating more opportunities as well. An objectives approach does not only create more local flexibility, but takes the context of the environmental conflict into account as well. Of course it can't be expected that the objectives approach can replace the standard approach including its positive aspects. But it might help defining a more comprehensive policy where both approaches might function next to each other, supplementing each other or replacing each other, depending on the local situation.

6. CONCLUSIONS

Policy dilemmas in Dutch compact cities do occur because enormous housing programs within the city limits are not easy to work out while taking environmental restrictions into account. The dilemmas are (partly) a conflict between spatial and environmental policy making. The environmental policy system in The Netherlands is highly central oriented, with a strong focus on environmental standards. Therefore spatial developments have to take place within a strict environmental framework. Unfortunately strict and generally used environmental standards are not always the best instruments to solve environmental conflicts in urban and compact areas. The local and often unique character of these conflicts have to be taken into account, especially when the complexity of the conflict is high. More responsibility should be given to local authorities:

- to benefit from local knowledge when solving environmental conflicts;
- to find solutions geared to local and unique situations;
- to have the possibility to include the conflict in a local context;

- to be able to define quality targets which can replace environmental quality targets if proven unfeasible.

The number of cases are limited in which the environmental standard approach will have to be replaced by a locally defined objectives approach to get a better outcome for conflicts to be solved. In other cases a locally defined objectives approach might prove to have a surplus value next to or supplementing the traditional and central based standard approach. If local authorities choose to work on the basis of an objectives approach undoubtedly they will have to put far more attention to strategic planning than they were used to when working with a centrally based standard approach. Strategic planning for spatial and environmental policy on a local level means local protection through local planning and development, which might very well lead to new developments in dealing with dilemmas of the compact city.

REFERENCES

1. Amsterdam (municipality) (1994) Ontwerp structuurplan (Concept structure plan) Physical Planning Department, Amsterdam.
2. Amsterdam (municipality) (1994b) Ontwerp beleidsnota Ruimtelijke Ordening en Milieu (Concept policy report Physical Planning and Environment), Physical Planning Department and Environmental Department, Amsterdam.
3. Amsterdam (municipality) (1995) Stedebouwkundig programma van eisen, Houthavens (Urban development program of requirements, Houthavens), The municipality, Amsterdam.
4. Bartelds, H., G. de Roo (1995) Dilemma's van de compacte stad, uitdagingen voor het beleid (Dilemmas of the compact city, challenges to policy developments), VUGA, The Hague.
5. Boei (1994) Arnhem, integrale milieuzonering op en rond het industrieterrein Arnhem-Noord (Arnhem, integrated environmental zoning in and around the industrial site Arnhem-Noord), In G. de Roo (ed) Kwaliteit van Norm en Zone (Quality by Standards and Zoning), Geo Pers, Groningen.
6. Borst, H., G. de Roo, H. van der Werf and H. Voogd (1995) Milieuzones in

Beweging (Environmental Zones on the move), Samsom H.D. Tjeenk Willink, Alphen aan den Rijn.

7. Faludi, A., A. van der Valk (1994) Rule and order: Dutch planning doctrine in the twentieth century, Kluwer Academic Publishers, Dordrecht.
8. Groningen (municipality) (1996) In Natura, milieubeleidsplan (By Nature, environmental policy plan), The municipality, Groningen.
9. Gun, V.E. Van der, G. de Roo (1994) An Integrated Environmental Approach to Land-use Zoning, In H. Voogd (ed) Issues in Environmental Planning, Pion, London.
10. Miller, D., G. de Roo (1996) Integrated Environmental Zoning; an innovative Dutch approach to measuring and managing environmental spillovers in urban regions, Journal of the American Planning Association (JAPA), pp 373-380.
11. Needham, B., T. Zwanikken, J. Mastop, A. Faludi, W. Korthals Altes (1994) Evaluatie van het VINEX-verstedelijkingsbeleid (Evaluation of the VINEX urbanization policy), Dutch Ministry of Housing, Physical Planning and Environment, The Hague.
12. NPPC (National Physical Planning Council) (1985) De compacte stad gewogen (the compact city evaluated), Study reports no. 27, Dutch Ministry of Housing, Physical Planning and Environment, The Hague.
13. VROM (Dutch Ministry of Housing, Physical Planning and Environment) (1983) Verstedelijkingsnota (Urbanization report), SDU-publishers, The Hague.
14. VROM (Dutch Ministry of Housing, Physical Planning and Environment) (1989) National Environmental Policy Plan I, To choose or to lose, SDU publishers, The Hague.
15. VROM (Dutch Ministry of Housing, Physical Planning and Environment) (1990) Vierde nota over de ruimtelijke ordening Extra deel 1 (Fourth physical planning report Extra part 1), SDU publishers, The Hague.
16. VROM (Dutch Ministry of Housing, Physical Planning and Environment) (1990b) Ministerial Manual for a Provisional System of Integrated Environmental Zoning, The Ministry, The Hague.
17. VROM (Dutch Ministry of Housing, Physical Planning and Environment) (1993) Vierde nota over de ruimtelijke ordening Extra deel 3 (Fourth physical planning report Extra part 3), SDU publishers, The Hague.
18. VROM (Dutch Ministry of Housing, Physical Planning and Environment) (1993b) Evaluation NEPP '90, The Ministry, The Hague.

19. VROM (Dutch Ministry of Housing, Physical Planning and Environment) Waar vele willen zijn is ook een weg (Many intentions will lead to a way out), The ministry, The Hague.
20. VM (Ministerie van Volksgezondheid en Milieuhygiëne) (1972) Urgentienota Milieuhygiëne (Urgency report environmental), Tweede Kamer, 1971-1972, 11906, nr. 2, The Hague.

European sustainable cities: the challenge of citylife: being exposed to an air polluted urban environment

Prof dr. R.A.F. Smook

Delft University of Technology, Faculty of Civil Engineering, Department of Building and Construction Management, Section Design and Construct Management,
Stevinweg 1, 2628 CN Delft, The Netherlands

1. AIR POLLUTION AND THE COMPREHENSIVE APPROACH

1.1. "Stadtluft macht Frei" is not quite true in our days

The city is a dangerous place to live in, because of the very complex pattern of environmental threats. Finding ways to reduce $C0_2$ emission can be a goal, but this achievement has to be imbedded in a more comprehensive concern about the environment.

1.2. The complexity of the problem makes a comprehensive approach necessary

The well-known slogan "Think globally, act locally" is very much true in matters of fighting air pollution.

The slogan specifically for the attitude towards the urban environment could be: "Think comprehensively, act sectorially". This with trying to improve the environment and also the air environment step by step. The slogan is hopefully followed by: "Plan globally, implement locally".

1.3. EU involvement in air pollution

Being here invited as one of the chairmen of the EU Expert Group on the Urban Environment I like to tell something about the work we just completed and to show something about the relation with the topic at stake in this workshop.

The Sustainable Cities (Policy) Report is prepared by the EU Expert Group on the Urban Environment. The Report rounds off a four year period of work. The presentation of the Report took place on a conference on Sustainable Cities in Lisbon, Portugal, September 1996. To show the attitude the EU chooses herewith I present the chapter arrangement in the Report:

1.4. Chapters of the (Policy) Report:

Positive Context for Sustainable Cities;
Sustainable Urban Management;
Sustainable Management of Natural Resources, Energy and Waste;
Socio-economic aspects of Sustainability;
Sustainable Accessibility;
Sustainable Spatial Planning;
Conclusion and Recommendations.

1.5. Comprehensive approach

The problem of environmental pollution of urban entities is studied following the categorization reflected in the chapter arrangement. It is clearly identified that all aspects are close coherent and very much interdependent.

The interdependency has to be understood, the comprehensive approach has to be followed, a sectorial strategy has to be defined.

1.6. The main problems to be identified

Cultural defined lack of awareness;
Fixation on economic growth, that causes $C0_2$ production;
Poor investment in waste disposal;
Over-estimation of mobility needs, in order to achieve $C0_2$ reduction;

Lack of willingness to cooperate.

Lack of performance of integrated policy making and management systems.

1.6.1. Cultural defined lack of awareness

All that goes wrong is due to the fact that we do not understand the real already existing situation and take the attitude that our generation, in our living environment will not be confronted with the ultimate failure of the system (selfishness).

1.6.2. Fixation on economic growth

The Report states very clearly that only an other attitude can bring about results. Mankind only can survive with an annual addition to our already outrageous style of living. The economic system is build on this principle.

Economic growth goes hand in hand with more pressure on the eco-system.

1.6.3. Poor investment in waste disposal

Looking back at the urban history in Europe we can evaluate that for instance sewerage systems were only introduced half a millennium after city life organized itself. It takes years before we are used to the idea that we have to deal properly with the waste we produce and invest in waste disposal on before hand.

This extra investment has to be balanced out against the economic growth.

1.6.4. Over-estimation of mobility needs

In terms of air pollution mobility seems to be a key issue. As long as a certain over-estimation of mobility needs is considered as being "quite harmless" very little progress can be made in the field of improvement of the urban environment.

In passenger and goods transportation we are used to an unsustainable level of quantity and quality. "Freedom of movement" is a commodity we can no longer afford. This is perhaps a key issue in the improvement of the air quality in cities.

1.6.5. Lack of willingness to cooperate

Since our choking in pollution is a shared and international phenomena it has to be understood that we have to refrain from our unwillingness to cooperate over national, cultural, sectorial and even personal borders.

1.6.6. Lack of performance of integrated policy making and management systems

Policy making in the field of environment is fragmented. The town planning and urban management instrument is underdeveloped and unfortunately so is the usage of the instrument. Problems can only be solved in the framework of comprehensive urban and regional planning: "Planned globally, implemented locally".

1.7. Four key principles and ways for problem solving are identified in the Urban Environment Expert Group Report:

1. Management; Get it organized in the best way.
2. Integration; See the cohesion of problems and solutions.
3. Ecosystems thinking; Understand that the environment is a system.
4. Cooperation; Work together in order to survive.

2. SOME CONCLUSIONS IN RESPECT TO AIR POLLUTION OF THE SUSTAINABLE CITIES REPORT 1996

2.1. Introduction

2.1.1. General approach of the problem of air pollution

The Sustainable Cities Report looks at the problem of air pollution as one of the major problems to be solved in improving the urban environment. The Report states that the problem can only be solved in the context of other important aspects of the urban environment. It identifies necessary linkages to aspects as economic growth, spatial planning in general and mobility.

Within the European context already lots of regulations and directives are in force. All of them aim at reduction of emission levels and "coordination" of legislation in this field in member states.

2.1.2. Management

The policy report of the Sustainable Cities Project is intended as a contribution, on the part of the Urban Environment Expert Group, to the growing debate about cities and sustainability in Europe. It draws together a wide range of thinking and practical experience in addressing questions of urban management for sustainability. Despite a growing raft of legislation, directives and regulations, European cities continue to face economic and social problems and environmental degradation. New ways of managing the urban environment need to be found so that European cities can both solve local problems and contribute to regional and global sustainability.

2.1.3. Diversity of cities

The report recognizes and celebrates the diversity of European cities. Clearly the legal and organizational basis for urban environmental action varies between Member States, in part reflecting differences in the responsibilities assigned to different tiers of local government. In addition, cities differ in their geographical circumstances and city administrations vary in terms of the sophistication of local responses, processes and techniques. Approaches to sustainable development are likely to be different in different cities.

2.1.4. Exchange of experience

The report, therefore, does not suggest blanket solutions or recipes for all cities. Instead it advocates the provision of supportive frameworks within which cities can explore innovative approaches appropriate to their local circumstances, capitalising on traditions of local democracy, good management and professional expertise. Whatever their responsibilities and competencies, local governments throughout Europe, through the many and varied roles which they perform, are now in a strong position to advance the goals of sustainability.

2.1.5. Addressing wider audience

The report and its conclusions are aimed at a wide audience. For whilst elected representatives in cities, city managers/administrators and urban environment professionals have key roles to play in sustainable urban management, successful progress depends upon the active involvement of local communities and the creation of partnerships with the private and voluntary sectors within the context of strong and supportive government frameworks at all levels. Political leadership and commitment are critical if progress is to be made.

The remainder of this part describes the principal approaches advocated in the report, makes conclusions and recommendations for policy, practice and research and sets out a continuing agenda for the Urban Environment Expert Group.

2.2. Issues of key concern

The report envisages the sustainable city in process terms rather than as an end point. Accordingly, it highlights policy processes as well as policy content. Both emphases are significant when it comes to the transfer of good practice from one locality to another. The city is seen as a complex system requiring a set of tools which can be applied in a range of settings. Although the system is complex, it is appropriate to seek simple solutions which solve more than one problem at a time, or several solutions that can be used in combination.

Four key principles underlie the solutions advocated in the report, and should form the basis for sustainable urban management, as follows:

1. Management:
2. Integration;
3. Ecosystems thinking;
4. Cooperation.

2.3. Management

2.3.1. The need of putting sustainability on the agenda

Sustainable development will only happen if it is explicitly planned for. Market forces or other unconscious and undirected phenomena cannot solve the serious problems of sustainability. Agenda 21 specifies a thorough process of considering a wide range of issues together, making explicit decisions about priorities, and creating long term frameworks of control, incentives and motivation, combined with quantitative, dated targets in order to achieve what has been decided. Sustainable urban management should be based on the above process.

2.3.2. Variation of tools working towards integration

The process of sustainable urban management requires a range of tools addressing environmental, social and economic concerns in order to provide the necessary basis for integration. There are various tools, some addressing environmental, social, or economic concerns of urban management separately, others attempting to combine these concerns. The Sustainable Cities Project focuses on the environmental tools available to urban management processes.

2.3.3. Five main tools

Five main groups of environmental tools are advocated. These are:

1. collaboration and partnership;
2. policy integration;
3. market mechanisms;
4. information management;
5. measuring and monitoring.

Each tool is considered as an element within an integrated system of sustainable urban management. There can be no prescriptions for how to use or combine these tools; there are many ways of moving towards sustainability. Institutional and environmental contexts are different in different Member States and in different cities, and each therefore requires a novel approach. The fundamental goal is to achieve an integrated urban management process, but the elements in that process will evolve through the interplay of different interests.

2.3.4. Active role of the government and the "super government"- question of subsidiarity

The approach to these tools implies a need for a broader and more active view of the role of government, especially municipal government, than has become current in parts of Europe. Management for sustainability is essentially a political process which has an impact on urban governance. The tools advocated this report are all means of modifying or constraining the operation of professions, performance monitoring, and markets within sustainability objectives set from outside. By applying these tools, urban policy making for sustainability can become much broader, more powerful and more ambitious than has hitherto been generally recognized.

2.3.5. Acceptance by the public: legitimation

The political process of democratic choice can legitimate both sustainability objectives and the means to achieve them - provided people are educated and accurately informed about the consequences of their choices. Many of the problems related to unsustainability are only soluble if the people accept limits on their freedoms. These limitations can only be acceptable if the people affected choose or at least consent to them. The 'social contract' model of politics, in which civil society is created through individuals voluntarily agreeing to collective limitations on their own actions in order to make them all better off, holds the solution to sustainable urban management.

2.4. Integration

2.4.1. 5th Action Program

The necessity for coordination and integration is emphasized in Chapter 8 of the Fifth Action Program on the Environment. This is to be achieved through the combination of the subsidiarity principle with the wider concept of shared responsibility. In setting out the recommendations which emerge from the Sustainable Cities Project the Experts Group is seeking to achieve both horizontal and vertical integration.

Horizontal integration is necessary so that further integration of social, environmental, and economic dimensions of sustainability will have synergetic effects and therefore strongly stimulates the process towards sustainability. Horizontal integration requires integration between the policy fields within municipalities, within regional and national authorities and within the European Union. This latter is required across the European Commission's activities as well as within each Directorate General.

2.4.2. Coordination of sectors

At local, regional and national level a movement towards integration between policy field or sectors has started. Projects, research programs etc. are, at least in some Member States, developed, stimulated and disseminated through horizontal structures in organizations.

Vertical integration across all levels at European Union, Member States and regional and local governments is equally important. Vertical integration might result in greater coherence of policy and action, so that the development of sustainability at local level is not undermined by decisions and actions by Member States governments and the EU.

2.5. Ecosystems thinking

2.5.1. Starting with hydrological systems/networks and infrastructure

The ecosystems approach regards energy, natural resources and waste production as flows or chains. Maintaining, restoring, stimulating and closing the flows or chains contribute to sustainable development. The regulation of traffic and transport is another element of the ecosystems strategy.

The dual network approach is based on these principles. This dual network approach provides a framework for urban development at a regional or local level. This framework consists of two networks: the hydrological network and the infrastructure network. This hydrological network structures the urban planning in order to improve the water quality and avoid fast run-off. The infrastructure network provides opportunities to minimize car mobility and to stimulate the use of public transport systems and walking or cycling. In the plan design process attention should be paid to:

- water quality and water quantity, main structures, ecological values etc.
- existing or new public transport, employment and amenities in relation to residential areas, integration of walk and cycle routes in residential areas etc.

The Report does not take in account the ecosystems thinking in the "air domain". This is an aspect yet to be elaborated.

Analyzing these aspects will result in basic principles for urban sustainability from a physical ecosystems point of view. However, the ecosystems approach contains a social dimension as well, which considers each city as a social ecosystem. Niches and diversity form the elements of this social ecosystem.

Finally, the ecosystems approach emphasizes the city, as a complex system which is characterized by continuous processes of change and development.

2.6. Cooperation

2.6.1. Learning and doing and exchange

Cooperation is an essential part of moving towards sustainability. Collaboration and partnerships between different levels, organizations, and interests is essential for two reasons. First, it reduces the tendency of individual organizations and agencies to pursue their own agendas in isolation from the broader public interest. Second, most problems can only be solved through coordinated action by a range of actors and agencies, in line with the principle of shared responsibility as advocated by the Fifth Action Program on the Environment.

The Sustainable Cities Project emphasizes the importance of 'learning by doing'. Involvement in decision making and management means that organizations and individuals engage in a process of mutual betterment. Viewing sustainable urban management as a learning process both reinforces the point made earlier about taking the first step towards sustainability and highlights the importance of experimentation.

Much can be learned from sharing experiences between cities. However, it must be acknowledged that transferring lessons on physical matters, such as river basin management and recycling initiatives, is currently easier than with spatial planning initiatives because of the extra complication imposed by the variety of legal and cultural issues on which planning systems are based. The possible emergence of a spatial development perspective for Europe over the coming years offers considerable potential for the application of sustainability approaches to spatial planning.

Two categories of collaboration are specifically promoted in the report. The first category is focused on the operations of local authorities and include professional education and training; cross-disciplinary working; and partnerships and networks. The second category is focused on the relationship between a local authority and its community and include community consultation and participation; and innovative educational mechanisms and awareness raising.

2.6.2. Cooperation and partnership

A key goal is to create the conditions that enable collaboration and partnership to take place. This is important for the above mentioned reasons, as well as because cooperation promotes equivalence between actors, rather than hierarchy, thus facilitating increased understanding and sense of responsibility among different actors.

2.7. Sustainable management of natural resources - Conclusions:

Chapter 4 in the Sustainable Cities Report identifies the problems of consumption of non-renewable or slowly renewable natural resources that exceeds the capacity of the natural system. It links this problem to the related waste accumulation that is characteristic of today's lifestyles in European cities.

The functioning of urban systems is compared to natural systems, where equilibrium is maintained by circulating resources and wastes internally. The difference between the functioning of the natural and the urban system lies in the way the latter is dependent on importing natural resources and energy into the city and exporting waste and pollution out to the surrounding areas. Instead of being closed systems, where natural resources are used in an economical way to provide energy, and any unused material is reused, recycled or processed for re-entering the circulation process, cities are highly dependent open systems. By depending on surrounding areas for the provision of natural resources and energy, and for the disposal of waste, cities impose their problems on these areas. Depletion of natural resources, pollution and environmental degradation with their resulting social, economic and environmental consequences affects the rural population as well as urban systems themselves.

The chapter emphasizes that a more sustainable functioning of urban systems requires a move towards management of cities that makes use of the lessons that nature can teach us about ecological and economical flow management.

An integrated approach to closing the cycles of natural resources, energy and waste should be adopted within cities. The objectives of such an approach should include minimizing consumption of natural resources, especially non-renewable and slowly renewable ones; minimizing production of waste by reusing and recycling wherever possible; minimizing pollution of air, soil and waters; and increasing the proportion of natural areas and biodiversity in cities. These objectives will be easier to work towards on a small scale, which is why local ecological cycles are the ideal basis for introducing more sustainable policies for urban systems.

2.8. Sustainable management of natural resources - Recommendations:

The issues of natural resources, energy and waste are closely interconnected. Cities are places of high energy intensity, and energy plays an increasingly important role in the operation of the urban systems. The more energy that is consumed, the higher the need for natural resources to support the energy production. Similarly, the higher the consumption of natural resources and energy, the more waste is accumulated.

Due to this inter-relationship it is logical that several of the relevant policy options have multiplier effects. So by addressing one particular problem, the policy options may simultaneously solve one or more other problems.

The key goal of sustainable management in relation to air is to ensure quality and supply. This can be achieved by reducing pollution sources and quantities, and by promoting air generation and filtering. Several technical measures and regulatory instruments can contribute to reducing pollution. The provision of more green elements in cities can increase the capacity for air generation and filtering. Green elements also serve to reduce noise pollution, assist the formation of suitable micro-climatic conditions, manage storm water, and provide recreational and aesthetic values. It is recommended that the measures designed to improve air quality and supply are developed within an overall framework of an action plan for air quality. This will be a requirement once the EU's 'Framework Directive on ambient air quality assessment and management' becomes operational.

The general aim in relation to soil, flora and fauna is to increase the proportion of natural and human made eco-systems within cities. Developing green corridors linking country-side to the various green elements within cities provides the best ecological frameworks for habitats, thus combining an increase in biodiversity with recreational value. A move away from mono-culture towards increased biodiversity is another important aspect in the sustainable management of cities. It is also recommended that green elements are used to provide a ground for education and awareness raising in relation to the way ecosystems function and how urban functions can be integrated into the natural system.

The principles of sustainable water management are related to water conservation and minimizing the impact of all water related functions on the natural system. Measures to green cities are also beneficial for the water system. Maximizing the use of penneable surfaces facilitates the infiltration and cleansing of storm water, while creating ponds, ditches and wetlands facilitates the retention of storm water, purifies the water and enriches the flora and fauna. The issue of water use efficiency is also crucial in sustainable water management. Taking into account the end use of the water in determining the required quality is a means that aids water conservation. Collecting storm water for secondary uses and recycling grey water are important measures.

Significant advances have been made in developing more environmentally friendly sewerage solutions. Biological treatment plants and passive water treatment methods based on ecological functions should be more widely utilized. These have clear environmental benefits, and some provide recreational and educational values too. The basic aim of sustainable energy management is concerned with energy conservation. The key to energy conservation lies in behavior of individuals and organizations, but also in energy production and distribution. Non-renewable energy sources should be replaced by renewable ones wherever possible. Both energy management and production should be decentralized. Developing local energy management systems and local energy production facilities increases the possibility for coordinating actors, for working actively towards reducing energy demand and increasing efficiency of production and distribution. The application of sustainable design principles play an important role in energy conservation. Densities, siting, layout, bio-climatic architectural design, materials, insulation, orientation of buildings, micro-climate, green elements etc. can make significant contributions to minimizing energy consumption.

The role that cities can play directly is also significant. Through an energy audit of both internal and external activities, and of the city's own building stock, cities can move towards adopting suitable energy efficiency measures, thus contributing to energy conservation while reducing running costs. Greening the city's activities in such a way, also provides a valuable example for other organizations and individuals to follow. It adds credibility to any awareness raising initiatives undertaken by cities.

Various solutions that utilize waste for energy production serve the dual purpose of conserving natural resources and making efficient use of waste products.

The ultimate aim of sustainable waste management is, however, to minimize production of waste. The reduction of packaging, and the increased use of reusable and recyclable packaging contribute towards this aim. Maximum separation at source and composting minimizes waste production, reduces the level of contamination of waste, and turns some of the waste into useful forms, such as topsoil or biogas. Other policy options for waste management coincide with those for water management, especially those for liquid waste management.

Finally, influencing behavior through education, information and practical evidence is a key factor in achieving more sustainable urban systems. The relationship between influencing behavior and sustainable management of natural resources is particularly evident. It is an area where people's behavior affects the level of sustainability very directly. Often it is also an area where people can see the results of changed behavior in a very transparent way.

3. PAY-OFF

The work of the EU Expert Group on the Urban Environment shows clearly its commitment to tackle problem in the environment in a comprehensive way. So the solving of the air pollution problem has to be looked upon.

Air quality and urban planning: towards an integrative approach

G. Knaap [a] and T.J. Kim [b]

[a] Gerrit Knaap is associate professor of Urban Planning at the University of Illinois at Arbana-Champaign, Visiting Fellow at the Center for Urban Policy and the Environment at Indiana University and Senior Research Fellow at the American Planning Association

[b] T. John Kim is professor of Urban Planning and civil Engineering at the University of Illinois at Urbana-Champaign.

AIR QUALITY AND LAND USE PLANNING: TOWARDS AN INTEGRATIVE APPROACH

Integrated environmental management (IEM) is a concept that now holds great currency at the national, state, and local levels. Although definitions vary, IEM is generally considered to convey an interorganizational, coordinated approach to environmental management that takes into account social, political, economic, and institutional human actions in a particular environmental system (Margerum and Born 1995). At the federal level the U.S. Environmental Protection Agency, the Forest Service, and the Bureau of Land Management have recently launched 'ecosystem' approaches to carrying out their mandates. And at the state level, new institutional relationships have been formed to address land and water resource management. According to the Director of the Wisconsin Department of Natural Resources, a multi disciplinary, integrated approach to environmental stewardship may represent the most important scientifically and philosophically based management principle yet developed (Besadny, 1991, as quoted in

Born and Sonzogni 1995).

Despite its growing popularity, IEM as an approach to managing urban air quality is relatively new. Until recently, U.S. air policy has focused on technological approaches to emissions reductions rather than on the interrelationships between urban land use, transportation, and air quality. In recent years, however, that has changed. Since the passage of the Clean Air Act Amendments (CAAA) of 1990 and the Intermodal Surface Transportation Efficiency Act (ISTEA) of 1991, U.S. policy makers have increasingly viewed transportation and land use as integral issues in the management of urban air quality.

In this paper we discuss integrated land use approaches to urban air management. We begin with a review the CAA and ISTEA focusing on air quality standards and land use and transportation planning as approaches to air quality management. Next we describe policy instruments that can be used to implement such an approach and obstacles to successful implementation. These obstacles include the limitations of urban reconfiguration as an approach to air quality management, institutional impediments to the transmission of policy from the federal to the local level, and market resistance to policy implementation. We conclude with suggestions for further research and institutional development.

1. THE CLEAN AIR ACT AMENDMENTS AND THE INTERMODAL SURFACE TRANSPORTATION EFFICIENCY ACT

The 1990 CAAA and the 1991 ISTEA provide the institutional foundation for an integrated approach to air quality management. In brief, the CAAA sets new standards for air quality, offers guidelines for meeting those standards, and authorizes incentives and sanctions the EPA can use to motivate states that fail to meet federal standards. ISTEA delegates responsibility for transportation planning to states and metropolitan planning organizations (MPO's), requires states and MPO's to develop short- and long-term integrated transportation plans, and requires such plans to further the goals of the CAAA.

1.1. Transportation and Land Use Implications of the CAAA

Under the 1970 and 1977 amendments to the original 1955 Clean Air Act, states were required to prepare state implementation plans (SIPs) to meet and maintain national ambient air quality standards (NAAQS). Although the 1955 act and the 1963 amendments did not mandate transportation planning, the 1970 amendments gave states the authority to adopt transportation control measures and the 1977 amendments explicitly encouraged greater integration of transportation and air quality management (Garrett and Wachs 1996). These amendments also enabled the EPA to impose a federal air quality plan or to withhold federal highway funds if air quality standards were not met by 1987. By 1987, however, few urban areas had met the standards, despite continuing reductions in emissions from cleaner cars and fuels, federal limits on fuel volatility, and unusually favorable weather patterns. The failure to meet these standards, in part, stemmed from continuing increases in vehicle miles traveled (VHT). (See Figure 1.)

Figure 1. Growth in Vehicle-Miles Traveled (VMT) in the U.S.

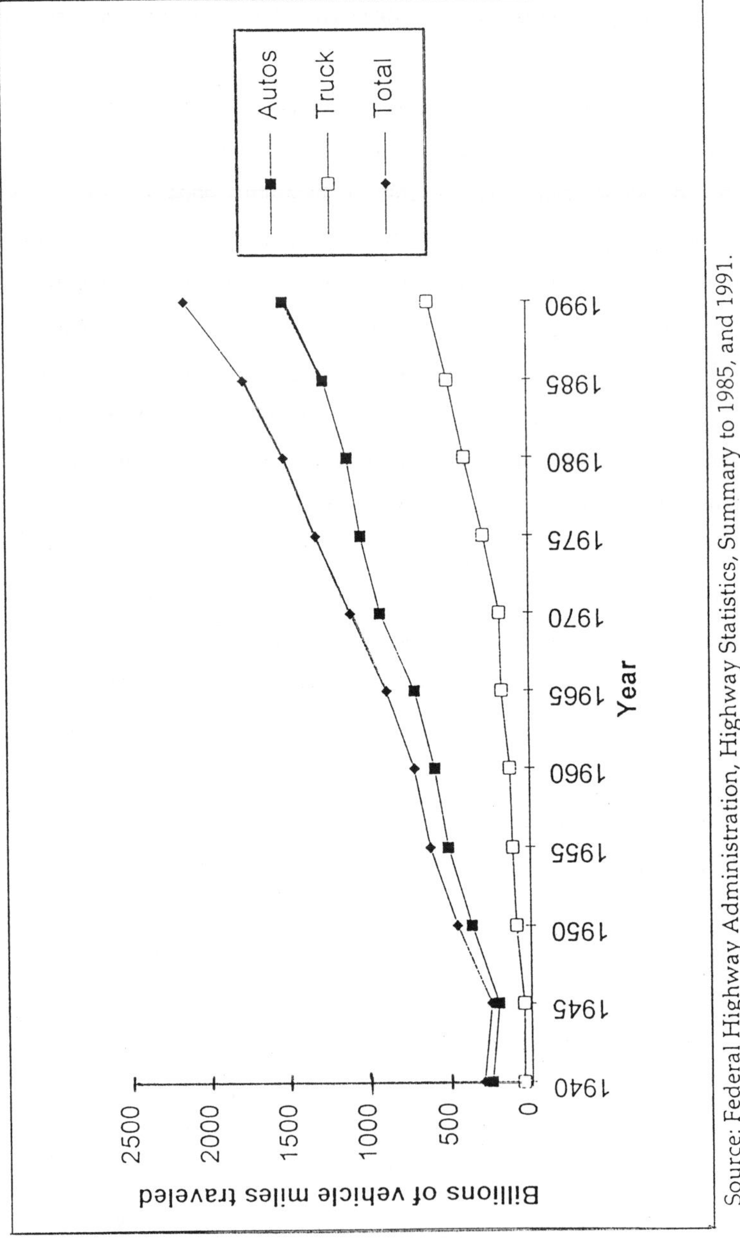

Source: Federal Highway Administration, Highway Statistics, Summary to 1985, and 1991.

The 1990 CAAA further altered the relationship between air quality management and transportation planning. Besides establishing longer, more realistic schedules for meeting air quality standards, the amendments included a classification system for metropolitan areas, based on the degree to which metropolitan areas were in compliance with national air quality standards. Metropolitan areas with the most serious air quality problems, for example, had to develop the most comprehensive strategies for reducing growth in VMT's. In addition, the CAAA identifies transportation control measures (TCM's) that states can use in their SIP's. These include public transit options, high-occupancy vehicle lanes, employer-based transportation programs, trip reduction ordinances, traffic flow improvements, mandatory no driving days, and gasoline rationing.[1] (See Table 1.)

The CAAA also contained new provisions for conformity.[2] These provisions made it the affirmative responsibility of the Federal agency supporting an action to ensure that its activities conform to an approved or promulgated air quality implementation plan (Shrouds 1995, p.200). Most pertinently, transportation plans and programs pursuant to Title 23, US Code or the Urban Mass Transportation Act, as amended, have to conform with the mobile source emission reduction targets in the SIP. The effect of these conformity requirements is that federal funds for transportation improvement projects will only be available for those projects that conform with the purpose of the state SIP--reducing the severity and number of NAAQS violations and the attainment of the standards. These requirements significantly altered the scope of metropolitan transportation plans. Whereas transportation plans have traditionally focused on accessibility and congestion mitigation, plans developed after the CAAA must also consider the effects of plans on VMT and the contribution of VMT's to air quality.

---

[1] Transportation control measures include both incentives and regulatory controls. For more on TCM's, see Shrouds (1995).

[2] For more on conformity, see Shrouds (1992)

Table 1. TCM's Listed in the 1990 Clean Air Act Amendment

- Programs for improved public transit
- Restriction of construction of certain lanes or roads for use by buses of HOV's
- Employer-based transportation management programs, including incentives
- Trip reduction ordinances
- Traffic flow improvement programs that achieve emissions reductions
- Fringe and corridor parking facilities serving HOV's and transit
- Programs to limit or restrict vehicle use downtown or in other areas of emission concentration, particularly during peaks
- HOV/ridesharing service program
- Time or place restrictions of road surfaces or area to bikes and pedestrians
- Bike storage, lanes and other facilities, public and private
- Programs to control extended vehicle idling
- Programs to reduce extreme cold start emissions
- Employer-sponsored programs to permit flexible work schedules
- Localities' SOV trip reduction planning and development programs for special events and major activity centers, including shopping centers
- Pedestrian and non-motorized transport facility construction and reconstruction
- Programs for voluntary removal of pre-1980 vehicles

1.2. The Air Quality Implications of ISTEA

The passage of ISTEA marked the beginning of a new era of transportation planning in the United States. The planning provisions of ISTEA require MPO's, in cooperation with state and local transit operators, to develop both a long-term regional transportation plan (RTP) and a short-term transportation improvement program (TIP) that includes projects consistent with the RTP. Fifteen factors must be addressed in preparing these plans. (See Table 2.) The RTP must accommodate transportation demands for 20 years. To serve as an implementation vehicle for the RTP, the TIP sets project funding priorities for three years and must be updated every two years. In addition ISTEA requires states to prepare a statewide RTP and TIP. As a part of the statewide plan, each state must develop traffic congestion management systems for all areas with a population greater than 200,000, known as transportation management areas (TMA's). All transportation plans and programs within a TMA must be prepared and implemented by the MPO with local transit operators. Newly designated MPO's must include officials of major local transit agencies and state and local elected officials. The planning process must be certified at least every three years by the U.S. Department of Transportation (DOT).[3]

Table 2.

THE PLANNING FACTORS TO BE CONSIDERED IN THE METROPOLITAN PLAN

1. Preserve and enhance *existing transportation systems*
2. Conserve *energy*
3. Relieve and prevent *congestion*
4. Integrate transportation policies with *land use* and development policies (demand management, growth management, APFO)
5. Fund *"enhancements"*
6. Include *all transportation projects* (not just federally funded ones)
7. Make major connections *(connectivities)* with:
   - international borders
   - ports and airports
   - freight routes to modes
   - intermodal facilities
   - recreational, historic and military destinations
8. Ensure *connectivity* of metro and non-metro roads
9. Meet the needs identified through *management systems* prepared by SDOT's and MPO's
10. Preserve *right-of-way* for future projects
11. Provide for the efficient movement of *freight*
12. Use *life-cycle costing* analysis of proposed investments (and prepare major multimodel investment analysis: other modes treated like transit 'alternatives analysis' to facilitate comparative analysis across modes)
13. Transportation *impact analyses*: social, economic and environmental (include linkages between housing, jobs and transportation)
14. Enhance *transit services*
15. Enhance *transit security*

[3] For an analysis of the certification process see U.S. GAO (1996).

ISTEA enhances the role of MPO's and elevates the role of air quality as a transportation planning goal. For TMA's that contain non-attainment areas for ozone and carbon monoxide, all highway projects that increase highway capacity must be addressed in the congestion management plan. Funding for projects that contribute to NAAQS attainment is provided by the Act's Congestion Mitigation and Air Quality Improvement (CMAQ) program. To further enhance their flexibility, ISTEA provides funding for projects that include wetland mitigation banking, brownfield cleanup, intelligent transit, research and planning, congestion pricing, habitat mitigation and banking, highway runoff mitigation, pedestrian greenways, and bicycle programs.

Combined, the CAAA and ISTEA provide the institutional framework for a new era in land use, transportation, and air quality planning and management. The CAAA contains specific standards, deadlines, and sanctions. Metropolitan areas that do not meet these standards must take actions to reduce VMT's through transportation planning. Transportation plans must be developed by MPO's in cooperation with state and local transportation agencies and governments, certified by the DOT, and conform with SIPs submitted to the EPA.

2. LAND USE AND TRANSPORTATION PLANNING

Concern for air quality in land use planning is not new. For many years a fundamental concern of land use planners has been the separation of uses to minimize nuisances. Evidence that air pollution represents such a nuisance has been confirmed by research demonstrating that air pollution significantly reduces land values (Knaap 1997). Recognition of the relationship between transportation and land use is also not new. For decades, transportation planners have followed the 4-step process illustrated in Figure 2.

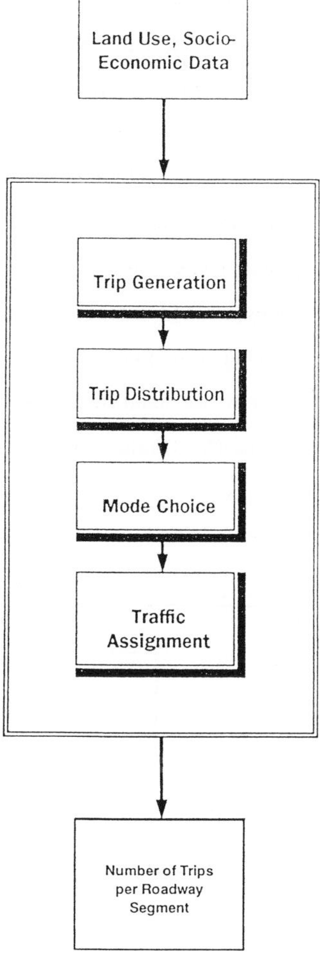

Figure 2.

In this process, land use is taken as given and the transportation system is designed to accommodate transportation demands from one location to another.

Recently, however, land use has been viewed as less of a given and more as a factor determined in part by transportation policy (Moore and Thorsness 1994). The notion that transportation systems affect land use, however, is also not new. Quoting the comments of Charles Gordon at the 1939 National Conference on Planning, Segoe (1941, P.244-5) writes: "From the planners point of view, it is important .... to stress the fact that the influence of transportation is such that it directly affects the nature and direction of urban growth and development. It is not only an important urban facility to be planned, but is in itself a tool of planning. ... through the planning of transportation the planner may exercise a direct and potent influence upon the other physical aspects of urban development."

What is relatively new, and what has been highlighted by the CAAA and ISTEA, is the notion that transportation and land use planning can serve to further air quality goals. Financed in part by federal grants, new models have been developed that include feedbacks from transportation policy to land use. These models incorporate the complex dynamic interrelationships among land use, transportation, and air quality. What's more, these models provide the foundation for land use planning as an approach towards meeting national air quality standards. (See Figures 3 and 4.)

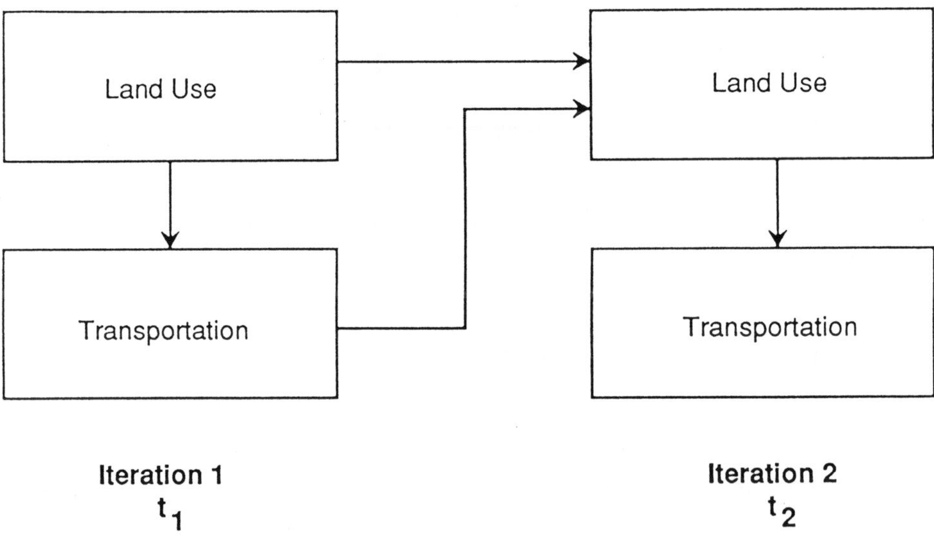

Figure 3. Iterative Land use / Transportation Analysis

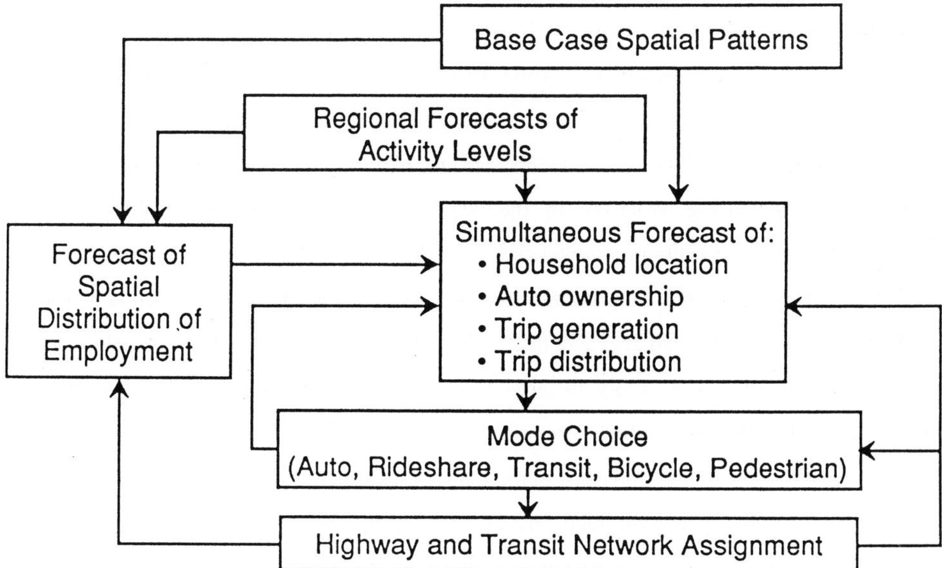

Figure 4. Simulation of Land use / Transportation Interaction

3. LAND USE POLICY INSTRUMENTS FOR AIR QUALITY MANAGEMENT

Like transportation plans, land use plans have little impact without implementation vehicles. Transportation plans are implemented primarily though direct investments in public transportation infrastructure. Because most urban land is privately owned, however, land use plans must be implemented indirectly using a variety of policy instruments. Land use policy instruments that can be used by local governments to further air quality goals include urban growth boundaries, zoning, parking regulations, design review, jobs-housing balance programs, property taxes, impact fees, and public expenditure programs.

3.1. Urban Growth Boundaries

Urban growth boundaries (UGB's) are policy instruments used for a variety of purposes that include urban spatial containment, increasing urban densities, guiding the extension of urban infrastructure, and increasing certainty in the regulatory environment--all of which could serve to improve air quality. Urban containment could lead to shorter trip lengths. Increasing urban density could encourage transit as a modal choice. Guiding the extension of urban infrastructure could lead to better integrated urban transportation systems. And increasing certainty in the regulatory environment could guide private investments toward transit-complementary uses. In the Portland, Oregon, metropolitan area, for example, UGB's are being used to contain urban sprawl, facilitate transit ridership, reduce automobile trip lengths, and enhance urban air quality. (See Figure 5.) The efficacy of UGB's as tools towards these ends, however, remains unproven, and requires considerable cooperation among local governments within a metropolitan area. Further, to have significant impacts on the density and spatial extent of the metropolitan area, UGB's are likely to raise land and housing prices. Whether UGB's can withstand the political pressures created by rising land and housing prices remains to be seen.

Region 2040 at a glance...

CONCEPT A

What we did:

- **Urban Form:** Significant expansion of the UGB. New growth at urban edge develops mostly in the form of housing.
- **Major Roads:** 10,190 lane-miles.
- **Transit:** 12,322 daily service-hours.

What happened:

- **Congestion:** Worst of the three growth concepts, with nearly 12% of roadways having significant peak-hour congestion.
- **Transit ridership:** 372,390 daily riders.
- **Trip length:** Total vehicle miles traveled (VMT) more than double 1990 levels; VMT per capita would increase 5.2% over 1990.

CONCEPT B

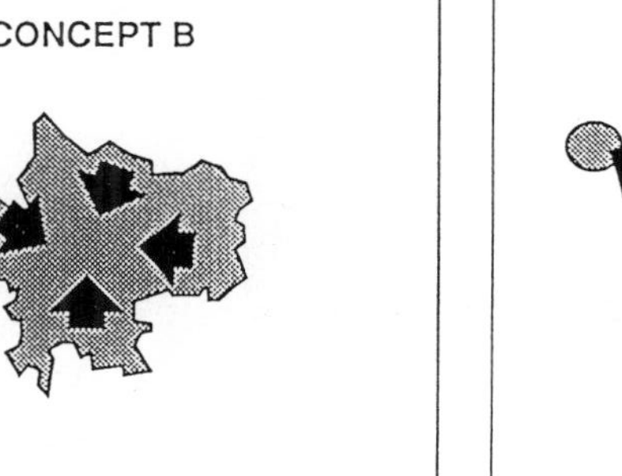

What we did:

- **Urban Form:** No expansion of the UGB; growth accommodated through development of existing land and infill.
- **Major Roads:** 9,820 lane-miles.
- **Transit:** 13,192 daily service-hours.

What happened:

- **Congestion:** Slightly less than Concept A, with significant congestion on more than 11% of major roadways
- **Transit ridership:** Highest of the three concepts, with 527,758 daily riders.
- **Trip length:** Greatest reduction in VMT per capita, dropping 12.4% from 1990.

CONCEPT C

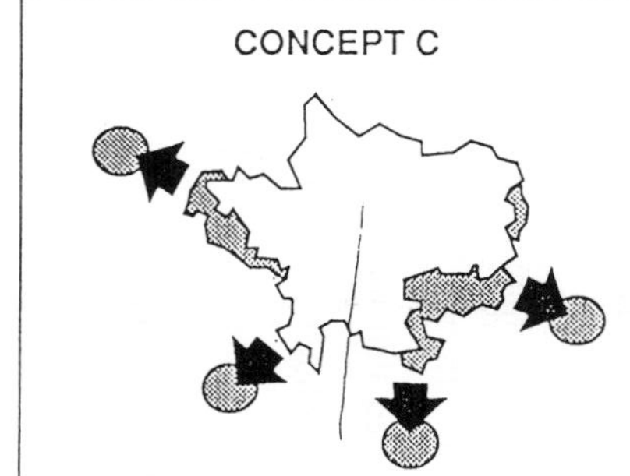

What we did:

- **Urban Form:** Slight expansion of the UGB, with growth encouraged in regional centers and neighboring cities.
- **Major Roads:** 10,327 lane-miles.
- **Transit:** 12,553 daily service-hours.

What happened:

- **Congestion:** Least of the three concepts, with slightly over 8% of roadways having significant peak-hour congestion.
- **Transit ridership:** 437,178 daily riders.
- **Trip length:** Slight reduction in VMT per capita, with a decrease of 3.8% over 1990.

**Population and employment were held constant among the growth concepts, with an assumption of 2.67 million residents and 1.63 million jobs.*

Figure 5.

3.2. Zoning

Zoning is perhaps the most widely used instrument to control the type and intensity of urban land uses. Most often, zoning has been used to separate uses and to limit residential density, which increases trip lengths and encourages automobile use. Relaxing such zoning constraints might have the opposite effects. Under the label the 'new urbanism' plans for entire cities have been developed that include regional town centers, local town centers, and transit focused neighborhood centers (Katz 1994). New comprehensive plans for the cities of Portland and Seattle, for example, rely extensively on mixed use and transit focused zoning. (See Figure 6.)

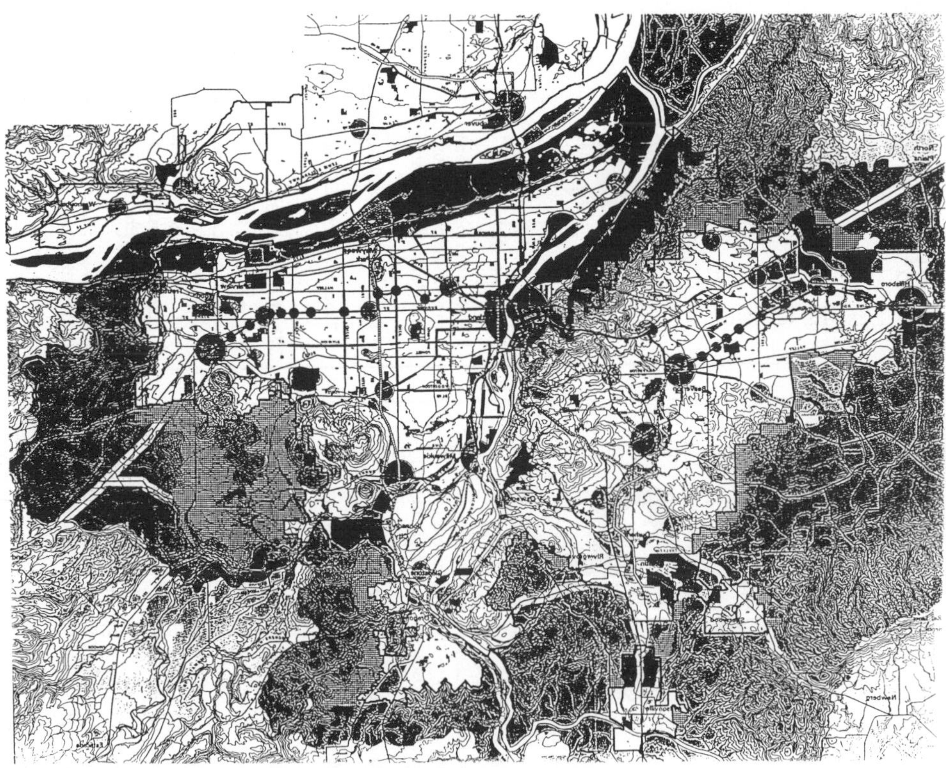

Figure 6.

Proponents of the new urbanism argue that mixed use zoning can reduce automobile travel (Bartholomew 1995, Morris 1997). These proponents claims, for example, that zoning which allows vertical separation of retail and residential uses in the same building, can reduce home-based travel; zoning that allows banks, day care, restaurants, and dry cleaning near suburban employment centers can reduce work-based travel; and zoning that requires transit-supportive uses in transit station areas encourages transit use. Finally, proponents argue, zoning that permits high residential densities enables public transit routes to serve a larger number of points, can lower the per-person cost of ridership, and can increase the frequency of service. Research has shown, for example, that feeder bus service is viable at a density of seven units per acre and a high-frequency bus service is viable at 15 units per acre, yet current zoning regulations typically limit residential densities below five units per acre (Morris 1997).

The extent to which traditional zoning regulations limit densities and separate uses, however, remains a matter of dispute. Although early studies found that allowable use and density zoning affected land values and thus, by inference, lowered density and separated uses, more recent studies have found that zoning tends to follow that market and that actual development densities fall short of zoned densities (Knaap 1997). These results cast doubt on the extent to which land uses would change if zoning constraints were removed. Further, the extent to which mixed use zoning results in a mixture of land uses remains untested.

3.3. Parking Regulations

Parking is an inherent by-product of automobile transportation and a large consumer of urban space; parking regulations thus have a potentially significant influence on transportation mode choice and the spatial structure of urban areas. To encourage walking, bicycling, and transit ridership, parking regulations can be used to alter the location, supply, and design of parking space (Morris 1997). Parking space located between buildings and a public street, for example, give automobile users greater building access than pedestrians. To encourage pedestrian travel, parking could be prohibited between buildings and the street, building set backs could be limited, and parking could be located behind buildings. (See Figure 7.). Parking regulations that require excessive parking supplies for commercial or residential development lower the overall density of development, facilitate automobile use, and can create large spaces that impede pedestrian and bicycle travel. To counter these effects, communities can lower minimum parking requirements or impose parking space maximums. Finally, parking regulations can be used to alter the design of parking lots. Design requirements might include buffering and interior landscaping, retail uses on the lower levels of parking structures, and dedicated spaces for bicycles.

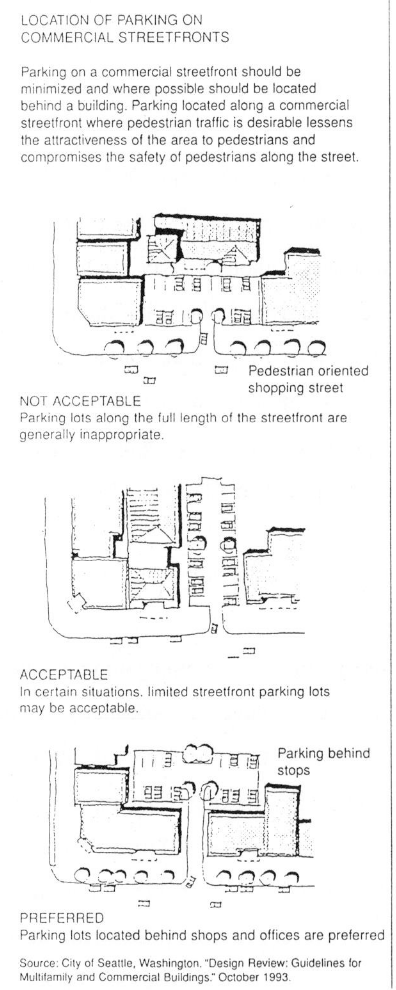

Figure 7. Parking lot guidelines

3.4. Design Review

In recent years, urban designers have designed transit-oriented developments (TOD) that maximize transit friendliness (Calthrope 1993). Such TOD's incorporate mixed land uses, high density residential development, and street patterns that facilitate public transit use. (See Figure 8.)

Transit-oriented development, such as this community designed by Peter Calthorpe Associates, is one component of the LUTRAQ alternative to a bypass freeway west of Portland. Such site plans could generate significantly fewer auto trips than conventional development.

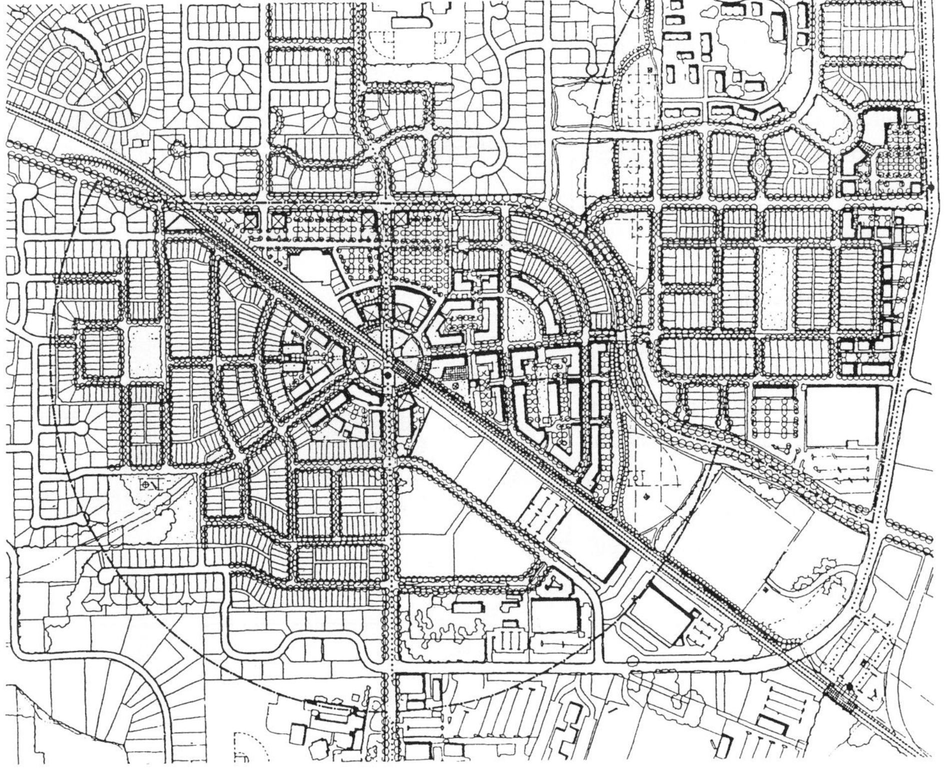

Figure 8.

The market for TOD's in the United States remains largely untested, but new developments could become marginally more transit friendly--perhaps though design review. Design review is a process through which a proposed development or subdivision must pass before receiving a building permit or subdivision approval. Design reviews could be conducted, for example, to assure that subdivisions and developments provide a continuous, direct and convenient transportation linkages; provide a pedestrian-friendly environment; offer interesting and attractive spaces; and conserve open spaces for public use (Morris 1997). (See Figure 9.)

Figure 9.A. Disconnected Streets

Recent practice has emphasized discontinuous streets, such as loops and cul-de sacs, in order to discourage through traffic.
Unfortunately, such streets also make it impossible for buses to pass through these areas.
Transit service is convenient to most residents in the development.

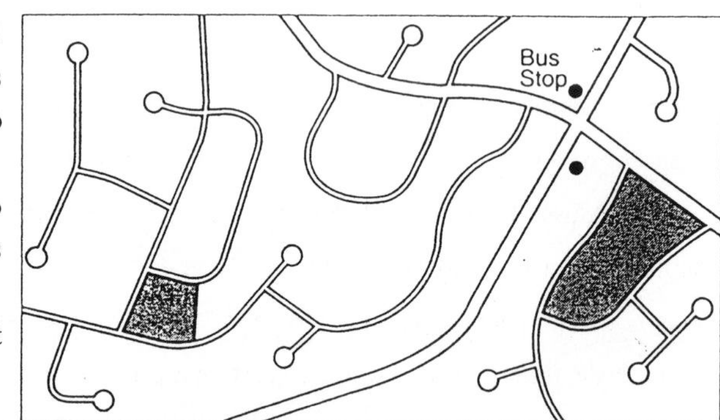

Figure 9.B. Interconnected Streets

Interconnected streets gives pedestrians many alternative walking paths and help shorten walking distances. When streets are connected in this way, auto drivers have many routes to follow as well. This disperses traffic and reduce the volume of cars on any one street in the network.

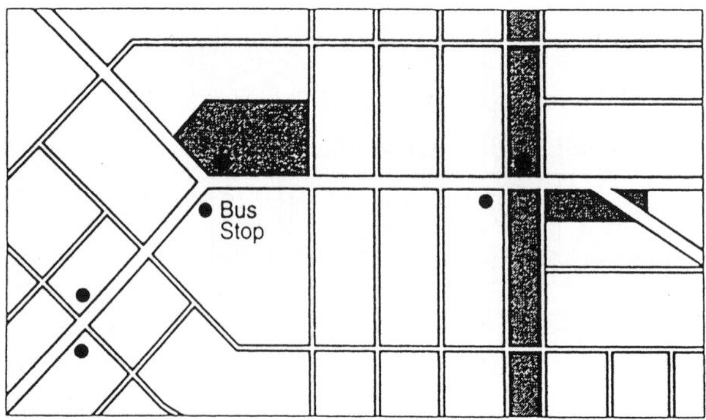

Source: Ontario Ministry of Transportation, *Transit-Supportive Land Use Planning Guidelines*, April 1992.

3.5. Jobs-Housing Balance

The separation of jobs from residences has created what some call a jobs-housing imbalance. This imbalance, according to this perspective, causes excessive commuting between job rich and housing rich communities. Evidence of a jobs-housing imbalance in the Chicago and San Francisco metropolitan areas is offered by Cervaro (1989, 1996). To correct the imbalance Cervaro (1989) offers a number of policy alternatives, including growth phasing (in which building permits are regulated to assure a balance of jobs and housing units), office-housing linkages (which require office developers to provide or fund housing construction), jobs-housing negotiations in the development review process, and state and regional housing allocation programs.

The role of public policy in balancing jobs and housing among communities, however, remains highly disputed. Gordon and Richardson (1988), for example, argue that journey-to-work trips represent a minority of VMT, especially during peak-hour travel, and that jobs-housing imbalances are self correcting as jobs follow housing to the suburbs. Further, Moore and Thorsnes (1994, P.108) claim policies relating to a jobs-housing balance will get lip service but, for lack of an ability to define, implement, or enforce them, little more. At best, concerns about having housing near employment opportunities will manifest themselves as mixed-use zoning or performance standards, which could be more sensible than numerical quotas for a specific mix of jobs and residences. Finally, after reexamining the issue in the San Francisco area, Cervaro (1996, P. 508) concludes: "To the degree that it exists, any problem of jobs-housing imbalance is fundamentally one of barriers to the production of suitable housing in job-rich cities and subregions .... Thus, one of the many policy challenges to planners in coming years will be to break down barriers to residential mobility, such as NIMBY resistance, large-lot zoning, and other exclusionary policies."

3.6. Tax Abatement, Impact Fees, and Public Expenditures

Tax abatement and public expenditures have long been popular policy instruments used to influence private investment and urban form. The widespread use of impact fees is relatively new. Traffic impact fees are a potentially effective tool for shaping urban developments and transportation choices (Morris 1997). By assessing traffic impact fees, local governments can force developers to bear the costs they impose on the transportation system. When properly assessed, such fees can foster a more efficient and equitable system of infrastructure finance. What's more, by assessing such fees based on the extent of traffic impacts, such fees will automatically encourage a more transit-friendly mix, density, and design of urban development. (See Table 3.)

Table 3.

TRAFFIC IMPACT FEE REDUCTION INCENTIVES	
Action	TIF Reduction
Development within the Transit Overlay District*	2%
Construction of on-site but off road internal walk/bike network	12%
Construction of direct walkway connections to the nearest arterial for non-abutting development	3%
Commercial development which would be occupied by an employer subject to and complying with, section	4%
Direct walk/bikeway connection to destination activity (such as a commercial/retail facility, park or school) if residential development or to origin activity (such as a residential area) if commercial/retail facility	2%
Installation of on-site sheltered bus-stop (with current of planned service or bus stop within 1/4 mile of site with adequate walkways if approved by C-TRAN)	1%
Installation of one secure bike parking space per 10 vehicular parking stalls	1%
Connection to existing or future regional bike trail (either 1% directly, or by existing, safe access)	1%
Voluntary compliance with Commute Trip Reduction Ordinance by non-regulated employers	5%
Designation of 10 percent of all non-residential parking as carpool/vanpool parking facilities if located in a manner maximizing accessibility subject to ADA requirements**	1%
Total if all strategies were implemented	22%
* Automatic reduction for developing within Transit Overlay District and compliance with the provisions of this Ordinance. ** Requires regular maintenance.	

Tax abatement and public expenditures have often been used to foster economic development, though often for dubious reasons. Such instruments could also be used, however, to further air-quality management goals. In the Portland metropolitan area, for example, financial incentives are now used to encourage transit-supportive land use development in light-rail station areas. In such station areas, developers are offered tax incentives and public subsidies if their proposed development meets certain transit-oriented criteria. Tri-Met, the operator of the metropolitan transit system, has also published adds in periodicals such as Urban Land to attract the attention of appropriate developers.

4. DILEMMAS IN PLANNING FOR AIR QUALITY

Although a variety of land use policy instruments have been developed and are now being used to further air quality goals, dilemmas remain. These include the uncertain influence of land use on air quality, institutional impediments to policy transmission, and market resistance to plan implementation.

4.1. The Influence of Land Use on Air Quality

Despite the growing popularity of land use planning as an approach to air quality improvement, the influence of land use change on air quality remains in considerable doubt. To address this question, transportation planning models have been developed to incorporate the interrelationship between land use and transportation and the effects of transportation choices on ambient air quality. To date, however, such models have produced mixed results. Preliminary results from an ongoing research project in the Portland, Oregon, metropolitan area, led by 1000 Friends of Oregon, suggest that transit-oriented land use development can lower automobile ownership rates, encourage walking, biking and transit ridership, reduce VMT, and by implication, improve urban air quality (Bartholomew 1995). Although the projected changes in transportation patterns are not dramatic, they are expected to increase over time. The results of this work has been examined by the Oregon Department of Transportation as one of five alternatives to a proposed freeway in an environmental impact statement. In a less well known study, Scheuernstuhl and May (1992) examined the relationship between land use and air quality in the Denver metropolitan area. Although they found that increased densities led to increase public transit ridership, they concluded that land use change had little or no effect on air quality except perhaps over a very long period.

4.2. Institutional Impediments to Policy Transmission

The CAAA and ISTEA provide the foundation for a land use approach to air quality management. The CAA sets national air quality standards and requires states to consider transportation and land use approaches for meeting those standards. ISTEA requires states to prepare integrated transportation plans and requires that those plans conform with CAAA implementation plans. Most importantly, federal legislation includes financial support to prepare and implement such plans.

Federal legislation also establishes institutional linkages from the state to the metropolitan level. ISTEA requires all states to prepare statewide transportation plans and to designate MPO's to prepare metropolitan plans that conform with the statewide plans. Therefore, because MPO's are creations of the state without home rule powers, there is both an institutional and procedural link between state and metropolitan transportation planning.

Within most metropolitan areas, however, the institutional linkages between MPO's and local governments are relatively weak. Although MPO's were required to maintain the 3-C (comprehensive, continuous, and cooperative) planning process, the extent of coordination between MPO's and local governments varies widely. In Chicago, for example, the MPO is the Chicago Area Transportation Study, with complex links to local governments. (See Figure 10.) In many metropolitan areas, the MPO is a Council of Government, a voluntary association of local governments with no statutory power. In Portland, the MPO is Metro, a regional government with a directly-elected board and statutory authority to require local land use plans to conform with metropolitan plans. In Portland, therefore, it is likely that regional land use approaches to air quality problems can be successfully transmitted to local governments.[4] In other metropolitan areas, the likelihood is considerably less.

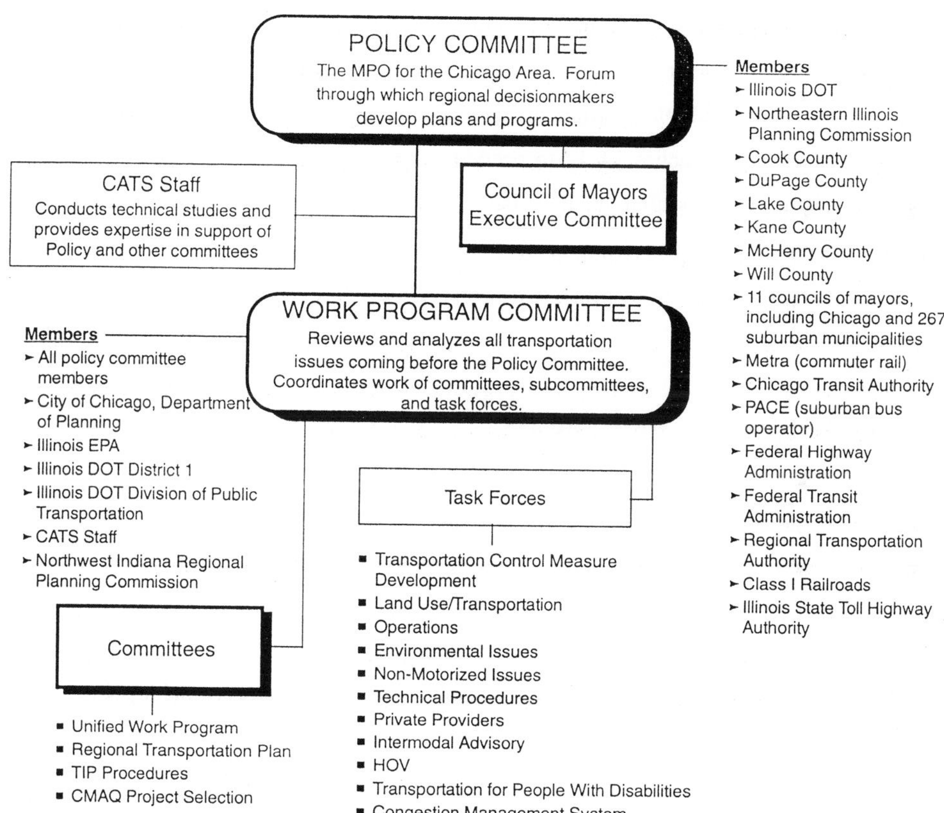

Figure 10. Organization and Membership of the Chicago Area Transportation Study (CATS)

[4] The transmission of metropolitan land use policies to local governments in Oregon is strengthened by the statewide land use program. See Knaap and Nelsen (1992).

4.3. Market Response to Land Use Instruments

Finally, even if all local government adopt land use plans designed to further air quality goals, the influence of land use policy instruments remain in doubt. Gordon and Richardson (1997), for example, argue that U.S. metropolitan areas would have looked much the same without the construction of the interstate highway system. Further, the influence of urban growth boundaries, zoning, design review, parking regulations, and jobs-balance strategies on land use and transportation patterns remain untested and highly disputed. Even supporters of land use approaches to transportation and air quality problems acknowledge that no single land use instrument is likely to have much effect, and that only when multiple instruments are use in a complementary and coordinated fashion are significant changes likely to occur (Moore and Thorsnes 1994). Clearly what is needed is empirical research on how land markets respond to transportation-oriented land use instruments, and how transportation behavior and air quality respond to land use change. Fortunately, significant research on both of these issues is now underway.[5]

5. SUMMARY AND CONCLUSIONS

Despite its uncertain efficacy, land use planning to address transportation and air quality problems has been institutionalized as a national policy approach. The mandates imposed and the funds made available by the CAAA and ISTEA assure that such an approach will receive considerable attention by local governments over the next few years as well. To implement such an approach a variety of policy instruments are available, some new and some old but redirected. The efficacy of these policy instruments, however, remains in dispute and will probably remain so for years if not decades. In light of this uncertainty, therefore, it seems inappropriate for the federal government to prescribe specific policy approaches or instruments. Land use and transportation policy choices

---

[5] One example of a research project designed to address these issues is underway at the University of Illinois. Titled "Does Planning Matter" this project uses dynamic geographic information systems to monitor the land development process in metropolitan Portland Oregon and to analyze the efficacy of land use policy instruments. See Ding, Hopkins and Knaap (1997).

involve difficult value-laden choices on which the residents of different metropolitan areas might strongly disagree. Over the next several years, therefore, mandates for compliance with national standards should remain, but local experimentation with alternative policy instruments should be allowed to continue. Only though continued policy experimentation, empirical research, and institutional development can a nation can learn what contributions integrated land use, transportation, and air quality management can make toward air quality improvement, and what sacrifices must be made to achieve such a result.

REFERENCES

1. Bartholomew, K.A., 1995, A Tale of Two Cities, Transportation, 22: 273-93.
2. Born, S.M. and W.C. Sonzogni, 1995, Integrated Environmental Management: Strengthening the Conceptualization, Environmental Management, 19,2: 167-81.
3. Calthorpe, P., 1993, The Next American Metropolis--Ecology, Community, and the American Dream, New York,NY: Princeton Architectural Press.
4. Cervaro, R., 1996, Jobs-Housing Balance Revisited: Trends and Impacts in the San Francisco Bay Area, Journal of the American Planning Association, 62,4: 492-511.
5. Cervaro, R., 1989, Jobs-Housing Balancing and Regional Mobility, Journal of the American Planning Association, 55,2: 136-50.
6. Ding, C., L. Hopkins, and G. Knaap, 1997, Does Planning Matter? Visual Examination of Urban Development Events, Landlines, 9,1: 4-5.
7. Environmental Planning and Economics, Inc, 1992, Clearing the AIR, Choosing the Future: Reducing Highway Vehicle Emissions in the Chicago Ozone Non-attainment Area, Final Report, Springfield, IL: the Illinois Department of Energy and Natural Resources Office of Research and Planning.
8. Ewing, R., 1997, Is Los Angeles-style Sprawl Desirable?, Journal of the American Planning Association, 63,1: 107-26.
9. Garrett, M. and M. Wachs, 1996, Transportation Planning on Trail: The Clean Air Act and Travel Forecasting, Thousand Oaks, CA: Sage.
10. Gordon, P. And H.W. Richardson, 1997, Are Compact Cities a Desirable Planning Goal?, Journal of the American Planning Association, 63,1: 95-106.
11. Gordon, P., A. Kumar, and H.W. Richardson, 1988, Beyond the Journey to Work,

Transportation Research-A, 22,6: 419-26.

12. Katz, P., 1994, The New Urbanism--Toward an Architecture of Community, New York, NY: McGraw Hill.
13. Kessler, J. and W. Schroeer, 1995, Meeting Mobility and Air Quality Goals: Strategies that Work, Transportation, 22: 241-272.
14. Knaap, G.J., 1997, The Determinants of Metropolitan Land Values: Implications for Regional Planning, unpublished working paper, Department of Urban and Regional Planning, University of Illinois.
15. Knaap, G.J. and A.C. Nelson, 1992, The Regulated Landscape: Lessons on State Land Use Planning from Oregon, Cambridge, MA: Lincoln Institute of Land Use Policy.
16. Lyons, W. M., 1995, Policy Innovations of the US Intermodal Surface Transportation Efficiency Act and the Clean Air Act Amendments, Transportation, 22: 217-40.
17. Moore, T. and P. Thorsness, 1994, The Transportation/ Land Use Connection, PAS report # 448/339, Chicago, IL: American planning Association.
18. Morris, M., 1996, Creating Transit-Supportive Land-Use Regulations, PAS report # 468, Chicago, IL: American Planning Association.
19. Scheuernstuhl, G.J. and J.H. May, 1992, Land Use, Transportation, and Air Quality Relationship, in R.L. Wayson, ed., Transportation Planning and Air Quality, p. 90-99, New York, NY: American Society of Civil Engineers.
20. Shrouds, J.M., 1992, Transportation Planning Requirements of the Federal Clean Air Act: A highway Perspective, in R.L. Wayson, ed., Transportation Planning and Air Quality, p. 14-29, New York, NY: American Society of Civil Engineers
21. Shrouds, J.M., Challenges and Opportunities for Transportation: Implementation of the Clean Art Act Amendments of 1990 and The Intermodal Surface Transportation Efficiency Act of 1991, Transportation, 22: 193-215.
22. United States Advisory Commission on Intergovernmental Relations, 1995, MPO Capacity; Improving the Capacity of metropolitan Planning Organizations to Help Implement National Transportation Policies, Washington, DC: ACIR.
23. United States General Accounting Office, 1996, Urban Transportation: Metropolitan Planning Organizations' Efforts to Meet Federal Planning Requirements, GAO/RCED-96-200, Washington, DC: USGAO.

GENERAL CHAIRMAN

G. van der Slikke	Ministry of Housing, Spatial Planning and the Environment, The Netherlands
L.D. Grant	National Center for Environmental Assessment, US Environmental Protection Agency

ORGANIZING COMMITTEE

United States of America

S.D. Lee	National Center for Environmental Assessment, US EPA
J. Bachmann/D. Evarts	Office of Air Quality Planning and Standards, US EPA
L. Folinsbee	National Center for Environmental Assessment, US EPA
L.D. Grant	National Center for Environmental Assessment, US EPA
J. Graham/T. Hartlage	National Exposure Research Laboratory, US EPA
E.T. Oppelt/A. Miller	National Risk Management Research Laboratory, US EPA
J. Vandenberg	National Health and Environmental Effects Research Laboratory, US EPA
W. Wilson	National Center for Environmental Assessment, US EPA

The Netherlands

T. Schneider	Chairman
K. Krijgsheld	Ministry of Housing, Spatial Planning and the Environment, The Netherlands
J. van Ham	Scientific secretary, TNO Institute of Environmental Sciences, Energy Research and Process Innovation
O. van Steenis	Logistics and registration, National Institute of Public Health and the Environment

PARTNERS PROGRAM

M. Schneider-Ferrageau de St. Amand

ADVISORY/PROGRAM COMMITTEE

United States of America

W.H. Farland	National Center for Environmental Assessment, US EPA
G.J. Foley	National Exposure Research Laboratory, US EPA
W. Nitze/K. Jacobs-Mudd	Office of International Activities, US EPA
E.T. Oppelt	National Risk Management Research Laboratory, US EPA
L.W. Reiter	National Health and Environmental Effects Research Laboratory, US EPA
J.S. Seitz	Office of Air Quality Planning and Standard, Office of Air and Radiation, US EPA

The Netherlands

H.P. Baars	TNO Institute of Environmental Sciences, Energy Research and Process Innovation
B.J. Heij	National Institute of Public Health and the Environment
K. Krijgsheld	Ministry of Housing, Spatial Planning and the Environment
P. Rombout	National Institute of Public Health and the Environment
B. Weenink	Ministry of Housing, Spatial Planning and the Environment

LIST OF PARTICIPANTS

Richard Ackermann
International Finance Corp. World Bank
1850 I (Eye) Street, NW
WASHINGTON DC 20433
USA

Tel: 1-202-4732606
Fax: 1-202-4730986
Email:

Joris Al
Ministry of Housing, Spatial Planning and the Environment
PO Box 30945
2500 GX DEN HAAG
The Netherlands
Tel: 31-70-3394852
Fax: 31-70-3394254
Email:

Markus Amann
IIASA
A-2361 LAXENBURG

Austria

Tel: 43-2236-432236807/432
Fax: 43-2236-71313
Email: amann@iiasa.ec.eg

Jan van Amsterdam
RIVM
PO Box 1
3720 BA BILTHOVEN
The Netherlands

Tel: 31-30-2742888
Fax: 31-30-2744446
Email: j.van.amsterdam@rivm.nl

Lucie Audette
OAR, USEPA
401 M Street SW
WASHINGTON DC 20460
USA

Tel: 1-313-7417850
Fax: 1-313-6684531
Email:

Hans-Peter Baars
TNO, Institute of Environmental Science, Energy Research and Process Innovation
PO Box 342
7300 AH APELDOORN
The Netherlands
Tel: 31-55-5493693
Fax: 31-55-5493252
Email:

Dick Bakker
TNO, Institute of Environmental Science, Energy Research and Process Innovation
PO Box 57
1780 AB DEN HELDER
The Netherlands
Tel: 31-223-638805
Fax: 31-223-630687
Email: d.j.bakker@mep.tno.nl

Ron Barnes
CONCAWE
Madouplein 1
1201 BRUSSELS
Belgium

Tel: 32-2-2203111
Fax: 32-2-2194646
Email:

Annemarie Bastrup-Birk
NERI, Department of Terrestrial Ecology
Vejlsoevej 25
DK-8600 SILKEBORG
Denmark

Tel: 45-89201400
Fax: 45-89201413
Email:

Judith Bates
AEA Technology, ITSU
Harwell
DIDCOT OX11 ORA
UK

Tel: 44-1235-432522
Fax: 44-1235-432662
Email:

H.J. Belois
Province of Gelderland
PO Box 9090
6800 GX ARNHEM
The Netherlands

Tel: 31-26-3598797
Fax: 31-26-3599480
Email:

Jan Berdowski
TNO-MEP
PO Box 342
7300 AH APELDOORN
The Netherlands

Tel: 31-55-5493171
Fax: 31-55-5493252
Email:

Chris Bernabo
Science & Policy Associates
The West Tower, Suite 400
1333 H Street, NW
WASHINGTON DC 20005
USA
Tel: 1-202-7891201
Fax: 1-202-7891206
Email: cbernabo@scipol.com

Lars Björkbom
Swedish Environmental Protection Agency
Blekholmsterrassen 36
S-10648 STOCKHOLM
Sweden

Tel: 46-8-6981063
Fax: 46-8-6981504
Email:

Bart van Bleek
National Spatial Planning Agency
PO Box 30940
2500 GX DEN HAAG
The Netherlands

Tel: 31-70-3393344
Fax: 31-70-3393052
Email: VanBleek@rop.rpd.minvrom.nl

Arie Bleyenberg
Deputy Director Centre for Energy Conservation a
Environmental Technology
Oude Delft 180
2611 HH DELFT
The Netherlands
Tel: 31-15-2150150
Fax: 31-15-2150151
Email:

Henk Bloemen
RIVM
PO Box 1
3720 BA BILTHOVEN
The Netherlands

Tel: 31-30-2742389
Fax: 31-30-2287531
Email:

Andries Blommers
Pollution Prevention Consulting BV
Margrietstraat 22
2555 PW DEN HAAG
The Netherlands

Tel: 31-70-3688382
Fax: 31-70-3234193
Email:

Margaretha de Boer
Minister, Ministry of Housing, Spatial Planning and the Environment
PO Box 30945
3500 GX DEN HAAG
The Netherlands
Tel:
Fax:
Email:

Yvo de Boer
Ministry of Housing, Spatial Planning and the Environment
PO Box 30945
3500 GX DEN HAAG
The Netherlands
Tel: 31-70-3394386
Fax: 31-70-3391313
Email:

Cor van den Bogaard
Inspectorate for the Environment
PO Box 30945
DEN HAAG
The Netherlands

Tel: 31-70-3394610
Fax: 31-70-3391298
Email: C.j.m.VanDenBogaard@HIMH.D

Brian Brangan
Department of the Environment
Custom House
DUBLIN 1
Ireland

Tel: 353-1-6793377
Fax: 353-1-8742423
Email: brian.brangan@environ.irlgov.ie

Josef Brechler
Dep. of Meteolology & Environ- mental Protection, Charles University
V Holesovickach 2
18000 PRAGUE
Czech REPUBLIC
Tel: 42-2-21912549
Fax: 42-2-21912533
Email:

Leendert van Bree
RIVM
PO Box 1
3720 BA BILTHOVEN
The Netherlands

Tel: 31-30-2742843
Fax: 31-30-2744448
Email:

Frans de Bree
Buro Blauw BV
Vadaring 96
6702 EB WAGENINGEN
The Netherlands

Tel: 31-317-425200
Fax: 31-317-426111
Email:

Bert Brunekreef
Wageningen Agricultural University
Vakgroep Humane Epidemiology
PO Box 238
6700 AE WAGENINGEN
The Netherlands
Tel: 31-317-483305
Fax: 31-317-482782
Email:

Eltjo Buringh
RIVM
PO Box 1
3720 BA BILTHOVEN
The Netherlands

Tel: 31-30-2742187
Fax: 31-30-2744448
Email: e.buringh@rivm.nl

Morris Burton Snipes
Lovelace Respiratory Research Institute
P.O. Box 5890
ALBUQUERGUE, NEW MEXICO 87185-5890
USA

Tel: 1-505-8451021
Fax: 1-505-8451198
Email: bsnipes@lucy.lrri.org

Flemming Cassee
National Institute of Public Health and the Environment
PO Box 1
3720 BA BILTHOVEN
The Netherlands
Tel: 31-30-2742406
Fax: 31-30-2744448
Email: fr.cassee@rivm.nl

Jancy Cohen
The Bureau of National Affairs
1231-25th ST.NW suite S-348
WASHINGTON, DC 20037
USA

Tel: 1-202-452-4438
Fax: 1-202-452-4150
Email: jcohen@bna.com

Dan Costa
Pulmonary Toxicology Branch, NHEERL, USEPA
(MD-68)
RESEARCH TRIANGLE PARK NC 27711
USA

Tel: 1-919-5412531
Fax: 1-919-5410026
Email:

Ellis Cowling
SOS, North Carolina State University
1307 Glenwood Avenue Suite 157
RALEIGH, NC 27605
USA

Tel: 1-919-5157564
Fax: 1-919-5151700
Email: cowling@ncsu.edu

Jacqueline Cramer
TNO-STB / University of Amsterdam
PO Box 541
7300 AM APELDOORN
The Netherlands

Tel: 31-55-5493500
Fax: 31-55-5421458
Email: STB@stb.tno.nl

José van Daalen
20 Maple Avenue
HU10 GPF WILLERBY EAST YORKSHIRE
UK

Tel: 44-1482-652702
Fax: 44-1482-652702
Email:

Catrien van Dam
Provincie Zuid-Holland
PO Box 90602
2509 LP DEN HAAG
The Netherlands

Tel: 31-70-4417647
Fax: 31-70-4417826
Email:

Arthur Davidson
Davidson & Associates, Environmental Consultin;
4814 Sommerset Drive SE
BELLEVIEW, WA 98006
USA

Tel: 1-206-8659386
Fax: 1-206-5621085
Email: art-d@msn.com

Walter Debruyn
Flemish Institute for Technological Research (VITO)
Boeretang 200
B-2400 MOL
Belgium

Tel: 32-14-335868
Fax: 32-14-321185
Email: dbruyn@vito.be

Dick Derwent
Meteorological Office
London Road
Bracknell
BERKSHIRE RG 12 2SZ
UK
Tel: 44-1344-854624
Fax: 44-1344-854493
Email:

Robert Devlin
N HEERL US EPA
RESEARCH TRIANGLE PARK NC 27711
USA

Tel: 1-919-9666255
Fax: 1-919-
Email:

Wouter van Dieren
IMSA
Van Eeghenstraat 77
1071 EX AMSTERDAM
The Netherlands

Tel: 31-20-5787600
Fax: 31-20-6622336
Email: imsa@euronet.nl

Jan Dormans
RIVM
PO Box 1
3720 BA BILTHOVEN
The Netherlands

Tel: 31-30-2742879
Fax: 31-30-2744437
Email: j.dormans@rivm.nl

A. Dumez
ESSO
PO Box 1
4803 AA BREDA
The Netherlands

Tel: 31-76-5291899
Fax: 31-76-5229545
Email:

Elizabeth Dutrow
OAR, USEPA
401 M Street SW
WASHINGTON DC 20460
USA

Tel: 1-202-2339061
Fax: 1-202-2339575
Email:

Hans Eerens
RIVM
PO Box 1
3720 BA BILTHOVEN
The Netherlands

Tel: 31-30-2743012
Fax: 31-30-2287531
Email:

Klaas van Egmond
RIVM
PO Box 1
3720 BA BILTHOVEN
The Netherlands

Tel: 31-30-2742045
Fax: 31-30-2744411
Email:

Marius Enthoven
Commission of the European Union, DG XI
200, Rue de la Loi
B-1049 BRUSSELS
Belgium

Tel:
Fax:
Email:

Hans Erbrink
KEMA
PO Box 9035
6800 ET ARNHEM
The Netherlands

Tel: 31-26-3562545
Fax: 31-26-3515022
Email: erbrink@mta6.kema.nl

William Farland
NCEA, USEPA (MD-52)
401 M Street SW (8601)
WASHINGTON DC 20460
USA

Tel: 1-202-2607316
Fax: 1-202-4012492
Email:

Terence Fitz-Simons
OAQPS, OAR, USEPA (MD-14)
RESEARCH TRIANGLE PARK NC 27711
USA

Tel: 1-919-5410889
Fax: 1-919-5411903
Email: fitz.simons.terence@epamail.epa.

Arthur FitzGerald
International Finance Corp., Technical and Env. Division
1850 I Eye Street NW
Room no I 10-002
DC 20433 WASHINGTON
Tel:
Fax:
Email:

Larry Folinsbee
Environmental Media Assessment Group, USEPA (MD-52)
RESEARCH TRIANGLE PARK NC 27711
USA

Tel: 1-919-5412229
Fax: 1-919-5411818
Email: Folinsbee.Lawrence@epamail.epa

Maarten van der Gaag
Ministry of Housing, Spatial Planning and the Environment
DGM/DWL/MLG IPC 630
PO Box 30945
2500 GX DEN HAAG
Tel: 31-70-3394945
Fax: 31-70-3391313
Email:

Judith Graham
National Exposure Research Laboratory (NERL)
USEPA (MD-75)
Research Triangle Park
NORTH CAROLINA 27711
USA
Tel: 1-919-541-0349
Fax: 1-919-541-3615
Email: jgraham@epamail.epa.gov

Lester Grant
NCEA/RTP USEPA (MD-52)
RESEARCH TRIANGLE PARK NORTH CAROLINA 27711
USA

Tel: 1-919-5414173
Fax: 1-919-5415078
Email: grant.lester@epamail.epa.gov

Bronno Haan
RIVM
PO Box 1

3720 BA BILTHOVEN
The Netherlands
Tel: 31-30-2743080
Fax: 31-30-2744435
Email: Bronno.de.Haan@rivm.nl

Tomas Halenka
Department of Meteolology & Environmental Protection, Charles University
V Holesovickach 2
18000 PRAGUE
CZECH REPUBLIC
Tel: 42-2-21912514
Fax: 42-2-21912533
Email:

Joop van Ham
TNO Institute of Environmental Sciences, Energy Research and Process Innovation
PO Box 6013
2600 JA DELFT
The Netherlands
Tel: 31-15-2696877
Fax: 31-15-2613186
Email: J.vanHam@mep.tno.nl

Pieter Hamelink
Ministry of Housing, Spatial Planning and the environment
PO Box 30945
2500 GX DEN HAAG
The Netherlands
Tel: 31-70-3394838
Fax: 31-70-3391313
Email:

Frans Harren
University of Nijmegen
Department of Molecular and Laserphysics
Toernooiveld
6525 EO NIJMEGEN
The Netherlands
Tel: 31-24-3652128
Fax: 31-24-3653311
Email: fransh@sci.kun.nl

Ineke van Harten
National Research Programme on Global Air Pollution and Climate Change (RIVM)
PO Box 1
3720 BA BILTHOVEN
The Netherlands
Tel: 31-30-2743211
Fax: 31-30-2744436
Email:

David Hawkins
Natural Resources Defense Fund
1200 New York Avenue, NW
WASHINGTON DC 20005
USA

Tel: 1-202-2896868
Fax: 1-202-2891060
Email:

Stephen Hedley
South East Institute of Public Health
Broomhill House, David Salomons Estate, Broom Road
TUNBRIDGE WELLS, KENT
England
Tel: 44-1892-515153
Fax: 44-1892-516344
Email:

BertJan Heij
National Research Programme on Global Air Pollution and Climate Change (RIVM)
PO Box 1

3720 BA BILTHOVEN
Tel: 31-30-2743108
Fax: 31-30-2744436
Email: bertjan.heij@rivm.nl

Kees Hendriks
DLO Winand Staring Centre
PO Box 125
6700 AC WAGENINGEN
The Netherlands

Tel: 31-317-474235
Fax: 31-317-424812
Email: c.m.a.hendriks@sc.dlo.nl

Bill Hogsett
NHEERL, USEPA, Ozone Research Group
200 Southwest 25th Street
CORVALLIS, OREGON 97333
USA

Tel: 1-541-7544632
Fax: 1-541-7544799
Email: bill@hert.cor.epa.gov

Mike Holland
ETSU
Rue de Trèves 48
1040 BRUSSELS
Belgium

Tel: 32-2-2387845
Fax: 32-2-2387709
Email: mike.holland@ocat.co.uk

Nico Hoogervorst
RIVM
PO Box 1
3720 BA BILTHOVEN
The Netherlands

Tel: 31-30-2743653
Fax: 31-30-2744416
Email:

Oystein Hov
NILU
Norwegian Institute for Air Research
PO Box 100
N-2007 KJELLER
Norway
Tel: 47-63898000
Fax: 47-63898050
Email:

Theodore Hullar
Department of Environmental Toxicology, University of California
DAVIS, CALIFORNIA 95616
USA

Tel: 1-916-7549288
Fax: 1-916-7523394
Email: tlhuller@ucdavis.edu

Leon Janssen
RIVM
PO Box 1
3720 BA BILTHOVEN
The Netherlands

Tel: 31-30-2742771
Fax: 31-30-2287531
Email:

Wouter de Jong
Ministry of Housing, Spatial Planning and the Environment
PO Box 30945
2500 GX DEN HAAG
The Netherlands
Tel: 31-70-3393042
Fax: 31-70-3393087
Email:

Lex de Jonge
VROM
PO Box 30945
2500 GX DEN HAAG
The Netherlands

Tel: 31-70-3394693
Fax: 31-70-3391313
Email:

Til de Jonge-van Swaay
De Jonge Water en Milieu
Badhuisweg 77
2587 DEN HAAG
The Netherlands

Tel: 31-70-3584363
Fax: 31-70-3585138
Email: dejonge@bart.nl

Menno Keuken
TNO
Schoemakerstraat 97
2600 JA DELFT
The Netherlands

Tel: 31-15-2696227
Fax: 31-15-2617217
Email: keuken@mep.tno.nl

Ray Klicius
Environment Canada
Environment Protection Service
351 St. Joseph Blvd.
18th floor
HULL (QUEBEC) K1A OH3
Tel:
Fax: 1-819-9534705
Email:

Gerrit Knaap
University of Illinois
311 Coble Hall, 801 S. Wright Street
CHAMPAIGN, ILLINOIS 61820
USA

Tel: 1-217-2445369
Fax: 1-217-2441717
Email: g-knaap@uiuc.edu

Eric Kreileman
RIVM
PO Box 1
3720 BA BILTHOVEN
The Netherlands

Tel: 31-30-2743554
Fax: 31-30-2744427
Email: eric.kreileman@rivm.nl

Klaas Krijgsheld
Ministry of Housing, Spatial Planning and the Environment
PO Box 30945
3500 GX DEN HAAG
The Netherlands
Tel: 31-70-3394391
Fax: 31-70-3391313
Email:

Oskar de Kuijer
DTO
PO Box 6063
2600 JA DELFT
The Netherlands

Tel: 31-15-2697543
Fax: 31-15-2697547
Email: o.d.kuijer@worldaccess.nl

Lester Lave
Carnegie-Mellon University
PITTSBURG PENNSYLVANIA 15213
USA

Tel: 1-412-2688837
Fax: 1-412-2687357
Email: Lester.Lave@andrew.cmu.edu

Erik Lebret
RIVM
PO Box 1
3720 BA BILTHOVEN
The Netherlands

Tel: 31-30-2742777
Fax: 31-30-2744407
Email:

Si Duk Lee
NCEA, USEPA (MD-52)
RESEARCH TRIANGLE PARK NC 27711
USA

Tel: 1-919-5414477
Fax: 1-919-5410245
Email: Lee.siduk@epamail.epa.gov

Rolaf van Leeuwen
WHO, European Centre for Environment and Health
PO Box 10
3730 AA DE BILT
The Netherlands

Tel: 31-30-2295307
Fax: 31-30-2294252
Email: rle@who.nl

Walter Lehr
Hoechst Schering AgrEvo GmbH
Werk Höchst, Tor Süd K 607
D-65926 FRANKFURT AM MAIN
Germany

Tel: 49-69-3052939
Fax: 49-69-3053826
Email:

Jip Lenstra
Ministry of Housing, Spatial Planning and the Environment
PO Box 30945
2500 GX DEN HAAG
The Netherlands
Tel: 31-70-3394412
Fax: 31-70-3391313
Email:

Leo de Leu
Gemeentewerken Rotterdam Ir buro Milieu
PO Box 6633
3002 AP ROTTERDAM
The Netherlands

Tel: 31-10-4894956
Fax: 31-10-4894500
Email:

William Linak
Air Pollution Prevention and Control Division, USEPA (MD-65)
RESEARCH TRIANGLE PARK NC 27711
USA

Tel: 1-919-5415792
Fax: 1-919-5410554
Email:

Rob van Lint
Ministry of Housing, Spatial Planning and the environment
PO Box 30945
2500 GX DEN HAAG
The Netherlands
Tel: 31-70-3392250
Fax: 31-70-3391289
Email:

Frederick Lipfert
Consultant
23 Carill Court

NORTHPORT, NY 11768
USA
Tel: 1-516-261-5735
Fax: 1-516-344-7867
Email:

Morton Lippmann
New York University Institute of Environmental Medicine
TUXEDO, NY 10987
USA

Tel: 1-914-3512396
Fax: 1-914-3515472
Email: lippmann@charlotte.med.nyu.edu

Hans Luttikholt
HASKONING
PO Box 151
6500 AD NIJMEGEN
The Netherlands

Tel: 31-24-3284284
Fax: 31-24-3239346
Email: JLU@HASKONING.NL

Petra Mahrenholz
Federal Environmental Agency
Bismarckplatz 1
14193 BERLIN
Germany

Tel: 49-30-89032084
Fax: 49-30-89032285
Email:

Petra Mahrenholz
Federal Environmental Agency
Bismarckplatz 1
14193 BERLIN
Germany

Tel: 49-30-89032084
Fax: 49-30-89032285
Email:

Gordon McInnes
European Environment Agency
Kongens Nytov 6
DK 1050 COPENHAGEN
Denmark

Tel: 45-33367100
Fax: 45-33367199
Email:

Michelle McKeever
OAQPS, USEPA (MD-15)
RESEARCH TRIANGLE PARK NC 27711
USA

Tel: 1-919-5415488
Fax: 1-919-5410839
Email:

Richard Minard
Center for Economy and the Environment, Nation
Academy of Public Administration
1120 G Street NW Suite 850
WASHINGTON DC 20005-3821
USA
Tel: 1-202-3473190
Fax: 1-202-3930993
Email: minard@tmn.com

Richard Morgenstern
OPPE, USEPA
1616 P Street NW
WASHINGTON DC 20036
USA

Tel: 1-202-3285037
Fax: 1-202-9393460
Email: morgenst@rff.org

Jan Mulderink
DTO
Valkenierspad 6
6871 CH ROZENDAAL
The Netherlands

Tel: 31-26-3635113
Fax: 31-26-3648489
Email:

Samuel Napolitano
Office of Atmospheric Program, USEPA (6201)
401 M Street SW
WASHINGTON DC 20460
USA

Tel: 1-202-2339751
Fax: 1-202-2339854
Email:

Frans de Neve
3M Netherlands
PO Box 3235
4800 DE BREDA
The Netherlands

Tel: 31-76-5301292
Fax: 31-76-5301355
Email:

Hans Nieuwenhuis
Ministry of Housing, Spatial Planning and the Environment
PO Box 30945
3500 GX DEN HAAG
The Netherlands
Tel: 31-70-3394408
Fax: 31-70-3391313
Email:

Jan Nilsson
MISTRA
G. Brogatan 36
11120 STOCKHOLM
Sweden

Tel: 46-08-7911022
Fax: 46-08-7911089
Email:

Timothy Oppelt
NRMRL, USEPA
26 Martin Luther King Drive
CINCINNATI OHIO 45268
USA

Tel: 1-513-5697418
Fax: 1-513-5697276
Email:

Joseph Paisie
OAQPS, OAR, USEPA (MD-15)
RESEARCH TRIANGLE PARK NC 27711
USA

Tel: 1-919-5415556
Fax: 1-919-5415489
Email: paisie.joe@epamail.epa.gov

Randy Pasek
Atmospheric Processes Research Section, Air Resources Board
P.O. Box 2815
SACRAMENTO, CALIFORNIA 95812
USA
Tel: 1-916-3248496
Fax: 1-916-3224357
Email: rpasek@arb.ca.gov

J. Peeters
Ministry of Housing, Spatial Planning and the Environment
PO Box 30945
2500 GX DEN HAAG
The Netherlands
Tel: 31-70-3394529
Fax: 31-70-3391280
Email: peeters@dgv.dgm.minvrom.nl

Caroline Petti
Office of Policy Analysis and Review, USEPA (mail code 6103)
401 M Street SW
WASHINGTON DC 20460
USA
Tel: 1-202-2603832
Fax: 1-202-2600253
Email: petti.caroline@epamail.epa.gov

Per Pinstrup Andersen
International Food Policy Institute
1200 17th Street N.W.
WASHINGTON DC 20036
USA

Tel: 1-202-8625600
Fax: 1-202-4674439
Email: P.Pinstrup-Andersen@cgnet.com

Frank Princiotta
Air Pollution Prevention and Control Division,
USEPA (MD-60)
RESEARCH TRIANGLE PARK NC 27711
USA

Tel: 1-919-5412821
Fax: 1-919-5415227
Email: Princiotta.

Wim Quick
VNCI
PO Box 443
2260 AK LEIDSCHENDAM
The Netherlands

Tel: 31-70-3378738
Fax: 31-70-3209418
Email:

Knut Rauchfuss
Landesumweltamt - NRW
Wallneyerstrasse 6
D-45133 ESSEN
Germany

Tel: 49-201-7995165
Fax: 49-201-7995446
Email: k.rauchfuss@link-do.soli.de

John Rea
Department of the Environment
Romney House
43, Marsham Street
LONDON SW1P 3PY
UK
Tel: 44-171-2768155
Fax: 44-171-2768299
Email:

Lawrence Reiter
NHEERL, USEPA (MD-51)
RESEARCH TRIANGLE PARK NC 27711
USA

Tel: 1-919-5412281
Fax: 1-919-5410245
Email:

Michiel Roemer
TNO-MEP
PO Box 342
7300 AH APELDOORN
The Netherlands

Tel: 31-55-5493787
Fax: 31-55-5493252
Email:

Peter Rombout
RIVM
PO Box 1
3720 BA BILTHOVEN
The Netherlands

Tel: 31-30-2742238
Fax: 31-30-2744448
Email:

Dennis Rondinelli
University of North Carolina Kenan-Flagler
Business School
CHAPEL HILL NORTH CAROLINA 27599-344
USA

Tel: 1-919-9628201
Fax: 1-919-9628202
Email: dennis.rondinelli@vne.edu

Gert de Roo
University of Groningen
PO Box 800
9700 AV GRONINGEN
The Netherlands

Tel: 31-50-3633895
Fax: 31-50-3633901
Email:

Benjamin Santer
PCMDI
Lawrence Livermore National Laboratory

PO Box 808, Mail Stop L-264
LIVERMORE, CA 94550
Tel: 1-510-4234249
Fax: 1-510-4227675
Email: bsanter@pcmid.llnl.gov

Erik Schmieman
Wageningen Agricultural University
Hollandseweg 1
6706 KN WAGENINGEN
The Netherlands

Tel: 31-317-483809
Fax: 31-317-484933
Email: erik.schmieman@alg.shhk.wau.nl

Toni Schneider
RIVM
PO Box 1

3720 BA BILTHOVEN
The Netherlands
Tel: 31-30-2742696
Fax: 31-30-2744436
Email: toni_s@euronet.nl

Mini Schneider
Nassauplantsoen 7
3761 BH SOEST
The Netherlands

Tel:
Fax:
Email:

Emile Schols
DSM, Research
PO Box 18
6160 MD GELEEN
The Netherlands

Tel: 31-46-4761344
Fax: 31-46-4760700
Email:

Elly Schreur
Gemeentewerken, Afdeling Milieubeleid
PO Box 6633
3002 AP ROTTERDAM
The Netherlands

Tel: 31-10-4896223
Fax: 31-10-4896231
Email:

Peter Segaar
Municipality of Utrecht / DSO
PO Box 8406
3503 RK UTRECHT
The Netherlands

Tel: 31-30-2864864
Fax: 31-30-2946634
Email: BSMilieu@knoware.nl

Henrik Selin
Department of Water and Environmental Studies,
Linkoping Univ.
58183 LINKOPING

Sweden
Tel: 46-13-282996
Fax: 46-13-133630
Email:

Gert van der Slikke
Ministry of Housing, Spatial Planning and the
Environment
PO Box 30945
3500 GX DEN HAAG
The Netherlands
Tel: 31-70-3394381
Fax: 31-70-3391311
Email: slikke@dle.dgm.minvrom.nl

Roger Smook
TU Delft
Fac. Civiele Techniek
PO Box 5048
2600 GA DELFT
The Netherlands
Tel: 31-15-2786636
Fax: 31-15-2787683
Email:

Jan Paul van Soest
Centre for Energy Conservation and Environmenta
Technology
PO Box 180
2611 HH DELFT
The Netherlands
Tel: 31-15-2150150
Fax: 31-15-2150151
Email:

Ottelien van Steenis
National Research Programme on Global Air
Pollution and Climate Change (RIVM)
PO Box 1

3720 BA BILTHOVEN
Tel: 31-30-2742970
Fax: 31-30-2744436
Email: ottelien.van.steenis@rivm.nl

Björn Stigson

Tel:
Fax:
Email:

P.A.J. Thomassen
Olie Contact Commissie
PO Box 29764
2502 LT DEN HAAG
The Netherlands

Tel: 31-70-3382369
Fax: 31-70-3382363
Email:

Richard Tol
Free University
De Boelelaan 1115
1081 HV AMSTERDAM
The Netherlands

Tel: 31-20-4449503
Fax: 31-20-4449553
Email:

Alfred Tonneijck
AB-DLO
PO Box 14
6700 AA WAGENINGEN
The Netherlands

Tel: 31-317-475908
Fax: 31-317-423110
Email: a.e.g.tonneijck@ab.dlo.nl

Gene Tucker
Air Pollution Prevention and Control Division,
USEPA (MD-65)
RESEARCH TRIANGLE PARK NC 27711
USA

Tel: 1-919-5412746
Fax: 1-919-5412157
Email: Tucker.Gene@epamail.epa.gov

John Vandenberg
NHEERL, USEPA (MD-51)
RESEARCH TRIANGLE PARK NC 27711
USA

Tel: 1-919-5414527
Fax: 1-919-5410642
Email:

Guus Velders
RIVM
PO Box 1
3720 BA BILTHOVEN
The Netherlands

Tel: 31-30-2742331
Fax: 31-30-2287531
Email: Guus.Velders@rivm.nl

Harm Verbruggen
Free University
De Boelelaan 1115
1081 HV AMSTERDAM
The Netherlands

Tel: 31-20-4449555
Fax: 31-20-4449553
Email:

Pieter Verkerk
Ministry of Housing, Spatial Planning and the Environment
PO Box 30945
2500 GX DEN HAAG
The Netherlands
Tel: 31-70-3394621
Fax: 31-70-3391307
Email:

Hans Visser
KEMA
Afdeling Milieu Services
PO Box 9035
6800 ET ARNHEM
The Netherlands
Tel: 31-026-3563012
Fax: 31-026-3515022
Email: visser@mta6.kema.nl

Johan Vollenbroek
MOB
W. Pyzmontsingel 18
6521 BC NIJMEGEN
The Netherlands

Tel: 31-24-3230491
Fax: 31-24-3238469
Email:

Jaro Vostal
EHAC International
6360 Hills Drive
BLOOMFIELD HILLS, MI 48301
USA

Tel: 1-810-6446527
Fax: 1-810-6446527
Email:

Jan Tjalling van der Wal
RIVM
PO Box 1
3720 BA BILTHOVEN
The Netherlands

Tel: 31-30-2742388
Fax: 31-30-2287531
Email: jan.van.der.wal@rivm.nl

Mike Walsh
3105 North Dinwiddie Street
ARLINGTON, VIRGINIA 22207
USA

Tel: 1-703-2411297
Fax: 1-703-2411418
Email: mpwalsh@ipg.apc.org

Bert van Wee
RIVM
PO Box 1
3720 BA BILTHOVEN
The Netherlands

Tel: 31-30-2749111
Fax: 31-30-2742971
Email: Bert.Van.Wee@RIVM.NL

Bas Weenink
Ministry of Housing, Spatial Planning and the Environment
PO Box 30945
3500 GX DEN HAAG
The Netherlands
Tel: 31-70-3394408
Fax: 31-70-3391313
Email:

Steven Weiner
Battelle Pacific Northwest National Laboratory
901 D Street, SW Suite 900
WASHINGTON DC 20024-2115
USA
Tel: 1-202-6467870
Fax: 1-202-6465020
Email: sc_weiner@pnl.gov

Ronald Whitfield
Argonne National Laboratory
ARGONNE, ILLINOIS 60439-4382
USA
Tel: 1-630-2528430
Fax: 1-630-2526073
Email:

Peter Wicks
European Commission DGXI
Rue de la Loi 200
B-1049 BRUXELLES
Belgium
Tel:
Fax: 32-2-2991067
Email:

Jeanine van de Wiel
Gezondheidsraad
PO Box 90517
2509 LM DEN HAAG
The Netherlands
Tel: 31-70-3407520
Fax: 31-70-3407523
Email:

Arnold van der Wielen
Ministry of Housing, Spatial Planning and the environment
PO Box 30945
2500 GX DEN HAAG
The Netherlands
Tel: 31-70-3394896
Fax: 31-70-3391297
Email:

Martin Williams
Department of the Environment Ait Quality Division
Romney House
43 Marsham
LONDON SW1P 3PY
UK
Tel: 44-171-2768881
Fax: 44-171-2768821
Email:

William Wilson
National Exposure Research Laboratory USEPA
(MD-52)
RESEARCH TRIANGLE PARK NC 27711
USA
Tel: 1-919-5412551
Fax: 1-919-5410245
Email: wilson.william@epamail.epa.gov

Gerard Wolters
Ministry of Housing, Spatial Planning and the Environment
PO Box 30945
2500 GX DEN HAAG
The Netherlands
Tel: 31-70-3394651
Fax: 31-70-3391308
Email:

Remco Ybema
Energy Research Foundation
PO Box 1
1755 ZG PETTEN
The Netherlands
Tel: 31-224-564428
Fax: 31-224-563338
Email:

Kees Zoeteman
Ministry of Housing, Spatial Planning and the Environment
PO Box 30945
2500 GX DEN HAAG
The Netherlands
Tel: 31-70-3394017
Fax: 31-70-3391308
Email:

Job van Zorge
Ministry of Housing, Spatial Planning and the Environment
PO Box 30945
2500 GX DEN HAAG
The Netherlands
Tel: 31-70-3394947
Fax: 31-70-3391314
Email:

Thijs Zuidema
Univeristy of Groningen
PO Box 800
9700 AV GRONINGEN
The Netherlands

Tel: 31-50-3633757
Fax: 31-50-3637337
Email:

Henk van Zuylen
Transport Research Centre (AVV)
PO Box 1031
3000 BA ROTTERDAM
The Netherlands

Tel: 31-10-2825732
Fax: 31-10-2825643
Email: h.j.vzuylen@avv.rws.minvenw.nl